# Hot Carriers in Semiconductors

# Hot Carriers in Semiconductors

Edited by

## Karl Hess
## Jean-Pierre Leburton
and
## Umberto Ravaioli

*University of Illinois at Urbana – Champaign*
*Urbana, Illinois*

Plenum Press • New York and London

Library of Congress Cataloging-in-Publication Data

On file

Proceedings of the Ninth International Conference on Hot Carriers in Semiconductors,
held July 31 – August 4, 1995, in Chicago, Illinois

ISBN 0-306-45366-5

# PREFACE

This volume contains invited and contributed papers of the Ninth International Conference on Hot Carriers in Semiconductors (HCIS-9), held July 31-August 4, 1995 in Chicago, Illinois. In all, the conference featured 15 invited oral presentations, 60 contributed oral presentations, and 105 poster presentations, and an international contingent of 170 scientists. As in recent conferences, the main themes of the conference were related to nonlinear transport in semiconductor heterojunctions and included Bloch oscillations, laser diode structures, and femtosecond spectroscopy. Interesting questions related to nonlinear transport, size quantization, and intersubband scattering were addressed that are relevant to the new quantum cascade laser. Many lectures were geared toward quantum wires and dots and toward nanostructures and mesoscopic systems in general. It is expected that such research will open new horizons to nonlinear transport studies.

An attempt was made by the program committee to increase the number of presentations related directly to devices. The richness of nonlocal hot electron effects that were discussed as a result, in our opinion, suggests that future conferences should further encourage reports on such device research.

On behalf of the Program and International Advisory Committees, we thank the participants, who made the conference a successful and pleasant experience, and the support of the Army Research Office, the Office of Naval Research, and the Beckman Institute of the University of Illinois at Urbana-Champaign. We are also indebted to Mrs. Sara Starkey and Mrs. Carol Willms for their invaluable assistance in the conference organization, administration, and the preparation of these proceedings, and to P. Matagne, S. Nagaraja, A. Pacelli and T. Singh for their contributions to the administration of the conference.

Karl Hess, Chair
J.P. Leburton, Co-chair
U. Ravaioli, Co-chair

# The Ninth International Conference on Hot Carriers in Semiconductors (HCIS-9)
## Chicago, Illinois, July 31 - August 4, 1995

**Conference Committee**

| | | |
|---|---|---|
| K. Hess | Chairman | University of Illinois |
| J. P. Leburton | Co-chair | University of Illinois |
| U. Ravaioli | Co-chair | University of Illinois |

**Progam Committee**

| | |
|---|---|
| A. Andronov (Russia) | S. Laux (USA) |
| J. R. Barker (UK) | P. Lugli (Italy) |
| G. Bauer (Austria) | D. N. Mirlin (Russia) |
| J. Devreese (Belgium) | J. P. Nougier (France) |
| D. K. Ferry (USA) | W. Porod (USA) |
| E. O. Göbel (Germany) | J. F. Ryan (UK) |
| Z. S. Gribnikov (Ukraine) | J. Shah (USA) |
| P. Hawrylak (Canada) | C. Stanton (USA) |
| I. Kizilyalli (USA) | K. Taniguchi (Japan) |

**International Advisory Committee**

| | |
|---|---|
| J. R. Barker (UK) | K. Hess (USA) |
| G. Bauer (Austria) | M. Inoue (Japan) |
| L. R. Cooper (USA) | P. Lugli (Italy) |
| T. Elsaesser (Germany) | S. Luryi (USA) |
| D. K. Ferry (USA) | L. Reggani (Italy) |
| E. O. Göbel (Germany) | J. F. Ryan (UK) |
| S. Goodnick (USA) | J. Shah (USA) |
| E. Gornick (Austria) | C. Snowden (UK) |
| H. Grubin (USA) | P. Vogl (Germany) |
| C. Hamaguchi (Japan) | K. Yokoyama (Japan) |

# CONTENTS

## 1. Hot Carrier Luminescence and Femtosecond Spectroscopy

## 2. Bloch Oscillations and Fast Coherent Processes in Semiconductors

## 3. Hot Carriers in Nanostructures and Low-Dimensional Systems

# 4. High Field Transport and Impact Ionization

## 5. Hot Electrons in Devices

1. Hot Carrier Luminescence and Femtosecond Spectroscopy

# FIELD-INDUCED EXCITON IONIZATION
# STUDIED BY FOUR-WAVE MIXING

M.Koch,[1,2] G. von Plessen,[1,3] T. Meier,[1] J. Feldmann,[1] S.W. Koch,[1]
P. Thomas [1], E.O. Göbel,[1,4] K.W. Goossen,[2] J.M. Kuo,[2] and R.F. Kopf [2]


[1]Department of Physics and Material Sciences Center,
  Philipps University of Marburg, 35032 Marburg, Germany

[2]AT&T Bell Laboratories, Holmdel, New Jersey 07733, USA

[3]Clarendon Laboratory, University of Oxford, Oxford OX1 3PU, UK

[4]Physikalisch Technische Bundesanstalt, 38116 Braunschweig, Germany

## INTRODUCTION

The dynamics of carriers induced by electric fields in semiconductor superlattices has received much interest in recent years. Phenomena like Bloch oscillations and negative differential velocity in these structures have been studied using a variety of experimental techniques [1-5]. One particular point of interest has been the transition of the miniband regime to the Bloch oscillation regime with increasing electric field [6,7]. Cw spectra of strongly-coupled superlattices have evidenced that this transition is concomitant with a rapid field-induced ionization of the zero-field miniband exciton [5,8]. While it has been pointed out from the theoretical side that the field-induced ionization process strongly influences the coherent dynamics in the transition region [9], no systematic experimental investigation of the ionization process has been carried out so far. One interesting question that such an investigation could help to answer is to what extent this ionization process differs from the well-documented case of field-induced exciton ionization in bulk semiconductors.

A frequently employed technique for time-resolving ultrafast processes in semiconductors is transient four-wave mixing (FWM). FWM has been used to investigate phase destroying processes, which limit the coherent lifetime of excitons [10], and to study interference phenomena, e.g. the dynamics of excitonic wave packets (EWP) composed of bound and continuum edge states of quantum well excitons [11,12]. Recently, we have used FWM to study the field-induced ionization dynamics of excitons in a multiple-quantum well structure [13]. Under zero applied field, the decay of the FWM signal reflects the dephasing of the excitonic transition due to phase-breaking scattering processes; application of an electric field results in an additional enhancement of the signal decay rate due to the field-induced dissociation of the Coulomb-correlated electron-hole pair. Here, we apply this technique to study the electric-field-induced dynamics of the superlattice exciton in a strongly-coupled GaAs/Al$_{0.3}$Ga$_{0.7}$As superlattice in the low-field (miniband) regime.

## EXPERIMENTAL

The sample investigated is a GaAs/Al$_{0.3}$Ga$_{0.7}$As superlattice consisting of 100 periods of 3nm GaAs wells and 3nm Al$_{0.3}$Ga$_{0.7}$As barriers, embedded into the intrinsic region of a p-i-n diode. Kronig-Penney calculations yield 62 meV for the electronic and 5 meV for the heavy-hole miniband widths, respectively. We note that for a similar superlattice structure, the binding energy of the heavy-hole exciton has been determined to be 4.5 meV. [14], which is close to the GaAs bulk value. DC electric fields can be applied by varying the external voltage applied to the diode. The FWM experiments are performed in the so called self-diffraction geometry, where two laser pulses are focused onto the sample under a small angle and the diffracted FWM signal is monitored as a function of the time delay between the optical pulses. While the first pulse sets up a macroscopic optical polarization in the material, the second delayed pulse probes the polarization decay caused by scattering events and the field induced exciton dissociation. For details of the FWM technique and set up see, e.g. [10]. As an excitation source we have used a Kerr-lens mode-locked Ti:sapphire laser giving either 60 fs or 120 fs laser pulses. All experiments are performed at sample temperatures below 10 K.

## RESULTS AND DISCUSSION

First, we discuss the FWM transients obtained with the 60 fs pulses. Although we detune the center frequency of the laser pulses to 13 meV below the heavy-hole resonance, due to their large spectral width (approx. 40 meV) the pulses still excite all bound excitonic states and part of the continuum states. FWM transients obtained under these conditions for different applied voltages are depicted in the semi-logarithmic plot in Fig. 1. The upmost curve in Fig. 1, which is obtained for flat band conditions, shows after a first signal maximum around zero time delay a pronounced minimum followed by a monotonic signal decay. This minimum, which stems from a destructive interference of the 1s resonance on the one hand, and a coherent superposition of the higher-lying bound states and a part of the continuum states on the other hand, is a clear indication for the existance of an EWP [11]. We have also performed FWM measurements with time-resolved detection (not shown) which reveal a further typical signature of an EWP, namely an unstructured echo-like real-time signal which gets broader with increasing time delay between the two exciting pulses [12].

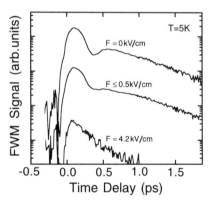

**Figure 1.** Time-integrated FWM transients for different electric fields F obtained with 60 fs pulses. The curves are displaced for clarity.

When a small field (F ≤ 0.5kV/cm) is applied to the superlattice, the modulation depth associated with the interference is drastically reduced. This means, that the continuum components of the excitonic wave packet, which give rise to the interference with the 1s resonance, undergo a strong field-induced dephasing at very low electric fields. A possible explanation for this pronounced field sensitivity of the continuum components is that they involve unbound electron-hole pair states, which will require very low fields to undergo significant dephasing. In contrast the long-lived signal component associated with the tightly-bound 1s resonance is not measurably affected by the weak field. However, at higher fields, the probability of the 1s resonance for tunneling out of the Coulomb potential well is considerably enhanced. Consequently, the corresponding signal component experiences a drastically enhanced decay rate, as can be seen from the lowest curve in Fig. 1.

To futher explore the effect of the field on the 1s resonance we perform additional FWM experiments using 120 fs pulses. By choosing a detuning of only 3 meV below resonance, we predominantly excite the 1s exciton only. The excitonic continuum states are only weakly excited under this conditions, which is due to the narrower spectrum of the 120 fs-pulses (approx. 20 meV). Consequently, the interference modulation becomes much less pronounced as can be seen from the upper curve in Fig. 2a (dots). The FWM signal reveals a mainly plateaulike profile for time delays between +0.1 ps and +0.7 ps and a slow decay beyond +0.7 ps [15]. As seen before, the signal decay becomes faster when an electric field is applied (Fig. 2, diamonds,squares).

In order to obtain ionization times from the experimental data, we perform a fit procedure on the FWM transients using theoretical model calculations. These calculations are performed by solving the semiconductor Bloch equations, which have been extented to include the external electric field [12]. The simple superlattice model which we use here has cosine miniband dispersions [6]; the miniband widths are as given above. The basic idea of the fit procedure is as follows: First, the theoretical FWM curve for F = 0 is adjusted to fit the corresponding experimental transient, using the inhomogeneous broadening, $\Gamma_{inhom}$, of the optical transition and the excitonic dephasing time at zero field, $T_2(F = 0)$, as fit parameters. Good agreement between theory and experiment is found for $\Gamma_{inhom}$ = 3.5 meV and $T_2(F = 0)$ = 1.5 ps. Assuming that the scattering rates and $\Gamma_{inhom}$ remain unaffected by the field, we then compute the theoretical curves for nonzero field. A comparison with the experimental transients shows that good agreement is obtained if a constant voltage offset of 1.2 V at the diode due to screening of the applied field by accumulated carriers is allowed for. We note that a similar voltage offset has recently been found in FWM measurements on Schottky diode structures [7]. An important application of the calculated curves lies in the fact that they extrapolate the measured transients towards long decay times, where the combined effects

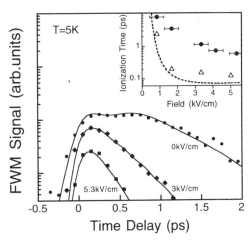

**Figure 2.** Time-integrated FWM transients for different applied fields obtained with 120 fs pulses (symbols). The solid lines show model calculations based on the semiconductor Bloch equations.
Inset: Exciton ionization times as a function of the applied field. Ionization times determined for the superlattice are shown as dots. Ionization times calculated for bulk GaAs are shown as triangles (solutions of the semiconductor Bloch equations), and as a dotted line (results of Eq. A7 in Ref. [19]).

of inhomogeneous broadening and Coulomb many-body interaction cease to affect the signal shape, and the signal decay becomes monoexponential [17]. This enables us to extract the field-dependent dephasing times $T_2(F > 0)$ from the signal decays, $\tau_D$ (F > 0), according to the usual relation $T_2 = 4\tau_D$ for inhomogeneously broadened transitions [18]. With the reasonable assumption that the dephasing time for nonzero field is given by $T_2^{-1}(F) = T_2^{-1}(F = 0) + T_{ion}^{-1}(F)$, we finally obtain the field dependent ionization time $T_{ion}(F)$ (see Ref.[15] for details).

The dots in the inset of Fig. 2 show the field dependence of the ionization time. As expected, we see a decrease of the ionization time with increasing field. It is interesting to compare these results to the ionization times, which can be expected for a 1s exciton in bulk GaAs. While the dashed line in the inset is a result of a WBK calculation (Eq. A7 in Ref. [19]), the triangles are obtained from solving the semiconductor Bloch equations in the same way as before, now assuming parabolic band dispersions in growth direction. Obviously, the values for $T_{ion}$ calculated for a bulk exciton are much smaller than the ones for the superlattice exciton. This discrepancy is remarkable since the binding energies of the two excitons are almost equal (within less than 10%), which should result in a similar field dependence. We interpret the difference in the ionization times as a consequence of the nonparabolicity in the miniband dispersion. While the effective mass at the miniband minimum of our superlattice structure is approximately equal to that of the bulk case, the carriers photogenerated near the miniband edge become heavier as they are ballistically accelerated by the electric field [4], thus making it harder for the field to dissociate the Coulomb-correlated electron-hole pairs than in the bulk case. This effect will tend to impede the field-induced ionization of the superlattice exciton in comparison with that of bulk excitons of similar binding energies.

## SUMMARY

In summary, we have performed time-domian studies of electric-field-induced exciton ionization in a strongly coupled semiconductor superlattice. We find the excitonic continuum states to be more sensitive to low electric fields than the 1s-state. By comparing our experimental results to numerical model calculations we are able to extract ionization times for the 1s exciton. The ionization times are considerably longer than those expected for a bulk exciton, which can be explained as a consequence of the nonparabolicity in the miniband dispersion.

## REFERENCES

1. J. Feldmann, et al., *Phys. Rev. B* 46: 7252 (1992).
2. C. Waschke, et al. *Phys. Rev. Lett.* 70: 3319 (1993).
3. T. Dekorsy, et al., *Phys. Rev. B* 50: 8106 (1994).
4. A. Sibille, et al., *Phys. Rev. Lett.* 64:52 (1990).
5. G. Cohen, and I. Bar-Joseph, *Phys. Rev. B.* 46:9857 (1992).
6. G. von Plessen, et al., *Phys. Rev. B* 49: 14058 (1994).
7. P. Leisching, et al., *Phys. Rev. B* 50: 14389 (1994).
8. E.E. Mendez, F. Agullo-Rueda, and J.M.Hong, *Phys. Rev. Lett.* 60: 2426 (1988).
9. T. Meier, et al., *Phys. Rev. B.* 51: 14490 (1995).
10. J. Kuhl, et al., *in*: "Festkörperprobleme, Advances in Solid State Physics, Vol. 29" U. Rössler ed., Vieweg, Braunschweig (1989).
11. J. Feldmann, et al., *Phys. Rev. Lett.* 70:3027 (1993).
12. F. Jahnke, et al., *Phys. Rev. B* 50: 8114 (1994).
13. G. von Plessen, et al., *Appl.Phys. Lett.* 63: 2372 (1993).
14. A. Chomette, et al., *Europhys. Lett.* 4: 461 (1987).
15. G. v. Plessen, et al. (unpublished).
16. T. Meier, et al., *Phys. Rev. Lett..* 73: 902 (1994).
17. M. Wegener, et al., *Phys. Rev. A* 42: 5675 (1990).
18. T. Yajima and Y. Taira, *J. Phys. Soc. Jpn.* 47:1620 (1979).
19. D.A.B. Miller, et al., *Phys. Rev. B* 32:1043 (1985).

# Dephasing of Excitons in Multiple Quantum Well Bragg Structures

M. Hübner[1], E. J. Mayer[1], N. Pelekanos[1], J. Kuhl[1], T. Stroucken[2], A. Knorr[2], P.Thomas[2], S.W. Koch[2], R. Hey[3], K. Ploog[3], Y. Merle D'Aubigne[4], A. Wasiela[4], and H. Mariette[4]

[1]Max-Planck-Institut für Festkörperforschung , D-70569 Stuttgart, Germany
[2]Fachbereich Physik und Zentrum für Materialwissenschaften, Philipps University, D-35032 Marburg, Germany
[3]Paul Drude Institut für Festkörperelektronik, D-10177 Berlin, Germany
[4]Laboratoire de Spectrometrie Physique, University J . Fourier et CNRS, Grenoble, France

The recent progress in the development of ultrashort laser pulses enables observations of the coherent dynamics of excitons in quantum wells (QW) with less than 100 fs time resolution by transient degenerate-four-wave-mixing (DFWM). Such investigations have attracted rapidly growing attention as a powerful technique to explore nonlinear processes in semiconductors which may have enormous potential for future semiconductor device applications. Only very recently the role of many-body exciton-exciton interaction effects for the magnitude of the nonlinear optical response as well as for the optical dephasing rate has been studied in detail. So far, the discussion has been confined, however, to the coupling of excitons via Coulomb fields[1]. For GaAs multiple QW (MQW) samples, this interaction mechanism seems to be negligible for excitons excited in different wells if the barrier thickness exceeds values of 10-15 nm so that tunneling processes of carriers between the different wells can be excluded.

Here we present dephasing measurements of excitons in special MQW structures with a varying number of equidistantly separated QWs grown by MBE on semi-insulating substrates. The number N of the QWs was varied between 1 and 10 and the interwell distance d was equal to either $\lambda/2$ or $\lambda/4$, where $\lambda$ is the wavelength of the exciton resonance in the material.

Previous linear reflectivity studies on  CdTe samples containing 10 QWs with a period $P=\lambda/2$ or $\lambda/4$ showed huge differences of the amplitude reflectivity at the exciton transition for $P=\lambda/2$ or $\lambda/4$ which can be explained by interference due to multiple reflections. Moreover, the photon confinement provided by such resonant Bragg ($\lambda/2$) and anti-Bragg ($\lambda/4$) structures is expected to change the exciton-photon coupling and thus the radiative contribution to exciton dephasing[2,3]. The much broader linewidth of the reflectivity spectrum observed for $P= \lambda/2$ as compared to $P=\lambda/4$ has been attributed to strong radiative broadening of the excitonic state following Ivchenko's prediction[3] that in the case of the

Bragg condition a set of N QW's behaves as a single QW with a radiative coupling equal to N times the coupling of a single QW. This enhanced radiative coupling should be potentially observed in time-domain dephasing measurements provided that the enhancement is sufficiently large to make the radiative contribution to dephasing comparable to competing scattering rates due to interface roughness, defects or interactions with acoustic phonons. In very low excitation (incident laser intensity a few $Wcm^{-2}$ ) DFWM experiments on GaAs MQWs, however, we observed no significant differences in the dephasing rates between Bragg ($\lambda/2$) and anti-Bragg ($\lambda/4$) structures, contrary to expectations based on the cw reflectivity measurements of ref.2. On the other hand our finding is in good agreement with existing data for the dephasing time and the radiative lifetime of excitons in GaAs QWs. The dephasing time associated with nonradiative scattering processes amounts to several picoseconds, whereas the radiative contribution to the dephasing time $T_2^{rad}$ is approximately 40-50 ps. Thus a substantial shortening of the radiative lifetime is required before the radiative contribution to dephasing becomes detectable.

The transmission, reflection, absorption or diffraction of resonant light in such periodic structures are strongly modified as compared to the optical properties of bulk samples by two effects  to : (i) interference phenomena resulting from the superposition of the phase-coupled polarisation contributions of the individual wells and (ii) coupling of excitons excited in different QWs via the resonant electromagnetic field transmitted through the wells and partially reflected at their surfaces. Since both, the interference and the radiative coupling are sensitive to the QW period, pronounced variations of the signal amplitude and shape for the two types of samples as well as for the forward and backward diffraction configuration  are expected. Employing a tranfer matrix technique, we have calculated the transmitted ($I_t$), reflected ($I_r$) and absorbed intensity ($I_a= I_0 - I_t- I_r$ ; $I_0$= incident intensity ) in dependence on N and d. These calculations yield the following surprising results:(i) the intensity absorbed in the sample cannot be simply related to the absorption coefficient usually defined as $\alpha= -1/L \ln ( I_t/ I_0 )$, where L is the absorbing sample length. (ii) the decay rates of the DFWM signals diffracted in the forward and backward direction for the $\lambda/4$ structure are predicted to be significantly different whereas they should be the same for the sample with the $\lambda/2$ period.

We have investigated two different sets of QWs: (i) 20-nm-GaAs QW's cladded by $Al_xGa_{1-x}As$ ( x =0.3) barriers and 10-nm-CdTe QWs embedded between thick $Cd_{1-x}Zn_xTe$ (x=0.13) barriers. The samples were kept at 8 K in a Helium cryostat and excited by two pulses of  700 fs duration from a Kerr-lens mode-locked Ti: Sapphire laser which were focused to a spot size of 70 μm diameter. The cocircularly polarized pulses have equal intensity and wavevectors $k_1$, $k_2$. The DFWM-signal is monitored either in the forward or backward diffraction geometry in the direction $2k_2- k_1$ as a function of the delay between the two excitation pulses.

Figure 1 compares DFWM signals measured in the backward diffraction geometry on GaAs MQW structures with N=10 for a modest excitation intensity of approximately  20 $Wcm^{-2}$. In accordance with destructive and constructive interference occurring in the $\lambda/4$ and $\lambda/2$ configuration, respectively, the signals differ by almost one order of magnitude. Our theoretical analysis indicates that the optical resonances in a $\lambda/4$ structure are split as a consequence of the light-induced coupling. This splitting is the origin of the strong modulations detected for the $\lambda/4$ structure.Their presence prevents an accurate comparison of the two decay rates but the data reveal a remarkably faster dephasing in the $\lambda/2$ sample.

Signals detected in the forward and backward diffraction for a $\lambda/4$ structure consisting of 10 CdTe QWs deposited on a transparent substrate reveal significant differences of the  decay times. The decay rate of the signal for the $\lambda/4$ sample in the forward direction is about 15 % larger at low excitation intensities, but this difference increases to almost a factor of 2 at intensities of 50 $Wcm^{-2}$. In contrast,  the two rates are the same within the experimental error of < 10 % for the $\lambda/2$ sample for all densities in this regime.

8

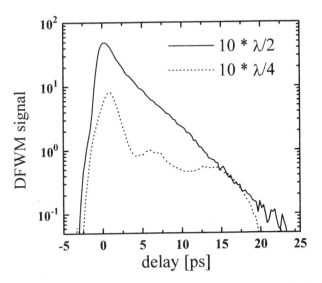

Fig.1: Two-pulse DFWM signal on the hh exciton transition of 20-nm-GaAs MQW Bragg
and anti-Bragg structures with N=10 measured in the backward diffraction geometry.
Excitation intensity 20 Wcm$^{-2}$.

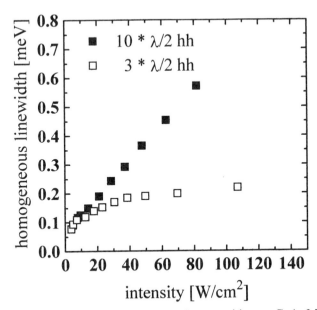

Fig.2: Homogeneous linewidth of the hh-exciton transition on GaAs MQW Bragg
structures with N=3 and 10 as a function of excitation intensity.

Using the backward diffraction geometry, we have analyzed the time-integrated DFWM signal for the hh exciton for 20-nm-GaAs QWs with N=1, 3 and 10 in dependence on the excitation density. The FWHM of the PL line of our samples increases from 300 μeV to approximately 400 μeV if N is increased from 1 to 10. This slight increase of the linewidth demonstrates that inhomogeneous line broadening originating from width fluctuations of the different wells in the MQW structures is very small. It should be noticed that the DFWM signal intensity scales as $N^2$ for both the backward and forward diffraction geometry in a Bragg structure contrary to DFWM on thicker QW's or bulk samples[5]. In the latter the forward signal is proportional to $L^2$ whereas its backward counterpart oscillates with L and reaches its first maximum for L=20-25 nm. Figure 2 depicts homogeneous linewidth data measured on $\lambda/2$ samples with N = 3 and 10 as a function of excitation density n. At low intensities ( < 10 Wcm$^{-2}$ ), the dephasing time is 10 ps independent on N, proving that the radiative contribution to dephasing of excitons in our GaAs QWs Bragg structure is negligible. The absence of effective radiative broadening at low excitation derived from the time domain spectra is also confirmed by the spectral linewidth of reflectivity spectra.

At higher densities n, we found a much faster rise of the dephasing rate with n for samples with the larger number N of quantum wells. This observation reflects the rapidly increasing superradiant coupling between excitons located in the different wells at higher densities. This interpretation is confirmed by experiments performed on the lh exciton. The latter show no significant variation of the dephasing time as a function of N even at excitation intensities as high as 100 Wcm$^{-2}$ in accordance with the lower oscillator strength of the lh exciton transition and the slight detuning of the Bragg resonance. A rise of the density dependent contribution to $T_2$ with N has been observed in the presence of large populations of either coherent or incoherent excitons in the wells. However, the effect seems to be remarkably stronger for coherent excitons. Incoherent excitons were created within the excitation volume by a third stronger pump pulse incident onto the sample 20 ps in advance of the first pulse of the self diffraction experiment. This timing ensures that the additional excitons have completely lost their phase coherence and that excitation density dependent coherent coupling of excitons in the different QWs is suppressed.

In summary, we have observed remarkable variations of the signal amplitude and shape for TI-DFWM performed on GaAs and CdTe MQW Bragg and anti-Bragg structures in dependence on the period of the structure, the number of QWs and the excitation density. These findings are attributed to interference effects and excitation density dependent interwell coupling of excitons via confined resonant electromagnetic fields. The interwell coupling is negligible for lh-excitons up to intensities of 100 Wcm$^{-2}$ in accordance with their small oscillator strength. For hh excitons it is remarkably stronger for phase coherent as compared to incoherent excitons. Different decay times observed for hh excitons on a N= 10 CdTe QWs anti-Bragg structure in the forward and backward direction are in qualitative agreement with theoretical model calculations.

References:

1.  see e.g.: K. Bott et al., Phys Rev. **B 48**, 17418 (1993); E.J. Mayer et al., Phys. Rev. **B 50,** 14733 (1994) , E. J. Mayer et al.,  Phys. Rev **B 51**, 10909 (1995)
2.  Y. Merle d'Aubigne, A Wasiela , H. Mariette and A Shen, 22nd Internat. Conference on "The Physics of Semiconductors", Vol.2 pg. 1201, ed. D.J. Lockwood, World Scientific Singapore. 1995
3.  E.L. Ivchenko and A.I. Nesvizhskii, and S. Jorda, Phys. Sol. State **36,**1156 (1994) and Superlattices and Microstructures **16**, 17 (1994)
4.  T. Stroucken et al., Phys. Rev. Lett. **74**, 2391 (1995)
5.  A. Honold, L. Schultheis, J. Kuhl, and C.W. Tu , Appl. Phys. Lett. **52**, 2105 (1988)

# TEMPERATURE DEPENDENCE OF PHOTOLUMINESCENCE IN InGaAsP/InP STRAINED MQW HETEROSTRUCTURES

O. Y. Raisky, W. B. Wang and R. R. Alfano

New York State Center of Advanced Technology for Ultrafast Photonic Materials and Applications, Physics Department, The City College of New York, New York, NY 10031

C. L. Reynolds, Jr. and V. Swaminathan

AT&T Bell Laboratories, 9999 Hamilton Blvd., Breinigsville, PA 18031

Multiple quantum well (MQW) InGaAsP/InP heterostructure systems have been drawn considerable research interest in recent years due to its suitability for long wavelength opto-electronic devices[1]. The performance of such devices is strongly affected by peculiarities of recombination processes in the quantum wells[2] (QW).

The goal of this study was to investigate the effect of barrier width on the radiative recombination of carriers. In our study, the photoluminescence spectra from InGaAsP/InP MQW double heterostructures have been measured in the 77-290 K temperature range with different excitation intensities.

A set of samples with fixed well width and various barrier thickness were grown on $n^+$-type InP substrates in a low-pressure metalorganic chemical vapor deposition (MOCVD) reactor. The undoped active MQW region consisting of 9x70 A strained wells of InGaAsP ( band gap value $\lambda \cong 1.35$ $\mu$) and 8x80 A, 100 A and 150 A lattice-matched barriers of InGaAsP ($\lambda \cong 1.12$ $\mu$) (samples #382, #380 and #381 correspondingly) was confined by 500 A undoped InGaAsP ($\lambda \cong 1.12$ $\mu$) separate confinement layers and embedded between $n$- and $p$-cladding InP layers. A mode-locked $Nd^{3+}$:YAG laser with second harmonic generation system (1064 nm and 532 nm) was used as an excitation source in all our measurements and PL spectra were recorded on 0.25 m SPEX spectrometer using standard lock-in technique.

The photo-generated carriers in MQW region can effectively screen the built-in field and thus cause change in the measured spectra. To account for this effect, a series of PL spectra at different excitation intensities were measured. Within 2.5 decades of pump intensities we did not observe any significant line shape or peak position changes, indicating that excited carrier densities were relatively low.

PL spectra of the samples at 77 K are shown in Fig. 1 ( $\lambda_{exc}$ = 532 nm). The peaks at 1.23 $\mu$ are assigned to radiative recombination of carriers in the QW, while the 1.07 $\mu$ peaks are due to the emission from barrier layers and confined layers.

Fig. 2 shows the temperature dependence of the integrated peak intensities of the QW. We found that the decrease rate of the QW PL intensity varied with the sample structure, being the highest for the sample with 80 A barriers and the lowest for 150 A, which may be caused by some barrier-dependent non-radiative recombination mechanism. For comparison, we evaluated carrier tunneling and thermionic lifetimes for each sample, based on transfer matrix calculations and thermionic emission theory[3] ( see Table 1, material parameters from Adachi[4]). The estimated average electric field was 35-45 kV/cm. These calculations predicted that 100 A barrier sample PL data should be placed closer to that of 80 A sample, rather than 150 A sample, which seemed not agree with our measured results. Therefore a model including non-radiative recombination processes and possible space charge effects is required to more adequately describe the experimental data.

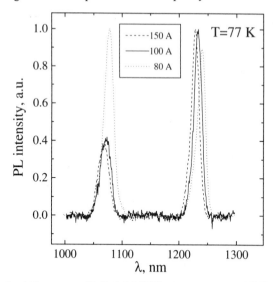

**Figure 1.** Normalized PL spectra of InGaAsP MQW *p-i-n* structures at 77 K: solid line - barrier width = 100 A, dotted line - 80 A, dashed line - 150 A. Excitation wavelength 532 nm, 30 W/cm².

**Table 1.** Calculated tunneling and thermionic escape times. Numbers in ( ) give the value at 77 K.

| sample | barrier width, A | tunneling time | | | thermionic time, T=290 K(77 K) | | |
|--------|------------------|----------------|------------|------------|----------|------------|------------|
| | | electron | light hole | heavy hole | electron | light hole | heavy hole |
| #381 | 150 | 0.6 ps | 17 ps | 6.6 sec | 0.2 ps (8 ps) | 0.78 ps (820 ps) | 8 ps (0.4 μs) |
| #380 | 100 | 0.12 ps | 2 ps | 0.2 msec | 0.17 ps (6 ps) | 0.6 ps (560 ps) | 7 ps (0.2 μs) |
| #382 | 80 | 0.06 ps | 1 ps | 4 μs | 0.16 ps (5 ps) | 0.6 ps (480 ps) | 6.7 ps (9 μs) |

We found that the PL intensity is proportional to excitation intensity for 0.3 - 45 W/cm² range, without evidence of saturation. It may indicate the fast escape of the carriers from the wells.

We compared our PL spectra with absorption measurements and found that the QW PL peak positions of all samples were shifted by 8-10 meV up on energy scale in respect to the corresponding excitonic absorption peaks. This suggests band-to-band transitions as a dominant radiative recombination process. In addition, we found that QW PL peak position shows essentially the same temperature dependence as that of band gap of bulk InGaAsP alloy. This is similar to the results reported[5] for InGaAs/InP MQW structures.

In conclusion, the temperature and intensity dependencies of PL were investigated in InGaAsP/InP MQW double heterostructures with different barrier thickness. The radiative efficiency at 293 K was found to be smaller than that at 77K by a factor of 60-95 % , depending on the barrier width. The decrease rate of the QW PL intensity for 80 A barrier sample was observed to be higher than that for 100 A and 150 A ones.

This research at CUNY is supported by NASA and the New York State Technology Foundation.

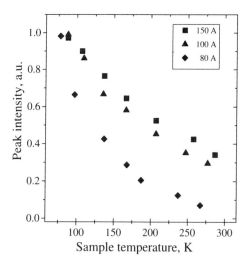

**Figure 2**. - integrated QW PL peak intensities as a function of the sample temperature. Squares - barrier width = 150 A , triangles - 100 A, diamonds - 80 A.

# REFERENCES

1. B.W. Takasaki, J.S. Preston, J.D. Evans, and J.G.Simmons, *Can. J. Phys.* **70**,1017 (1992)
2. A.M. Fox, D.A.B. Miller, G. Livesku, J.E.Cunningham, and W.Y.Jan, *IEEE J. Quantum Electron* **27**(10), 2281(1991)
3. H. Schneider and K. V. Klitzing, *Phys. Rev. B* **38**(9), 6160 (1988)
4. S. Adachi, *J.Appl.Phys* **53**(12), 8775(1982)
5. H. Temkin, D. Gershoni, and M.B. Panish, Optical properties of GaInAs/InP quantum wells, *in:* "Epitaxial Microstructures," A.C. Gossard, ed., Academic Press, New York, (1994)

# SPIN-FLIP DYNAMICS OF EXCITONS IN GaAs QUANTUM WELLS

A. Lohner,[1] D. Snoke,[2] W.W. Rühle[1], and K. Köhler[3]

[1]Max-Planck-Institut für Festkörperforschung
Heisenbergstraße 1
70569 Stuttgart, Germany

[2]University of Pittsburgh, Department of Physics and Astrophysics
3941 O'Hara St.
Pittsburgh PA, USA

[3]Fraunhofer-Institut für Angewandte Festkörperphysik
Tullastr.72
79108 Freiburg, Germany

## INTRODUCTION

Several recent studies [1-3] have attempted to determine how excitons in quantum wells can flip their spin. In GaAs quantum wells, excitons in the ground state can have either total spin J=1 or J=2, since they are comprised of a spin-$\frac{1}{2}$ conduction electron and a spin-$\frac{3}{2}$ hole. While a number of studies have examined the flip from +1 to -1 spin states, little experimental work has directly addressed the processes by which J=2, or "dark," excitons, can convert to J=1 excitons. Since J=2 excitons do not couple to a single photon, in most experiments they can not be observed, and therefore unverifiable assumptions are made about their behavior.

In this paper we report results from a direct method for measuring the conversion rate of J=2 to J=1 excitons. The experiment is basically as follows: first, J=2 excitons are created via two-photon (infrared) excitation. Following the generation of the excitons, the single-photon-recombination luminescence (visible or near-infrared) from the J=1 excitons is detected with a streak camera with an S20 photocathode, which is completely insensitive in the infrared. Since the streak camera does not respond to the exciting laser light, the J=2 excitons can be created by resonant excitation and observed immediately thereafter, without unwanted background from the laser light.

This experiment relies on the fact that just as single-photon emission from J=2 states is forbidden, so two-photon absorption by J=1 excitons is forbidden and two-photon absorption by J=2 excitons is allowed. Previous studies [4] have shown that two-photon absorption into the 1s heavy-hole exciton state in GaAs quantum wells is comparable to two-photon absorption into the 2p state. Therefore, in these experiments

*Hot Carriers in Semiconductors*
Edited by K. Hess *et al.*, Plenum Press, New York, 1996

excitons can be created directly in the J=2 ground state of the quantum well.

## EXPERIMENT

An Optical Parametric Oscillator (OPO) sychronously pumped by a Ti:Sapphire laser was used to generate 100 fs infrared pulses, which could be tuned in the wavelength range 1400-1550 nm to create excitons in a GaAs quantum well sample via two-photon absorption. RG1000 filters in the exciting laser beam path ensured that no light from the pump laser reached the sample. Single-photon-recombination luminescence from the sample was detected in the wavelength range 700-800 nm by a Hamamatsu streak camera with time resolution of about 10 ps. Several samples with Al.3Ga.7As barriers and varying GaAs well width were examined. For a 3 nm well width, heavy-hole excitons in the 1s state are generated by laser light at approximately 1460 nm.

Three pitfalls must be avoided in these experiments. First, since the 100 fs exciting laser has a full width at half maximum (FWHM) of 20 meV, when the laser is tuned to the 1s resonance, it is also possible to generate excitons in the 2p states, which can then drop down into 1s exciton states and give a rise time of J=1 luminescence unrelated to the J=2/J=1 conversion time. To avoid this problem, the exciting laser was tuned to 10 meV *below* the 1s resonance. Since the number of 2p excitons depends on the *square* of the laser intensity resonant with the 2p state, the contribution due to down conversion from 2p excitons can be made negligible.

Second, at high powers it is also possible to generate highly excited excitons via *three*-photon excitation from the substrate or from wider quantum wells in a multiple-quantum-well structure (two photons create an exciton, which is then excited over a barrier into a higher-lying quantum-well states.) This was checked in these experiments via observation of the 1s luminescence when the exciting laser photon energy was *well below* the ground state; i.e. when no excitons could be created directly by two-photon excitation at all. In this case, weak but measureable luminescence from the 1s excitons still occurred, with total intensity proportional to the laser power to the third power. This could only come about due to relaxation of excitons excited over the barrier layers by absorption of a photon. Using low laser power substantially reduces this effect, but in general, some contribution from this effect always occurs. Since this small contribution does not depend strongly on laser wavelength, the three-photon signal, taken at laser photon energies well below resonance, can simply be subtracted from the signal when the laser is near resonance.

Third, since only excitons with energy less than the homogenous linewidth can recombine, at high lattice temperature excitons will be excited into higher-energy, non-radiating states, substantially complicating the analysis of the conversion times. To prevent this, the sample was held at 2 K via immersion in a liquid helium bath.

## RESULTS

Figure 1 shows luminescence at 730 nm, as a function of time, from a 3 nm quantum well at 2K, excited by circularly polarized OPO light at 1471 nm (i.e., about 10 meV below the 1s heavy-hole resonance.) The data is reasonably well fit by the simple model

$$\frac{\partial n_1}{\partial t} = -\frac{n_1}{\tau_1} + \frac{n_2}{\tau_1} - \frac{n_1}{\tau_2} \tag{1}$$

$$\frac{\partial n_2}{\partial t} = -\frac{n_2}{\tau_1} + \frac{n_1}{\tau_1} - \frac{n_2}{\tau_2} \tag{2}$$

where $n_1$ is the population of J=1 excitons, and $n_2$ is the population of J=2 excitons. Since almost no excitons are created in higher states in this experiment, these should be the only relevant populations. A single coupling constant, $\tau_1$, is assumed between the two populations, and both populations are assumed to decay with the same lifetime, $\tau_2$. Solution of these equations gives $n(t) \propto e^{-t/\tau_2}(1 - e^{-2t/\tau_1})$; i.e., a rise time of $\tau_1/2$. For the fit shown in Fig. 1, the values $\tau_1 = 104$ ps and $\tau_2 = 103$ ps are obtained. The long lifetime measured here is assumed to arise from the effect of localization on radiative recombination [5]. This will only be seen at very low lattice temperature, e.g. 2 K as in these experiments, because at higher temperatures thermalization to higher, nonradiative states will give an initially fast decay of the luminescence.

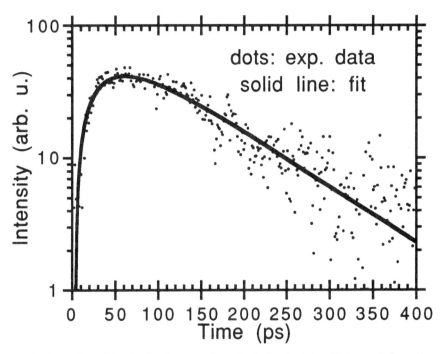

**Figure 1.** Dots: Recombination luminescence from the J=1 spin state of 1s heavy-hole excitons in a 3 nm GaAs quantum well, following generation in the J=2 spin state by a short (100 fs) laser pulse at 1471 nm, slightly below the resonance for two-photon generation. Solid line: a fit of the rate equations discussed in the text.

Since the excitons are created by circularly polarized light, there could, in general, be two different time scales for the coupling to $m_J = +1$ and $m_J = -1$ excitons. Because the mechanism for the conversion process is presumed to involve flip of the spin of either the electron or the hole, two different matrix elements can be written for conversion to different J=1 states [6]. In these experiments, however, no appreciable difference in the data is seen when the visible J=1 luminescence is analyzed for linear polarization or for circular polarization of either handedness.

Similar experiments on quantum wells with different widths show that the hole spin flip time decreases with increasing well widths. Finally, a fast initial rise of luminescence, faster than our time resolution, is observed if the laser two-photon energy is tuned in resonance with the 1s-state. We tentatively attribute this fast transient to the fast transformation of 2p into 1s excitons, since the 2p excitons are also excited in this case due to the spectrally broad laser pulses. A fast transition from 2p into 1s excitons is possible without spin-flip.

## CONCLUSIONS

This experiment seems to provide the first direct measurement of the rate of conversion of J=2 excitons to J=1 excitons. The time scale for the spin-flip process is of the order of 50 ps. The conversion rate seems to be the same for transitions from the $m_J = +2$ spin state to either the $m_J = +1$ or the $m_J = -1$ spin states.

## REFERENCES

1. A. Vinattieri et al., Phys. Rev. B **50**, 10868 (1994).
2. A. Frommer et al., Phys. Rev. B **50**, 11833 (1994).
3. E. Blackwood et al., Phys. Rev. B **50**, 14246 (1994).
4. M. Nithisoontorn et al., Phys. Rev. Lett. **62**, 3078 (1989).
5. D.S. Citrin, Phys. Rev. B **47**, 3832 (1993).
6. M.Z. Maialle et al., Phys. Rev. B **47**, 15776 (1993).

# DYNAMICS OF EXCITON FORMATION AND RELAXATION IN SEMICONDUCTORS

P.E. Selbmann,[1] M. Gulia,[1] F. Rossi,[2]
E. Molinari,[1] and P. Lugli[3]

[1]Istituto Nazionale di Fisica della Materia (INFM) and
Università di Modena, 41100 Modena, Italy
[2]Fachbereich Physik und Zentrum für Materialwissenschaften
Philipps-Universität Marburg, 35032 Marburg, Germany
[3]II Università di Roma 'Tor Vergata', 00173 Roma, Italy

## INTRODUCTION

Optical excitation above the band gap of a semiconductor generates hot carrier distributions which subsequently relax towards an equilibrium with the crystal lattice. Simultaneously, electron-hole (e-h) pairs may undergo assisted transitions to excitonic bound states. These free excitons are created with large center-of-mass wave vectors, $K$: to become optically active they have to be scattered into states with $K \approx 0$ and $s$-symmetry by inelastic collisions. The time-resolved measurement of the resulting luminescence[1] provides information about these intrinsic processes of exciton formation and relaxation. However, the interpretation of the experiments is difficult due to the various competing interactions of free carriers and excitons and a kinetic model of the system is needed.

In this paper, we present a kinetic description of the incoherent relaxation processes in semiconductors after ultrashort photoexcitation including exciton formation and dissociation. Our approach is based on an extension of the system of Boltzmann equations for the free carrier and exciton distributions by scattering integrals for phonon-assisted exciton reactions. It is valid for low enough particle densities where exciton screening and band gap renormalization can be neglected. The kinetic equations are solved by means of a new Ensemble Monte Carlo (EMC) method. Numerical results for bulk GaAs demonstrate the dominant role of LO-phonons for the creation of excitons in the initial stage after excitation with short $(500 fs)$ laser pulses. It is shown that the resonant structure of the transition probability for this reaction results in an oscillatory dependence of the density of created excitons and the luminescence rise time on the energy of the exciting laser, similar to that found for narrow quantum wells[2].

*Hot Carriers in Semiconductors*
Edited by K. Hess *et al.*, Plenum Press, New York, 1996

# KINETIC MODEL AND MONTE CARLO SOLUTION

The incoherent relaxation processes following ultrafast photoexcitation of semiconductors are usually well described by coupled semiclassical Boltzmann equations for electrons and holes. The interaction of the particles with lattice imperfections (phonons, impurities, etc.) and among each other are accounted for by suitable nonlinear collision integrals: the kinetic equations are Markovian Master equations in k-space. Free excitons may be treated in a similar way. To include, however, excitonic binding and dissociation one has to complement the Boltzmann equations with corresponding scattering terms coupling the electron $(f_e)$, hole $(f_h)$ and exciton $(f_X)$ populations, respectively. For simplicity we will consider 1s excitons only. To lowest order in density, the formation of excitons is assumed to be an assisted bimolecular mechanism in which a free electron-hole pair interacts with a phonon of wave vector, $\mathbf{q}$, and the final state is that of a free exciton with center-of-mass wave vector, $\mathbf{K}$. Denoting the microscopic probability for such a transition with $W^B$ and that for the reverse event of exciton dissociation with $W^D$, we may set up the scattering integrals for exciton reactions as:

$$\frac{\partial}{\partial t} f_X(\mathbf{K})\,|_{Reaction} = \sum_{\mathbf{k}_i,\mathbf{k}_j,\mathbf{q}} \{W_{\mathbf{q}}^B(\mathbf{k}_i,\mathbf{k}_j\mid\mathbf{K})f_i(\mathbf{k}_i)f_j(\mathbf{k}_j) - W_{\mathbf{q}}^D(\mathbf{K}\mid\mathbf{k}_i,\mathbf{k}_j)f_X(\mathbf{K})\},$$

$$\frac{\partial}{\partial t} f_i(\mathbf{k}_i)\,|_{Reaction} = \sum_{\mathbf{k}_j,\mathbf{K},\mathbf{q}} \{W_{\mathbf{q}}^D(\mathbf{K}\mid\mathbf{k}_i,\mathbf{k}_j)f_X(\mathbf{K}) - W_{\mathbf{q}}^B(\mathbf{k}_i,\mathbf{k}_j\mid\mathbf{K})f_i(\mathbf{k}_i)f_j(\mathbf{k}_j)\},$$

$$(i,j = e,h; i \neq j). \tag{1}$$

Obviously, this *ansatz* ensures conservation of the number of e-h pairs. The binding contributions are independent from $f_X$ since excitons can be treated as bosons for low densities; in the same sense we can omit the Pauli factors in the dissociation terms. The probability functions $W^{B/D}$ are calculated from the exciton-phonon interaction Hamiltonian[3] using Fermi's Golden rule. Accordingly, the sum of the kinetic energies of the free e-h-pair, the exciton and the phonon is conserved. The same holds for the center-of-mass momenta, but not for the relative momentum of the e-h pair since the transition involves a bound state of the relative motion.

The kinetic equations with the reaction terms (1) are solved by an extension of the EMC technique of Ref. 4. The k-space is subdivided into cells and stochastic particle trajectories are generated on this grid using the known microscopic transition probabilities. A detailed description of the simulation of the bimolecular exciton reactions (which resembles in some sense that of carrier-carrier scattering) will be given elsewhere[5]. The optical generation of free e-h pairs is included in terms of a semiclassical generation rate[4].

# NUMERICAL RESULTS AND DISCUSSION

Simulations have been performed for a simplified bulk GaAs model[4] with spherical parabolic bands for electrons $(m_e = 0.063m_0)$ and holes $(m_h = 0.450m_0)$, respectively, and a gap energy of $E_g = 1.519eV$ (at $T = 0$). Photoexcitation takes place by Gaussian laser pulses of $500fs$ and different intensities adjusted to the desired final pair densities. The laser energy, $E_{las}$, has been varied between $1.54eV$ and $1.64eV$.

The excess energy of the incident photon is shared such that the initial electron distribution is centered around $(E_{las} - E_g)/(1 + m_e/m_h)$. If allowed, emission of LO-phonons is the most efficient scattering process and leads, within few $ps$, to the appearance of satellite peaks in the distribution functions at integer multiples of the

LO-phonon energy, $\hbar\omega = 36.4 meV$, below the initial peak. These sharp features are broadened by carrier-carrier scattering. Deformation potential scattering by acoustic phonons provides a much slower relaxation mechanism for low energies. The initial correlation of the photocreated e-h pairs is rapidly destroyed by the various collisions.

Fig. 1 displays the transition probability for exciton binding under LO-phonon emission, $W_{LO}^{B}$, integrated over all possible phonon and exciton wave vectors. The pronounced structure of this function results from the conservation laws and, in particular, from the exciton form factor[3]. Scattering into bound states is favoured for electrons just above the threshold given by the LO-phonon energy minus the exciton binding energy of $4.5 meV$. Since the holes are always generated well below this threshold, the high probabilities for transitions with large initial hole energies are of no practical importance. For the same reason the holes are not subject to LO-phonon scattering and the slow variation of their distribution has little effect on the exciton formation.

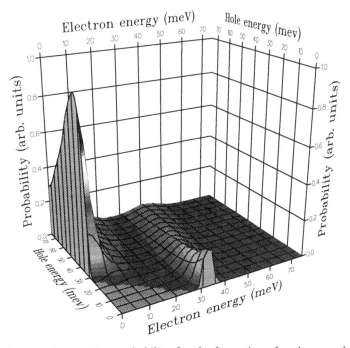

**Figure 1.** Integrated transition probability for the formation of excitons under emission of one LO-phonon as a function of the particle energies.

e-h pairing turns out to be an effective process with rates up to $\approx 10 ps^{-1}$, whenever the electron distribution peaks in the region of high exciton formation probability which can be achieved by tuning the laser energy. In fig. 2a) we show the temporal build-up of the exciton population for different $E_{las}$. For curve 1) the electron excess energy is in resonance with the maximum of the transition probability, for curves 3) and 5) it is one and two LO-phonon energies above. Following a fast increase, the density saturates for times $> 100 ps$. The final densities for larger $E_{las}$ are somewhat smaller since the distribution function have broadened during the relaxation. For off-resonant excitation (curves 2) and 4)) the exciton creation is much slower and leads to smaller densities. It should be noted that our model includes acoustic phonon mediated reactions, too. These processes are slower than those under participation of LO-phonons by at least one order of magnitude and act at low energies only.

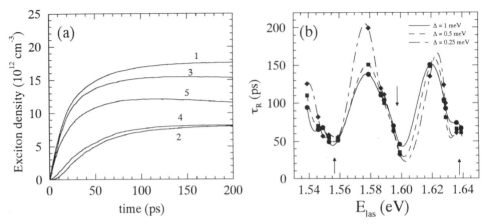

**Figure 2.** (a) Time dependence of the exciton density for different excitation energies: 1.553 (1), 1.587 (2), 1.595 (3), 1.628 (4), and $1.636eV$(5). (b) Fitted rise time of luminescence in dependence on $E_{las}$ for three different linewidths, $\Delta$. Arrows indicate the resonance energies. The total density of generated e-h pairs is $10^{14}cm^{-3}$ and the temperature $T = 4.2K$.

To contribute to the luminescence, the initial exciton distributions have to relax by deformation potential scattering to energies within the homogeneous linewidth, $\Delta$, around the $K = 0$ state. Unlike carrier-carrier scattering, the present calculations do not contain exciton-exciton and exciton-free carrier interactions.

We have fitted the time dependence of the fraction of 'radiative' excitons for different assumed $\Delta$ to an exponential with characteristic time $\tau_R$, which is interpreted as the rise time of luminescence. As shown in fig. 2b), $\tau_R$ exhibits an oscillatory dependence on excitation energy with period $\hbar\omega(1 + m_e/m_h)$. The origin of this behaviour are the sharp LO-phonon related features in the hot electron distribution, $f_e$, and the sensitivity of the exciton formation to the initial electron energy. The argument of 'sensitive spots' has been used recently for the explanation of experimentally observed rise time oscillations in thin quantum wells[2]. It has to be expected that any additional broadening of the peaks in $f_e$ will tend to reduce the effect. Indeed, our simulations reveal that it is weakened for shorter laser pulses.

## ACKNOWLEDGMENT

This work has been supported in part by the EU network 'Ultrafast'.

## REFERENCES

1. T.C. Damen, J. Shah, D.Y. Oberli, D.S. Chemla, J.E. Cunningham, and J.M. Kuo, Dynamics of exciton formation and relaxation in GaAs quantum wells, Phys. Rev. B42: 7434 (1990).
2. P.W.M. Blom, P.J. van Hall, C. Smit, J.P. Cuypers, and J.H. Wolter, Selective exciton formation in thin $GaAs/Al_xGa_{1-x}As$ quantum wells, Phys. Rev. Lett. 71: 3878 (1993).
3. For a review see: J. Singh, The dynamics of excitons, Solid State Phys. 38: 295 (1984).
4. T. Kuhn and F. Rossi, Monte Carlo simulation of ultrafast processes in photoexcited semiconductors: Coherent and incoherent dynamics, Phys. Rev. B46: 7496 (1992).
5. P.E. Selbmann, M. Gulia, F. Rossi, E. Molinari, and P. Lugli, to be published.

# ULTRAFAST OPTICAL ABSORPTION MEASUREMENTS OF ELECTRON-PHONON SCATTERING IN GaAs QUANTUM WELLS

K. Turner, L. Rota, R.A. Taylor, and J.F. Ryan

Clarendon Laboratory, Department of Physics, Oxford University, Parks Road, Oxford OX1 3PU, U.K.

## INTRODUCTION

A knowlege of intra- and intersubband electron-phonon scattering rates in quantum wells is central to the understanding and manipulation of hot carrier dynamics in nanostructures. A significant amount of both experimental and theoretical work has been reported on this issue, but variations over two orders of magnitude exist in the measured scattering rates. Our purpose here is to provide a joint experimental and theoretical investigation which avoids many of the problems and clarifies many of the inconsistencies encountered by previous reports.

Time-resolved optical absorption in GaAs/AlAs quantum wells has been measured at low carrier density and the experiments have been analysed using a multi-subband Monte Carlo simulation which contains all the important scattering mechanisms. Electron-phonon scattering rates obtained from dielectric continuum theory have been used, and in contrast to some earlier reports, we find excellent agreement with the experiment. The dominance of interface phonons in intrasubband relaxation is confirmed.

The dynamics of spatially-confined, nonequilibrium electrons in semiconductor nanostructures depends strongly on their interaction with optical phonons. This interaction involves predominantly those phonons which are confined to the same region of the structure, except that as the size is reduced below $\sim$ 10nm the role of interface (IF) phonons becomes important: long wavelength IF modes produce long range potentials which can induce both inter- and intrasubband electron scattering. Dielectric continuum theory predicts scattering times which are typically <1ps[1-3] but experimental measurements between 160fs and 14ps have been published[4-9]. Many experiments have been performed at high density ($> 10^{11}$cm$^{-2}$) and high excitation energy resulting in carrier-carrier (CC) scattering, nonequilibrium phonons and degeneracy effects dominating the signal, preventing accurate determination of the electron-phonon scattering rates[10]. The complexity of the initial photoexcited state in such experiments, which may involve several populated valence and conduction subbands, invariably means that quantitative interpretation of the experiment is not straightforward, and consequently a full numerical simulation is required in order to obtain an accurate quantitative analysis of the data[11-13]. Here we report the lowest density measurements to date ($9 \times 10^9$cm$^{-2}$). Considering carefully where the carriers are pumped within the energy bands and all the interactions they can undergo we obtain results in very good agreement with the dielectric continuum theory.

*Hot Carriers in Semiconductors*
Edited by K. Hess *et al.*, Plenum Press, New York, 1996

## EXPERIMENTAL RESULTS

The technique we employed was that of time-resolved optical absorption using the conventional pump-probe configuration. A tunable modelocked Ti:sapphire laser producing 125 fs pulses was used to excite carriers preferentially into different subbands of the quantum well. A time-delayed probe pulse, derived from the same laser pulse as the pump, subsequently monitored the induced transmission. In order to achieve high detection sensitivity, the beams were amplitude modulated at different frequencies, $\nu_1$ and $\nu_2$, both of the order 1kHz, and the transmitted probe intensity occurring at the difference frequency ($\nu_1$-$\nu_2$) was measured using a lock-in amplifier. Transmission changes of $\sim 1{:}10^5$ could be detected, which allowed very low photoexcited carrier densities to be employed. The probe transmission signal obtained at large negative time delays was subtracted from the measured intensity, to yield the differential transmission $\mathcal{T}$.

The sample was a 60-period 7.6nm/7.6nm GaAs/AlAs symmetric superlattice grown by MBE on a GaAs substrate. The substrate was removed by mechanical polishing and chemical etching. The electron subband separation $E_{12}$ is $\sim 150$meV, so LO phonon emission processes will dominate the intersubband dynamics. Our time-resolved experiment comprised two distinct parts. First, the sample was excited at low energy, 1.592 eV so that carriers were excited from the n=1 light and heavy hole bands into the n=1 conduction subband and only *intrasubband* electron relaxation occurs. Second, the sample was excited at somewhat higher energy, 1.687 eV so that a population of electrons was excited also into the n=2 subband, but close to the minimum. In this case *intersubband* scattering by both phonon emission and absorption are now permitted in addition to the intrasubband processes. However, from simple effective mass consideration we expect that approximately only 35% of the electrons are created in the n=2 subband, the remainder being excited into the n=1 subband, so that intrasubband relaxation processes are expected to continue to have a strong influence on the induced transmission. In some work, this aspect is not correctly considered. Furthermore, as $\mathcal{T} \propto \sum_n \int dk(f_n^e + f_n^{lh,hh})$, where $f_n^e$, $f_n^{lh}$, and $f_n^{hh}$ are the occupancies of optically-coupled electron-hole states in the $n$th subband, it is then clear that the dynamics of the light and heavy holes have a strong influence on the induced transmission, so that the measured decay times cannot simply be assigned to electron scattering times.

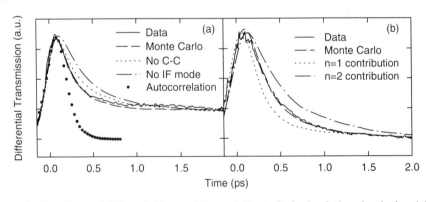

**Figure 1.** Experimental differential transmission and Monte Carlo simulations for the low (a) and high (b) energy excitation.

The time-dependent differential transmission measured at room temperature is presented in Fig. 1(a) and (b) (solid line) respectively for the low-energy and high-energy experiments. The autocorrelation of the laser pulse is also shown for comparison in Fig. 1(a) (solid dots). The carrier density excited in both cases was estimated to be $\sim 9 \times 10^9$ cm$^{-2}$. In the former case the transmission increases rapidly, closely following the rising edge of the pump pulse. The signal then decays to reach a plateau after about 500 fs, and remains relatively constant at later times. The initial decay is also fast: when the laser pulse temporal profile is deconvoluted from $\mathcal{T}$, we obtain a characteristic decay time of $\sim 145$fs. The intrasubband electron scattering rate given by the dielectric continuum model[1] predicts a relaxation time of 90 fs for electrons in the initial state. When the n=2 subband is excited (Fig. 1(b)) the decay profile is significantly different. The measured decay time is now 335 fs, the increase reflecting the influence of slower intersubband processes. For comparison the predicted intersubband n=2 $\rightarrow$ 1 scattering time for electrons in the initial state is $\sim 520$ fs.[1]

## THEORETICAL MODEL

In order to extract accurate electron-phonon scattering rates from the data we have made a three-band ensemble Monte Carlo simulation of the experiments: it includes conduction band, heavy and light hole bands, but neglects higher conduction bands and the spin-split valence band which are not excited in the experiments. Electron and hole wavefunctions have been derived from an envelope function calculation of the band structure within the effective mass approximation. LO confined modes and IF modes are included in the simulation within the dielectric continuum model.[1] Interband and intraband hole scattering by bulk TO phonons are also included. Although the carrier density is low, we have included CC scattering.

The results of the simulations of the two experiments are shown by the dashed curves in Fig. 1. The excellent agreement between theory and experiment gives strong confirmation of the theoretical model. The simulation also demonstrates quite clearly the marked effect of IF phonon scattering: the dashed line of Fig. 1 shows the results obtained when IF mode scattering is switched off, and clearly there are significant departures from the experimental data. In the intrasubband case, the IF mode emission process contributes a factor of $5.5 \times 10^{12}$ s$^{-1}$ to the overall rate of $9.5 \times 10^{12}$ s$^{-1}$.

From our data we can evaluate the separate contributions to the differential transmission from the n=1 $\rightarrow$ 1 and the n=2 $\rightarrow$ 1 processes. Although only 35% of electrons are excited into the n=2 subband, they produce an effect of similar magnitude on the probe transmission as those excited into the lower subband; this is due to the latter being scattered out of the initial state quite rapidly, on the same timescale as the excitation pulse, whereas the former electrons remain in the initial state for times considerably longer than the duration of the probe pulse[13] (See dotted and chain curves of Fig. 1(b)). We find that the intersubband scattering rate due to phonon emission is $1.6 \times 10^{12}$ s$^{-1}$, with the confined mode contribution being about a factor of two stronger than the IF mode contribution. Even in our almost ideal experimental condition we can see that a direct determination of the scattering time from the slope of the transmission signal results in a relatively inaccurate result, 335 fs against the real value obtained by the Monte Carlo fitting of 520 fs (accounting for both emission and absorption processes).

With our model we can now see if it is possible to reconcile many of the previously published results. The fastest intersubband scattering time reported in literature in Ref. 8 is actually obtained in conditions very similar to our intrasubband experiment as, by exciting resonantly with the second subband exciton only a small fraction of carriers ($\sim 15$-20%) are generated in the second subband. The fast decay then reflects mainly the intrasubband scattering time. On the other hand, an intrasubband scattering of several picoseconds has been recently reported through luminescence measurements.[9] To verify if this long time is somehow consistent with our experiment we performed a simulation under the same conditions for two different quantum well widths, both corresponding to a subband separation larger than the LO phonon energy. Our simulation is shown in Fig. 2. Although the physical model is the same and the intersubband scattering time is still 520 fs, we can see how the decay of the simulated luminescence is very slow. Looking more deeply into the simulation we are able to verify that the long decay is not representative of intersubband scattering but rather is determined by hot carrier cooling, the first two subbands being equally populated at the same total energy.

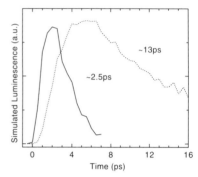

**Figure 2.** MC simulation of the luminescence from the energy in the region of n=2 subband for a 100 Å(solid line) and a 180 Å(dashed line) quantum well.

The results presented here provide strong confirmation of the electron-optical phonon scattering rates in quantum well structures predicted by dielectric continuum theory. Intrasubband scattering rates are found to be intermediate between those of the bulk constituent materials, in the present case GaAs and AlAs. The n=2 population lifetime measured here agrees well with that measured by time-resolved Raman spectroscopy[14], where the subband separation was $\sim$ 50 meV. The intrasubband time agrees well with time-resolved Raman experiments performed on the same sample[15], where an upper limit of $\sim$ 250fs was measured. As we have demonstrated, caution must be taken to ensure the scattering time measured is not obscured by other effects, especially at high density or excitation energy.

We wish to acknowledge financial support from the European Commission through the NANOPT and ULTRAFAST programmes.

## REFERENCES

1 See e.g. H. Rücker, E. Molinari, and P. Lugli, Phys. Rev. **B44**, 3463 (1991); Phys. Rev. **B45**, 6747 (1992); and references therein.

2 N. Mori and T. Ando, Phys. Rev. **B40**, 6175 (1989).

3 S. Rudin and T.L. Reinecke, Phys. Rev. **B41**, 7713 (1990); P.A. Knipp and T.L. Reinecke, Phys. Rev. **B48**, 18037 (1993).

4 A. Seilmeier, H.J. Hübner, G. Abstreiter, G. Weimann, and W. Schlapp, Phys. Rev. Lett. **59**, 1345 (1987).

5 D.Y. Oberli, D.R. Wake, M.V. Klein, J. Klem, T. Henderson, and H. Morkoç, Phys. Rev. Lett. **59**, 696 (1987).

6 K.T. Tsen, K.R. Wald, T. Ruf, P.Y. Yu and H. Morkoç, Phys. Rev. Lett. **67**, 2557 (1991).

7 Levenson J.A., Dolique G., Oudar J.L., Abram I., Phys. Rev. B **48**, 3688 (1990)

8 Hunsche S., Leo K., Kurtz H., Köhler K., Phys. Rev. B **50**, 5791 (1994)

9 Morris D., Houdé D., Deveaud B., Regreny A., Superlatt. and Microstr. **15**, 309 (1994)

10 See e.g. P. Lugli and S.M. Goodnick in *Hot Carriers in Semiconductor Nanostructures: Physics and Applications*, edited by J. Shah, Academic Press, New York (1992).

11 J. Shah, B. Deveaud, T.C. Damen, W.T. Tsang, A.C. Gossard, and P. Lugli, Phys. Rev. Lett. **59**, 2222 (1987).

12 L. Rota, P. Lugli. T. Elsaesser, and J. Shah, Phys. Rev. **B47**, 4226 (1993)

13 L. Rota and D.K. Ferry, Appl. Phys. Lett. **62**, 2883 (1993).

14 M.C. Tatham, J.F. Ryan, and C.T. Foxon, Phys. Rev. Lett. **63**, 1637 (1989).

15 Maciel A.C., Mayhew N., Turner K., Ryan J.F., Gulia M., Molinari E., Lugli P., To be published.

# HOT PHOTOLUMINESCENCE AND ELECTRON-PHONON INTERACTION IN GaAs/AlAs MQW STRUCTURES

D.N.Mirlin, B.P.Zakharchenya, P.S.Kop'ev, I.I.Reshina,
A.V.Rodina, V.F.Sapega, A.A.Sirenko, and V.M.Ustinov

A.F.Ioffe Physical Technical Institute, St.Petersburg 194021,Russia

Energy relaxation of hot electrons by electron-phonon interaction in 2D structures has received much attention in recent years. Not only the confinement of electrons but also the modification of the phonon spectrum, the appearance of the confined and the interface phonons is of great importance in these structures.

Experimental work by time-resolved optical technique provided a lot of valuable information on ultrafast carrier dynamics but showed also very much controversy in electron-phonon scattering rates. Under excitation with high power laser pulses distorting factors such as screening and especially hot phonon distributions strongly influence the experimental results.

In experiments on anti-Stokes Raman scattering by Tatham and Ryan[1] and Tsen et al.[2] a tendency of the increasing role of the IF AlAs-like phonons in 2D electron scattering in narrow GaAs/AlAs MQW's was demonstrated. However, the total scattering rate was several times underestimated in [2], probably because the phonons probed in anti-Stokes Raman scattering have a restricted range of in-plane wave vectors ($q_\parallel$) as compared to phonons emitted during energy relaxation of hot electrons.

We have used a new approach in studying electron-phonon interaction in GaAs/AlAs MQWs and superlattices based on hot band-to-acceptor photoluminescence[3]. In our experiments the hot electrons are excited by continuous wave lasers at sufficiently low power density ($\sim100$ W/cm$^2$). Thus no complicating factors such as phonon heating and screening influence the experiment. The structures are lightly p-doped with Be in the central parts of the wells (to a concentration $1\text{-}3\times10^{17}$cm$^{-3}$). The photoexcited electrons recombine with the acceptor-bound holes both from the initial excited state and after sequential emission of optical phonons. From energy losses revealed in the spectra of hot photoluminescence we could determine directly for the first time what kind of phonons play the dominant role in the scattering process[4].

Measurements of the magnetic depolarization of hot luminescence allow us to find the hot electron scattering rate.

We studied a series of GaAs/AlAs MQW's with the well widths in the range 40-160 Å and a fixed barrier width of 100 Å. Figure 1 shows how the separation between the peaks in the spectrum changes with the well width. It is equal to the LO-

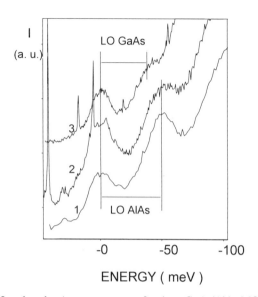

I
(a. u.)

LO GaAs

3

2

1    LO AlAs

-0          -50          -100

ENERGY ( meV )

**Figure 1**. Hot photoluminescence spectra for three GaAs/AlAs MQW structures at T=4K. 1-40 Å/100 Å, 2-85 Å/100 Å, 3-130 Å/100 Å.
" Zero-phonon" peaks in all spectra are adjusted to the zero of the energy scale.

phonon energy of GaAs ( 37 meV) for the well width greater than 120 Å. On the other hand, for narrow wells the separation between maxima is 50 meV, that value being very close to the energy of the LO-phonon in AlAs.

To analyze these results we compare them with the calculations of intraband scattering rate for different phonon types. It was shown recently by Rücker *et al* [5] that the dielectric continuum model ( DCM ) that uses electromagnetic boundary conditions at the interfaces predicts the displacements and potentials of the phonon modes and their symmetries that are very similar to the results of microscopic calculations for GaAs/AlAs systems for large $q_{\parallel}$ that are essential in hot electron scattering . Deviations from microscopic results are limited to a small region near the interface where the electron wave functions of the first subbands are small anyway. Therefore we performed the calculations of the scattering rate for intra- and interband scattering for different type of phonons within the framework of the DCM following the treatment by Mori and Ando [6].

Figure 2 shows the calculated dependence of the intraband scattering rate of a 200 meV electron in the first subband as a function of the well width for different type of phonons. For MQW's with narrow wells smaller than 100 Å the main contribution to the scattering rate comes from the AlAs-like IF modes. The potential of the IF phonons decreases away from the interface as exp-($q_{\parallel}$ z) and the main contribution is made by IF phonons with minimal $q_{\parallel}$ values. In narrow wells such phonons have large overlap integrals with the electron wave functions and also large values of electron-phonon coupling constants . The frequencies of these IF phonons are very close to the frequencies of AlAs LO phonons as can be seen from their dispersion curves. These results are in agreement with the spectra of narrow wells where separation between phonon peaks was about 50 meV. In narrow wells scattering by IF AlAs-like phonons dominates at all energies.

For larger wells contribution of IF phonons decreases and the confined GaAs phonons start to dominate the scattering process. The contribution of IF GaAs-like phonons is relatively small. The barrier phonons make extremely small contribution because their overlap integrals with electron wave functions in the barrier are very small.

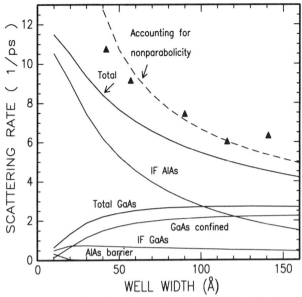

**Figure 2**. Calculated intraband scattering rate of a 200 meV -electron
vs. well width .The curves refer to interaction with the different types
of phonons. Experimental results for the total scattering rate are shown
by symbols.

The agreement between the calculated total scattering rate and the measured
values     is  quite  satisfactory.  It  becomes  even  better  if  some  account  of
nonparabolicity of the conduction band is made in the calculations, as is shown by
the dashed curve in Fig.2.

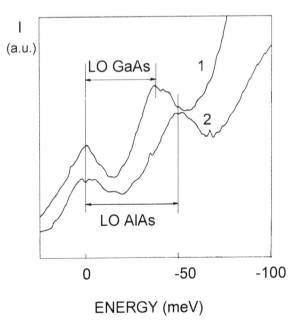

**Figure 3**. Comparison of the spectra :
(1) - GaAs/Al$_{0.32}$Ga$_{0.68}$As MQW 46 Å/100 Å
(2) - GaAs/AlAs MQW 40 Å /100 Å

We also studied experimentally the influence made by the barrier composition. Figure 3 shows that in a case of the mixed $Al_{0.32} Ga_{0.68}As$ barrier the dominant contribution is made by confined GaAs LO phonons even in narrow wells, in contrast to a MQW with an AlAs barrier where the scattering is by AlAs-like IF phonons. For a mixed barrier the total scattering rate is two thirds of the rate in case of the AlAs barrier.

Recently the same method was used in the study of electron-phonon interaction in GaAs/AlAs superlattices as a function of the barrier width [7]. In superlattices with a well width of 40 Å and barrier width smaller than 10 Å the GaAs confined phonons dominate the scattering. As the barrier width increases, the contribution of the IF AlAs-like phonons increases and for a 20 Å barrier they already make the main contribution.

**Acknowledgments**

We acknowledge the support of International Science Foundation, grant R 47000 and Russian Fundamental Research Foundation, grant 94-02-04821.

**REFERENCES**

1. M.C.Tatham and J.F.Ryan, Inter- and intra-subband relaxation of hot carriers in quantum wells probed by time-resolved Raman spectroscopy, *Semicond. Sci. Technol.* 7: B 102 (1992).
2. K.T.Tsen, R.Joshi and H.Morkoç Direct measurements of electron-optical phonon scattering rates in ultrathin GaAs-AlAs multiple quantum well structures, *Appl. Phys. Lett.* 62: 2075 (1993).
3. D.N.Mirlin, I.Ja.Karlik, L.P.Nikitin, I.I.Reshina, and V.F.Sapega, Hot electron photoluminescence in GaAs Crystals *Solid State Commun.* 37:757 (1981).
4. D.N.Mirlin, P.S.Kop'ev, I.I.Reshina, A.V.Rodina, V.F.Sapega, A.A.Sirenko, and V.M.Ustinov, Hot electron luminescence and Raman scattering in MQW structures at high magnetic field, in: "22 International Conference on the Physics of Semiconductors" ,D.J.Lockwood, ed., World Scientific, Singapore (1995).
5. H.Rücker, E.Molinari, and P.Lugli, Microscopic calculation of the electron-phonon interaction in quantum wells, *Phys. Rev.* B 45:6747 (1992).
6. N.Mori and T.Ando, Electron-optical-phonon interaction in single and double heterostructures, *Phys. Rev.* B 40: 6175 (1989).
7. V.F.Sapega, M.P.Chamberlain, T.Ruf, M.Cardona, D.N.Mirlin, A.Fischer, K.Tötemeyer, and K.Eberl, Optical phonon emission in GaAs/AlAs MQW's determined by hot-electron luminescence, *Phys.Rev.* B, in press.

# TIME-RESOLVED STUDIES OF INTERSUBBAND RELAXATION USING THE FREE ELECTRON LASER

B. N. Murdin[1], W. Heiss[2], C. R. Pidgeon[3], E. Gornik[2], S-C. Lee[3], I. Galbraith[3], C. J. G. M. Langerak[1,3], H. Hertle[4] and M.Helm[5]

[1]FOM-Institute for Plasma Physics, 'Rijnhuizen', P. O. Box 1207, 3430 BE Nieuwegein, The Netherlands;
[2]Institut für Festkorperelektronik, T.U. Wien, A1040 Wien, Austria;
[3]Department of Physics, Heriot Watt University, Edinburgh, UK;
[4]Walter Schottky Institut, T.U.Munchen, Germany;
[5]Institut für Halbleiterphysik, T.U.Linz, A-4040 Linz, Austria.

## INTRODUCTION

The knowledge of intersubband lifetimes is important for the design of intersubband based emitters and detectors. In wide GaAs/AlGaAs quantum wells, where the intersubband spacing is smaller than the optical phonon energy, rather long lifetimes are expected, consistent with intersubband relaxation by acoustic phonon scattering and Auger processes. However, this has not always been found to be the case, and lifetimes ranging from 20ps[1] to 1.6ns[2] have been determined using several different experimental techniques[1-4]. From a saturation experiment a lifetime of 600ps[2] was estimated in a quantum well of width 400 Å, while with a pump/probe technique we have directly measured a lifetime of 40ps in a sample of well width 270 Å [3]. In experiments where the carriers were produced by photoexcitation, lifetimes of 20ps[1] and 325ps[4] were determined in different experiments on samples of well widths about 210 Å.

In the present work we report measurement of subband lifetimes in two $Si/Si_{1-x}Ge_x$ quantum well samples. We have determined $T_1$ by a time resolved pump and probe experiment using the far infrared ps free electron laser (FEL) source FELIX at Rijnhuizen, the Netherlands. Preliminary results were reported elsewhere[5]. We discuss optical and acoustic phonon scattering as possible limiting processes for the lifetimes observed in GaAs/AlGaAs and Si/SiGe quantum wells.

It is possible radically to alter the situation by applying an external magnetic field, which suppresses the in-plane $k_x$ and $k_y$ motion. In high purity GaAs/AlGaAs heterostructures under high magnetic fields, cyclotron resonance absorption has been observed to undergo a true saturation, showing that the electron temperature, $T_e$, is relatively unaffected by the laser intensity[6,7]. We show in the second part of the paper that this is true only for sample concentrations below a certain electron concentration; above this critical value the polaron non-parabolicity (see below) required for saturation is nullified and heating becomes the dominant effect. In order to investigate the role of electronic heating in the lifetime determinations above, we intend to perform further lifetime measurements in quantum well samples under an applied magnetic field.

## INTERSUBBAND LIFETIMES IN Si/SiGe AND GaAs/AlGaAs QUANTUM WELLS

The n-type Si/SiGe multiple quantum well samples were grown at Daimler-Benz by molecular beam epitaxy (MBE)[5] on a (100)-oriented silicon wafer. They consist of a 400 Å Si buffer, a 3000 angstrom strain relaxed $Si_{0.7}Ge_{0.3}$ buffer, a multiple quantum well structure formed by five periods of silicon wells between 250 Å barriers wherein the central 50 Å are antimony doped, and a 100 Å Si capping layer. The well width, carrier density per well, and mobility is 50 Å, $2.1x10^{12}cm^{-2}$, 12,600 cm²/Vs for sample C0660 and 75 Å, $1.1x10^{12}cm^{-2}$ and 16,900 cm²/Vs for sample C0663 respectively. A multipass waveguide geometry was used to observe an enhanced intersubband absorption. For all experiments the samples were mounted on a cold finger within the insulation vacuum of a helium bath cryostat. The samples were characterized with a Fourier transform spectrometer[8], from which the absorption cross-section was deduced. The absorption peaks were at 34.7meV (C0660) and 19.8meV (C0663) respectively.

The pump and probe lifetime measurements were performed with FELIX delivering pulses which varied between 8ps and 2ps, with a spacing of 1ns. To deduce the lifetime, $T_1$, the change in transmittance of the probe, $\Delta T$, as the absorption recovers, was measured as a function of the delay, $t_d$, with a sensitive liquid helium cooled Ge:Ga detector. The earlier experiments on a grating -coupled stack of GaAs/AlGaAs quantum wells designed to have an intersubband absorption peak at 18meV were conducted in the same way, and have been described in detail elsewhere[3].

For the analysis we use the rate equation for a two level, homogeneously-broadened system[3]:

$$\frac{dn_1}{dt} = \frac{\sigma I_{pump}(t)}{\hbar\omega}[n_1 - n_2] - \frac{n_2}{T_1} \tag{1}$$

with $n = n_1 + n_2$. Here, $n_1$ and $n_2$ are the carrier concentrations in the first and the second subbands, $\sigma$ is the absorption cross-section and $T_1$ is the lifetime of the electrons in the second subband. The time dependence of the carrier concentration is calculated by integrating this formula over the intensity of the pump beam, $I_{pump}(t)$, which is modeled by a train of Gaussian pulses with a duration $t_p$ and a spacing $\delta t = 1ns$. The absorption is calculated by :

$$A = \int_{-\infty}^{\infty} \frac{\sigma}{\hbar\omega} I_{probe}(t - t_d)[n_1 - n_2]dt \tag{2}$$

where $I_{probe}$ denotes the intensity of the probe pulse, which is shifted by the delay time $t_d$.

The results of the earlier pump/probe experiment for GaAs/AlGaAs[3] and the present one for Si/SiGe (sample C0660) are shown in Fig. 1 a) and Fig. 1b), respectively. From the best fit of the experimental data a lifetime of 40ps was obtained for the wide wells of GaAs/AlGaAs, and a value of 30ps for the Si/SiGe sample C0660 and 20ps for sample C0663.

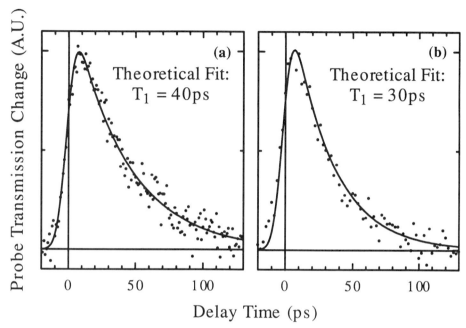

**Figure 1.** **(a)** Normalized change in probe transmission versus time between excite and probe pulses for wide wells of GaAs/AlGaAs[3], taken with the laser centered on the subband resonance at 69μm; **(b)** normalized change in probe transmission, as in (a), for wide wells of Si/SiGe (sample C0660)[5], taken with the laser centered on the subband resonance at 35.7μm

We have performed calculations of the intrasubband and intersubband scattering rates in an infinite square quantum well due to the emission and the absorption of LO phonons in wide GaAs/AlGaAs quantum wells[9]. The occupation is taken into account by assuming a thermal equilibrium over the three lowest subbands and a Fermi distribution with an electron temperature $T_e$. The electron temperature is allowed to exceed the lattice temperature which determines the phonon occupation number. An electron temperature higher than the lattice temperature is plausible for the following reason. After the excitation of the electrons by the laser pulse the carriers have a non-equilibrium distribution which thermalizes through rapid electron-electron scattering to a Fermi distribution with $T_e$ higher than the lattice temperature. It is shown that intrasubband scattering by LO phonon emission has lifetimes which become very long at low $T_e$, thus limiting the cooling of the electron plasma to around 50K. Using the theoretical results all the experimental lifetimes given in the literature[1-4] for wide GaAs/AlGaAs quantum wells ranging between 20ps and 1.6ns can be explained by choosing a certain $T_e$ between 30K and 80K.

Up to now there have been no theoretical predictions of lifetimes in Si/SiGe quantum well structures. The lattice relaxation of the carriers in the excited subband is governed by inelastic scattering processes which are described by three basic mechanisms: polar optical scattering (which is not present in Si), optical and acoustic deformation potential scattering. In our Si/SiGe quantum well samples the intersubband transition is taking place well below the energy of the optical phonons (63meV in bulk Si), so that the optical deformation potential scattering should not be the limiting process. Furthermore, from transport measurements in rather low doped n-Si samples[10] and from Monte Carlo simulations[11] an inelastic scattering time due to interactions with acoustic phonons of about 10ps can be deduced for bulk n-Si at liquid helium temperatures, which is comparable to the value we measure. However, we have performed a similar calculation to that above for the optical deformation potential scattering, and find lifetimes that are comparable to that we measure, again for electron temperatures below 100K. Thus, in this case both processes appear to be important.

## LANDAU LEVEL LIFETIMES IN GaAs/AlGaAs HETEROSTRUCTURES

There have been contradictory reports about the possibility of achieving saturation of the 2-DEG cyclotron resonance (CR) absorption in GaAs/AlGaAs heterostructures. Earlier reports showed either partial saturation of the CR absorption[12] or electron heating at higher laser intensities[13]. More recently saturation was reported in ultra-pure heterostructures[6] with a sheet density, $N_s$, lower than $7 \times 10^{10} cm^{-2}$. Saturation of CR was also observed in coupled quantum well samples[14] with carrier concentrations lower than $1 \times 10^{11} cm^{-2}$ and recently by us[7] in samples with concentrations below $2 \times 10^{11} cm^{-2}$. Here we report measurements to higher concentrations, showing conclusively that the resulting screening of the polaron non-parabolicity removes the saturation and results in electron heating. Thus, provided the sample concentration and the experimental magnetic field range are chosen appropriately, then true saturation of CR is possible.

In a ladder of equidistant Landau levels (LL's) no saturation can be achieved. In such a system the increase of the laser intensity leads to the population of an increasing number of excited LL's. In GaAs/AlGaAs heterostructures, a non-equidistant spacing between the LL's is produced principally by the polaron effect. The resonant polaron effect leads to a significant increase of the effective mass when a LL is close to a virtual state, which is formed by the n = 0 LL plus the excitation of the optical phonons (at $E_{opt} = 36meV$)[15,16]. For the n = 1 LL the resonant polaron effect is observed at a magnetic field around 22T, while for the n = 2 LL the effective mass increase occurs around 11T. Thus at 11T the n = 2 LL is significantly shifted, and the intensity dependence of the (n = 0 to n = 1) CR absorption can be modeled within a two level system. This procedure is described in detail elsewhere[6,7], where the solution of the rate equations is fitted to the experimental saturation CR data in terms of two adjustable parameters, $\tau$ (the electron momentum scattering time) and T (the upper state lifetime).

The samples were mounted at the center of a superconducting magnet and immersed in liquid helium. The laser beam was directed onto the sample by an oversized waveguide. The intensity was controlled as described above, and the transmitted radiation detected by the broadband free electron photoconductivity in bulk InSb. Two detectors were mounted, one in the waveguide in front of the sample as a reference and the other just behind the sample to measure the transmitted signal. We have investigated two modulation doped n-type GaAs/AlGaAs single heterostructure samples of sheet densities $N_s = 2.2 \times 10^{11} cm^{-2}$ and $4 \times 10^{11} cm^{-2}$. In Fig. 2a) the CR absorption is shown for the low concentration sample measured at a wavelength of 70μm. The solid lines are best fits to the experimental data. At an intensity of 270 W/cm$^2$ a single Lorentzian shaped absorption line is observed with a transmission minimum at 10.8T. The FWHM of the transition is about 0.3T. When the intensity is increased to 2700W/cm$^2$ the transmission minimum is shifted to a higher magnetic field value. This is a result of the band nonparabolicity. In addition the amplitude of the absorption line is reduced and the FWHM is enlarged. However, the linewidth is increased in such a way that the absorption curve at high intensity lies within the absorption at low intensity. The same behavior is observed with carrier concentrations <u>lower</u> than this by ourselves and other authors[6,7].

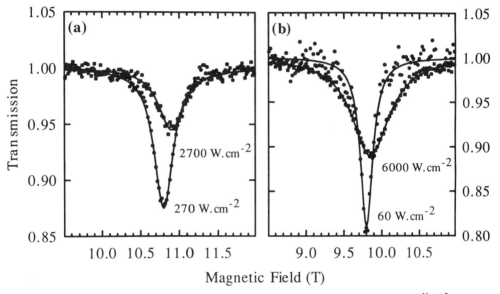

**Figure 2.** (a) Experimentally determined CR transmission for a sheet density of $2.2 \times 10^{11} cm^{-2}$ at two intensities. The solid lines are the best fitted theoretical curves. A homogeneous saturation behavior is observed; (b) CR transmission for a sheet density of $4 \times 10^{11} cm^{-2}$ at two intensities. The cyclotron resonance transition clearly shows strong thermal broadening with increasing laser intensity.

The intensity dependent CR transmission for the high concentration sample, measured at 77.3μm, is shown in Fig. 2b). At low intensities a narrow CR absorption line is observed. When the intensity is increased the transmission minimum is again shifted to higher magnetic fields. In this sample, by contrast, increasing the intensity by a factor of 100 produces a reduction of the absorption of only 50%. Furthermore, with increasing intensity a dramatic increase of the linewidth is observed, characteristic of heating. At an intensity of 6000W/cm$^2$ the FWHM amounts to 0.8T in comparison to a linewidth of 0.2T at 60W/cm$^2$. The wings of the CR curve at the higher intensity overlap the absorption line at lower intensity. Thus, we conclude that in heterostructures with carrier concentrations higher than about $2 \times 10^{11} cm^{-2}$ electron heating is produced by increasing the laser intensity, preventing true saturation being observed at any intensity for the n = 0 to n = 1 LL transition at 11T.

As mentioned above, the resonant polaron effect is essential for the formation of a two level system within a ladder of LL's. The density dependence of the polaron effect was studied previously in GaAs/AlGaAs heterostructures[16]. From reflectivity measurements within the Reststrahlen band a strongly reduced polaron effect was observed for samples with $N_s > 3.4 \times 10^{11} cm^{-2}$. The reason is that at these values of $N_s$ the lowest LL is almost completely filled, thereby suppressing the resonant part of the $n = 1$ LL polaron effect by removing the final states from the virtual transition; the inclusion of the occupation probabilities of the LL's in a "many-polaron memory-function" calculation shows that a combination of occupation and screening effects provides almost complete suppression of the polaron non-parabolicity at this carrier density[17]. In our case, where we are working at half the resonant polaron field to establish our two level CR system, this would correspond to an electron concentration of about $2 \times 10^{11} cm^{-2}$, which is in good accord with the observations described in Figs. 2(a) and 2(b) above. For the pure sample of Fig. 2(a) we determine an intensity-independent scattering time value of $\tau = 15ps$.

This work is partially supported by the "FWF", Austria (P9301 TEC). It was also performed as a part of the research program at the "FOM" with financial support from the "NWO", Netherlands. One of us (CJGML) is grateful for a research assistantship from SERC, UK.

# REFERENCES

1.  J. A. Levenson, G. Dolique, J. L. Oudar, I. Abram, Phys. Rev. B41, 3688 (1990).
2.  K. Craig, C. L. Felix, J. N. Heyman, A. G. Markelz, M. S. Sherwin, K. L. Campman, P. F. Hopkins, A. C. Gossard, Semicond. Sci. Techno. 7, 627 (1994).
3.  B. N. Murdin, G. M. H. Knippels, C. J. G. M. Langerak, A. F. G. Van der Meer, C. R. Pidgeon, M. Helm, W. Heiss, K. Unterrainer, E. Gornik, K. K. Geerink, N. J. Hovenyer, W. Th Wenckebach, Semic. Sci. Techno. 9, 1554 (1994).
4.  D. Y. Oberli, D. R. Wake, M. V. Klein, J. Klem, T. Henderson, H. Morkoc, Phys. Rev. Lett. 59, 696 (1987).
5.  W. Heiss, E Gornik, H. Hertle, B. Murdin, G. M. H. Knippels, C. J. G. M. Langerak, F. Schaffler and C. R. Pidgeon, Appl. Phys. Lett. 66, 3313 (1995).
6.  I. Maran, W. Seidenbusch, E. Gornik, G. Weimann, M. Shayegan, Semicond. Sci. Technol. 9, 700 (1994).
7.  W. Heiss, P. Auer, E. Gornik, C. R. Pidgeon, C. J. G. M. Langerak, B. N. Murdin, G. Weimann and M. Heiblum, Appl. Phys. Lett., to be published (1995).
8.  H. Hertle, G. Schuberth, E. Gornik, G. Abstreiter, F. Schaffler, Appl. Phys. Lett. 59, 2977 (1991).
9.  S. C. Lee, I. Galbraith and C. R. Pidgeon, Phys. Rev. B52, 1874 (1995).
10. P. Norton, T. Braggins and H. Levinstein, Phys. Rev. B8, 5632 (1973).
11. C. Jacoboni and L. Reggiani, Rev. Mod. Phys. 55, 645 (1983).
12. M. Helm, E. Gornik, A. Black, G. R. Allan, C. R. Pidgeon, K. Mitchell and G. Weimann, Physica 134B, 323 (1985).
13. G. A. Rodriguez, R. M. Hart, A. J. Sievers, F. Keilmann, Z. Schlesinger, S. L. Wright and W. I. Wang, Appl. Phys. Lett. 49, 458 (1986).
14. W. J. Li, B. D. McCombe, J. P. Kaminski, S. J. Allen, M. I. Stockman, L. S. Muratov, L. N. Pandey, T. F. George and W. J. Schaff, Semicond. Sci. Technol. 9, 630 (1994).
15. G. Lindemann, R. Lassnig, W. Seidenbusch and E. Gornik, Phys. Rev. B28, 4693 (1983).
16. C. J. G. M. Langerak, J. Singleton, P. J. Van der Wel, J. A. A. Perenboom, D. J. Barnes, R. J. Nicholas, M. A. Hopkins and C. T. B. Foxon, Phys. Rev. B38, 13133 (1988).
17. F. M. Peeters, Wu Xiaoguang and J. T. Devreese, Surf. Sci. 196, 437 (1988).

# TIME-RESOLVED FEMTOSECOND INTERSUBBAND RELAXATIONS AND POPULATION INVERSION IN STEPPED QUANTUM WELLS

C.Y. Sung, A. Afzali-Kushaa , T.B. Norris, X. Zhang and G.I. Haddad

Center for Ultrafast Optical Science,
The University of Michigan, Ann Arbor, MI 48109

## INTRODUCTION

The generation of coherent far-infrared radiation in superlattices or multiple-quantum-well (MQW) structures has been a goal for many years. The use of intersubband transitions to generate coherent mid-infrared radiation was first successfully demonstrated by J. Faist et al. [1]. However, new structures are needed to generate radiation in the 30-300 mm regime (1-10 THz). In this paper, we demonstrate how a stepped quantum well structure can modify the intersubband relaxation rates, allowing a population inversion between subbands to be achieved. We measured the intersubband relaxation rates in the stepped quantum well with femtosecond differential transmission spectroscopy and experimentally observed the population inversion.

The basic idea is to design the structure so that it can behave as a 4-level laser system (the inset of Fig. 1.), with a population inversion between levels 3 and 2. A pump laser (CO2 laser or other IR sources) would be used to excite a doped QW, pumping carriers from subband 1 to subband 4 or higher subbands. The energy separation between subband 4 and subband 3 is designed to be greater than the LO phonon energy; thus the excited carriers will relax very fast (within 500 fs) to subband 3 [2-4]. The dominant relaxation mechanism for relaxation is polar LO phonon scattering, which is proportional to the wavefunction overlap and inversely proportional to the square of the phonon wavenumber involved. [4] Thus, relaxation rate to subband 3 will be faster than to the other subbands 2 or 1. If we design the QW with the energy separation between subband 2 and 1 to be larger than the LO-phonon energy, while the 3-2 separation is less than the LO-phonon energy, the carriers in subband 2 will be depopulated very fast to level 1; however, the 3 to 2 scattering rate will be significantly reduced. [5,6] Furthermore, a larger wavevector is required for intersubband scattering from level 3 to level 1 than from 2 to 1. Thus carrier relaxation via LO-phonon emission from n=2 to 1 will be much faster than from 3 to 1.

## EXPERIMENTAL RESULTS

From calculations of the intersubband relaxation rates including LO and LA phonon scattering at room temperature, we get $\tau_{31} \approx 15$ ps, $\tau_{32} \approx 1$ ns, $\tau_{21} \approx 500$ fs. To measure these rates experimentally, we used differential transmission spectroscopy with femtosecond resolution, pumping and probing across the band gap [5]. Two white-light continuum pulses were generated using a 250 kHz, 3.5 $\mu$J, 85 fs Ti : sapphire amplifer. A fraction from one of the continua, ranging from 1.4 eV to 1.65 eV was used as a broad band probe pulse. The dispersion of the broad-band probe was compensated by double-pass prism pairs so that transmission spectra of the entire near-band-edge region could be obtained with 120-fs resolution. A 10 nm bandwidth filter was used to select the pump pulse wavelength from the other continuum pulse. An optical multichannel analyzer (OMA) was used to measure differential transmission spectra (DTS).

The stepped quantum well sample consisted of 20 periods of 100-Å GaAs wells which are surrounded by 150-Å $Al_{0.15}Ga_{0.85}As$ step layers and 100-Å $Al_{0.25}Ga_{0.75}As$ cladding layers. The substrate was removed for transmission measurements. The calculated subband splitting between n1 and n2 is about 68 meV, almost equal to two times the LO-phonon energy. The splitting of n2 and n3 is about 28 meV, smaller than the LO-phonon energy. The positions of the excitonic peaks agree well with calculated values. We performed the measurement at room temperature without doping or bias. In the case of low optical density $(-\Delta\alpha d \ll 1)$, the normalized transmission changes $\Delta T / T_0$ were approximately equal to $-\Delta\alpha d$. The changes of the reflectivity which are smaller than 5 percent of the DTS signal were not included. [7] In [8], a simulation based on the DTS proportional to the total carrier density was used to fit the measured time evolution of the DTS signal almost perfectly, while the integrated DTS signals increase linearly with pumping density. As plotted in the inset of Fig. 4, we measured the bleaching signal which depends almost linearly on pumping carrier density, thus, the integrated DTS can be used as a direct measure of the subband population.

In the valence band, Hopfel et al. have measured a very long relaxation time from HH2 to HH1 about 130 ps when the heavy hole subband splitting is smaller than the LO phonon energy. [6] The hole relaxation times should be around a hundred picoseconds in our case, due to the energy separation between HH2 and HH1 being 29 meV. Furthermore, the larger density of states in the valence band will also reduce the contribution the DTS bleaching by holes. Thus, electronic relaxation in the conduction band will dominate the time evolution of the DTS signal.

In Figure 1, we show a series of differential transmission spectra (DTS), where we resonantly pump the E2HH2 transition. The E1HH1 exciton peak also includes a small contribution from E1LH1. A total carrier density $5 \times 10^{11}$ cm$^{-2}$ was estimated from the pump power and spot size. The DTS show a peak at E2HH2 which has a fast partial decay as the E1HH1 peak rises. The spectrally integrated peak amplitudes are shown in Fig. 2. The 250 fs

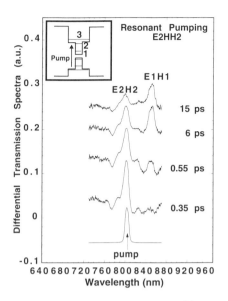

**Figure 1** Differential transmission spectra with resonant E2HH2 excitation at time delays of 0.35,0.55, 6 and 15 ps. The peaks are exciton transitions bleached by pump induced carriers.

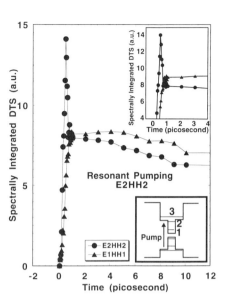

**Figure 2** Time evolution of the spectrally integrated DTS changes at E2HH2 and E1HH1 excitons from Figure 1. The inset shows the data at early times.

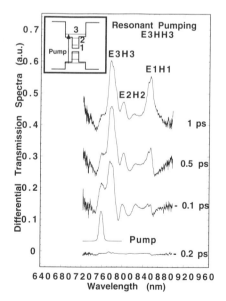

**igure 3** Differential transmission spectra with ~sonant E3HH3 excitation taken at time delays of ).2, -0.1, 0.5, 1 ps. The peaks are exciton transitions leached by pump induced carriers.

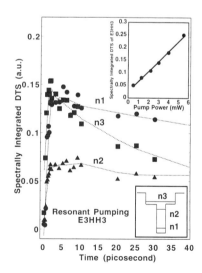

**Figure 4** Time evolution of the spectrally integrated DTS changes at E3HH3, E2HH2 and E1HH1 transitions from Figure 3. The inset shows spectrally integrated DTS values of E3HH3 transition vs. pump density.

decay of E2HH2 and associated rise of E1HH1 are attributed to the electronic relaxation between the first and second subband in the conduction band. [5] This value is very close to the calculated LO-phonon-mediated scattering time. The 250 fs fast decay is only observed when pump energy equal to E2HH2 transition and coherent artifacts will be shorter than 120 fs, thus the pump-probe transient interactions are not the major reasons for the fast relaxation. After the fast intersubband relaxation, both E2HH2 and E1HH1 peaks show slow decays (about 80 ps) due to carrier recombination because of poor interface quality. The nonzero amplitude of the E2HH2 transition after the initial decay we attribute to the residual hole population in the HH2 valence band level. In figure 3, we show a series of DTS where we resonantly pump E3HH3 transition. Initially the E3HH3 signal rises with the integral of the pump pulse. Due to the reduced intersubband scattering rate from level 3 to level 1, the E3HH3 peak shows a much slower decay. The spectrally integrated DTS are shown in Fig. 4. The E3HH3 decay is on the order of 50 ps (15 ps from calculation), significantly longer than the E2HH2 decay. Our calculations show a very long decay time $\tau_{32} \approx 1$ ns due to the subband splitting (28 meV) between 3 and 2 being smaller than the LO phonon energy, thus the n3->n1 relaxation dominates the E3HH3 decay process. From the slow E3HH3 and fast E2HH2 decay rates shown by DTS, we demonstrate that the intersubband relaxation rates can be modified in the stepped QW structure.

The calculated relative oscillator strengths for E3HH3, E2HH2 and E1HH1 as 1, 0.97 and 0.94 respectively are roughly equal. Since the E3HH3 peak is much larger than that of E2HH2 and the E3HH3 decay time (50 ps) is much longer than that of E2HH2 (250 fs), we can confirm the presence of a population inversion between the n=3 and n=2 levels due to spectrally integrated DTS signals proportional to each subband population. In fact, the population inversion is maintained during the entire carrier decay. This implies that a significant population inversion between levels 3 and 2 can be maintained by CW pumping carriers from the n1 to n$\geq$3 in a doped QW sample.

## SUMMARY

We have investigated intersubband relaxation rates in a stepped quantum well structure at room temperature using differential transmission spectroscopy with subpicosecond time resolution. Due to the reduced wavefunction overlap and larger wavevector required for intersubband scattering, the intersubband relaxation rates can be modified from a few hundred femtosecond to a few tens of picoseconds in a stepped QW. Our data clearly show a fast electronic intersubband relaxation time of 250 fs from level 2 to level 1 and a slow relaxation time 50 ps from level 3 to level 1. Those measured time constants are consistent with our calculated values. A population inversion between levels n=3 and 2 separated by 7 THz has been observed for the first time to our knowledge. Optical pumping ($CO_2$ or other IR sources) from n=1 to n$\geq$3 in doped structure should be able to generate FIR radiation in this stepped quantum well structure.

# REFERENCES

[1] J. Faist, F. Capasso, D.L. Sivco, C. Sirori, A. Hutchinson, and A.Y. Cho, Science **264**, 553 (1994).

[2] S. Hunsche, K. Leo, H. Kurz, and K. Kohler, Phys. Rev. B **50**, 5791(1994).

[3] R.Ferreira and G. Bastard, Phys. Rev. B **40**, 1074 (1989)

[4] J. A. Brum, T. Weil, J. Nagle, and B. Vinter, Phys. Rev. B **34**, 2381 (1986)

[5] D. Y. Oberli, D. R. Wake, M. V. Klein, J. Klem and H.Morkoc, Phys. Rev. Lett. **59**, 696 (1987).

[6] R. A. Hopfel, R. Rodrigues, Y. Iimura, T. Yasui, Y. Segawa, Y. Aoyagi, S. M. Goodnick, and M. C. Tatham, Phys. Rev. B. **47**, 10943 (1993).

[7] S. Hunsche, H. Heesel, H. Kurz, Optics Communications, **109**, 258, 1994.

[8] S. Hunsche, K. Leo, and H. Kurz, Phys. Rev. B **49**, 16565 (1994).]

# FAR-INFRARED MODULATED PHOTOCURRENT IN GaAs/AlGaAs COUPLED QUANTUM WELL TUNNELLING STRUCTURES

R.J. Stone[†], J.G. Michels[†], S.L. Wong[†], C.T. Foxon[*], R.J. Nicholas[†] and A.M. Fox[†]

[†]Clarendon Laboratory, Physics Department, University of Oxford, Parks Road, Oxford, OX1 3PU, UK
[*]Department of Physics, University of Nottingham, University Park, Nottingham, NG7 2RD, UK.

## INTRODUCTION

Interest in the high frequency response of quantum well tunnelling structures has prompted a number of studies by far-infrared (FIR) spectroscopy either using free electron lasers[1], or more conventional sources[2]. In principle, these FIR studies are able to determine the fundamental limits of the high frequency response, since there is a transition from classical rectification to quantum effects when the FIR frequency exceeds the intrinsic time constants of the device[3]. As yet, there have been no studies into the effect of FIR on the tunnelling current generated in quantum well photodiodes. In this work, we report studies of the response of a GaAs/ $Al_{0.33}Ga_{0.67}As$ coupled quantum well p-i-n diode to FIR radiation in the 50-300µm range. We find that the FIR induces a strong modulation of the tunnelling photocurrent, especially between 57 and 90µm. The origin of this new effect is not fully understood, but is thought to be associated with carrier heating processes. The modulation effect we observe could prove to be useful for the detection of FIR radiation, in a similar way that intersubband transitions are used to detect infrared radiation at shorter wavelengths[4].

## SAMPLE AND TECHNIQUE

The GaAs/$Al_{0.33}Ga_{0.67}As$ p-i-n photodiodes studied contained 25 double quantum well (DQW) pairs, located in the 1µm intrinsic region of the diode. The repeat unit of the DQW consisted of a wide (158.3Å) well, a thin (17Å) barrier and a narrow (79.1Å) well, separated from the next DQW by a thick (152.6Å) barrier. The diode was grown on an $n^+$ substrate, with a 0.24µm thick short period superlattice and a 1µm thick $Al_{0.33}Ga_{0.67}As$ buffer layer between the intrinsic region and the substrate. The n-dopant used was Si, with a doping level within the buffer of $1.3 \times 10^{18}$ cm$^{-3}$. The top contact consisted of a 0.75µm $Al_{0.33}Ga_{0.67}As$ layer doped with Be at $1.3 \times 10^{18}$ cm$^{-3}$

*Hot Carriers in Semiconductors*
Edited by K. Hess *et al.*, Plenum Press, New York, 1996

with a $0.1\mu m$ $p^+$ GaAs capping layer. The device was processed into a $4mm^2$ mesas which had a breakdown voltage under reverse bias of -3.5V.

Figure 1 shows a schematic diagram of the experimental arrangement. The sample was mounted in a He cryostat at 2.2K. Carriers were excited selectively into the wells and not the barrier material using a 670nm diode laser. The power was kept below 200nW to reduce space charge effects, resulting in a well photogenerated carrier density of $<10^6 cm^{-2}$. The FIR was provided by a CW molecular gas laser, which gives discrete lines between $30\mu m$ and $700\mu m$. The FIR was focused onto the device with a waveguide cone.

The FIR modulated photocurrent (FIRPC) was measured with a chain of two lock-in amplifiers (see fig. 1). The visible and FIR beams were chopped at frequencies of $f_1$ and $f_2$. The first lock-in amplifier was synchronised to $f_1$ and measured the photocurrent (PC), with a small time constant. The output of the first lock-in amplifier was fed to the second lock-in amplifier synchronised to $f_2$, thus measuring FIRPC. This technique is analogous to optically detected cyclotron resonance[5].

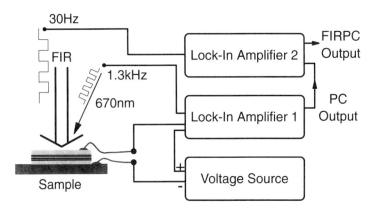

**Figure 1.** Schematic of double modulation experiment used to measure FIRPC.


**RESULTS AND DISCUSSION**

Figure 2a shows the voltage dependence of the photocurrent with no FIR incident upon the sample. We observe peaks at -1.8V and -1.3V corresponding to resonant electron tunnelling between adjacent DQWs[6]. The resonance at -1.8V arises from tunnelling from the narrow well (NW) $n = 1$ state to the wide well (WW) $n = 3$ state, while at the -1.3V peak we observe (NW $n = 1$) to (NW $n = 2$) tunnelling. Figure 2b shows the first and second derivatives of the photocurrent, used later in the analysis.

Figure 3 shows the FIRPC response to an incident wavelength ($\lambda$) of $70\mu m$. We observe that the FIRPC signal closely resembles the negative first derivative of the PC, with peaks at -1.5V and -0.95V. This indicates that the FIR has an influence on the electron tunnelling dynamics. We do not observe any second derivative component of the PC in the FIRPC, which demonstrates that we do not have optical rectification. The inset to figure 3 shows the wavelength dependence of the magnitude of the FIRPC at the -1.8V resonance normalised to the FIR laser power. We observe that the normalised response falls off with $\lambda$ until at 305µm the signal is barley resolvable.

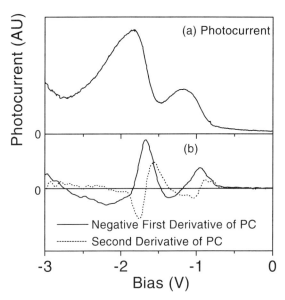

**Figure 2. (a)** Photocurrent response of sample at 2.2K. **(b)** First and second derivatives of photocurrent.

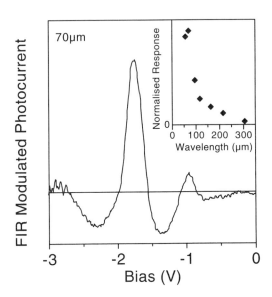

**Figure 3.** FIR modulated photocurrent signal at 70 µm. Inset: FIR power normalized response at −1.8V vs. wavelength.

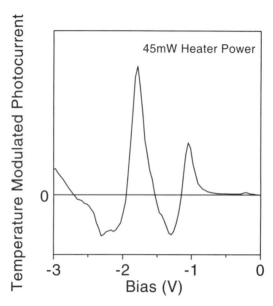

**Figure 4.** Temperature modulated photocurrent signal with 45mW heater power, modulation frequency 0.1Hz.

The most likely explanation of the observed behaviour is carrier heating. It is known that the photocurrent in these samples is extremely sensitive to temperature[7]. Fig. 4 shows the result of an experiment in which the temperature of the device was cycled up and down at 0.1Hz from 6K, using a heater coil. The temperature modulated photocurrent (TPC) was recorded in a similar manner to the FIRPC. We notice that the TPC has a first derivative form, which is comparable to the FIRPC signal, and is thus strong evidence that the FIRPC is a heating effect.

The mechanism by which temperature variations (ΔT) change the photocurrent is understood as follows: increasing the carrier temperature causes a redistribution of the carriers to higher energy within the subband. Since the tunnelling current is primarily a function of the difference in energy between the states in adjacent DQWs, this increase in the carrier energy is analogous to increasing the bias across the sample, thus leading to a derivative signal. Other studies on the effect of THz radiation on semiconductor tunnelling devices[8] report a rectified response proportional to the second derivative of the current. As noted above, this is not observed in our data, which suggests that the FIRPC signal must be due to other effects. The fact that we do not observe rectification suggests that the tunnelling time through the thick barrier in our sample is slower than the lowest FIR frequency used (5.3Thz).

The FIR could cause carrier heating either by dissipating power directly in the quantum wells, or by being absorbed in the p-type top contact. In the former case, the mechanism could be either intersubband or intraband transitions, or simply ohmic heating. In the latter case, the absorption mechanism could be either free carrier, or intervalence band absorption, the heat being transferred to the quantum wells by thermal conduction through the lattice. None of these possibilities fit the data convincingly. Intraband absorption in the wells should increase as $\lambda^2$, which is not observed. Intrasubband or ohmic heating require photons polarised along the growth ($z$) direction: coupling could occur via aerial effects[8] in the contact wires, but would be expected to roll off as $\omega^4$, which is again contrary to our data. The intervalence band absorption does predict a heating effect with the observed $\lambda$ - dependence, but the p-density of $5 \times 10^{17}$ cm$^{-3}$ required to give a peak at $\sim 70\mu m$ is unrealistically large for Be doped layers at 2K.

It is worth noting that because the carrier density in the wells is very small, it is not necessary to absorb much power to change the carrier temperature significantly. A simple calculation using the known energy loss rates[9] shows that $\Delta T$ can be as large as $\sim 10K$ with only 1nW of power absorbed. Such increases in carrier temperature due to FIR radiation have been observed in quantum well structures with $\Delta T$ as large as $\sim 100K$[10].

## CONCLUSION

We report a new FIR modulated photocurrent effect in GaAs/Al$_{0.33}$Ga$_{0.67}$As DQW tunnelling p-i-n diodes under normal incidence. We find that the modulation due to the FIR is large (1%, or a responsivity of $10\mu A/W$ which could possibly be increased by using a higher optically generated carrier density), which indicates our device may be of use as a detector, especially considering the normal incidence geometry and the high impedance associated with photodiodes. The exact details of the processes involved are not yet fully understood due to the complexity of the system, but we consider carrier heating to be the most likely cause.

## ACKNOWLEDGEMENTS
This work was partly supported by the EPSRC (grant H32995) and The Royal Society. The samples were grown at Philips Research Laboratories, and processed by F. Stride at University College London.

## REFERENCES
1. J.P. Kaminski et al, Nucl. Instr. and Meth. in Phys. Res. A, **341** 169 (1994)
2. V.A. Chitta, R.E.M. de Bekker, J.C.Maan, S.J. Hawksworth, J.M. Chamberlain, H. Henini and G.Hill, Semicond. Sci. Technol., **7** 432 (1992)
3. H.C. Liu, "Resonant Tunnelling: Physics and Applications", Plenum, NewYork
4. B.F. Levine, J. App. Phys, **4** R1 (1993)
5. J.G. Michels, R.J. Warburton, R.J. Nicholas and C.R. Stanley, Semicond. Sci. Technol. **9** 198 (1994)
6. R.G. Ispasoiu, A.M. Fox, C.T. Foxon, J.E. Cunningham and W.Y. Yan, Semicond. Sci. Technol., **9** 545 (1994)
7. R.G. Ispasoiu, A.M. Fox, C.T. Foxon, J.E. Cunningham and W.Y. Yan, Appl. Phys. Lett. **63** 2917 (1993)
8. J.S. Scott, J.P. Kaminski, M. Wanke, S.J. Allen, D.H. Chow, M. Lui and T.Y.Liu, Appl. Phys. Lett., **64** 1995 (1994)
9. D.R. Leadley, R.J. Nicholas, J.J. Harris and C.T. Foxon, Semicond. Sci. Technol., **4** 879 (1989)
10. P.C. van Son, J. Cerne, M.S. Sherwin, S.J. Allen Jr., M. Sundaram, I.-H. Tan, D. Bimberg, Nucl. Instr. and Meth. in Phys. Res. A, **341**, 174 (1994)

# DYNAMICS OF EXCITONS IN A CdSe-ZnSe MULTIPLE QUANTUM WELL

F.Yang, G.R.Hayes, R.T.Phillips and K.P.O'Donnell*

Cavendish Laboratory
University of Cambridge
Madingley Road
Cambridge
England, CB3 0HE

The optical spectrum of II-VI quantum wells (QWs) is, in general, dominated by inhomogeneous broadening, which is caused by random fluctuations of QW layer thickness. In these QWs, hot carriers have not only kinetic energy but also potential energy which varies following the in-plane bandgap fluctuations. Apart from hot-exciton formation and relaxation as in the intrinsic QW, hot carrier relaxation in these QWs also includes the localisation of excitons and the migration of localised excitons between localisation sites. In this paper, we present a study of the exciton relaxation in a 100-period 10Å CdSe-50Å ZnSe multiple quantum well by using time-resolved photoluminescence spectroscopy. The sample has a ±1 monolayer fluctuation in QW thickness. The experimental setup[1] includes a frequency-doubled modelocked Ti:Sapphire laser, a liquid helium cooled cryostat, and a time-resolved luminescence detection system using the up-conversion technique. The overall temporal resolution was about 300 fs and the spectral resolution was 18 meV.

Figure 1 shows the luminescence spectra at different times after excitation. For the first 1 to 10 ps, the luminescence spectra have almost the same shape and peak position. From 10 to 100ps, the luminescence peak redshifts and narrows by about 30meV. For times greater than 100ps, both the peak position and the width of the spectrum show little further change.

The time evolution of the luminescence intensity was measured at 2.30, 2.34, 2.42, 2.48, 2.52eV, as shown as symbols in Figure 2. The rise of the luminescence at the peak position 2.42eV is also measured from -5 to 50ps with a smaller temporal step. The common feature of these decay curves is that the luminescence intensity follows a rapid rise during the first few picoseconds. After the rapid rise, the luminescence intensity at different energies change differently. For the data measured at 2.52eV, the intensity decreases exponentially. For other data, as the photon energy decreases the rapid rise of the luminescence turns into a slower rise before it falls off exponentially. The slow rise time increases with decreasing photon energy.

The above experimental results can be explained by an exciton relaxation model including the migration of localised excitons[2]. The key points are: first, photo-excited hot carriers lose kinetic energy and form localised excitons at a much faster rate than the exciton migration or recombination rate. The formation rate of localised excitons is independent on

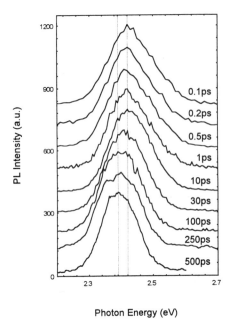

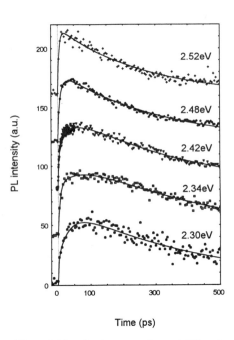

Figure 1. Time-resolved photoluminescence spectra of the CdSe-ZnSe MQWs at 10K.

Figure 2. Photoluminescence decay at different detecting energies at 10 K.

energy. Second, the luminescence is due to the recombination of localised excitons, which has a *single* time constant across the whole inhomogeneously broadened luminescence spectrum. Such a model can be formalised into a rate equation for the time-dependent distribution of localised excitons $f(E,t)$,

$$\frac{df(E,t)}{dt} = G(E,t) + \int_{E}^{E_{max}} \frac{f(\varepsilon,t)}{\tau(\varepsilon,E,t)}d\varepsilon - \frac{f(E,t)}{\tau_r} - \int_{E_{min}}^{E} \frac{f(E,t)}{\tau(E,\varepsilon,t)}d\varepsilon \qquad (1)$$

where $f(E,0)$ is the steady-state distribution of localised states[3], $E_{min}$ and $E_{max}$ are the lower and upper limit of the localised exciton distribution. The density of the exciton population at $t$ and within the energy interval $\Delta E$ can then be written as $n_0 f(E,t)\Delta E$, where $n_0$ is the density of photo-excited electron-hole pairs at $t=0$. $G(E,t)$ describes the generation rate of localised excitons through hot carrier and/or hot-exciton cooling. $1/\tau(E_1,E_2,t)$, describes the migration rate of a localised exciton at time $t$ from a state of energy $E_1$ to $E_2$. $\tau_r$ is the life-time of localised excitons including radiative and non-radiative recombination. In order to compare this rate equation with our experimental results in a more explicit way, (1) can be further simplified into a set of equations as,

$$\frac{dn_h}{dt} = -\frac{n_h}{\tau_l}$$

$$\frac{d\langle n_1 \rangle}{dt} = F(E)\frac{n_h}{\tau_l} - \frac{\langle n_1 \rangle}{\tau_r} - \frac{\langle n_1 \rangle}{\tau_{ml}} \qquad (2)$$

$$\frac{dn(E,t)}{dt} = f(E,0)\frac{n_h}{\tau_l} + \frac{\langle n_1 \rangle}{\tau_{mi}} - n(E,t)(\frac{1}{\tau_r} + \frac{1}{\tau_{mo}})$$

where at t=0, $n_h$ is the density of the electron-hole pair created by photo-excitation, $\langle n_1 \rangle = n_0 \cdot \int_E^{E_{max}} f(E,t)dE$ is the density of localised excitons from E to $E_{max}$; $F(E) = \int_E^{E_{max}} f(E,0)dE$ ; n(E,t) is the same as in Eq(1). The different time constants are $\tau_l$ the localisation time of excitons, $\tau_r$ the recombination time of localised excitons, $\tau_{mo}$ the migration-out time of localised excitons to lower energy states, $\tau_{mi}$ the migration-in time from higher energy states. $1/\tau_{ml} = \int_E^{E_{max}} \frac{1}{\tau(\varepsilon,E,t)}dE$ is the total migration rate of $\langle n_1 \rangle$ to lower energy states. (2) can be solved analytically by assuming $\tau_l \ll \tau_{ml} \ll \tau_r$,

$$I_{lum} \propto n(E,t)/\tau_r \propto$$

$$[(\exp(-t/\tau_r - t/\tau_{mo}) - \exp(-t/\tau_l)) + C(\exp(-t/\tau_r - t/\tau_{mo}) - \exp(-t/\tau_r - t/\tau_{ml}))] \quad (3)$$

where C is a function of $f(E,0)$, F(E), $\tau_l$ and $\tau_{mi}$.

The least square fits of (3) to the experimental data are shown in Fig.2 as solid lines, and the fitting parameters are listed in Table 1. A constant localisation time $\tau_l$ of ~4ps for all detection energies agrees well the prediction of our model and confirms that the hot-excitons (-carriers) cooling and localising rate is independent of energy. The recombination time of localised excitons is a constant (of ~470ps). This also agrees with our model. The migration-out time increases with decreasing energy. This is because at lower energy, there are fewer available states for localised excitons to migrate to.

Table2.　Time constants obtained using (5).

| Photon energy (eV) | Localisation time $\tau_l$(ps) | Decay time $\tau_r$(ps) | Migration-out time $\tau_{mo}$*(ps) | Migration time $\tau_{ml}$(ps) |
|---|---|---|---|---|
| 2.30 | 4.0 | 466 | 9262 | 40.4 |
| 2.34 | 4.0# | 468 | 3858 | 76.6 |
| 2.42 | 4.0 | 476 | 2030 | 43.9 |
| 2.48 | 4.0 | 494 | 939 | 12.3 |
| 2.52 | 4.2 | 472 | 610 | 20.3 |

#This value is fixed in the fitting. By varying $\tau_l$ the least square fit gives $\tau_l$ 6.6ps and $\tau_r$ is 405ps, which is plotted by a dotted line in Fig.2.

# REFERENCES

1. G.R.Hayes, I.D.W.Samuel and R.T.Phillips, Exciton dynamics in electroluminescent polymers studied by femtosecond time-resolved photoluminescence spectroscopy, Phys. Rev. B (to be published)
2. F.Yang, G.R.Hayes, R.T.Phillips, and K.P.O'Donnell, Exciton relaxation in a thin CdSe-ZnSe multiple quantum wells, Phys.Rev. B (submited)
3. F.Yang, M.Wilkinson, E.J.Austin, and K.P.O'Donnell, Origin of the Stokes shift: a Geometrical model of exciton spectra in 2D semiconductors, Phys.Rev.Lett. **70**, 323 (1993).

# LOW TEMPERATURE ANTI-STOKES LUMINESCENCE IN CdTe/(Cd,Mg)Te QUANTUM WELL STRUCTURES

R. Hellmann[1], A. Euteneuer[1], S.G. Hense[1], J. Feldmann[1], P. Thomas[1], E.O. Göbel[1,*], A. Waag[2], and G. Landwehr[2]

[1] Fachbereich Physik and Zentrum für Materialwissenschaften, Philipps-Universität, Renthof 5, D-35032 Marburg, Germany
[2] Physikal. Institut der Universität Würzburg, D-97074 Würzburg, Germany
[*] present address: Physikalisch Technische Bundesanstalt, Bundesallee 100, D-38116 Braunschweig, Germany

## INTRODUCTION

Over the last decades, anti-Stokes photoluminescence (ASPL), i.e., emission that occurs at a wavelength shorter than the wavelength of the excitation source, has been studied in a variety of systems, such as atoms and molecules[1], polymers[2] and amorphous semiconductors[3]. The main issue addressed in these studies was the identification of the underlying microscopic mechanisms as, e.g., multi-photon processes, Auger recombination or thermal excitation. In the case of crystalline bulk semiconductors and semiconductor heterostructures ASPL excited by a coherent two-photon absorption process, using photon energies well below the optical band-gap is a well known phenomenon[4]. The radiation then exhibits a superlinear dependence on excitation intensity. Recently, however, ASPL from an asymmetric double quantum well (QW) showing a linear intensity dependence has been reported[5] and was attributed to a dipole-dipole transfer.

In this contribution, we report on low temperature photoluminescence (PL) studies on semi-conductor multiple quantum wells (MQW), which provide evidence for a transfer of excitons from confined QW states to above-barrier states. We find an efficient ASPL from the barrier when the excitation energy is tuned to energies equal to or higher than the spectral position of the lowest heavy-hole (hh) excitonic n=1 QW transition. Energy, intensity and temperature dependent measurements reveal that the ASPL is due to a zero-phonon two step absorption process, if we assume that localized and/or impurity bound QW-excitons act as saturable intermediate states.

## EXPERIMENTAL

We have studied a set of nominally undoped $Cd_{1-x}Mg_xTe/Cd_{1-y}Mg_yTe$ MQWs grown by MBE. The structures consist of 7.5 nm $Cd_{1-x}Mg_xTe$ wells (x=0 and 0.075) and of $Cd_{0.60}Mg_{0.40}Te$ barriers with thicknesses $L_B$ = 7.5, 76, and 144 nm, respectively. PL is excited either using a pulsed dye laser or a pulsed, frequency doubled Ti:sapphire laser and is monitored using a photomultiplier or a diode array. All experiments are performed in a temperature variable helium flow cryostat.

## RESULTS AND DISCUSSION

Figure 1 shows a PL spectrum of the MQW with $L_B$ = 144 nm excited at the light-hole (lh) n=1 exciton resonance of the QW. It is composed of two high energy peaks, which stem from the recombination of localized and bound excitons in the QW and of two minor peaks together with a long wavelength tail arising from the substrate. Surprisingly, however, the spectrum also shows luminescence at wavelengths shorter than the excitation wavelength, the spectral position and shape of which is identical to the barrier PL under above barrier excitation.

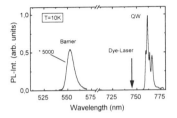

Fig.1: Time-integrated PL spectrum of the MQW with a barrier thickness of 144 nm excited below the barrier (T=10K).

The dependence of the ASPL on excitation energy, i.e., the AS-PLE is shown in the inset of Fig. 2 together with the conventional PLE spectrum of the MQW. Clearly both PLE spectra are essentially the same displaying the spectral features of the 1s hh and lh resonances of the QW at 759.4 nm and 745 nm, respectively. In addition, no ASPL is detectable for excitation energies less than the 1s hh resonance implying that the excitation scenario of the anti-Stokes barrier PL must involve real QW states. We therefore exclude coherent two-photon absorption via below band-gap states to contribute considerably to the ASPL.

The intensity of the spectrally integrated ASPL versus excitation density is shown in Fig. 2. The double logarithmic plot reveals two straight lines having slopes of 2.1 and 1.08, i.e., quadratic and linear dependences, for densities below and above $2 \times 10^8$ cm$^{-2}$, respectively. Since both the QW and barrier PL under direct excitation reveal linear density dependences this two-fold density dependence displays the intrinsic behavior of the anti-Stokes process and is not related to a saturation of nonradiative traps. In addition, the change in slopes indicates saturation effects in the anti-Stokes transfer. Based on these observations both coherent two-photon absorption and most likely Auger recombination can be ruled out, since they would exhibit quadratic or cubic dependences on excitation intensity in the entire intensity range.

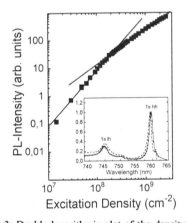
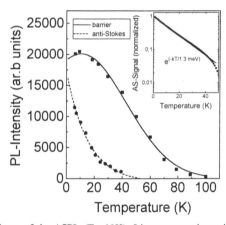

Fig.2: Double logarithmic plot of the density dependence of the ASPL (T= 10K). Linear regressions yield slopes of 2.1 and 1.08 for densities below and above $2 \times 10^8$ cm$^{-2}$, respectively. The inset shows the QW PLE spectrum (dashed) and the anti-Stokes PLE spectrum (dots and solid line as a guide to the eye)

Fig.3: Temperature dependence of the anti-Stokes (dots) and barrier (squares) PL. Dashed and solid lines are fits to the experimental data. The inset shows the ASPL normalized to the barrier PL dependence.

Figure 3 shows the temperature dependences of the anti-Stokes and barrier PL intensity, as measured at a density higher than the saturation density (cf. Fig.2). While it might be expected that the ASPL displays the same temperature dependence as the barrier PL, the results, on the contrary, clearly show that the ASPL is much more sensitive to temperature than the barrier PL. In particular, while the ASPL decreases much faster with temperature and is not detectable for T > 50K, the barrier luminescence intensity is nearly constant up to 20K and is detectable even up to 100K. Hence, the temperature dependence of the ASPL reflects predominantly the temperature dependence of the anti-Stokes excitation process. For a further analysis, however, we have normalized the temperature dependence of the ASPL with respect to the barrier PL dependence (inset of Fig.3) in order to exclude contributions due to an enhanced carrier trapping. Note that in this temperature range the QW PL exhibits nearly no temperature dependence. As the ASPL occurs at low temperatures, i.e., kT is much less than the energy spacing between the lowest n=1 QW level and the band gap of the barrier, and since it displays an exponential decrease with temperature rather than an activation type behavior, thermal excitation can be excluded as being the anti-Stokes excitation process.

In the following we show that our results are in full agreement with the assumption of a direct zero-phonon two-step absorption process (TSA), if we assume that localized and/or impurity bound QW-excitons act as saturable intermediate states. In this scenario the anti-Stokes process starts with the excitation of electron-hole pairs, which relax in energy, form excitons and, finally, reach localized or impurity bound states. Since these localized or bound exciton states are composed of a broad spectrum of $k$-states, excitation into high-energetic electron hole pair states with a well defined wave-vector $k$ by direct photon absorption without the participation of phonons (which could account for momentum conservation) is possible. This `hot' electron-hole pair density then relaxes in energy and the carriers either reach the lowest barrier states or are trapped into the QW. Those carriers which reach localized states in the ternary barrier material then recombine radiatively giving rise to the observed ASPL.

This excitation scenario consistently explains our observations: (i) The AS-PLE follows the QW-PLE, i.e., the QW absorption, since in this TSA process localized or bound states in the well act as intermediate states (after relaxation). (ii) The ASPL intensity is proportional to the number of photons responsible for the second absorption step and the population of intermediate states. While the former has previously been shown to be proportional to the excitation intensity[6], the latter is, at low excitation intensities where the intermediate states are not saturated, also propotional to the excitation intensity. As a result, at low excitation intensities the ASPL displays a quadratic and, at higher intensities where the intermediate state population is saturated, a linear dependence on excitation density. Consequently, the change of the slope of the density dependence gives a rough estimate of the density $N_{loc}$ of the intermediate exciton states. From our data we obtain $N_{loc} = 2 \times 10^8$ cm$^{-2}$ and $2.3 \times 10^8$ cm$^{-2}$ for the MQWs with 144nm and 77 nm barrier thickness, respectively. In the $Cd_{0.925}Mg_{0.075}Te/Cd_{0.60}Mg_{0.40}Te$ structure, where presumably compositional disorder due to alloy fluctuations increases the density of localized states in the QW, we obtain $N_{loc} = 1.1 \times 10^9$ cm$^{-2}$. (iii) The temperature dependence reflects the thermal depopulation of intermediate states. In this respect, the observed exponential decrease of the ASPL intensity, described by $\exp(-kT/1.3 \text{ meV})$ gives a measure of the energetic depth of the density of intermediate states. We note that an indirect phonon-assisted two-step absorption process cannot account for our observations, as it transfers both localized *and* delocalized QW excitons. Hence, there are no saturable intermediate states leading to a purely quadratic intensity dependence.

## SUMMARY

In summary, we have presented detailed photoluminescence studies of low temperature anti-Stokes luminescence occuring in II/VI semiconductor quantum wells. The underlying microscopic process is identified as being zero-phonon two-step absorption involving localized or impurity bound exciton states in the quantum well, which act as saturable intermediate states.

## ACKNOWLEDGMENTS

The work at the University of Marburg has been supported by the Deutsche Forschungsgemeinschaft (DFG) through the Leibniz Förderpreis and the SFB 383. R.H. acknowledges support by the DFG through the Graduierten Kolleg `Optoelektronik Mesoskopischer Halbleiter'.

## REFERENCES

[1] see, e.g., Y.R. Shen, The Principles of Nonlinear Optics, John Wiley & Sons, New York, (1984)
[2] Z.G. Soos, and R.G. Kepler, Phys. Rev. B **43**: 11908 (1992)
[3] S.Q. Gu, M.E. Raikh, and P. C. Taylor, Phys. Rev. Lett. **69**: 2697 (1992)
[4] see, e.g., R. Cingolani, and K. Ploog, in: *Advances in Physics* **40**, 535 (1992); and references therein
[5] A. Tomita et al., Int. Conf. on Quantum Electronics, Technical Digest **9**: 102 (1992)
[6] R. Hellmann et al., Phys. Rev. B **51**: 18053 (1995)

# FAR-INFRARED EMISSION AND MAGNETO TRANSPORT
# OF 2D HOT HOLES

Yu.L.Ivanov, G.V.Churakov,V.M.Ustinov, A.E.Zhukov

A.F.Ioffe Physico-Technical Institute, Russian Academy of Sciences
Politekhnicheskaya 26, St Petersburg, 194021, Russia

Streaming of hot carriers in bulk materials such as p-Ge is well known. A lot of interesting galvanomagnetic and FAR emission phenomena is connected with it [1]. However streaming in quantum well structures has not been studied experimentally [2]. We report new experimental results of a galvanomagnetic and FIR emission investigation in 2DHG in a strong electric field **E** parallel and magnetic field **H** perpendicular to the 2DHG layers.

GaAs/AlGaAs quantum well structures grown by MBE in [001] direction are Be modulation doped and the width of wells vary from 60 to 400 Å. The samples are supplied by ohmic contacts which length is much large than the space between them to short Hall electric field. The pulse electric field used is up to 8 kV/cm and the magnetic field is up to 6.4 T. The Ge(Ga) detector for the integral FIR emission is used. The temperature of measurements is 4.2 K.

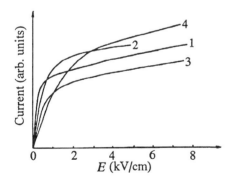

**Fig. 1.** Current-voltage characteristics of the GaAs/AlGaAs MQWs structures. Curve 1—the 400 Å well width structure; Curves 2 and 3—the 200 Å well width structures; Curve 4—the 68 Å well width structure.

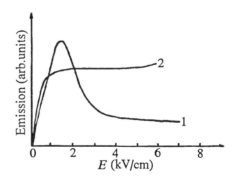

**Fig. 2.** FIR emission intensity dependencies on electric field. The radiation propagates perpendicular to the plane of the MQWs structures. Curve 1—for the 400 Å well width structure; Curve 2—for the 200 Å well width structure.

The current-voltage characteristics show a clear saturation of the current at strong fields E (see Fig. 1). It is connected with strong interaction between holes and LO-phonons (streaming state) and the hole distribution function in velocity space becomes strip-shaped along the E direction. Note that the narrower quantum well width the stronger is the electric field at which the carriers come to the streaming state that we explain by the influence of interface on the carrier mobility.

FIR emission intensity vs electric field for the samples of 400 and 200 Å is shown in Fig 2. Curve 1 presents the 400 Å well width sample, where the effect of confinement for holes is rather small. Therefore the hole behavior in a strong electric field is almost the same as it is in some bulk materials. The experimental dependence we have obtained is possibly connected with "the bulk" light-to-heavy hole radiative transitions that agrees well with the theory. In this case the emission intensity has a maximum. We explain the decrease of intensity and further independence on electric field by decreasing of the light - heavy holes concentration ratio which takes place in the streaming state as compared to equilibrium state.

Fig.2, curve 2 presents the 200 Å well width sample, where the effect of confinement is stronger. The calculations and photoluminescence experiments indicate that the equilibrium holes occupy only the ground quantum subband. The holes being heated by the electric field occupy the upper quantum subbands and come back to the ground state, in particular, with radiative transitions. In the upper subbands the heavy and the light hole states are mixed that allows the FIR emission which is connected with intersubband radiative transitions to propagate both perpendicular and parallel to the quantum layers. In this case the subband occupation is not practically changed in the streaming state, therefore FIR emission intensity is also not changed with increas of electric field.

Fig.3 shows the FIR emission intensity dependences vs electric fields for the 68 Å well width samples. We calculate that in such quantum wells only two quantum levels are below the LO-phonon energy. The emission intensities we observe have different characters in two directions of emission propagation. One of them, obtained in the direction along the plane of QWs, has a clear maximum and increases at higher E. The other, obtained perpendicular to the plane through the special diaphragm ( which shuts the edges of the structure), is much weaker than in the first case and increases steadily with slow saturation.

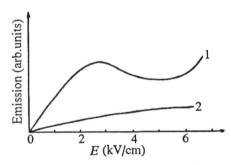

**Fig. 3.** FIR emission intensity dependencies on electric field for the 68 Å well width structure. Curve 1—the radiation propagates along to the layers. Curve 2— the radiation propagates perpendicular to the layers.

The former we connect with the radiative transitions between the two quantum hole subbands lying below the LO-phonon energy, when the electric wave vector is directed orthogonal to QW layers and orthogonal to the electric field direction. At low electric fields the hole states are diffusive that means that the holes are able to be scattered elastically by the acoustic phonons in to the first quantum subband and then emitting photon transit to the ground subband. The stronger the electric field the more hot holes occupy the first subband. Therefore the emission intensity increases. Then the electric field becomes stronger and the holes begin to reach the streaming state. During the time of LO-phonon emission the hot holes are able to gain a surplus of energy above $\hbar\omega$, but it is still small for the holes occupying the first subband after LO-phonon emission. The number of the streaming holes increases with the electric field, the hole population of the first subband decreases and therefore the FIR emission intensity goes down. Further increase of the electric field appears to be enough to gain the hot holes for the first subband occupation and the emission intensity rises up again.

In the other case the emission perpendicular to the quantum layers is basically connected with the light-to-heavy hole state transitions, which radiation is able to propagate both orthogonal and along the quantum layers. In the 68 Å QW width structure the ground light hole subband is very close to the LO-phonon energy, therefore only the hot holes from the tail of the distribution function are able to occupy it. The number of these holes increases with the electric field and the radiation increases too.

Fig. 4 shows the current and FIR emission dependencies vs magnetic field for the sample of 200 Å. There is a clear maximum when the applied field **E** is along [110] crystallographic direction. This can be considered as a manifestation of the negative magneto resistance (see Fig.4a, curve 1). Also the magnetoresistance has been obtained as a positive when the field **E** is oriented along the [100] crystallographic direction. This experimental data can be explained by warping of a constant energy surface (for instance at the optical phonon energy) in velocity space. In the [110] direction this surface has a hollow and the hole velocity in the streaming condition at the optical phonon energy is not so big as it is in the [100] direction (see Fig.4b, curve 1, point A). The magnetic field turns the hot hole trajectory in the velocity space and the hole velocity at the optical phonon energy becomes bigger (see Fig 4b, curve 2, point B). It makes the dissipative current bigger too. Note that theory predicts a sharp drop of the dissipative current at high magnetic fields in the streaming condition. It is not observed in the experiments. Apparently a noticeable part of nondissipative current flow through samples at high magnetic fields. However, above stated explanation of the negative magneto resistance is valid too.

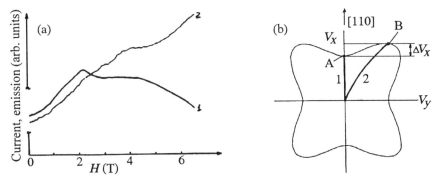

**Fig. 4.** (a) Dependences of dissipative current at E = 1.1 kV/cm (1) and FIR emission at E = 1.65 kV/cm (2) vs magnetic fields. The increase of current is up to 5% of the magnitude. The increase of FIR radiation is up to 30% of the magnitude. (b) Constant energy surface in velocity space (schematically) at the optical phonon energy. Curve 1 shows streaming state at H=0, curve 2 shows streaming state deformed by a magnetic field.

The experiment shows that magnetic field can raise the FIR emission connected with intersubband transitions. However, the mechanism of the FIR emission in a magnetic field is not absolutely clear and requires detailed theoretical study as well as further experiments.

The work is supported by The Russian Science Foundation ( Grant No. 93-02-2838).

1. *Optical and Quantum Electronics.*, v.23, No 2, (1991) Special Issue on FIR Lasers
2. Xu W., Peeters F.M. and Devreese J.T. Diffusion-to-streaming transition in a two-dimensional electron system in a polar semiconductor. *Phys. Rev.* B, 43,17, 14134 (1991)

# OSCILLATORY TRANSPORT OF ELECTRONS IN GaAs SURFACE-SPACE-CHARGE FIELDS

Wolfgang Fischler,[1] Günther Zandler,[2] and Ralph A. Höpfel[1]

[1]Institut für Experimentalphysik, Universität Innsbruck
Techniker-Straße 25, A-6020 Innsbruck, Austria
[2]Walter Schottky Institut, Technische Universität München
Am Coulombwall, D-85747 Garching, Germany

Injection of electron-hole pairs into the surface-space-charge layer of doped semiconductors leads to an acceleration of the generated carriers following the electrostatic field in the surface depletion region. The electrons and holes move into opposite directions, thus reducing the driving force by building up an opposite electric field. As a consequence of the carrier seperation, the sign of the resulting electric field in the surface depletion zone changes and the carriers are driven back until the sign changes again. This process of an altering resulting electric field gives rise to a coherent oscillation of the carriers within the time scale of momentum relaxation (ballistic regime). The oscillation amplitude of the holes should be much smaller than that of the electrons because of their larger mass. This scenario is expected only for weak scattering, momentum relaxation of the carriers leads to damping of the oscillation.

Theoretical ensemble Monte-Carlo simulations support this view of the time development of the described process. Figure 1 shows the electron density below the surface as a function of time delay after excitation with three different carrier densities. The calculated system is a Schottky-diode ($N_D = 3 \cdot 10^{16}$ cm$^{-3}$) described in the next paragraph. Excitation of electron-hole pairs occurs with a penetration depth of ~1.5 μm at the Γ-point with 100 fs FWHM pulses. The results of the space and time-dependent Monte-Carlo simulations, taking into account carrier-phonon interaction, impurity scattering (no Pauli exclusion principle and no carrier-carrier scattering) demonstrate that under appropriate conditions coherent oscillations of the electric field and the electron density within the surface depletion region are present. The oscillation frequency is close to $\omega_p^2 = ne^2/\varepsilon\varepsilon_0 m^*$ and increases with increasing excitation density. The decrease of the calculated curves to a quasistationary level after about 2.5 ps is a consequence of the weighting procedure (carrier drift into the bulk due to the surface-space-charge field).

In our femtosecond time-resolved experiments we measured the transient change of reflectivity using a combined fast scan and lock-in technique. The samples are n-GaAs Schottky-diodes with a semitransparent Cr-gate with a thickness of 10 nm. The doping

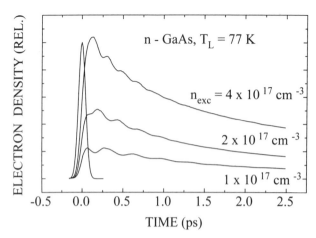

**Figure 1.** Theory: The electron density of the surface depletion region as a function of the time delay after excitation for a Schottky-diode ($N_D = 3 \cdot 10^{16}$ cm$^{-3}$). The density is weighted by a factor exp $(x/\lambda)$, where x is the distance from the surface and $\lambda$ is chosen as 50 nm. The 100 fs pump pulse is centered around 0 fs.

concentrations of the samples vary from $3 \cdot 10^{16}$ cm$^{-3}$ to $3 \cdot 10^{17}$ cm$^{-3}$ which gives rise to built-in electrical fields of several 10 kV/cm and depletion lengths of about 90 nm to 180 nm. The temperature during the measurements is kept at 100 K, in order to operate near the maximum of the electron mobility.

The photon energy of the infrared laser pulses ($\tau$ = 90 fs) of a Ti:Sapphire laser is tuned around the band gap energy (~1.5 eV corresponding to $\lambda$ = 828 nm) with a spectral width of 20 meV. It is split into an intense pump and a weak probe pulse which are polarized perpendicular to each other. The probe beam is oriented along the (110) crystal direction. A shaker periodically delays the probe beam with a frequency of 1 Hz, the pump beam is chopped with a frequency of 4 kHz. The focus of the probe beam at the sample is intentionally kept smaller than that of the pump beam which has a diameter of about 60 μm. A large area silicon pin-photodiode followed by a lock-in amplifier and a signal averager records the reflected probe beam intensity.

In Figure 2 the relative change of reflectivity is plotted as a function of the time delay between pump and probe pulse. The center of the spectral distribution of the laser beam was tuned from $\lambda$ = 830 nm to $\lambda$ = 813 nm. The doping concentration of the n-GaAs is $3 \cdot 10^{16}$ cm$^{-3}$, the average power of the pump pulse was about 400 mW. All signals exhibit a fast initial rise during the absorption of the pump pulse. Depending on the chosen spectral distribution a minimum and a second variably delayed maximum occur followed by a decrease to a quasistationary level on a picosecond time scale. At wavelengths shorter than $\lambda$ = 813 nm no oscillations are observed. The time between the two observed maxima is plotted in the inset of Figure 3 as a function of the central wavelength of the spectrum. The oscillation period varies from about 180 fs to 480 fs.

In Figure 3 the measured excitation density is obtained from the measured oscillation period $\tau_P$ using the relation $(2\pi/\tau_P)^2 = ne^2/\varepsilon\varepsilon_0 m^*$. For the calculation of the theoretical density a decrease of the absorption coefficient due to band filling effects[1] is taken into account.

From the qualitative and quantitative agreement of the experiment with the theoretical predictions we conclude that the observed oscillations can be interpreted as coherent plasma

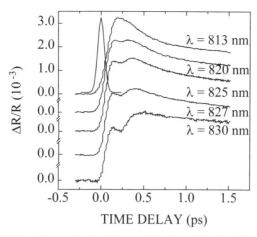

**Figure 2.** Experiment: Relative change of reflectivity as a function of the time delay between pump and probe pulse varying the central wavelength of the spectrum. The pump pulse is centered around 0 ps.

oscillations of the photoexcited carriers. This interpretation is supported by the observed dependencies on excitation density and wavelength, which is a variation of excitation density as well as carrier energy. Tuning the broad spectrum (FWHM approximately 11 nm) from below band gap energies ($\lambda$ = 833 nm) to higher energies causes an increase of the excitation density since both the part of the spectrum above the band gap energy and the joint density-of-states increase. Higher excitation densities lead according to $\omega_p^2 = ne^2/\varepsilon\varepsilon_0 m^*$ to higher oscillation frequencies of the coherent plasmons. Tuning to the high energy range ($\lambda$ > 813 nm) no oscillation is observed any more. We attribute this effect to a strongly increaesed electron-phonon scattering process when the kinetic energy of a larger fraction of electrons exceeds the LO-phonon energy (36 meV). Additionally, excitation at higher energies causes higher excitation densities for constant pump power. At higher carrier concentrations electron-hole scattering becomes more important, too. Both scattering processes shorten the momentum relaxation time of the single electrons leading to strong damping of the plasma oscillations.

The measured signals are insensitive to rotation of the polarization of pump and probe beam if their polarizations are kept perpendicular to each other. The oscillatory behaviour of the reflectivity therefore is not due to the electrooptic effect but rather to the change of the electron density (phase-space-filling) in the surface-space-charge region. The observed changes of the reflectivity ($\Delta R/R$ up to $3 \cdot 10^{-3}$) are in qualitative agreement with calculations based on band-filling by the photoexcited carrier concentrations. The influence of electro-absorption[2,3] is discussed in detail in reference 2, where the first observation of coherent plasmons by time-resolved electroabsorption has been reported.

As pointed out in references 3 and 4 an undesirable problem in reflective pump-probe experiments arises from the inhomogeneity of the excited density distribution due to the beam profile of the laser pulses. The signal read out by the probe beam is an average over different densities across the excited spot. This explains on the one hand the different detected reflectivity signals depending on the position of the focus of the probe beam within that of the pump beam and on the other hand why only one oscillation period is observed. A second problem which concerns measurements of the surface depletion layer is the fact that the probe reflection is determined by a larger region than the depth of the surface depletion region.

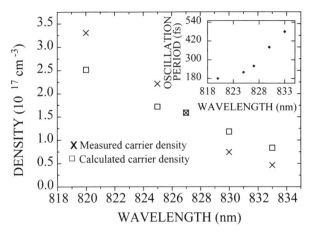

**Figure 3.** Calculated and measured density as a function of the central wavelength of the spectrum.

The measured reflectivity signals of Figure 2 uniformly reveal a broadening of the second maximum compared to the first one. This broadening indicates a slowdown of the oscillation frequency. This would cause a decrease of the electron density in the surface-depletion-region. The reason is not determined yet, but it could be a consequence of a fast hole diffusion into the Cr-layer since the Fermi energy in the metal is far above the edge of the valence band at the interface of the semiconductor-metal junction. Also scattering of hot accelerated electrons certainly plays an important role in the scenario.

In our measurements we do not observe coherent LO-phonons[5] which are generated and detected in a similar way in the surface-space-charge field of bulk GaAs. Strong oscillations of the generated holes are not expected since about 2/3 of the holes are in the heavy hole band and so their effective mass is much larger and their mobility, therefore, lower.

Further experiments will be necessary to explain open questions such as the contribution of the electro-reflectance to the measured signals. Moreover a comprehensive investigation of coupled plasmon-phonon modes[4] is an outstanding problem.

This work has been supported by the "Fonds zur Förderung der wissenschaftlichen Forschung", Austria (project No. 10065).

1. Brian R. Bennett, Richard A. Soref, and Jesús A. del Alamo, "Carrier-induced change in refractive index of InP, GaAs and InGaAsP", *IEEE J. Quantum Electron.*, 26:113 (1990).
2. W. Sha, Arthur L. Smirl, and W. F. Tseng, "Coherent plasma oscillations in bulk semiconductors", *Phys. Rev. Lett.*, 74:4273 (1995).
3. G. C. Cho, *private communications*.
4. V. Kuznetsov, and C. J. Stanton, "Coherent phonon oscillations in GaAs", *Phys. Rev. B*, 51:7555 (1995).
5. G.C. Cho, W. Kütt, and H. Kurz, "Subpicosecond time-resolved coherent-phonon oscillation in GaAs", *Phys. Rev. Lett.* 65:764 (1990).

# TEMPERATURE DEPENDENCE OF THE INTERSUBBAND HOLE RELAXATION TIME IN P-TYPE QUANTUM WELLS

Z. Xu,[1] G. W. Wicks,[2] C. W. Rella,[3] H. A. Schwettman,[3] and P. M. Fauchet[1,2,4]

[1]Department of Physics & Astronomy, University of Rochester
 Rochester, NY 14627
[2]The Institute of Optics, University of Rochester, Rochester, NY 14627
[3]Stanford Picosecond FEL Center, Stanford University, Stanford, CA 94305
[4]Department of Electrical Engineering, University of Rochester
 Rochester, NY 14627

## INTRODUCTION

Relaxation times ranging from 1 to 10 ps have been measured with infrared bleaching techniques in n-type quantum wells (QWs) when the inter-subband transition energy is larger than the LO-phonon energy.[1-3] These relaxation times have been attributed to electron-LO-phonon scattering. However, they are significantly longer than the theoretical predictions and the discrepancy is attributed to LO-phonon screening, intervalley scattering, and hot-phonons.[4] Methods that employ a lower excitation levels are a better tool for exploring the intraband relaxation. For example, sub-picosecond time-resolved Raman-scattering measurements have revealed an inter-subband relaxation time of less than 1 ps, which is consistent with the theoretical predictions.[5] For p-type QWs, the density of states and the band structure are different and it would be interesting to determine the role of different scattering mechanisms, such as LO-phonon scattering and deformation-potential scattering. In our recent work, we have shown that pump-probe measurements are suitable for p-type QWs.[6] In this paper, we report the temperature dependence of the relaxation time and present some details of the theoretical calculations.

## SAMPLE DESCRIPTION

The samples used in our experiments are strained quantum wells grown by molecular-beam epitaxy (MBE). The QW layer consists of 50 $In_{0.5}Ga_{0.5}As/Al_{0.5}Ga_{0.5}As$ periods grown on top of a 2 μm thick $In_xGa_{1-x}As$ graded layer to adjust the strain. The InGaAs well is 4 nm wide, and the AlGaAs barrier is 8 nm wide. The sample is uniformly doped with Be. The doping concentration is $10^{19}$ $cm^{-3}$ and the sheet density per QW is $1.2 \times 10^{13}$ $cm^{-2}$ if all the carriers are transferred to the wells. An absorption peak is observed

*Hot Carriers in Semiconductors*
Edited by K. Hess *et al.*, Plenum Press, New York, 1996

at 5.25 µm (220 meV) with a full width at half maximum (FWHM) of 50 meV.[6] This energy is consistent with the transition energy from the n = 1 heavy-hole (HH1) subband to the n = 1 light-hole (LH1) subband obtained from a simple effective-mass calculation including the energy splitting of 150 meV induced by the strain.[7] The broad absorption peak results from hole-hole scattering[8] and the difference in dispersion between the two subbands. We do not detect the transition involving the HH2 which is located approximately 100 meV away from HH1 because the Fermi level is below the HH2 subband edge.

## EXPERIMENTAL RESULTS AND MODELING

The pump-probe experimental setup was discussed in our previous paper.[6] We measure the pump-induced transmission change as a function of the time delay between the pump pulse and the probe pulse, both of which are tuned in resonance with the infrared absorption near 5 µm. In all the experiments, we find an increase of transmission following the pump pulse. The recovery can be described by a single exponential relaxation time of the order of 1 picosecond. The maximum change in transmission is linear with pump intensity below 1 GW/cm$^2$ and saturates to ~3% with a saturation intensity $I_{sat}$ of ~3 GW/cm$^2$. As the saturation regime is entered, the relaxation time increases from 0.8 ps to 1.8 ps.

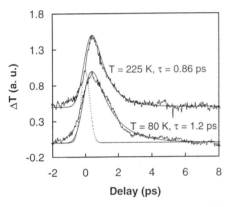

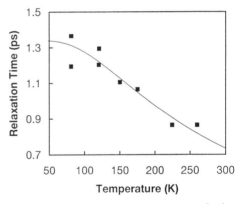

**Figure 1.** Pump induced transmission change measured at 5 µm for a pump intensity of ~1 GW/cm$^2$. The dashed line is the laser pulse autocorrelation (pulse duration of 0.7 ps). The traces correspond to different temperatures as labelled.

**Figure 2.** Temperature dependence of the relaxation time. The solid line assumes that the relaxation rate is proportional to (2n+1), where n is the Bose-Einstein occupation number.

In the present work, we perform temperature (T) dependence measurements at the wavelength of 5 µm and the pump intensity of ~1 GW/cm$^2$. Typical data at T = 120 K and T = 225 K are shown in Figure 1, where the solid lines are the best fits. The relaxation increases from 0.8 ps to 1.3 ps as T decreases from 260 K to 80 K, as shown in Figure 2, where the solid line is the calculated temperature dependence assuming that the relaxation rate is proportional to the summation of phonon emission and phonon absorption (2n+1), where n is the phonon occupation number. Finally, when the laser is tuned through the absorption band, the magnitude of the signal changes but its temporal behavior does not change within the accuracy of the measurements.

## DISCUSSION

In p-type QWs, $I_{sat}$ is much higher than in n-type QWs in large part because the absorption cross-section is one order of magnitude lower in p-type QWs. The wavelength independence of the relaxation time is consistent with the assumption of a homogeneously broadened two-level system. An important question is whether the measured relaxation time is the dwell time (i.e., the time it takes for an energy-nonconserving scattering event to empty the excited state) or the cooling time (here, the excited carriers are assumed to form a distribution on a time scale faster than the pulse duration). Within the second model, the increase in relaxation time with increasing intensity might be attributed to hot phonons. However, as shown in Figure 2, the increase of the relaxation time with decreasing temperature is fitted very well assuming that the total scattering rate is due to the summation of phonon absorption and phonon emission by the photoexcited holes. Thus, we conclude that at low excitation intensity, we measure the dwell time. At high intensity, the depletion of the cold hole reservoir in the ground state (HH1) begins to play a significant role and slows the overall relaxation time. The scattering processes are shown in Figure 3.

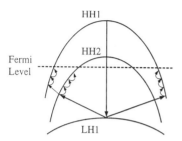

**Figure 3.** Schematic diagram of the absorption and scattering processes between two subbands in $k_{//}$ space. The carriers excited to the LH1 by the pump beam first scatter through deformation potential scattering, and then thermalize with the cold hole reservoir very quickly through hole-hole scattering. Finally the entire hole distribution cools down through optical- and acoustic-phonon emission.

Numerical calculations of the phonon scattering rate were performed within the effective-mass approximation and assuming a parabolic band structure. We modified the theoretical framework proposed by B. K. Ridley,[9,10] who calculated the electron-phonon scattering. Following the envelope wavefunction approximation and assuming an infinite well, the interaction matrix element has the following expression:[9]

$$\left| \langle k' | H_{ep} | k \rangle \right|^2 = \frac{\hbar}{2NM\omega_Q} I^2(k,k')C^2(Q)\left( n(\omega_Q) + \frac{1}{2} \pm \frac{1}{2} \right) \delta_q G^2(q_z) \qquad (1)$$

where k and k' are the wavevectors of the initial state and the final state respectively, q and $q_z$ are the in-plane and growth-direction components of the phonon wavevector respectively, C(Q) is the coupling strength which depends on the scattering process, $G(q_z)$ sets the selection rule for $q_z$, and $I^2(k,k')$ is the overlap integral over a unit cell. The major difference between electron and hole scattering results from $I^2(k,k')$. In our calculation $I^2(k,k')$ is equal to $(1+3\cos^2\theta_k)/4$ for HH-HH scattering and to $3/4\sin^2\theta_k$ for LH-HH scattering, where $\theta_k$ is defined as the angle between k and k'.[11] We also used a guided mode phonon in our calculation. $G(q_z) \neq 0$ only for $q_z = |k_z \pm k_z'|$. For the guided mode, there is no $q_z = 0$ mode. However, when we calculate the LH1 to HH1 scattering rate for an infinite-well, the scattering rate has a contribution only from the $q_z = k_z + k_z'$ term because $k_z = k_z'$. For a finite-well, $k_z \neq k_z'$ because of the different effective masses of the LH and HH subbands. As a result, there must be contribution from $q_z = |k_z-k_z'|$. In order to take this correction into account, we added a contribution from the $q_z = 0$ mode for this scattering process while we excluded this term for the intra-subband scattering calculations. The mixing between the HH and LH subbands was ignored, which changes the overlap term $I^2(k,k')$ and the effective masses at large wavevector as well. In our sample, the HH and LH subbands may be decoupled efficiently due to the large strain-induced energy splitting. Thus the HH subbands should have an in-plane effective mass close to that of an ideal decoupled two-dimensional system, a value which is much smaller than that of the bulk material. Finally, the screening length for the LO-phonon scattering was estimated using the formula given by Asada.[8]

The calculations show that in all cases the LO-phonon-scattering rate is much smaller than the deformation potential scattering rate due to screening. The total inter-subband scattering yields a dwell time of ~0.6 ps for the carriers in LH1: the LH1-HH1 scattering rate is ~0.5 ps$^{-1}$ and the LH1-HH2 scattering rate is ~1 ps$^{-1}$. The calculated dwell time agrees very well with the relaxation times observed at low excitation level. These numbers are significantly smaller than those obtained by others using bulk-like effective masses.[12] At low excitation intensity, the hot holes scattered from LH1 are thermalized with the cold hole reservoir very quickly through hole-hole scattering.[13] The temperature of the hole reservoir changes very little. As a result, the cooling does not affect the results. The relaxation time we measured is thus the dwell time.[13] However, at high excitation intensity, the entire hole reservoir may be heated significantly and the cooling need to be considered.

In conclusion, we have measured and modeled the relaxation of hot holes excited by inter-subband pumping with a mid-IR sub-picosecond FEL laser. We find good agreement between the theoretical calculation and the experimental results.

## ACKNOWLEDGMENT

We acknowledge support from the US Office of Naval Research through grants N00014-92-J-4063 and N00014-91-C-0170.

## REFERENCE

1. A. Seilmeier, H. -J. Hubner, G. Abstreiter, G. Weimann, and W. Schlapp, *Phys. Rev. Lett.* **59**, 1345 (1987)
2. F. H. Julien, J. -M. Lourtioz, N. Herschkorn, D. Delacourt, J. P. Pocholle, M. Papuchon, R. Planel, and G. Le Roux, *Appl. Phys. Lett.* **53**, 116 (1988)
3. T. Elsaesser, R. J. Bauerle, W. Kaiser, H. Lobentanzer, W. Stolz, and K. Ploog, *Appl. Phys. Lett.* **45**, 256 (1989)
4. F. F. Sizov and A. Rogalski, *Prog. Quant. Electr.* **17**, 93 (1993)
5. M. C. Tatham, J. R. Ryan, and C. T. Foxon, *Phys. Rev. Lett.* **63**, 1637 (1989)
6. Z. Xu, P. M. Fauchet, C. W. Rella, B. A. Richman, H. A. Schwettman, and G. W. Wicks, *Phys. Rev.* **B51**, 10631 (1995)
7. X. Z. Lu, R. Garuthara, S. Lee, and R. R. Alfano, *Appl. Phys. Lett.* **52**, 93 (1988)
8. M. Asada, *IEEE J. Quantum Electron.* **QE-25**, 2019 (1989)
9. B. K. Ridley, *J. Phys. C:Solid State Phys.* **15**, 5899 (1982)
10. B. K. Ridley, *Phys. Rev.* **B39**, 5282 (1989)
11. J. D. Wiley, *Phys. Rev.* **B4**, 2485 (1971)
12. S. Seki and K. Yokoyama, *J. Appl. Phys.* **76**, 7399 (1994)
13. J. F. Young, T. Gong, P. M. Fauchet, and P. J. Kelly, *Phys. Rev.* **B50**, 2208 (1994)

# HOT PHONON EMISSION
# IN A DOUBLE BARRIER HETEROSTRUCTURE

V. V. Mitin, G. Paulavičius, and N. A. Bannov

Department of Electrical and Computer Engineering,
Wayne State University Detroit, MI 48202

## INTRODUCTION

Hot acoustic phonon emission represents one of the major channels for thermal energy removal from heterostructures; in addition, detection of phonons emitted by hot electrons provides a valuable tool for investigation of electron-phonon interactions in heterostructures.[1] We have studied the acoustic phonon emission by hot electrons in double barrier heterostructures. We have solved the electron kinetic equation to obtain the electron distribution function, which has been used to solve the kinetic equation for phonons and to determine the radiation and absorption patterns for the energy carried by acoustic phonons. The radiation and absorption patterns have highly pronounced maxima inside the solid angle close to the normal to the quantum well direction (z-direction). These orientational dependencies are related to the quantum confinement of electrons and uncertainty in z-component of their wave vectors.

## BASIC EQUATIONS

Our basic approach is very close to one, used in Refs.[2,3] However, in contrast to these papers, we take into account the effect of hot electrons on phonon emission and absorption; in addition, we allow for both spontaneous and stimulated phonon processes. The kinetic equation for a one particle phonon density matrix, $\sigma_{\mathbf{q},\mathbf{q}'}$, has the following form

$$\partial \sigma_{\mathbf{q},\mathbf{q}'}/\partial t + i(\omega_{\mathbf{q}} - \omega_{\mathbf{q}'})\sigma_{\mathbf{q},\mathbf{q}'} = I_R \qquad (1)$$

$$-2i \sum_{n,n',\mathbf{k}_{\|}} \frac{|<n|\Gamma_{\mathbf{q}'}|n'>|^2}{\varepsilon_{n,\mathbf{k}_{\|}} - \varepsilon_{n',\mathbf{k}_{\|}+\mathbf{q}'_{\|}} + \omega_{\mathbf{q}} - i\lambda}[(1 - f_{n,\mathbf{k}_{\|}})f_{n',\mathbf{k}_{\|}+\mathbf{q}'_{\|}}\delta_{\mathbf{q},\mathbf{q}'} + (f_{n',\mathbf{k}_{\|}+\mathbf{q}'_{\|}} - f_{n,\mathbf{k}_{\|}})\sigma_{\mathbf{q},\mathbf{q}'}]$$

$$+2i \sum_{n,n',\mathbf{k}_{\|}} \frac{|<n|\Gamma_{\mathbf{q}}|n'>|^2}{\varepsilon_{n,\mathbf{k}_{\|}} - \varepsilon_{n',\mathbf{k}_{\|}+\mathbf{q}_{\|}} + \omega_{\mathbf{q}'} + i\lambda}[(1 - f_{n,\mathbf{k}_{\|}})f_{n',\mathbf{k}_{\|}+\mathbf{q}_{\|}}\delta_{\mathbf{q},\mathbf{q}'} + (f_{n',\mathbf{k}_{\|}+\mathbf{q}_{\|}} - f_{n,\mathbf{k}_{\|}})\sigma_{\mathbf{q},\mathbf{q}'}] .$$

$I_R$ is the phonon relaxation term due to interactions other than electron scattering. The electron states are determined by subband number $n$ and electron in-plane wave vectors $\mathbf{k}_{\|}$, the phonon states are determined by wave vectors $\mathbf{q} = (\mathbf{q}_{\|}, q_z)$, where $\mathbf{q}_{\|}$ is the phonon in-plane wave vector, axis $z$ is normal to the quantum well. The matrix

*Hot Carriers in Semiconductors*
Edited by K. Hess *et al.*, Plenum Press, New York, 1996

elements in Eq. (1) are given by the formulae

$$|< n|\Gamma_{\mathbf{q}}|n' >|^2 = \frac{E_a^2 q^2}{2\rho V_{pr}\omega(q)}\, \mathcal{G}(n',n,q_z)\,,$$

where the overlap integral is determined by the formula

$$\mathcal{G}(n',n,q_z) = |< n'|\exp(iq_z z)|n >|^2\,,$$

$E_a$ is the deformation potential constant, $V_{pr}$ is the principal volume, $\omega(q)$ is the phonon dispersion relation, and we use such units, that $\hbar = 1$.

Because our system is translationally invariant in the plane of the quantum well we will use the following Wigner function

$$N_{\mathbf{q}_\parallel}(q_z,z) = \sum_{\Delta q_z} \sigma_{(\mathbf{q}_\parallel,q_z+0.5\Delta q_z);(\mathbf{q}_\parallel,q_z-0.5\Delta q_z)}\, \exp(i\Delta q_z z)\,,$$

which satisfies the kinetic equation

$$\partial N_{\mathbf{q}_\parallel}(q_z,z)\,/\,\partial t + S_z(\mathbf{q})\,\partial N_{\mathbf{q}_\parallel}(q_z,z)\,/\,\partial z = \mathcal{I}_R + \mathcal{I}_{PE} \qquad (2)$$

In Eq. (2), $S_z(\mathbf{q})$ is the $z-$ component of the phonon velocity, $S_{\mathbf{q}}$, $\mathcal{I}_{PE}$ is the collision integral due to phonon interaction with electrons. At distances from quantum well $d_z$ which are much larger than the inverse z-component of phonon wave vector, the collision integral $\mathcal{I}_{PE}$ is equal to zero. Thus, it may be taken into account through the boundary conditions for function $N_{\mathbf{q}_\parallel}(q_z,z)$:

$$S_z\left(N_{\mathbf{q}_\parallel}(q_z,+d_z) - N_{\mathbf{q}_\parallel}(q_z,-d_z)\right) = \sum_{n,n'}\int d\,\mathbf{k}_\parallel\, \frac{E_a^2 q^2}{2\pi\rho\omega(q)}\, \mathcal{G}(n',n,q_z) \times \qquad (3)$$

$$[(1 - f_{n,\mathbf{k}_\parallel})\, f_{n',\mathbf{k}_\parallel+\mathbf{q}_\parallel}\, + (f_{n',\mathbf{k}_\parallel+\mathbf{q}_\parallel} - f_{n,\mathbf{k}_\parallel})\, N_{\mathbf{q}_\parallel}(q_z,z)\,]\,\delta(\varepsilon_{n,\mathbf{k}_\parallel} - \varepsilon_{n',\mathbf{k}_\parallel+\mathbf{q}_\parallel} + \omega_{\mathbf{q}})\,.$$

In Eqs. (1), (2) and (3) we have assumed that the electron subsystem is homogeneous in the $x-y$ plane and described by a diagonal density matrix (distribution function) $f_{n,\mathbf{k}_\parallel}$. For the function $f_{n,\mathbf{k}_\parallel}$ we have the kinetic equation which allows for a strong (in-plane) electric field and electron scattering by optical and acoustic phonons. We have used equilibrium distributions for phonons in the collision integrals of electrons with phonons.


## RADIATION PATTERNS

The kinetic equation for electrons and Eq. (2) have been solved by the Monte Carlo technique. We have obtained the electron distribution function $f_{n,\mathbf{k}_\parallel}$ and the phonon Wigner function $N_{\mathbf{q}_\parallel}(q_z,z)$. The primary objects of our interest in this report are the fluxes of energy carried by acoustic phonons. We will use the three densities of energy fluxes which are defined as follows

$$\mathcal{N}(\Omega,\omega) = \frac{L_x L_y}{8\pi^3 \Delta\Omega\Delta\omega} \int_{\Delta\Omega\Delta\omega} d\mathbf{q}\, S_{\mathbf{q}}\, \omega_{\mathbf{q}}\, N_{\mathbf{q}_\parallel}(q_z,z)$$

$\mathcal{N}_\Omega(\omega) = \int \mathcal{N}(\Omega,\omega)d\Omega$ , $\mathcal{N}_\omega(\Omega) = \int \mathcal{N}(\Omega,\omega)d\omega$ , where $\Omega$ is the solid angle, $\omega$ is the phonon frequency, $L_x L_y$ is the area of the quantum well. $\mathcal{N}(\Omega,\omega)$ is the energy flux due to emitted (and/or absorbed) acoustic phonons per unit energy range, per unit solid angle (dimension is $1/(\text{s srad})$). We restrict our consideration to a nondegenerate electron gas, then, in accordance with Eqs.(1) and (2), the densities of phonon energy fluxes are proportional to the electron concentration and we normalize these quantities per one electron.

We have calculated functions $\mathcal{N}(\Omega,\omega)$, $\mathcal{N}_{\Omega}(\omega)$, and $\mathcal{N}_{\omega}(\Omega)$ for a wide range of external parameters. To demonstrate the results we will use the following coordinate system: axis $x$ is going in the direction of the average electron velocity, axis $z$ is perpendicular to the quantum well, axis $y$ augments the axis $x - z$ to a right basis. We will also use the spherical angles $\theta, \varphi$. Fig. 1 and 2 demonstrate radiation patterns in two electric fields, $E = 10V/cm$ and $E = 1000V/cm$. In the case $E = 10V/cm$ electrons are mostly scattered by acoustic phonons, in the electric field $E = 1000V/cm$ electrons loose most of theeir energy by emitting the optical phonons.

It is worth mentioning that the ordinates in Figs. 1 and 2 have much larger scales that the abscissas. Therefore, acoustic phonons carry energy basically always in directions close to the $z-$direction. This results from the fact, that the $z-$ component of the phonon wave vector is conserved only within an accuracy $\pi/a$ (where $a$ is the width of the quantum well) and this quantity is usually larger than the electron (and phonon) in-plane wave vector. At the same time, the flux of energy exactly along the $z-$direction is equal to zero. This is merely a consequence of the conservation laws. The in-plane distribution of the energy flux has a good pronounced dependence on angle $\varphi$ and has higher values for small $\varphi$ $(0 < \varphi < \pi)$ because the electron distribution function is shifted in the $x-$ direction. The large bend on the curve 1 in the Fig. 2 for $\varphi = 5°$ for $\theta$ close to $45°$ is due to much larger scale in the $z-$direction on Fig. 2 and due to stretching the electron distribution function in $x-$direction.

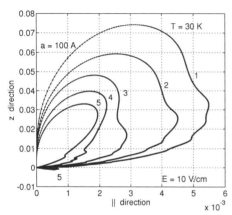

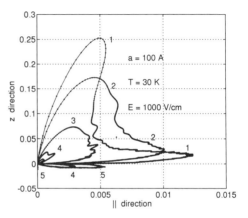

**Figure 1.** Radiation pattern of acoustic phonons, $\mathcal{N}_{\Omega}(\omega)$, for azimuthal angles $\varphi = 5°(1)$, $55°(2)$, $85°(3)$, $115°(4)$, $175°(5)$ in units: meV/(s srad electron). GaAs quantum well of width $100\text{Å}$, electric field $10\ V/cm$, lattice temperature T=30$K$. Negative fluxes (below the abscissa) correspond to prevailing phonon absorption.

**Figure 2.** Radiation pattern of acoustic phonons, $\mathcal{N}_{\Omega}(\omega)$, for azimuthal angles $\varphi = 5°(1)$, $55°(2)$, $85°(3)$, $115°(4)$, $175°(5)$ in units: meV/(s srad electron). GaAs quantum well of width $100\text{Å}$, electric field $1000\ V/cm$, lattice temperature T=30$K$. Negative fluxes (below the abscissa) correspond to prevailing phonon absorption.

*Acknowledgment* - This work was supported by ARO.

[1] J. K. Wigmore, M. Erol, M. Sahraoui-Tahar, M. Ari, C. Wilkinson, J. Daviest, M. Holland, C. Stanley, Measurement of energy-loss from a hemt structure by observing emitted phonons, *Semicond.Sci.Technol.* 8:322 (1993).
[2] F. T. Vasko, Emission of acoustic phonons by two-dimensional electrons, *Sov.Phys.Solid State* 30:1207 (1989).
[3] W. Xu, J. Mahanty, Acoustic phonon emission angle in selectively doped AlGaAs-GaAs-AlGaAs single quantum well, *J.Phys.: Condens.Matter* 6:6265 (1994).

# INTRA- AND INTERSUBBAND TRANSITION RATES DUE TO EMISSION OF OPTIC PHONONS IN QUANTUM WELLS: EFFECTS OF THE SUBBAND NONPARABOLICITY

Augusto M. Alcalde and Gerald Weber

Instituto de Física, Universidade Estadual de Campinas,
Caixa Postal 6165, 13083-970 Campinas SP, Brazil

Electron-phonon interaction in polar semiconductor quantum wells attracted a great amount of interest over the last years due to its importance for electronic properties. Particular interest was directed towards LO-phonon confinement which affects significantly the scattering rates in quantum wells. In general, the use of dielectric continuum models is considered as well established[1] and scattering rates calculated with such models compare successfully with experimental results.[2] From the calculation of capture times[3] it became evident that capture processes with large kinetic energy influence the overall capture time. For those large kinetic energies, implying in large momenta, the parabolic-band approximation becomes less justified even for GaAs-AlGaAs structures where in general nonparabolicity effects can be safely neglected. A question which has not yet been properly addressed is how strongly the subband nonparabolicity affects the intra- and intersubband scattering rates in quantum wells.

In the present work we study the influence of subband nonparabolicity on intra- and intersubband scattering rates and compare these results to those obtained with the parabolic-band approximation. This enables us to determine the situations for which the subband nonparabolicity is negligible or not. For GaAs-Al$_{0.3}$Ga$_{0.7}$As quantum wells we find that the subband nonparabolicity is important for transitions involving higher energy states and for structures with well thicknesses smaller that $\approx$150 Å. Also, we find that the nonparabolicity influences differently transition rates due to confined and interface phonon mode.

Several theoretical models were proposed, in various degrees of complexity, to take into account the effect of nonparabolicity in the electronic band structure (e. g. Ref. 4 and references therein). We chose an energy dispersion relation which is believed to adequately represent the subband nonparabolicity of GaAs-GaAlAs quantum wells,[4]

$$\mathcal{E} = \frac{\hbar^2 k^2}{2m^*}(1 - \gamma k^2),$$

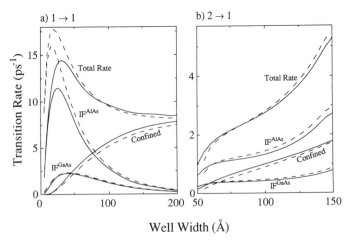

**Figure 1.** Transition rates for 1→1 and 2→1 transitions as a function of the well width, due to confined and AlAs-like and GaAs-like interface (IF) phonon modes. Solid and dashed lines are for nonparabolic and parabolic subbands, respectively.

where $\mathcal{E}$ is the total electron energy, $m^*$ is the band edge mass and $\gamma$ the nonparabolicity parameter. The electron–LO-confined phonon scattering rates are obtained from the Fermi golden rule and with the electron–phonon (Fröhlich interaction) interaction Hamiltonian. The scattering rates take the form

$$W_{np}^{(i)} = \frac{m^* \lambda^2 L}{\hbar^3} \sum_n |G_n|^2 (N_{LO} + 1) \frac{1}{\alpha} \left\{ \frac{a_n L^2}{2\gamma} \left[ (1 - 2\gamma k_z^2) - \alpha \right] + b_n \right\}^{-1}, \tag{1}$$

with $\alpha = \left[ (2\gamma k_z^2 - 1)^2 + 4\gamma Q^2 \right]^{1/2}$ where $k_z$ is the $z$-component of the initial (final) electron wave vector for intrasubband (intersubband) transitions, and $G_n$ is the electron-phonon overlap integral. For the description of the confined phonon modes we used the corrected slab model,[5] whereas for the interface phonon modes we use the Hamiltonian described by Mori and Ando.[6] For the calculation of the nonparabolic energy states we used the simple model proposed by Nag.[4] This model has the advantage that the overlap functions have the same analytic functional form of the parabolic-band case, thus it allows a straightforward inclusion of the nonparabolicty effects into existing parabolic-band calculations.[7] For the calculations presented in this work we used the material parameters for GaAs-Al$_{0.3}$Ga$_{0.7}$As given by Adachi.[8]

In Figs. 1 and 2 we show the calculated scattering rates for several transitions due to the emission of confined and interface phonon modes as a function of the quantum well width. We observe that the effect of the subband nonparabolicty on the scattering rates increases as transitions involve higher energy states, but the scattering rates are not qualitatively very different from those in the parabolic-band approximation. The enhancement of the scattering rates observed for confined phonons with the inclusion of subband nonparabolicity results mainly from a larger electron-phonon overlap as well as from a larger density of final electron states. This enhancement occurs despite of the fact that the emitted confined phonons have larger wave vectors due to the nonparabolicity of the electron subband. Thefore simple reasoning used for for bulk, where a larger wave vector implies in a smaller Fröhlich coupling factor and smaller scattering rates, is not applicable for quantum wells. Differently from the confined

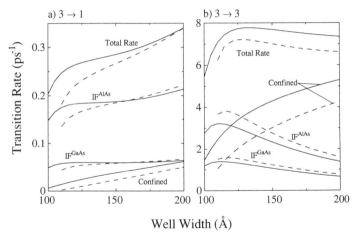

a) 3 → 1    b) 3 → 3

Well Width (Å)

**Figure 2.** Transition rates for 3→1 and 3→3 transitions as a function of the well width, due to confined and AlAs-like and GaAs-like interface (IF) phonon modes. Solid and dashed lines are for nonparabolic and parabolic subbands, respectively.

modes, the transition rates due to interface phonon modes are in general reduced by the nonparabolicity. This can be explained by a smaller electron-phonon overlap, as the electron wave function moves away from the interfaces where these phonon modes have their maxima. However, except for some specific well widths, the reduction of the rates due to interface phonon modes is not compensated by the increase due to confined phonon modes as shown by the total rates in Figs. 1 and 2.

In conclusion, we have calculated the scattering rates for intrasubband and intersubband transitions due to electron–phonon interaction in quantum wells where we included the subband nonparabolicity. We find that for higher subbands and larger electron confinement the nonparabolicity effects becomes more important.

Financial support by CNPq is acknowledged.

## REFERENCES

1. H. Rücker, E. Molinari, and P. Lugli, "Microscopic calculation of the electron–phonon interaction in quantum wells," *Phys. Rev. B* 45:6747 (1992).
2. A. M. de Paula and G. Weber, "Γ to $X_z$ electron transfer times in type-II superlattices due to emission of confined phonons," *Appl. Phys. Lett.* 65:1281 (1994).
3. G. Weber and A. M. de Paula, "Carrier capture processes in GaAs-AlGaAs quantum wells due to emission of confined phonons," *Appl. Phys. Lett.* 63:3026 (1993).
4. B. R. Nag and S. Mukhopadhyay, "Energy levels in quantum wells of nonparabolic semiconductors," *Phys. Stat. Sol. (b)* 175:103 (1993).
5. G. Weber, "Electron–confined-phonon interaction in quantum wells: reformulation of the slab model," *Phys. Rev. B* 46:16171 (1992).
6. N. Mori and T. Ando, "Electron–optical-phonon interaction in single and double heterostructures," *Phys. Rev. B* 40:6175 (1989).
7. G. Weber, A. M. de Paula, and J. F. Ryan, "Electron–LO-phonon scattering rates in GaAs-Al$_x$Ga$_{1-x}$As quantum wells," *Semicond. Sci. Technol.* 6:397 (1991).
8. S. Adachi, "GaAs, AlAs, and Al$_x$Ga$_{1-x}$As: Material parameters for use in research device aplications," *J. Appl. Phys.* 58:R1 (1985).

# HOT-PHONON EFFECTS IN A QUANTUM KINETIC MODEL FOR THE RELAXATION OF PHOTOEXCITED CARRIERS

Jürgen Schilp[1] and Tilmann Kuhn[2]

[1] Institut für Theoretische Physik und Synergetik,
   Universität Stuttgart, 70550 Stuttgart, Germany
[2] Lehrstuhl für Theoretische Physik, Brandenburgische
   Technische Universität Cottbus, 03013 Cottbus, Germany

## INTRODUCTION

In solid state physics the microscopic quasiparticle dynamics are usually described by the semiclassical Boltzmann equation. There, by using the completed collision approximation, scattering processes are treated as instantaneous in space and time and the single-particle energy is conserved in each process. On short length and time scales quantum effects enter the description of the dynamics. A well-known example is the generation of electron-hole pairs as described by the semiconductor Bloch equation. Within this concept the correlation between electrons and holes is described by the interband polarization and coherent effects are taken into account.[1]

In quantum kinetic theory characteristic quantum effects enter the description: (i) Energy-time uncertainty which leads to a time-dependent broadening of transitions, (ii) renormalization effects due to the correlation between different interactions, and (iii) the collisional broadening as an influence of higher order perturbation theory. In this paper we present a quantum kinetic analysis of the coupled carrier-phonon dynamics in a polar semiconductor based on the density matrix approach. By comparing the results with the semiclassical dynamics both for a one-band model and a two-band model, we investigate the role of nonequilibrium phonons for the various quantum effects.

## COUPLED ELECTRON-PHONON SYSTEM

First we consider a homogeneous bulk semiconductor with one electron band with energy $\epsilon_{\mathbf{k}}^e = \hbar^2 k^2 / 2m_e$, wave vector $\mathbf{k}$, and effective mass $m_e$. The energy of the LO-phonons $\hbar\omega_{op}$ is taken to be constant. The Hamiltonian contains the unperturbed part of the free motion of electrons and phonons and the Fröhlich interaction with coupling matrix element $g_{\mathbf{q}}^e$.

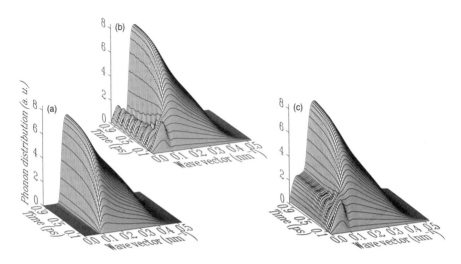

**Figure 1.** Nonequilibrium phonon distribution $N_q = q^2 n_q$ (a) in the semiclassical case, (b) in the quantum kinetic case without and (c) in the quantum kinetic case with collisional broadening.

The basic variables are the distribution function of electrons (phonons) $f_k^e = \langle c_k^\dagger c_k \rangle$ ($n_q = \langle b_q^\dagger b_q \rangle$) with creation $c_k^\dagger$ ($b_q^\dagger$) and annihilation $c_k$ ($b_q$) operators. Their dynamics is given by

$$\frac{d}{dt} f_k^e = 2 \sum_q \left[ \text{Re}\left\{ s_{k+q,k}^e \right\} - \text{Re}\left\{ s_{k,k-q}^e \right\} \right] \quad , \quad \frac{d}{dt} n_q = 2 \sum_k \text{Re}\left\{ s_{k+q,k}^e \right\} . \quad (1)$$

The equations of motion (1) for the expectation values of two operators contain expectation values of three operators, the so-called phonon-assisted density-matrices (PADMs) defined as $s_{k+q,k}^e = \frac{i}{\hbar} g_q^e \langle c_{k+q}^\dagger b_q c_k \rangle$. In the next step the equations of motion for the PADMs have to be evaluated and we get expectation values of four operators, i.e., two-particle density-matrices. The resulting hierarchy is closed by factorizing the two-particle density-matrices in two one-particle density-matrices and one obtains

$$\frac{d}{dt} s_{k+q,k}^e = i\Omega_{k+q,k}^e s_{k+q,k}^e + \frac{1}{\hbar^2} \left| g_q^e \right|^2 \left[ f_{k+q}^e \left( 1 - f_k^e \right) \left( n_q + 1 \right) - \left( 1 - f_{k+q}^e \right) f_k^e n_q \right] \quad (2)$$

with the oscillation frequency $\Omega_{k+q,k}^e = (\epsilon_{k+q}^e - \epsilon_k^e - \hbar\omega_{op})/\hbar$. The set of equations (1), (2) describes non-Markovian dynamics with respect to the distribution functions because the knowledge of $f_k^e$ and $n_q$ at a given time is not sufficient to specify the state of the system.

It turns out that within second order perturbation theory the above system shows unphysical results for certain parameter regimes.[2,3] This can be avoided by calculating the deviation from the factorization of the two-particle density-matrices in the next order and when performing a random phase and Markov approximation, this leads to a microscopic model for the complex polaron self-energy resulting in a damping due to the imaginary part $\Gamma_k^e$,

$$\Gamma_k^e = \pi \sum_{q',\pm} \left| g_{q'}^e \right|^2 \delta(\epsilon_k^e - \epsilon_{k+q'}^e \pm \hbar\omega_{op}) \left[ \left( n_{q'} + \tfrac{1}{2} \pm \tfrac{1}{2} \right) f_{k+q'}^e + \left( n_{q'} + \tfrac{1}{2} \mp \tfrac{1}{2} \right) \left( 1 - f_{k+q'}^e \right) \right] ,$$

which depends only on distribution functions and we get a complex frequency $\Omega_{k+q,k}^e = (\epsilon_{k+q}^e + i\Gamma_{k+q}^e - \epsilon_k^e + i\Gamma_k^e - \hbar\omega_{op})/\hbar$ in Eq. (2). These self-energy terms give rise to the

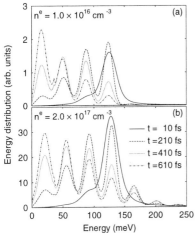
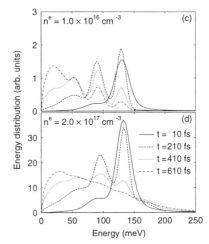

**Figure 2.** Energy distribution of electrons in (a), (b) the semiclassical and (c), (d) the quantum kinetic case at two different densities.

collisional broadening. It should be noted that mainly due to the random phase assumption an overestimation of the collisional broadening is expected.[4]

In Fig. 1 we have studied the relaxation from an initial Gaussian electron distribution with excess energy 60 meV and a width corresponding to a laser pulse of 80 fs duration. The nonequilibrium phonon distribution is plotted for three different models. In the semiclassical case [Fig. 1(a)] the fixed energy of the initial and final state leads to a minimum and maximun wave vector of emitted phonons. The phonon distribution shows an increase to smaller wave vectors due to the $\frac{1}{q}$-dependence of the Fröhlich coupling. In the quantum kinetic cases without [Fig. 1(b)] and with [Fig. 1(c)] damping these sharp limits are smeared out on short time scales due to the energy-time uncertainty. With increasing time, in the semiclassically allowed region it approaches the semiclassical shape. In the small wave vector region, however, we find a pronounced difference. In the quantum kinetic cases we observe an oscillating phonon population. This is due to the build up of the polaron from an initially uncorrelated electron state. The oscillations are damped by the collisional broadening.

## TWO-BAND MODEL

The drawback in a one-band model is the uncorrelated initial condition which typically leads to an overestimation of quantum effects at short times. Therefore, in this section we will investigate a two-band semiconductor coupled to the LO-phonon system by the polar Fröhlich interaction. Additionally we consider carrier-light and Coulomb interaction within Hartree-Fock approximation. The dynamical quantum effects like time-dependent broadening and renormalizations are modified by hot phonon effects. The dynamics of the quasiparticles is described by the equations of motion for the distribution functions of electrons $f_k^e$, holes $f_k^h$, phonons $n_q$ and the interband polarization $p_k$. In addition to the intraband PADMs $s_{k',k}^e$ and $s_{k',k}^h$ also interband PADMs, e.g., $t_{k',k}^{(+)} = \frac{i}{\hbar} g_q^e \langle d_{k'} b_q c_k \rangle$, $d_k$ denoting an annihilation operator for a hole, have to be included. The corresponding set of equations of motion can be found in Ref. [4,5].

In the following simulations we have applied a short laser pulse with excess energy 140

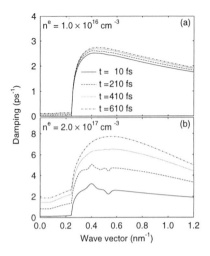

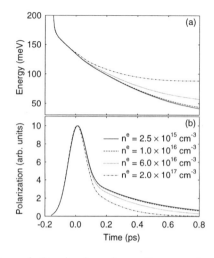

**Figure 3.** The imaginary part of the self-energy at two different densities.

**Figure 4.** Density-dependence of energy relaxation and dephasing in the quantum kinetic case.

meV and a duration of 80 fs. In Fig. 2 we compare the generation and relaxation dynamics in the semiclassical and quantum kinetic case for two densities. In the semiclassical case [Fig. 2(a), (b)] we clearly see the phonon replicas at both densities, at the high density phonon reabsorption leads to replicas also above the generated peak. In the quantum kinetic case at low densities [Fig. 2(c)] we find an initial broadening of the replicas due to energy-time uncertainty and a subsequent narrowing due to the build-up of energy conservation. At high densities [Fig. 2(d)], however, this narrowing is completely absent resulting in a broad, structureless distribution. The reason can be understood from the damping rate $\Gamma_k$ plotted in Fig. 3 for the two densities. At high densities $\Gamma_k$ strongly increases due to nonequilibrium phonons resulting in a much stronger collisional broadening. Fig. 4 shows the density-dependence of energy relaxation and dephasing in the quantum kinetic case with collisional broadening. The energy relaxation is reduced due to phonon reabsorption and the dephasing is increased due to the higher scattering efficiency. The qualitative behavior is the same as in the semiclassical case but due to the increase of the collisional broadening the density-dependent effects are enhanced.

## CONCLUSION

We have presented a quantum kinetic analysis of hot phonon effects in photoexcited semiconductors. Phonons are generated in semiclassically forbidden regions of phase space due to energy-time uncertainty and the build-up of polarons. Nonequilibrium phonons strongly influence the collisional broadening which leads to an increase in the density-dependence of energy relaxation and dephasing.

## REFERENCES

1. T. Kuhn and F. Rossi, Phys. Rev. B **46**, 7496 (1992).
2. D. B. Tran Thoai and H. Haug, Phys. Rev. B **47**, 3574 (1993).
3. R. Zimmermann and J. Wauer, J. Lumin. **58**, 271 (1994).
4. J. Schilp, T. Kuhn, and G. Mahler, phys. stat. sol (b)**188**, 369 (1995).
5. J. Schilp, T. Kuhn, and G. Mahler, Phys. Rev. B**50**, 5435 (1994).

# SIMULATION OF RAMAN SCATTERING FROM NONEQUILIBRIUM PHONONS IN InP AND InAs

D. K. Ferry[1], E. D. Grann[2], and K. T. Tsen[2]

[1]Department of Electrical Engineering
[2]Department of Physics and Astronomy
Arizona State University, Tempe AZ 85287

## INTRODUCTION

Since the original studies of picosecond Raman scattering in semiconductors (von der Linde *et al.*, 1980; Kash *et al.*, 1985), it has become fairly well established that the cooling of the excited electron-hole plasma is dominated by the emission of optical phonons (on the picosecond time scale) and that the deviation of the phonon population from equilibrium is quite important (Pötz and Kocevar, 1980; Shah *et al.*, 1985; Lugli and Goodnick, 1987; Das Sarma *et al.*, 1988; Tsen *et al.*, 1989; Joshi and Ferry, 1989; Joshi *et al.*, 1990). Most of these studies have been carried out in the GaAs and/or GaAlAs system, and the dominant conclusion of these studies is that the decay (or cooling) time of the hot electron-hole plasma is limited by the lifetime of the hot, nonequilibrium polar optical phonons in the system.

In this paper, we discuss nonequilibrium phonon populations in InP and InAs and measured through Raman scattering. The two materials used here provide different qualitative behaviors. The decay of the hot plasma in InP corresponds to a decay of the hot phonons, as in GaAs, and the lifetime of this decay is determined primarily by the lifetime of the hot phonons, which we find to be 2.0-2.3 ps. In the case of InAs, however, the decay is dominated by the long storage times of the carriers in the satellite valleys of InAs.

## APPROACH AND RESULTS

The undoped, 2 μm thick InAs and InP samples were grown by molecular beam epitaxy on (001)-oriented undoped InAs and InP subtrates, respectively. 900 fs, 50 mW pulses at 1.952 eV were generated by a DCM double-jet dye laser synchronously pumped by the second harmonic of a cw mode-locked Nd:YAG laser, with a 76 Mhz repetition rate. These were split into two beams of equal intensity but different polarization. The anti-Stokes Raman signal was collected and analyzed by a computer-controlled CCD Raman system. The experiments were carried out in a backscattering geometry and at room temperature.

From the power density of the pump pulse and the absorption coefficients of the samples, the average photoexcited carrier density was estimated to be $\approx 3 \times 10^{15}$ cm$^{-3}$ for both InAs and InP.

Simulations of the laser excited plasma were carried out by the ensemble Monte Carlo technique (Ferry *et al.*, 1991a). Only the electrons were considered, as the population of the polar modes was of primary interest. Nonparabolicity was assumed for the various conduction bands, and all normal scattering processes were included. For this reason, we treat the Coulomb interaction via a real-space molecular dynamics approach (Ferry *et al.*, 1991b), including the role of the exchange energy (Kann *et al.*, 1990; Kriman *et al.*, 1990)..

Modeling of the nonequilibrium phonons is handled within the ensemble Monte Carlo procedure by a secondary self-scattering and rejection process pioneered by Lugli *et al.* (1987). The buildup of the phonon population (stored on a grid in momentum space) through emission and absorption processes is monitored throughout the simulation. The difference between the instantaneous value, for a given momentum wave vector, and some prescribed maximum value is used for the rejection technique. The presence of the nonequilibrium phonons slows the energy decay of the hot carriers.

### Indium Phosphide

Excitation at 1.952 eV is assumed to create an electron-hole plasma with an approximate density of $3 \times 10^{15}$ cm$^{-3}$. This density is taken to be uniform in the excitation volume for the thin samples studied here. Carriers are excited from all three valence bands. The satellite valleys in InP lie considerably higher than those in the case of GaAs, and here it is assumed that the L and X valleys are 0.5 eV and 0.98 eV, respectively, above the $\Gamma$ minimum (Ferry, 1991). At the high energy of the X valleys, there is no excitation into these valleys from the photo-excited electrons. Only a small number of electrons are scattering into the L valleys. By the time at which the peak of the phonon distribution, essentially all of the carriers have returned to the central valley of the conduction band. This process is quite fast and a value of the $\Gamma$–L coupling constant of $7 \times 10^8$ eV/cm was used. Varying this value, as well as raising the separation energy between these two valleys, had no significant effect on the simulated Raman scattering. Once the carriers are in the central valley of the conduction band, cooling is dominated by a cascade of LO phonon emission processes.

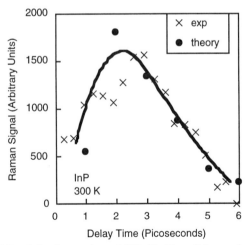

**Figure 1.** Response of InP to 0.6 ps laser pulse at 1.952 eV. The solid curve is a guide to the eye for the theoretical fit to the data. The phonon lifetime is 2.0 ps.

In Fig. 1, we plot the computed population of the polar LO phonons (with a wave vector appropriate to the range of wave vectors in the Raman scattering experiment) and the Raman signal that is measured for InP. The amplitudes are arbitrary, but the peak in the phonon population corresponds to about a 4-5% increase in the density of phonons. It is noted that the peak in phonon occupation is delayed relative to the start of the laser pulse, and this corresponds to the need for the phonon cascade to build up the population. A lifetime of the nonequilibrium phonons in the Monte Carlo of 2.0-2.3 picoseconds leads to a good agreement between the experiment and the simulation.

## Indium Arsenide

The model that we use for InAs assumes that the L and X valleys are located approximately 0.72 eV and 0.98 eV, respectively, above the $\Gamma$ minimum. Since the band gap of InAs is so small (0.36 eV at 300 K), the electrons are excited well up into the conduction band, and a very large fraction of these are scattered into the X valleys. The latter, however, are scattered to the L valleys quite rapidly, so that the main relaxation dynamics is dominated by the $\Gamma$ and L valleys. In contrast to InP, however, the coupling between the $\Gamma$ and L valleys is such that the very low effective mass of the former leads to a quite slow return of the carriers from the L valley to the $\Gamma$ valley.

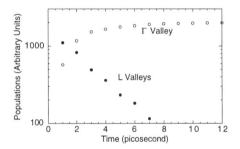

**Figure 2.** The populations of the satellite L valleys and the central $\Gamma$ valley, as a function of time after initiation of the laser pulse. The pulse half-width is 0.6 ps.

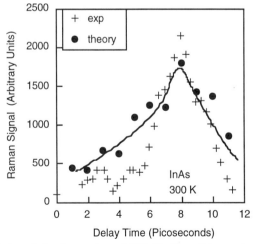

**Figure 3.** Response of InAs to 0.6 ps laser pulse at 1.952 eV. The solid curve is a guide to the eye for the theoretical fit to the data. The phonon lifetime is 0.5 ps.

In Fig. 2, we plot the population of the $\Gamma$ and L valleys. From the figure, it may be seen that the population of the L valleys decays with a time constant of approximately 6 ps. This may be compared with the simulated and measured population of nonequilibrium phonons in Fig. 3. For the latter, a short phonon lifetime of 0.5 ps has had to be assumed, in order to shorten the simulated decay to that found in the experiment. The storage of the carriers in the L valleys leads to a significant continued generation of nonequilibrium phonons at long times, and the overall decay of the optical phonons replicates the decay of the population of the satellite valleys.

## DISCUSSION

The lifetimes of the phonons found in InP and InAs are considerably smaller than that found in GaAs. In fact, InP is softer than GaAs, and InAs is softer than InP. The trend GaAs-InP-InAs is one moving toward softer materials and toward smaller phonon lifetimes. In fact, Weinreich (1965) has pointed out that the anharmonic coefficients should be proportional to the elastic compliance constants (the reciprocals of the normal stiffness constants), so that these should scale with the lattice constant as $d^5$ (Harrison, 1980). This alone, however, does not account for the size of the change, and one must also consider that the phonon frequencies are becoming smaller across this ordering. All of this supports the general trend in the lifetimes observed. More extensive studies will be published elsewhere.

### Acknowledgements

This work was supported by the U.S. Office of Naval Research (DKF) and the U.S. National Science Foundation (KTT).

## REFERENCES

Das Sarma, S., Jain, J. K., and Jalabert, R., 1988, *Phys. Rev. B* 37:1228.

Ferry, D. K., 1991, "Semiconductors," Macmillan, New York.

Ferry, D. K., Kriman, A. M., Kann, M.-J., and Joshi, R., 1991a, in "Monte Carlo Device Simulation: Full Band and Beyond," K. Hess, Ed., Kluwer Acad. Pub., Norwall, MA, pp. 99-121.

Ferry, D. K., Kriman, A. M., Kann, M.-J., and Joshi, R. P., 1991b, *Comp. Phys. Commun.* 67:119.

Harrison, W. A., 1980, "Electronic Structure and the Properties of Solids," W. H. Freeman, San Francisco.

Joshi, R. P., and Ferry, D. K., 1989, *Phys. Rev. B* 39:1180.

Joshi, R. P., Tsen, K. T., and Ferry, D. K., 1990, *Phys. Rev. B* 41:9899.

Kann, M. J., Kriman, A. M., and Ferry, D. K., 1990, *Phys. Rev. B* 41:12659.

Kash, J. A., Tsang, J. C., and Hvam, J. M., 1985, *Phys. Rev. Lett.* 54:2151.

Kriman, A. M., Kann, M.-J., Ferry, D. K., and Joshi, R., 1990, *Phys. Rev. Lett.* 65:1619.

Lugli, P., and Goodnick, S. M., 1987, *Phys. Rev. Lett.* 59:716.

Lugli, P., Jacoboni, C., Reggiani, L., and Kocevar, P., 1987, *Appl. Phys. Lett.* 50:1251.

W. Pötz, W. and Kocevar, P., 1980, *Phys. Rev. B* 28:7040.

Shah, J., Pinczuk, A., Gossard, A. C., and Wiegmann, W., 1985, *Phys. Rev. Lett.* 54:2045.

Tsen, K. T., Joshi, R. P., Ferry, D. K., and Morkoc, H., 1989, *Phys. Rev. B* 39:1446.

von der Linde, D., Kuhl, J., and Klingenburg, H., 1980, *Phys. Rev. B* 37:1228.

Weinreich, G., 1965, "Solids: Elementary Theory for Advanced Students," John Wiley, New York.

# HOT ELECTRONS AND NONEQUILIBRIUM PHONONS IN MULTIPLE δ-DOPED GaAs

M. Asche[1], P. Kleinert[1], R. Hey[1], H. Kostial[1],
B. Danilchenko[2], A. Klimashov[2], and S. Roshko[2]

[1] Paul-Drude-Institut für Festkörperelektronik,
Hausvogteiplatz 5-7, 10117 Berlin, Germany
[2] Institute of Physics of the Ukrainian Academy of Sciences,
Prospect Nauki 46, 252650 Kiev, Ukraine

## INTRODUCTION

The hot carrier intersubband transfer in V-shaped multiple quantum wells (VMQW) is investigated as a function of electric field strength by time of flight spectra of phonons emitted by hot carriers, differential conductivity, and time resolved changes of the current induced by nonequilibrium phonons generated by a heater. Since the mobilities in the various subbands differ significantly, carrier heating and consequently the phonon emission by the heated carriers differ, too. Therefore, specific features can be observed, which reflect the change of subband population by applied fields.

## MULTIPLE δ-LAYERS

For this study GaAs containing 7 δ-layers separated one from another by 100 nm and doped with $5 \cdot 10^{11}$ cm$^{-2}$ Si atoms per layer was grown by MBE on a 3.5 mm thick substrate[1] chosen large enough to distinguish between the arrival of longitudinal and transverse acoustical phonons (LA and TA).

The VMQW is described by 3 subbands: the ground state (0), the first excited subband (1), which present 86 and 14 % of the carriers in the thermodynamic equilibrium at 4.2 K, and a second excited subband (2) extending at the top of the barriers. As experimentally shown [1] it will be populated by carrier heating already in the weak field limit. Since the electron mobility in subband 1 is estimated to be twice as high as in the subband 0 the heating of the carriers in subband 1 is stronger. Consequently, we assume carrier transfer mainly from the first to the second subband to be effective from the very beginning of carrier heating. Another argument in favour of its dominating role in this region is the fact that the overlap of the wavefunctions of the subbands 1 and 2 exceeds that of subbands 0 and 2. In spite of the smaller number of carriers in subband 1 in comparison to subband 0 we conclude that in the weak field limit the transfer 1-2 (between the subbands 1 and 2) prevails over 0-2 and 0-1, although the latter is assumed to be present from the low field limit, too.

## PHONON EMISSION BY HOT CARRIERS

The spectra of the phonons emitted by hot carriersas of the current changes induced by the ballistically arriving phonons generated by the heater as functions of time are integrated over the intervals according to the arrival of the LA and TA phonon contributions. The long tail observed for higher fields in the time of flight spectra is ascribed to the results of repeated decay and conversion processes of initially created optical phonons and is integrated from the end of the falltime of the ballistical phonon signal up to the limit, at which the signal vanishs in the noise. The time integrated fluxes of the phonons emitted by the electron gas are shown in Fig.1 for the VMQW at 3.7 K.

*Hot Carriers in Semiconductors*
Edited by K. Hess *et al.*, Plenum Press, New York, 1996

**Optical Phonons**

Curve 1 presents the products of the cascade of decay and conversion processes of initially emitted LO phonons (e.g.[3]). They set in above a clearly distinguished threshold field of 6 V/cm. An obtained threshold field strength of some V/cm is in agreement with numerical simulations of the contributions of optical phonons to the energy loss of the heated carriers in a simplified two-level MQW system [4]. Such a simulation show the important role of electron transitions from the confined to the extended states above the barrier in the presence of a heating field, which lead to a strong emission of optical phonons due to the high mobility in these interlayer states. The observed flux increases with input power into the δ-layers up to six orders of magnitude. In the high field region the time integrated flux per input power remains constant, reflecting that almost the total energy introduced into the electron gas is dissipated by optical phonons .

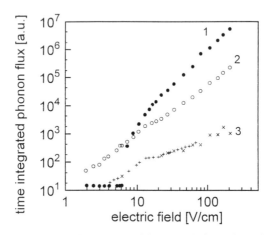

**Figure 1.** Time integrated phonon fluxes versus field strength (1-products from LO phonons, 2-TA, and 3-LA phonons)

**Acoustical Phonons**

Concerning the TA and LA phonons (curve 2 and 3, respectively) the time integrated fluxes per input power are generally decreasing with increasing field. One possible explanation is the higher probability that the carriers in the δ-layers emit acoustical phonons characterized by wave vectors with an increasing component along the direction of the heating field. Such an assumption should be confirmed by angular measurements as performed for GaAs/AlGaAs[5,6] for instance.
As usual the TA phonon signal is clearly pronounced and can be observed from the lowest field strength in the investigated region, whereas LA phonons can be detected only above 4 V/cm due to

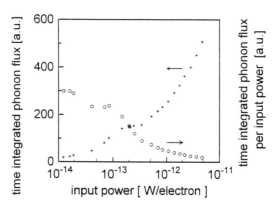

**Figure 2.** Time integrated LA flux and time integrated flux per power versus input power

the small signal on account of the focusing factor < 1. Both spectra exhibit specific features at 7 and 12 V/cm. In spite of the stronger signals the TA fluxes reflect those features less pronounced than the LA fluxes. This peculiarity is still better reflected in the time integrated signal per input power shown in Fig.2.

In order to explain the kink at 7 V/cm we study the differential conductivity (Fig. 3). It shows an increase at about 6 V/cm. We assume that already a significant depletion of subband 1 manifests itself in the change of the slope. If mainly the 0-2 transfer dominates the differential conductivity now in contrast to the weak field limit, the enhanced ratio of the drift velocities between subband 2 and the ground state in comparison to such a ratio between the subbands 2 and 1 could explain an increase. As a consequence of the partial depletion of the first excited subband, transitions from the ground state to this subband become enhanced. The intersubband transfer processes are assumed to be mainly assisted by phonons via deformation potential interaction. Therefore, a more pronounced increase in the LA flux than in the TA flux is observed.

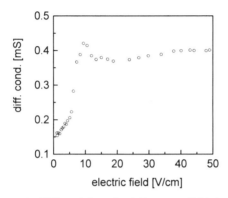

**Figure 3.** Differential conductivity versus field strength

Another specific feature is indicated above 12 V/cm. It is to be seen that in the TA characteristic (Fig.1) the increase diminishes, while there is even a region of constant flux up to 20 V/cm in the LA behaviour. In the field region between 12 and 18 V/cm the differential conductivity (cf. Fig.3) exhibits a hump indicating an additional electron transfer to a state with higher mobility. Especially the very pronounced effect in the LA spectrum that the flux does not rise but remains constant indicates a possible origin. We assume that phonons are reabsorbed to assist interband transitions of the electrons between subbands 0 and 1. The observed limitation to a restricted field region can be explained by the required conservation of momentum and energy as well as the changing population of the states in the involved subbands with increasing field strength.


## PHONON INDUCED CURRENT CHANGE

### Nonequilibrium phonons created by heat pulses

In order to investigate the questions of the involved transitions furtheron the current changes induced by ballistically arriving phonons generated by heat pulses are measured. The energy distributions of these phonons have a Planck spectrum, which is cut off by isotope scattering during their propagation through the substrate. Since the upper values of their energies are 2.75 and 2.4 meV for LA and TA phonons, respectively, they are comparable to the distance between the Fermi level of the ground state as well as the first excited state in the quantum wells and the lowest energy of the extended states above the barriers. Therefore we apply such phonon pulses for the study of intersubband transfer.

### Spectra induced by phonon absorption

The phonon induced current changes (PIC) are always positive, i.e. the transfer effect dominates over the additional scattering processes. After integration over the arrival times of LA and TA phonons the results are shown in Fig.4a and b as functions of the field strength applied to the $\delta$-layers. Again the interaction of electrons with LA phonons exhibits a more distinct structure. As already mentioned the requirements of energy and momentum conservation as well as the population of the subbands determine the lower as well as upper limits of field regions for the absorption of phonons connected with intersubband transitions. Because the change in current induced by phonons in comparison to the current without such an excitation is only about 0.1%, the subband populations are regarded the same as in the absence of arrival of nonequilibrium phonons.

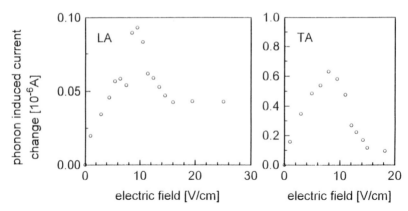

**Figure 4.** PIC integrated over the arrival time of LA (a) and TA (b) phonons versus field

The peculiarity especially in the LA PIC characteristic at about 7 V/cm can again be seen in context with the change of the slope of the differential conductivity at 6 V/cm, i.e. it could be connected with the higher difference in drift velocities in the involved subbands achieved by intersubband transfer between 0 and 2 in comparison to 1 and 2.

With increasing field strength above 11 V/cm a shoulder is seen in the LA PIC characteristic. Since this effect resembles the peculiarity in the spectra of emitted phonons and the appearance of a small increase in the differential conductivity, the origin should be the same. The similar limits of 16 V/cm for this region in the PIC and of 18 V/cm in the differential conductivity support this assumption.

## SUMMARY

Hot electron transfer between the subbands 0, 1, and 2 of the V-shaped mutiple quantum well is reflected in time resolved emission and absorption spectroscopy of ballistically propagating phonons. The specific features of the LA phonon flux are mainly due to the 0-1 transition assisted by emission as well as reabsorption of phonons. The differences in LA and TA phonon induced current changes support the role of deformation potential interaction for intersubband transitions. Differential conductivity measurements distinctly reflect the repopulation of the subbands.

## ACKNOWLEDGEMENT

We want to express our thanks to O. G. Sarbey for stimulating discussions of the problems.

## REFERENCES

1. H. Kostial, T.Ihn, P. Kleinert, R. Hey, M. Asche, and F.Koch, Phys. Rev. **B 47**, 4485 (1993)
2. M. Asche, R. Hey, H. Kostial, B. Danilchenko, A. Klimashov, and S. Roshko, Phys. Rev. **B 12**, 7966 (1995)
3. B. Danilchenko, A. Klimashov, S. Roshko, M. Asche, R. Hey, and H. Kostial, J. Phys.: Condens. Matter 6, 7955 (1995)
4. P. Kleinert and M. Asche, Phys. Rev. **50**, 11022 (1994)
5. J. K. Wigmore, A. J. Kent, O. H. Hughes, and L. J. Challis, Semicond. Sci. Technol. **7 B**, 29 (1992)
6. P. Hawker, A. J. Kent, N. Hauser, and C. Jagadish, Semicond. Sci. Technol. **10**, 601 (1995)

# PHONON OSCILLATIONS IN A SPECTRUM OF REVERSIBLE BLEACHING OF GALLIUM ARSENIDE UNDER INTERBAND ABSORPTION OF A HIGH-POWER PICOSECOND LIGHT PULSE

I.L.Bronevoi,[1] A.N.Krivonosov,[1] and V.I.Perel'[2]

[1]Institute of Radioengineering and Electronics,
Russian Academy of Sciences, 103907, Moscow,
GSP-3, Mokhovaya st., 11, Russia
[2]A.F.Ioffe Physicotechnical Institute, Russian
Academy of Sciences, 194021, St.-Petersburg,
Politekhnicheskaya st., 26, Russia

This text describes the results of studies with picosecond resolution of the optical transparency T spectra of a thin layer of a high-purity GaAs exposed to a high-power light pulse having the photon energy $\hbar\omega_{ex}$, which is somewhat larger than the forbidden gap width $E_g$. Experiments were carried out at room temperature. The duration of both the exciting and probing pulses was about 14 ps. Figure 1 shows results of the sample transparency measure-

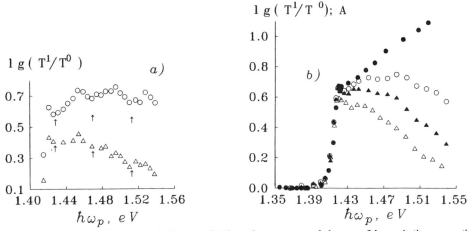

**Figure 1.** Transparency T variation (indices 1 and 0 show the presence and absence of the excitation, respectively) of GaAs as a function of the energy of a photon of the probing pulse at $\hbar\omega_{ex}$ =1.558 eV. a) The delay time of the probing pulse with respect to the exciting one $\tau_d$= -3 ps: o- $W_{ex}$ =1.0 a.u.; ▵- $W_{ex}$ = 0.19 a.u. b) $W_{ex}$= 1.0 a.u.: o -$\tau_d$ = 13 ps; ▴- $\tau_d$ = 31 ps; ▵- $\tau_d$ = 80 ps. The dependence A = f($\hbar\omega_p$) - ● (See explanations in the text).

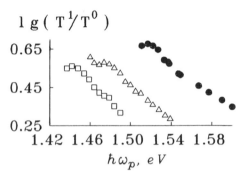

$$lg(T^{1/2}/T^0)$$

**Figure 2.** GaAs transparency variation as a function of the energy of a photon of the probing pulse at $W_{ex} = 1.0$ a.u.: □- $\hbar\omega_{ex} = 1.415$ eV, $\tau_d = 5$ ps; △- $\hbar\omega_{ex} = 1.439$ eV, $\tau_d = 5$ ps; •- $\hbar\omega_{ex} = 1.492$ eV, $\tau_d = 0$.

ments as a function of the energy of a photon of the probing pulse $\hbar\omega_p$ (the bleaching spectrum) within the $\hbar\omega_p < \hbar\omega_{ex}$ range. In this spectrum, which has been measured nearly at the moment of the maximum of the exciting pulse intensity, were found local minima. An oscillating character of the bleaching has been observed for a variety of values of $\hbar\omega_{ex}$[1] and the exciting pulse integral energy $W_{ex}$, the spectral localization of minima being retained. However at the drop of excitation and after its cessation the oscillations smoothened and then disappeared (See Figure 1(b)) despite the bleaching degree being nearly equal to that shown in Figure 1(a) (Figure 1(b) also shows the experimental dependence $A = lg[T^0(\hbar\omega_p = 1.379$ eV)/ $T^0(\hbar\omega_p)]$ which describes the spectrum of optical density of non-excited sample $A_0 \approx A \ln 10$). Oscillations have also been far less strong at $\hbar\omega_p > \hbar\omega_{ex}$, Figure 2. This behavior of oscillations, as well as the fact that there is an antireflective coating on the sample surface, do not permit to attribute the oscillations to the interference in the sample.

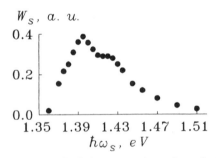

$$W_s, \; a. \; u.$$

**Figure 3.** Integral spectrum of radiation propagating at the angle $\Theta = 75°$ to the normal to sample surface at $\hbar\omega_{ex} = 1.558$ eV, $W_{ex} = 1.0$ a.u. (radiation absorption within the non-excited range does not make it possible, regrettably, to observe radiation that propagates along the GaAs layer).

It should be noted that during the excitation pulse there appears a recombinative edge superluminescence from the sample [2,3] (See Figure 3). Recombination should result in the appearance of a carrier flow in the energy space (partially due to emitting by electrons optical phonons having the energy $\hbar\Omega_0$) towards the level at which the carrier recombination occurs. The recombination must lead to a dip on the curve of electron distribution at a certain energy $E_0$ near the conduction band bottom. This dip is inevitably followed by another dip when $E_1 = E_0 + \hbar\Omega_0$ since the energy transitions from $E_1$ to $E_0$ under phonon emission are more frequent than those in the opposite direction occurring with phonon absorption (in the absence of the dip at $E_0$ the transition rates should have been equal) and so on. This must lead to minima in the bleaching spectrum that are spaced by energy intervals $\hbar\Omega_0 (1+m_e/m_h)$, where $m_e$ and $m_h$ are the masses of electrons and holes, respectively. In Figure 1(a), the arrows indicate the positions of the supposed minima spaced by 42 meV. It is seen that they all agree well with minima observed experimentally thus proving the model suggested.

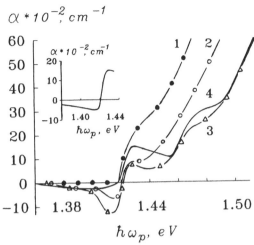

**Figure 4.** The dependence of the absorption coefficient $\alpha$ of the sample on $\hbar\omega_p$ at $\hbar\omega_{ex} = 1.558$ eV; $W_{ex} = 1.0$ a.u.: 1 - $\tau_d = 80$ ps; 2 - $\tau_d = 31$ ps; 3 - $\tau_d = 13$ ps; 4 - $\tau_d = -3$ ps (the insert gives the same dependence 4 but with regard of a correction to a finite duration of the probing pulse).

In the range of energies in which recombination takes place the absorption spectra are presented more clear in Figure 4. These spectra have been obtained from comparison of blea-ching spectra with the experimental dependence $A = f(\hbar\omega_p)$ given in Figure 1(b). It is seen that as long as the residual bleaching has not been reached, in the spectral range $\hbar\omega < E_g^0$ ($E_g^0$ being the width of the band gap of a non-excited sample) a small amplification appears. This amplification should result in arising the recombination superluminescence, that develops, preferentially, along the GaAs layer.[3] The recombination superluminescence must lead to a deviation of the carrier distribution from the quasi-equilibrium one thus causing the distribution depletion in the spectral range of the amplification. This should to distort the absorption spectrum shape as compared to the "smooth" spectrum peculiar to quasi-equilibri-um distribution, which, apparently, explains in this way the appearance of a local absorp-tion maximum at $\hbar\omega \approx 1.428$ eV that corresponds to the above mentioned electron distribu-tion dip reflecting in phonon oscillations.

Acknowledgements - The research described in this publication was made possible in part by Grant No. 95-02-05871-a from the Russian Foundation of Fundamental Research and Grant No. M3S300 from the International Science Foundation.

1. I.L.Bronevoi, A.N.Krivonosov, and V.I.Perel', Phonon oscillations in the spectrum of the reversible bleaching of gallium arsenide under interband absorption of a high-power picosecond light pulse, Solid State Communications. 94:805 (1995).
2. N.N.Ageeva, I.L.Bronevoi, E.G.Dyadyushkin, and B.S.Yavich, Anomalous emission from gallium arsenide during interband absorption of intense picosecond light pulses, JETP Lett. 48:276 (1988).
3. N.N.Ageeva, I.L.Bronevoi, E.G.Dyadyushkin, V.A.Mironov, S.E.Kumekov, and V.I.Perel', Superluminescence and brightening of gallium arsenide under interband absorption of picosecond light pulses, Solid State Communications. 72: 625 (1989).

# LO Phonon Emission and Femtosecond Non-Equilibrium Dynamics of Hot Electrons in GaAs

A. Leitenstorfer[1], C. Fürst[1], G. Tränkle[2], G. Weimann[2], and A. Laubereau[1]

[1]Physik Department E 11, Technische Universität München
[2]Walter-Schottky-Institut, Technische Universität München
D-85748  Garching, Germany

**Abstract.** We present the first spectrally <u>and</u> temporally resolved studies of hot electron dynamics in the low density regime ($N \approx 10^{15} \mathrm{cm}^{-3}$) where LO phonon emission is found to be the dominant process for energy relaxation.

Ultrafast optical spectroscopy is a major tool for the investigation of hot carrier dynamics since the strong coupling of the elementary excitations leads to a rapid relaxation typically occuring on a sub-picosecond timescale. In intrinsic semiconductors two important processes are responsible for this ultrafast temporal evolution: carrier-carrier (cc) and carrier-phonon (cp) scattering. Valuable information on these basic interactions is extracted from experiments where the different contributions to the kinetics can be clearly separated. However, time and energy resolved data under conditions where cp scattering is the dominating effect for the first relaxation steps, are difficult to obtain because of the limited sensitivity of ultrafast measurements at low carrier concentrations.

In previous work on hot carrier distributions with femtosecond pulses, time and energy resolved transient absorption measurements [1-3] and luminescence upconversion studies [4,5] were carried out at excitation densities above $5 \times 10^{16} \mathrm{cm}^{-3}$. In this regime, the rapid carrier-carrier collisions lead to a thermalization of the carriers on a sub-100 fs-timescale. Non-equilibrium dynamics of hot electrons with an experimental time resolution of a few picoseconds has been studied employing a sensitive streak camera [6].

Energy exchange of the hot carriers with the crystal lattice could not be observed independently in these earlier investigations. In particular, the dominant process of electron-phonon scattering in polar semiconductors, the Fröhlich interaction with longitudinal optical (LO) phonons, is too slow to be directly observed at the high carrier densities of previous femtosecond measurements, while on the other hand the process is too fast for the time resolution of Ref. [6].

*Hot Carriers in Semiconductors*
Edited by K. Hess *et al.*, Plenum Press, New York, 1996

Non-equilibrium features of highly energetic electron distributions owing to LO phonon emission in GaAs and InP have been seen in spectrally resolved hot luminescence studies with cw-, ps- and fs-excitation [7-9]. However, the time-integrated nature of these experiments provides only limited information on the carrier dynamics.

In this paper we present transient absorption changes in GaAs measured with the help of a novel two-color modelocked Ti:sapphire laser [10]. The high repetition rate (76 MHz) together with the spectral variability of our system enables us to observe for the first time the femtosecond dynamics of the energetic distributions of hot carriers at excitation densities as low as $10^{15} \mathrm{cm}^{-3}$.

In our experiments, a 500 nm thick specimen of high purity GaAs grown by molecular beam epitaxy is studied. The sample is anti-reflection coated and held at a lattice temperature of $T_L$=15 K. Electron-hole pairs are excited by transform-limited Gaussian pump pulses of a duration of 80 fs and a central photon energy of 1.63 eV. Exciting GaAs with this energy generates electrons starting at the heavy hole and light hole bands. The electrons originating from the the heavy hole band possess an initial excess energy of 103 meV in the conduction band, substantially above the threshold for LO phonon emission (36 meV). The corresponding heavy hole distribution is centered around a kinetic energy of 13 meV. Excitation out of the light hole band into the conduction band yields electrons of an excess energy of 67 meV and light holes of 47 meV. The transmission changes induced in the sample are monitored with weak probe pulses of a duration of 15 to 22 fs at central photon energies between 1.55 and 1.6 eV. Pump and probe pulse trains are perfectly synchronized [10] with linear polarizations perpendicular to each other. Energy information is obtained by analyzing the transmitted broadband probe pulses with a double monochromator of spectral resolution set to 4 meV.

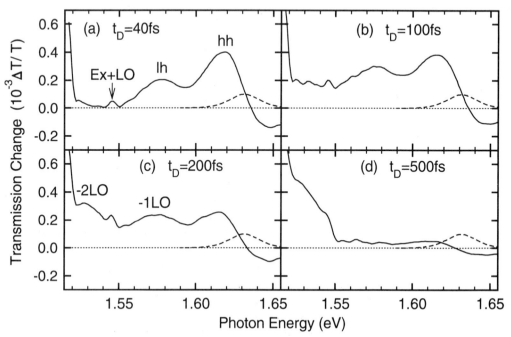

**Figure 1.** (a) to (d) Spectrally resolved transmission changes recorded at various probe delays $t_D$ at an excitation density of $N = 1.6 \times 10^{15} \mathrm{cm}^{-3}$ and a lattice temperature of $T_L$=15 K. The dashed lines represent the spectrum of the excitation pulses.

In Fig. 1 we present energy and time resolved transmission changes recorded within the first 500 fs after excitation of $1.6 \times 10^{15}$ electron-hole pairs per $cm^3$. The spectra are clearly dominated by the bleaching of the heavy hole to conduction band transition due to the hot electron population. At a delay time of 40 fs with respect to the maximum of the pump pulse (Fig. 1a), the most prominent peak appears at a photon energy of 1.62 eV (hh), representing the electrons originating from the heavy hole band. A less pronounced maximum is situated near 1.58 eV (lh). This peak is due to the electrons excited from the light hole band. The relative magnitude of the two peaks is in good agreement with the ratio of 5:2 that is expected for excitation from the two valence bands. After 100 fs (Fig. 1b) the maximum at 1.62 eV has decreased, while the peak at 1.58 eV is increasing. This effect is due to the electrons excited from the heavy hole band which have emitted one LO phonon of an energy of 36 meV, thus scattering in the energy region originally populated by the light hole electrons. Of special interest is the observation of a third maximum arising near 1.53 eV at a time delay of 200 fs (Fig. 1c). This peak is due to electrons from the light and heavy hole bands which have emitted one resp. two LO phonons. At a time delay of 500 fs (Fig. 1c), most electrons have relaxed to the bottom of the conduction band where no LO phonon emission is possible ($E_{gap} = 1.52$ eV in GaAs at 15 K).

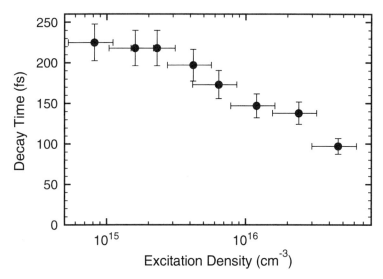

**Figure 2.**   Initial decay constant of the bleaching due to hot electrons at a probing energy of 1.62 eV versus excitation density.

The steep increase in $\Delta T / T$ observed near 1.52 eV is connected to the screening of the 2s exciton by the generated charge carriers. At all positive time delays, we observe a narrow band near 1.55 eV (Ex+LO), one optical phonon energy above the 1s exciton in different samples. We attribute this feature to the bleaching of a two quantum absorption process where a photon creates an exciton together with an optical phonon. To our knowledge this absorption line is reported here for the first time and was not found in a linear absorption spectrum. It appears in our transient measurement because of the high nonlinearity of the exciton screened by the excited carriers.

Another interesting feature is observed on the high energy side of the spectra (see Fig. 1): In sharp contrast to the expected bleaching by the carrier population, a pronounced induced absorption appears above the maximum of the pump spectrum. This phenomenon is caused by an increase of the coulomb enhancement factor of the absorption due to the non-equilibrium distribution of the electrons [11] and is closely related to the Fermi edge singularities observed in metals and highly doped semiconductors [12]. The resulting slight redshift of the bleaching peaks has already been found at higher excitation densities [2,13]. We emphasize that at our low carrier densities the induced absorption survives at delay times much longer than the excitation pulses.

In order to gain better insight into the dynamics of electrons governed by LO phonon emission and to learn more about the increasing influence of cc scattering with excitation density, we perform a series of time resolved measurements at a probing energy of 1.62 eV. The recorded data represent the carrier population probed within an energy interval of 4 meV and its relaxation by the rapid scattering events. In Fig. 2, the initial decay time of the bleaching is depicted versus excitation density. The fast initial dynamics is density independent up to $3 \times 10^{15} \text{cm}^{-3}$ and shows a time constant of $220\pm10$ fs, in reasonable agreement with LO phonon scattering times reported in Refs. [14,15]. This decay is closely related to the emission of LO phonons by the hot electrons. At higher carrier densities the dynamics becomes substantially faster due to the additional energy relaxation by cc collisions, resulting in a doubling of the decay rate near approximately $3 \times 10^{16} \text{cm}^{-3}$.

More detailed data, in particular our observations on the hole distributions will be published elsewhere.

## References

[1]     W.Z. Lin, R.W. Schoenlein, J.G. Fujimoto, and E.P. Ippen, IEEE J. Quant. Electron. **24**, 267 (1988)

[2]     J.-P. Foing, D. Hulin, M. Joffre, M.K. Jackson, J.-L. Oudar, C. Tanguy, and M. Combescot, Phys. Rev. Lett.**68**, 110 (1992)

[3]     S. Hunsche, H. Heesel, A. Ewertz, H. Kurz, and J.H. Collet, Phys. Rev. B **48**, 17818 (1993)

[4]     T. Elsaesser, J. Shah, L. Rota, and P. Lugli, Phys. Rev. Lett. **66**, 1757 (1991)

[5]     U. Hohenester, P. Supancic, P. Kocevar, X.Q. Zhou, W. Kütt, and H. Kurz, Phys. Ref. B **47**, 13233 (1993)

[6]     D.W. Snoke, W.W. Rühle, Y.-C. Lu, and E. Bauser, Phys. Rev. Lett. **68**, 990 (1992); Phys. Rev. B **45**, 10979 (1992)

[7]     D.N. Mirlin, I.Ja. Karlik, L.P. Nikitin, I.I. Reshina, and V.F. Sapega, Solid State Comm. **37**, 757 (1981)

[8]     R.G. Ulbrich, J.A. Kash, and J.C. Tsang, Phys. Rev. Lett. **62**, 949 (1989); J.A. Kash, Phys. Rev. B **40**, 3455 (1989); B **47**, 1221 (1993)

[9]     A. Leitenstorfer, A. Lohner, T. Elsaesser, S. Haas, F. Rossi, T. Kuhn, W. Klein G. Boehm, G. Traenkle, and G. Weimann, Phys. Rev. Lett. **73**, 1687 (1994)

[10]    A. Leitenstorfer, C. Fürst, and A. Laubereau, Opt. Lett. **20**, 916 (1995)

[11]    R. Zimmermann, Phys. Stat. Sol. B **146**, 371 (1988)

[12]    C. Tanguy and M. Combescot, Phys. Rev. Lett. **68**, 1935 (1992)

[13]    J.H. Collet, S. Hunsche, H. Heesel, and H. Kurz, Phys. Rev. B **50**, 10649 (1994)

[14]    J.A. Kash, and J.C. Tsang, Solid State Electron. **31**, 419 (1988)

[15]    M. Heiblum, D. Galbi, and M. Weckwerth, Phys. Rev. Lett. **62**, 1057 (1989)

# HOT CARRIER THERMALIZATION DYNAMICS IN LOW-TEMPERATURE-GROWN III-V SEMICONDUCTORS

A.I. Lobad,[1,2] Y. Kostoulas,[1,2] G.W. Wicks[3] and P.M. Fauchet[1,2,3,4]

[1]Laboratory for Laser Energetics
[2]Department of Physics and Astronomy
[3]The Institute of Optics
[4]Department of Electrical Engineering
University of Rochester, Rochester, NY 14623

The presence of point defects is expected to influence the properties of free carriers in semiconductors. Low-temperature-grown (LT) GaAs is a material with a high density of point defects and a unique combination of properties: sub(picosecond) mobile carrier lifetime, good carrier mobility and very high dark resistivity.[1] The ultrashort mobile carrier lifetime in LT GaAs and other III-Vs has been extensively investigated by many groups and found to be due to trapping.[2-6] The purpose of the present paper is to report results on the ultrafast thermalization dynamics of hot carriers in LT grown III-Vs.

Measurements were performed on LT $Ga_{0.51}In_{0.49}P$, InP and $In_{0.53}Ga_{0.47}As$. The 0.5 and 1μm thick LT GaInP films were grown at 500°C and 200°C on GaAs wafers, respectively. The GaAs wafers were then selectively etched. The 200°C film was cleaved into two pieces, one of which was subjected to a 10 min annealing step at 450°C. The 0.2 μm thick LT InP films were grown at 200°C and 300°C, and then separated from their substrate by the lift-off technique.[2] The LT $In_{0.53}Ga_{0.47}As$ films were grown at 200°C. Some films were subjected to an annealing step at 450°C for 10 min and others intentionally doped with Be. Experiments were performed at room temperature using either an amplified colliding-pulse modelocked (CPM) dye laser producing 100 fs white-light continuum pulses centered at 2 eV or an amplified additive-pulse modelocked (APM) color center laser producing 200 fs white light continuum pulses centered at 0.8 eV. We have used the techniques of time resolved luminescence up-conversion and pump-probe spectroscopy to study the hot carrier dynamics.

In figure 1 we show the amplitude of the absorption bleaching in GaInP at the peak of the excitation pulse normalized by the linear absorption ($\Delta\alpha(\tau = 0)/\alpha_o$) as a function of the probe wavelength for three samples: the 500°C grown and the 200°C as-grown and annealed layers. All samples show a spectral hole burning peak at 640 nm instead of 620 nm which is the excitation energy. This is due to significant carrier-carrier scattering at these high injected carrier densities which leads to a spreading of the carrier distribution during photo-excitation. The sample grown at 500°C shows a peak bleaching of 60% but when the growth temperature is lowered to 200°C, the peak bleaching is reduced to less than 10%. This can be attributed to either an ultrafast carrier scattering mechanism, such as impact ionization of defects, which spreads the carrier distribution within our time resolution or an ultrafast (< 100 fs) trapping time. An ultrafast trapping time has been ruled out by subgap measurements.[7] After annealing, the density of point defects is greatly reduced, leading to a partial recovery of the spectral hole peak bleaching to ~ 40%.

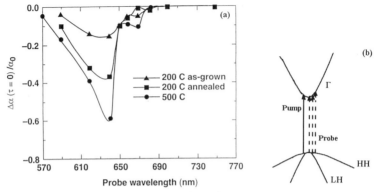

**Figure 1** (a) Spectral hole burning in InGaP grown at 500°C, 200°C and 200°C + annealing, (b) InGaP band diagram.

A similar ultrafast spreading of the as-excited carrier distribution has also been observed in the case of 200°C grown LT-InP, as can be seen from the instantaneous rise time of the bandgap luminescence in figure 2. A large number of electrons (holes) reach the bottom (top) of the conduction (valence) band within 100 fs or less, despite the fact that the carriers are injected high in the band (~650 meV excess energy). In normally grown III-Vs, the rise time of the photoluminescence can be as long as several ps, because inelastic scattering of carriers with phonons takes place on a time scale of 150 fs. In normally grown InP, the band edge luminescence risetime is 3 ps[8] which is shorter than in GaAs because intervalley scattering is not energetically allowed for the photoexcited electrons in InP.

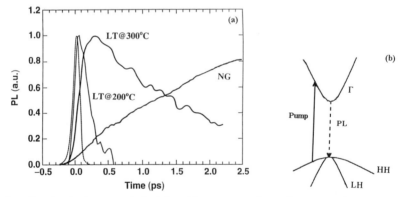

**Figure 2** (a) Transient luminescence obtained from InP grown at 300°C, 200°C (Ref.6) and NG-InP (Ref. 8). The dashed line shows the pump pulse duration, (b) InP band diagram.

To investigate the thermalization dynamics of hot carrier in LT InGaAs, we use a high energy pump of 1.58 eV, obtained by frequency doubling the amplified color center laser, and a near band edge probe of 0.79 eV. We estimate the maximum injected carrier density to be $10^{19}$ cm$^{-3}$. Excitation at 1.58 eV produces three distributions of carriers in the conduction band, which originate from the heavy hole, the light hole and the spin split-off valence bands. The heavy hole-electron distribution has an excess energy of 0.78 eV, 0.23 eV above the threshold for intervalley scattering to the L valley. This leads to a preferential and efficient scattering of this distribution into the L valley. The two other distributions are below the threshold for intervalley scattering and thermalize within the $\Gamma$

valley. In this experiment the onset of bleaching at the probe wavelength is a measure of how fast the carrier distribution spreads to the bottom of the conduction band. Figure 3 compares the response of NG and LT grown InGaAs. The NG layer shows an initial increase of absorption attributed to band-gap renormalization,[9] which is absent in the 200°C grown LT layer. In the LT layer, the hot carriers have thermalized and reached the band edge within the time resolution thus canceling the effect of band-gap renormalization. The subsequent and gradual decrease of absorption on a picosecond time scale indicates the return of electrons from the L valley. Both layers exhibit a similar time constant of ~ 3.5 ps for the L to Γ scattering process.

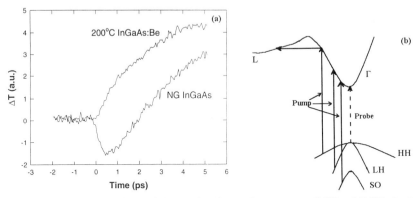

**Figure 3** (a) Temporal evolution of transmission obtained from InGaAs grown at 500°C and 200°C +Be doping, (b) InGaAs band diagram.

The fact that point defects have a strong influence on the properties of free carriers is not surprising when one considers that the presence of a large density of traps has the same effect in silicon. In a-Si:H, a material that contains ~ $10^{19}$ cm$^{-3}$ bandtail states and ~ $10^{16}$ cm$^{-3}$ midgap states, the initial scattering time is of the order of 1 fs[10] and the energy relaxation time rate is approximately 10 times faster than in crystalline silicon.[11] These results are consistent with the results of this paper: (i) an ultrafast initial scattering time spreads the carrier distribution throughout the band and eliminates spectral hole burning, and (ii) an ultrafast energy relaxation time pushes the hot carriers towards the bottom of the band and leads to immediate near band-gap bleaching and luminescence.

## ACKNOWLEDGMENT

This work was supported in part by the Office of Naval Research. A.I. Lobad was the recipient of the Frank Horton fellowship.

## REFERENCES

1. F.W. Smith, Mat. Res. Soc. Symp. Proc. **241**, 3 (1992).
2. Y. Kostoulas et al, in *Ultrafast Phenomena in Semiconductors*, SPIE **2142**, 100 (1994).
3. E.S. Harmon et al, Appl. Phys. Lett. **63**, 2248 (1993).
4. B.C. Tousley et al, J. of Electronic Materials **22**, 1477 (1993).
5. S. Gupta et al, J. of Electronic Materials **22**, 1449 (1993).
6. Y. Kostoulas et al, Appl. Phys. Lett. **66**, 1821 (1995).
7. Y. Kostoulas et al, submitted for publication.
8. J. Shah, IEEE J. of Quantum Electronics **24**, 276 (1988).
9. T. Gong et al, Phys. Rev. B **44**, 6542 (1991).
10. P.M. Fauchet et al, J. Non-Cryst. Solids **141**, 76 (1992).
11. D. Hulin et al, J. Non-Cryst. Solids **137&138**, 527 (1991).

# HOT CARRIER DYNAMICS IN THE X VALLEY
# IN Si AND Ge MEASURED BY PUMP-IR-PROBE
# ABSORPTION SPECTROSCOPY

W. B. Wang, M. A. Cavicchia, and R. R. Alfano

Institute for Ultrafast Spectroscopy and Lasers, and New York State Center
for Advanced Technology for Ultrafast Photonic Materials and Applications,
The City College of New York, Convent Avenue at 138th St, NY, NY 10031

Si is the semiconductor of choice for nanoelectronic roadmap into the next century for computer and other nanodevices. With growing interest in Si, Ge, and $Si_mGe_n$ strained superlattices, knowledge of the carrier relaxation processes in these materials and structures has become increasingly important.[1] The limited time resolution for earlier studies[2,3] of carrier dynamics in Ge and Si, performed using Nd:glass lasers, was not sufficient to observe the fast cooling processes.

In this paper, we present a direct measurement of hot carrier dynamics in the satellite X valley in Si and Ge by time-resolved infrared(IR) absorption spectroscopy, and show the potential of our technique to identify whether the X valley is the lowest conduction valley in semiconductor materials and structures.

The details of the femtosecond visible-pump and IR-probe absorption set up has been reported elsewhere.[4] In this measurement, ~585 nm, ~400 fs pulses, obtained from the output of a synchronously pumped dye laser with a pulse-dye-amplifier, was used as a pump, and IR pulses tuned from 2.5 μm to 5.5 μm, generated in a $LiIO_3$ crystal by difference frequency method, was used to monitor the photoinduced IR absorption.

The measured change of the induced IR absorption at $\lambda_{probe}$ =3.3 μm as a function of delay time between pump and probe pulses for intrinsic Si and Ge samples at room temperature are displayed in Figs.1 and 2, respectively. The salient features of the curves are the following: although the curve for Si is characterized by a long flat decay within the measured time range of ~30 ps, the curve for Ge has a short decay of ~6 ps followed by a long flat decay.

The measurements were also extended to other probe wavelengths of 2.8 μm, 4.5 μm and 5.2 μm for both Si and Ge. The time evolution of the induced IR absorption at those probe wavelengths was found to be similar as that for $\lambda_{probe}$ =3.3 μm: a single flat decay appears with Si, and a short decay followed by a flat decay accompanies with Ge. It is clear that the different temporal behavior of the induced IR absorption for Si and Ge reflects their different carrier cooling processes and arises from their different band structures.

The induced total IR absorption is attributed to interconduction band absorption (ICA) by electrons from the lower X valley to the higher X valley separated by an energy difference in IR; free carrier absorption (FCA) by holes in the valence bands and by electrons in the conduction bands; and inter-valence band absorption (IVA) by electrons from the split-off to the heavy and light hole bands. Since the calculated values[3,5] of OD for IVA at the probe wavelengths are much smaller than the measured total OD, IVA can be neglected in our

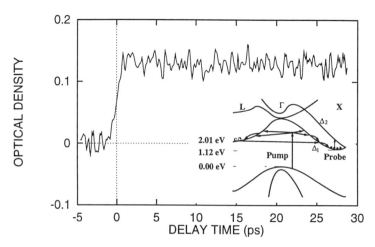

**Figure 1.** The measured change of the induced IR absorption as a function of delay time at $\lambda_{probe}=3.3$ μm in Si. The energy band structure of Si is shown in the inset.

consideration. The decay of FCA is flat because the recombination times of electrons and holes in Ge and Si are several order of magnitude longer than our experimental time scale. The long flat decay of FCA has been previously observed.[2-4]

The temporal characteristics of ICA in Si and Ge depends on their energy band structures. In Si, the lowest conduction band energy (1.12 eV measured from the top of the $\Gamma$ valence band) is located at the X ($\Delta_1$, $\Delta_2$) valley[6] as shown in the inset of Fig.1. Hot carriers produced by 585 nm pump photons through an indirect transition can obtain sufficient energy to reach both the X and L valleys, from which they can scatter to each other as well as scatter inside each of the X and L valleys. Since the energy of the X valley minimum is lower than that of the L valley (2.01 eV)[7], after a few times of the X-L back-and-forward intervalley scattering, almost all of the electrons will finally scatter into the X valley. Those X valley electrons then decay to the bottom of the X valley through intravalley scattering, and stay there until they combine with holes in a few hundred nanoseconds.[8] Therefore, the induced ICA by X valley electrons from $\Delta_1$ to $\Delta_2$ (see the inset of Fig.1) has a long flat decay.

In contrast to Si, the lowest conduction band energy in Ge is located at the L valley[6] (0.66 eV) as shown in the inset of Fig.2. Electrons pumped by 585 nm photons through direct transition in the $\Gamma$ valley obtain enough energy to undergo intervalley scattering from the $\Gamma$ valley to the L and X valleys. The population of electrons in the X valley increases at first to its maximum through the $\Gamma$,L->X intervalley scattering. Since the minimum energy of the X valley (0.86 eV) is higher than that of the $\Gamma$ and L valleys (0.80 eV and 0.66 eV, respectively), electrons stay in the X valley for only a short time, and then decay back to the $\Gamma$ and L valleys. Therefore, the ICA in the X valley should have a short decay.

In order to quantitatively understand the change of the ICA with band structures, a parameter $f_X$, defined as the fraction of carrier density in the X valley relative to the total carrier density at the delay time corresponding to the flat absorption region, should be discussed. Since electrons in the conduction bands at the flat absorption region may be assumed to be in thermal equilibrium and characterized by a Boltzmann distribution, $f_X$ can be written[9] as:

$$f_X = 1/[1+(t_X/t_L)(m_L/m_X)^{3/2}\exp(\Delta E_{XL}/K_BT)+(t_X/t_\Gamma)(m_\Gamma/m_X)^{3/2}\exp(\Delta E_{X\Gamma}/K_BT)] \quad (1)$$

where $t_i$ and $m_i$ are the lifetime and density-of-state effective mass of electrons in the ith (i=$\Gamma$, L and X) conduction valley, respectively; $\Delta E_{Xi} =(E_g)^X-(E_g)^i$ is energy difference between the X and ith vallys (i=$\Gamma$, L) at their minima; $K_B$ is the Boltzmann constant and T is

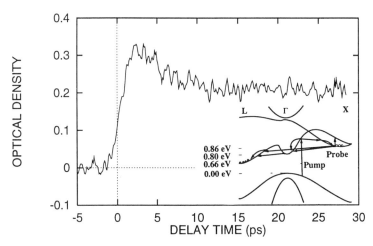

**Figure 2.** The measured change of the induced IR absorption as a function of delay time at $\lambda_{probe}$=3.3 $\mu$m in Ge. The energy band structure of Ge is shown in the inset.

the effective temperature of relaxed electrons (T=293 K). For Si, substituting $\Delta E_{XL}$=-890 meV and other parameters[6,8] of $t_X$, $t_L$, $m_X$, and $m_L$ into eqn.(1), $f_X$ was calculated to be $[f_X]_{Si} \sim 1$, which indicates that almost all of the electrons will decay and stay at the bottom of the X valley. Therefore, the ICA for Si is flat. For Ge, substituting values of $\Delta E_{XL}$=200 meV and $\Delta E_{X\Gamma}$=60 meV, and other parameters[6,8] of $t_i$, $m_i$, (i=L, $\Gamma$) into eqn.(1), $f_X$ was calculated to be $[f_X]_{Ge} \sim 0$, which means that the X valley holds almost no electrons at the long delay time. Since electrons stay in the X valley only for a short time, the ICA for Ge has a short decay.

The temporal behavior of the photoinduced IR absorption in a semiconductor can be used as a criteria to determine whether the X valley is the lowest conduction band valley. According to eqn.(1), materials with a conduction band energy minimum at the X valley ($\Delta E_{Xi}$<0 for i=$\Gamma$ and L, therefore, $f_X$~1) should have a single flat decay, while materials with a conduction band valley minimum not at the X valley ($\Delta E_{Xi}$>0 for i=$\Gamma$ or L, therefore, $f_X$~0) should have a short decay followed by a flat decay if the pump and probe wavelengths used can make the transition for ICA in the X valley. Over past several years, it has become a hot and controversial topic that certain $Si_mGe_n$ strained superlattices may be converted from the indirect band gap of Si to a direct band gap via Brillouin zone folding.[1] The pump-IR-probe absorption technique presented here will be particularly useful for identifying this fundamental change of the lowest conduction band from the X valley to the $\Gamma$ valley.

This research is supported by NASA and the NYS Technology Foundation.

**References**

1. H.Presting, H.Kibbel, M.Jaros, R.M.Turton, U.Menczigar, G.Abstreiter, and H.G. Grinmeiss, Semicond. Sci. Technol., 7, 1127 (1992).
2. A.L.Smirl, J.Ryanlindle, and S.C.Moss, Phys. Rev., B18, 5489 (1978).
3. N.Ockman, R.Dorsinville, W.B.Wang and R.R.Alfano, IEEE QE-23, 2008 (1987).
4. W.B.Wang, K.Shum, R.R.Alfano, D.Szmyd, and A.J.Nozik, Phys. Rev. Lett., 68, 662 (1992).
5. R.Braustein, Phys. Rev., 130, 869 (1963).
6. O.Madelung, in "Semiconductors: Group IV and III-V Compounds, in Data in Science and Technology", edited by R. Poerschke (Springer, Berlin, 1991).
7. J.Weber and M.I.Alonso, Phys. Rev., B40, 5683 (1989-I).
8. B.R.Nag, in "Semiconductors probed by ultrafast laser spectroscopy", edited by R. R. Alfano (Academic, Orlando, 1984), Vol.I, Ch. 1.
9. W.B.Wang, R.R.Alfano, D.Szmyd, and A.J.Nozik, Phys. Rev., B46, 15828 (1992).

# ULTRAFAST HOLE HEATING IN INTRINSIC GaAs

F. Vallée, P. Langot, and R. Tommasi

Laboratoire d'Optique Quantique
du Centre National de la Recherche Scientifique
Ecole Polytechnique, 91128 Palaiseau cédex, France

Ultrafast interactions of free carriers between themselves and with their environment is a central problem of semiconductor physics, both from a fundamental point of view and for their technological implications. Ultrafast carrier dynamics has thus been extensively studied in semiconductors using time resolved femtosecond techniques, yielding important information on carrier scattering processes[1-3]. In direct gap semiconductors, most of the investigations have focused on electron relaxation dynamics and although electron scattering processes are relatively well characterized, little is known about hole relaxation dynamics. Because of their larger occupation number close to the band edge and/or of their generally slower thermalization dynamics, electrons usually dominate the probed transient semiconductor properties and experimental conditions have to be specifically chosen in order to access nonequilibrium hole dynamics.

Recently, this problem has been addressed using time resolved luminescence in $n$-doped bulk GaAs and GaAs/AlGaAs quantum wells[4-6]. In these experiments, the high-density electron plasma introduced by doping acts as a reservoir for low density nonequilibrium hole recombination making the measurements possible but also strongly modifying the hole environment. In particular, at low temperature, it has been shown that hole cooling is essentially due to hole-thermal electron scattering in these samples[5,6]. In contrast to time-resolved luminescence, the electron and hole contributions are intrinsically separated in absorption saturation measurements and, taking advantage of the higher k-space localization of the electrons than the holes, the heavy-hole (HH) dynamics can be selectively analyzed in intrinsic bulk GaAs using a two-wavelength femtosecond technique[7].

Because of the larger HH mass compared to the electron one, for a fully thermalized system the two distributions are partly separated in k-space and, in particular, the HH occupation number is larger than the electron one for large wave vector ($k \geq 5 \times 10^6$ cm$^{-1}$). The same situation can be realized for nonequilibrium distributions by photoexciting electrons with an average energy smaller then the thermal one. Consequently, using a short pulse to photoexcite small wave vector carriers and probing large wave vector carrier states (i.e., probing with higher energy photons) the sample transient transmission change $\Delta T/T$ predominantly contains information on filling of the HH states. This is illustrated in Figure 1, showing the relative contributions of electrons (EL), heavy-holes and light-holes (LH) to the normalized absorption change $\Delta \alpha / \alpha$ (only band filling effects were considered).

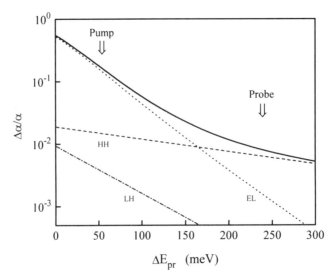

**Figure 1.** Total normalized absorption change, $\Delta\alpha/\alpha$ (full line) due to band filling as a function of the probe photon excess energy, for an electron-hole plasma with a density of $2\times10^{17}$ cm$^{-3}$ at T=295 K. The individual contributions due to the filling of conduction (EL), heavy- (HH) and light-hole (LH) bands are also shown.

In order to better characterize the transient contributions due to the different photoexcited carriers we have numerically calculated the time evolution of $\Delta T/T$ for pump and probe wavelengths of respectively 840 nm and 750 nm and a carrier density of $2\times10^{17}$ cm$^{-3}$ in GaAs at room temperature (Figure 2). Here, the electrons are photoexcited with an excess energy close to their final thermal energy to limit spurious effects due to cold-electron heating[8]. The simulations are based on a numerical solution of the coupled carrier-phonon Boltzmann's equations[9] for parabolic electron, light-hole and heavy-hole bands with static screening for hole interactions. Only intraband processes are taken into account for carrier-carrier scattering and both intraband and interband processes are included for carrier-phonon scattering. Polar and nonpolar hole-optical phonon scattering were considered. The many body effects: band gap renormalization (BGR) and modification of the Coulomb enhancement factor (CEF), are also introduced[10].

The calculated individual contributions due to BF of the electron and hole (HH and LH) states and to many body effects are shown in Figure 2. Because of the large momentum difference between the probed and photoexcited states, BF effects are delayed and the short time delay signal is mainly due to BGR and CEF reduction. Actually, the former is found to dominate resulting in an almost instantaneous decrease of the sample transmission close to the band gap. This contribution essentially depends on carrier density and is only weakly modified for longer time delay. The subsequent rise of $\Delta T/T$ is mainly due to filling of the probed HH states as the initially cold distribution thermalizes. Direct contribution due to filling of the LH states is negligible but their inclusion into the calculation is essential as LH contain most of the initial excess energy injected into the hole system. The calculated dynamics indicates that holes reach thermal equilibrium with the lattice in less than 1 ps.

Experiments were performed using a femtosecond Ti:Al$_2$O$_3$ laser operated at 800 nm with 1.6 W average power. The laser beam, after passing a Faraday isolator, is coupled into a single mode optical fiber (5.5 cm long, 3.9 μm core diameter) where it is spectrally broadened by self-phase modulation to 130 nm (FWHM). The fiber output is split into two synchronized, independently tunable (in the range 720-890 nm), 70 fs pulses using two identical systems consisting of a grating pair compressor and a slit placed out of the Fourier

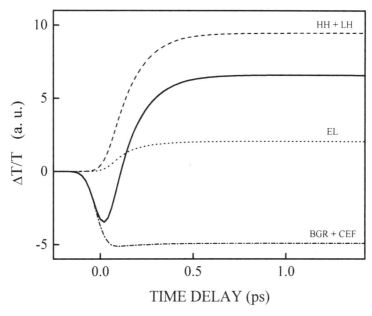

**Figure 2.** Total calculated transmission change, $\Delta T/T$ (full line) as a function of the probe time delay, for an electron-hole plasma density of $2\times10^{17}$ cm$^{-3}$ in GaAs at T=295 K. The pump and probe wavelengths are 840 nm and 750 nm, respectively. The individual contributions due to the filling of conduction (EL), heavy (HH) and light-hole (LH) bands together with the contribution of BGR and CEF reduction are also shown.

plane. The two beams are cross-polarized and sent into a standard pump-probe set-up. The high repetition rate (76 MHz) and stability of our system permit a very high sensitivity with a noise level for $\Delta T/T$ measurements of $\sim10^{-6}$. The pump and probe beams were focused to a focal spot 15 μm in diameter using a 10x microscope objective. All of the experiments were performed at room temperature in a 0.2 μm thick intrinsic GaAs sample with $Al_{0.6}Ga_{0.4}As$ cladding layers.

The measured $\Delta T/T$ is shown in Figure 3 in the same conditions as the numerical simulation described above. The measured temporal behavior is in good agreement with the theoretical predictions with, in particular, a transient decrease of $\Delta T/T$ due to BGR and a subsequent fast rise due to HH state filling. Nearly complete thermalization is observed after $\sim700$ fs. The normalized time evolution of the difference between the maximum differential transmission $(\Delta T/T)_M$ and the transient one $\Delta T/T$ is reported in the inset of Figure 3, on a logarithmic scale. $(\Delta T/T)_M$ has been determined experimentally for a probe time delay of 5 ps. The transient differential transmission approaches its quasi-equilibrium value with an estimated time constant of $\tau_{th}\sim125$ fs.

As the initial hole energy is lower than the thermal energy at T=295 K, holes must gain energy to thermalize with the lattice. The energy source might be both the lattice and the photoexcited electrons. Measurements performed as a function of carrier density (from $2\times10^{16}$ cm$^{-3}$ to $7\times10^{17}$ cm$^{-3}$) show little modifications of the thermalization time (from 160 fs to 120 fs) indicating that hole heating is dominated by hole-phonon interactions. This is in agreement with our numerical calculations and the hole cooling measurements in n-doped samples at room temperature[4]. Although the excitation conditions are different, the minor role played by electrons in our measurements is also consistent with the longer hole cooling times measured in *n*-doped samples at low-temperatures ($\tau_{th}\sim560$ fs for a donor density of $6\times10^{17}$ cm$^{-3}$).[5]

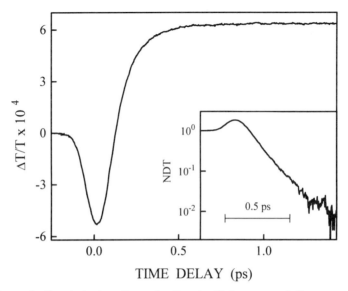

**Figure 3.** Transient absorption saturation in GaAs measured for a pump wavelength of 840 nm and a probe wavelength of 750 nm at T=295K. The photoexcited carrier density is $2 \times 10^{17} \text{cm}^{-3}$. The inset shows, on a logarithmic scale, the experimental time evolution of NDT=$[(\Delta T/T)_M - \Delta T/T] / (\Delta T/T)_M$.

These first selective measurements of hole thermalization in an intrinsic direct gap semiconductor open up many possibilities for the investigation of hole interactions with their environment in these systems. In particular, temperature dependent measurements should give a better insight into the relative importance of hole-electron and hole-optical phonon interactions by strongly modifying the optical phonon absorption probabilities. Hole-hole interaction could also be investigated in p-doped systems or after photoexcitation of an electron-hole plasma.

## REFERENCES

1. C.L. Tang and D.J. Erskine, Femtosecond relaxation of photoexcited nonequilibrium carriers in $Al_xGa_{1-x}As$, *Phys. Rev. Lett.* 51:840 (1983).
2. T. Elsaesser, J. Shah, L. Rota, and P. Lugli, Initial thermalization of photoexcited carriers in GaAs studied by femtosecond luminescence spectroscopy, *Phys. Rev. Lett.* 66:1757 (1991).
3. L.H. Acioli, M. Ulman, F. Vallée, and J. Fujimoto, Femtosecond carrier dynamics in the presence of a cold plasma in GaAs and AlGaAs, *Appl. Phys. Lett.* 63:666 (1993).
4. X.Q. Zhou, K. Leo, and H. Kurtz, Ultrafast relaxation of photoexcited holes in *n*-doped III-V compounds studied by femtosecond luminescence, *Phys. Rev. B* 45:3886 (1992).
5. A. Chébira, J. Chesnoy, and G.M. Gale, Femtosecond relaxation of photoexcited holes in bulk gallium arsenide, *Phys. Rev. B* 46:4559 (1992).
6. A. Tomita, J. Shah., J.E. Cunningham, S.M. Goodnick, P. Lugli, and S.L. Chuang, Femtosecond hole relaxation in *n*-type modulation-doped quantum wells, *Phys. Rev. B* 48:5708 (1993).
7. R. Tommasi, P. Langot, and F. Vallée, Femtosecond hole thermalization in bulk GaAs, *Appl. Phys. Lett.* 66:1361 (1995).
8. P. Langot, R. Tommasi and F. Vallée, Cold-phonon effect on electron heating in GaAs, to be published.
9. J.H. Collet and T. Amand, Model calculation of the laser-semiconductor interaction in subpicosecond regime, *J. Phys. Chem. Solids* 47:153 (1986).
10. H. Haug and W. Koch, in "Quantum Theory of the Optical and Electronic Properties of Semiconductors," World Scientific, Singapore (1993).

# GREEN'S FUNCTIONS VERSUS DENSITY MATRICES: ASPECTS OF FREE-CARRIER SCREENING IN HIGHLY EXCITED SEMICONDUCTORS

U. Hohenester[1] and W. Pötz[2]

[1] Inst. f. Theoretische Physik, Karl–Franzens– Universität
A–8010 Graz, Austria
[2] University of Illinois at Chicago, Physics Department,
Chicago, Il 60607

**Abstract.** The question of how to treat free-carrier screening in the description of highly nonequilibrium carrier plasmas is still widely disputed. To further inquire into this problem we compare different techniques to derive generalized Boltzmann Bloch equations (BBEs) for optically excited semiconductor systems which account for both the coherent and incoherent dynamics. Mostly such BBEs have been obtained either within the framework of non-equilibrium Green's functions (NEGFs) or by using the density matrix approach (DMA). A loose end of the latter approach has been the lack of a consistent way to treat free–carrier screening, and the conventional procedure has been to replace the bare Coulomb potential in the basic Hamiltonian by an appropriately screened one. We show that from the equation of motion (EoM) for the two–particle NEGF one can derive a simple prescription of how to treat free-carrier screening within the DMA. As an example we obtain the semiconductor Bloch equations for a homogeneous bulk semiconductor and we show that the ad-hoc implementation of screening within the DMA introduces errors.

Over the last few years, the development of optics and modern techniques for sample growth have made it possible to study coherence phenomena on a sub–picosecond time–scale. The derivation of BBEs which are capable of describing such phenomena as well as the incoherent dynamics in semiconductors are usually based on either the NEGF formalism[1] or the DMA.[2,3,4] The latter approach seems to be more attractive and transparent, since the quantities are more directly related to physical observables. However, it has been an open question how to consistently treat free-carrier screening within the DMA, and we will address this problem in the following.

We will restrict ourselves to a simplified model, where only the interactions of the carriers with an external field and the Coulomb interaction between the carriers will be taken into account. The full Hamiltonian of the problem is then given by $H = H_o + H_e + H_i$, where $H_o$ describes the noninteracting carriers. Introducing the

*Hot Carriers in Semiconductors*
Edited by K. Hess *et al.*, Plenum Press, New York, 1996

shorthand notation $1 \equiv (\mathbf{r}_1, \alpha_1)$, with the real-space coordinate $\mathbf{r}_1$ and the band index $\alpha_1$ we can write

$$H_o = \sum_1 \psi^\dagger(1)\varepsilon(1)\psi(1)$$
$$H_e = \tfrac{1}{2}\sum_{11'}(T(11',t)\psi^\dagger(1')\psi(1) + \text{h.c.})$$
$$H_i = \tfrac{1}{2}\sum_{11'}v(11')\psi^\dagger(1)\psi^\dagger(1')\psi(1')\psi(1),$$

with the kinetic energy $\varepsilon$, the transition matrix $T$ and the unscreened Coulomb matrix element $v$. The field operators $\psi$ obey the usual anticommutation rules $[\psi(1), \psi^\dagger(1')]_+ = \delta(11')$ and zero otherwise. Next we define the NEGF $g(11') = -i\langle T_c \psi(1)\psi^\dagger(1')\rangle$, with the contour ordering operator $T_c$ (we set $\hbar = 1$). The time arguments have been extendend to the Keldysh contour[5] and we have included the contour time $t_1$ into $1 \equiv (\mathbf{r}_1, \alpha_1, t_1)$. A perturbative expansion for the NEGFs can be obtained on the basis of the functional–derivative method. Following Refs. 6 and 7 we introduce an additional coupling to an auxiliary field $u$, $H_u = \sum_1 u(1,t)\psi^\dagger(1)\psi(1)$, and we get the EoM

$$(i\partial_{t_1} - \varepsilon(1) - u(1))g(11';u) = \delta(11') + \sum_2 (T(12)g(21';u) - iv(12)g_2(12, 1'2^+;u))\Big|_{t_2 \to t_1},$$

with the two-particle NEGF $g_2(12, 1'2^+;u) = (g(22^+;u) - \frac{\delta}{\delta u(2)}))g(11';u)$ and $2^+$ denoting an infinitesimally increased time argument. As discussed in some detail in Ref. 6, in Coulombic systems a perturbative expansion converges only with respect to the screened auxiliary field $u_s(1) = u(1) - i\sum_2 v(12)(g(22^+;u) - g(22^+))$. The lowest order approximation is to treat the interaction with the external field only in first order and to neglect $\frac{\delta \Sigma'}{\delta u_s}$, where $\Sigma'$ denotes the self–energy without the Hartree term. We then get the screened exchange contribution $\Sigma'_{SX}(11') = iv_s(11')g(11')$; here the screened Coulomb potential obeys a Dyson–like equation $v_s(11') = v(11') + \int d2d\bar{2}\, v(12)\Pi(2\bar{2})v_s(\bar{2}1')$ and the polarization in random–phase approximation (RPA) is given by $\Pi(2\bar{2}) = -ig(2\bar{2})g(\bar{2}2)$. The corresponding approximation for $g_2$ then reads

$$\bar{g}_2(12, 1'2^+) = i\int d3d4\, g(13)g(31')v_s(24)g(42)g(24)$$
$$= i\int d3d4\, v(34)g(13)g(31')(g(24)g(42) - \bar{g}_2(24, 2^+4^+)), \qquad (1)$$

with $\bar{g}_2(12, 1'2^+) = (g_2(12, 1'2^+) - g(11')g(22^+) + g(12)g(21'))_{u=0}$. In order to find the connection with the DMA we have to operate with $g^{-1}$ on Eq. (1) and we finally have to make the limit $t_{1'} \to t_2 \to t_1$. Within the Hartree approximation for $g^{-1}$ we obtain with $\varepsilon_H(1) = \varepsilon(1) - i\sum_2 v(12)g(22^+)$, the density matrix $\rho(1'1) = \langle \psi^\dagger(1')\psi(1)\rangle$ and the two particle correlation function $c_2(1'2', 12) = \langle \psi^\dagger(1')\psi^\dagger(2')\psi(2)\psi(1)\rangle_{t_{1'}=t_{2'}=t_2=t_1} - \rho(11')\rho(22') + \rho(12')\rho(2'1)$ the final result

$$\left(i\partial_t - (\varepsilon_H(1) - \varepsilon_H(1') + \varepsilon_H(2) - \varepsilon_H(2'))\right)c_2(1'2', 12)\Big|_{2' \to 2} =$$
$$(v(12) - v(1'2))\rho(1'1)\rho(22) + \sum_3(v(13) - v(1'3))\rho(1'1)(c_2(23, 23) - \rho(23)\rho(32)). \qquad (2)$$

This EoM has been obtained within the DMA by various authors[3,4] and has been solved by (i) neglecting the correlation transfers (CT) on the rhs of Eq. (2) and (ii) performing a Markov and an adiabatic approximation. However, the Coulomb potential had to be screened "by hand". As an important result of this work we have shown that (i) screening is indeed contained within the DMA (through the CT in (2)) and (ii) the screened exchange approximation together with RPA screening is closely related to a truncation of the BBGKY hierarchy at the level of the four point functions and keeping only terms as indicated by Eq. (2).

The solution of Eq. (2) within the DMA for a one–component plasma can be found in Ref. 8 while the extension to many bands including also coherence effects will be subject of a future publication. In the following we will describe how to derive a BBE within the NEGF formalism for the multi–component case. As a starting point we use the differential Dyson equation[5,7,9]

$$\left(i\partial_t - (\varepsilon(1) - \varepsilon(1'))\right)\rho^<(11') = i(\Sigma^< \otimes g^a - g^r \otimes \Sigma^< - g^< \otimes \Sigma^a + \Sigma^r \otimes g^<)(11')\Big|_{t_1=t_{1'}},$$
(3)

with $\rho^<(11') = ig^<(11')|_{t_1=t_{1'}}$. Here we have introduced the shorthand notation $(a \otimes b)(11') = \int d2\, a(12)b(21')$ (note that the time arguments for the $<$, retarded and advanced components of $g$ and $\Sigma$ are defined on the time axis running from $-\infty$ to $+\infty$). The standard procedure to solve Eq. (3) is to perform a gradient expansion,[5] keeping only the lowest-order contributions, and to make a generalized Kadanoff–Baym ansatz for $g^<$.[1,9] In the gradient expansion it is important to explicitly take into account the fast center–of–mass time dependence of the interband polarizations. Details of the rather lengthy calculations can be found in Ref. 10. We finally make the further (bulk–material) approximation $v(12) = v(|\mathbf{r}_1 - \mathbf{r}_2|)$. The correctly screened contributions to the semiconductor Bloch equations due to $\Sigma'_{SX}$ then read

$$\dot{\rho}^<_{\alpha\alpha',\mathbf{k}}\Big|_{SX} = -i\sum_\mathbf{q}((V_\mathbf{q}(\omega_{\alpha'}) - V^*_\mathbf{q}(\omega_\alpha))\rho^<_{\alpha\alpha',\mathbf{k}} + (V^*_{-\mathbf{q}}(-\omega_\alpha) - V_{-\mathbf{q}}(-\omega_{\alpha'}))\rho^<_{\alpha\alpha',\mathbf{k+q}}+$$
$$i\sum_{\beta,\mathbf{q}}(v^{s\,r}_\mathbf{q}(\omega_\beta)\rho^<_{\alpha\beta,\mathbf{k+q}}\rho^>_{\beta\alpha',\mathbf{k}} - v^{s\,a}_\mathbf{q}(\omega_\beta)\rho^>_{\alpha\beta,\mathbf{k}}\rho^<_{\beta\alpha',\mathbf{k+q}}),$$
(4)

with $V_\mathbf{q}(\omega) = -2\sum_{\beta\beta',\mathbf{k}'} v^{s\,r}_\mathbf{q}(\omega'_\beta)v^{s\,a}_\mathbf{q}(\omega'_{\beta'})\rho^<_{\beta\beta',\mathbf{k}'+\mathbf{q}}\rho^>_{\beta'\beta,\mathbf{k}'}R_+(\omega - \omega'_{\beta'})$ and the screened Coulomb potentials $v^{s\,r,a}_\mathbf{q}(\omega) = v_\mathbf{q}/\epsilon^{r,a}(\mathbf{q},\omega)$. The dielectric function in RPA is given by $\epsilon^{r,a}(\mathbf{q},\omega) = 1 - v_\mathbf{q}\Pi^{r,a}_\mathbf{q}(\omega)$, where the polarization reads $\Pi^{r,a}_\mathbf{q}(\omega) = -2\sum_{\beta,\mathbf{k}'}(\rho^<_{\beta\beta,\mathbf{k}'} - \rho^<_{\beta\beta,\mathbf{k}'+\mathbf{q}})R_\pm(\omega - \omega'_\beta)$. Here we have introduced the abbreviations $\rho^<_{\alpha\alpha',\mathbf{k}} = -\langle a^\dagger_{\alpha',\mathbf{k}}a_{\alpha,\mathbf{k}}\rangle$, $\rho^>_{\alpha\alpha',\mathbf{k}} = \delta_{\alpha\alpha'} + \rho^<_{\alpha\alpha',\mathbf{k}}$; moreover $\omega_\alpha = \varepsilon_{\alpha,\mathbf{k+q}} - \varepsilon_{\alpha,\mathbf{k}}$, $\omega'_\beta = \varepsilon_{\beta,\mathbf{k}'+\mathbf{q}} - \varepsilon_{\beta,\mathbf{k}'}$ and $R_\pm(\omega) = (\omega \pm i0^+)^{-1}$. In order to recover the equations obtained within the DMA by neglect of the correlation transfers,[3] we split up $v^{s\,r,a}_\mathbf{q}$ in the last sum on the rhs. of Eq. (4) into real and imaginary parts and we finally replace the screened Coulomb potentials by the bare ones. One easily observes that only the imaginary parts of $R_\pm$ in the polarizations $\Pi^{r,a}$ contribute to the imaginary parts of $v^{s\,r,a}_\mathbf{q}$. Comparing these results to the DMA results we find that in the latter case also the real parts of $R_\pm$ appear. This shows that the DMA based on an ad hoc screened Hamiltonian can lead to spurious contributions to BBEs.

## REFERENCES

[1] A. Kuznetsov, Phys. Rev. B **44**, 8721 (1991).
[2] T. Kuhn and F. Rossi, Phys. Rev. B **46**, 7496 (1992).
[3] M. Lindberg and S. W. Koch, Phys. Rev. B **38**, 3342 (1988).
[4] F. Adler, G. F. Kuras, and P. Kočevar, SPIE Proc. **2142**, 206 (1994).
[5] J. Rammer and H. Smith, Rev. Mod. Phys. **58**, 323 (1986).
[6] L. P. Kadanoff and G. Baym, *Quantum Statistical Mechanics* (Benjamin, New York, 1962).
[7] H. C. Tso and N. J. .M. Horing, Phys. Rev. B **44**, 8886 (1991).
[8] R. Balescu, *Statistical Mechanics of Charged Particles* (Interscience, London, 1963).
[9] P. Lipavský, V. Špička, and B. Velický, Phys. Rev. B **34**, 6933 (1986).
[10] W. Pötz and U. Hohenester, to be published.

# ULTRAFAST PHOTOLUMINESCENCE DECAY IN GaAs GROWN BY LOW-TEMPERATURE MOLECULAR-BEAM-EPITAXY

A. Krotkus,[1] S. Marcinkevičius,[2] and R. Viselga[1]

[1]Semiconductor Physics Institute, 2600,
A. Goštauto 11, Vilnius, Lithuania
[2]Department Physics II,
Royal Institute of Technology,
S-10044, Stockholm, Sweden

GaAs layers grown by molecular-beam-epitaxy at low (~200°C) substrate temperatures (LT-GaAs) are characterized by very short, subpicosecond duration, carrier lifetimes. Because these layers have, at the same time, reasonably high electron mobility and large resistivity and breakdown field, the number of their applications in ultrafast optoelectronic and microwave electronic devices is steadily growing[1]. On the other hand, LT-GaAs could be an interesting system for investigating the dynamics of hot optically-excited carriers in the very early moments of their evolution. Ultrafast recombination in this material with a perfect crystallinity could be exploited as an intrinsic, "built-in" sampling gate allowing to reach temporal resolution unlimited by the laser pulse duration.

In the present work we have investigated hot electron dynamics in LT-GaAs by using a temporally-resolved photoluminescence (PL) technique. Transient PL is a more straightforward way to study carrier dynamics in LT-GaAs than the most frequently used transient photocurrent and transient optical reflectivity measurements, because PL intensity gives a direct information on carrier distribution functions and their kinetics.

Temporally-resolved PL measurements were performed at T=300°C using an up-conversion set-up described previously[2]. Self-mode-locked Ti:sapphire laser with the pulse duration of 100 fs and the central wavelength of 770 nm was employed in the set-up. Photoexcited carrier density was $10^{18}$ cm$^{-3}$, PL decay was monitored at different emission wavelengths.

LT-GaAs layers were grown by molecular-beam-epitaxy on semi-insulating (100) GaAs substrates at the temperature of about 200°C and under arsenic supersaturation. The growth rate was 2.5 μm/h; as-grown layer contained up to 1% of excess arsenic. Both as-grown layers and layers annealed for 30 min at 600°C at As overpressure were investigated.

*Hot Carriers in Semiconductors*
Edited by K. Hess *et al.*, Plenum Press, New York, 1996

Figure 1 shows PL dynamics of annealed LT-GaAs layer measured at three different emitted photon energies. Characteristic times of the PL transients are about 1.0-1.1 ps at

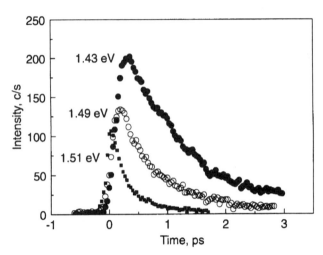

**Figure 1.** Photoluminescence dynamics of annealed LT-GaAs layer measured at different emitted photon energies

energy corresponding to the transitions between the conduction and valence band edges, and decrease when the emission energies are higher. The rising part of the PL signal also depends on the energy at which the signal is monitored, becoming steeper with increase of the emission energy. These features have been explained by comparing the experimental PL

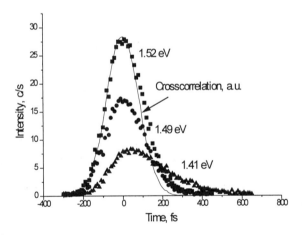

**Figure 2** Photoluminescence dynamics of as-grown LT-GaAs at different emitted photon energies.

transients with the results of the Monte-Carlo simulations in terms of electron redistribution during their thermalization. Experimental and theoretical results fit best together for a photoexcited carrier lifetime of 400 fs.

PL dynamics of as-grown, nonannealed LT-GaAs is shown in Figure 2. PL intensity in this case is considerably lower than for annealed material. Transient PL traces are symmetrical, their width is close to the width of the laser pump and probe pulse crosscorrelation. This indicates that carrier lifetime in as-grown LT-GaAs is much shorter than it becomes after the annealing. The estimation of that parameter was made by comparing the integral PL intensities over whole spectra measured on both types of the samples. The value for carrier lifetime in as-grown LT-GaAs was found to be less than 70 fs, which, to our knowledge, is the shortest carrier lifetime ever measured in semiconductors.

PL spectra of as-grown and annealed LT-GaAs show striking differences, too. These spectra are presented in Figure 3 and can be, essentially, considered as photoexcited electron

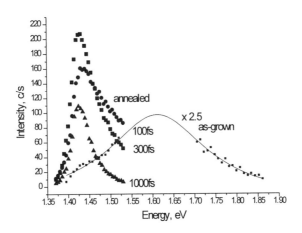

**Figure 3.** Photoluminescence spectra of as- grown and annealed (at different times after photoexitation) LT-GaAs. The PL spectrum for as-grown LT-GaAs is approximated by a Lorentzian function centered at the energy of the excitation quanta.

distribution functions measured at different moments after their excitation. Measurements on annealed samples show that the electron distribution is thermalized already at times as short as 100 fs after illumination of the sample. On the other hand, PL spectrum of as-grown sample has a strongly inverted, nonthermal shape. Remarkable PL signal was detected even at the photon energies larger than the excitation energy. The PL spectrum for as-grown LT-GaAs is best approximated by a Lorentzian function centered at the energy of the excitation quanta. It evidences that in this case we are observing a very early moment of the hot optical electron evolution, probably, earlier than the estimated 70 fs.

## REFERENCES

1. G.L.Witt: in Semi- insulating III- V Materials, ed. M.Godlewski, World Scientific, Singapure (1994).
2. A.Krotkus, S.Marcinkevičius, V.Pašiškevičius, U.Olin, *Semicond.Sc.Technol.*, **9**:1382 (1994).

# SUB-PICOSECOND LUMINESCENCE SPECTRA OF

# PHOTOEXCITED ELECTRONS RELAXATION IN p-GaAs

Mohamed A. Osman and Novat Nintunze

School of Electrical Engineering & Computer Science
Washington State University
Pullman, WA 99164-2752

## INTRODUCTION

Femtosecond lasers have allowed closer examination of the dynamics of photoexcited carriers in semiconductors during the first tens of femtosecond after laser excitation[1-3]. In particular, the ultrafast thermalization and relaxation of minority photoexcited carriers in p-GaAs have been a subject of intensive experimental investigations. For example, Furuta and Yoshii[2] used time resolved photoluminescence to examine minority electron relaxation over a wide range of acceptor concentrations and demonstrated the importance of electron-hole interaction in the relaxation of the photoexcited minority electrons at high acceptor concentrations. However, because the energy used in the photoexcitation excitation process (1.9 eV) generated electrons with energies above the intervalley transfer threshold, energy loss through intervalley phonon emission contributed to the overall energy and momentum relaxation process. More recently, Rodrigus and coworkers[3] also examined the relaxation of photoexcited minority electrons in p-GaAs using shorter laser pulses with energies below the intervalley transfer process in order to minimize the role of intervalley phonons and examine the role of e-h in the cooling process. They also compared the measured electron temperatures with Monte Carlo simulations and obtained good agreement between the measured and simulated data.

In this contribution we use ensemble Monte Carlo (EMC) approach to simulate the photoluminescence spectra using the same laser excitation conditions reported by Furuta and Yoshii[2]. The coherent coupling and other quantum effects[4] are neglected in this simulation because the high doping levels result in strong carrier-carrier interaction rates, the high lattice temperature (300K), and the fast momentum randomizing effects of the intervalley phonons. Also, the very fast interband scattering of excited holes from the light to the heavy hole band results in further damping of the coherent effects. Earlier photon echo measurements by Becker and coworkers[5] with 2eV laser pulse showed polarization de-phasing times of 14fs and 44fs at carrier densities of $7 \times 10^{18}$ cm$^{-3}$ and $1.5 \times 10^{17}$ cm$^{-3}$, respectively. At lower carrier densities dephasing times were longer due to the weakening of the role of carrier-carreir interactions. In this investigation, the dynamic screening of carrier-carrier interaction is taken into account in the EMC simulation and the electron temperatures were calculated from the luminescence spectra at times longer than 200 fs after laser excitation which are significantly longer than the above polarization dephasing times. The details of the Monte Carlo model are briefly discussed next followed by discussion of the results and summary.

*Hot Carriers in Semiconductors*
Edited by K. Hess *et al.*, Plenum Press, New York, 1996

## THE MONTE CARLO COMPUTATIONAL MODEL

The ensemble Monte Carlo program includes three nonparabolic conduction band $\Gamma$, L, and X valleys, warped heavy and light hole bands, and spherical parabolic split-off band. The scattering mechanisms include acoustic phonon, screened hot polar optical phonons, screened carrier-carrier scattering (electron-electron, hole-hole, and electron-hole), intervalley phonon scattering, and intra and interband optical phonons. The screening of the c-c scattering is treated dynamically by using a frequency and momentum dependent dielectric function. The details of the implementation of the dynamic c-c scattering and its comparison with static screening models were reported earlier[6]. The simulations were performed at lattice temperature of 300K for $10^{16}$ cm$^{-3}$ excitation density and acceptor concentration ranging from $10^{16}$ cm$^{-3}$ to $10^{19}$ cm$^{-3}$. The laser pulse used in the simulation assumed a photon energy of 1.9 eV and 130fs duration which excites electrons from the heavy, light, and split-off bands.

## RESULTS AND DISCUSSION

The photoluminescence spectra for band to band transitions at an observation energy of 1.45 eV is shown in figure 1 for three acceptor doping levels with a fitted continuous line drawn as an eye guide. During the first 0.4ps, the photoluminescence intensity exhibits the sharpest rise at acceptor concentration of $10^{19}$ cm$^{-3}$ due to the strength of the e-h energy loss channel. However, the duration of this rapid rise region is also the shortest because the electrons relax faster to the bottom of the conduction band. The subsequent slow rise is due to carrier-phonon interaction. Notice that the slowest and longest rise occurs in the intrinsic case where the energy loss is dominated by carrier-phonon interactions. To further examine, the nature of energy loss at high acceptor concentrations, the transient photoluminescence spectra were calculated at two photon energies and are plotted in figure 2. For the near band edge case of 1.45eV observation energy, the photoluminescence intensity exhibits a continuous rise, though with different time constants. However, at the higher photon energy of 1.64eV, the transient photoluminescence intensity undergoes a rapid rise during the first 180fs followed by a steep decay of about the same duration and slow decay at longer times. The rapid rise at 1.64eV is due to the increase of the number of electrons scattering into in these energy states as they cascade down the central valley after losing their energy to the cold hole plasma. The rapid decay indicates a fast scattering out of the electrons from 1.64 eV to lower energy states as they rapidly lose more energy mainly to the cold hole plasma.

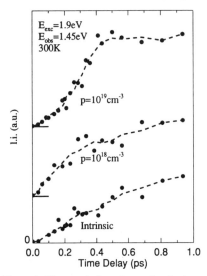

Figure 1: Time evolution of near band edge photoluminescence intensity in p-GaAs

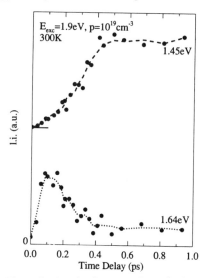

Figure 2: photoluminescence intensity in p-GaAs at two photon energies.

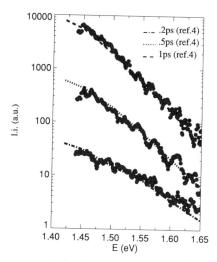

Figure 3: Time-resolved photoluminescence spectra in p-GaAs for three different time delays after excitation.

In figure 3, the experimental time resolved photoluminescence spectra reported by Furuta and Yoshii (dashed lines) are plotted together with the Monte Carlo simulation results (solid dots) at three different times after excitation for acceptor concentration of $10^{19}$ cm$^{-3}$. The measured and simulated curves show good agreement specially at 1.0ps. The effective electron temperatures could be obtained from the slope of the high energy regions. This amounts to assuming a Boltzmann distribution which is acceptable for longer times due to the strong carrier-carrier scattering. The effective carrier temperatures from EMC simulation at 0.5 and 1.0ps reach 561K and 430K, respectively. Furuta and Yoshii determined 540K and 420K from their experimental spectra. The relatively larger deviation at 0.5 ps is due to the fact that the electron population is not fully thermalized(i.e. the assumption of Boltzmann distribution is not good) due to the week e-e interaction and delayed return of electrons from the upper valleys.

## SUMMARY

The subpicosecond photoluminescence spectra of photoexcited electrons in p-doped GaAs have been simulated using ensemble Monte Carlo approach and showgood agreement with measured spectra. The calculations confirm the importance of e-h scattering as an energy loss channel in p-doped semiconductors.

## REFERENCES

1. L. Rota, P. Lugli, T. Elsaesser, and J. Shah, "Ultrafast Thermalization of photoexcited carriers in Polar Semiconductors," *Phys. Rev. B* 47:4226 (1993).
2. T. Furuta and A. Yoshii, "Ultrafast Energy Relaxation Phenomena of Photoexcited Minority Electrons in p-GaAs," *Appl. Phys. Lett.* 59:3607 (1991).
3. R. Rodrigues, M. Sailor, P. Buchberger, R.A. Hopfel, N. Nintunze, and M.A. Osman, "Ultrafast energy loss of electrons in p-GaAs," *Appl. Phys. Lett.* 67:264 (1995).
4. T. Kuhn and F. Rossi, "Monte Carlo Simulation of Ultrafast Processes in Photoexcited Semiconductors: Coherent and Incoherent Dynamics," *Phys. Rev. B* 46:7496 (1992).
5. P.C. Becker, H.L. Fragnito, C.H. Brito Cruz, R.L. Fork, J.E. Cunningham, J.E. Henry, and C.V. Shank, "Femtosecond Photon Echoes from Band-to-Band Transitions in GaAs," *Phys. Rev. Lett.* 61:1647 (1988).
6. N. Nintunze & M.A. Osman, "Ultrafast relaxation of highly photoexcited carriers in p-doped and intrinsic GaAs," *Phys. Rev. B* 50:10706 (1994).

# ROLE OF THE ELECTRON-HOLE INTERACTION ON FEMTOSECOND HOLE RELAXATION IN InP

G.R. Allan and H.M. van Driel

Department of Physics, University of Toronto
60 St. George St. Toronto, Canada  M5S 1A7

## INTRODUCTION

Investigations of carrier relaxation in optically excited polar semiconductors are often frustrated by the inability to separate electron and hole relaxation processes. In general, photo-luminescence experiments measure the product of electron $(f_e)$ and hole $(f_h)$ occupation functions, whereas absorption saturation experiments measure their sum. Since holes relax much faster than electrons due to a stronger lattice-coupling and larger effective mass, $e$-$h$ plasma dynamics are often dominated by electron dynamics. As a result, hole-relaxation mechanisms are less well understood than those involved in electron relaxation. Doped semiconductors have been used to provide initial conditions that isolate the minority-carrier dynamics. In particular, Zhou $et$ $al$[1] observed hole relaxation in n-GaAs and n-InP by time-resolved luminescence. With excitation at 2.0 eV, the relaxation time was determined to be 400 fsec for n-InP and independent of doping density in the range $5 \times 10^{17}$ to $2.5 \times 10^{18}$ $cm^{-3}$. The dominant hole-relaxation mechanism was identified as deformation-potential coupling to optical phonons. However, electrons are degenerate in these doped materials and, therefore, hole relaxation via electron-hole scattering is impeded[2]. We report on experiments performed in $intrinsic$ InP with $non$-$degenerate$ plasmas that are sensitive to hole relaxation via energy transfer from hot holes to low-energy electrons.

Hole-to-electron energy transfer is most efficient for low-energy electrons and hot holes. In order to be most sensitive to $e$-$h$ scattering, we measure the relaxation dynamics of low-energy electrons excited from the split-off band in the presence of hot holes. We exploit the tunability of our femtosecond laser to vary the excitation conditions and compare the electron dynamics for two situations: where hot holes are photo-excited with and without low-energy electrons. Thus, we are able to determine the role of $e$-$h$ energy transfer on the cooling of holes in a non-degenerate plasma. We use InP (instead of GaAs) to avoid complications due to inter-valley scattering of electrons.

The energy distribution of carriers in a semiconductor is known to have a significant impact on the dynamics of carrier relaxation. Two distinct effects can arise due to a change in the energy distribution: (i) new scattering mechanisms can become important and (ii) many-body interactions can be modified. For example, (i) inter-valley scattering plays a dominant role in highly-excited GaAs and (ii) low-energy carriers are more efficient at screening the Coulomb interaction than high-energy carriers. The experiments reported here focus on how the introduction of a new scattering mechanism may influence the carrier relaxation in InP.

## EXPERIMENT

A series of degenerate pump-probe experiments were performed to measure the transmission and reflection of the probe pulse as a function of probe delay over a range of wavelengths. A self-modelocked Ti:Sapphire laser was used as a source for all the experiments. The wavelength is tunable from 840 to 780 nm ($1.48 < \hbar\omega < 1.59eV$) with a 90 fsec pulsewidth and a 83 MHz repetition rate. Experiments were performed at three different wavelengths for each carrier density, which were $4 \times 10^{16}$ $cm^{-3}$ and $1 \times 10^{17}$ $cm^{-3}$ (peak density). The sample used is a 1 $\mu$m thin epitaxial layer of InP obtained by a two-step etch procedure[3] and Van der Waals bonded to a sapphire window. Photoluminescence data indicate that the sample was free of strain. The sample temperature was held at 77 K for all experiments.

## RESULTS

Figure 1 shows the pump-induced transmission change for a range of photon energies that spans the SO-band edge. The transmission change is interpreted directly as the absorption change, $\log(T/T_o) = -\Delta\alpha L$ where $T_o$ is the transmission of the unexcited material. Curve (a) shows data for the case of excitation below the SO-band edge. Hot electrons are injected from the HH and LH bands. The absorption is bleached by both $f_e$ and $f_h$ during the pump pulse but recovers rapidly due to the short dwell time of electrons and holes in the high-energy states. The transmission decreases at longer times due to bandgap renormalization (BGR). After 0.5 psec, $f_e$ and $f_h$ at the LH and HH transitions are negligible since there is no time dependence to the transmission. Curve (b) shows data for excitation above the SO-band edge, where low-energy electrons are injected from the SO band along with hot electrons and hot holes. During the pump pulse, the HH, LH and SO transitions are bleached by $f_e$ and $f_h$; whereas for longer delays, bleaching of the SO transition by only $f_e$ dominates the absorption change. Thermalization of the electrons injected at high energy causes the band-edge population to increase rapidly. Subsequent cooling of the electrons further increases the band-edge population on a slower timescale. Curve (c) shows the transmission change for a photon energy of 1.495 eV, at which electrons are injected from the LH and HH bands but not from the SO band. BGR causes a portion of the pulse spectrum to probe the SO transition. The increase of the transmission is qualitatively similar to curve (b), although the magnitude of the increase is smaller since only a small fraction of the spectrum probes the SO transition. The three cases can be summarized as probes of (a) HH and LH transitions (b) primarily SO transition and (c) equal weights of HH, LH and SO transitions. A comparison between case (b) and (c) illustrates the effect of e-h energy transfer on the dynamics of low-energy electrons.

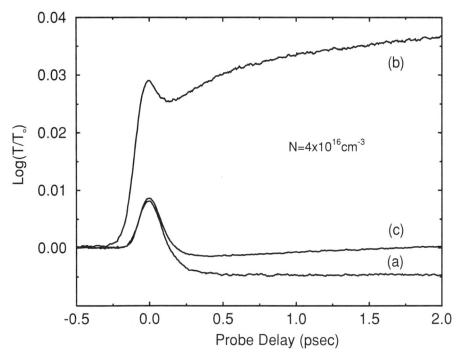

Figure 1: Time-resolved transmission change for photon energies (a) below (1.480 eV) (b) above (1.510 eV) and (c) slightly below (1.495 eV) the split-off bandgap of InP. Carrier denisty is $4 \times 10^{16} cm^{-3}$.

## DISCUSSION

We isolate the contribution from the SO transition in curves (b) and (c) of Fig. 1 and interpret it as proportional to the band-edge electron population. Contributions from the LH and HH transitions are assumed to be the same at all three photon energies and are removed by subtraction of curve (a) from curves (b) and (c). The results of the subtraction are shown in Fig. 2. The data for curve (b) has been scaled so that both curves approach the same level at long delay. As well, the data up to 200 fsec are not shown because in this interval the data is convolved with the pulse shape.

Hole-to-electron energy transfer should be more efficient for the case where low-energy electrons are injected from the SO band (curve a) as compared to the case where no electrons are injected from the SO band (curve b). The hole distributions are nearly the same in both cases. We observe no distinguishable difference between the rates of change of the band-edge population for the two cases during the time when the holes are hot ($\sim$ 500 fsec). Therefore, we conclude that hole relaxation must proceed with negligible energy transfer to cold electrons. Although we have no direct evidence, we assume that hole relaxation is dominated by lattice-coupling. In order to establish an upper limit on the energy transfer rate, $W_{eh}$, we fit the transmission data to a simple model based on energy transfer between electrons, holes and the lattice in the relaxation-time approximation. The results, shown as solid lines in Fig. 2, give adequate agreement without including $e$-$h$ energy transfer. As an illustration, the dashed line in Fig. 2 shows the results for an initial energy-transfer rate $W_{eh}$ = 30 meV/psec (this rate decreases as the holes cool). Based on the scatter in the data, we estimate that $W_{eh}$ is less than 20 meV/psec.

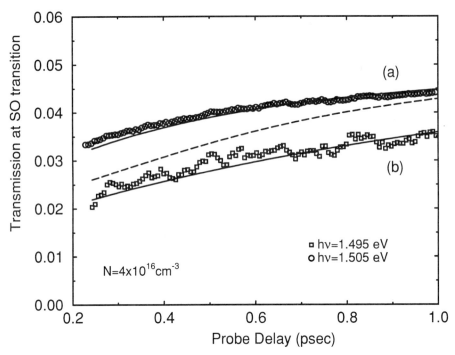

Figure 2: Transmission at the SO band transition versus probe delay. This is proportional to the band-edge electron population. Data for excitation (a) above and (b) below the split-off bandgap of InP are shown as symbols. The solid lines are results of a calculation without *e-h* energy transfer. The dashed line is based on an *e-h* energy transfer rate $W_{eh} = 30$ meV/psec.

## CONCLUSIONS

We have performed experiments in *intrinsic* InP to determine the role of the *e-h* interaction in the dynamics of *non-degenerate* low-energy electrons. We use the low-energy electrons as an indirect probe of hole relaxation. By comparing the dynamics of the band-edge electron occupancy for two excitation conditions, we determine that the *e-h* interaction is not an efficient channel of energy loss for holes. Therefore, we suggest that hole relaxation must be dominated by mechanisms associated with lattice coupling.

## ACKNOWLEDGMENTS

We gratefully acknowledge the financial support of NSERC and The Premier of Ontario's Technology Fund.

## REFERENCES

1. X.Q. Zhou, K. Leo and H. Kurz, *Phys Rev B*. 45: 3886 (1992).
2. M. Combescot and R. Combescot, *Phys Rev B*. 35: 7986 (1987).
3. G. Augustine, N.M. Jokerst and A. Rohatgi, *Appl Phys Lett*. 61: 1429 (1992).

# FIELD INDUCED ELECTRON CAPTURE BY METASTABLE CENTERS IN PLANAR-DOPED GaAs:Si

R.Stasch[1], M.Asche[1], M.Giehler[1], R.Hey[1], and O.G.Sarbey[2]

[1] Paul-Drude-Institut für Festkörperelektronik
Hausvogteiplatz 5-7, D-10117 Berlin, Germany
[2] Institute of Physics of the Ukrainian Academy of Sciences
Prospekt Nauki 46, 252650 Kiev, Ukraine

## INTRODUCTION

Capture of hot carriers by deep traps above the Fermi level in GaAs highly planar-doped with Si by molecular beam epitaxy is investigated at 77 K. Usually carrier trapping in such centers is studied by applying pressure[1-5], thus reducing the energy difference between the conducting $\Gamma$-states and the centers. In contrast to those measurements performed in thermodynamic equilibrium, in our experiments the carriers gain the desired energy by heating in an applied electric field.

## DETERMINATION OF THE THRESHOLD FIELD

Concerning MBE growth and sample preparation we refer to Stasch et al.[6]. The measurements are performed by immersing the sample in liquid nitrogen and applying rectangular voltage pulses of varying duration. In Fig.1 we demonstrate current voltage characteristics for GaAs planar-doped with $7 \cdot 10^{12}$ and $2 \cdot 10^{13} \mathrm{cm}^{-2}$ Si atoms. They show a saturation behaviour for fields above 6 and 9 kV/cm, respectively. This shift is expected for the $\Gamma$-L transfer (Gunn effect) in dependence on carrier heating as a function of scattering on ionized impurities. The upper curves correspond to the initial state, when the sample is cooled down in the dark and only single voltage pulses are applied for each field strength. The lower curves reflect the characteristics after saturation of the charge transfer (SCT) in the high field limit achieved after many repeated pulses[6]. The reduction of the saturation current is stable after switching off the applied voltage pulses. Under illumination with a red LED (0.66 $\mu$m) or at sufficiently high lattice temperature the initial state is restored. Therefore, it is proven that the carriers are trapped by metastable centers.

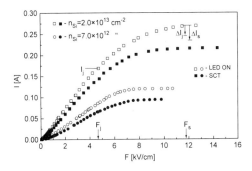

**Figure 1.** Current as a function of applied field at 77 K for two samples with different Si concentrations. Open symbols – initial curve or under illumination with a red LED, solid symbols – after saturation of charge transfer (SCT). The current changes marked by arrows are explained in the text.

The field-induced current change is about 20% for doping densities between $7 \cdot 10^{12}$ and $1.4 \cdot 10^{14}$ cm$^{-2}$ Si atoms per layer, but is less than 5% for $5 \cdot 10^{12}$ cm$^{-2}$ and is not observed for $2 \cdot 10^{12}$ cm$^{-2}$. For a given field strength the current decrease scales with the total time of the applied field (number of pulses × pulse duration), i.e., it is determined by the total current passed through the sample.

In order to demonstrate the field dependence of the capture process we compare equivalent charge flows through the sample as a function of the electric field. The initial condition is implemented by illumination with the red LED. Then, the LED is switched off and for a fixed field strength $F_s$ in the saturation region of the characteristic a certain current change $\Delta I_s$ is introduced during a definite time $\Delta t_s$ by a sequence of pulses (see Fig.1). At an other field strength $F_j < F_s$ the initial current $I_j$ (LED on) is measured and the time $\Delta t_j$ is determined, in which the same amount of carriers should be trapped as during $\Delta t_s$ in the saturation region of the I-U characteristics at $F_s$ (cf. arrows in Fig.1.) The LED is switched off again and the equivalent charge flow through the sample is established at $F_j$. Then at $F_s$ the corresponding current change $\Delta I_j(F_s)$ implemented at $F_j$ during $\Delta t_j$ is measured in a single pulse measurement. The ratio of this current change $\Delta I_j(F_s)$ to the initially chosen $\Delta I_s$ is shown in Fig.2.

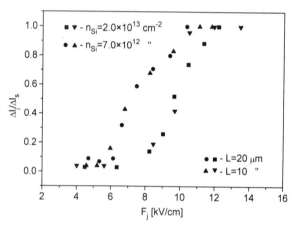

**Figure 2.** Ratio of the changes in current determined by the procedure described in the text versus applied field for both doping concentrations (cf. Fig.1) and two different sample lengths.

The results demonstrate that the current change appears only for field strengths above a certain threshold and that it rises in a steep way , i.e., the charging of the traps demands a critical excitation energy. The threshold shifts to higher fields with increasing impurity density of the samples. This is expected on account of the weaker carrier heating and is in accordance with the observed shift of the threshold field for current saturation caused by the $\Gamma$-L transfer .

It should be emphasized that the threshold field for trapping is evidently below the set-in of current saturation caused by negative differential conductivity (ndc). The latter becomes realized at a redistribution of about 75% according to the Monte Carlo simulation[7]. However, a remarkable transfer is present already at lower field strengths. Therefore, the field induced capture is already possible at a lower energy than the high barrier of 550 meV as determined in [8] necessary for capture in the DX center. Our result can be explained when the trapping proceeds either via the L minima with a subsequent lattice relaxation or directly from the $\Gamma$ minimum to the center.

## DISCUSSION OF THE METASTABLE CENTER

Both proposed processes are phonon assisted. In our case the heated carriers are able to emit such phonons. From our results it is not understood why the carriers are transferred over a barrier of 550 meV[8]. In spite of the strong transverse fields the DLTS measurements are performed in thermodynamic equilibrium. In the temperature region used for DLTS investigations one could expect that carriers in the tail of the distribution function could be transferred to the center by assistance of high energy phonons more easily than across a considerably higher barrier. Further on, it is not clear why only current changes of about 20% are observed. A possible explanation, however, is a strong change of mobility accompanying the trapping[4,9,10] , then a significantly higher percentage of trapped carriers may be involved.

As mentioned above illumination with a red LED (or light of shorter wavelength) restores the initial state of the center. Studies of the release mechanism in dependence on the wavelength[11] yield an optical ionization threshold of about 0.9 eV. This is in agreement with our observation that a LED with 1.4 $\mu$m is sufficient for carrier release. The carriers can also be released by heating the sample above 140 K. The high energy difference necessary for both types of ionization is in agreement with the well known negative U-model discussed for $DX^-$ centers, but does also not contradict any metastable center with small lattice relaxation provided it forms a deep state[12]. Furthermore, with respect to photoionization of metastable centers in AlGaAs sometimes [13,14] a lower ionization threshold at 0.25 eV is found besides the one at 0.95 eV indicating the existence of two types of traps[15]. Therefore, we tried to change the number of filled traps by irradiation with a halogen lamp with an edge filter for $\lambda \geq 2.2\mu$m, however, we did not observe a carrier release. By field induced transfer at still higher fields it can also not be decided, whether there are two centers present or not, since at essentially higher field strengths the percentage of carriers in the L minima would become reduced on behalf of scattering in the X minima[7].

As already mentioned from the present current voltage measurements only the combined change of carrier density and mobility can be determined. In the case of the $DX^-$ center two electrons are trapped, i.e., the number of charged impurity centers remains the same. However, in spite of this circumstance the mobility could be increased as observed by correlation of shallow and deep donors[4,9,10,16]. Because in our $\delta$-doped GaAs the trapping exists for highly doped layers, only correlation effects could be involved.

## SUMMARY

We established that in highly planar-doped GaAs metastable centers above the Fermi level are present. These levels are charged by hot electrons above a certain threshold field, which is lower than the field necessary for current saturation on behalf of $\Gamma$-L transfer of the electrons. Both critical fields shift to higher values with lower carrier density in more highly doped samples. The existence of the threshold indicates a significantly lower carrier energy for trapping than reported for trapping in thermodynamic equilibrium. Because only one photoionization threshold was observed at 0.9 eV.our results are consistent with the presence of the DX center.

## ACKNOWLEDGEMENT

We thank P. Krispin for helpful discussions.

## REFERENCES

1. M. Mizuta, M. Tachikawa, H. Kukimoto, and S. Minomura, Jap. Journ. Appl. Phys., **24**, L143 (1985)
2. T. Suski, R. Piotrzkowski, P. Wisniewski, E. Litwin–Staszewska, L. Dmowski, Phys. Rev. **B40**, 4021, (1989)
3. W. Jantsch, G. Ostermayer, G. Brunthaler, G. Stoeger, J. Woeckinger, and Z. Wilamowski, "The Physics of Semiconductors", (Ed. E. M. Anastassakis and J. D. Joannopouls, World Scientific, Singapore) p. 485 (1990)
4. D. K. Maude, J. C. Portal. L. Dmowski, T. J. Foster, L. Eaves, M. Nathan, M. Heiblum, J. J. Harris, and R. B. Beall, Phys. Rev. Lett. **59**, 815, (1987)
5. M. F. Li, P. Y. Yu, E. R. Weber, and W. L. Hansen, Appl. Phys. Lett. **51** 349 (1987)
6. R. Stasch, M. Asche, L. Däweritz, R. Hey, H. Kostial, and M. Ramsteiner, Journ. Appl. Phys., **77**, 4463 (1995)
7. J. Pozela and A. Reklaitis, Sol State Electron. **23**, 927 (1980)
8. T. N. Theis, P. M. Mooney, and S. A. Wright, Phys. Rev. Lett **60**, 361 (1988)
9. P. J. van der Weel, M. J. Anders, L. J. Giling, and J. Kossut, Semicond. Sci. Technol. **8**, 211 (1993)
10. E. Buks, M. Heiblum, Y. Levinson, and H. Shtrikman, Semicond. Sci. Technol **9**, 2031 (1994)
11. B. Danilchenko, A. Klimachov, and S. Roshko, private commun.
12. M. Lannoo, *Physics of DX centers in GaAs Alloys (Sci. Tech.Publications, ed. J.C.Bourgoin)*, Solid State Phenomena **10**, 209 (1990)
13. J. C. M. Henning and J. P. M. Ansems, Phys. Rev. **B 38**, 5772 (1988)
14. R. Legros, P. M. Mooney, and S. L. Wright, Phys.Rev. **B 35** ,7505 (1987)
15. J. C. M. Henning, E. A. Montie, and J. P. M. Ansems,*Materials Science Forum (Trans Tech Pub, Switzerland)*, 1085 (1889)
16. D. J. Chadi, K. J. Chang, and W. Walukiewicz, Phys. Rev. Lett. **62**, 1923 (1989)

# FEMTOSECOND TIME-RESOLVED PHOTOLUMINESCENCE MEASUREMENTS OF PPV, MEH-PPV AND CN-PPV

G.R.Hayes, I.D.W.Samuel and R.T.Phillips

Cavendish Laboratory
University of Cambridge
Madingley Road
Cambridge
England, CB3 0HE

Conjugated polymers based on poly(arylenevinylene)s are of potential technological importance due to their electroluminescence properties[1,2]. In poly($p$-phenylenevinylene) [PPV] and its soluble derivative poly(2-methoxy,5-(2'ethyl-hexyloxy)-$p$-phenylenevinylene) [MEH-PPV] the luminescence produced after photoexcitation or charge injection is attributed to the radiative recombination of the same singlet excited state, namely the neutral polaron-exciton. Obtaining a detailed understanding of the dynamics of excitons in PPV and its derivatives prior to recombination is therefore highly desirable for both fundamental and practical reasons. The method employed in our experiment was that of time-resolved photoluminescence spectroscopy. Optically thick films of the polymers were excited, under a vacuum of $8\times10^{-6}$ mbar, by 200 fs pulses of 3.06 eV photons from the frequency-doubled output from a modelocked Ti:Sapphire laser. The energy of the photons corresponded to the peak of the absorption spectrum for PPV. The luminescence was temporally and spectrally resolved using the upconversion technique[3]. The overall temporal resolution was about 300 fs and the spectral resolution was 18 meV.

Figure 1 shows the room temperature photoluminescence spectrum of PPV, MEH-PPV and CN-PPV, which is a cyano substituted PPV derivative[2], at various times after photoexcitation. The continuous-wave (cw) emission spectra are displayed in the inset. There are three main features of the time-resolved data for PPV and MEH-PPV that we wish to highlight. The first is the extremely rapid rise in the luminescence extending across the entire spectral region. After this a rapid decay of the high-energy luminescence tail is observed within a few hundred femtoseconds and a slower redshift and narrowing of the luminescence peaks, which occurs on a picosecond timescale. Thirdly, the overall decay of the luminescence can be seen.

The ultrafast luminescence rise is attributed to the formation of excitons followed by their subsequent vibrational relaxation onto the lowest vibrational level of the first excited state. Our results show that this formation and vibrational relaxation occurs in less than 300 fs. Conformational defects or impurities cause the effective conjugation length of the polymer to be much less than the actual chain length. This limits the extent along the chain

*Hot Carriers in Semiconductors*
Edited by K. Hess *et al.*, Plenum Press, New York, 1996

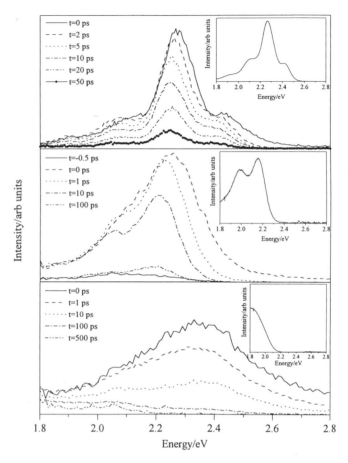

Figure 1. Time-resolved photoluminescence spectra of PPV, MEH-PPV and CN-PPV after excitation with 200 fs pulses at 3.06 eV. Inset:cw emission spectra at room temperature.

to which the exciton can be delocalised and thus raises the exciton's energy. The broad high-energy luminescence tail that exists at very short times after photoexcitation is believed to be caused by the recombination of excitons from short-conjugated chain-segments. The rapid removal of this high-energy luminescence tail is explained by exciton migration from higher-energy sites that occurs on a sub-picosecond timescale due to the presence of a large density of lower-lying sites to which the exciton can hop. This migration then becomes a slower diffusion because as the excitons lose more and more energy there are fewer and fewer energetically available sites to which the exciton can migrate. Figure 2(a) shows the redshift of the luminescence peak of PPV and MEH-PPV versus time. This clearly shows that the redshift is larger in MEH-PPV than PPV. This result suggests that MEH-PPV is less-conjugated than PPV because the change in energy of an exciton that hops from a site with an effective conjugation length of B units to one of B+1 units is larger if B is smaller. We attribute this feature to an increased concentration of defects in the MEH-PPV film compared to the PPV film due to the reduced crystallinity and hence lower degree of order of MEH-PPV compared to PPV. In figure 2(b) the decay of the luminescence at 2.25 eV in PPV and 2.2 eV in MEH-PPV is shown. These energies correspond to the cw luminescence peaks. The decay time was measured to be 330±30 ps in PPV and 145±20 ps in MEH-PPV.

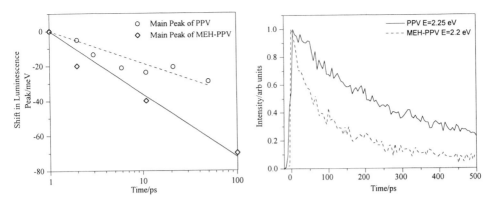

Figure 2. (a) Position of the peak of the main emission band of PPV and MEH-PPV versus time. The solid and dashed lines are guides to the eye that highlight the increased redshift seen in MEH-PPV. t=0 has been offset by 1 ps for clarity. (b) Evolution of the photoluminescence in PPV monitored at 2.25 eV and in MEH-PPV at 2.2 eV.

If we assume that the radiative lifetime of the two polymers is of the order 1.2 ns which is what is observed in *trans,trans*-distyrlbenzene which is a model oligomer of PPV[4] then this suggests that more efficient non-radiative decay channels, caused by defects, are present in MEH-PPV films. The photoluminescence efficiency of PPV films prepared by the same method as our samples is 0.27±0.02 and for MEH-PPV films is 0.10-0.15[5]. These results are fully consistent with the decay times that we observe.

The CN-PPV film shows markedly different behaviour from that of PPV and MEH-PPV. A high-energy peak exists at short times after photoexcitation with a decay time of 6±1 ps. The low-energy peak seen at longer times has a measured decay time of greater than 2 ns and corresponds to the cw luminescence peak. The photoluminescence efficiency of CN-PPV is 0.35±0.03[5]. Our results suggest that the low-energy CN-PPV emission is from an inter-chain excitation such as an excimer[6,7]. The long decay time may be caused by the low mobility of an inter-chain excitation which reduces the possibility of non-radiative decay at quenching sites. This has important implications for the design of highly luminescent conjugated polymers.

## REFERENCES

1. J.H.Burroughes, D.D.C.Bradley, A.R.Brown, R.N.Marks, K.Mackay, R.H.Friend, P.L.Burn and A.B.Holmes, Light-emitting diodes based on conjugated polymers, Nature 347:539 (1990)
2. N.C.Greenham, S.C.Moratti, D.D.C.Bradley, R.H.Friend and A.B.Holmes, Efficient light-emitting diodes based on polymers with high electron affinities, Nature 365:628 (1993)
3. G.R.Hayes, I.D.W.Samuel and R.T.Phillips, Exciton dynamics in electroluminescent polymers studied by femtosecond time-resolved photoluminescence spectroscopy, accepted by Phys. Rev. B
4. I.B.Berlman, Handbook of Fluorescence Spectra of Aromatic Molecules, Academic, New York (1967)
5. N.C.Greenham, I.D.W.Samuel, G.R.Hayes, R.T.Phillips, Y.A.R.R.Kessener, S.C.Moratti, A.B.Holmes and R.H.Friend, Measurement of absolute photoluminescence quantum efficiencies in conjugated polymers, Chem. Phys. Lett. 241:89 (1995)
6. I.D.W.Samuel, G.Rumbles, C.J.Collison, B.Crystall, S.C.Moratti and A.B.Holmes, Luminescence efficiency and time-dependence in a high electron affinity conjugated polymer, Synth. Met. (submitted)
7. G.R.Hayes, I.D.W.Samuel and R.T.Phillips, Excitations in a high electron affinity conjugated polymer studied by time-resolved photoluminescence spectroscopy, (to be submitted)

2. Bloch Oscillations and Fast Coherent Processes in Semiconductors

# STRONG TERAHERTZ-PHOTOCURRENT RESONANCES
# IN MINIBAND SUPERLATTICES AT THE BLOCH FREQUENCY

K. Unterrainer[1], B.J. Keay[1], M.C. Wanke[1], S.J. Allen[1], D. Leonard[2],
G. Medeiros-Ribeiro[2], U. Bhattacharya[2], and M.J.W. Rodwell[2]

[1]Center for Free -Electron Laser Studies,
[2]Materials and ECE Department
  University of California, Santa Barbara
  Santa Barbara, CA 93106

## ABSTRACT

We have observed resonant changes in the current-voltage characteristics of miniband semiconductor superlattices when the Bloch frequency is resonant with a terahertz field and its harmonics. This corresponds to absorption and gain of THz radiation in the Wannier Stark ladder of the superlattice. The resonant feature consists of a peak in the current which grows with increasing laser intensity accompanied by a decrease of the current at the low bias side. When the intensity is increased further the first peak starts to decrease and a second peak at about twice the voltage of the first peak is observed due to a two-photon resonance. At the highest intensities we observe up to a four-photon resonance.

## INTRODUCTION

In strongly coupled superlattices minibands are formed, and electrons accelerated by an electric field should show negative differential velocity due to the negative effective mass /1/. Electrons accelerated up to the zone boundary of a miniband are Bragg reflected. This repetitive motion of acceleration and Bragg reflection is called Bloch oscillation characterized by the Bloch frequency $\omega_B = eEd/\hbar$, where d is the superlattice period. THz emission from coherent Bloch oscillations in narrow miniband semiconductor superlattices should be possible, since the Bloch frequency can exceed the scattering rate $1/\tau$. However, it took more than two decades before the first observation of Bloch oscillations was made in a transient THz emission experiment of optically excited electrons /2/. Continuous Bloch emission from electrically injected carriers in a superlattice has not yet been observed.

The formation of Bloch oscillations in a DC biased superlattice also influences the DC conductivity. At low bias when the Bloch frequency is not high enough to overcome scattering ($\omega_B \cdot \tau < 1$) the electrons never reach the inflection point and the current in the superlattice direction is given by the Drude conductivity. When the bias is high enough to enable coherent Bloch oscillations, the electrons start to oscillate in space and can not contribute to the DC current anymore, which leads to negative differential conductivity.

In this work we have studied the influence of strong THz radiation on the DC conductivity of the superlattice. The problem of an electron in a superlattice with a dispersion relation $\varepsilon(k) = E_1 + \Delta/2(1 - cos(kd))$ driven by an electric field $E(t) = E^{DC} + E^{AC}cos(\omega t)$ can be treated in a Boltzmann equation approach /3,4/. These calculations show a strong modulation of the current at resonances of the Bloch frequency $\omega_B$ with the drive frequency $\omega$ or its higher harmonics.

For $\omega_B \to 0$ the current is given by $J_0^2(eE^{AC}d/\hbar\omega)$. At the roots of $J_0$ the DC current is completely quenched by the AC field. This AC localisation was also predicted by fully quantum mechanical calculations of the energy levels of a superlattice in an intense periodic electric field /5/.

**EXPERIMENT**

The samples used in this study are GaAs/$Al_xGa_{1-x}As$ (x = 0.3) superlattices grown on a semi-insulating GaAs substrate by molecular beam epitaxy. The superlattice structure consists of 40 periods of 80Å wide GaAs wells and 20Å thick AlGaAs barriers (mini band width $\Delta$ = 22 meV). The superlattice is homogeneously Si doped with a concentration of n=$3 \cdot 10^{15}$cm$^{-3}$. The superlattice is embedded in 3000Å thick GaAs layers with a carrier concentration of n=$2 \cdot 10^{18}$cm$^{-3}$ which serve as contact regions. The highly-doped contact layers are separated from the superlattice by 80Å GaAs lightly-doped setback layers.

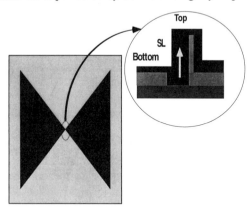

Fig. 1: Coplanar antenna structure and superlattice device geometry

Superlattice mesas with an area of 8µm$^2$ are formed by dry etching through the top n$^+$ region and the superlattice followed by an ion (H$^+$) implantation-isolation process. Ohmic Au/Ge/Ni contacts are fabricated on the top and the bottom of the mesas. The THz radiation is coupled to the superlattice by a co-planar broad band bow-tie antenna. This antenna technique ensures that the polarisation of the AC electric field inside the mesa is parallel to the growth direction independently from the polarization of the incident field. A high-purity Si hemisphere is mounted on the substrate side of the sample acting as a lens and focusing the radiation onto the area of the antenna. Gold wires are bonded to pads at the bow-tie antennas to allow for conductivity measurements. The UCSB free electron lasers were used as a THz radiation source, giving pulses of intense THz radiation with a pulse length of several microseconds.

## RESULTS and DISCUSSION

Fig. 2 shows the current voltage (IV) curve of the superlattice device without external radiation as a function of temperature. At 300 K the current is linear for small bias voltages and becomes sublinear at higher voltages. With decreasing temperature this nonlinearity gets more pronounced. At about T=100K the current shows a plateau for voltages larger than 20mV and starts to increase again above 200mV. At lower temperatures a region of negative differential conductance (NDC) forms in this voltage range. The current is linear for bias voltages below 20mV. The NDC-region starts at a bias of 20mV which corresponds to a critical electric field of 500V/cm. Assuming that the onset of the NDC region is due to Esaki-Tsu type localisation when $\omega_B \cdot \tau > 1$ we find a scattering time $\tau=1.3$ps. The maximum current density is about 100 A/cm$^2$.

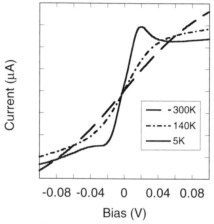

Fig. 2: DC current voltage characteristics of the superlattice without external radiation

The influence of an external harmonic electric field on the superlattice current is shown in Fig. 3a for a FEL frequency of 0.6 THz. The DC-current is measured during the FEL pulse and the voltage bias is changed. The curves in Fig. 3a are shown for increasing AC field strength (the curves are shifted down for increasing intensity). At low intensities an additional peak emerges in the NDC region. When the intensity is increased further the first peak starts to decrease and a second peak at about twice the voltage of the first peak is observed due to a two photon resonance. At the highest intensities we observe up to a four photon resonance. The AC localisation can be seen form the strong decrease of the initial current peak of the DC IV with increasing intensity. The position of the peaks does not change with intensity of the FEL. We attribute the first additional peak to a resonance of the external laser field where the Bloch frequency $\omega_B = \omega$. The other peaks are higher resonances where $\omega_B = n \cdot \omega$.

In Fig. 3b the results for a laser frequency of 1.5 THz are shown. The peaks are shifted to higher voltages and are much more pronounced. However, only the fundamental and the second harmonic is observed. In addition we observe a suppression of the current value in between the peaks. The peaks show a clear asymmetry with a steeper slope on the high voltage side. This asymmetry is different from the shape of the peak of original DC IV which shows a steeper slope at the low voltage side.

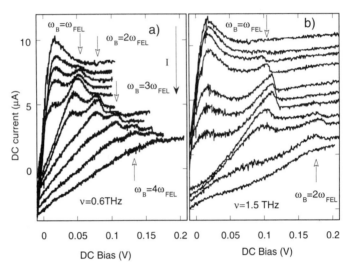

Fig. 3: DC current-voltage curve for increasing FEL intensity (The curves are shifted downwards for increasing laser intensity). The FEL frequency was fixed to 0.6 THz (Fig. 3a) and 1.5 THz (Fig. 3b).

The relationship between laser frequency and applied bias is linear and the slopes of the n-th harmonic are n-times the slope of the one photon resonance. The magnitude of the slope shows that the electric field is higher than expected from a linear voltage drop across the whole superlattice. The data indicate that a high electric field domain is formed which extends over one third of the superlattice. The electric field in the low field domain is assumed to be below the critical field for localisation and puts this part of the superlattice in the high-conductive miniband transport regime.

Our results show clearly that external radiation can coherently couple to Bloch oscillations in contrary to theoretical suggestions that THz radiation would not couple to a uniform Wannier-Stark ladder /6/. We conclude that this result is intimately related to dissipation and line broadening of the otherwise identical states in the ladder: absorption appears above the Wannier-Stark splitting ($\omega_B < \omega$) and gain below ($\omega_B > \omega$).

The authors would like to thank the staff at the Center for Free-Electron Laser studies: J.R. Allen, D. Enyeart, J. Kaminski, G. Ramian and D. White. Funding for the Center for Free-Electron Laser studies is provided by the Office of Naval Research. One of us (K.U.) was supported by a Schrödinger fellowship of the Austrian Science foundation.

**REFERENCES**

/1/     L. Esaki, R. Tsu, IBM J. Res. Dev. <u>14</u>, 61 (1970).

/2/     C. Wascke, H.G. Roskos, R. Schwedler, K. Leo, H. Kurz, PRL <u>70</u>, 3319 (1993).

/3/     V.V. Pavlovich, E.M. Epshtein, Sov. Phys. Semicond. <u>10</u>, 1196 (1976).

/4/     A.A. Ignatov, K.F. Renk, E.P. Dodin, PRL <u>70</u>, 1996 (1993).

/5/     M. Holthaus, PRL <u>69</u>, 351 (1992).

/6/     G. Bastard, R. Ferreira, C.R. Acad. Sci. <u>312</u>, 971 (1991).

# PHOTON-ASSISTED ELECTRIC FIELD DOMAINS AND MULTIPHOTON-ASSISTED TUNNELING IN ANTENNA COUPLED SEMICONDUCTOR SUPERLATTICES

B.J. Keay [1], S.J.Allen Jr.,[1] J. Galán[2,†], K..L. Campman[3],
A.C.Gossard[3], U. Bhattacharya[4], M.J.W. Rodwell[4]

[1]Center for Free-electron Laser Studies
[3]Materials Department
[4]Department of Electrical and Computer Engineering
UCSB, Santa Barbara, CA 93106

[2]Physics Department
Ohio State University, Columbus, OH 43210-1106

## INTRODUCTION

Photon-assisted tunneling (PAT), long restricted to transport in superconducting junctions[1], has recently emerged in transport in semiconductor superlattices and nanostructures in the presence of high frequency fields[2,3,4,5,6,7,8] The full character of PAT, however, has not been experimentally established.

Electric field inhomogeneities in multi-quantum well superlattices have also been studied extensively in the last twenty years[9]. Here, we show that the PAT channels can also support high and low field domains if the terahertz field is strong enough. By using bow tie antennas we have improved the coupling of terahertz radiation into semiconductor superlattices, thereby enabling us to explore PAT induced inhomogeneities as well as multiphoton assisted tunneling and the terahertz electric field dependence of the PAT channels. An extension of the model of Bonilla et al.[10] is put forward that can account in a quantitative way for the detailed I-V characteristics of these systems in strong high frequency fields, including both PAT channels and their effect on the electric field domain distribution.

## EXPERIMENT AND RESULTS

The experiments were performed on a superlattice material consisting of 300 nm of GaAs doped at $n+ = 2 \times 10^{18} cm^{-3}$, followed by a 50 nm GaAs spacer layer and ten 33 nm GaAs quantum wells and eleven 4 nm $GaAs/Al_{.30}Ga_{.70}As$ barriers and capped by another 50 nm GaAs spacer layer and 300 nm of $n+ = 2 \times 10^{18} cm^{-3}$ doped GaAs. The substrate was semi-insulating and the superlattice and spacer layers were n doped to 3 ×

*Hot Carriers in Semiconductors*
Edited by K. Hess *et al.*, Plenum Press, New York, 1996

$10^{15}$ cm$^{-3}$. The active area of the superlattice was 8 $\mu$m$^2$. Lithographically defined and processed broad band bow-tie antennas were integrated with the superlattice device and electrically connected to the top and bottom of the superlattice. The devices were then glued onto a high resistivity hemispherical silicon lens and gold wires were bonded to the two gold bows. The experiments were performed over a temperature range of 6-12 K with the sample mounted in a temperature controlled flow cryostat with Z-cut quartz windows. The radiation was incident on the curved part of the Si lens with the polarization parallel to the axis connecting the two gold bows of the bow-tie antenna.

The measured I-V characteristic without radiation ($E_{AC} = 0$), dotted line in Fig. 1c, shows a series of steps in current separated by saw tooth oscillations associated with sequential resonant tunneling in the presence of high and low electric field domains[9,10]. At low bias the current through the sample occurs via ground state to ground state tunneling. As the bias is increased, the current approaches the critical current, the maximum current that the ground state to ground state tunneling can support, and a quantum well breaks off forming a high field domain. The high field domain is characterized by the alignment of the ground state in one well aligned with the excited state in the "down hill" well. As the bias is increased still further one well after another breaks off into the high field domain, resulting in the saw tooth, negative differential conductance (NDC) structure, until the entire sample is encompassed by the high field domain. When this occurs the electric field is again uniform and this defines the beginning of a step.

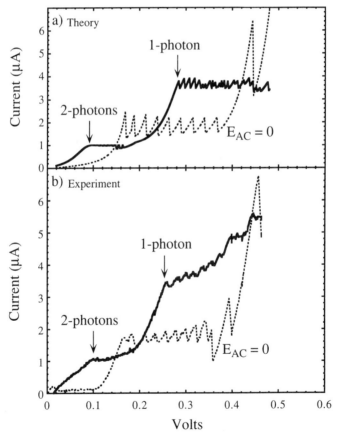

**Figure 1.** The current-voltage characteristics of a superlattice with 3.42 THz radiation (solid lines) and without terahertz radiation (dotted lines): (a) The calculated I-V characteristic with electric field domains. (b) The measured I-V.

In the presence of the intense terahertz radiation striking new structure appears in the I-V characteristic. The solid line in Fig. 1b shows the measured I-V characteristic when the superlattice is driven at 3.42 THz. In analogy to our description of the I-V characteristic in the absence of the terahertz radiation, the beginning of the new steps correspond to a uniform alignment of the 1 or 2-photon sidebands, or "virtual" states, associated with the ground state with the third level in the "down hill" well. Transport through the ground state (0-photons) and the structure associated with the I-V without radiation has been suppressed.

Remarkably, the saw tooth NDC structure observed on the 1-photon step is similar to the structure the I-V without terahertz radiation. This implies that on the plateau associated with the PAT step the superlattice supports high and low electric field domains, but sustained by the PAT channels. That is to say, transport occurs via tunneling between the virtual state associated with the ground state, and the third energy level in the low field domain and via the real ground state and the third energy level tunneling in the high field domain (Fig. 2).

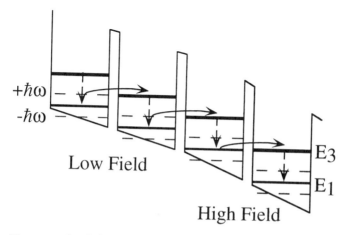

**Figure 2.** The energy level diagram, with ±1-photon sidebands, showing the state alignment corresponding to the electric field domain structure on the 1-photon step in Fig. 1. The second energy level, $E_2$, and the $\pm 2\hbar\omega$ sidebands are not shown for clarity.

To theoretically model the behavior described here, we have extended the model introduced by Bonilla et al.[10]. The key assumption in that model is the existence of a tunneling probability between neighboring quantum wells in the presence of an external electric field that exhibits a region of NDC. The origin of this NDC is physically very clear, alignment and misalignment of subband states in an applied DC electric field ($E_{DC}$).

Following Tien and Gordon[1], the action of the high frequency field is included by introducing new quantum well virtual states separated from the static level by $\pm n\hbar\omega$ (Fig. 2), with a state density proportional to $J_n^2(edE_{AC}/\hbar\omega)$, where $J_n$ is the nth order Bessel function, d is the superlattice period, $E_{AC}$ is the magnitude of the terahertz electric field and $\hbar\omega$ is the photon energy. In this model these new virtual states only appear in the well from which the electron tunnels. The calculated current-voltage characteristic, with electric field domains, is shown in Fig. 1a and compares well with the experimental results (Fig. 1b).

In conclusion, we report the first observation of electric field domain structure defined by tunneling between virtual states and the real quantum well states as well as multiphoton-assisted tunneling and a non-monotonic dependence of the multiphoton assisted tunneling induced currents on terahertz field strength in semiconductor multi-quantum well superlattices. We have successfully extended a recent model for current voltage characteristics of sequential resonant tunneling superlattices in the presence of electric field domains by inserting new subband states in the spirit of Tien and Gordon[1].

## ACKNOWLEDGMENTS.

We would like to thank help of the staff at the Center for Free-Electron Laser Studies, J.R. Allen, D. Enyeart, J. Kaminski, G. Ramian and D. White. The Center for Free-Electron Laser Studies is supported by the Office of Naval Research. Funding for this project was provided by the Army Research Office, the NSF, and the Air Force Office of Scientific Research. J.G. wishes to acknowledge financial support from MEC-Fulbright and ONR.

†Permanent address: Universidad Carlos III Madrid. Leganés 28911, Spain.

## References.

[1] P.K. Tien and J.P. Gordon, *Phys. Rev.* **129,** 647 (1963).

[2] P.S.S. Guimaraes, B.J. Keay, J.P. Kaminski, S.J. Allen Jr., P.F. Hopkins, A.C. Gossard, L.T. Florez, J.P. Harbison, *Phys. Rev. Lett.* **70,** 3792 (1993).

[3] B.J. Keay, P.S.S. Guimaraes, J.P. Kaminski, S.J. Allen Jr., P.F. Hopkins, A.C. Gossard, L.T. Florez, J.P. Harbison, *Surface Science* **305**, 385 (1994) .

[4] B.J. Keay, S.J. Allen Jr., J.P. Kaminski, K.L. Campman, A.C. Gossard, U. Bhattacharya, M.J.W. Rodwell, J. Galán, *The Physics of Semiconductors: Proceedings of the 22nd International Conference*, vol **2**, 1055, D.J. Lockwood ed., Vancouver, Canada (World Scientific, 1995).

[5] B.J. Keay, S.J. Allen Jr., J.P. Kaminski, J. Galán, K.L. Campman, A.C. Gossard, U. Bhattacharya, M.J.W. Rodwell, (preprint).

[6] S. Verghese, R.A. Wyss, Th. Schäpers, A. Förster, M.J. Rooks, Q. Hu, (preprint).

[7] L.P. Kouwenhoven, S. Jauhar, J. Orenstein, P.L. McEuen, *Phys. Rev. Lett.* **73**, 3443 (1994).

[8] J. Faist, F. Capasso, D.L. Sivco, C. Sirtori, A.L. Hutchinson, A.Y. Cho, *Science* **264**, 553-556 (1994).

[9] K.K. Choi, B.F. Levine, R.J. Malik, J. Walker, and C.G. Bethea, *Phys. Rev. B* **35**, 4172 (1987).

[10] L.L. Bonilla, J. Galán, J. A. Cuesta, F. C. Martínez and J. M. Molera, *Phys. Rev. B*, **50**, 8644 (1994).

# WANNIER-STARK RESONANCES IN DC TRANSPORT AND ELECTRICALLY DRIVEN BLOCH OSCILLATIONS

A. Di Carlo,[1] C. Hamaguchi,[2] M. Yamaguchi,[2] H. Nagasawa,[2] M. Morifuji,[2] P. Vogl,[1] G. Böhm,[1] G. Tränkle,[1] G. Weimann,[1] Y. Nishikawa[3] and S. Muto[3]

[1]Walter Schottky Institute, Technical University of Munich, D-85748 Garching, Germany
[2]Department of Electronic Engineering, Faculty of Engineering, Osaka University, Suita City 565, Japan
[3]Fujitsu Laboratories Ltd., 10-1 Morinosato-Wakamiya, Atsugi City 243-01, Japan

## INTRODUCTION

Only recently one has been able to realize the pioneering proposal of Esaki and Tsu from 1970 to utilize Bloch oscillations in superlattices as a source for tunable coherent THz emission.[1] This has been achieved by using coherent optical excitations of Bloch wave packets in superlattices and experiments indicate that the emission is superradiant.[2-4]

In this paper we show that coherent Bloch oscillations can be generated all electrically by using Zener (i.e., interband) tunneling as injection mechanism. In particular, we show that Bloch oscillations can maintain their coherence over many periods of oscillations in spite of the presence of strong Zener tunneling.

The stability of Bloch oscillations in the presence of interband tunneling has been a controversial subject for many years (see, e.g., Refs. 5-9). In the first part of this paper, we present both experimental and theoretical results that show that localized Wannier-Stark resonances can form in multi-quantum well structures in the presence of extremely high fields and correspondingly strong Zener tunneling. A clear evidence of Wannier-Stark resonances in carrier transport has been missing so far. Field induced tunneling between localized hole and electron states has been observed previously.[10] In the second part of this paper, we utilize the stability of Wannier-Stark resonances in the interband tunneling regime to predict electrically driven Bloch oscillations in superlattices.

## WANNIER-STARK RESONANCES WITH ZENER-TUNNELING

Experimentally, we have studied the current-voltage characteristics and electro-reflectance of highly doped p-i-n diodes with multi-quantum well structures in the intrinsic region at low temperatures (10 K). The intrinsic zone contains 10 superlattice periods, each period consisting of 2 atomic layers of AlAs and 5 atomic layers of GaAs. The samples were processed into $150 \times 150 \ \mu m^2$ mesa structures. For an applied reverse bias voltage of 0.95 V, the self-consistent energy band profile is shown in Fig. 1(a). The corresponding electric field in the intrinsic region is ~1 MeV/cm which is very high indeed but still well below the onset of impact ionization.

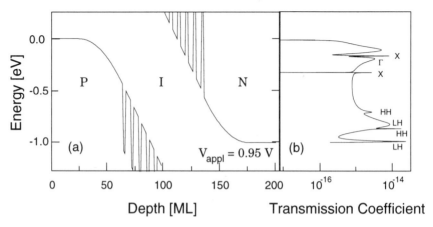

Figure 1. (a) Calculated energy band profile of p-i-n structure with 10 $(GaAs)_5/(AlAs)_2$ quantum wells in the 26 nm intrinsic region and an applied reverse bias voltage of 0.95 V. (b) Calculated transmission coefficient for tunneling across the band gap for this structure as a function of energy of the initial state.

We have calculated the high field tunneling current from the p-bulk valence into the n-bulk conduction band under reverse bias, using a recently developed multi-band multi-channel scattering approach that includes the self-consistent electric field profile in the entire p-i-n structure non-perturbatively and properly accounts for the field induced interband mixing effects between valence and conduction bands.[11,12]

The crucial result is that the Zener tunneling current across the band gap exhibits pronounced resonances whenever the energy of an incoming valence electron and an outgoing conduction electron state is aligned with one of the Wannier-Stark resonances associated with the multi-quantum well structure in the intrinsic zone. In Fig. 1(b), we show the calculated transmission coefficient as a function of energy for the multi-quantum well structure and the field profile as displayed in Fig. 1(a). There are sharp resonances due to the quantum well states derived from the light and heavy hole and GaAs-$\Gamma$ and AlAs-X conduction band states.

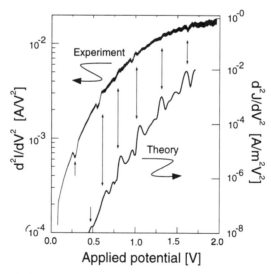

Figure 2. Comparison between experimental second derivative of the Zener tunneling current with the theoretical current density in the p-i-n diode of Fig. 1 as a function of reverse bias.

Figure 2 compares the calculated current density with the measured current. The observed oscillations in the I-V characteristics appear reproducibly in approximately 100 diodes prepared from two different wafers with different multi-quantum wells. Importantly, the positions and the separation between the extrema, including their field dependence, are in good agreement with the calculated results. The absolute magnitude of the measured current, however, is significantly higher than the calculated one, probably due to parasitic side currents that have a fairly smooth voltage dependence.

In order to further check these findings, we have also performed electroreflectance measurements and were able to clearly identify the dominant optical transitions between states in the same or adjacent GaAs quantum wells, again in good agreement with the theory and consistent with the peak separations in the DC current, shown in Fig. 2.[13] Altogether, these data show that well localized Wannier-Stark resonances form in the presence of Zener tunneling, in contrast to what has been widely believed so far.

## BLOCH OSCILLATIONS

We now consider a unipolar n-i-n+ diode with a GaAs:AlGaAs multi-quantum well structure in the intrinsic region and calculate the time-dependent evolution of an electronic wave packet that has been generated in the n-zone and traverses the intrinsic region. A wave packet with spatial extensions $\Delta x \lesssim 0.1$ μm may be generated electrically with a Gunn diode that is attached to the n-i-n+ structure, as will be described in detail elsewhere. The key point is that any wave packet that propagates towards the intrinsic region can enter the central region of the diode by Zener tunneling but *still perform coherent Bloch oscillations for many periods* and consequently give superradiant emission.

We take an i-zone with 20 unit cells of GaAs:GaAlAs with a barrier height of 0.4 eV, a GaAs and GaAlAs well width of 100 Å and 8 Å, respectively and apply an electric field that gives a total potential drop of 0.4 eV. We consider a Gaussian wave packet with $\Delta x = 400$ Å and solve the time-dependent Schrödinger equation in effective mass approximation. The wave packet starts in the the left n region and has an average energy of 0.16 eV.

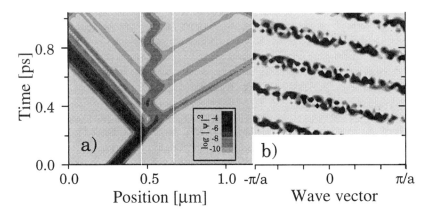

Figure 3. (a) Time development of Bloch wave packet in n-i-n+ superlattice in a high electric field, depicted in real space. The potential drops mostly in the middle intrinsic zone. The n and n+ region are on the left and right side, respectively. The wave packet starts at time t=0 on the n-side and propagates towards the i-zone. (b) Time development of this Bloch wave packet, projected onto zero field Bloch states of the 3rd miniband of the superlattice.

In Fig. 3(a), we show the time development of the wave packet across the structure in real space. When the packet reaches the middle region, most of the packet gets reflected, whereas some portion of it tunnels into the Wannier-Stark resonances associated with the 3rd mini-band. The portion of the wave packet that stays in the middle region performs an oscillatory motion in position space with a frequency given by the Bloch oscillation frequency (4.83 THz). A small part of the wave packet gets transmitted into the right $n^+$ region. The k-space projection on zero field Bloch states of the 3rd superlattice miniband in Fig. 3(b) shows that indeed the wave packet cycles through all k-states of the mini-band and changes its average k-value linearly in time, in accord with the semiclassical Bloch oscillation picture.

The oscillating electron density is given by $n(t) = n_0 \exp(-t/\tau)$ where $\tau = -\tau_B /\ln (1-T_L-T_R+T_L T_R)$ with $T_L(T_R)$ the tunneling coefficient from the middle region to the right (left) bulk and $\tau_B$ is the Bloch oscillation period. Our key finding is that there is a wide range of field strengths with $\tau \gg \tau_B$ and, correspondingly, a coherent oscillation of the wave packet in the i-zone.

This work was supported by a Grant-in-Aid for Scientific Research on Priority Area, `Quantum Coherence Electronics: Physics and Technology' from the Ministry of Education, Science and Culture, Japan, and by the Deutsche Forschungsgemeinschaft (SFB 348), Germany.

REFERENCES

1. L.Esaki and R. Tsui, *IBM J. Res. Develop.* 14:61 (1970).
2. C. Waschke, H. G. Roskos, R. Schwedler, K. Leo, H. Kurz, K. Köhler, *Phys. Rev. Lett.* 70:3319 (1993).
3. T. Dekorsy, R. Ott, H. Kurz, K. Köhler, *Phys. Rev.* B 51:17275 (1995).
4. H.G. Roskos, *Festkörperprobleme* 34, in press.
5. J. Zak, *Phys. Rev.* B 38:6322 (1988); ibid., B 43:4519 (1991).
6. J.B. Krieger and G.J. Iafrate, *Phys. Rev.* B 38:6324 (1988).
7. P.N. Argyres and S. Sfiat, *J. Phys.* 2:7089 (1990).
8. E. Avron, J. Nemirovsky, *Phys. Rev. Lett.* 68:2212 (1992).
9. J. Rotvig, Antti-Pekka Jauho, H. Smith, *Phys. Rev. Lett.* 74:1831 (1995).
10. J. Allam, F. Beltram, F. Capasso, and A. Y. Cho, *Appl. Phys. Lett.* 51:575 (1987).
11. M. Morifuji, Y. Nishikawa, C. Hamaguchi and T. Fujii, *Semicond. Sci. Technol.* 7:1047 (1992).
12. A. Di Carlo, W. Pötz, and P. Vogl, *Phys. Rev.* B 50:8358 (1994).
13. M. Morifuji, A. Di Carlo, M. Yamaguchi, C. Hamaguchi, H. Nagasawa, P. Vogl, G. Böhm, G. Tränkle, G. Weimann, to be published.

# QUANTUM PROPERTIES OF BLOCH ELECTRONS IN SPATIALLY HOMOGENEOUS ELECTRIC FIELDS

Jun He,[1] Gerald J. Iafrate[2]

[1]Department of Electric and Computer Engineering
North Carolina State University, NC 27695-7911
[2]U.S. Army Research Office
Research Triangle Park, NC 27709-2211

## INTRODUCTION

The existence of the Wannier Stark (WS) ladder and Bloch Oscillations (BO) have been the subjects of great controversy [1] until recent experimental demonstrations of their existence in semiconductor superlattices. The photoluminescence and photocurrent measurements of the biased GaAs/GaAlAs superlattices [2] provided the earliest experiment evidence of the electric field induced WS ladder energy levels. Recent optical experiments on excitonic emission from double wells [3], four-wave mixing from superlattices [4], and optical absorption studies clearly indicated oscillatory electron dynamics and the manifestation of concomitant Stark-ladder transitions when optical probing is invoked. Since all these recent experiments for observing WS ladder and BO involve semiconductor superlattices in electric fields, it is very timely to investigate the Bloch electron dynamics in a homogeneous electric field.

In this paper, the theory of the Bloch electron dynamics in homogenous electric field of arbitrary strength is presented. From the theory, a complete analysis of single-band and multiband processes are discussed for a Bloch electron in the presence of a uniform electric field as well as in the presence of a superimposed uniform and oscillatory electric field. In the formalism, the electric field is described through the use of the vector potential, and a basis set of localized, electric field dependent Wannier, and related envelope function are developed and utilized to describe the dynamics [5-7]. For the case of uniform electric field, Bloch oscillations in the single-band model and Zener tunneling in the multiband model are discussed. For the case of the superimposed uniform and oscillatory electric field, the single-band velocity for the Bloch electron is calculated, and the DC components of the velocities are examined for some special

*Hot Carriers in Semiconductors*
Edited by K. Hess *et al.*, Plenum Press, New York, 1996

tuning conditions. Further, in the multiband analysis, the transition processes for the Bloch electron transition out of the initially occupied band are investigated, and the transition rate is calculated as a function of the electric field, band parameters and initial conditions. In the derivation, the interband coupling is treated in the Wigner-Weisskopf (WW) approximation [8], a classic approach for describing the time decay from an occupied quasi-stationary state; the WW approximation allows for the analysis of the long-time, time-dependent tunneling characteristics of an electron transition out of an initially occupied band due to the power absorbed by the electric field while preserving conservation of total transition probability over the complete set of excited bands.

The Hamiltonian for a single electron in a periodic crystal potential subject to a electric field of arbitrary time dependence and strength is

$$
H = \frac{1}{2m}(\vec{P} - \frac{e}{c}\vec{A}(t))^2 + V_c(\vec{x}) ,
$$

(1)

where $V_c(\vec{x})$ is the periodic crystal potential, and $\vec{A}(t)$ is the vector potential for the electric field. For a homogeneous electric field $\vec{E}(t)$, $\vec{A}(t) = -c\int_0^t \vec{E}(t')dt'$. In the localized Wannier representation, the wave function of the electron is $\Psi(\vec{x},t) = \sum_n \sum_{\vec{l}} f_n(\vec{l},t) W_n(\vec{x} - \vec{l},t)$, where $W_n(\vec{x} - \vec{l},t)$ is the time-dependent Wannier function, and "$n$" indexes the band. The envelope function in the Wannier representation $f_n(\vec{l},t)$, for the Bloch electron in a homogeneous electric field of arbitrary time-dependence, satisfy the differential equation [7]

$$
i\hbar\frac{\partial f_n(\vec{r},t)}{\partial t} = \epsilon_n(-i\vec{\nabla} - \frac{e}{\hbar c}\vec{A})f_n(\vec{r},t) - e\vec{E}(t) \cdot \sum_{n'} \vec{R}_{nn'}(-i\vec{\nabla} - \frac{e}{\hbar c}\vec{A})f_{n'}(\vec{r},t) ,
$$

(2)

where "$\vec{r}$" represents an arbitrary lattice position within the crystal, $\epsilon_n(\vec{K})$ is the n-th Bloch energy band with crystal momentum $\vec{K}$, and $\vec{R}_{nn'}$ is the interband coupling parameter.

Equation (2) depicts the multiband Wannier envelope function for a Bloch electron in a homogeneous electric field. It is clear that the solution to Eq.(2) for $f_n(\vec{r},t)$ can be written as

$$
f_n(\vec{r},t) = \frac{1}{\sqrt{N}} \sum_{\vec{K}} e^{-\frac{i}{\hbar}\int_0^t \epsilon_n(\vec{K} - \frac{e}{\hbar c}\vec{A})dt'} e^{i\vec{K}\cdot\vec{r}} A_n(\vec{K},t) ,
$$

(3)

where the expansion coefficients, $A_n(\vec{K},t)$, satisfy the equations

$$
i\hbar\frac{\partial}{\partial t}A_n(\vec{K},t) = - \sum_{n'\neq n} B_{nn'}(\vec{K},t)A_{n'}(\vec{K},t) ,
$$

(4)

and where "$n$", "$n'$" index the complete set of energy bands for the crystal. In Eq.(4), the time-dependent interband matrix elements, $B_{nn'}(\vec{K},t)$, are given by

$$
B_{nn'}(\vec{K},t) = e\vec{E}(t) \cdot \vec{R}_{nn'}(\vec{K} - \frac{e}{\hbar c}\vec{A})e^{-\frac{i}{\hbar}\int_0^t (\epsilon_{n'}-\epsilon_n)dt'} ,
$$

(5)

where $\epsilon_n$, $\epsilon_{n'}$ are explicit function of $\vec{K} - \frac{e}{\hbar c}\vec{A}$ .

148

## UNIFORM ELECTRIC FIELD

For the case of the uniform electric field, $\vec{E}(t) = \vec{E}_o$, and the vector potential $\vec{A}(t) = A_o = -c\vec{E}_o t$. The single band envelope function is connected to the initial values of the envelope function by a time evolution kernel as $f_n^{SB}(\vec{l}, t) = \sum_{l'} K_n(\vec{l} - \vec{l'}, t) f_n(\vec{l'}, 0)$, where the time evolution kernel is

$$K_n(\vec{l} - \vec{l'}, t) = \frac{1}{N} \sum_{\vec{K'}} e^{-i/\hbar \int_0^t \epsilon_n[\vec{K} - (e/\hbar c)\vec{A}_o]dt'} e^{i\vec{K}\cdot(\vec{l} - \vec{l'})} \ . \tag{6}$$

The general oscillatory nature of $f_n^{SB}(\vec{l}, t)$ in the direction of electric field is delineated explicitly by invoking the $\vec{K}$-space periodicity of $\epsilon_n(\vec{K})$, namely $\epsilon_n(\vec{K}) = \epsilon_n(\vec{K} + \vec{G}_j)$, where $| \vec{G}_j | = j2\pi/a$ are the appropriate reciprocal-lattice vectors. It can be shown that the single-band envelope function $f_n^{SB}(\vec{l}, t)$ oscillates with frequency $\omega_B$ ($\omega_B = eE_o a/\hbar$) in the direction of the electric field, and is diffusive in the direction perpendicular to the electric field. In the case of one dimension, it is clearly seen that $| f_n^{SB}(\vec{l}, s\tau_B) | = | f_n(\vec{l}, 0) |$, where "s" refer to a positive integer, and $\tau_B = 2\pi/\omega_B$ is a Bloch period. Further, in the single-band one-dimensional model, with a given $\epsilon_n(\vec{K})$, the total wave function is periodic in time with Bloch frequency up to a constant phase factor in the direction of the field.

In the multiband analysis, the envelope function, as given by Eq.(3) with band coupling dependence reflected in the coefficients, $A_n(\vec{K}, t)$, given by Eq.(4). Since the set of equations given in Eq.(4) are not solvable exactly, with most approximate analytical method addressing the short time behavior, we use the Wigner-Weisskopf approximation to establish the long-time behavior of $A_n(\vec{K}, t)$ in Eq.(4). The Wigner and Weisskopf approximation couples the state of interest "n" to all other states of the system "n'" while including only direct reflective feedback from the states "n'" to the state of interest, thereby ignoring multiple probability reflections. In addition, the above approximation guarantees conservation of probability. In a recent paper [8], we derived the multiband solution for a Bloch electron in a homogeneous constant electric field through the use of Wigner-Weisskopf approximation. It is found that $| A_n(\vec{K}, t) |^2 = | A_n(\vec{K}, 0) |^2 \exp[-\gamma_n(\vec{K}_\perp)t]$, with the decay rate given by

$$\gamma_n(\vec{K}_\perp) = \frac{\pi}{8} \omega_B \sum_{n' \neq n} e^{-2q_{nn'}(\vec{K}_\perp)a[\epsilon_{nn'}(\vec{K}_\perp)/eE_o a]} \tag{7}$$

where $q_{nn'} = | q |$, with "q" determined by continuation of the energy band function between band "n" and "n'": $\epsilon_n(q, \vec{K}_\perp) = \epsilon_{n'}(q, \vec{K}_\perp)$, and $\epsilon_{nn'}(\vec{K}_\perp)$ is the average value of the energy-band difference between states "n" and "n'" taken over the direction $K_x$ in the Brillouin zone.

The total probability for an electron to be in its initial band "n" after time "t" is $\rho_n(t) = \sum_{\vec{K}} | A_n(\vec{K}, t) |^2 = \sum_{\vec{K}} | A_n(\vec{K}, 0) |^2 e^{-\gamma_n(\vec{K}_\perp)t}$, where the exponential terms indicate the irreversible decay out of n-th band into the upper bands of the crystal due to the power absorbed by the electric field. Thus, the total transition rate for an electron in initial band "n" is $\Gamma_n(t) \approx \Gamma_n(0) = \sum_{\vec{K}} | A_n(\vec{K}_\perp, 0) |^2 \gamma_n(\vec{K})$. This transition rate is the total transition probability per unit time for an electron to tunnel out of state "n" while coupled to all other bands. Thus, the time of Zener tunneling, $\tau_Z$, is determined by $\tau_Z \equiv \frac{1}{\Gamma_n(0)}$.

It is interesting to note that the total transition rate depends not only on the decay rate $\gamma_n(\vec{K})$, but also on the initial conditions $A_n(\vec{K}, 0)$ of the system. For initial conditions based on Wannier states, $\Gamma_n(0) = \frac{1}{N} \sum_{\vec{K}} \gamma_n(\vec{K}_\perp)$, the average value of

$\gamma_n(\vec{K}_\perp)$ over the entire $\vec{K}$-space. However, for initial conditions based on Bloch states, $\Gamma_n(0) = \sum_{\vec{K}} \delta_{\vec{K},\vec{K}_o} \gamma_n(\vec{K}_\perp) = \gamma_n(\vec{K}_{o\perp})$, the decay rate at $\vec{K}_\perp = \vec{K}_{o\perp}$.

## UNIFORM AND OSCILLATORY ELECTRIC FIELD

For the case of superimposed uniform and oscillatory electric field, $\vec{E}(t) = \vec{E}_o + \vec{E}_1 \cos \omega t$, and the vector potential $\vec{A}(t) = -c\vec{E}_o t - \frac{c}{\omega}\vec{E}_1 \sin \omega t$. In the single-band model, $B_{nn'}(\vec{K}, t) = 0$ for all "$n$" and "$n'$"; so that $A_n(\vec{K}, t)$ is a constant in time; the velocity for a Bloch electron can be calculated from its initial condition $A_n(\vec{K}, 0)$, and the field-dependent energy band function $\epsilon_n(\vec{K} - \frac{e}{\hbar c}\vec{A})$ as

$$< \vec{v}_n >= \frac{1}{\hbar} \sum_{\vec{K}} | A_n(\vec{K}, 0) |^2 \vec{\nabla}_{\vec{K}} \epsilon_n(\vec{K} - \frac{e}{\hbar c}\vec{A}) . \tag{8}$$

It is clear that for initial Wannier state, $| A_n(\vec{K}, 0) |^2 = \frac{1}{N}$, $< \vec{v}_n >= 0$; and for initial Bloch state, $A_n(\vec{K}, 0) = \delta_{\vec{K},\vec{K}_o}$, the velocity is $< \vec{v}_n >= \frac{1}{\hbar}\vec{\nabla}_{\vec{K}} \epsilon_n(\vec{K} - \frac{e}{\hbar c}\vec{A}) |_{\vec{K}=\vec{K}_o}$, which is electric field and time dependent.

By expressing the energy band function $\epsilon_n(\vec{K} - \frac{e}{\hbar c}\vec{A})$ in Fourier series expansion, the velocity for an Bloch electron of initial Bloch state can be written in a series expansion with explicit time-dependence,

$$< \vec{v}_n >= \frac{i}{\hbar} \sum_{\vec{l}} \varepsilon_n(\vec{l})\vec{l}e^{i\vec{K}_o \cdot \vec{l}} \sum_m J_m(l_x \alpha)e^{i(l_x \omega_B + m\omega)t} . \tag{9}$$

where $\varepsilon_n(\vec{l})$ are the Fourier components of the energy band function, the parameter $\alpha = \frac{eE_1 a}{\hbar \omega}$, and $J_m$ is the Bessel function of the first kind with integer order "$m$".

At some special tuning conditions, the velocity has time-independent component in the direction of electric field, i.e., DC velocity. If the frequency of the AC field $\omega$ is a multiple of the Bloch frequency $\omega_B$, $\omega = N_o\omega_B$; the Bloch frequency $\omega_B$ becomes the driven frequency of the system, i.e., the system oscillates at frequency $\omega_B$ since the energy band function and the velocity of the Bloch electron oscillate in time with the Bloch frequency. There are time-independent terms in the velocity, these terms contributes to the DC component of the velocity,

$$< \vec{v}_n >_x^{DC} = -\frac{2a}{\hbar} \sum_{m>0} \epsilon_n(mN_o a, \vec{K}_{o\perp})(mN_o)(-)^m J_m(mN_o \alpha) \sin mN_o K_{ox}a , \tag{10}$$

where $\epsilon_n(l_x, \vec{K}_\perp)$ is defined by $\epsilon_n(l_x, \vec{K}_\perp) = \sum_{\vec{l}_\perp} \varepsilon_n(\vec{l})e^{i\vec{K}_\perp \cdot \vec{l}_\perp}$. It can easily be proved that $< \vec{v}_n >_x^{DC}$ of Eq.(10) equals time average of $< v_n >_x$ in one Bloch period.

For weak AC field strength, i.e., for the AC field parameter $\alpha = eE_1 a/\hbar \omega \ll 1$, the time evolution kernel shows the probability for the Bloch electron to stay in the lattice sites with the same x-coordinate and to hop forward or backward in x-direction by $N_o$ lattice spacing, and the envelope function for the one dimensional case is

$$f_n(l, s_{TB}) = e^{-\frac{i}{\hbar}s_{TB}\varepsilon_N(0)}\{f_n(l, 0) + \frac{1}{\hbar}\frac{N_o\alpha\varepsilon_n(\alpha N_o)s_{TB}}{2}[f_n(l-N_o, 0) + f_n(l+N_o, 0)]\} . \tag{11}$$

For the tuning condition which assume that the Bloch frequency is a multiple of the frequency of the AC field, $\omega_B = M_o \omega$; the AC field frequency $\omega$ is the driven frequency of the system, *i.e.*, the system oscillates with frequency $\omega$. The time-independent terms exist in the single band velocity. These time independent terms contribute to the DC component of the velocity,

$$< \vec{v}_n >_x^{DC} = -\frac{2a}{\hbar} \sum_{l_x > 0} (-)^{M_o l_x} \epsilon_n(l_x, \vec{K}_{o\perp}) J_{l_x M_o}(l_x \alpha) l_x \sin l_x K_{ox} a . \tag{12}$$

It is very illustrative to examine the effects of a strong AC field with no DC field on one-dimensional crystal in the nearest neighbor tight-binding approximation, which is the model studied by M. Holthaus *et. al.* [9] for far-infrared irradiated superlattices. When the bias field is turned off, $E_o = 0$, $\omega_B = 0$, the deformed energy band becomes $\tilde{\epsilon}_n(K) = \varepsilon_n(0) + 2\varepsilon_n(a)J_o(\alpha) \cos Ka$, where the band width of the deformed band $\tilde{W}_n = 4\varepsilon_n(a)J_o(\alpha)$. For $\alpha$ equals the roots of $J_o$, $\tilde{W}_n = 0$, which is the condition for the collapse of minibands in superlattices given in Ref.[9]. Furthermore, the DC velocity of the Bloch electron for this model is $< v_n >^{DC} = -\frac{2a\varepsilon_n(a)}{\hbar} J_o(\alpha) \sin K_o a$, when radiation field is such that $J_o(\alpha) = 0$, $< v_n >^{DC} = 0$, the electron oscillates about the lattice site with frequency $\omega$ without developing of DC current.

In the multiband analysis, we again use the Wigner-Weisskopf approximation to solve the set of differential equation for $A_n(\vec{K}, t)$, the solution can be written as

$$A_n(\vec{K}, t) = A_n(\vec{K}, 0) e^{i\Delta\omega_n(\vec{K})t - \frac{1}{2}\gamma_n(\vec{K})t} , \tag{13}$$

with a $\vec{K}$-dependent frequency shift $\Delta\omega_n(\vec{K})$, and a $\vec{K}$-dependent decay rate $\gamma_n(\vec{K})$. The decay rate is

$$\begin{aligned}
\gamma_n(\vec{K}) &= 2\pi \sum_{n' \neq n} \{| D_{nn'}(0, \vec{K}_\perp) |^2 [(\frac{eE_o}{\hbar})^2 \delta(\tilde{\omega}_{nn'}(\vec{K})) + \frac{1}{4}(\frac{eE_1}{\hbar})^2 \delta(\tilde{\omega}_{nn'}(\vec{K}) + \omega) \\
&\quad + \frac{1}{4}(\frac{eE_1}{\hbar})^2 \delta(\tilde{\omega}_{nn'}(\vec{K}) - \omega)] \\
&\quad + \sum_{l_x \neq 0} \sum_m | D_{nn'}(l_x, \vec{K}_\perp) |^2 J_m^2(l_x\alpha)(\frac{eE_o}{\hbar} + \frac{eE_1}{\hbar}\frac{m}{l_x\alpha})^2 \delta(\tilde{\omega}_{nn'}(\vec{K}) - l_x\omega_B - m\omega)\}
\end{aligned} \tag{14}$$

where $D_{nn'}(l_x, \vec{K}_\perp) = \frac{i}{4} a e^{-q_{nn'}a|l_x|}$, with $q_{nn'}$ defined earlier; and $\tilde{\omega}_{nn'}(\vec{K}) = \frac{1}{\hbar}[\tilde{\epsilon}_{n'}(\vec{K}) - \tilde{\epsilon}_n(\vec{K})]$, where $\tilde{\epsilon}_n(\vec{K})$ is the equivalent energy band function for the deformed band.

In general, $\Delta\omega_n$ and $\gamma_n$ are $\vec{K}$-dependent, so that $| A_n(\vec{K}, t) |^2 = | A_n(\vec{K}, 0) |^2 \exp(-\gamma_n(\vec{K})t)$, $| A_n(\vec{K}, t) |^2$ decays exponentially with a decay rate of $\gamma_n(\vec{K})$. The total transition rate for an electron in initial band "$n$" can be calculated from $\Gamma_n \approx \sum_{\vec{K}} | A_n(\vec{K}, 0) |^2 \gamma_n(\vec{K})$ for given initial condition. It is noted that $\gamma_n(\vec{K})$ consists of several $\delta$-functions, which indicate the possible physical processes for Bloch electron in homogeneous, superimposed uniform and oscillatory electric field. In $\gamma_n(\vec{K})$, the terms $| D_{nn'}(0, \vec{K}_\perp) |^2 (\frac{eE_o}{\hbar})^2 \delta(\tilde{\omega}_{nn'}(\vec{K}))$ are for transitions between band "$n$" and "$n'$" due to Zener tunneling between the deformed bands, which are proportional to $E_o^2$; the terms $\frac{1}{4} | D_{nn'}(0, \vec{K}_\perp) |^2 (\frac{eE_1}{\hbar})^2 \delta(\tilde{\omega}_{nn'}(\vec{K}) \pm \omega)$ indicate the transitions between the two deformed bands through absorption/emission of a photon, which are proportional to $E_1^2$, and the terms $| D_{nn'}(l_x, \vec{K}_\perp) |^2 J_m^2(l_x\alpha)(\frac{eE_o}{\hbar} + \frac{eE_1}{\hbar}\frac{m}{l_x\alpha})^2 \delta(\tilde{\omega}_{nn'}(\vec{K}) - l_x\omega_B - m\omega)$

suggest transitions between the deformed bands via Wannier Stark transitions, as well as single photon or multiphoton absorption/emission.

## SUMMARY

In this paper, we have presented a theory of Bloch electron dynamics in a homogeneous, uniform field, as well as superimposed uniform and oscillatory electric field of arbitrary strength. In the theory, the envelope function in the Wannier representation is used to describe the Bloch electron, and the vector potential is used to described the electric field. For the case of uniform electric field, the envelope function in Wannier representation is analyzed, and the Bloch oscillations are shown in the single-band model; from multiband model, the Zener tunneling process is studied, and Zener tunneling rate is derived as a function of initial condition, band parameter and the electric field. For the case of the superimposed uniform and oscillatory electric field, the single-band velocities are calculated. The results show that for the tuning of the AC field frequency to some multiple of the Bloch frequency results in the resonant displacement of the electron from a given site to the tuned multiple of lattice sites along with a concomitant development of a DC velocity. Further, when the Bloch frequency equals multiple of AC field frequency, a development of DC velocity is also observed. In the multiband model, the general transition rate for an electron transition out of the initially occupied band is derived. The multi-band analysis shows all possible transitions including Zener tunneling, photon assisted tunneling, multiphoton emission/absorption, as well as transitions between Wannier-Stark ladders.

## ACKNOWLEDGEMENT

This work was supported by Office of Naval Research and U.S. Army Research Office.

## REFERENCES

1. see G. Nenciu, *Rev. Mod. Phys.*, 63, 91 (1991), and references therein.
2. E.E. Mendez, F.Agulló-Rueda, and J.M. Hong, *Phys. Rev. Lett.*, 60, 2426(1988).
3. J. Feldmann, K. Leo, J. Shah, D.A.B. Miller, J.E. Cunningham, T. Meier, G. Von Plessen, A. Schulze, P. Thomas, and S. Schmitt-Rink, *Phys. Rev.* B46, 7252 (1992).
4. C. Waschke, H.G. Roskos, R. Schwedler, K. Leo, H. Kurz, and K. Köhler, *Phys. Rev. Lett.* 70, 3319 (1993).
5. J.B. Krieger and G.J. Iafrate, *Phys. Rev.*B33,5494 (1986).
6. J.B. Krieger and G.J. Iafrate, *Phys. Rev.* B35, 9644 (1987).
7. G.J. Iafrate and J.B. Krieger, *Phys. Rev.* B40, 6144 (1989).
8. J. He and G. J. Iafrate, *Phys. Rev.* B50, 7553 (1994).
9. M. Holthaus and D. W. Hone, *Phys. Rev.* B49, 16605 (1994), M. Holthaus, *Phys. Rev. Lett.* 69, 351 (1992).

# BLOCH OSCILLATIONS AND DYNAMIC NDC UNDER

# A MULTIMINIBAND TRANSPORT IN SUPERLATTICES

A.Andronov and I.Nefedov

Institute for Physics of Microstructures, Russian Academy of Sciences
Nizhny Novgorod 603600, Russia

## INTRODUCTION

Works on superlattice (SL) transport treat mostly the case of strong SL potential
with narrow minibands - A in Fig.1 - providing transport in single miniband.

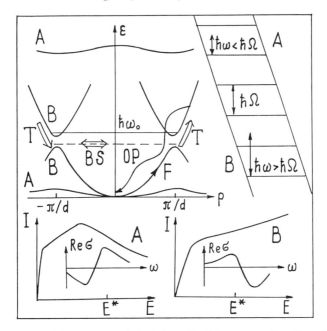

**Figure 1** Scheme of wide ($A$) and narrow ($B$) minigap SL with corresponding dispersion laws, transport processes, $I - V$ curves, differential conductivity versus frequency and Wannier-Stark transitions

Here at high electric field $E$ the Bloch oscillations (BO) occur and current is a decreasing function of the field [1] - i.e. static NDC takes place - provided the BO frequency $\Omega = eEd/\hbar$ (d is SL period) is higher than appropriate relaxation rate. The NDC occupies frequency region from $\omega = 0$ up to $\omega \simeq \Omega$ - Fig.1, A [2]. The NDC at $\omega = 0$ leads to formation of domains which hinder direct observation and utilization of the BO.

Here we discuss low temperature high field transport for weak potential SL with *narrow minigaps* (B in Fig.1) with first minigap just at an optical phonon energy $\hbar\omega_0$ (Fig.1, B). We discover peculiar carrier distribution ("roding") and show that though there is no static NDC nearby resonance (at $\omega \geq \Omega$) the system can exhibit (dynamic) NDC (Fig 1,B). Due to absence of the domains the dynamic NDC gives a unique possibilities for observation of BO and for development of the BO-based Terahertz oscillator.

## MODEL AND IDEALIZED CALCULATIONS

We suppose that weak 1-d SL potential has equal Fourier harmonics with value $u_0$ which produce the Bragg scattering (BS) in SL Brillouin zone (BZ) in supposed parabolic band (effective mass $m^*$) leading to minigaps $\epsilon_{gn}$ at $p_z = \pm np_B$, $p_B = \pi * \hbar/d$, $\vec{z}_0$ is SL axis. In electric field $E$ tunneling (T) through minigaps occurs. For $u_0 \to 0$ $\epsilon_{gn} = \epsilon_g = 2u_0$ and tunneling probability $P_{tn}$ and the BS probability $P_{Bn}$ are

$$P_{tn} = \exp(-E_t/nE), \qquad E_t = \frac{\pi^2 \epsilon_g^2}{4\epsilon_B ed}, \qquad P_{Bn} = 1 - P_{tn} \qquad (1)$$

where $\epsilon_B = p_B^2/2m^*$. We suppose that $p_B \simeq p_0$, $p_0 = \sqrt{2m^*\hbar\omega_0}$,, temperature T is low: $kT \ll \hbar\omega_0$ and impurity scattering is low so that at carrier energy $\epsilon > \hbar\omega_0$ scattering is mainly due to OP emission with high scattering rate $\nu \simeq \nu_0 = eE_0/p_0$, $E_0$ is characteristic field while at $\epsilon < \hbar\omega_0$ the rate is low: $\nu \simeq \nu^* \ll \nu_0$. High field transport in the bulk in this case involves "streaming". With the SL potential we have additionally the following rates: $\nu_B = \Omega/2\pi = (2p_B/eE)^{-1}$ - inverse BO period; $\nu_t = \delta\Omega/2\pi$, $\delta = \exp(-E_t/E)$ - the characteristic decay tunneling rate. We will suppose the following hierarchy of the rates:

$$\nu^* \ll \nu_B\delta, \quad \nu_B \leq \nu_0 \qquad (2)$$

In this case a carrier performs many BO in the first miniband and tunnels to the second miniband (before being scattered in the first miniband); then it quickly return to the first miniband due to OP emission before it is able to perform even single BO in the second miniband. Below we will use the Boltzman kinetic equation with both the BS and the T considered as ordinary scattering processes. This is justified because "true" scattering - OP emission - randomize phases of the carrier wave functions.

Consider first the idealized case $\nu^* \to 0$, $\nu_0 \to \infty$, which can be treated in one dimensional picture: distribution function $f(\vec{p}) = g(p_z)\delta(p_x)\delta(p_y)$. We put also $p_B = p_0$ ($p_B = p_0 - 0$); then the Boltzman equation for the static carrier distribution function $g_d$ in d.c electric field $E_d$ and boundary condition at BZ boundary are:

$$eE_d \frac{dg_d}{dp_z} = I_d\delta(p_z), \qquad g_d(-p_B) = g_d(p_B)(1 - \beta), \qquad (3)$$

Value of ($I_d$) should be found from the carrier density $N$: $\int_{-p_B}^{p_B} g_d(p_z)dp_z = N$. Similarly the linearized Boltzman equation for amplitude $g_{a0}$ of small alternative part of the distribution function in d.c. field $E_d$ and small a.c. electric field $E_a = E_{a0}\exp(i\omega t)$ and the linearized boundary condition are:

$$i\omega g_{a0} + F\frac{dg_{a0}}{dp_z} = -eE_{a0}\frac{dg_d}{dp_z} + I_a\delta(p_z) = \hat{I}_a\delta(p_z), \qquad (4)$$

$$g_{a0}(-p_B) = g_{a0}(p_B)(1-\beta) - g_d(p_B)E_{a0}E_t\beta/E^2 \qquad (5)$$

Value of $\hat{I}_a$ should be found now from $\int_{-p_B}^{p_B} g_{a0}(p_z)dp_z = 0$. Solution of the equations (3 - 5) is straightforward and one gets:

$$j_d = \frac{eNp_0\beta}{2m^*(2-\beta)}, \quad g_d(p_z) = \frac{N(1-\beta+\beta\theta(p_z))}{p_0(2-\beta)}, \qquad (6)$$

$$\sigma_1 = \sigma_{0E} \times \frac{\beta}{(\beta-1)^2 + 2(1-\beta)\cos\varphi + 1} \times \frac{\sin\varphi}{\varphi} \qquad (7)$$

Here $j_d$ is current, $\sigma_1$ is the real part of differential conductivity $\theta(p_z) = 1$, $p_z \geq 0$; $\theta(p_z) = 0$, $p_z < 0$; $\sigma_{0E} = eNp_0E_t/m^*E^2$, $\varphi = \pi\omega/\Omega$. At $\beta \geq 1$ tunneling is high; here carrier distribution is close to (ideal) streaming with $g_d \simeq 0$ at $p_z < 0$. At low $\beta$ the distribution is almost symmetrical with respect to $p_z = 0$ due to the intense BS. The distribution looks like a rod in momentum space and may be called here *roding*. Conductivity $\sigma < 0$ at $\omega > \Omega$ only - i.e. there is *dynamic* NDC. It is a result of carrier bunching in the BZ due to the two factors: strong electric field dependent interminiband tunneling and the BO.

## MONTE-CARLO SIMULATION

In the simulation for n-GaAs with weak SL potential we use simple two valley model ("old" $\Gamma - X$ model). The most important scattering is OP scattering

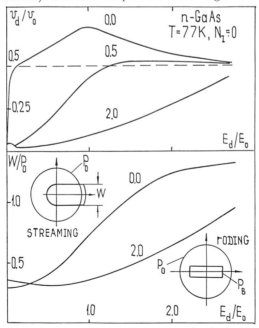

**Figure 2** Drift velocity $v_d$ ($v_0 = p_0/m^*$) and transverse distribution width $W$ for n-GaAs versus $E_d/E_0$; numbers are $E_t/E_0$.

with characteristic field $E_0 \approx 6.5$ kV/cm in n-GaAs. Figure 2 presents d.c. characteristics. We see that with an increase of $E_t$ the static NDC at high electric field (which is due to intervalley transfer) disappears. On the other hand in the low field region there appears new static NDC. This NDC is ordinary SL NDC arising from BS: in this low field region tunneling is negligible and transport takes place in the first miniband. We see also that at high enough $E_t$ *roding* with $W \ll p_d$ (where $p_d = m^* * v_d$ is the drift momentum) occurs. Simulations of $\sigma_1$ for $E = 0.5 - 1.0E_0$ , $E_t \simeq 0.25 - 1.0E_0$, T = 77K and impurity concentration $N_I \leq 10^{15} cm^{-3}$ show that the overall behaviors of the conductivity is similar to that in the idealized model with value of $\sigma_1$ an order of magnitude lower. It is likely that the suppression is due to the fact that at these fields ($E \simeq E_0$) electron spend substantial part of BO period in the second miniband before returning back to the first miniband being as a result "out of phase" and suppresses th NDC. The NDC survives for $E_t = 0.75E_0$ at T = 77K even for $N_I \simeq 10^{15} cm^{-3}$ demonstrating relaxed enough condition for observation of the NDC.

## CONCLUSION

From the quantum mechanical point of view the BO correspond to transitions between the Wannier-Stark levels (see Fig.1). Because under transport the SL is spatially homogenous system the levels are equally populated - i.e. there is *no population inversion* between the levels and conductivity should vanishes. However the present results for the narrow minigap SL together with earlier ones [2] for the wide minigap SL demonstrate that the conductivity around the Bloch frequency appears due to transport processes and crucially depends on the type of the process involved.

The SL parameters needed for observation of the effects may be met in GaAlAs - based SL, consisting of GaAs wells and few monolayer Al (or AlAs) barriers while a possible device could be just the Gunn-like diode where instead of pure GaAs the above SL with 500 or more periods will be provided. Such diode without domain could be an ideal system for observation of spontaneous BO emission. At the same time in the diode made say in the form of a ring (a strip-line resonator) the Terahertz field could also grow up as a result of the dynamic NDC.

**Acknowledgments.** One of us (AA) takes this opportunity to thank E.Gornik and the Technical University of Vienna for the invitation and the hospitality during his stay in Vienna where a part of this work was performed. We also acknowledge many valuable discussion with V.Aleshkin and I.Shereshevsky. This work is supported by Russian Foundation for Fundamental Research, Russian Nanophysics Program and by INTAS.

## REFERENCES

1. Esaki L.,Tsu R. "Superlattices and Negative Differential Conductivity in Semiconductors", IBM Jour. Res. Dev. 1970,V.14, P.61-65.
2. Ktitorov S.A., Simin G.S. and Sandolovski V.Ya. "Bragg reflection and the high-frequency conductivity of electronic solid-state plasma", Sov.Phys. - Solid State, 1971, V.13, N.8, P.1872-1874.

# PHONON-INDUCED SUPPRESSION OF
# BLOCH OSCILLATIONS IN SEMICONDUCTOR SUPERLATTICES:
# A MONTE CARLO INVESTIGATION

F. Rossi,[1,2] T. Meier,[1] P. Thomas,[1] S.W. Koch,[1]
P.E. Selbmann,[2,3] and E. Molinari[2,3]

[1]Fachbereich Physik und Zentrum für Materialwissenschaften
Philipps-Universität Marburg, 35032 Marburg, Germany
[2]Istituto Nazionale di Fisica della Materia (INFM), Italy
[3]Dipartimento di Fisica, Università di Modena, 41100 Modena, Italy

## INTRODUCTION

The progress in the generation of ultrashort laser pulses, together with the development of spectroscopies on this time-scale, has led to a series of experiments which give new insight into the microscopic carrier dynamics in semiconductors. In particular, the energy relaxation of photoexcited carriers has been widely investigated. At the same time, recent progress in the fabrication and characterization of semiconductor heterostructures and superlattices allows a detailed study of a new class of phenomena induced by an applied electric field, such as Bloch oscillations[1,2]. Both classes of phenomena typically occur on a pico- or femtosecond time-scale, where the coupling between coherent and incoherent phenomena is known to play a dominant role[3]. Therefore, an adequate theoretical model of the ultrafast dynamics on this time-scale must account for both coherent and incoherent effects on the same kinetic level.

In this contribution we study the energy relaxation of photoexcited carriers and their vertical transport. The energy-relaxation analysis shows the existence of different relaxation times corresponding to the different inter- and intraminiband scattering mechanisms. On the other hand, the vertical transport analysis clearly shows the dominant role of carrier-polar optical phonon interaction in determining the carrier dynamics; for superlattice structures characterized by a miniband width smaller than the optical-phonon energy and for laser excitations close to the band gap, the carrier dynamics exhibits Bloch oscillations. On the contrary, in agreement with experimental results[4], for laser excitations far from the band gap, the carrier-phonon scattering results in a strong damping of such charge oscillations. In addition to this scattering-induced suppression of the Bloch oscillations, our theoretical approach allows a systematic investigation of coherent phenomena such as excitonic and field-induced effects.

## THEORETICAL APPROACH AND NUMERICAL PROCEDURE

The starting point of the proposed approach are the well known semiconductor Bloch equations (SBE)[5]. Here, these equations are generalized to a three-dimensional multiband description of anisotropic semiconductors. In addition to including the Coulomb interaction in the time-dependent Hartree-Fock approximation, they also include the intraband drift due to an applied electric field[2]; and on the same kinetic level also incoherent processes, such as carrier-phonon scattering[6].

More specifically, by denoting with $f_{i,\mathbf{k}}^e = \langle c_{i,\mathbf{k}}^\dagger c_{i,\mathbf{k}} \rangle$ $(f_{j,\mathbf{k}}^h = \langle d_{j,\mathbf{k}}^\dagger d_{j,\mathbf{k}} \rangle)$ the electron (hole) distribution function in band $i$ ($j$) and with $p_{ji,\mathbf{k}} = \langle d_{j,-\mathbf{k}} c_{i,\mathbf{k}} \rangle$ the corresponding interband polarization, our generalized SBE are given by

$$
\left(\frac{\partial}{\partial t} + \frac{e\mathbf{F}}{\hbar}\cdot\nabla_\mathbf{k}\right) f_{i,\mathbf{k}}^e = \frac{1}{i\hbar}\sum_{j'}(\mathcal{U}_{ij',\mathbf{k}}p_{j'i,\mathbf{k}}^* - \mathcal{U}_{ij',\mathbf{k}}^* p_{j'i,\mathbf{k}}) + \left.\frac{\partial f_{i,\mathbf{k}}^e}{\partial t}\right|_{inco}
$$

$$
\left(\frac{\partial}{\partial t} + \frac{e\mathbf{F}}{\hbar}\cdot\nabla_\mathbf{k}\right) f_{j,-\mathbf{k}}^h = \frac{1}{i\hbar}\sum_{i'}(\mathcal{U}_{i'j,\mathbf{k}}p_{ji',\mathbf{k}}^* - \mathcal{U}_{i'j,\mathbf{k}}^* p_{ji',\mathbf{k}}) + \left.\frac{\partial f_{j,-\mathbf{k}}^h}{\partial t}\right|_{inco}
$$

$$
\left(\frac{\partial}{\partial t} + \frac{e\mathbf{F}}{\hbar}\cdot\nabla_\mathbf{k}\right) p_{ji,\mathbf{k}} = \frac{1}{i\hbar}\sum_{i'j'}(\mathcal{E}_{ii',\mathbf{k}}^e\delta_{jj'} + \mathcal{E}_{jj',-\mathbf{k}}^h\delta_{ii'})p_{j'i',\mathbf{k}}
$$

$$
+ \frac{1}{i\hbar}\mathcal{U}_{ij,\mathbf{k}}(1 - f_{i,\mathbf{k}}^e - f_{j,-\mathbf{k}}^h) + \left.\frac{\partial p_{ji,\mathbf{k}}}{\partial t}\right|_{inco}, \tag{1}
$$

where

$$
\mathcal{U}_{ij,\mathbf{k}} = \mu_{ij,\mathbf{k}}E(t) - \sum_{i'j',\mathbf{k}'} V\left(\begin{smallmatrix}\mathbf{k};&-\mathbf{k}';&-\mathbf{k};&\mathbf{k}'\\i;&j';&j;&i'\end{smallmatrix}\right)p_{j'i',\mathbf{k}'} \tag{2}
$$

and

$$
\mathcal{E}_{ii',\mathbf{k}}^e = \epsilon_{i,\mathbf{k}}^e\delta_{ii'} - \sum_{i'',\mathbf{k}''} V\left(\begin{smallmatrix}\mathbf{k};&\mathbf{k}'';&\mathbf{k};&\mathbf{k}''\\i;&i'';&i';&i''\end{smallmatrix}\right)f_{i'',\mathbf{k}''}^e
$$

$$
\mathcal{E}_{jj',\mathbf{k}}^h = \epsilon_{j,\mathbf{k}}^h\delta_{jj'} - \sum_{j'',\mathbf{k}''} V\left(\begin{smallmatrix}\mathbf{k};&\mathbf{k}'';&\mathbf{k};&\mathbf{k}''\\j';&j'';&j;&j''\end{smallmatrix}\right)f_{j'',\mathbf{k}''}^h \tag{3}
$$

are, respectively, the renormalized Rabi-frequencies and electron and hole energies[5]. Here, $V$ is the Coulomb matrix element, $\mu$ is the optical dipole matrix element between conduction and valence-band states, $\epsilon^{e,h}$ are the single-particle energies, and $\mathbf{F}$ is the applied electric field.

The accelerating (intraband) part of the electric field has been treated as described in Ref. 3. The last terms on the right hand sides of Eqs. (1) refer to incoherent contributions, i.e. various scattering processes. As for the case of the bulk[3], within the Markov approximation they have the structure of the usual Boltzmann collision term.

The set of SBE (1) is numerically solved using a generalized Monte Carlo approach recently proposed and applied to the case of bulk semiconductors[3]: the coherent contributions in Eqs. (1) are evaluated by means of a direct numerical integration while the incoherent ones are "sampled" by means of a conventional Monte Carlo simulation in the three-dimensional k-space.

The above theoretical scheme has been applied to a semiconductor superlattice. As described in Ref. 4, the dispersion and the corresponding wavefunctions along the growth direction ($z$) are computed within the well-known Kronig-Penney model, while for the in-plane direction an effective-mass model has been used. Starting from these three-dimensional wavefunctions, the carrier-phonon and the Coulomb matrix elements are numerically computed. The resulting matrix elements are functions of the various miniband indices and depend separately on $k_z$ and $k_{xy}$, thus reflecting the anisotropy of the superlattice structure.

## RESULTS

The above theoretical approach has been applied to the analysis of Terahertz (THz) radiation generated by the ultrafast dynamics of photoexcited carriers in the presence of a static electric field. We have investigated a superlattice structure with 111ÅGaAs wells and 17ÅAl$_{0.3}$Ga$_{0.7}$As barriers. In our simulations low temperature (T = 10$K$) and excitations by Gaussian laser pulses of 100$fs$ have been considered. The strength of the applied electric field is assumed to be 4$kV/cm$. For such a structure there has been recent experimental evidence for THz-emission from Bloch oscillations associated with resonant excitation of the second miniband[4]. However, from these experiments it remains unclear which carriers contribute to the measured transients.

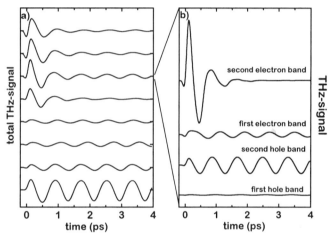

**Figure 1.** (a) Total THz-signals for eight different spectral positions of the exciting laser pulse: 1540, 1560, 1580, 1600, 1620, 1640, 1660, 1680$meV$ (from bottom to top). (b) Individual THz-signal of the electrons and holes in the different bands for central spectral position of the laser pulse of 1640$meV$.

To clarify this point, we have performed a series of "simulative experiments" corresponding to 8 different spectral positions of the laser excitation; from these simulations the THz-signal (proportional to the time-derivative of the current) has been directly computed. The general behavior of the magnitude of this signal (see Fig. 1), the oscillations and their damping are in very good agreement with the experimentally measured signals[4]. For resonant excitation of the lowest exciton connected to the optical transition between the first electron and hole minibands ($\hbar\omega_L = 1540meV$), we find a strong THz-signal. In this case, the signal is undamped because the bandwidth of the first miniband (13$meV$) is smaller than the LO-phonon energy (36.4$meV$) and, therefore, no LO-phonon emission is possible for these excitation conditions[6]. The amplitude of the signal decreases when the excitation energy is increased; there are also some small changes in the phase of these oscillations, which are induced by the electron-LO phonon scattering. When the laser energy comes into resonance with the transitions between the second electron and hole minibands ($\hbar\omega_L \approx 1625 meV$) the amplitude of the THz-signal increases again. The corresponding THz-transients show an initial part, which is strongly damped and some oscillations for higher times that are much less damped. For a better understanding of these results, we show in Fig. 1b the individual THz-signals, originating from the 2 electron and 2 heavy-hole minibands for the excitation with $\hbar\omega = 1640 meV$. The BO performed by the electrons within the second miniband are strongly damped. This damping is due to intra- and interminiband LO-phonon scattering processes. Since the width of this miniband ($45 meV$) is somewhat larger than the LO-phonon energy, also intraminiband scattering is possible, when the electrons are accelerated into the high-energy region of the miniband. The THz-signal originating from electrons within the first miniband shows an oscillatory behavior, with a small amplitude and a phase, which is determined by the time the electrons need to relax down to the bottom of the band. On the contrary, the holes in both minibands exhibit undamped BO, since the minibands are so narrow in energy that for these excitation conditions no LO-phonon emission can occur; the amplitude of the signal due to the holes within the first miniband is quite small, while the corresponding signal due to the holes of the second miniband is a little larger than the amplitude of the signal originating from the electrons within the first miniband. The phase-shift between the signals from the holes and the electrons within the first miniband is determined by the typical energy-relaxation times.

## CONCLUSIONS

We have presented a microscopic approach for the analysis of optical and transport properties of anisotropic semiconductors. Applying this theory to semiconductor superlattices, we have analyzed the THz radiation for different excitation conditions and have shown the different contributions to the measured signal. Our results clearly show, that at early times the THz radiation is mainly determined by the electrons within the second miniband, at later times the observed signal is due both to the electrons within the first miniband and to the heavy holes.

## REFERENCES

1. C. Waschke, H.G. Roskos, R. Schwedler, K. Leo, H. Kurz, and K. Köhler, Coherent submillimeter-wave emission from Bloch oscillations in a semiconductor superlattice, Phys. Rev. Lett. 70: 3319 (1993).
2. T. Meier, G. von Plessen, P. Thomas, and S.W. Koch, Coherent electric-field effects in semiconductors, Phys. Rev. Lett. 73: 902 (1994).
3. F. Rossi, S. Haas, and T. Kuhn, Ultrafast relaxation of photoexcited carriers: The role of coherence in the generation process Phys. Rev. Lett. 72: 152 (1994).
4. H.G. Roskos, C. Waschke, R. Schwedler, P. Leisching, Y. Dhaibi, H. Kurz, and K. Köhler, Bloch oscillations in GaAs/AlGaAs superlattices after excitation well above the bandgap, Superlattices and Microstructures 15: 281 (1994).
5. For a textbook discussion see, H. Haug and S.W. Koch, "Quantum Theory of the Optical and Electronic Properties of Semiconductors", 3rd ed., World Scientific, Singapore (1994).
6. F. Rossi, T. Meier, P. Thomas, S.W. Koch, P.E. Selbmann, and E. Molinari, Ultrafast carrier relaxation and vertical-transport phenomena in semiconductor superlattices: A Monte Carlo analysis, Phys. Rev. B 51: 16943 (1995).

# THIRD HARMONIC GENERATION IN A GaAs/AlGaAs SUPERLATTICE IN THE BLOCH OSCILLATOR REGIME

M.C. Wanke,[1] A.G. Markelz,[1] K. Unterrainer,[1] S.J. Allen,[1] R. Bhatt[2]

[1]Center for Free Electron Laser Studies, UCSB, Santa Barbara, CA 93106
[2]Bellcore, Red Bank, NJ

## ABSTRACT

A GaAs/AlGaAs superlattice driven at 600 GHz exhibits strong, power dependent third harmonic generation. Conversion efficiency up to 0.1% was observed. The power dependence agrees semi-quantitatively with a simple model describing Bloch oscillations driven by a strong THz electric field.

## INTRODUCTION

While the conventional view of Bloch oscillation involves transport in superlattices in the presence of strong DC electric fields, its experimental realization is complicated by the formation of electric field inhomogenieties (domains).[1] Although difficult to realize, strong THz fields can drive Bloch oscillation without domain formation. This regime is characterized by self-induced transparency, negative absolute resistance, and non-monotonic harmonic generation.[2,3,4] Here we focus on the THz field dependence of the harmonic generation, which is predicted to display unique features directly related to Bloch oscillation.

## EXPERIMENTAL METHOD

The measurements were performed on two identical superlattice (SL) samples grown by MOCVD on a GaAs semi-insulating substrate. The SL consists of 600 periods of 80Å GaAs wells separated by 20Å $Al_{0.3}Ga_{0.7}As$ barriers with the entire structure n-doped near $5 \times 10^{15}/cm^3$. Aluminum was evaporated on the top of the SL. In one sample the aluminum covered the entire 7x8mm sample, whereas in the other, the aluminum covered a 3mm diameter circle with the uncovered SL etched away. The samples were then mounted on an ultra-pure silicon prism ($10^4$ ohm-cm).

The samples, held at 77K, were illuminated through the prism and substrate with 2.5 µs pulses from the UCSB Free Electron Laser (FEL) tuned to 600 GHz and with peak powers reaching 5 KW (See Fig. 1a). The silicon prism allows us to couple p-polarized radiation to the growth direction of the sample with minimum loss; the aluminum surface serves to confine the field in the growth direction.

The power incident on the sample was controlled by a set of calibrated, crossed polarizers between the FEL and the sample/prism. The output power of the third harmonic was measured with a Ge bolometer using a 58cm-1 high pass filter to block the fundamental and any second harmonic generated in the sample. The measured output was shown to be third harmonic by use of a Fabry-Perot spectrometer.

*Hot Carriers in Semiconductors*
Edited by K. Hess *et al.*, Plenum Press, New York, 1996

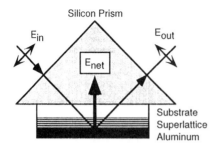
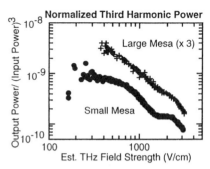

**Figure 1.** (a) The sample is aluminized and affixed to the face of a Si prism. (b) Third harmonic power normalized by the cube of the incident power as a function of estimated incident field strength displays the THz field dependence of the third order nonlinearity.

## RESULTS

The experimental results for the two samples are shown in Fig. 1b. The output power is divided by the cube of the input power to emphasize the departure from a cubic power law. In the smaller sample the results approach zero slope for low field strengths, which would indicate a purely cubic dependence. At higher powers the dependence dramatically deviates from cubic, dropping nearly an order of magnitude and developing well defined features. The larger sample shows similar features which are not as well defined.

## MODEL AND DISCUSSION

To model our results, we follow Pavlovitch and Epshtein[2] and use an analytic expression for the current generated in a superlattice in the presence of both a strong DC and AC field. For zero DC field, only odd harmonics are generated which have very strong field dependencies. Fig. 2 shows the calculated fundamental and third harmonic currents for the SL described above driven at 600 GHz.

The peak in the fundamental current, in the regime where $\omega\tau > 1$, occurs when $\omega_B/\omega \approx 1$, where $\omega_B$ is the AC Bloch frequency equal to $eE^{AC}d/\hbar$; e is the electron charge, $E^{AC}$ is the drive field strength, d is the superlattice period, and $\omega$ is the drive frequency. In the absence of scattering, the nth order current, where the fundamental is n=1, is proportional to $J_0(eE^{AC}d/\hbar\omega)J_n(eE^{AC}d/\hbar\omega)$.

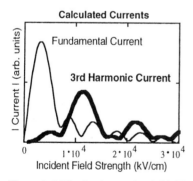
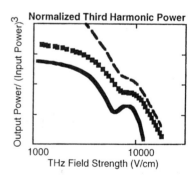

**Figure 2.** (a)The fundamental and third harmonic currents induced in the superlattice by Bloch oscillation in the presence of an 600 GHz AC electric field (including scattering). (b) Normalizing the radiated power by the cube of the incident power displays the electric field dependence of the third order susceptibility. The electric field variation over the gaussian beam profile averages the electric field dependence over the sample size. Two curves (+,- -) take into account a gaussian beam profile with a HWHM of 1mm with a mesa size of 3mm and 6 mm respectively. The curves are shifted for clarity.

Since the beam is gaussian, the incident electric field is not uniform. To compare with the experimental results, the harmonic current was averaged over the sample assuming a gaussian electric field variation. Fig. 2b shows the calculated output power normalized by the input power cubed, where the output power is assumed to be proportional to the current squared. The solid curve represents a uniform field across the sample, whereas the + and - - curves represent mesas 1.5 and 3 times larger than the spot size respectively. It is clear that the non-uniform field smears the feature but the overall reduction in third harmonic power at high fields remains the same.

The predicted normalized harmonic power, taking the non-uniform field in to account, strongly resembles the experimental results. As expected the smaller mesa has much sharper structure. At low fields the curves become horizontal, indicating a cubic power dependence. At the highest field a power conversion effeciency of 0.1% was observed.

Although the results strongly suggest that we are observing the onset of the non-linear response of Bloch oscillating carriers, more work needs to be done. The estimated THz electric field at which features appear is a factor of 4 too low; a possible explanation of this might be field enhancement from plasma oscillations. A sample replacing the SL with equivalently doped GaAs would confirm that the features seen are due to the presence of the SL. Finally, larger field strengths should allow us to explore more of the rich THz electric field dependence predicted in Fig 2a. This work is in progress.

## CONCLUSION

A striking power dependence of third harmonic generation in a semiconductor superlattice in the mini-band limit has been observed. The result agrees in a semiquantitative way with predictions of Bloch oscillation in the presence of strong AC fields.

## ACKNOWLEDGMENTS

The UCSB Center for Free-electron Laser Studies is supported by the Office of Naval Research. This project is also supported by the Army Research Office and the National Science Foundation.

## REFERENCES

1. A. Sibille, Observation of Esaki-Tsu negative differential velocity in GaAs/AlAs superlattices, *Phys. Rev. Lett.* 64, 52 (1990).
2. V.V. Pavlovitch and E.M. Epshtein, Conductivity of a superlattice semiconductor in strong electric fields, *Sov. Phys. Semicond.* 10:1196 (1976).
3. A.A. Ignatov et. al., Self-induced transparency in semiconductors with superlattices, *Sov. Phys. Sol. State.* 17:2216 (1975).
4. M. Holthaus, The quantum theory of an ideal superlattice responding to far-infrared laser radiation, *Z. Phys. B. Condensed Matter* 89:251 (1992).

# ULTRAFAST DYNAMICS OF ELECTRONIC WAVEPACKETS IN QUANTUM WELLS

Gary D. Sanders,[1,2] Alex V. Kuznetsov[2] and Christopher J. Stanton[1,2]

[1]Mikroelektronik Centret
Danmarks Tekniske Universitet
DK-2800 Lyngby, Denmark

[2]Department of Physics
University of Florida
Gainesville, Florida 32611

Ultrafast optical excitation of semiconductor quantum wells (QW) biased by a dc electric field has been the subject of much theoretical[1,2] and experimental[3,4] work in recent years. Such excitation can lead to generation of *ultrashort electric current pulses* through a mechanism known as *virtual photoconductivity*[5,6]: since the transport photocurrent is suppressed by the barriers in a QW system, the only mechanism of charge transport is believed to be the displacement current that is caused by generation of polarized electron-hole pairs[3] and is therefore present only during the optical pulse.

This behavior is qualitatively different from the photocurrent dynamics in bulk systems where the current does not vanish after the pulse is over. By gradually increasing the width of the wells, one can expect a transition from fast displacement photocurrent that adiabatically follows the excitation, to a slower bulk-like transport photocurrent. In this paper we investigate this dimensional crossover within the framework of a density matrix approach. Our aim is to establish quantitative criteria that can predict which of the possible types of photocurrent response will be present for a given system and excitation conditions.

We describe the carrier dynamics in a quantum well in terms of the time-dependent *density matrix*:

$$N_{nm,\vec{k}}^{\alpha\beta}(t) \equiv \langle a_{n\vec{k}}^{\alpha\dagger}(t) \, a_{m\vec{k}}^{\beta}(t) \rangle \tag{1}$$

where $a_{n,\vec{k}}^{\alpha\dagger}(t)$ is a Heisenberg operator that creates a carrier in a subband $n$ of the band $\alpha$ (we include conduction, heavy and light hole bands) with the wavevector $\vec{k}$ in the plane of the well. The wavefunctions for different subbands, $\varphi_{n,\vec{k}}^{\alpha}(\vec{r})$, are determined by numerically solving the Schroedinger equation for an infinite well in the external dc electric field.

The optical excitation is described by a time-dependent electric field of the laser pulse, $E^{opt}(t)$. The equations of motion for the density matrix (1) are obtained as a generalization of optical Bloch equations:[2,7]

$$\frac{\partial N^{\alpha\beta}_{nm,\vec{k}}}{\partial t} = i\left(\varepsilon^\alpha_n(k) - \varepsilon^\beta_m(k)\right) N^{\alpha\beta}_{nm,\vec{k}} + iE^{opt}(t)\sum_{\gamma j}\left(N^{\alpha\gamma}_{nj,\vec{k}}d^{\gamma\beta}_{jm} - d^{\alpha\gamma}_{nj}N^{\gamma\beta}_{jm,\vec{k}}\right) \tag{2}$$

where $\varepsilon^\alpha_n(k)$ is the energy dispersion of the subbands, and

$$d^{\alpha\beta}_{nm,\vec{k}} = \int d\vec{r}\,\varphi^\alpha_{n\vec{k}}(\vec{r})\,ex\,\varphi^\beta_{m\vec{k}}(\vec{r}) \tag{3}$$

are the dipole matrix elements between different subbands. The generalized Bloch equations (2) are solved numerically, and the density matrix is used to determine the time dependence of the observables such as charge density and total dielectric polarization of the well:

$$D(t) = \sum_{\alpha\beta knm} N^{\alpha\beta}_{nm,\vec{k}}(t) \cdot d^{\alpha\beta}_{nm,\vec{k}}\ . \tag{4}$$

The time derivative of the polarization (4) gives the photocurrent, and its second time derivative is proportional to the radiated signal in terahertz radiation experiments.[8–10] The effects of relaxation processes are modeled by adding a phenomenogical decay term to eq. (2) and the formalism is described in a forthcoming paper.[11]

Depending on the well width and the excitation pulse duration, we find three distinct regimes for the photocurrent dynamics. The first (quantum confined) regime corresponds to the conventional "polarized pair" displacement current[3,4] that adiabatically follows the excitation. It is achieved for narrow wells and/or long pulses such that the spectral width of the excitation is smaller than intersubband separation. In this case, each subband is excited independently. For shorter pulses/wider wells the excitation spectrum starts to cover several subbands which leads to *quantum beats* between the subbands.[2,3] The beats are seen in the time domain as oscillatory contribution to the polarization (4) which no longer follows the pulse adiabatically. Finally, for very wide wells or very short pulses the excitation spectrum begins to cover a large number of subbands. We demonstrate[11] that in this regime the carriers are created as *wavepackets* formed by a coherent superposition of many subbands. The characteristic size of these wavepackets (the *coherence length*)[9,11] is much smaller than the well width in this regime. Consequently, the carriers move inside the well as an ensemble of *classical* particles and create a bulk-like transport photocurrent. The different possible regimes are summarized in Fig. 1.

Our findings indicate that the dimensional crossover from a quantum-confined 2D behavior to the bulk-like 3D photocurrent response can be achieved not only by increasing the well width, but also by using shorter pulses. Consequently, it is possible to have a 100 Å quantum well behave like a 3D system if the excitation pulses are short (such as 10 fs pulses).[12] Therefore, using shorter pulses does not always make the current transient shorter. For short pulses or wide wells one has to include the *intraband coherence* between different subbands in the description, which is physically equivalent to treating carriers as wavepackets formed by superposition of different eigenstates. This effect can drastically change the dynamics of photocurrent and the terahertz radiation from quantum well systems.

**Acknowledgments**

This work was supported by U.S. Office of Naval Research through Grant N00091–JJ-1956 and by the NSF grant DMR 8957382. CJS and GDS wish to thank the Danish Research Academy and Mikroelektronik Centret at DTU for support during part of this work.

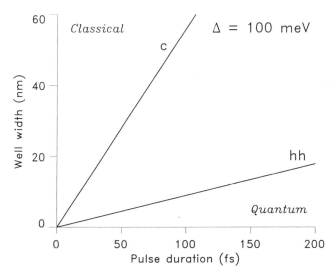

**Figure 1.** "Phase diagram" for different photocurrent regimes for excitation 100 meV above the band gap. The upper left corner (wide wells/short pulses) corresponds to a classical regime, the lower right corner (narrow wells/long pulses) is the quantum-confined regime. The intermediate quantum beats regime corresponds to the vicinity of the lines. Note that the criteria are different for electrons and holes — for heavy holes (hh) the quantum confined regime is very difficult to achieve.

## REFERENCES

[1] S. Schmitt-Rink, D.S. Chemla, and D.A.B. Miller, Adv. Phys. **38**, 89 (1989)

[2] T. Kuhn, E. Binder, F. Rossi, A. Lohner, K. Rick, P. Leisching, A. Leitenstorfer, T. Elsaesser, and W. Stolz, in: *Coherent Optical Interactions in Semiconductors*, ed. R.T. Phillips (Plenum, New York, 1994), p. 33.

[3] P.C.M. Planken, M.C. Nuss, W.H. Knox, D.A.B. Miller, and K.W. Goossen, Appl. Phys. Lett. **61**, 2009 (1992); P.C.M. Planken, M.C. Nuss, I. Brener, K.W. Goossen, M.S.C. Luo, S.L. Chuang, and L. Pfeiffer, Phys. Rev. Lett. **69**, 3800 (1992).

[4] M.S.C. Luo, S.L. Chuang, P.C.M. Planken, I. Brener, H.G. Roskos, and M.C. Nuss, IEEE J. Quantum Electron. **QE-30**, 1478 (1994); M.C. Nuss, P.C.M. Planken, I. Brener, H. Roskos, M.S. Luo, and S.L. Chuang, Appl. Phys. **B 58**, 2216 (1994).

[5] Y. Yafet and E. Yablonovich, Phys. Rev. **B 43**, 12480 (1991); E. Yablonovich, J.P. Heritage, D.E. Aspnes, and Y. Yafet, Phys. Rev. Lett. **63**, 976 (1989)

[6] D.S. Chemla, D.A.B. Miller, and S. Schmitt-Rink, Phys. Rev. Letters **59**, 1018 (1987); M. Yamanishi, *ibid* 1014 (1987).

[7] M. Lindberg, R. Binder, and S. W. Koch, Phys. Rev. **A 45**, 1865 (1992); Y.Z. Hu, R. Binder, and S.W. Koch, Phys. Rev. **B 47**, 15679 (1993).

[8] S.L. Chuang, S. Schmitt-Rink, B.I. Greene, P.N. Saeta, and A.F.J. Levi, Phys. Rev. Lett. **68**, 102 (1992).

[9] A.V. Kuznetsov and C.J. Stanton, Phys. Rev. **B 48**, 10828 (1993); B.B. Hu, A.S. Weling, D.H. Auston, A.V. Kuznetsov, and C.J. Stanton, Phys. Rev. **B 49**, 2234 (1994).

[10] X.-C. Zhang, B.B. Hu, J.T. Darrow, and D.H. Auston, Appl. Phys. Lett. **56**, 1011 (1990); B.B. Hu. X.-C. Zhang, and D.H. Auston, Phys. Rev. Lett. **67**, 2709 (1991).

[11] G.D. Sanders, A.V. Kuznetsov, and C.J. Stanton, to appear in Phys. Rev. **B**

[12] B.B. Hu, E.A. de Souza, W.H. Knox, J.E. Cunningham, M.C. Nuss, A.V. Kuznetsov, and S.L. Chuang, Phys. Rev. Letters **74**, 1689 (1995).

# PROSPECTS FOR MANIPULATING BLOCH OSCILLATIONS IN BIASED SUPERLATTICES THROUGH FEMTOSECOND PULSE SHAPING

A. M. Weiner, Y. Liu

Purdue University
School of Electrical Engineering
West Lafayette, IN 47907-1285

## INTRODUCTION

Coherent subpicosecond charge oscillations have been observed in several recent experiments performed on GaAs/GaAlAs quantum well [1] and superlattice [2] samples. In such experiments a single femtosecond laser pulse photoexcites from the valence band to a superposition state consisting of two or more eigenfunctions in the conduction band. The time-varying interference between these wave functions gives rise to an ultrafast wave packet motion. This wave packet motion manifests itself as periodic tunneling between the two wells in a coupled quantum well system or as a Bloch oscillation in a biased superlattice.

In addition, experiments have also been performed using specially shaped femtosecond waveforms (either pulse pairs or terahertz-repetition-rate optical pulse trains) to manipulate charge oscillations in coupled quantum wells [3]. These studies provide valuable insight into the role of the optical phase in determining the charge oscillation dynamics. However, the use of specially shaped pulses to control Bloch oscillation dynamics in superlattices has not been reported. In this paper we discuss the possibility of manipulating Bloch oscillations through femtosecond pulse shaping. Our analysis indicates that substantial control over Bloch oscillations may be possible by appropriately engineering the femtosecond photoexcitation pulse.

## THEORY

Our analysis is based on the simple, well-known Stark ladder wave functions neglecting electron-hole interactions and dephasing. In this simple approximation, the periodic electron wave packet can be described by a superposition of the Stark-ladder electronic eigenstates [4]:

$$\Psi(z,t) = \sum_p C_p(t) e^{-i2\pi pt/\tau_b} \chi_p^e(z) \tag{1}$$

where $\tau_b$ is the Bloch oscillation period, $\chi_p^e(z)$ is the conduction-band Stark-ladder wave function centered at the $p^{th}$ well with respect to the photoexcited hole [4], and from first-order perturbation theory,

*Hot Carriers in Semiconductors*
Edited by K. Hess *et al.*, Plenum Press, New York, 1996

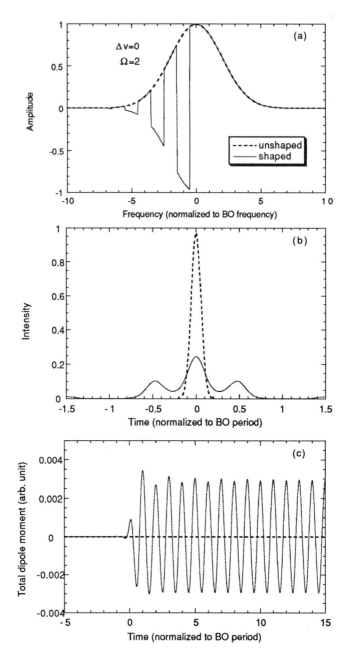

**Figure 1.** (a) Spectrum of input optical pulses; (b) Pulse shape of input optical pulses; (c) Calculated Bloch oscillating dipole moment; for shaped (solid lines) and unshaped (dashed lines) pulses.

$$C_p(t) \propto J_{-p}(\theta) \int^t dt' \, e^{i2\pi pt'/\tau_b} e_L(t') \tag{2}$$

where $J_p(\theta)$ is a Bessel function of the first kind, $\theta$ is the delocalization parameter [4] and $e_L(t)$ is the exciting optical field.

As pointed out in [4], photoexcitation using either a very short pulse or a pulse with zero frequency detuning results in this simple model in an electron wave packet which executes a breathing mode. Since this motion results in zero macroscopic dipole moment, it would not be evident in terahertz radiation or electro-optic sampling experiments. We find that by appropriately reshaping the input pulse, one should be able to generate a wave packet which really should oscillate in space and time. Fig. 1(a) and Fig. 1(b) show our input laser pulses in the frequency and time domains respectively. The dashed and solid lines are respectively our unshaped Gaussian pulses (with zero frequency detunings at center well absorption $\Delta v=0$ with a laser spectrum bandwidth $\Omega=2$, see ref. [4] for definitions) and shaped optical pulses. We have normalized time or frequency in units of a single-particle Bloch period or Bloch frequency. From Eq. (1) and (2), we calculated time-dependent wave packet motion and macroscopic dipole moments based on these two pulses. The results of our calculation for the time-dependent dipole moment are shown in Fig. 1(c). With unshaped pulses, there is zero macroscopic dipole moment because the electron wave packet moves in the breathing mode. Appropriately shaped pulses result in a nonzero oscillating dipole moment which builds up rapidly in one oscillation period and then slightly overshoots before relaxing back to equilibrium. This time-varying dipole is expected to be detectable experimentally.

## DISCUSSION

We note that the shaped input pulses are fully compatible with proven programmable femtosecond pulse shaping technology [5]. We also note that although dephasing is not yet explicitly taken into account, the calculated shaped input pulses have durations on the order of one Bloch oscillation period. With the initial rapid build up of the Bloch oscillation and overshoot, the shaped input pulses should prove relatively robust against dephasing.

The above calculation is based on a single-particle picture [4] without considering excitonic effects, which involve the Coulomb interaction between electrons and holes. In real experiments, excitonic effects such as nonuniform energy spacings and nonsymmetric absorption strengths will play an important role [4]. We have begun to extend our calculations to include the excitonic effect. Although the excitonic wave packets are more complicated due to the complexity of the excitonic energy levels and wave functions, nevertheless coherent control of wave packet motion through femtosecond pulse shaping should still be feasible.

## ACKNOWLEDGMENT

This work was supported by the National Science Foundation under grant No. 9404677-PHY.

## REFERENCES

1. K. Leo, et al., Phys. Rev. Lett. 66, 201 (1991); H. G. Roskos, et al., Phys. Rev. Lett. 68, 2216 (1992).
2. J. Feldman, et al., Phys. Rev. B46, 7252 (1992); C. Waschke, et al., Phys. Rev. Lett. 63, 2213 (1993).

3. P. C. M. Planken, et al., Phys. Rev. B48, 4903 (1993); I. Brener, et al., Appl. Phys. Lett. 63, 2213 (1993).
4. M. Dignam, J. E. Sipe, J. Shah, Phys. Rev. B49, 10502 (1994).
5. A. M. Weiner, D. E. Leaird, J. S. Patel, J. R. Wullert, IEEE J. Quantum Electron. 28, 908-920 (1992).

# MASSIVE INTERMINIBAND TRANSITIONS TRIGGERED BY STRONG STATIC ELECTRIC FIELDS IN SUPERLATTICES

J.P. Reynolds and Marshall Luban

Ames Laboratory and Department of Physics and Astronomy
Iowa State University, Ames, Iowa 50011

## INTRODUCTION

The recent experimental observations of Bloch oscillations and the emission of electromagnetic radiation[1] in the terahertz regime in semiconductor superlattices subject to static uniform electric fields, suggest that these systems may be developed as useful THz generators. The Bloch period is given by $\tau_B = h/(eFa)$, where $e$ and $F$ denote the magnitude of the electron charge and the electric field, $h$ is Planck's constant, and $a$ is the spatial period of the superlattice. Modeling these systems in terms of independent electrons in an idealized one-dimensional superlattice potential in the effective mass approximation, and using high-accuracy numerical methods to solve the time-dependent Schrödinger equation, the electron wave function has been found[2-4] to exhibit long-lived Bloch oscillations for electric fields in the range used in experiment. We here present results for this model, obtained using the methods of Ref. 2-4, for the regime where the electrons are subject to very strong static electric fields, so that they exhibit characteristics of acceleration as well as some residual but faint signs of Bloch oscillations. Some preliminary results for this regime have been given in Sec. IIID of Ref. 3, while a very detailed analysis is presented in Ref. 4.

## RESULTS

In Fig. 1 we display results for the system GaAs/$Al_{0.2}Ga_{0.8}$As with wells of width $w = 200$Å and barriers of width $b = 5$Å so that $a = w+b = 205$Å, and for the field strength $F = 3.9$ kV/cm, so that $eFa = 8$ meV and $\tau_B = 0.52$ psec. For this system the potential barrier height is $V_0 = 162$ meV. The initial wave function selected is normalized to unity and it is localized within five successive cells of the superlattice. More specifically it is a linear combination of five Wannier functions constructed from Bloch eigenstates of the lowest field-free miniband, and which are associated with a small interval of wavevectors within the Brillouin zone. The raster plot shown in Fig. 1 provides the electron probability density as a function of position and time, the latter measured in units of the Bloch period. The most striking feature of the plot is the interwoven mesh pattern of wispy parabolic line segments, with intersections that occur in the immediate vicinity of integer multiples of $\tau_B/2$. In view of the very thin AlGaAs layers of this superlattice, the appearance of parabolic segments that propagate towards large negative values of $z$ are to be expected, reminiscent of the behavior of a free quantum (or classical) particle subject to a static uniform electric field. The perhaps surprising feature is the recurrence of an additional such "parabola" every $\tau_B$, and second, the occurrence of a corresponding family of recurring "parabolas" describing *deceleration* in the direction of the field.

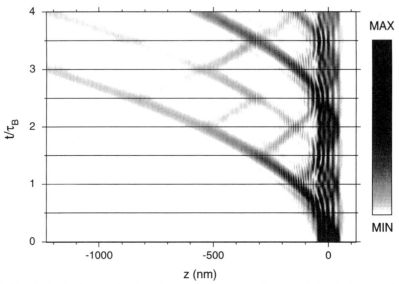

**Figure 1.** Probability density as a function of $z$ and $t/\tau_B$ for the superlattice and electric field described in the text. Note the interwoven mesh of parabolas describing acceleration and deceleration.

However, all of these features can readily be explained[4] upon expanding the computed time-dependent wave function in terms of the basis of *field-free* Bloch eigenstates $\phi_l(k,z)$ for the given superlattice, where $k$ and $l$ denote the wave vector and the miniband index, respectively. The resulting time-dependent expansion coefficients can be identified as the rigorous solutions of the full set of Houston equations.[5] Those equations incorporate an intraminiband process which may be described as a displacement of the wave packet with time, in an extended zone scheme, that proceeds at a *fixed rate* set by the electric field. They also include coupling terms that trigger interminiband transitions, allowing the exchange of probability between minibands at the same value of $k$. The interminiband coupling terms are linear in $F$ and are roughly inversely proportional to the energy difference between the minibands for given wavevector, so that in practice the transitions essentially occur only between adjacent minibands. For the superlattice under discussion, the minibands are relatively wide and are separated by small gaps (see Fig. 2), so that *the interminiband coupling terms play a significant role in the extended zone only in the vicinity of integral multiples of $\pi/a$*. Hence, for initial wave functions which are well localized in the variable $k$, interminiband transitions occur rather abruptly in time, at half-integral and integral multiples of the Bloch period. For these special times we find that a portion of the wave packet undergoes a rapid transition to a higher (or lower) adjacent miniband, whereas for intermediate times the remaining portions propagate within their own minibands, i.e., without appreciable interminiband transitions. The amount of probability exchanged between successive minibands depends on the strength of the electric field, the specific form of the minibands, and on the details of the initial wave function. We find that, in general, transitions to successively higher field-free minibands do not proceed monotonically; transitions from higher to lower minibands occur as well. These results are illustrated schematically in Fig. 2 with the aid of the arrows and filled circles. The transitions to successively higher minibands are associated with acceleration anti-parallel to the field, while successive transitions to lower minibands correspond to deceleration in the field direction. In short, the interwoven parabolic segments seen in Fig. 1 are associated with the recurring yet intermittent interminiband transitions of the electrons.

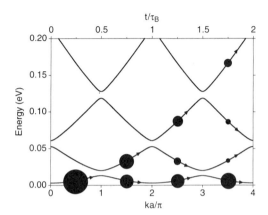

**Figure 2.** Miniband energies for the superlattice considered in the text. The circles and arrows summarize in a schematic manner results obtained when the computed wave function is expanded in terms of the field-free Bloch eigenstates of the superlattice. The radii of the filled circles are proportional to the instantaneous occupancy of each miniband.

If the initial wave packet spans a relatively wide range of values of $k$ the transitions between minibands are spread out over larger time intervals and the features of the mesh of interwoven parabolas can be significantly less distinct. Similarly, for a superlattice with relatively narrow bands and wide gaps (e.g., superlattices with thicker AlGaAs barriers), interminiband transitions occur at roughly the same rate at all times and the effects described in the previous paragraph are less distinct, even for initial wave functions which span a narrow range of $k$-values.

We have also explored in detail[4] the impact of Zener resonances on the dynamics of Wannier-Stark electrons in superlattices. These resonances, occur for a selected set of values of the electric field, as an electron occupying an orbital localized about a particular unit cell of the superlattice becomes degenerate with, and thus makes a transition to, an excited orbital localized about a different unit cell.

## REFERENCES

1. See, for example, P. Leisching, P. Haring Bolivar, W. Beck, Y. Dhaibi, F. Brüggemann, R. Schwedler, H. Kurz, K. Leo, and K. Köhler, Phys. Rev. B50, 14389 (1994) and numerous references listed in that article.
2. A.M. Bouchard and M. Luban, Phys. Rev. B **47**, 6815 (1993).
3. A.M. Bouchard and M. Luban, Phys. Rev. B (in press).
4. J.P. Reynolds and M. Luban (unpublished).
5. W.V. Houston, Phys. Rev. **57**, 184 (1940).

# OSCILLATORY INSTABILITIES AND FIELD DOMAIN FORMATION IN IMPERFECT SUPERLATTICES

E. Schöll, G. Schwarz, M. Patra, F. Prengel, and A. Wacker

Institut für Theoretische Physik, Technische Universität Berlin,
Hardenbergstraße 36, 10623 Berlin, Germany

We theoretically investigate vertical high-field transport in semiconductor superlattices, which exhibit self-generated current oscillations and the formation of stable stationary electric field domains depending on the available carrier density. We demonstrate that this behavior is strongly affected by growth-related imperfections like fluctuations of the doping density, the well and the barrier widths. We propose to use this as a novel noninvasive method to detect growth-related disorder in superlattices.

## INTRODUCTION

We consider a semiconductor superlattice where electric field domains form in the growth direction under high-field conditions if the superlattice is sufficiently doped or optically excited[1-4]. Previous studies have shown that the current-voltage characteristic consists of a sequence of branches (their number being roughly equal to the number of quantum wells), which arise from different locations of the domain boundary. These branches overlap in a certain range of the voltage, leading to multistability and different curves for sweep-up and sweep-down of the voltage[5]. Recently, time-dependent features like transient[6] and persistent[7] oscillations have also been observed and reproduced by different models of structurally "perfect" superlattices[8,9].

The model used here is an extension of the approach presented in Refs.[10,8] for a "perfect" super-lattice consisting of $N$ GaAs quantum wells separated by $N - 1$ AlAs barriers. Here we study imperfections associated with frozen-in fluctuations of the doping and the well and barrier widths in the growth direction only. We denote by $b_i$ the width of the $i^{th}$ barrier, which is located between the $i^{th}$ and $(i - 1)^{st}$ well of widths $l_i$ and $l_{i-1}$, respectively. The wells are n-doped with a doping concentration (per unit volume) $N_D^{(i)}$ in the $i^{th}$ well. We define a "local" lattice constant $d_i := b_i + (l_i + l_{i-1})/2$ to describe the vertical transport across the $i^{th}$ barrier. For simplicity we consider only the two lowest subbands, $k = 1, 2$. The rate of change of the carrier densities $n_k^{(i)}$ (per unit area) in the $k^{th}$ subband of the $i^{th}$ well is given by

$$\dot{n}_1^{(i)} = n_2^{(i)}/\tau_{21} + R_1^{(i)} n_1^{(i-1)} - R_1^{(i+1)} n_1^{(i)} - n_1^{(i)}(X_r^{(i+1)} + X_l^{(i)}) + n_2^{(i+1)} Y_l^{(i+1)} + n_2^{(i-1)} Y_r^{(i)} \quad (1)$$

$$\dot{n}_2^{(i)} = -n_2^{(i)}/\tau_{21} + R_2^{(i)} n_2^{(i-1)} - R_2^{(i+1)} n_2^{(i)} - n_2^{(i)}(Y_l^{(i)} + Y_r^{(i+1)}) + n_1^{(i-1)} X_r^{(i)} + n_1^{(i+1)} X_l^{(i+1)} \quad (2)$$

where $\tau_{21} = 1$ ps is the intersubband relaxation time, $R_k^{(i)}$ is the rate of electrons crossing the $i^{th}$ barrier between equivalent subbands $k$ of two neighbouring wells modelled by miniband conduction[10].

The tunnelling coefficients $X_r^{(i)}$, $X_l^{(i)}$, $Y_r^{(i)}$, and $Y_l^{(i)}$ for transitions between different subbands of neighbouring wells depend on the field $F^{(i)}$; the subscripts $r$ and $l$ denote resonant tunnelling to the right and left, respectively. $X$ stands for transitions from the first to the second subband, and $Y$ for the reverse process. They are calculated from perturbation theory[10], but using local energy levels and barrier widths. $X_r^{(i)}$ and $Y_l^{(i)}$ exhibit a distinct maximum for large electric fields where the first and the second subband of adjacent wells are in resonance. The electric field $F^{(i)}$ can be calculated from Poisson's law $\epsilon(F^{(i+1)} - F^{(i)}) = e(n_1^{(i)} + n_2^{(i)} - l_i N_D^{(i)})$, where $\epsilon$ is the permittivity of GaAs. The fields have to satisfy $\sum_{i=1}^{N+1} d_i F^{(i)} = U$, where $U$ is the voltage applied to the sample. The sample contacts are treated as two additional "virtual" wells denoted by 0 and $N + 1$, for which the boundary conditions $n_i^{(0)} = n_i^{(1)}$ and $n_i^{(N+1)} = n_i^{(N)}$ are assumed.

## SIMULATIONS

For uniform electric fields the current-voltage characteristic following from (1), (2) exhibits a two-peak-structure with a sharp maximum due to resonant tunneling. However, this characteristic with a regime of negative differential conductivity (NDC) is stable only at low doping. At higher doping, spatio-temporal instabilities lead to self-oscillations of the current associated with the build-up of space-charges[8,11]. At the highest doping densities, the number of available carriers is sufficient to provide the space charge necessary to form a stable boundary between a low-field and a high-field domain. Stationary domains are then found. This behavior is summarized in Fig. 1 for a "perfect" superlattice. The inset depicts a bifurcation scenario for fixed $N_D$ where limit cycle oscillations are generated from the inhomogeneous branch by a supercritical Hopf bifurcation $H_{super}$. The amplitude of the current oscillations is indicated by the hatched area. There is a small regime of bistability between the inhomogeneous steady state and the oscillations beyond the subcritical Hopf bifurcation $H_{sub}$. $T$ denotes transcritical bifurcations of various steady states. Fig. 2 (a) shows the periodic oscillations for intermediate doping. The homogeneous field distribution breaks up into a low- and a high-field domain. The latter shrinks while at the same time its field grows rapidly forming a steep but unstable domain wall. This leads to a sharp rise of the current. When the current has reached a certain value the high-field domain collapses, resulting in the original quasi-homogeneous field distribution. This process is repeated periodically. When a small amount of doping fluctuations is introduced (Fig. 2 (b)) the spatially homogeneous phase of the field distribution decreases resulting in more sinusoidal oscillations of the current

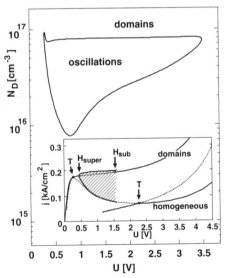

Figure 1: Phase diagram of spatio-temporal instabilities as a function of doping density $N_D$ and bias voltage $U$ for a perfect superlattice of $N = 40$ periods of GaAs/AlAs layers with $l = 90$ Å, $b = 15$ Å. The inset shows a bifurcation scenario for fixed $N_D = 2 \cdot 10^{16}$ cm$^{-3}$ (full lines: stable steady states, dotted lines: unstable steady states, hatched area: limit cycle oscillations)

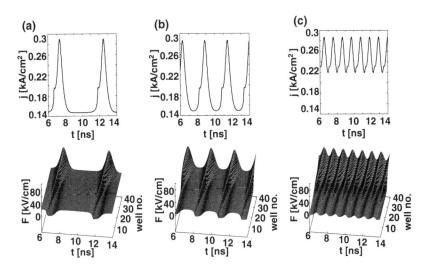

Figure 2: Self-oscillations with $N_D = 3 \cdot 10^{16}$ cm$^{-3}$ at $U = 1$ V (a) for a perfect superlattice, (b) for doping fluctuations of $\alpha = 0.1\%$, (c) for $\alpha = 3\%$, where $N_D^{(i)} = N_D(1 + \alpha e_i)$ with a random set of $N$ values $e_i$ from the interval $[-1, 1]$. The current density $j$ versus time $t$ and the evolution of the field distribution $F(x, t)$ are shown.

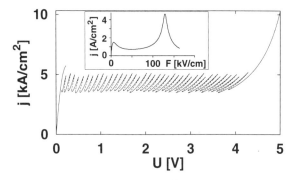

Figure 3: Current density $j$ vs. voltage $U$ for $N_D = 6.7 \cdot 10^{17}$ cm$^{-3}$ and doping fluctuations of $\alpha = 3\%$. Both stable (full) and unstable (dashed) domain states are shown. The inset depicts the current-voltage characteristics for uniform fields.

and a higher frequency. For larger fluctuations (Fig. 2 (c)) the homogeneous part vanishes completely, since the presence of irregularities supports the formation of charge accumulations. Furthermore, at doping concentrations $N_D$ for which the perfect superlattice would exhibit damped oscillations which asymptotically tend to a stable domain field distribution, we find persistent self-generated oscillations above a certain threshold of disorder. The actual shape of the boundary between both regimes, however, depends sensitively on the individual sequence of the irregularities. It is even possible that a sample with a particular spatial sequence of fluctuations shows stable domains, while the inverted sequence (corresponding to reversed bias) leads to self-oscillations.

In Fig. 3 the current-voltage characteristic is shown at a higher mean doping density where stable stationary high-field domains form at the anode in the NDC regime (cf. inset). In contrast to earlier work we have displayed here the full connected current-voltage characteristic. Along the characteristic stable and unstable parts alternate. They correspond to a continuous shift of the domain boundary across the superlattice from the anode ($i = N$) to the cathode ($i = 1$) with on average increasing bias. On the $i^{th}$ stable part (with rising voltage) the domain boundary is pinned at the $(N - i)^{th}$ well, while along the unstable parts (with falling voltage) the boundary is shifted to the next well. For neighbouring stable branches the domain boundary is thus displaced by one superlattice period. The irregularly varying length of the different branches is due to the doping fluctuations which determine the maximum and minimum current. Upon voltage sweep-up or sweep-down only parts of the stable branches can be reached (Fig. 4 (a)). With increasing degree of disorder, the irregularities are enhanced, and some stable branches are missed out altogether, as a result of their reduced length. The characteristics are in good agreement with the experiments[5] and allow even an estimate of the range of doping fluctuations between 3–10%.

In Fig. 4 (b) monolayer fluctuations of well and barrier widths are studied. In ($\alpha$) the $18^{th}$ well is chosen to be larger by one monolayer, while in ($\beta$) two wells are larger (the $11^{th}$ and the $32^{nd}$), and two wells are smaller (the $12^{th}$ and the $18^{th}$) by one monolayer. In ($\gamma$) there are four larger and four smaller wells. Finally, in ($\delta$) the $31^{st}$ barrier is wider by one monolayer. For increasing disorder ($\alpha$)–($\gamma$), we find the sequence of branches to exhibit a more and more irregular behavior. When only a small number of irregularities is present, it is even possible to determine their location within the superlattice structure. When at higher voltages a high-field domain forms near the anode (well no. 40) its influence becomes visible in the current-voltage characteristic only if the domain boundary crosses the barriers close to the perturbation. In this case the transition of the associated charge accumulation layer from one well into the next one occurs at a smaller or a larger applied voltage compared to the case of the perfect superlattice. This results in an extension or reduction, respectively, of the length of that current branch. It is thus possible to determine the location of a single irregularity of the well width from the current-voltage characteristic by enumerating the branches. (Fig. 4 (b)). If two perturbations are well

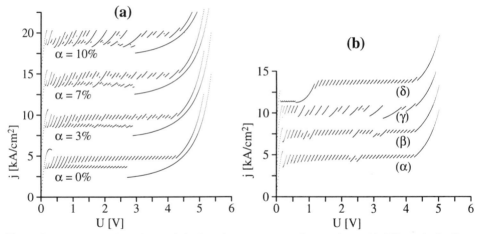

Figure 4: (a) Current-voltage characteristics for voltage sweep-up and sweep-down with different doping fluctuations. (b) Voltage sweep-up with different imperfections of wells and barriers. ($N_D = 6.7 \cdot 10^{17}$ cm$^{-3}$; the vertical scale is shifted for each curve)

separated within the superlattice structure so that the regions where they affect the field distribution do not overlap, their influences can still be distinguished. If the number of irregularities increases, their interaction leads to a more complex behavior in the corresponding part of the current-voltage characteristic ($\gamma$). The widths of the barriers can also severely affect the current-voltage characteristic, which can be seen in ($\delta$) where the $31^{st}$ barrier is wider by one monolayer. The voltage at which the domain boundary crosses the perturbation (i. e. the location of the perturbation within the superlattice) can be easily detected from the characteristic.

In conclusion, we propose to use macroscopic nonlinear transport properties far from equilibrium as a novel, noninvasive method of probing growth-related disorder and imperfections in superlattices. By simple global macroscopic electric measurements, in combination with model calculations, microscopic structural features can thus be investigated.

We thank L. Bonilla, H. T. Grahn, and J. Kastrup for enlightening discussions. This work was supported by DFG in the framework of Sfb 296.

# REFERENCES

1. L. Esaki and L. L. Chang, Phys. Rev. Lett. **33**, 495 (1974).
2. K. K. Choi, B. F. Levine, R. J. Malik, J. Walker, and C. G. Bethea, Phys. Rev. B **35**, 4172 (1987).
3. H. T. Grahn, R. J. Haug, W. Müller, and K. Ploog, Phys. Rev. Lett. **67**, 1618 (1991).
4. Y. Zhang, X. Yang, W. Liu, P. Zhang, and D. Jiang, Appl. Phys. Lett. **65**, 1148 (1994).
5. J. Kastrup, H. T. Grahn, K. Ploog, F. Prengel, A. Wacker, and E. Schöll, Appl. Phys. Lett. **65**, 1808 (1994).
6. R. Merlin, S. H. Kwok, T. B. Norris, H. T. Grahn, K. Ploog, L. L. Bonilla, J. Galán, J. A. Cuesta, F. C. Martínez, and J. M. Molera, in *Proc. 22nd Int. Conf. Phys. Semicond.*, edited by D. J. Lockwood (World Scientific, Singapore, 1995), p. 1039.
7. H. T. Grahn, J. Kastrup, K. Ploog, L. L. Bonilla, J. Galán, M. Kindelan, and M. Moscoso, Jpn. J. Appl. Phys. (1995), in press.
8. A. Wacker, F. Prengel, and E. Schöll, in *Proc. 22nd Int. Conf. Phys. Semicond.*, edited by D. J. Lockwood (World Scientific, Singapore, 1995), p. 1075.
9. L. L. Bonilla, J. Galán, J. A. Cuesta, F. C. Martínez, and J. M. Molera, Phys. Rev. B **50**, 8644 (1994).
10. F. Prengel, A. Wacker, and E. Schöll, Phys. Rev. B **50**, 1705 (1994).
11. L. L. Bonilla, in *Nonlinear Dynamics and Pattern Formation in Semiconductors*, edited by F. J. Niedernostheide (Springer, Berlin, 1995), Chap. 1.

# BALLISTIC ELECTRON CURRENT IN SYSTEM OF ASYMMETRICAL DOUBLE QUANTUM WELLS AND ITS TERAHERTZ OSCILLATIONS

N.Z. Vagidov, Z.S. Gribnikov, and A.N. Korshak

Institute of Semiconductor Physics,
Ukrainian National Academy of Sciences,
Kiev, 252650, Ukraine

## INTRODUCTION

In this paper a ballistic transport of current carriers with special dispersion relation $\varepsilon(p_x)$ across a thin doped base of the $n^+nn^+$- or $p^+pp^+$- diode is considered ($p_x$ is a component of crystal momentum $\mathbf{p}$ in the current direction). A distinguished feature of the $\varepsilon(p_x)$- relation consists in the presence of some region in the $\mathbf{p}$- space where effective mass of the carriers $(\partial^2\varepsilon/\partial p_x^2)^{-1}$ is negative.

Among possible mechanisms of realization of the required dispersion relation we turn attention to such of them which can be described by matrix Hamiltonian:

$$H_{11} = \varepsilon^{(1)}(\mathbf{p}), \qquad H_{22} = \varepsilon^{(2)}(\mathbf{p}), \qquad H_{12} = H_{21} = \delta, \qquad (1)$$

where $\mathbf{p} = -i\hbar\partial/\partial\mathbf{r}$, $\mathbf{r} = (x, \mathbf{r}_\perp)$ is a radius vector, $\varepsilon^{(1)}(\mathbf{p}) = p^2/2m$, $\varepsilon^{(2)}(\mathbf{p}) = \varepsilon_0 + p^2/2M$, $\varepsilon_0 > 0$; $\delta$ is an effective potential of interaction. Here $m$, $M$ are light and heavy masses; in the following we shall need $M > 2m$. Hamiltonian (1) describes an electron motion both in asymmetric heterostructural double quantum wells (DQW) (Fig.1,a) and in composite (two-layer) $\Gamma X$-quantum wells (Fig.1,b). It results in the dispersion relation:

$$\varepsilon_{1,2}(p) = 1/2\left(\varepsilon^{(1)} + \varepsilon^{(2)} \pm [(\varepsilon^{(1)} - \varepsilon^{(1)})^2 + \delta^2]^{1/2}\right), \quad p = |\mathbf{p}|. \qquad (2)$$

The branch $\varepsilon_1(p)$ (see, Fig.1,c) contains required part with negative effective mass (NEM).

For the values of current density near $j_c = \lambda_c n_0$ and voltage interval $(V_c, V_k)$ (see, Fig.1,c; $n_0$ is a base doping, $\lambda_c$ is a slope of the tangent line in the point $p = p_c$) stationary concentration and potential distributions are characterized by an extensive quasi-neutral region with drifting NEM-carriers (NEM-region), it is located between two space charge regions, one of them (usual) is situated near cathode and the other (unusual) is near anode[1,2]. Current saturation near $j_c$ takes place in this voltage interval. Increase of the voltage $V_a$ across the base shortens the NEM-region and extends the anode space charge region. Small exceeding $j$ over $j_c$ eliminates the NEM-region completely and unites together both charged regions separated before.

To abandon the effective-cathode approximation in the numerical simulations[3] 1D-Poisson equation is solved self-consistently with stationary collisionless kinetic equation for 3D-electrons with the dispersion relation $\varepsilon_1(p)$ in form (2) and other relations close to $\varepsilon_1(p)$. In the absence of a current, the electrons were assumed to have an equilibrium Fermi-Dirac distribution which served as boundary condition for the incident electrons. Thus an ideal ballistic electron transport across the base is considered. Stationary electron concentration distributions are shown in Fig.2 for $j \cong j_c$ and different voltages. Both charge regions have dipole structure; they include enrichment and depletion regions.

## OSCILLATIONS

It is evident that vast NEM-region is unstable plasma medium. It can be shown[1,2] that convective instability takes place in such medium with increment which is equal to plasma frequency of the NEM-electrons with the concentration $n_0$ and differential effective mass in the point $p=p_c$. If the ballistic diode base is long enough this instability regenerates in global (for the short-circuit case) and stationary values of concentration, potential and current become unstable in the interval $(V_c, V_k)$. Therefore we have researched non-stationary behaviour of the diodes on the basis of the self-consistent solution of Poisson equation together with non-stationary collisionless kinetic equation. In the stable current flowing regime stationary distributions are reached as result of decaying transient oscillation process. But in the unstable regime oscillations do not decay completely and the transient process finishes by the quasi-stationary current oscillations. The most typical picture is the oscillating regime close to sinusoidal with comparatively small amplitude $J_m$ (it does not exceed 6-10% of the mean current). Despite the smallness of the oscillation amplitude the above described picture is not identical to the following one: "stationary value + small oscillations". In particular mean current value does not coincide with the stationary value from previous calculations. Mean current slowly increases in the interval $(V_c, V_k)$ and mean electron concentrations differs noticeably from the stationary distributions in numerous details. There is not distinct division of the base into the NEM-region and the charged layers. The concentration oscillations (Fig.3) are space charge waves with alternation of enrichment and depletion layers. A number of these drifting layers increases with $l$. The noticeable difference between mean and stationary values proves the existence of new solution which does not connect with the stationary one directly. Frequency of quasi-harmonic oscillations with small amplitude is a fundamental parameter of the diode structure which depends on the base length $l$, voltage across the diode $V_a$, concentration of doping impurities $n_0$, boundary condition and dispersion relation parameters.

To calculate thoroughly frequency $\omega$ and amplitude $J_m$ as a function of $V_a$ the method of computer characteriograph[4] (CC) is applied. This method consists in calculation of the non-stationary currents for continuously increasing voltage $V_a$: $V_a(t)=V_a(0)+V_a'\cdot t$. This approach holds true if obtained results for different values of rate $V_a'$ are similar.

Some dependencies $j(t)$ calculated with the CC are shown in Fig.4. Side by side with real time $t$, voltage $V_a(t)$ is plotted on the horizontal axis. Oscillation region takes place if base thickness exceeds some minimum length $l_c$ for determined $\varepsilon(p)$-relation, doping and boundary conditions. Interval of the current oscillations widens with increasing $l$. For small exceeding $l$ over $l_c$ the oscillation amplitude increases monotonically from the interval boundaries to its center. The following increasing of the base length leads to crucial changes of $J_m(V_a)$-dependence: an essentially nonlinear regime replace quasi-linear (Fig.4).

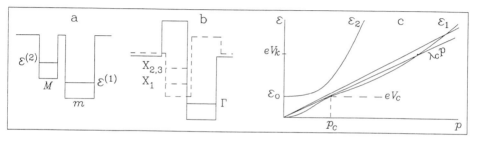

**Figure 1.** Examples of the structures $(a,b)$ with dispersion relation of electrons containing the NEM-part $(c)$: $a$-asymmetric DQW; $b$-composite ΓX-quantum well; $c$-the $\varepsilon_{1,2}(p)$-relation with the model parameters $m=0.085m_0$, $M=0.54\,m_0$, $\varepsilon_0=0.1\,eV$, $\delta=0.02\,eV$.

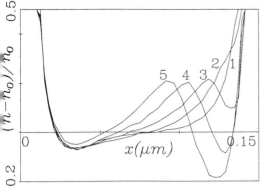

**Figure 2.** Stationary distributions of electron concentration $(n(x)-n_0)/n_0$ for different voltages: $0.15\ V(1)$, $0.2\ V(2)$, $0.25\ V(3)$, $0.3\ V(4)$, $0.35\ V(5)$. The parameters of $\varepsilon_1(p)$-relation are the same as in Fig. 1,c: $l=0.15\ \mu m$, $n_0=5\cdot10^{17}\ cm^{-3}$, equilibrium concentration $n(0)=10^{19}cm^{-3}$.

**Figure 3.** Concentration waves $(n(x,t)-n_0)/n_0$ for the same set of parameters as in Fig.2, $Va=0.25V$, time in $ps$.

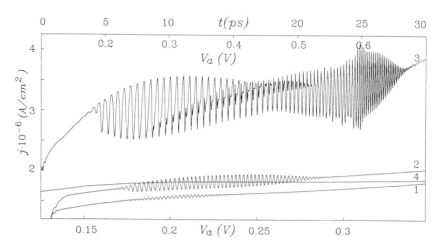

**Figure 4.** Dependencies $j(t)$ calculated with the help of the CC. Below is plotted voltage range with $V_a'$ $= 7.4mV/ps$ for curves 1,2 and overhead with $V_a'=20mV/ps$ for curve 3. Curves 1 and 2 correspond to the set of parameters from Fig.2 with $l=0.13\mu m(1)$ and $l=0.15\mu m(2)$. Curve 3 is distinguished by $\varepsilon_0=0.2eV$, $\delta=0.002eV$ and $l=0.2\mu m$. Curve 4 represents the stationary IV-characteristic which corresponds to the distributions in Fig.2. Curve 2 is shifted down by $0.2\times10^6\ A/cm^2$.

Monotone increase of the frequency $\omega$ with increasing $V_a$ for fixed $l$ and with shortening $l$ for fixed $V_a$ make possible soft electrical tuning of the generator's frequency in sufficiently wide range (up to 1.5 times).

## REALIZATIONS OF REQUIRED DISPERSION RELATION

There are two main ways of obtaining dispersion relations $\varepsilon(p_x)$ with NEM-region: (i) anisotropic deformation of $p$-type semiconductor with degenerate valence bands at $\mathbf{p}=0$; (ii) heterostructural hybridization of different dispersion relations. The first of these ways is considered in[5]. The most obvious method of hybridization of relations $\varepsilon^{(1)}$ and $\varepsilon^{(2)}$ that leads to Hamiltonian (1) is asymmetric DQW (Fig.1,a). DQW is formed by two original quantum wells with dispersion relations $\varepsilon^{(1)}$ and $\varepsilon^{(2)}$ in them, and separating tunnel-transparent barrier. Parameters of the barrier determine the small potential $\delta$ in Hamiltonian (1). Naturally electron gas in DQW is 2D-gas and in order to the current oscillations be established we have to use a system of these wells separated by isolating barriers of width $w$. Their thickness must be small enough to form quasi-homogeneous medium for 1D-ballistic transport. It is convenient to use separating isolating barriers between DQWs for modulation donor doping that allows to decrease impurity scattering and thus to satisfy the condition of ballistic transport.

Condition $M>2m$ is difficult to fulfill in asymmetric DQW. In the popular structure $In_xGa_{1-x}As$ (well 1)/$In_yAl_{1-y}As$ (barrier)/$InP$ (well 2) (which is convenient because the conduction band offsets in system $In_xGa_{1-x}As/InP$ are smaller than valence band offsets) $M{\approx}2m$ is satisfied approximately for isomorphic compound with $x=0.53$, so for all pseudomorphic compounds with $x>0.53$ the required condition $M>2m$ is hold. For $x{\rightarrow}1$ we have $M/m{\approx}3.5$ ($M=0.076-0.081m_0$; $m=0.021-0.024m_0$).

Another heterostructure defined by Hamiltonian (1) is a composite (two-layer) $\Gamma X$-well which consists of $\Gamma$- and X-wells (Fig.1,b). An example of such structure is $GaAs$($\Gamma$-well)/$AlAs$(X-well)/$Al_xGa_{1-x}As$(barrier), optimum values of $x$ are $0.45-0.5$. Exchange potential $\delta$ in Hamiltonian (1) is determined by the exchange rate in the $\Gamma X$-junction[6]. The mass of $\Gamma$-electrons in $GaAs$ ($0.066m_0$) is $m$, and the transverse mass of X-electrons in $AlAs$ ($0.19m_0$) is $M$ (i.e. ratio $M/m$ here does not exceed 3). Another disadvantage of composite $\Gamma X$-well is difficulty to control the $\Gamma X$-coupling (i.e. $\delta$).

In fact, most of our conclusions are independent of the assumption about real electron-electron interaction.

## REFERENCES

1. Z.S. Gribnikov, and A.N. Korshak, Ballistic injection of electrons with negative effective masses, Fiz. Techn. Poluprov. 28: 1445 (1994).
2. Z.S. Gribnikov, and A.N. Korshak, Space-charge-limited ballistic transport in short heterostructures with complicated electron dispersion relations, In: "Quantum Confinement. Physics and Application," Electrochem. Soc. Inc., San-Francisco (1994).
3. N.Z. Vagidov, Z.S. Gribnikov, and A.N. Korshak, Space charge of injected ballistic electrons with negative effective masses in terahertz range, Fiz. Techn. Poluprov. 29: in press (1995).
4. N.Z. Vagidov, Z.S. Gribnikov, and A.N. Korshak, Terahertz-range oscillations of the ballistic current of electrons with a negative effective mass, Pis'ma ZhETF. 61: 38 (1995).
5. A.N. Korshak, Z.S. Gribnikov, and N.Z. Vagidov, Subterahertz and terahertz oscillations of ballistic hole current in uniaxially-compressed narrow bases, this volume.
6. I.L. Aleiner, and E.L. Ivchenko, Electron minizones in $(GaAs)_N(AlAs)_M$ superlattices for even and odd $M$, Fiz. Techn. Poluprov. 27: 594 (1993).

# SUBTERAHERTZ AND TERAHERTZ
# OSCILLATIONS OF BALLISTIC HOLE CURRENT
# IN UNIAXIALLY COMPRESSED NARROW BASES

A.N. Korshak, Z.S. Gribnikov, and N.Z. Vagidov

Institute of Semiconductor Physics,
Ukrainian National Academy of Sciences,
Kiev, 252650, Ukraine

Mostly $\Gamma$-electrons in direct gap semiconductors with zinc-blend crystal lattice are used for experimental researches of the ballistic transport. Mean free paths of both heavy and light holes in these materials (as well as in the diamond-like semiconductors) are small enough[1] because of predominant scattering into heavy-hole states. Situation changes crucially at an uniaxial compression which eliminates the degeneracy of valence bands at the point $\mathbf{k}=0$ and shifts this point along the compression direction. In the energy range $\delta E_{12}(0)=E_2(0)-E_1(0)$ (where $1,2$ are numbers of the mixed valence bands) there are only light holes with substantially anisotropic dispersion relations $E_1(\mathbf{k})$. Thus their mean free path and time increase substantially.

Dispersion relations of heavy and light holes can be determined by two valence band parameters $A$ and $B$ (we assume the third of them $C$ is equal to 0 for simplicity)

$$E_{2,1}(k)=(A\pm B)k^2, \qquad A=\hbar^2/4(1/m+1/M), \qquad B=\hbar^2/4(1/m-1/M), \qquad (1)$$

where $m$ and $M$ are masses of light and heavy holes. (We assume that $A$, $B$ and $E_{1,2}(k)$ are positive). Compression along $x$-axis changes relations (1) in the following manner:

$$E_{2,1}(\mathbf{k})=Ak^2+a\varepsilon_\Sigma\pm[B^2k^4+b^2\varepsilon_\Delta^2+Bb\varepsilon_\Delta(2k_x^2-k_\perp^2)]^{1/2}, \qquad (2)$$

where $k^2=k_x^2+k_\perp^2$, $\varepsilon_\Sigma=\varepsilon_{xx}+2\varepsilon_\perp$, $\varepsilon_\Delta=\varepsilon_{xx}-\varepsilon_\perp$, $\varepsilon_\perp=\varepsilon_{yy}=\varepsilon_{zz}$, $\varepsilon_{ij}$ are components of deformation tensor; $a$ and $b$ are deformation potential constants[2]. Described above splitting takes place both at the uniaxial compression along $x$-axis and at isotropic extension of semiconductor layer in the $yz$-plane (as a result of pseudomorphic growth).

The dispersion relation $E_1(\mathbf{k})$ contains a negative effective mass $((\partial^2 E_1/\partial k_x^2)^{-1}<0)$ (*NEM*) part, in the neighbourhood of the new degeneracy points $k_x^2=k_c^2=-b\varepsilon_\Delta/B$, $k_\perp^2=0$. Constant energy contours $E_{1,2}(k_x,k_\perp)=E$ are shown in Fig. 1,a; the degeneracy point is seen clearly. The dispersion relations $E_{1,2}(k_x,k_\perp)$ for different $k_\perp$ are plotted in Fig. 1,b, they are

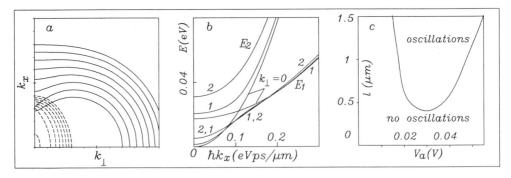

**Figure 1.** Constant energy contours (*a*), dispersion relations $E_{1,2}(k_x, k_\perp)$ (*b*) ($k_\perp(1) < k_\perp(2)$), and domain of existanse of the oscillation region (*c*) for *p-Ge* sample with $n_0 = 3 \times 10^{15} cm^{-3}$, $\varepsilon_\Delta = 0.35\%$, $T = 4.2K$.

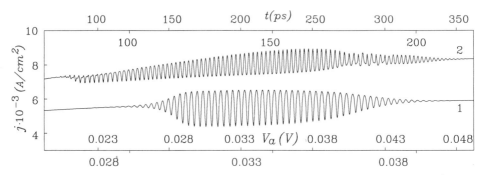

**Figure 2.** Dependencies $j(t, V_a)$ calculated with the help of the CC for *p-Ge*-samples from Fig. 1 with $l = 0.5\mu m$ (1) and $l = 1\mu m$ (2). Abscissae: below is plotted voltage range and overhead is the time range (lower scale is for curve 1, upper one is for curve 2). Curve 1 is shifted down by $3 \times 10^3 A/cm^2$.

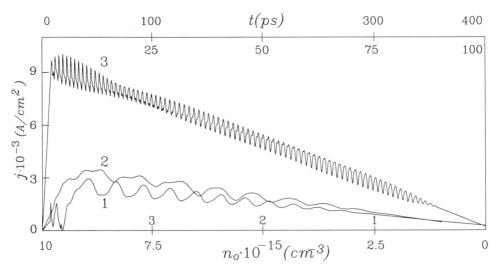

**Figure 3.** Dependencies $j(t, n_0)$ calculated with the help of the CC for: (1) *p-InAs* ($l = 0.9\mu m$, $V_a = 0.038V$, $\varepsilon_\Delta = 0.74\%$); (2) *p-InSb* ($l = 1.5\mu m$, $V_a = 0.027V$, $\varepsilon_\Delta = 0.56\%$); (3), *p-Ge* ($l = 0.9\mu m$, $V_a = 0.035V$, $\varepsilon_\Delta = 0.44\%$). Abscissae: below is plotted concentration range and overhead is the time range (lower scale is for curve 1,2, upper one is for curve 3). Curve 1 is shifted down by $3 \times 10^3 A/cm^2$.

similar to the analogous ones in[3], but contrary to[3] parameter similar to the interaction potential $\delta$ is not constant, and is proportional to $k_\perp$. Therefore the gap $\delta E_{12}(k_c)$ is equal to zero in $k_\perp=0$ and increases linearly with $k_\perp$. As it is shown in[4] the existence of NEM-part in the dispersion relation of the carriers leads to generation of current oscillations in ballistic diodes with sufficiently long doped base. A particular advantage of the considered relation $E_1(k)$ is the large mass ratio $M/m$ (in p-Ge we have $M/m= 0.34/0.043=7.9$; in p-InAs - $M/m=0.4/0.026=15.4$) which can not be obtained in the case of layered heterostructures[3]. It allows to obtain large voltage interval $(V_c, V_k)$ with effective generation. To begin generation a voltage across the diode has to exceed the critical value $V_c$ and base length $l$ has to exceed the width $l_c$ of the cathode-adjacent dipole layer at the voltage $V_c$:

$$V_c=b|\varepsilon_\Delta|(1+A/B)/e, \qquad l_c=\pi v_c/2\omega_p, \qquad \omega_p^2=e^2 n_0/\varepsilon_d m, \qquad (3)$$

where $v_c=\hbar k_c/m$, $n_0$ is doping concentration in the base, $\varepsilon_d$ is a dielectric constant.

Current oscillations calculated for two model p-Ge samples by the computer characteriograph[3,4] (CC) are shown in Fig. 2. For shorter sample quasi-linear harmonic oscillations with frequencies $f\sim 500GHz$ take place and for longer one non-linear oscillations in the wider interval $(V_c, V_k)$ with frequency $f\sim 300GHz$ are found. Fig. 1,c presents the dependencies $V_c(l)$ and $V_k(l)$. The critical length $l_c$, saturation of the voltage $V_c$ and increasing $V_k$ with $l$ are seen clearly. The voltage $V_k$ saturates at greater lengths $l$. We use CC-method to investigate current oscillations as a function of $n_0$ for fixed $l$ and $V_a$. Concentration $n_0$ varies with time slowly so that variation of $n_0$ during one period of the oscillations is relatively small. Fig.3 shows some of such dependencies.

If we were free to choose all the parameters which determine the oscillation frequency and amplitude we could change this frequency in the range from 100 GHz to 10 THz (and greater) and amplitudes up to 20 % of mean current. But we are restricted by the ballistic length. Mean free path of the holes with $k=k_c$ in the NEM-region (see[3] ) $l_f=v_c\tau(k_c)/2$ has to exceed the critical length $l_c$ with a sufficient reserve. $\tau(k_c)$ is a mean free time of the holes with $k_x=k_c$, $k_\perp=0$, and $v_c/2$ is mean velocity of the holes in the NEM-region, that is $\omega_p\tau(k_c)>\pi$. This condition we have to complement with the coherence condition $\omega\tau(k_c)>1$ where $\omega$ is the current oscillation frequency.

Analysis of the $\tau(k_c)$-dependencies leads to the following conclusion: inequality $\omega_p\tau(k_c)>\pi$ holds in two regions of $k_c$ The first region is found at low temperatures (for example, 4,2K) for $E_1(k_c)<\hbar\omega_0$ ($\omega_0$ is an optical phonon frequency). This region is characterized by the light doping ($\sim 10^{15}cm^{-3}$), long bases ($l\sim 1\mu m$) and comparatively low oscillation frequencies ($f=\omega/2\pi\approx 0.5THz$). The second region is characterized by very large $k_c$ with the base lengths $l$ shorter than the scattering length for optical phonon emission and absorption. The oscillation frequencies in this region attain and exceed 10THz. We have to insure high values of doping concentration $n_0$ ($\sim 10^{18}cm^{-3}$) and decrease ionized impurity scattering simultaneously. Usually a multilayer structure of the base with longitudinal hole transport in the conducting layers and with modulation doping of the barriers between them is used. The multilayer bases with modulation doping are analogous to base structures, described in[3].

1. M. Heiblum, K. Seo, H.P. Meier, T.W. Hickmott, Observation of ballistic holes, Phys.Rev. Letters, 60: 828 (1988)
2. G.L. Bir, G.E. Pikus, "Symmetry and Strain-Induced Effects in Semiconductors," J. Willey and Sons, NY (1974)
3. N.Z. Vagidov, Z.S. Gribnikov, A.N. Korshak, Ballistic electron current in system of asymmetrical double quantum wells and its terahertz oscillations, this volume (1995)
4. N.Z. Vagidov, Z.S. Gribnikov, A.N. Korshak, Terahertz-range oscillations of the ballistic current of electrons with a negative effective mass, JETP Lett., 61: 38 (1995)

# NEGATIVE DIFFERENTIAL MOBILITY AND CONVECTIVE INSTABILITY OF A SEMICONDUCTOR SUPERLATTICE

X. L. Lei,[1] N. J. M. Horing,[1] H. L. Cui,[1] and K. K. Thornber[2]

[1]Department of Physics and Engineering Physics
Stevens Institute of Technology, Hoboken, New Jersey 07030
[2]NEC Research Institute
4 Independence Way, Princeton, New Jersey 08540

Although narrow-miniband superlattices have an important transport inhomogeneity manifested in the formation of high-field domains and in periodic structure of the I-V characteristic when the applied voltage increases beyond a threshold,[1] the most interesting transport phenomena in wide-miniband systems are those related to Esaki-Tsu negative differential mobility (NDM) at relatively low electric field.[2] We report here on a careful examination of the frequency dependence of small-signal ac mobility and the possible fluctuation instability of a planar superlattice subject to negative differential miniband conductance. Our analysis is based on three-dimensional hydrodynamic non-linear balance equations with accurate microscopic treatment of phonon and impurity scatterings.

We consider an infinite GaAs-based planar superlattice in which electrons move freely in the transverse ($x$-$y$) plane and can travel along the growth axis ($z$-direction) through the (lowest) miniband. The electron energy dispersion $\varepsilon(\mathbf{k})$ can be written as the sum of a transverse energy $\varepsilon_{\mathbf{k}_{\parallel}} = k_{\parallel}^2/2m$ ($m$ being the band mass of the carrier in the bulk semiconductor), and a tight-binding miniband energy $\varepsilon(k_z) = (\Delta/2)(1 - \cos k_z d)$ related to the longitudinal motion.

In the spatially inhomogeneous case, we employ the average (per-carrier) center-of-mass momentum $\mathbf{p}_d$, the electron temperature $T_e$ and the ratio of the chemical potential to $T_e$, $\zeta$, as fundamental variables to describe the transport state of a many-electron system in an arbitrary energy band,[3-5] and treat $\mathbf{p}_d$, $T_e$ and $\zeta$, and correspondingly the carrier number density $n$, the average drift velocity $\mathbf{v}_d$ and all other quantities, as being time- and space-dependent.[6] Considering balances of the carrier number density, the acceleration and the energy in a small volume element around a spatial position, we obtain hydrodynamic balance equations of the following type for superlattice vertical transport, in which the electric field and the carrier drift are assumed in the $z$ direction: $\mathbf{E} = (0, 0, E)$, $\mathbf{p}_d = (0, 0, p_d)$ and $\mathbf{v}_d = (0, 0, v_d)$, with spatial inhomogeneity only along the superlattice $z$-axis.

$$\frac{\partial n}{\partial t} = -\frac{\partial}{\partial z}(nv_d), \tag{1}$$

$$\frac{\partial}{\partial t}(nv_d) = -\frac{\partial}{\partial z}(nB_z) + \frac{neE}{m_z^*} + nA, \tag{2}$$

*Hot Carriers in Semiconductors*
Edited by K. Hess *et al.*, Plenum Press, New York, 1996

$$\frac{\partial}{\partial t}(n\epsilon) = -\frac{\partial}{\partial z}(nS_z) + neEv_d - nW. \tag{3}$$

Here, $n$ is the average carrier density, $\epsilon$ the average carrier energy, $v_d$ the average carrier drift velocity, $1/m_z^*$ the $zz$ component of the inverse effective-mass tensor, $B_z$ the $zz$ component of the average velocity-velocity dyadic, and $S_z$ the $z$ component of the energy flux vector. They are functions of $z_d \equiv p_d d$, $T_e$, and $\zeta$, and thus are time and space dependent through these fundamental variables. In these equations, the frictional acceleration $A = A_i + A_p$ (due to impurity and phonon scatterings), and the energy loss rate (per particle) $W$ (due to phonon scattering), are given in Ref. 5.

The electric field relates to space charge through its divergence in accordance with the Gauss theorem (Poisson equation):

$$\partial E/\partial z = e(n - n_0)/\epsilon_0 \kappa, \tag{4}$$

where $e$ is the electron charge, $\kappa$ the dielectric constant, and $n_0$ the background density of positive charge which is assumed to be uniformly distributed over the whole superlattice.

To investigate the dynamic mobility and to proceed with mode analysis under the drifted condition, we consider small wavelike fluctuations: $\delta z_d$, $\delta T_e$, $\delta \zeta$, $\delta v_d$ and $\delta Q \sim \exp[i(kz - \omega t)]$ ($Q$ stands for $E$, $n$, $1/m_z^*$, $B_z$, $S_z$, or $\epsilon$), superimposed on the dc bias such that $z_d = z_0 + \delta z_d$, $T_e = T_0 + \delta T_e$, $\zeta = \zeta_0 + \delta\zeta$, $v_d = v_0 + \delta v_d$, and $Q = Q_0 + \delta Q$. In the zero-th order, the continuity equation (1) and the Poisson equation (4) are identities, and the other two equations are just the dc steady-state equations for the effective force and energy balance in the spatially homogeneous case:

$$0 = eE_0/m_{z0}^* + A_0, \tag{5}$$

$$0 = eE_0 v_0 - W_0. \tag{6}$$

The balance equations for linear order fluctuations can be written in terms of four variables: $\delta z_d$, $\delta T_e$, $\delta \zeta$ and $\delta E$, and form a set of four linear algebraic equations. The condition to have a nonzero solution yields

$$D(k, \omega) = 0, \tag{7}$$

which determines the dispersion relation between $k$ and $\omega$ for all the eigen modes.

In the spatially uniform case, $k = 0$, Eq. (7), $D(0, \omega) = 0$, is satisfied for arbitrary $\omega$. Therefore, we can, in principle, apply a small uniform ac field $\delta E$ of (real) frequency $\omega$ superimposed on a bias dc field $E_0$, and consider the velocity response $\delta v_d$ of the system to obtain the small-signal (complex) mobility defined as $\mu_\omega = \delta v_d/\delta E$. As an example, we consider a GaAs-based superlattice with period $d = 10\,\text{nm}$, miniband width $\Delta = 900\,\text{K}$, carrier sheet density $N_s = 1.5 \times 10^{15}\,\text{m}^{-2}$, and low-temperature linear dc mobility $\mu(0) = 1.0\,\text{m}^2/\text{Vs}$ at lattice temperature $T = 300\,\text{K}$. The real part of the calculated small-signal mobility in the space-uniform case, $\text{Re}\,\mu_\omega$, is shown in Fig. 1 as a function of frequency $\nu \equiv \omega/2\pi$ of the ac driving field at several different dc biases $E_0$.

In the inhomogeneous case, $k \neq 0$, Eq. (7) can generally be satisfied only for complex $k = k_1 + ik_2$ and/or complex $\omega = \omega_1 + i\omega_2$. We concentrate here on the most interesting drift-relaxational modes and related instabilities and examine the case of a real wavevector: $k = k_1$ and $k_2 = 0$. The real wavevector $k_1$ and the imaginary part of frequency, $\omega_2$, obtained from equation (7) as functions of real frequency $\nu = \omega_1/2\pi$ are shown in Fig. 2 at several dc bias fields $E_0$ from 2.0 to 7.0 kV/cm. When biased in the NDM regime, $\omega_2$ is positive at low $\omega_1$ (low $k$), and continues to be positive with increasing frequency until $\omega_1$ reaches a maximum value dependent on the bias. A positive $\omega_2$ implies a traveling wave

$$\exp[\omega_2 t + i(k_1 z - \omega_1 t)] \tag{8}$$

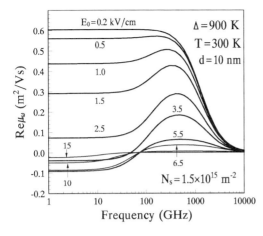

**Figure 1.** The real part of the small-signal mobility in the spatially uniform case, $\mathrm{Re}\mu_\omega$, is shown as a function of frequency $\nu \equiv \omega/2\pi$ at lattice temperature $T = 300\,\mathrm{K}$, for the GaAs-based superlattice described in the text, with the bias field $E_0$ ranging from 0.2 to $15\,\mathrm{kV/cm}$.

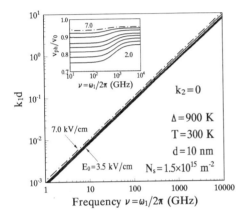
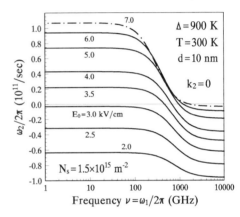

**Figure 2.** (a) Real wavevector parameter, $k_1 d$, and (b) imaginary frequency parameter, $\omega_2/2\pi$, in the case of $k_2 = 0$, are plotted as functions of the real frequency $\nu = \omega_1/2\pi$. The inset shows the phase velocity $v_{ph}$ in units of bias drift velocity $v_0$ as a function of $\nu$ with bias electric field $E_0$ equal to 2.0, 2.5, 3.0, 3.5, 4.0, 5.0, 6.0, and $7.0\,\mathrm{kV/cm}$.

propagating along the positive $z$ direction with its amplitude growing exponentially as a function of time. The phase velocity of this space-charge wave, $v_{ph} = \omega_1/k_1$, is shown in the inset of Fig. 2. Depending on the bias and frequency, $v_{ph}$ varies from $0.75v_0$ to $0.95v_0$, $v_0$ being the bias drift velocity. These results are significantly different from those based on the conventional drift-diffusion model or on one-dimensional hydrodynamic equations.[7,8]

The existence of unstable convective space-charge waves suggests that the Gunn-like phenomena due to Bragg-scattering-induced NDM may appear in wide-miniband superlattices. The full 3D hydrodynamic balance equations (1)-(3) introduced here provide a convenient and accurate tool to deal with these miniband effects.

The authors thank NEC Research Institute, Princeton, New Jersey, for partial support of this work.

## REFERENCES

1. L. Esaki and L.L. Chang, Phys. Rev. Lett. **33**, 495 (1974).
2. L. Esaki and R. Tsu, IBM J. Res. Dev. **14**, 61 (1970).
3. X.L. Lei and C.S. Ting, Phys. Rev. B **30**, 4809 (1984); **32**, 1112 (1985).
4. X.L. Lei, N.J.M. Horing and H.L. Cui, Phys. Rev. Lett. **66**, 3277, (1991).
5. X.L. Lei, Phys. Stat. Sol. (b), **170**, 519 (1992).
6. X.L. Lei, J. Cai, and L.M. Xie, Phys. Rev. B **38**, 1529 (1988).
7. A.A. Ignatov and V.I. Shashkin, Sov. Phys. JETP **66**, 526 (1987).
8. M. Büttiker and H. Thomas, Phys. Rev. Lett. **38**, 78 (1977).

# ON THE OBSERVABILITY OF DYNAMIC LOCALIZATION
# IN SEMICONDUCTOR SUPERLATTICES USING
# OPTICAL SPECTROSCOPY

T. Meier, F. Rossi, K.-C. Je, J. Hader, P. Thomas, and S.W. Koch

Department of Physics and Materials Sciences Center,
Philipps University, Renthof 5, D-35032 Marburg, Germany

## INTRODUCTION

During the last decade, a variety of field-induced coherent phenomena in semi-conductor superlattices has received considerable attention both from a theoretical and from an experimental point of view. For the case of a static applied electric field, the existence of Bloch oscillations has been predicted theoretically[1] and confirmed by means of different experimental techniques of nonlinear optical spectroscopy[2,3].

Moreover, if the semiconductor is subject to an oscillating electric field, the theo-retically predicted *dynamic localization*[4] is expected to occur. As for the case of Bloch-oscillations, it should be possible to obtain experimental evidence for this dynamic loca-lization using optical spectroscopy. Such an optical experiment will clearly be influenced by excitonic effects, which therefore have to be included in a theoretical investigation of this phenomenon[5,6].

We present a theoretical analysis of dynamic-localization effects in semiconductor superlattices. In particular we study the signatures of dynamic localization in the linear spectrum and in the Terahertz radiation.

## THEORY

Our theoretical approach is based on a generalized set of semiconductor Bloch equations[7], which include the Coulomb-interaction in time dependent Hartree-Fock approximation and the applied electric field[5]. Our approach is based on a full three-dimensional superlattice model. We factorize the superlattice wavefunctions as a pro-duct of a plane wave for the free particle motion in the $x - y$ plane and a superlattice periodic Bloch type function in the $z$ (growth) direction. To calculate the wavefunctions in the growth direction we use the Kronig-Penney model. A discussion of this model

*Hot Carriers in Semiconductors*
Edited by K. Hess *et al.*, Plenum Press, New York, 1996

for the superlattice wavefunctions is given in Ref. 8.

Within this basis we numerically calculate the dispersion, and the optical and Coulomb matrix elements, which enter into the semiconductor Bloch equations. Numerically solving these equations we calculate the optical signals, which are discussed in the following.

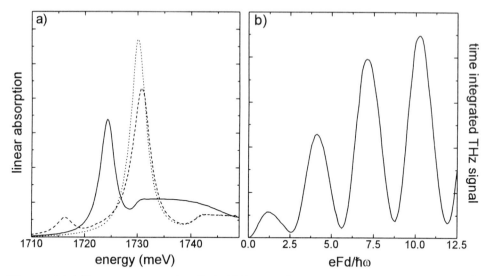

**Figure 1.**(a) Linear spectrum of a $GaAs/AlAs$ superlattice with well (barrier) width $34\mathring{A}$ ($17\mathring{A}$). Solid: without applied electric field, dashed: with localizing electric field ($\hbar\omega = 20meV$), and dotted: with localizing electric field ($\hbar\omega = 40meV$). (b) Time-integrated Terahertz signal as function of the ratio between the amplitude and the frequency of the oscillating electric field $eFd/\hbar\omega$, for $\hbar\omega = 20meV$.

## RESULTS

In the linear optical regime an oscillating applied electric field has strong influence on the spectrum. The full line in Fig. 1a shows the linear spectrum of a $GaAs/AlAs$ superlattice in the absence of an applied field. The spectrum exhibits an exciton resonance with a binding energy of about $7meV$. The wavefunction of this exciton is three-dimensional with anisotropic shape[9]. The high energy shoulder is related to the miniband dispersion. The dashed line in Fig. 1a shows the spectrum for the case of an oscillating applied electric field (with frequency $\omega = 20meV/\hbar$) which leads to dynamic localization, if the amplitude of the field is chosen to be $eFd/\hbar\omega = 2.4$, being the first root of the Bessel function $J_0$. According to analytical models this corresponds to the first occurrence of dynamic localization[10]. Due to the collapse of the miniband induced by the field the exciton line shifts towards higher energies. Associated with this shift is an increase in oscillator strength and binding energy, which is now about $13meV$. This behavior is due to the fact that the field suppresses the motion in $z$ direction and the exciton changes from an anisotropic three-dimensional to a basically two-dimensional one. This transition can also be analyzed directly based on the exciton wavefunction[9]. The significant absorption below the lowest exciton, which is shown by the dashed line of Fig. 1a, is similar to the below gap-absorption induced by a static field in the Wannier-Stark regime. For alternating electric field, the existence of these resonances is however very sensitive to the superlattice parameters and to the frequency of the applied electric field, see e.g. dotted line in Fig. 1a.

In this localization regime a decrease of the Terahertz radiation, which monitors the movement of electrons and holes in the growth direction, is expected[6]. In Fig. 1b the time-integrated squared Terahertz signal[6] is shown as function of the ratio $eFd/\hbar\omega$. As expected, the minima of this signal occur close to the zeros of the Bessel-function $J_0$. Since we have not used a simple tight-binding model, but a more realistic Kronig-Penney model for the description of the wavefunctions the dynamic localization is not complete[4], i.e. the Terahertz signal does not vanish completely.

## CONCLUSIONS

Within a realistic model for semiconductor superlattices we have demonstrated how dynamic localization can be observed experimentally. In particular it has been shown that the dynamic localization strongly modifies the dimensionality of the super-lattice exciton. This effect, which is associated with an increase of oscillator strength and binding energy of the exciton, should be observable in the linear spectrum. It has also been shown that dynamic localization leads to a strong decrease of the Terahertz signal, for appropriate values of field strength and frequency. By monitoring the ampli-tude of the Terahertz signal as function of the strength of an applied oscillating field, it should be possible to experimentally identify dynamic localization.

## REFERENCES

1. G. von Plessen and P. Thomas, Method for observing Bloch oscillations in the time domain, Phys. Rev. B 45:9185 (1992).
2. J. Feldmann, K. Leo, J. Shah, D.A.B. Miller, J.E. Cunningham, T. Meier, G. von Plessen, A. Schulze, P. Thomas, and S. Schmitt-Rink, Optical investigation of Bloch oscillations in a semiconductor superlattice, Phys. Rev. B 46:7252 (1992).
3. C. Waschke, H.G. Roskos, R. Schwedler, K. Leo, H. Kurz, and K. Köhler, Coherent submillimeter-wave emission from Bloch oscillations in a semiconductor superlattice, Phys. Rev. Lett. 70:3319 (1993).
4. D.H. Dunlap and V.M. Kenkre, Dynamic localization of a charged particle moving under the influence of an electric field, Phys. Rev. B 34:3625 (1986).
5. T. Meier, G. von Plessen, P. Thomas, and S.W. Koch, Coherent electric-field effects in semiconductors, Phys. Rev. Lett. 73:902 (1994).
6. T. Meier, G. von Plessen, P. Thomas, and S.W. Koch, Coherent effects induced by static and time-dependent electric fields in semiconductors, Phys. Rev. B 51:14490 (1995).
7. for a textbook discussion see, H. Haug and S.W. Koch, "Quantum Theory of the Optical and Electronic Properties of Semiconductors", 3rd ed., World Scientific, Singapore (1994).
8. F. Rossi, T. Meier, P. Thomas, S.W. Koch, P.E. Selbmann, and E. Molinari, Ultrafast carrier relaxation and vertical-transport phenomena in semiconductor superlattices: A Monte Carlo analysis, Phys. Rev. B 51:16943 (1995).
9. T. Meier, F. Rossi, P. Thomas, and S.W. Koch, Dynamic Localization in Anisotropic Coulomb Systems: Field induced Crossover of the Exciton Dimension, submitted for publication.
10. M. Holthaus, Collapse of minibands in far-infrared irradiated superlattices, Phys. Rev. Lett. 69:351 (1992).

# THE ROLE OF COHERENCE IN THE PHOTOGENERATION
# PROCESS OF HOT CARRIERS

T. Kuhn,[1] F. Rossi,[2] A. Leitenstorfer,[3] A. Lohner,[3] T. Elsaesser,[3]
W. Klein,[4] G. Boehm,[4] G. Traenkle,[4] and G. Weimann[4]

[1]Lehrstuhl für Theoretische Physik, Brandenburgische Technische
Universität, Karl-Marx-Straße 17, 03013 Cottbus, Germany
[2]Fachbereich Physik und Zentrum für Materialwissenschaften,
Philipps-Universität Marburg, 35032 Marburg, Germany
[3]Physik Department E11, Technische Universität München,
85748 Garching, Germany
[4]Walter-Schottky-Institut, Technische Universität München,
85748 Garching, Germany

## INTRODUCTION

The process of carrier generation is the basic step for all experiments analyzing the relaxation dynamics of photoexcited carriers. It is usually described by a semiclassical generation rate with spectral and temporal characteristics determined by the laser pulse. In that case, the fact that a coherent laser pulse induces coherence also in the carrier system is completely neglected. Such a coherence is expressed in terms of the interband polarization. The dynamics of the interband polarization is strongly influenced by carrier-carrier and carrier-phonon scattering processes resulting in a dephasing of the polarization which, because carrier generation involves the interaction of the polarization with the light field, leads to a strong density dependence of the generation rate.

In this contribution we present a theoretical and experimental analysis of the role played by a coherent modeling of the generation and recombination processes of nonequilibrium electrons in GaAs. Time-integrated band-to-acceptor luminescence spectra calculated from an approach including the coherent interband polarization are in good agreement with experimental results which demonstrate initial carrier distributions much broader than the spectrum of the femtosecond excitation pulse.

## CARRIER GENERATION PROCESS

On the semiclassical level the generation and scattering dynamics of photoexcited carriers is described by the Boltzmann equation (BE) for the distribution functions of electrons ($f_{\mathbf{k}}^{e}$)

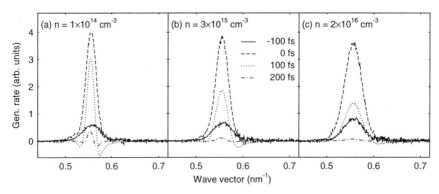

**Figure 1.** Generation rate of electrons as a function of $k$ at different times and densities.

according to

$$\frac{d}{dt} f_{\mathbf{k}}^{e} = g_{\mathbf{k}}(t) - \sum_{\mathbf{k'}} [W_{\mathbf{k'k}}^{e} f_{\mathbf{k}}^{e} (1 - f_{\mathbf{k'}}^{e}) - W_{\mathbf{kk'}}^{e} f_{\mathbf{k'}}^{e} (1 - f_{\mathbf{k}}^{e})] \tag{1}$$

and the analogous Boltzmann equation for the distribution function of holes ($f_{\mathbf{k}}^{h}$). Here, the generation rate $g_{\mathbf{k}}(t)$ and the scattering rates $W_{\mathbf{k'k}}^{e}$ are calculated from Fermis golden rule. The generation rate is, except for phase space filling effects, fully determined by the temporal and spectral characteristics of the laser pulse. On this level all effects related to the coherence of the laser pulse are neglected. The coherent dynamics in semiconductors, on the other hand, is described by the semiconductor Bloch equations (SBE) which involve, besides the distribution functions, also the interband polarization $p_{\mathbf{k}}$. Its dynamics is described by

$$\frac{d}{dt} p_{\mathbf{k}} = \frac{1}{i\hbar} \left[ \left( \epsilon_{\mathbf{k}}^{e} + \epsilon_{\mathbf{k}}^{h} \right) p_{\mathbf{k}} + M_{\mathbf{k}} E_0(t) e^{-i\omega t} \left( 1 - f_{\mathbf{k}}^{e} - f_{-\mathbf{k}}^{h} \right) \right]$$
$$- \sum_{\mathbf{k'}} \left[ W_{\mathbf{k'k}}^{p} p_{\mathbf{k}} - W_{\mathbf{kk'}}^{p} p_{\mathbf{k'}} \right] . \tag{2}$$

The dynamics of the distribution functions is again described by Eq. (1), the generation rate, however, being calculated according to

$$g_{\mathbf{k}} = \frac{1}{i\hbar} \left[ M_{\mathbf{k}} E_0(t) e^{-i\omega t} p_{\mathbf{k}}^{*} - \text{c.c.} \right] . \tag{3}$$

Here, $M_{\mathbf{k}}$ is the dipole matrix element, $E_0(t)$ is the temporal shape of the laser pulse with frequency $\omega$, $\epsilon_{\mathbf{k}}^{e,h}$ denote the single-particle energies of electrons and holes, and c.c. denotes complex conjugate.* The scattering part in Eq. (2) has the same structure as in Eq. (1) and the corresponding matrices are given by

$$W_{\mathbf{k'k}}^{p} = \frac{1}{2} \sum_{\nu=e,h} [W_{\mathbf{k'k}}^{\nu} (1 - f_{\mathbf{k'}}^{\nu}) + W_{\mathbf{k'k}}^{\nu} f_{\mathbf{k'}}^{\nu}] . \tag{4}$$

The generation rate (Eq. (3)) involves the interband polarization which is influenced by the density-dependent scattering processes entering Eq. (2) and therefore, in contrast to the semiclassical case, the generation rate becomes density dependent.

The second term in the second row of Eq. (2) which has the structure of an in-scattering term is often neglected on the basis of a random phase argument because it involves a summation

---

*Coulomb renormalizations of the carrier energies and the external field are taken into account in the calculations. However, they are not important under the conditions studied here.

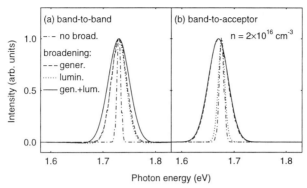

**Figure 2.** Luminescence linewidth of (a) band-to-band and (b) band-to-acceptor transitions calculated for a carrier distribution as generated by a 150 fs laser pulse.

over the complex polarization components. Then, the scattering part for $p_k$ reduces to the common structure with a k-dependent dephasing time $T_2$. However, it turns out that a correct modeling of the dephasing including the in-scattering term is crucial to obtain a physically reasonable density dependence of the dynamics.[1]

To compare the different models of carrier dynamics in photoexcited semiconductors the BE and the SBE have been solved numerically for GaAs excited by a 150 fs laser pulse. The BE has been solved by a standard ensemble Monte Carlo (EMC) simulation. The solution of the SBE is based on a combined technique including a generalized Monte Carlo simulation for the incoherent part of both distribution functions and polarization and a direct integration of the coherent part of the equations.

In Fig. 1 the generation rate at different times as obtained from the SBE is plotted as a function of wave vector for three different densities. At the lowest density the behavior is essentially the same as in the case without carrier-carrier scattering: Energy-time uncertainty leads to an initially very broad generation rate; with increasing time the line narrows and, in the tails, exhibits negative parts due to a stimulated recombination of carriers initially generated off-resonance.[2] After the pulse, the distribution function of the generated carriers is in good agreement with the BE result. Scattering processes destroy the coherence between electrons and holes which is necessary for the stimulated recombination processes. As a consequence, with increasing density the negative tails are strongly reduced and the generation remains broad for all times resulting in a much broader carrier distribution than in the BE case.

## LUMINESCENCE LINE SHAPE

Carriers are generated due to the interaction of the electronic system with the electro-magnetic field. The luminescence which is used to monitor the carrier dynamics is based on the same interaction mechanism. Therefore the radiative recombination process should be influenced by scattering processes in the same way as the generation process resulting in a density dependent broadening. In contrast to the generation by a laser pulse, where the light field is treated classically, the luminescence due to spontaneous emission requires a quantum mechanical treatment of the photon field. A quantum kinetic calculation can be performed along the same line as for carrier-phonon interaction[3] by introducing photon-assisted density matrices. The scattering part for these photon-assisted density matrices is the same as for the interband polarization resulting in the same broadening as for the generation process.

The situation changes, however, when looking at band-to-acceptor transitions. These transitions are described by an electron-acceptor polarization $p_k^a$. The quantum mechanical

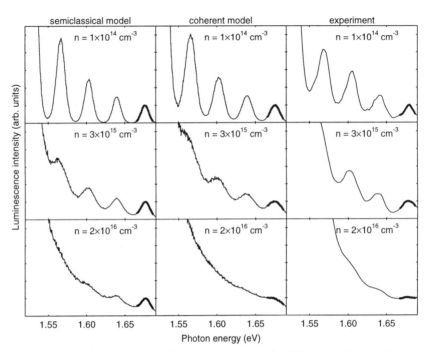

**Figure 3.** Calculated and measured band-to-acceptor spectra for different carrier densities.

calculation leads to a scattering part for this variable according to[4]

$$\frac{d}{dt}p_{\mathbf{k}}^a\bigg|_{\text{scat}} = -\sum_{\mathbf{k}'}\left[W_{\mathbf{k}'\mathbf{k}}^a p_{\mathbf{k}}^a - \beta_{|\mathbf{k}-\mathbf{k}'|}W_{\mathbf{k}\mathbf{k}'}^a p_{\mathbf{k}'}^a\right] \tag{5}$$

where $\beta_q = [1 + (qa_B/2)^2]^{-2}$ is the Fourier transform of the charge density in the acceptor state, $a_B$ being the acceptor Bohr radius. Since holes are not involved in the transitions, the scattering matrix is determined only by scattering processes of electrons, i.e.,

$$W_{\mathbf{k}'\mathbf{k}}^a = \frac{1}{2}\left[W_{\mathbf{k}'\mathbf{k}}^e\left(1 - f_{\mathbf{k}'}^e\right) + W_{\mathbf{k}'\mathbf{k}}^e f_{\mathbf{k}'}^e\right] . \tag{6}$$

Due to the complexity of the problem a full time-dependent quantum-kinetic calculation of the luminescence spectrum has not yet been performed. However, the importance of the broadening can be analyzed by looking at the spectrum produced by a given carrier distribution. In Fig. 2 band-to-band (BB) and band-to-acceptor (BA) spectra corresponding to an excitation by a 150 fs pulse and ignoring any relaxation of the carrier distributions after the pulse are shown for different models at a density of $2 \times 10^{16}$ cm$^{-3}$. In a semiclassical treatment generation and recombination exhibit no broadening and the width of the spectra is determined by the energetic width of the laser pulse (dash-dotted lines). If the broadening of the generation rate as shown in Fig. 1 is taken into account while still neglecting the broadening of the luminescence a strong broadening of the spectra is observed at this density (dashed lines). If only the broadening of the luminescence is taken into account we find the same broadening in the BB spectrum, however, the BA luminescence exhibits only a very small broadening (dotted lines). This behavior is due to the fact that the broadening in the BB transitions is mainly produced by scattering processes of the holes with their much higher density-of-states. As a consequence, when taking into account the broadening of generation and recombination processes, we find an additional broadening of the BB transition while

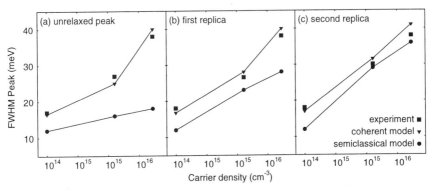

**Figure 4.** Full width at half maximum of the unrelaxed peak and the first and second phonon replica as a function of carrier density obtained from the calculated and measured spectra. The lines connecting the theoretical values are meant as guides to the eye.

for the BA spectrum the broadening of the luminescence has a negligible effect (solid lines). This result justifies the use of a semiclassical luminescence model in the dynamic calculations discussed in the next section.

## BAND-TO-ACCEPTOR LUMINESCENCE

Hot carrier luminescence has proven to be a powerful technique to study the ultrafast dynamics of photoexcited carriers in semiconductors.[5] In BA luminescence experiments, due to their high sensitivity, carrier densities as low as several $10^{13}$ cm$^{-3}$ have been reached. Thus, the transition between a dynamics dominated by carrier-phonon scattering to a dynamics dominated by carrier-carrier scattering with its consequences for the generation process discussed above is experimentally accessible.[6,7]

In the experiment a 3 $\mu$m thick GaAs layer doped with Be acceptors of a concentration of $3 \times 10^{16}$ cm$^{-3}$ is excited at a photon energy of 1.73 eV by transform limited 150 fs pulses from a mode-locked Ti:sapphire laser. Time-integrated luminescence spectra are recorded.[7] The right column of Fig. 3 shows measured BA luminescence spectra for three different values of the carrier density. At low density we observe an initial peak of the generated carrieres (marked bold) and pronounced replicas due to the emission of an integer number of optical phonons. With increasing density the peaks become broader and at the highest density only a slight structure related to the phonons is still visible.

These measured spectra are compared with the results of simulations based on the BE and on the SBE (left and centered column in Fig. 3). We find several pronounced differences between the models. In the semiclassical model the unrelaxed peak is clearly visible up to the highest density in contrast to the coherent model, where, in agreement with the experiment, this peak is strongly broadened. At the higher densities the semiclassical spectra exhibit an increase in the broadening of subsequent replicas which is not present in the coherent calculations and in the experimental results.

To demonstrate this difference quantitatively, in Fig. 4 we have plotted the full width at half maximum (FWHM) of the three highest peaks in the spectra. From the semiclassical model we obtain a strong increase in the density dependence of subsequent peaks. This behavior can be easily understood: Due to the emission time of an optical phonon of about 150 fs the carriers populate subsequent replicas at increasing times and, therefore, the efficiency of carrier-carrier scattering processes in broadening the peaks increases also. However, the measured spectra exhibit a different behavior which is quantitatively reproduced by the coherent model:

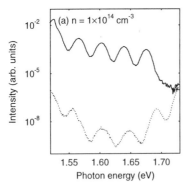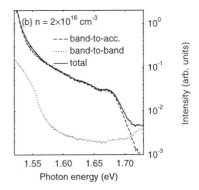

**Figure 5.** Band-to-band, band-to-acceptor, and total spectra for two densities obtained from the coherent model plotted on a logarithmic scale.

Already the unrelaxed peak exhibits a strong increase with increasing density and the density dependence of all replicas is approximately the same. The reason is that the broading of the generation process as discussed above obviously dominates over the broadening due to subsequent scattering processes. As has been shown in the previous section, this is due to the fact that the generation process is influenced by the scattering of electrons and holes, while the broadening during the relaxation processe is only due to scattering of the electrons.

Electrons do not only recombine with bound holes at the acceptors but also with free holes in the valence band. This results in a BB background in the spectra. In order to estimate the magnitude of this background, we have calculated also the BB luminescence spectrum in the coherent model. The results are shown in Fig. 5 on a logarithmic scale. We find that at the lowest density the BB spectrum in the relevant energy range is more the four orders of magnitude smaller than the BA spectrum. With increasing density this background increases, however, even at the highest density it is still one order of magnitude smaller.

## CONCLUSIONS

We have presented theoretical and experimental results demonstrating the importance of dephasing processes for the analysis of luminescence spectra. The density dependence of the spectra in the interesting region of a transition between carrier-phonon and carrier-carrier dominated dynamics can only be explained by a coherent modeling of the carrier generation including the dynamics of the interband polarization. The main mechanism determining the width of the peaks is the broadening of the generation rate. It turned out that for band-to-acceptor spectra the broadening of the recombination process is not important, however, for band-to-band spectra also this phenomenon should be taken into account. In the present case, the background due to band-to-band luminescence has been found to be of negligible importance.

## REFERENCES

1. T. Kuhn, S. Haas, and F. Rossi, phys. stat sol. (b) **188**, 417 (1995).
2. T. Kuhn and F. Rossi, Phys. Rev. B **46**, 7496 (1992).
3. J. Schilp, T. Kuhn, and G. Mahler, Phys. Rev. B **50**, 5435 (1994).
4. T. Kuhn, to be published.
5. T. Elsaesser *et al.*, Phys. Rev. Lett. **66**, 1757 (1991).
6. J. A. Kash, Phys. Rev. B **40**, 3455 (1989) and Phys. Rev. B **51**, 4680 (1995).
7. A. Leitenstorfer *et al.*, Phys. Rev. Lett. **73**, 1687 (1994).

# COHERENT DYNAMICS OF IMPACT IONIZATION IN BULK AND LOW-DIMENSIONAL SEMICONDUCTORS

H. Schröder,[1] H. Buss,[1] T. Kuhn,[1,2] and E. Schöll[1]

[1] Institut für Theoretische Physik, Technische Universität Berlin,
Hardenbergstraße 36, 10623 Berlin, Germany
[2] Lehrstuhl für Theoretische Physik, Brandenburgische Technische
Universität, Karl-Marx-Straße 17, 03013 Cottbus, Germany

## INTRODUCTION

In semiclassical kinetic theories scattering processes occur between well-defined single-particle states by conserving energy and momentum. This leads to a strong reduction of the phase-space for the final state in a scattering process. On short time-scales, however, due to energy-time uncertainty, the single-particle energy of a carrier is not yet a well-defined quantity which results in a time-dependent broadening of transitions as described by quantum kinetic theories. Deviations between semiclassical and quantum kinetics are particularly pronounced for cases where, due to phase-space restrictions, semiclassical scattering processes are completely inhibited in certain energy ranges. Coulomb scattering processes which in the semiclassical limit have a well-defined threshold energy occur typically in multi-band systems. In such systems two carriers initially in two bands $i$, $j$ may interact via the Coulomb potential ending up in the bands $m$, $n$. In a two-band system essentially two types of such processes with threshold occur: (i) A carrier in the higher band interacts with a carrier in the lower band and both carriers end up in the higher band. This is a typical impact ionization process.[1-3] (ii) The interaction between two carriers in the lower band leads to a transition of both carriers into the higher band. Such processes are not allowed between valence band and conduction band, however, they are possible between subbands in low-dimensional structures. In this contribution we present a quantum kinetic analysis of these two kinds of Coulomb processes based on a density-matrix approach. First, we investigate the intersubband dynamics in a quantum wire and then we study coherent effects for impact ionization processes in bulk semiconductors.

## QUANTUM KINETIC APPROACH

The analysis of the Coulomb quantum kinetics is based on a density matrix approach for a multi-band model. The Hamiltonian is given by

$$H = \sum_{i,\mathbf{k}} \epsilon_{i,\mathbf{k}} c_{i,\mathbf{k}}^\dagger c_{i,\mathbf{k}} + \frac{1}{2} \sum_{\substack{i,j,l,m \\ \mathbf{k},\mathbf{k}'\mathbf{q}}} V_{ijlm}(\mathbf{q}) c_{i,\mathbf{k}}^\dagger c_{j,\mathbf{k}'}^\dagger c_{l,\mathbf{k}'+\mathbf{q}} c_{m,\mathbf{k}-\mathbf{q}} \cdot \tag{1}$$

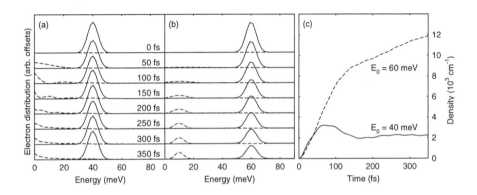

**Figure 1.** Distributions function of electrons in the lower (solid lines) and upper (dashed lines) subband for an initially Gaussian distribution in the lower subband, (a) below and (b) above the semiclassical threshold for intersubband transitions, and (c) electron densities in the upper subband as functions of time for both cases. In (a) the distribution functions in the upper subband have been multiplied by a factor of 10.

Here, $c^\dagger_{i,\mathbf{k}}$ ($c_{i,\mathbf{k}}$) denotes the creation (annihilation) operator of an electron in band $i$ with momentum $\mathbf{k}$, $\epsilon_{i,\mathbf{k}}$ is the single-particle energy (here: parabolic bands), and $V_{ijlm}(\mathbf{q})$ is the Coulomb matrix element. The basic kinetic variables are the single-particle density matrices $f_{ij}(\mathbf{k}) = \langle c^\dagger_{i,\mathbf{k}} c_{j,\mathbf{k}} \rangle$. The semiclassical limit is completely determined by the diagonal elements, the carrier distribution functions. A quantum kinetic description is characterized by additional variables, in particular the off-diagonal elements describing coherence between different bands and, in the case of Coulomb interaction, two-particle correlations $s_{ijlm}(\mathbf{k}, \mathbf{k}', \mathbf{q}) = \langle c^\dagger_{i,\mathbf{k}} c^\dagger_{j,\mathbf{k}'} c_{l,\mathbf{k}'+\mathbf{q}} c_{m,\mathbf{k}-\mathbf{q}} \rangle + f_{il}(\mathbf{k}) f_{jm}(\mathbf{k}') \delta_{\mathbf{k}',\mathbf{k}-\mathbf{q}}$. Neglecting higher order correlations due to higher-order terms in the Coulomb matrix element, a coupled set of equations of motion for the single-particle and two-particle variables is obtained.[3,4]

## INTERSUBBAND DYNAMICS IN QUANTUM WIRES

In low-dimensional systems where several subbands are involved in the dynamics two kinds of processes may occur which lead to transitions between the subbands. Matrix elements like $V_{2212}$ lead to processes where one electron changes from subband 1 to subband 2 while the other electron remains in subband 2. However, in many heterostructures matrix elements of this type vanish for symmetry reasons. Matrix elements like $V_{2211}$, on the other hand, are nonzero. They describe processes where both carriers make a transition from subband 1 to subband 2. In the semiclassical limit such processes have a threshold given by the condition $E_1 + E_2 = 2E_S$, where $E_1, E_2$ denote the initial energies of the carriers and $E_S$ is the subband splitting.

We have investigated the intersubband dynamics due to $V_{2211}$ in a quantum wire with $E_S = 50\,\mathrm{meV}$ based on the quantum kinetic approach for an initial Gaussian distribution of electrons in the lower subband. Fig. 1(a) shows the evolution of the distribution functions in the upper and lower subband for the case where the initial distribution is below the second subband. In the semiclassical limit no transitions are possible. In the quantum kinetic case we notice that, due to energy-time uncertainty, initially a broad occupation in the second subband builds up which narrows with increasing time. The total density in the second subband exhibits a decrease after the initial rise and, after about 100 fs, remains essentially constant (Fig. 1(c)). For an initial condition above the semiclassical threshold initially a strong broadening is observed also (Fig. 1(b)). In this case, however, the distribution in the second subband approaches the shape of the initial distribution and the density keeps growing.

## INTERBAND DYNAMICS IN BULK SEMICONDUCTORS

As a second example we have studied the impact ionization process of laser-pulse excited carriers in a bulk semiconductor. In this case it is convenient to work in an electron-hole picture using the

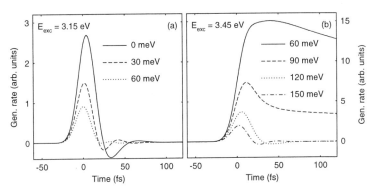

**Figure 2.** Generation rate of electrons as a function of time at different energies above the gap obtained from the coherent impact ionization model. The carriers are excited by a 20 fs laser pulse. In (a) the laser excitation is below, in (b) above the the semiclassical threshold.

indices $e$ and $h$. The interaction with the laser pulse is treated in terms of the well-known semiconductor Bloch equations[5] introducing an interband polarization $p_{\mathbf{k}}$. Impact ionization and Auger processes of electrons are then described by the matrix elements $V_{eehe}$ and $V_{ehee}$. The corresponding processes for holes are typically irrelevant because, due to the large hole mass, their semiclassical threshold energy is very high. The high dimensionality of two-particle correlations (which now depend on three 3D-vectors) require additional approximations in order to obtain a numerically tractable problem. By concentrating on the region close to the semiclassical threshold, we neglect the energy difference of the final states of the electrons. If, in addition, the momentum dependence of the matrix element in the relevant region is neglected (Keldysh model[6]), the two-particle correlations may be integrated over the angles resulting in functions of three scalar variables only.

Figure 2 shows the electron generation rate due to impact ionization processes at different energies above the gap for the case of excitation with a 20 fs laser pulse. For an excitation below the semiclassical threshold (Fig. 2(a)) we observe a transient impact ionization and, as in the case of intersubband transitions, a subsequent negative generation rate due to induced Auger recombination processes. For an excitation above threshold (Fig. 2(b)) the generation rate remains positive after the pulse at energies where semiclassically transitions are allowed while it approaches zero in the forbidden region exhibiting a similar oscillatory behavior as for an excitation below threshold.

## CONCLUSIONS

We have presented a quantum kinetic analysis of intersubband and interband Coulomb processes which change the carrier number in the bands. On short times energy-time uncertainty leads to transitions also in the parameter region which is semiclassically forbidden. Due to the coherence in the system, in this case the population is reduced at later times by induced inverse (Auger recombination) processes which is analogous to a negative generation rate in the interband dynamics described by the semiconductor Bloch equations.

This work has been supported by the Deutsche Forschungsgemeinschaft within the framework of the SFB 296.

## REFERENCES

1. N. Sano and A. Yoshi, Phys. Rev. B **45**, 4171 (1992).
2. J. Bude, K. Hess, and G. J. Iafrate, Phys. Rev. B **45**, 10958 (1992).
3. W. Quade, E. Schöll, F. Rossi, and C. Jacoboni, Phys. Rev. B **50**, 7398 (1994).
4. T. Kuhn, S. Haas, and F. Rossi, phys. stat sol. (b) **188**, 417 (1995).
5. T. Kuhn and F. Rossi, Phys. Rev. B **46**, 7496 (1992).
6. L. V. Keldysh, Sov. Phys.–JETP **10**, 509 (1960).

# ULTRAFAST COHERENT CARRIER CONTROL
# IN QUANTUM STRUCTURES

Jeremy J. Baumberg,[1] Albert P. Heberle,[1] K. Köhler,[2] and K. Ploog[3]

[1]Hitachi Cambridge Laboratory, Cavendish Labaratory, Cambridge, CB3 0HE, U.K.
[2]Fraunhofer-Institut für Angewandte Festkörperphysik, 79108 Freiburg, Germany.
[3]Paul-Drude-Institut für Festkörperelektronik, 10117 Berlin, Germany.

## INTRODUCTION

The advent of stable sources of ultrafast laser pulses has stimulated the investigation of coherent dynamics in solid-state systems.[1] Phase becomes an important factor in interactions with matter when the optical transitions have a phase relaxation time exceeding pulse separations. We show the relative phase between successive pulses in this case strongly influences the evolving dynamics as seen previously in molecular systems[2] and recently with THz downmixing in quantum wells.[3] Experiments have shown that many body interactions cause distinct differences between the optical transitions of semiconductors and ideal two-level systems.[4] Despite these complications,[5] we demonstrate the progression from exploration to the *manipulation* of coherent carrier dynamics in semiconductor quantum structures. We contrive the destruction of photoexcited carrier populations within a few hundred femtoseconds of their creation. This technique produces femtosecond optoelectronic nonlinearities which are faster than suggested by the transition linewidth.

## EXPERIMENTAL TECHNIQUE

Differential reflection ($\Delta R$) measurements are adopted to resolve the exciton population dynamics using the conventional pump-probe geometry in Figure 1(a). A mode-locked Ti:sapphire laser with dispersion precompensation for subsequent optics serves as the source of 100 fs transform-limited pulses. A computer-controlled Michelson interferometer produces two pump pulses of temporal separation, $\tau_{12}$. Accuracy of 0.01 $\lambda/c$ (0.03 fs) is achieved by active stabilization on the interference fringes from a HeNe laser beam copropagating with the pump pulses. An 80 nm step on one of the interferometer mirrors generates two sets of HeNe fringes with a $\pi/2$ phase difference to allow stabilization at any phase delay. Figure 1 shows the spectrum for one (b) and both (c) pulses centered at 804 nm when $\tau_{12}$ is 10 times the pulsewidth. Minima in the spectrum appear at wavelengths $\lambda$ for which the light taking different paths in the interferometer interferes destructively: $(n+1/2) \lambda = c \tau_{12}$ ($n$ integral). Translation by $\lambda/2$ ($\Delta\tau_{12}=1.34$ fs) inverts this interference pattern. The high contrast fringes

*Hot Carriers in Semiconductors*
Edited by K. Hess *et al.*, Plenum Press, New York, 1996

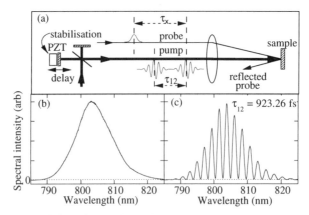

**Figure 1:** (a) Schematic experimental arrangement. (b) Spectrum of a single pump pulse. (c) Spectrum of the double pump pulse sequence at a separation of ten pulsewidths (923.26 fs). Curve (b) is scaled up by a factor of 4.

remain locked for any pump-pulse separation demonstrating the excellent phase stability of the arrangement. Pump and probe beams are separately acousto-optically modulated and focussed to a 200 μm spot on the sample. The pump-induced reflectivity change (ΔR) accesses the exciton population sampled at the probe delay time $\tau_x$. The pump and probe beams are polarized perpendicularly and have comparable intensities, each pulse exciting exciton densities of ~$10^{10}$ cm$^{-2}$ for the data shown here.

Two different samples were used for the experiments: sample A contains five 120 Å GaAs QWs separated by 120 Å Al$_{0.3}$Ga$_{0.7}$As barriers whilst sample B contains ten 250 Å GaAs QWs separated by barriers which consist of 8.5 Å AlAs, 300 Å Al$_{0.3}$Ga$_{0.7}$As and 8.5 Å AlAs. The luminescence line widths of 0.7 and 0.3 meV, respectively, demonstrate the high quality of the samples. Using a reflection geometry avoids strain problems which are inevitable in samples etched for transmission experiments and we take all measurements at a temperature of 4 K.

## PHASE SPECTROSCOPY

Figure 2 shows the phase-dependent effect of the second pump pulse on the total carrier density as a function of inter-pump pulse separation, $\tau_{12}$. Oscillations are visible in ΔR recorded by the probe pulse 10ps after the first pump pulse impinges on the sample, by which time most excitons have lost their phase coherence. Figure 2(a,b) shows selected sections of the several thousand fringes recorded on sample A when the laser was tuned to the heavy-hole exciton energy, $E_{hh}$, and a narrow pump spectrum (150 fs pulsewidth) optimised to minimise free-carrier excitation. The reflectivity change directly tracks the total exciton population and oscillates with a period $T_{hh}=h/E_{hh}$. Simultaneously-recorded interference of the pump pulses on a photodiode (open circles) demonstrates the strong influence of their relative phase long after they have ceased to overlap in time.

In the frequency domain these oscillations are easily understood as the direct overlap of the narrow exciton absorption line with the spectral fringes (Fig. 1c). As $\tau_{12}$ increases, the fringes move through the transition every $T_{hh}$ and as the separation of the fringes simultaneously narrows ($\propto 1/\tau_{12}$), their wings unavoidably excite the exciton degrading the interference minima. This picture suggests that the exciton responds only to the spectral band with which it overlaps implying a slow response limited by the inverse linewidth. However we will show that this is *not* the case and additional off-resonant excitation creates an ultrafast response.

From the oscillations, a slowly varying amplitude A(t) and phase φ(t) is extracted by fitting the oscillations piecewise in sections of ten periods. This technique is ideally suited to explore several current issues in exciton spectroscopy such as the effect of the local polarization field on the excitons and the microscopic nature of phase scattering outside the impact approximation.[6,7] Figure 2(c,d) shows results for sample B excited 0.9 meV above the heavy-hole exciton energy at 1525 meV. The laser interferogram was treated in the same way (dashed lines) and the lack of curvature in $\varphi_{laser}$ shows chirp is absent. Although the absolute phase value at $\tau_{12}=0$ is arbitrarily set, the slope of φ(t) dynamically tracks the evolution of the exciton energy. Strong heavy- light-hole quantum beating is observed with a period of $h/(4.2 \text{ meV})$ in agreement with the separation of heavy- and light-hole exciton lines seen in the linear reflectance spectrum, both of which are covered by the laser spectrum.

The quantum beats appear as steps in $\varphi_{AR}$ whose height is expected to be π only when equal electronic polarizations are generated. The bias of the laser towards the lower line and the different oscillator strengths of the two transitions imbalance the modulation here and reduce this step height. As time progresses, the steps increase in height suggesting that the phase of the light holes relaxes more *slowly* than the phase of the heavy holes. This effect is also visible in the increased modulation with time. Such non-intuitive results would imply that dephasing strongly depends on the angular momentum or the mass of the excitons. Further analysis of these results is hindered by the lack of theoretical models for the microscopic dephasing process.

A striking phase shift appears around $\tau_{12}=0$ which offsets the oscillations for $\tau_{12}<0$ and $\tau_{12}>0$. Although the quantum beats prevent direct determination of the energy shifts from the slope of $\varphi_{AR}$, they do reveal additional features. The phase shift indicates an instantaneous frequency excursion shown in Figure 2(d,inset), present only when the pump pulse hits the sample. Such an ultrafast transient would be expected from the optical Stark effect[8] (OSE) which cannot be measured in the small detuning, non-adiabatic regime using any existing technique. In particular there exist many substantial theoretical treatments of the OSE which show the dominating effect of Coulomb interactions at small detunings. As yet, experiments in quantum confined systems have probed only the phase-space filling component and not implicated exciton-exciton contributions. However such a phase shift might also be explained by excitation of free carriers and their subsequent ultrafast dephasing. Studies as a function of laser power and detuning to address these issues are planned.

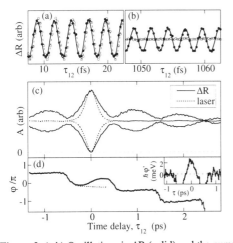

Figure 2: (a,b) Oscillations in ΔR (solid) and the pump interferogram (dashed) near zero delay and for $\tau_{12}\sim$1ps. (c,d) Amplitude A and phase φ of fit to oscillations. inset: differential of the phase near zero delay.

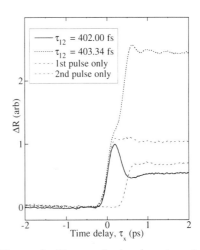

Figure 3: Time-resolved coherent control showing enhancement at 403.34fs and destruction at 402.00fs. The dotted curves record the effect of each pump pulse alone.

# TIME-RESOLVED COHERENT CONTROL

Destructive interference occurs when a minimum in the pump pair spectrum coincides with the exciton line. The first pulse alone, however, does not have interference minima in its spectrum and therefore must create excitons which are destroyed later by the second pulse. A similar effect has been proposed to enhance ultrafast carrier-carrier scattering.[9] Heisenberg's principle transiently broadens the initial spectral content of a transform limited laser pulse above the frequency width of the full pulse. Carriers created by this excess spectral content are removed during the later part of the pulse if there is no scattering. In the presence of scattering, these excess carriers cannot be reemitted coherently because they have lost their phase information. This effect is predicted to result in a much larger energy distribution of the carriers than would be expected from the laser pulsewidth alone although such conclusions remain contentious. With conventional techniques it is impossible to resolve the transiently-broad carrier distribution directly because the appropriate time resolution would degrade the spectral uncertainty. However, with the phase-locked double pulse we indeed resolve the transient carriers created by such an initial spectral distribution, which in our case is the spectrum of the first pulse alone (Fig. 1(b)). Correspondingly, the final spectral distribution here is the spectrum of the double pulse (Fig. 1(c)).

The time-resolved evolution of the population at a fixed inter-pump pulse separation demonstrates a transiently-excited exciton population. Figure 3 presents such evidence in $\Delta R(\tau_x)$ of sample A for fixed pump pulse separations ($\tau_{12}$) of 402.00 and 403.34 fs, equivalent to 150.00 and 150.50 $T_{hh}$. The dotted curves show the effect of each pump pulse alone, with the later pulse set to excite a smaller carrier density than the first one. The solid line (150.50 $T_{hh}$) resolves the creation of an initial exciton population followed by its subsequent destruction within a few hundred femtoseconds. The resulting exciton density collapses below that for either pump pulse alone, unambiguously verifying the coherent destruction of carriers. In the opposite extreme, if the second laser pulse arrives 150.00 $T_{hh}$. later (dashed), *more* carriers are generated than from the sum of the two pump pulses individually. The in-phase carriers that are already excited increase the subsequent probability of absorption if they remain coherent. As the time delay between the pump pulses increases, more phase scattering occurs and destruction is lost.

## CONCLUSION

Phase-locked femtosecond pulses can coherently manipulate the photoexcited carriers in quantum structures. Analysis of the phase and intensity of the observed oscillations in the carrier density reveal energy shifts underlying the coherent exciton dynamics. Resolving the destruction process of excitons involved in the interference process highlights ultrafast optical dynamics which is not limited by the optical transition line width. Such dynamical behaviour is promising for novel optoelectronic devices.

## REFERENCES

1. for recent reviews see special issue, phys.stat.sol.(b) 173: (1992) and references therein.
2. W.S. Warren, H. Rabitz, and M. Dahleh, Science 259:1581 (1993) and references therein.
3. P.C.M. Planken *et al.*, Phys. Rev. 48:4903 (1993); I. Brener *et al.*, J. Opt. Soc. Am. 11:2457 (1994).
4. S. Weiss *et al.*, Phys. Rev. Lett. 69:2685 (1992).
5. S.T. Cundiff *et al.*, Phys. Rev. Lett. 73:1178 (1994).
6. H. Wang *et al.*, Phys. Rev. Lett. 74:3065 (1995).
7. J.-Y. Bigot *et al.*, Phys. Rev. Lett. 70:3307 (1993).
8. S. Schmitt-Rink, D.S. Chemla and H. Haug, Phys. Rev. B 37:941 (1988).
9. A.V. Kuznetsov, Phys. Rev. 44:13381 (1991); T. Kuhn and F. Rossi, Phys. Rev. Lett. 69:977 (1992)

# COHERENT DYNAMICS AND DEPHASING OF INTER– AND INTRABAND–EXCITATIONS IN SEMICONDUCTOR HETEROSTRUCTURES

E. Binder and T. Kuhn

Lehrstuhl für Theoretische Physik, Brandenburgische Technische
Universität, Karl-Marx-Straße 17, 03013 Cottbus, Germany

## INTRODUCTION

The analysis of the coherent dynamics on ultrashort time-scales provides valuable information on basic material parameters and interaction mechanisms determining, e.g., energy levels, renormalizations, and dephasing rates. While in a spatially homogeneous bulk semiconductor a coherent coupling to a light field results in an interband polarization which can be measured in four-wave-mixing experiments, in heterostructures, due to the breaking of the translational invariance, additionally intraband (intersubband) polarizations can be excited which, if associated with an oscillating dipole moment, lead to the emission of an electromagnetic radiation in the terahertz range.[1] It has been found that the spectral details of this radiation crucially depend on the interband and intraband dephasing processes.[2] In this contribution we present an analysis of the interband and intraband dynamics in heterostructures based on a microscopic description for both the coherent and the scattering part.

## SINGLE-PARTICLE DENSITY MATRICES

The analysis of the dynamics is based on a density matrix approach for a multi-band model. The basic kinetic variables are the single-particle density-matrices, the interband polarizations $p_{ji,\mathbf{k}} = \langle d_{j,-\mathbf{k}} c_{i,\mathbf{k}} \rangle$, the electron intraband density-matrices $f_{ij,\mathbf{k}}^e = \langle c_{i,\mathbf{k}}^\dagger c_{j,\mathbf{k}} \rangle$, with the electron distribution functions as diagonal parts and the electron intraband polarizations as off-diagonal parts, and the corresponding hole intraband density-matrices $f_{ij,\mathbf{k}}^h = \langle d_{i,\mathbf{k}}^\dagger d_{j,\mathbf{k}} \rangle$. Here, $c_{i,\mathbf{k}}^\dagger$ ($c_{i,\mathbf{k}}$) denotes the creation (annihilation) operator of an electron in band $i$ with momentum $\mathbf{k}$, $\epsilon_{i,\mathbf{k}}^e$ is the single particle energy, and $d_{j,\mathbf{k}}^\dagger$, ($d_{j,\mathbf{k}}$) and $\epsilon_{j,\mathbf{k}}^h$ denote the corresponding quantities for holes.

By using the equations of motion for the operators, a coupled set of equations of motion for the single-particle variables is obtained, the generalized multi-subband semiconductor Bloch equations.[3,4] The coherent part of these equations, which, e.g., describes effects like generation and stimulated recombination of carriers, includes mean-field contributions of carrier-carrier interaction. This treatment and the resulting effects have been discussed previously.[2]

*Hot Carriers in Semiconductors*
Edited by K. Hess *et al.*, Plenum Press, New York, 1996

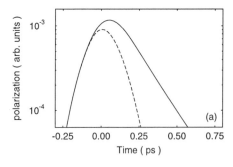 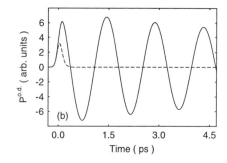

**Figure 1.** (a) Incoherently summed polarizations and (b) off-diagonal part of the intraband dipole-moment with in-scattering terms (solid lines) and neglecting in-scattering terms (dashed lines).

From the second order contribution in the Coulomb matrix element the scattering part of the equations is obtained. We find, e.g., for the intraband polarization:

$$\frac{d}{dt} f^e_{i_1 i_2, \mathbf{k}} = \sum_{i_5} [ - \Gamma^{ee,out}_{i_2 i_5}(\mathbf{k}) f^e_{i_1 i_5, \mathbf{k}} - \Gamma^{ee,out*}_{i_1 i_5}(\mathbf{k}) f^{e*}_{i_2 i_5, \mathbf{k}} \tag{1}$$
$$+ \; \Gamma^{ee,in}_{i_2 i_5}(\mathbf{k}) \left( \delta_{i_1, i_5} - f^e_{i_1 i_5, \mathbf{k}} \right) + \Gamma^{ee,in*}_{i_1 i_5}(\mathbf{k}) \left( \delta_{i_1, i_5} - f^{e*}_{i_2 i_5, \mathbf{k}} \right) ]$$

with the abbreviations

$$\Gamma^{ee,out}_{i_1 i_5, \mathbf{k}} = \sum_{\nu=e,h} \sum_{\substack{i_2, i_3, i_4 \\ i_6, i_7, i_8}} \sum_{\mathbf{k'}, \mathbf{q}} \frac{\pi}{\hbar} \delta \left( -\epsilon^e_{i_5, \mathbf{k}} - \epsilon^\nu_{i_6, \mathbf{k'}} + \epsilon^\nu_{i_7, \mathbf{k'+q}} + \epsilon^e_{i_8, \mathbf{k-q}} \right) \tag{2}$$
$$V^{e\nu}_{i_1 i_2 i_3 i_4}(\mathbf{q}) V^{e\nu}_{i_5 i_6 i_7 i_8}(\mathbf{q}) f^\nu_{i_2 i_6, \mathbf{k'}} \left( \delta_{i_7, i_3} - f^\nu_{i_7 i_3, \mathbf{k'+q}} \right) \left( \delta_{i_8, i_4} - f^e_{i_8 i_4, \mathbf{k-q}} \right) ,$$

and $\Gamma^{ee,in}$, which is obtained by exchanging $f_{ij}$ with $\delta_{ij} - f_{ij}$ in Eq. (2). $V^{e\nu}_{ijlm}(\mathbf{q})$ denotes the Coulomb matrix-element. We have neglected the effect of scattering at interband polarizations which can be justified by the strong damping of these quantities. If we neglect the off-diagonal terms (polarizations) on the r.h.s. in the equation of motion for the distribution functions, it reduces to the well-known Boltzmann transport equation. From the equations of motions for the polarizations we can see that there is not only dephasing, i.e. loss of coherence, but also relaxation, i.e. transport of the intraband-polarizations in $\mathbf{k}$-space. If we neglect terms which have the structure of "in-scattering" terms [second row on the r.h.s. of Eq. (1)] we arrive at a dephasing-time approximation with a $\mathbf{k}$-dependent dephasing-time $T_2$ for each of the polarizations. Equations with a similar structure are obtained for the interband polarization.

## DEPHASING OF THE POLARIZATIONS

We have performed calculations for an asymmetric double quantum-well structure under resonance conditions.[2] The coherent dynamics is induced by a 150 fs laser pulse with an excess energy of 20 meV. In Fig. 1(a) the incoherently summed interband polarizations $p^{inc} = \sum_{\mathbf{k}} \sum_{ij} |p_{ji, \mathbf{k}}|$ are plotted as functions of time. In Fig. 1(b) we show the off-diagonal parts of the intraband dipole-moment $P^{o.d.} = \sum_{\mathbf{k}} \sum_{i \neq j} M_{ij} f^e_{ij, \mathbf{k}}$, where $M_{ij}$ is the intraband dipole-matrix element, as functions of time. The dashed lines are results of a simulation where in-scattering terms have been neglected for both kinds of polarizations. In this case, the dephasing effects are due to removing polarization from the system. This yields similar dephasing rates for both kinds of polarizations, the remaining differences being due to details of the different interaction mechanisms. The dephasing is so strong that no oscillating dipole moment is built up. The solid lines are results of a simulation including in-scattering terms. The dephasing of the interband polarizations is reduced, but the change in the

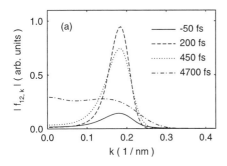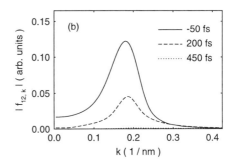

**Figure 2.** Modulus of the intraband polarization $|f_{12,\mathbf{k}}^e|$ as a function of $\mathbf{k}$ at different times. (a) with in-scattering, (b) in-scattering neglected.

behavior of the intraband polarizations is much more pronounced: dephasing is now very weak and a long-living oscillating dipole moment is clearly visible. The reason for this different behavior is that for the interband polarization the scattering occurs between k-components oscillating with different frequencies determined by the dispersion relation $\omega(\mathbf{k})$. Thus, a scattering process connects polarization components with different phases which yields a cancellation due to averaging over the frequencies in the in-scattering terms. Under the present excitation conditions with a carrier density of about $10^{10}$ cm$^{-2}$ screening is weak and small-angle scattering is strongly preferred in carrier-carrier interaction. Small angle-scattering, however, corresponds to the $q \rightarrow 0$ limit and connects regions in k-space with similar oscillation frequencies. Thus the dephasing effect for the interband polarizations is reduced by the in-scattering terms, but it is still present. The intraband polarizations, on the other hand, are oscillating with frequencies which are determined only by the subband-splitting and which thus are independent of k. Therefore, no cancellation due to random phases occurs and polarization is only transported between different regions of k-space. However, due to the different Coulomb matrix-elements in the in- and out-scattering terms in the equations of motion for the intraband polarizations, the symmetry between these terms, which is present in the Boltzmann transport equation for distribution functions resulting in the conservation of particle number, is exact only in the $q \rightarrow 0$ limit. Therefore, there is always a slight dephasing connected with the k-space dynamics. It turns out that this dephasing depends sensitively on the basis chosen for the dynamics.

Figure 2 shows the k-dependence of the intraband polarization at different times. In Fig. 2(a) (with inscattering terms) we clearly see the transport of the intraband polarization which is similar to the relaxation of a carrier distribution, in Fig. 2(b) (inscattering terms neglected) the intraband polarization is strongly damped and nearly vanishes after 450 fs.

## CONCLUSIONS

We have presented an analysis of the dephasing of inter- and intraband-excitations in semiconductor heterostructures based on the generalized multi-subband semiconductor Bloch-equations. It turns out that, as has been found in the bulk case,[3] in particular at low densities the concept of a dephasing rate obtained from the total scattering rate results in a strong overestimation of the dephasing due to the dominance of small-angle scattering. This effect is much more pronounced for the case of interband polarizations due to their k-independent oscillation frequencies.

## REFERENCES

1. H. G. Roskos *et al.*, Phys. Rev. Lett. **68**, 2216 (1992).
2. E. Binder, T. Kuhn, and G. Mahler, Phys. Rev. B **50**, 18319 (1994).
3. F. Rossi, S. Haas, and T. Kuhn, Phys. Rev. Lett. **72**, 152 (1994).
4. W. Pötz, M. Žiger, and P. Kočevar, Phys. Rev. B **52**, 1959 (1995).

# THEORY OF COHERENT PHONON OSCILLATIONS IN SEMICONDUCTORS

Alex V. Kuznetsov[1] and Christopher J. Stanton[1,2]

[1]Department of Physics
University of Florida
Gainesville, Florida 32611

[2]Mikroelektronik Centret
Danmarks Tekniske Universitet
DK-2800 Lyngby, Denmark

## INTRODUCTION

Recent experiments[1-6] have shown that ultrafast optical excitation of semiconductors can produce oscillating changes in the optical properties of the material. The frequency of the oscillations in transmission or reflection always matches one of the optical phonon modes, which indicates that the phonon mode becomes coherently excited. This phenomenon has been observed in GaAs,[1,2] Ge,[3] as well as in a variety of other less common materials.[4,5] Very recently, Dekorsy and co-workers[7] have also observed terahertz radiation from coherent phonons in Te.

While it is well known that photoexcited electrons and holes emit a large number of optical phonons in the course of their energy relaxation,[8] these phonons cannot contribute to the observed oscillations. Spontaneous emission of phonons is an *incoherent* process that populates the phonon modes and thus leads to quantum *fluctuations* of the displacement in each mode, but the expectation value of each mode's displacement is zero, and the fluctuations of different modes are uncorrelated. The experiments suggest that besides being incoherently populated, at least some of the phonon modes can also behave like classical oscillators and have non-zero time-dependent displacement. This effect is left out in the usual description of carrier relaxation based on the Boltzmann equation.[9] On a phenomenological level, it has been described in Ref. 10. In this paper, we describe a microscopic theory of coherent phonons,[11] and its application to the case of GaAs.[12]

## GENERAL THEORY

We define the *coherent amplitude* of **q**-th phonon mode in terms of Heisenberg phonon operators as:

$$D_{\mathbf{q}} \equiv \langle b_{\mathbf{q}} \rangle + \langle b^{\dagger}_{-\mathbf{q}} \rangle \equiv B_{\mathbf{q}} + B^{*}_{-\mathbf{q}} , \tag{1}$$

which is proportional to the displacement of the mode.[13] Operator averages in (1) are zero if the mode is in a Fock state with definite number of phonons in it. However, for states that are quantum-mechanical superpositions of different Fock states these expectation values are nonzero and oscillate as a function of time. In particular, $B_{\mathbf{q}}$ is nonzero for *coherent states* which describe an oscillator acted upon by a classical force.[14]

One can quite generally write down an equation of motion for the coherent amplitude (1) by commuting the phonon operators in (1) with the Hamiltonian:[11]

$$\frac{\partial^2}{\partial t^2} D_{\mathbf{q}} + \omega_{\mathbf{q}}^2 D_{\mathbf{q}} = -2\omega_{\mathbf{q}} \sum_{\alpha \mathbf{k}} M^{\alpha}_{\mathbf{kq}} \, n^{\alpha}_{\mathbf{k},\mathbf{k}+\mathbf{q}} , \tag{2}$$

where $\omega_{\mathbf{q}}$ is the mode frequency, $M^{\alpha}_{\mathbf{kq}}$ is the electron-phonon coupling in band $\alpha$ (conduction and valence), and $n^{\alpha}_{\mathbf{k},\mathbf{k}'}$ is the electronic density matrix.

Eq. (2) allows us to predict which modes are coherently driven by optical excitation. An optical pulse creates carriers in a macroscopically uniform state, so that the electronic density matrix is diagonal: $n^{\alpha}_{\mathbf{k},\mathbf{k}'} = f^{\alpha}_{\mathbf{k}}(t)\delta_{\mathbf{k}\mathbf{k}'}$ , where $f$ is the carrier distribution function. Consequently, the only phonon mode that is coherently driven by the optical excitation is **q=0** mode, which is also the only mode that can be probed by optical reflection or transmission measurements (modes with wavelengths smaller than the probe spot size will average to zero). For **q=0** mode, the total density of photocarriers, $N$, acts as a classical driving force in (2). Sudden changes in the carrier density caused by ultrafast optical excitation create a macroscopically large coherent amplitude of the **q=0** mode that is responsible for the observed oscillations.

The magnitude of the displacement of **q=0** mode can be expressed in terms of microscopic quantities:

$$U_o = \frac{C^v - C^c}{\rho \omega_0^2} \cdot \frac{N_o}{V} , \tag{3}$$

where $C^{\alpha}$ are the deformation potentials in the band $\alpha$, $\rho$ being the reduced density.[13] Note that this expression for the amplitude of the lattice displacement is purely classical and does not depend on the Planck constant.

For a typical set of parameters ($\rho = 5$ g/cm$^3$; $V = 1$ cm$^3$; $\omega_0 = 10$ THz; and $C^v - C^c = 10^9$ eV/cm ), we see that for the excitation density of $10^{19}$ cm$^{-3}$ the amplitude of coherent lattice displacement is about $3 \cdot 10^{-4}$ nm, which is $10^{-3}$ of a typical lattice constant. This value is consistent with the experimental data[4,7] and constitutes a huge lattice distortion. Under the same conditions, the dimensionless coherent amplitude $B$ will be about $3 \cdot 10^9 / \sqrt{V}$ , and its square equals the number of coherent phonons in the mode $\mathcal{N}_o^{coh} = 10^{19}$ cm$^{-3}$ . The fact that there is a macroscopically large number of phonons in a single quantum state ensures that this mode behaves classically for all practical purposes since the relative fluctuations of the displacement is proportional to $1/\mathcal{N}^{coh}$ and is thus extremely small.[11]

Therefore, we see that the coherent phonon oscillations are not caused by synchronized motion of many different modes whose phases are locked. Instead, the oscillations are produced by a single zero-wavevector mode that becomes occupied by a macroscopically large number of coherent phonons.

## COHERENT PHONONS IN GaAs

For cubic materials like GaAs, the above general theory has to be modified because for symmetry reasons the sum over **k**-space on the right-hand-side in (2) vanishes. In such materials, the excitation of coherent phonons is only possible if there is an externally created preferential direction in the system. Such a direction is provided by depletion field under the surface of bulk GaAs samples. It has been shown that the amplitude of the experimentally observed oscillations is directly related to the depletion field.[6,15]

In the presence of the external symmetry-breaking electric field $\mathbf{E}^{ext}$, the dynamics of electronic polarization $\mathbf{P}$ and the coherent displacement of the $\mathbf{q}=0$ LO mode, $W$, are governed by plasmon-phonon equations:[12,16]

$$\frac{\partial^2}{\partial t^2}\mathbf{P} + \gamma_{el}\frac{\partial}{\partial t}\mathbf{P} + \omega_{pl}^2\mathbf{P} = \frac{e^2 N(t)}{\varepsilon_\infty \mu}\left[\mathbf{E}^{ext} - 4\pi\gamma_{12}W\right], \tag{4}$$

$$\frac{\partial^2 W}{\partial t^2} + \gamma_{ph}\frac{\partial W}{\partial t} + \omega_L^2 W = \frac{\gamma_{12}}{\varepsilon_\infty}\left[\mathbf{E}^{ext} - 4\pi\mathbf{P}\right]. \tag{5}$$

In the absense of coupling, the polarization is expected to oscillate with the plasmon frequency $\omega_{pl} = \sqrt{4\pi e^2 N/\varepsilon_\infty \mu}$ ($N$ being the total density of carriers, $\mu$ — reduced electron-hole mass), while the frequency of lattice oscillations should be that of the longitudinal optical phonon $\omega_L$. The coupling of phonons to the electric field is characterized by $\gamma_{12} = \omega_T \cdot \sqrt{(\varepsilon_o - \varepsilon_\infty)/4\pi}$, where $\varepsilon_o$ and $\varepsilon_\infty$ are the low- and high-frequency dielectric constants, and $\omega_T$ is the transverse phonon frequency.[16]

After the ultrafast optical excitation the photocarriers will quickly screen the external field. The resulting fast field transient acts as a driving force for the lattice displacement in (5) which starts oscillating with the eigenfrequencies of (4) and (5) (plasmon-phonon modes).[12,16]

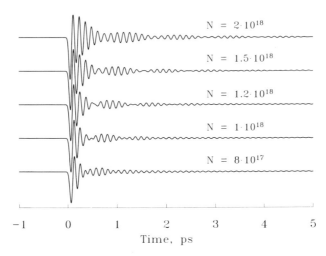

**Figure 1.** Time derivative of the lattice displacement at different excitation densities. The LO—TO beating is clearly visible. The oscillations with LO frequency originate at the low-density wings of the excited spot, while frequency of oscillations is reduced to TO frequency because of screening in the high-density center of the spot

In the experiment, the density is highly inhomogeneous across the excited spot, so one has to solve (4) and (5) for different excitation densities and then integrate the results over all densities present in the excited region.[12,15] As a result, the frequency-dependent modes such as plasmon modes are suppressed by the destructive interference from regions with different densities. The phonon-like branch of the spectrum is largely density-independent and leads to pronounced oscillations at the LO frequency in the observed response.[15] At higher densities, the strong screening causes the frequency of the phonon mode to approach that of the TO phonon. The resulting beating between LO and TO oscillations is shown in Fig. 1 and has been experimentally observed in Ref. 17.

In conclusion, we have developed a microscopic theory of coherent phonon oscillations in semiconductors. We identify the macroscopic coherent occupation of the zero-wavevector mode as the source of the observed oscillatory response. A modified version of the theory explains the experimental findings for GaAs, and can be expanded to treat the problem of terahertz emission from coherent phonons.[7,12]

## Acknowledgments

This work was supported by the NSF grant DMR 8957382. CJS wishes to thank the Danish Research Academy and the Mikroelektronik Centret at DTU for support during part of this work.

## REFERENCES

[1] G.C. Cho, W. Kutt, and H. Kurz, Phys. Rev. Lett. **65**, 764 (1990).

[2] W. Kutt, G.C. Cho, T. Pfeifer, and H. Kurz, Semicond. Sci. Technol. **7**, B77 (1992).

[3] T. Pfeifer, W. Kutt, H. Kurz, and H. Scholz, Phys. Rev. Lett. **69**, 3248 (1992).

[4] T.K. Cheng, J. Vidal, H.J. Zeiger, G. Dresselhaus, M.S. Dresselhaus, and E.P. Ippen, Appl. Phys. Lett. **59**, 1923 (1991).

[5] W. Albrecht, Th. Kruse, and H. Kurz, Phys. Rev. Lett. **69**, 1451 (1992).

[6] W. A. Kutt, W. Albrecht, and H. Kurz, IEEE J. Quantum Electronics **QE-28**, 2434 (1992).

[7] T. Dekorsy, H. Auer, C. Waschke, H.J. Bakker, H.G. Roskos, H. Kurz, V. Vagner, and P. Grosse, Phys. Rev. Lett. **74**, 738 (1995).

[8] J.A. Kash and J.C. Tsang, Solid State Electron. **31**, 419 (1989).

[9] See e.g. D.W. Bailey, C.J. Stanton and K. Hess, Phys. Rev. **B 42**, 3423 (1990) and references therein.

[10] H.J. Zeiger, J. Vidal, T.K. Cheng, E.P. Ippen, G. Dresselhaus and M.S. Dresselhaus, Phys. Rev. **B 45**, 768 (1992).

[11] A.V. Kuznetsov and C.J. Stanton, Phys. Rev. Lett. **73**, 3243 (1994).

[12] A.V. Kuznetsov and C.J. Stanton, Phys. Rev. **B51**, 7555 (1995).

[13] Ch. Kittel, *Quantum Theory of Solids*, (Wiley, New York, 1963).

[14] J.R. Klauder, E.C.G. Sudarshan, *Fundamentals of Quantum Optics* (Benjamin, New York, 1968).

[15] T. Dekorsy, T. Pfeifer, W. Kutt, and H. Kurz, Phys. Rev. **B 47**, 3842 (1993).

[16] A. Mooradian and A.L. McWhorter, Phys. Rev. Letters **19**, 849 (1967).

[17] T. Pfeifer, T. Dekorsy, W. Kutt, and H. Kurz, in *Phonon Scattering in Condensed Matter VII*, edited by M. Meissner and R.O. Pohl (Springer, Berlin, 1993), p. 110.

# A THEORETICAL ESTIMATE OF COHERENCE EFFECTS IN THE FEMTOSECOND SPECTROSCOPY OF BULK SEMICONDUCTORS

G. F. Kuras and P. Kočevar

Inst. f. Theoretische Physik, Karl–Franzens– Universität
A–8010 Graz, Austria

**Abstract.** A recently developed combined density–matrix and Monte–Carlo description of the femtosecond laser–pulse spectroscopy of bulk semiconductors is extended to explore the possibility of experimentally detectable coherence effects in the nonexcitonic regime at high excitation densities, and here in particular of Rabi–type density oscillations and corresponding oscillations in the time resolved degenerate pump–and–probe absorption spectroscopy. The analysis contains the first numerical evaluation of so–called higher–order polarisation scatterings. For GaAs as reference material these calculations allow to determine the essential experimental prerequisits for a demonstration of Rabi–type oscillatory effects in a laser–pulse excited bulk semiconductor.

## INTRODUCTION

The recent years have seen a growing interest in coherent contributions to the nonlinear electronic response of semiconductors to femtosecond laser pulses. Most of the experimental investigations were conducted in quantum–well systems, and here almost exclusively in the excitonic regime at low excitation densities to keep the dephasing action of scatterings as small as possible. Whether or not such coherence effects might be experimentally detectable in the scattering–dominated nonexcitonic regime, i.e. at higher excitation energies and densities, is still an open question and will be addressed in the following.

The first detailed numerical analysis of the interplay between the coherent and the incoherent carrier response used a combined density–matrix and Monte–Carlo approach [1], where, for a two–band model of the semiconductor, the heavy–hole valence band and the conduction band are modelled as an ensemble of coupled two–level oscillators in $\vec{k}$–space. The distribution functions $f_{\vec{k}}^e$ of electrons (e) and $f_{\vec{k}}^h$ of holes (h) and the laser–driven e–h amplitude $p_{\vec{k}}$ were chosen as the three basic components of the one–particle density matrix in $\vec{k}$–space, taken as diagonal in $\vec{k}$ by assuming spatial homogeneity. Their equations of motion, known as Semiconductor Bloch equations, were obtained through the Heisenberg equations of motion for the respective num-

*Hot Carriers in Semiconductors*
Edited by K. Hess *et al.*, Plenum Press, New York, 1996

ber and transition operators [1,2]. The Hamiltonian contained the conventional optic interband dipole excitation by the (classical) laser field, the free–particle terms and the carrier–carrier (cc) and carrier–phonon (cp) interactions [1]. The cc interaction $V_{\vec{q}}$ was taken as free–carrier–screened in the spirit of the screened–exchange approach for equilibrium Coulomb systems [1]. As usual it was assumed that the probe pulse does not produce any optical nonlinearities. The resulting BBGKY hierarchy was closed at the level of the three basic dynamical variables by appropriate factorisations and in conjunction with adiabatic and Markovian approximations at the four–point level [1,3,4].

While the first–order contributions in $V_{\vec{q}}$ led to mean–field renormalisations of the particle energies and the electric field, the leading dissipative contributions, the conventional Boltzmann collision intergrals, appeared at the second order in the cp interaction [1] and in $V_{\vec{q}}$ [3,4,5]. The numerical evaluation of these leading scattering contributions and of the coherent photogeneration dynamics to the time evolution of $f_{\vec{k}}^{e}$ and $f_{\vec{k}}^{h}$ was achieved by a proper extension of the well–established Ensemble–Monte–Carlo simulation of purely incoherent photo–excitation scenarios [1,3,4]. At the same time, $p_{\vec{k}}$, called interband polarisation because of its immediate relation to the macroscopic polarisation [1,2,5] and thereby to the optic susceptibility, was obtained by a direct parallel numerical integration of its equation of motion.

Using this combined density–matrix and Monte–Carlo description, we calculate the contributions of the so–called polarization scatterings, all again of second order in $V_{\vec{q}}$, to the damping of coherent Rabi–type charge oscillations. While only terms linear in the interband polarization $p$ had been included in earlier such estimates [4,6], the present work provides, to our knowledge, the first numerical evaluation of polarisation scatterings of second and third order in $p$, in particular the complete evaluation of all energy–conserving higher–order polarisation scatterings [2] (HOPS).

## THEORY AND RESULTS

Following our earlier estimates of coherence effects in bulk GaAs [4], we consider a degenerate pump–probe absorption measurement with a 1.7 eV / 50 fs excitation with excitation densities around $8 \cdot 10^{18}$ cm$^{-3}$ and a 15 fs probe pulse with spectral width given by Heisenberg's uncertainty relation. We first define the conventionally used theoretical 'standard scenario' [1,2,5] by assuming isotropic and parabolic bands within a two–band model, i.e. we consider holes from the upper valence band and $\Gamma$–valley electrons and neglect photoexcitation from secondary valence bands. Except for HOPS, this standard scenario contains all leading terms of 2nd order in $V_{\vec{q}}$ [4,6]. The calculated absorption spectrum describes the nonlinear response to the excitation pulse. It must be finally convoluted with the spectral and temporal shape of the probe pulse to account for the detection procedure in a pump–probe transmission/reflection experiment [3,4].

In the following the results obtained within the standard scenario will serve as reference for our estimate of the role of HOPS. The detailed expressions for the latter can be found in the literature [2]. In all three equations of motion we kept the terms of the general form of energy–conserving Boltzmann collision integrals, but neglected the non–conserving contributions of principle–value integrals [2].

At the earliest times of the excitation, the directly laser–driven polarisation $p_{\vec{k}}$ already evolves, while the carrier distribution functions $f_{\vec{k}}^{e}$ and $f_{\vec{k}}^{h}$ are still practically zero. Accordingly the (cubic) HOPS contributions to $\dot{p}_{\vec{k}}$, containing no distribution functions, strongly dominate at the very beginning of the pump pulse. Although a seemingly small effect, the corresponding initial broadening of the photogenerated polarisation and the ensuing band occupations have noticeable consequences. They strongly reduce the initial Pauli blocking of the carrier generation process, leading to a pronounced increase of the asymptotic excitation density, as shown in Fig.1. Note the non–monotonic time evolution of the carrier density, the direct trace of the Rabi character of the electronic response. It is caused by those short time intervals, during which coherence–induced stimulated e–h recombinations outweigh the generation [1]. The resulting modifications of the calculated relative absorption change ('bleaching') are also shown in Fig.1. The main consequence of HOPS is a reduction of the oscillation amplitudes, typically about a factor of 2, and follows from the smaller level occupations due to the increased broadening of the distribution functions.

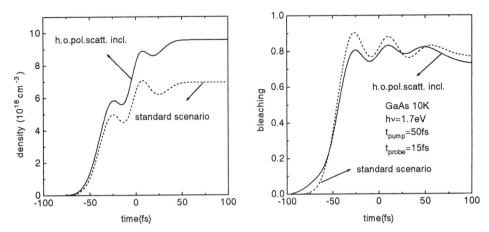

Figure 1: Left: total carrier density versus time; right: probe–pulse–convoluted relative bleaching versus time.

In this connection we shortly comment on our results obtained by also including the further broadening effect of the additional rigid band shifts through the so–called Coulomb hole [7]. As these renormalisations are directly proportional to the integral carrier density [7], they can be easily included in our simulations. It turns out [8] that the additional reduction of the Rabi amplitudes is small enough to leave the results of Fig.1 as a quantitative theoretical prediction of measurable Rabi–type coherence effects. The same still holds after the further inclusion of the light–hole excitation channel, of the secondary conduction–band minima and of additional hot–carrier effects like LO–phonon heating [8].

**REFERENCES**

[1] T. Kuhn and F. Rossi, Phys. Rev. B **46**, 7496 (1992).
[2] A. Kuznetsov, Phys. Rev. B **44**, 8721 (1991).
[3] F. J. Adler, P. Kočevar, J. Schilp, T. Kuhn, and F. Rossi, Semicond. Sci. Technol. **9**, 446 (1994).
[4] F. Adler, G. F. Kuras, and P. Kočevar, SPIE Proc. **2142**, 206 (1994).
[5] M. Lindberg and S. W. Koch, Phys. Rev. B **38**, 3342 (1988).
[6] F. Rossi, S. Haas, and T. Kuhn, Phys. Rev. Lett. **72**, 152 ( 1994).
[7] H. Haug and S. W. Koch, *Quantum Theory of the Optical and Electronic Properties of Semiconductors* (World Scientific, 1993).
[8] G. F. Kuras, Diploma Thesis, Karl–Franzens–Universiät Graz, 1995 (unpublished).

# AC LARGE SIGNAL CONDUCTION AND HARMONIC GENERATION OF A GaAs-AlGaAs SUPERLATTICE

X.L. Lei[1,2], N.J.M. Horing[2], H.L. Cui[2], and K.K. Thornber[3]

[1]State Key Laboratory of Functional Materials for Informatics
Shanghai Institute of Metallurgy, Chinese Academy of Sciences
865 Chang Ning Road, Shanghai, 200050, China
[2]Department of Physics and Engineering Physics
Stevens Institute of Technology
Hoboken, New Jersey 07030
[3]NEC Research Institute
4 Independence Way, Princeton, New Jersey 08540

The prospect of negative differential conductance in Bloch miniband transport predicted by Esaki and Tsu[1] brings into focus the need for an analysis of such features under dynamic conditions. In this paper, we discuss the current response of vertical superlattice transport when a strong ac signal of frequency $\omega$ is superposed on a strong dc bias field. The nonlinearity involved results in harmonic generation in conjunction with distortion of the periodic current response, and this is analyzed here using a quasi-analytical balance equation approach, which involves only a few percent of the computational effort of an earlier Monte Carlo simulation[2]. Moreover, our analysis treats random impurity and phonon (acoustic and polar optic) scatterings on a realistic three-dimensional microscopic basis, and is not constrained by the one-dimensional assumptions of other work[3].

We consider a model superlattice system in which electrons move along the axial-$z$-direction through the (lowest) miniband formed by periodically spaced potential wells and barriers of finite height. In the lateral/transverse ($x$-$y$) plane they are taken to move freely in this planar quantum well superlattice. The electron energy dispersion can be written as the sum of a transverse energy $\varepsilon_{\mathbf{k}_\parallel} = k_\parallel^2/2m$, related to the free lateral motion, and a tight-binding-type miniband energy $\varepsilon(k_z)$ related to the longitudinal motion of the electron:

$$\varepsilon(\mathbf{k}_\parallel, k_z) = \varepsilon_{\mathbf{k}_\parallel} + \varepsilon(k_z) \tag{1}$$

with

$$\varepsilon(k_z) = \frac{\Delta}{2}(1 - \cos k_z d), \tag{2}$$

where $d$ is the superlattice period along the $z$-direction, $-\pi/d < k_z \leq \pi/d$, and $\Delta$ is the miniband width.

*Hot Carriers in Semiconductors*
Edited by K. Hess *et al.*, Plenum Press, New York, 1996

Our analysis of nonlinear high frequency superlattice miniband transport and harmonic generation for this planar quantum well system is carried out within the framework of a balance equation approach to the description of the transport state in terms of the center-of-mass momentum $\vec{p}_d$ (per particle) and relative electron temperature $T_e$. For the planar quantum well superlattice under consideration, subject to a spatially-uniform time dependent electric field $E(t)$ in the $z$-direction of the superlattice axis (growth direction), there are two nonlinear balance equations, one representing momentum balance (Eq.3) and the other representing energy balance(Eq.4):

$$\frac{1}{v_m}\frac{dv_d}{dt} = eE(t)d\alpha(T_e)\cos(p_d d) + \frac{2}{\Delta d}(A_i + A_p), \tag{3}$$

$$\frac{2}{\Delta}\frac{dh_e}{dt} = eE(t)d\alpha(T_e)\sin(p_d d) - \frac{2W}{\Delta}. \tag{4}$$

Here, the average drift velocity $v_d$ is given by ($v_m = \Delta d/2$, $N$ is the total number of electrons)

$$v_d = \frac{2}{N}\sum_{\mathbf{k}_\parallel, k_z} \frac{d\varepsilon(k_z)}{dk_z} f(\varepsilon(\mathbf{k}_\parallel, k_z - p_d), T_e), \tag{5}$$

and the ensemble-averaged electron energy per carrier is

$$h_e = \frac{2}{N}\sum_{\mathbf{k}_\parallel, k_z} \varepsilon(\mathbf{k}_\parallel, k_z) f(\varepsilon(\mathbf{k}_\parallel, k_z - p_d), T_e), \tag{6}$$

where $\alpha(T_e)$ is a function of $T_e$, independent of $p_d$, as given by

$$\alpha(T_e) = \frac{2}{N}\sum_{\mathbf{k}_\parallel, k_z} \cos(k_z d) f(\varepsilon(\mathbf{k}_\parallel, k_z), T_e). \tag{7}$$

In these equations $f(\varepsilon, T_e)$ is the Fermi-Dirac distribution function at the electron temperature $T_e$, with chemical potential $\mu$. The expressions for the impurity- and phonon-induced frictional decelerations $A_i$, and $A_p$, and the energy-transfer rate $W$ from the electron system to the phonon system, have been determined in previous publications[4] on the basis of a realistic three-dimensional microscopic treatment of phonon (acoustic and polar-optic) and random impurity scatterings, jointly with dynamic, nonlocal electron-electron interactions. These expressions are too lengthy to include here.

We consider the electric field $E(t)$ to be composed of both a dc part $E_0$ and an ac part of frequency $\omega$ and amplitude $E_\omega$ applied such that

$$E(t) = E_0 + E_\omega \sin(\omega t). \tag{8}$$

On the basis of the balance equations (3) and (4), we calculate the nonlinear steady time-dependent response of the drift velocity $v_d(t)$ and the electron temperature $T_e(t)$ to this time-dependent electric field. When the high-frequency steady state is reached, the drift velocity and the electron temperature of the system are periodic time functions of period $T_\omega = 2\pi/\omega$. The drift velocity can then be written in the form

$$v_d(t) = v_0 + \sum_{n=1}^{\infty}[v_{n1}\sin(n\omega t) + v_{n2}\cos(n\omega t)], \tag{9}$$

where the dc component $v_0$ and the harmonic coefficients $v_{n1}$ and $v_{n2}$ are given by

$$v_0 = \frac{1}{T}\int_0^T v_d(t)dt, \quad v_{n1} = \frac{2}{T}\int_0^T v_d(t)\sin(n\omega t)dt, \quad v_{n2} = \frac{2}{T}\int_0^T v_d(t)\cos(n\omega t)dt. \tag{10}$$

We have carried out numerical calculations of the large-signal high-frequency current response for GaAs-based superlattices using the above equations. In this, realistic impurity, acoustic-phonon and polar-optic-phonon scatterings have been taken into consideration. The distortion of the periodic ac current response (drift velocity $v_d$) due to nonlinear harmonic generation is exhibited as a function of time in Figure 1, and an analysis of harmonic content is shown in Figure 2 in terms of the harmonic coefficients $v_{ij}$ as as functions of the period of the high-frequency field, $T_\omega$.

This work was partially supported by the Chinese National Natural Science Foundation and the China National and Shanghai Municipal Commissions of Science and Technology.

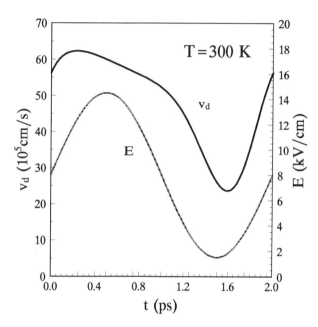

**Figure 1** Steady time dependent response (one period) of the drift velocity $v_d(t)$ (solid line) to an electric field $E(t) = E_0 + E_\omega \sin(2\pi t/T_\omega)$, with $T_\omega = 2.0\,\mathrm{ps}$, $E_0 = 8\,\mathrm{kV/cm}$ and $E_\omega = 6.5\,\mathrm{kV/cm}$ (dashed lines), in a planar quantum well superlattice of $d = 10\,\mathrm{nm}$, $\Delta = 900\,\mathrm{K}$, $N_s = 1.5 \times 10^{15}/\mathrm{m}^2$ and low-field, linear mobility $\mu(0) = 1.0\,\mathrm{m}^2/\mathrm{Vs}$ at lattice temperature $T = 300\,\mathrm{K}$.

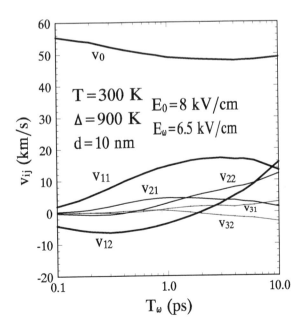

**Figure 2** Calculated dc component $v_0$ and the harmonic coefficients $v_{11}$, $v_{12}$, $v_{21}$, $v_{22}$, $v_{31}$ and $v_{32}$ of the drift velocity in the steady time-dependent response of a 3D-superlattice ($d = 10\,\mathrm{nm}$, $\Delta = 900\,\mathrm{K}$, $N_s = 1.5 \times 10^{15}/\mathrm{m}^2$ and $\mu(0) = 1.0\,\mathrm{m}^2/\mathrm{Vs}$ at lattice temperature $T = 300\,\mathrm{K}$) to the field $E(t) = E_0 + E_\omega \sin(2\pi t/T_\omega)$, with $E_0 = 8\,\mathrm{kV/cm}$ and $E_\omega = 6.5\,\mathrm{kV/cm}$, as functions of the period of the high-frequency field, $T_\omega$.

# REFERENCES

1. L. Esaki and R. Tsu, IBM J. Res. Dev. 14:61 (1970).
2. P.J. Price, in *Semiconductors and Semimetals*, Vol **14**, R.K. Willardson and A.C. Beer, ed., Academic Press, New York (1979).
3. A.A. Ignatov and V.I. Shashkin, Phys. Stat. Sol. (b) 110:K117 (1982); Phys. Lett. 94A:169 (1983).
4. X.L. Lei, N.J.M. Horing, and H.L. Cui, Phys. Rev. Lett. 66:3277 (1991); J. Phys: Condens. Matter, 4:9375 (1992).

# 3. Hot Carriers in Nanostructures and Low-Dimensional Systems

# HOT ELECTRON EFFECTS IN PHONON-ASSISTED

# LANDAUER RESISTANCE

V. L. Gurevich
Solid State Physics Division
A. F. Ioffe Institute
194021 Saint Petersburg, Russia

The advances in semiconductor nano-fabrication and material science in recent years have made available materials of great purity and crystalline perfection. The electrical conduction in such nanoscale structures has been a focus of numerous investigations, both theoretical and experimental, with a number of important discoveries and even patents.

The essence of the electrical conduction in these nanoscale structures is that the quantum nature of the electron leaves its distinct trace in a macroscopic measurement. Namely, electrons move through nanostructures phase coherently, while nanostructures act effectively as waveguides. This is possible since the largest dimension of the structure is smaller than the coherence breaking length in the problem (typically a few $\mu m$). Such nanoscale systems are characterized by low electron densities, which may be easily varied by means of electrical field (because of a large screening length). It also possible to make these nanostructures in such a way that the impurity scattering can be ignored. The transport of electrons in such a regime will be referred to as *ballistic* and is analogous to the 3D Sharvin's classical point contact conductance[1].

In this paper we wish to address three distinct yet related topics. ($i$) We investigate the influence of phonons on the the ballistic transport in semiconductor nanostructures. We also examine the onset of the collision-controlled transport and estimate the critical temperature interval which separates ballistic and fully collision-controlled transport for various values of electron concentration and sample dimensions. ($ii$) We investigate hot electron effects of a new type which are entirely peculiar to the nanostructures. They hold, provided that $eV \geq k_B T$, where $V$ the applied voltage bias.[1]. ($iii$) We predict resonances of a new type that can take place whenever the separation between a pair of levels of transverse quantization equals to the energy of an optical phonon.

We consider the simplest arrangement within the physical picture first discussed by Landauer, namely, there are two reservoirs (each in equilibrium with itself) which are connected by a uniform quantum wire. By this we mean a wire whose transverse

---

[1]In the classical regime, the nonlinear phenomena in the current-voltage characteristics of point contacts connecting normal metals were first observed and discussed by Yanson[2]

dimensions, $L_\perp$, are of the order of the de Broglie wavelength, $\lambda_F$, while the longitudinal dimension, $L_x$, is significantly bigger so that $L_x \gg \lambda_F$.

It is known (see [3] and the review [4]) that in the absence of scattering $G$ is a step-like function of the Fermi. Each step corresponds to the inclusion of a new mode of transverse quantization to the conduction process. The height of each step is equal to the quantum of conductance, $G_0 = 2e^2/h$ multiplied by a factor whose physical interpretation is that of a transmission probability (it is unity in our case).

The onset of phonon scattering results in exchange of energy and quasimomentum between electrons and phonons. However, the electronic transport may still be regarded as ballistic (and be treated as such to zeroth order) provided that the change in the conductance, $\Delta G$, due to phonon scattering is relatively small.

In order to calculate $\Delta G$ we have followed three different approaches: (i) iterations of the Boltzmann equation[5, 6], (ii) Matsubara diagrammatic techniques[7][2] (iii) Keldysh diagrammatic techniques[9]. In this paper we adopt the first approach due to its conceptual simplicity. In Eq. (2) (see below) we perform integration over the three components of the phonon wave vector. $\mathbf{q}_\perp$ indicates the two transverse wave vector components. The third component is given by $q_x = \pm(p - p')/\hbar$. Therefore the third integration is equivalent to the integration over the electron quasimomentum, $p'$, because of the conservation of quasimomentum.

The distribution function of the electrons, $f_n(p)$, in absence of the electron-phonon interaction[4] is $f_n^{(0)}(p) = f^{(F)}(\epsilon \mp eV/2 - \mu)$ where $f^{(F)}$ is the Fermi function, $V$ is the voltage bias across a nanostructure, $\epsilon = \epsilon_n(p) = \epsilon_n(0) + p^2/2m^*$ is the electron energy, $m^*$ is the electron effective mass, $p$ is the $x$-component of the electron quasimomentum, $n$ is a subband index, and the upper (lower) sign above is for $p > 0 (p < 0)$. To the lowest order in the electron-phonon coupling, the total current $\Delta J$ is a sum of currents carried by individual *channels currents*, $\Delta J_n$, which in turn are made up of *partial currents* $\Delta J_{nn'}$ for a particular inter-channel transitions[6, 7, 9]:

$$\Delta J = \sum_n \Delta J_n = \sum_{nn'} \Delta J_{nn'}. \tag{1}$$

respectively. The partial currents are given by[9]

$$\begin{aligned}
\Delta J_{nn'} &= -2eL_x \left[1 - \exp\left(-eV/k_B T\right)\right] \int \frac{d^d q_\perp}{(2\pi)^d} \int_0^\infty \frac{dp}{2\pi\hbar} \int_{-\infty}^0 \frac{dp'}{2\pi\hbar} \\
&\times C_{nn'} W_{\mathbf{q}} \left(\mathcal{A}_{nn'}^{(-)} + \mathcal{A}_{nn'}^{(+)}\right) \delta(\epsilon' - \epsilon - \hbar\omega_{\mathbf{q}}),
\end{aligned} \tag{2}$$

Here

$$\mathcal{A}^{(-)} = \left[1 - f^{(F)}\left(\epsilon' - \mu^{(-)}\right)\right]\left[f^{(F)}\left(\epsilon - \mu^{(+)}\right)\right] N_{\mathbf{q}}, \tag{3}$$

$$\mathcal{A}^{(+)} = \left[1 - f^{(F)}\left(\epsilon - \mu^{(-)}\right)\right]\left[f^{(F)}\left(\epsilon' - \mu^{(+)}\right)\right] (1 + N_{\mathbf{q}}),$$

where $C_{nn'} = |\langle n'| \exp(i\mathbf{q}_\perp \mathbf{r}_\perp)|n\rangle|^2$, $(d+1)$ is the dimension of the system, $L_x$ is the total length of the conductor, $N_{\mathbf{q}}$ is the average phonon occupation number (which we

---

[2] In this context we wish to mention work by Velický *et al.*[8] where the self-energy was calculated under the assumption that it is independent of the electron quasi-momentum. This assumption is in contradiction with the results obtained in Ref.[7] where the most essential features arise as a consequence of self-energy's explicit dependence on the electron quasi-momentum.

assume to be the equilibrium Bose function) and $\langle \mathbf{r}_\perp | n \rangle$ is the transverse part of the electron wavefunction. The terms proportional to $\mathcal{A}^{(+)}$ and $\mathcal{A}^{(-)}$ describe the phonon emission and absorption, respectively. This equation describes a contribution of a single phonon branch. As expected, $\Delta J$ diminishes the total current. One can show explicitly that only those phonons contribute that can *backscatter* the electrons. It is readily seen that this condition imposes a certain lower limit for the magnitude of the phonon wave vectors.

We consider only the scattering by three-dimensional extended phonons (although the phonons localized within the nanostructure or near its interface with the bulk could have been easily included into our scheme). For the acoustic phonons in the isotropic approximation, $W_\mathbf{q} = \pi Z^2 q^2 / \rho \omega_\mathbf{q}$. Here $Z$ is the deformation potential constant for the longitudinal phonons, and $\rho$ is the mass density. For the polar optical scattering, $W_\mathbf{q} = (2\pi e/q)^2 \omega_o (\varepsilon_0 - \varepsilon_\infty)/\varepsilon_0 \varepsilon_\infty$. We ignore the dispersion of the optical phonons, namely, $\omega_\mathbf{q} = \omega_o$.

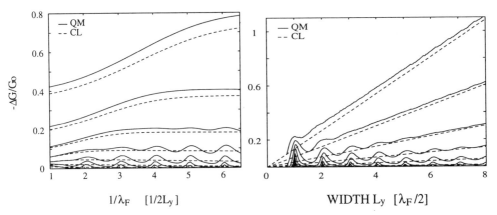

**Figure 1.** (a) $-\Delta G_{\mathrm{QM}}(\lambda_F)$ (solid) and $-\Delta G_{\mathrm{CL}}(\lambda_F)$ (dashed) (at $L_y=1000$ Å) increase with temperatures. $T_0=1.25$ Kelvin, (b) $-\Delta G_{\mathrm{QM}}(L_y)$ (solid) and $-\Delta G_{\mathrm{CL}}(L_y)$ (dashed) (at $\epsilon_F=14$ meV) increase with temperatures. $T_0=1.25$ Kelvin.

In Fig. 1a we present numerical results for the acoustic phonon scattering for quasi-1D GaAs wires, subject to a small bias, with $L_z=100$Å and $L_x=1\mu$m (the $x$-axis being along the propagation direction, while the effective width of the channel is $L_y$) as functions of Fermi wavelength, $\lambda_F \equiv h/\sqrt{2m^*\epsilon_F}$ (for a fixed $L_y=1000$Å). In Fig. 1b the results are given as functions of channel width $L_y$ (with fixed $\epsilon_F=14$meV corresponding to $\lambda_F=40$nm). The variation of both $L_y$ and $\lambda_F$ can be accomplished experimentally by varying the gate voltage. One can see from Fig. 1a and 1b that the condition $|\Delta G/G| \ll 1$ is fulfilled for a wide range of transverse wire dimensions and temperatures.

For the acoustic phonon scattering in the non-Ohmic regime, although $|\Delta \mathbf{J}|$ is somewhat smaller than the ballistic part of current $J^{(0)}$, its derivatives with respect to the applied voltage bias are larger than those of $J^{(0)}$. In Fig. 2 we show a typical hot electron effect, i.e. the dependence of $|\Delta \mathbf{J}|$ on the applied voltage.

For the optical phonon scattering we predict a new type of resonance. The phonon induced part of the current is proportional to the transition probability of the electron-phonon scattering. In turn, the latter is proportional to the product of initial and final densities of states, $1/\sqrt{\epsilon(\epsilon - R_{nm})}$, where $R_{nm} = \epsilon_m(0) - \epsilon_n(0) - \hbar\omega_o$. As long as $R_{nm} \neq 0$ the corresponding integral is well-behaved as a function of bias or gate voltage. However, when the resonant condition is met ($R_{nm} \rightarrow 0$), the product of the densities of states amounts to a $1/\epsilon$ singularity which renders energy integrals non-integrable (cf. with [10]). This leads to a logarithmic divergence (which is, of course, removed if one takes into consideration scattering events of higher order) and, consequently, to a signi-

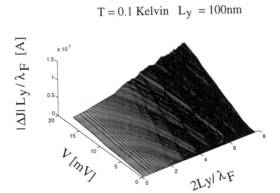

Figure 2 Acoustical phonon-assisted current as a function of bias voltage and $\lambda_F^{-1}$. $L_y = 100nm$ while $T=1$ Kelvin.

ficant increase in the resistance. Since no such singularities are present in 2,3D cases for an otherwise equivalent physical systems, no EPR in our sense of the word is to be expected there either.

Let us investigate a resonant behavior of a partial current, say, $\Delta J_{12}$. In Fig. 3 we show schematically the transitions that correspond to $\Delta J_{12}$. If the position of the Fermi level, $\mu$, is exactly halfway between the bottoms of the subbands 1 and 2 and the applied voltage bias slightly exceeds the optical phonon energy (see Fig. 3a) the transition indicated by an arrow is allowed. Such a transition would corresponds to the electro-phonon resonance.

Indeed, in this case the right-hand part ($p > 0$) of band 2, near its bottom is occupied by electrons whereas the left-hand part ($p < 0$) of band 1 is empty, so the transitions where an electron is backscattered by an optical phonon from the bottom of band 2 to the bottom of band 1 are allowed. If, on the other hand, the position of the Fermi level is shifted by, say, $\Delta\mu$ upward from the midpoint between the bottoms of subbands of 1 and 2 (see Fig. 3b) then in general transitions between the subbands accompanied by optical phonon emission are still possible (as is indicated by the arrow). However, they would not be resonant ones. Indeed, at $eV = \hbar\omega_o$ both

the right-hand part of band 2 and the left-hand part of band 1 are occupied so that the resonant transitions are forbidden. In Fig. 3c, however, the electrostatic potential exceeds the value of that is shown in Fig. 3b, to the extent that $\mu^{(-)} \leq \epsilon_1(0)$. The resonances are allowed but at a voltage bias that is higher than in Fig. 3a.

A number of cusps appear in Fig. 4a, each of these corresponds to a current drop due to electron-phonon interaction. Their sharpness depends on the magnitude of $\delta R$, the amount by which EPR condition is mismatched, in the limit of $\delta R \rightarrow 0$, the sharpness of real resonances is approached. The signature of EPR is particularly pronounced for

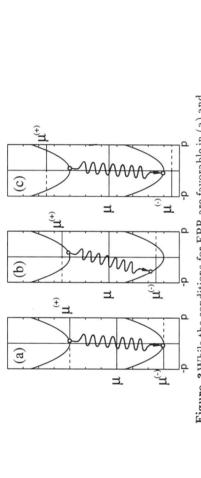

**Figure 3** While the conditions for EPR are favorable in (a) and (c), no EPR ' is possible for (b).

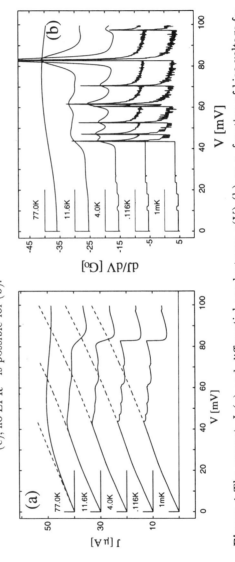

**Figure 4** The current $J$ (a) and differential conductance $g(V)$ (b) as a function of bias voltage for various temperatures.

the differential conductance $g(V)$ and may prove to be instrumental in the experimental investigation of this effect (see Fig. 4b). The resonant drops in current are a consequence of the blockage of the uppermost current carrying channel that would be open in the absence of collisions with phonons at the voltage bias values corresponding to the positions of cusps. As in the case of magnetophonon resonance[10], the existence of EPR is determined only by the singularities of the electron DOS and the existence of a nonvanishing phonon limiting frequency for $q \to 0$ and is independent of any other details of the electron and phonon spectrum and their interaction.

Kulik and Shekhter[11] treated theoretically classical diffusive electron transport through a narrow semiconductor wire and predicted periodic singularities in the $I - V$ characteristics resulting from the phonon-emission events. Such multiple emissions could take place in our case as well, although, the quantum nature of the electron transport in nanowires will bring about new features into this phenomenon. One can also compare our results with Monte Carlo simulations in Ref.[12] (in Q1D wires) and the results in Ref.[13] (in Q2D semiconductor structures) where a different type of nonmonotonous dependence of the conductivity on the gate voltage which is due to the optical phonons has been reported. The essential difference between the physics considered in this paper and in Refs. [12] and [13], as we see it, is that the condition $eV > \hbar\omega_o$ plays a crucial role in our case.

The hot electron effects, and EPR in particular, can be used as an experimental probe for the investigation of nanosystems. One can investigate the character of the electrons and phonon interactions, or the details of the electron band structure and the actual positions of the levels of transverse quantization, or the role of nonequilibrium phonons in the transport phenomena. We also encourage experimentalists to think of a physical situation where a blockage of a single channel due to EPR may play a crucial role so that one can separate a one-channel contribution experimentally.

# References

[1] Yu. V. Sharvin, Zh. Exp. Teor. Fiz. **48**, 984 (1965) [Sov. Phys. JETP **21**, 655 (1965)].

[2] I. K. Yanson, Zh. Exp. Teor. Fiz. **66**, 1035 (1974) [Sov. Phys. — JETP **39**, 506 (1974)]; I. O. Kulik, A. N. Omelyanchouk and R. I. Shekhter, Sov. Phys. J. Low Temp. Phys. **3**, 740 (1977).

[3] R. Landauer, IBM J. Res. Develop. **1**, 233 (1957); **32**(3), 306 (1989).

[4] See, for instance, C. W. J. Beenakker and H. van Houten, *Quantum Transport in Semiconductor Nanostructures* in Solid State Physics, **44**, Academic Press, 1991, edited by H. Ehrenreich and D. Turnbull.

[5] V. L. Gurevich, V. B. Pevzner and K. Hess, in *Quantum Transport in Ultrasubmicron Devices*, edited by D. K. Ferry, H. L. Grubin, C. Jacoboni and A.-P. Jauho (Plenum Press, New York, 1995).

[6] V. L. Gurevich, V. B. Pevzner and K. Hess, Phys. Rev. B **51**(8), 5219 (1995); J. Phys.: Condens. Matter **6**, 8363 (1994).

[7] V. B. Pevzner, V. L. Gurevich and E. W. Fenton, Phys. Rev. B **51**(15), 9465 (1995); E. W. Fenton, V. L. Gurevich, V. B. Pevzner and K. Hess, *Proc. 22nd Int. Conf. on Physics of Semiconductors*, 1994, Vancouver, Canada (World Scientific Publishing).

[8] B. Velický, V. Špička and J. Mašek, Sol. Stat. Comm., **72**(10), 981 (1989).

[9] V. L. Gurevich, V. B. Pevzner and G. J. Iafrate, Phys. Rev. Lett. (1995); to be published (1995).

[10] Yu. A. Firsov, V. L. Gurevich, R. V. Parfeniev and I. M. Tsidil'kovskii, *Landau level spectroscopy* in Modern Problems in Condensed Matter Sciences, vol. 27.2, pp. 1180-1308, edited by G. Landwehr and E. I. Rashba (Elsevier Science Publishers, New York, 1991).

[11] I. O. Kulik and R. I. Shekhter, Phys. Lett. **98A**, 132 (1983).

[12] D. Jovanovic *et al.*, Appl. Phys. Lett. **62**(22), 2824 (1993).

[13] W. Xu, F. M. Peeters and J. T. Devreese, Phys. Rev. B **48**, 1562 (1993).

# PHONON GENERATION IN NANOWIRES AND NON-OHMIC

# PHONON-ASSISTED LANDAUER RESISTANCE

V. B. Pevzner[1], V. L. Gurevich[2], K. Hess[3] and G. J. Iafrate[4]

[1]Electrical Engineering Department
North Carolina State University, Raleigh, NC 27695
[2]Solid State Physics Division, A. F. Ioffe Institute
194021 Saint Petersburg, Russia
[3]Beckman Institute, University of Illinois, Urbana, IL 61801
[4]Army Research Office, RTP, NC 27709

We investigate the phonon generation under the large bias condition as a function of voltage bias, as well as the relationship between the rate of energy absorption by the phonon bath and the current in such a non-Ohmic regime.

We consider two reservoirs (each in equilibrium with oneself) connected by a uniform nanowire, the simplest arrangement within the physical picture first discussed by Landauer[1]. The wire's transverse dimensions, $L_\perp$, are of the order of the de Broglie wavelength, $\lambda_F$, while the longitudinal dimension, $L_x$, is significantly bigger so that $L_x \gg \lambda_F$.

In the linear regime and in the absence of inelastic processes in the wire, $G$ is a step-like function of the Fermi level (see for instance the review [2]), with each step being equal to $G_0 = 2e^2/h$, which corresponds to the inclusion of a new mode of transverse quantization to the conduction process. The onset of phonon scattering results in exchange of energy and quasimomentum between electrons and phonons. However, electronic transport may still be regarded as ballistic (and be treated as such to zeroth order) provided that the change in the conductance, $\Delta G$, due to phonon scattering is sufficiently small.

In order to calculate the rate of phonon generation we have followed three different approaches: (i) successive iteration of the Boltzmann equation[3, 5], and the diagrammatic techniques of (ii) Matsubara[5] or (iii) Keldysh[6]. In this paper we adopt the first approach due to its conceptual simplicity. As usual we integrate in Eq. (1) (see below) over the three components of the phonon wave vector and over the electron quasimomentum, $p'$, employing the conservation of quasimomentum along $x$ direction (the orientation of a wire).

We consider only the scattering by three-dimensional extended phonons (although, the phonons localized within the nanostructure or near its interface with the bulk could have been easily included into our scheme). For the acoustic phonons in the isotropic approximation, $W_q = \pi Z^2 q^2/\rho\omega_q$. Here $Z$ is the deformation potential constant for the longitudinal phonons, and $\rho$ is the mass density. For the polar optical scattering, $W_q = (2\pi e/q)^2\omega_o(\varepsilon_0-\varepsilon_\infty)/\varepsilon_0\varepsilon_\infty$. Furthermore, we ignore the dispersion of the optical phonons, namely, $\omega_q = \omega_o$.

To the lowest order in electro-phonon coupling the rate of energy transfer $Q$ from electrons to the phonon system[6] is

$$
\begin{aligned}
Q &= \sum_q \hbar\omega_q \left[\frac{\partial N_q}{\partial t}\right]_{\text{coll}} = 2L_x \left[1 - \exp\left(-\frac{eV}{k_B T}\right)\right]^2 \int \frac{d^3 q}{(2\pi)^3} \\
&\times \sum_{nn'} \int \frac{dp'}{2\pi\hbar} \hbar\omega_q W_q C_{nn'} f(\epsilon' - \mu^{(+)}) \left[1 - f(\epsilon' - \mu^{(-)})\right] \\
&\times \left[1 - f(\epsilon - \mu^{(-)})\right] f(\epsilon - \mu^{(+)})\delta(\epsilon' - \epsilon - \hbar\omega_q)
\end{aligned}
\tag{1}
$$

here $f = f^{(F)}(\epsilon - \mu)$ is the Fermi function, $C_{nn'} = |\langle n'| \exp(i q_\perp r_\perp)|n\rangle|^2$, $L_x$ is the total length of the conductor, $N_q$ is the average phonon occupation number (which we assume to be the equilibrium Bose function) and $\langle r_\perp|n\rangle$ is the transverse part of the electron wavefunction, and where by $\epsilon$ we mean $\epsilon_n(p' - \hbar q_x)$ and by $\epsilon'$ we mean $\epsilon_{n'}(p')$. The summation over the various phonon branches is implied.

For the acoustic phonon scattering, although the phonon-controlled part of the current $|\Delta J|$ is somewhat smaller than the ballistic part of current $J^{(0)}$, its derivatives with respect to the applied voltage bias are larger than those of $J^{(0)}$. In Fig. 1 we show a typical hot electron controlled acoustical phonon generation effect, i.e. the dependence of $Q$ on the applied voltage and Fermi wavelength.

For the optical phonons (i.e. $\omega_q = \omega_o$), Eq. (1) takes the following form:

$$
\begin{aligned}
Q &= \frac{m^* L_x}{h^2} \left[1 - \exp\left(-\frac{eV}{k_B T}\right)\right]^2 \sum_{nn'} \int \frac{d^2 q_\perp}{(2\pi)^2} \\
&\times \int_0^\infty d\epsilon \, \hbar\omega_o W_q C_{nn'} \frac{\Theta(\epsilon - R_{nn'})}{\sqrt{\epsilon(\epsilon - R_{nn'})}} f(\epsilon + \hbar\omega_o - \mu_n^{(+)}) \\
&\times \left[1 - f(\epsilon - \mu_n^{(-)})\right] f(\epsilon - \mu_n^{(+)}) \left[1 - f(\epsilon + \hbar\omega_o - \mu_n^{(-)})\right].
\end{aligned}
\tag{2}
$$

As is seen in Fig. 2, the rate is quite sensitive to both voltage bias and gate voltage, and depends on an interplay of these parameters.

It then follows that

$$
\frac{Q}{|\Delta J| V} = \frac{\hbar\omega_o}{eV}.
\tag{3}
$$

One can easily visualize this result in the following way: $|\Delta J|/e$ is the variation of the electron flux due to the electron-phonon scattering. The scattered electrons reverse their quasimomenta and thus no longer contribute to the total current. However, since each electron that belongs to the blocked part of the flux emits just one phonon, one gets the following ratio: $Q = \hbar\omega_o|\Delta J|/e$. The rate of energy loss to the optical phonons is particularly significant, as compared to the total power, if EPR takes place near the generation threshold of $eV = \hbar\omega_o$ (to within $k_B T$). In nonhomogeneous wires, the generation is most likely to start in the section that meets EPR condition first. It is important to mention that only those phonons are generated which can *backscatter* the electrons. It is readily seen that this condition imposes a certain lower limit for the magnitude of the phonon wave vectors[4].

This situation is quite unlike the usual Landauer transport, where the electron-phonon (and the electron-electron) interactions are restricted to the contacts and hence all the heat is released in the contacts only. In our case some energy is transferred to the phonons and is released as heat in the vicinity of a nanowire. The rest of heat is released in the reservoirs. These two mechanisms of heat release are physically quite different.

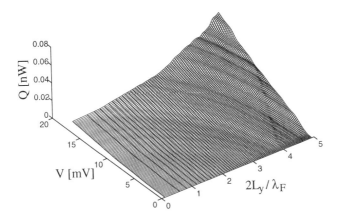

**Figure 1.** Rate of energy loss to the acoustical phonons as a function of voltage bias and Fermi wavelength. $L_y$=100nm and $T$=1 Kelvin.

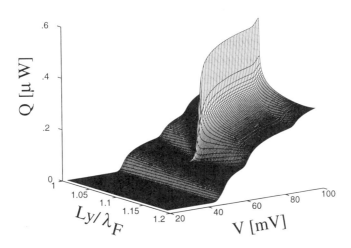

**Figure 2.** Rate of energy loss to the optical phonons as a function of voltage bias and channel width $L_y$. $\lambda_F$=40nm and $T$=4.2 Kelvin.

The singularities in Eq. (2) are due to the initial and final densities of states. As long as $R_{nn'} \neq 0$ they are integrable and the integral (2) is well-behaved as a function of bias or gate voltage. However, when the resonant condition is met ($R_{nn'} \rightarrow 0$), the denominator in Eq.(2) amounts to a $1/\epsilon$ singularity which renders (2) non-integrable (cf. with [7]). This leads to a logarithmic divergence (which is, of course, removed if one takes into consideration scattering events of higher order) and, consequently, to a significant increase in the resistance. Since no such singularities are present in 2,3D cases for an otherwise equivalent physical systems, no EPR in our sense of the word is to be expected there either.

One can see that actually, exactly as in the case of magnetophonon resonance[7], the existence of EPR is determined only by the singularities of the electron DOS and the existence of a nonvanishing phonon limiting frequency for $q \rightarrow 0$ and is independent of any other details of the electron and phonon spectrum and their interaction. Thus the existence of the 1D channels rather than their exact structure is sufficient for EPR to exist.

Kulik and Shekhter[8] predicted current oscillations which should be accompanied by the phonon generation. In contrast to the ballistic quantum transport investigated here, they have examined the classical diffusive electron transport in narrow wires.

We believe that EPR can be used as an experimental probe for the investigation of various physical aspects of nanosystems. To name a few, for instance, one can investigate the character of the actual interactions between the electrons and optical phonons, or the details of the electron band structure and the actual positions of the levels of transverse quantization, or the role of nonequilibrium phonons in the transport phenomena.

# References

[1] R. Landauer, IBM J. Res. Develop. **1**, 233 (1957); **32**(3), 306 (1989).

[2] See, for instance, C. W. J. Beenakker and H. van Houten, *Quantum Transport in Semiconductor Nanostructures* in Solid State Physics, **44**, Academic Press, 1991, edited by H. Ehrenreich and D. Turnbull.

[3] V. L. Gurevich, V. B. Pevzner and K. Hess, in *Quantum Transport in Ultrasubmicron Devices*, edited by D. K. Ferry, H. L. Grubin, C. Jacoboni and A.-P. Jauho (Plenum Press, New York, 1995).

[4] V. L. Gurevich, V. B. Pevzner and K. Hess, Phys. Rev. B **51**(8), 5219 (1995); Condens. Matter **6**, 8363 (1994).

[5] E. W. Fenton, V. L. Gurevich, V. B. Pevzner and K. Hess, *Proc. 22nd Int. Conf. on Physics of Semiconductors*, 1994, Vancouver, Canada (World Scientific Publishing); V. B. Pevzner, V. L. Gurevich and E. W. Fenton, Phys. Rev. B **51**(15), 9465 (1995).

[6] V. L. Gurevich, V. B. Pevzner and G. J. Iafrate Phys. Rev. Lett. (1995); to be published (1995).

[7] Yu. A. Firsov, V. L. Gurevich, R. V. Parfeniev and I. M. Tsidil'kovskii, *Landau level spectroscopy* in Modern Problems in Condensed Matter Sciences, vol. 27.2, pp. 1180-1308, edited by G. Landwehr and E. I. Rashba (Elsevier Science Publishers, New York, 1991).

[8] I. O. Kulik and R. I. Shekhter, Phys. Lett. **98A**, 132 (1983).

# ELECTRON TRANSPORT IN QUANTUM WIRES AT HIGH TEMPERATURE

T. Ezaki, N. Mori, H. Momose, K. Taniguchi and C. Hamaguchi

Department of Electronic Engineering, Osaka University,
2–1, Yamada-oka, Suita City, Osaka 565, Japan

## INTRODUCTION

There is an increasing interest in the study of low-dimensional electronic systems. It has been pointed out that the elimination of small-angle scattering by impurities can considerably enhances the mobility in quantum wires (QWs) at low temperature[1]. At high temperature, magnetophonon resonance peaks have been observed in QWs fabricated on a GaAs/AlGaAs heterostructure[2] and a conductance minimum, which is attributed to the electron–optical-phonon resonance, has been reported[3]. In the present work, we are interested in the electron transport in quasi-one-dimensional (Q1D) systems at high temperature. We solve two-dimensional (2D) Poisson and Schrödinger equations self-consistently to evaluate the electronic states in a QW and calculate the applied gate voltage dependence of conductance by using the Monte Carlo method.

## FORM FACTORS IN QUANTUM WIRE

Figure 1 shows a schematic diagram of an Al-gated QW fabricated on a GaAs/AlGaAs single hetero-structure. The device consists of an unintentionally $p$-doped GaAs substrate ($N_A = 10^{14}\,\mathrm{cm}^{-3}$), followed by an undoped AlGaAs spacer layer (30 nm) and an $n$-doped AlGaAs cap layer ($N_D = 1.2 \times 10^{18}\,\mathrm{cm}^{-3}$, 25 nm). We define that the $y$- and $z$-axes are parallel and normal to the GaAs/AlGaAs interface, respectively.

In order to obtain electronic states in the QW with good accuracy, we employed self-consistent calculation. Then we calculate 1D form factors $G_{ii'}$ between states $i$ and $i'$:

$$G_{ii'}(q_x) = \int d\vec{\rho}_1 \int d\vec{\rho}_2 \, \mathrm{K}_0(q_x|\vec{\rho}_1 - \vec{\rho}_2|)\Psi_{i'}^*(\vec{\rho}_1)\Psi_i(\vec{\rho}_1)\Psi_i^*(\vec{\rho}_2)\Psi_{i'}(\vec{\rho}_2) \tag{1}$$

where $\Psi_i$ is the $i$-th wave function, $\vec{\rho}$ is the 2D position vector in a plane of parallel to the hetero-interface and $\mathrm{K}_0$ is the modified Bessel function of the second kind. The subband index $i$ can be expressed by the node number $n$ of wave functions in $y$-direction and $m$ for $z$-direction, $i.e.$ $i \equiv (n, m)$.

*Hot Carriers in Semiconductors*
Edited by K. Hess *et al.*, Plenum Press, New York, 1996

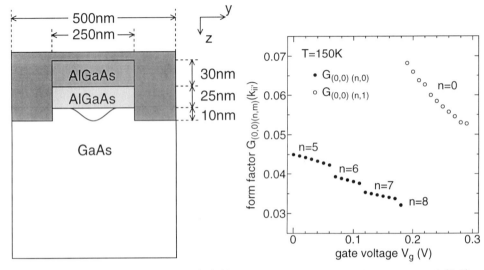

**Figure 1.** Cross-section of a mesa etched Al-gated QW structure. The device geometry: an unintentionally $p$-doped GaAs substrate ($N_A = 10^{14} \text{cm}^{-3}$) followed by an undoped AlGaAs spacer layer and an $n$-doped AlGaAs cap layer ($N_D = 1.2 \times 10^{18} \text{cm}^{-3}$).

**Figure 2.** Form factors between ground $(0, 0)$ and the excited $(n, m)$ states that almost satisfies the resonance condition $(E_{(n,m)} - E_{(0,0)} \lesssim \hbar\omega_0)$. $k_{ii'}$ is the wave number that satisfies the energy conservation. Solid circles show $G_{(0,0)(n,0)}$ and open circles $G_{(0,0)(n,1)}$.

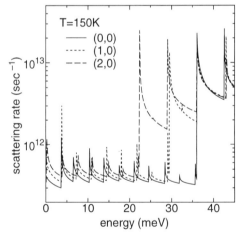

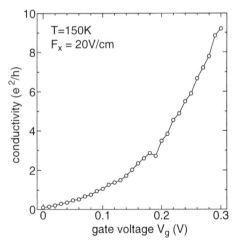

**Figure 3.** The total scattering rate as a function of electron kinetic energy at $T = 150$ K. Solid line shows the scattering rate for the ground state electrons, dotted line for the first excited states, and dashed line for the second excited states.

**Figure 4.** Calculated conductance as a function of applied gate voltage $V_g$ at $T = 150$ K. The applied longitudinal electric field $F_x$ is set to be 20 V/cm. A conductance minimum is clearly seen at around $V_g = 0.19$ V.

Figure 2 shows calculated results of the form factors between the ground state $(0, 0)$ and the excited states $(n, m)$ near the resonance condition, *i.e.* $E_{(n,m)} - E_{(0,0)} \lesssim \hbar\omega_0$ with $\hbar\omega_0$ being the longitudinal-optical (LO) phonon energy (36.2 meV). As seen in the figure, we can expect a strong resonance at around $V_g = 0.19$ V where the resonance state $(n, m)$ changes its node number $m$[4, 5].

## MONTE CARLO SIMULATION

Since the electron–optical-phonon scattering is the most dominant process in a GaAs/Al-GaAs QW at high temperature, we took only the electron–optical-phonon interaction into account. The scattering rate of electrons in the state $(i, \varepsilon)$ takes the form:

$$W_i(\varepsilon) = \sum_{i'} \alpha\omega_0 \sqrt{\frac{\hbar\omega_0}{\varepsilon - E_{i'} + E_i \pm \hbar\omega_0}} \left(N_0 + \tfrac{1}{2} \pm \tfrac{1}{2}\right) \{G_{ii'}(k - \gamma) + G_{ii'}(k + \gamma)\} \quad (2)$$

where $\alpha = 0.07$ is the Fröhlich's coupling constant, $N_0$ the LO phonon occupation number, $\gamma = \{2m^*(\varepsilon - E_{i'} + E_i \pm \hbar\omega_0)\}^{1/2}/\hbar$ and $m^* = 0.067m_0$. The $\pm$ sign corresponds to the emission and the absorption of LO phonon, respectively.

Figure 3 shows the total scattering rates as a function of electron kinetic energy $\varepsilon$. As seen in the figure, there are several peaks in scattering rates which arise from the nature of 1D density of state. These singularities in the Q1D systems make usual constant-$\Gamma$ algorithm inefficient due to the large percentage of self-scattering events. In order to reduce the self-scattering events, direct integration of the Monte Carlo free-flight equation

$$-\log r = \int_0^t dt' \, W_i(k(t')) \quad (3)$$

was applied[6, 7], where $r$ is a uniform random number. The Monte Carlo result of conductance as a function of gate voltage for $T = 150$ K is shown in Fig. 4. A conductance minimum at around $V_g = 0.19$ V is cleanly seen in the figure. This feature can be explained by the strong interaction between the ground $(0, 0)$ and the excited $(0, 1)$ states[4, 5].

## CONCLUSION

In conclusion, we performed Monte Carlo calculation to obtain applied gate voltage dependence of conductance in a Q1D system by using the eigen states determined by the self-consistent calculation. A conductance minimum due to the electron–optical-phonon interaction is found to occur at around $V_g = 0.19$ V for the device considered in this study.

## REFERENCES

1. H. Sakaki, Jpn. J. Appl. Phys. **19**, (1980) L735.
2. G. Berthold, J. Smoliner, C. Wirner, E. Gornik, G. Böhm, G. Weimann, M. Hauser, C. Hamaguchi, N. Mori and H. Momose, Semicond. Sci. Technol. **8**, 735 (1993).
3. K. Ismail, in *Science and Technology of Mesoscopic Structures*, edited by S. Namba, C. Hamaguchi and T. Ando, (Springer-Verlag, Tokyo, 1992) p.135.
4. D. Jovanovic, J. P. Leburton, K. Ismail, J. M. Bigelow and M. H. Degani, Appl. Phys. Lett. **62**, 2824 (1993).
5. N. Mori and C. Hamaguchi, Proc. Int. Workshop on Computational Electronics,University of Illinois at Urbana-Champaign, 1992, p.261.

6. C. Jacoboni and L. Reggiani, Rev. Mod. Phys, **55**, 645 (1983).
7. D. Jovanovic and J.P. Leburton, in *Monte Carlo Device Simulation: Full Band and Beyond*, edited by K. Hess, (Kluwer Academic Publishers, 1991) p.191.

# MONTE CARLO SIMULATION OF
# HOT ELECTRONS IN QUANTUM WIRES

H. Momose[1], N. Mori[2], K. Taniguchi[2], C. Hamaguchi[1,2]

[1] Low Temperature Center, Osaka University
[2] Department of Electronic Engineering, Osaka University
2-1 Yamada-oka, Suita, Osaka 565, Japan

## INTRODUCTION

One of the method of analyzing the electron transport in semiconductors is the Monte Carlo (MC) simulation that is able to solve the semi-classical Boltzmann transport equation. The conventional MC simulation, however, takes no accounts of quantum effects, such as collisional broadening, intra-collisional field effect and so on. We have proposed a new model for the MC simulation which can handle effects of broadening of the spectral density function, and calculated transport properties of 1DEGs where only the ground subband was taken into account[1]. In the present study, we extend this model to take into account of multi-subband effects and evaluate the transport response of 1DEGs. Throughout this paper we set $\hbar = 1$.

## MODEL FOR MONTE CARLO SIMULATION

We use the quantum Boltzmann transport equation for non-degenerate electrons which is derived by using the non-equilibrium Green's function method [2]:

$$\left\{\frac{\partial}{\partial t} - eE\frac{\partial}{\partial k}\right\} f(n, k, \omega, t) = \sum_{n'}\sum_{k'} \int \frac{d\omega'}{2\pi} W(n', k', \omega'; n, k, \omega) f(n', k', \omega', t)$$
$$- \sum_{n'}\sum_{k'} \int \frac{d\omega'}{2\pi} W(n, k, \omega; n', k', \omega') f(n, k, \omega, t) \quad (1)$$

where $n$ is the subband index, $k$ the 1D wave vector, $\omega$ the energy, and $f(n, k, \omega, t)$ a Wigner distribution function (WDF). Once the WDF is known, the average electron energy, the drift velocity or other physical quantities are readily calculated. The transition probability $W$ for electron–optical-phonon interaction is given by

$$W(n, k, \omega; n', k', \omega')$$
$$= 2\pi A(n', k', \omega') \sum_{\eta = \pm 1}\sum_{q} M_{n,n'}^2(q)(N_0 + \tfrac{1}{2} + \tfrac{1}{2}\eta)\delta(\omega' + \eta\omega_0 - \omega)\delta_{k', k+q} \quad (2)$$

where $A(n, k, \omega)$ is the spectral density function, $N_0$ the optical-phonon occupation, $\omega_0$ the optical-phonon energy, and $M_{n,n'}(q)$ the matrix element of electron–optical-phonon interaction. The effect of broadening is expressed by the spectral density function $A(n, k, \omega)$.

*Hot Carriers in Semiconductors*
Edited by K. Hess *et al.*, Plenum Press, New York, 1996

The spectral density function could be calculated by using a suitable model for the system in principle. In the following analysis, however, we assume $A(n, k, \omega)$ as a Gaussian function characterized by the constant level width $\Gamma$

$$A(n, k, \omega) = \frac{2\sqrt{\pi}}{\Gamma} \exp\left[-\left\{\frac{2(\omega - \varepsilon_{n,k})}{\Gamma}\right\}^2\right] \tag{3}$$

for simplicity, where $\varepsilon_{n,k}$ is the dispersion relation.

## MODEL OF A QUANTUM WIRE

We employ a simple model for a quantum wire in which a two-dimensional electron gas in $xy$-plane is confined by narrow gates or split gates, and electrons are free along only $x$-direction. The confinement in the $y$-direction is assumed to be characterized by a parabolic potential of frequency $\Omega$:

$$V(y) = \tfrac{1}{2}m\Omega^2 y^2. \tag{4}$$

The transition probability, therefore, is reduced to

$$
\begin{aligned}
&W(n, k, \omega; n', k', \omega') \\
&= 4\pi\alpha A(n', k', \omega') \sum_{\eta=\pm 1} (N_0 + \tfrac{1}{2} + \tfrac{1}{2}\eta) G(k' - k)\delta(\omega' + \eta\omega_0 - \omega)
\end{aligned} \tag{5}
$$

where

$$G(q) = \int dp \, \frac{1}{2\sqrt{q^2 + p^2}} |J_{n,n'}(v)|^2, \tag{6}$$

$$|J_{n,n'}(v)|^2 = \frac{n_0!}{n_1!} v^{n_1 - n_0} e^{-v} |L_{n_0}^{n_1 - n_0}(v)|^2, \tag{7}$$

$v = \tfrac{1}{2}(p/K)^2$, $K = (2m\Omega)^{1/2}$, $n_0 = \min(n, n')$, $n_1 = \max(n, n')$, $\alpha$ is Fröhlich coupling constant, and $L_p^q(v)$ are the associated Laguerre polynomials.

## TRANSPORT PROPERTIES IN A QUANTUM WIRE

Figure 1 shows calculated results of the scattering probability in a quantum wire of $\Omega = 20\,\mathrm{meV}$ at $T = 300\,\mathrm{K}$ for $\Gamma = 5\,\mathrm{meV}$. Since we consider momentum and energy as independent variables, the scattering probability is a function of $(k, \omega)$. The peaks at $\omega = \omega_0$, $\omega_0 + \Omega$, $\omega_0 + 2\Omega$, $\cdots$ are due to the resonant phonon coupling, and the peaks at around $(k, \omega) = (k, \varepsilon_k + \omega_0)$, $(k, \varepsilon_k - \omega_0)$ correspond to the phonon emission and absorption processes.

Using this scattering probability, we perform the MC simulation. Figure 2 shows the average electron energy in the steady state. We take into account lowest 5 subbands. As seen in the figure, the average electron energy increases monotonously with applied electric field showing no anomalous carrier cooling.

Steady state drift velocity as a function of the applied electric field is also shown in Fig. 3. We find that the electron mobility in low electric field is estimated at $5.8 \times 10^3 \mathrm{cm}^2/\mathrm{V\cdot sec}$ although the mobility estimated with the single-subband approximation is $7.2 \times 10^3 \mathrm{cm}^2/\mathrm{V\cdot sec}$ [1].

## ACKNOWLEDGMENTS

This work is supported in part by the Grant-in-Aid for Scientific Research on Priority Area "Electron Wave Interference Effects in Mesoscopic Structures" from the Ministry of Education, Science, Sports and Culture, Japan. We acknowledge the support by the Grant-in-Aid for Encouragement of Young Scientists to H. M. from the Ministry of Education, Science and Culture, Japan.

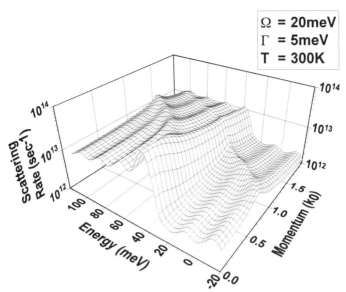

**Figure 1.** Scattering probability of electrons in a GaAs quantum wire of $\Omega = 20$ meV at room temperature. $k$ is normalized by $k_0 = (2m\omega_0)^{1/2}$.

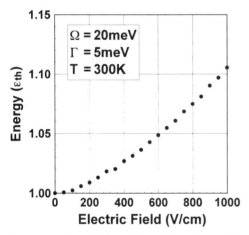

**Figure 2.** Applied electric field dependence of average electron energy in a GaAs quantum wire. The average electron energy is normalized by the thermal energy. $\Omega = 20$ meV and $T = 300$ K.

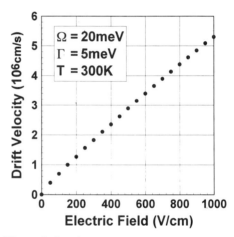

**Figure 3.** Steady state electron drift velocity as a function of applied electric field in a GaAs quantum wire of $\Omega = 20$ meV at room temperature.

## REFERENCES

1. H. Momose, N. Mori, K. Taniguchi and C. Hamaguchi: Semicond. Sci. Technol. 9 (1994) 958.
2. A.P. Jauho, in *Quantum Transport in Semiconductors*, edited by D.K. Ferry and C. Jacoboni, (Plenum, New York, 1992), Chapter 7.

# QUASI-ONE-DIMENSIONAL TRANSPORT AND HOT ELECTRON EFFECTS IN InAs MESOSCOPIC STRUCTURES

Masataka Inoue,[1] Shin-ichi Osako,[2] Shigehiko Sasa,[1] Kazuyuki Tada,[1] Tsuyoshi Sugihara,[1] Satoshi Izumiya,[1] Yoshitaka Yamamoto,[1] and Chihiro Hamaguchi[2]

[1]Department of Electrical Engineering, Osaka Institute of Technology
Omiya 5-16-1, Asahi-ku, Osaka 535, Japan
[2]Faculty of Engineering, Osaka University
Yamada-oka 2-1, Suita, Osaka 588, Japan

## INTRODUCTION

The study of quasi-one-dimensional (1D) electron transport in mesoscopic structures has become a very attractive research field during the past decade. These mesoscopic structures have shown a wide variety of new phenomena such as universal conductance fluctuations and quantum interferences.[1] A significant enhancement of 1D electron mobility due to the suppression of ionized impurity scattering has been predicted.[2] Electron-phonon and electron-electron scatterings in quantum wires were also studied to predict the significant difference from that in the two-dimensional electrons.[3] Meanwhile, the clear evidence of 1D density of states and mobility modulation of 1D electrons has been demonstrated in GaAs/AlGaAs quantum wires by Ismail et al.[4]

So far mesoscopic or 1D devices have been mostly studied by using modulation-doped GaAs/AlGaAs. For further development of these devices, however, the operating temperature is of the center of interest. The one-dimensional quantized conductance has been observed at around 77K for the first time in InAs/AlGaSb split-gate devices.[5] The low effective mass and strong confinement of electrons in the InAs quantum well are the major advantage for relatively high temperature operations over the conventional GaAs/AlGaAs devices.[6]

In this paper, we report the investigation of 1D transport in multiple quantum wires fabricated on InAs/AlGaSb heterostructures. The magnetoresistance measurements of the 1D wires with different width have been performed at 4.2K. The high-field transport properties have been also measured to analyze hot electron effects of 1D electrons with and without magnetic field, since the fields introduce additional features on the 1D electron system. Device characteristics of InAs quantum wire transistors with gate electrodes are also reported.

*Hot Carriers in Semiconductors*
Edited by K. Hess *et al.*, Plenum Press, New York, 1996

## FABRICATION OF QUANTUM WIRES

InAs/Al$_{0.5}$Ga$_{0.5}$Sb heterostructures grown by molecular beam epitaxy (MBE) have been fabricated by using conventional photolithography and wet chemical etching.[6] The structure consists of a GaAs buffer layer (250nm), an AlSb buffer layer (1.5μm), an AlGaSb barrier layer (200nm), an InAs channel layer (15nm), an AlGaSb upper layer (15nm), and a GaSb cap layer (10nm). The structure was grown on a semi-insulating GaAs substrate and was nominally undoped. The substrate temperature was set to 640℃ during the growth of the GaAs buffer, 600℃ and 450℃, for the AlSb buffer layer and the InAs layer or above, respectively. The measured mobility and density of two-dimensional (2D) electrons were ranging 0.7-1.8×10$^5$cm$^2$/V·s and 4-9×10$^{11}$cm$^{-2}$ at 77K, respectively.

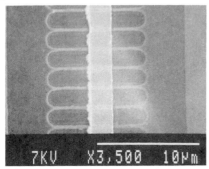 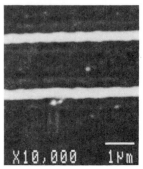

Fig.1 SEM image of wires and QWT.

In order to fabricate lateral surface superlattices (LSSLs) or quantum wire transistors (QWTs), we used the phosphoric acid based etchant which gave fairly isotropic mesa structures in contrast with the case for GaAs/AlGaAs. The fabrication process of the QWTs began with the definition of narrow wires, and was followed by the formation of non-alloyed ohmic contacts at both ends of the wires. Then, we loaded the wafer back into the MBE chamber, and grew 20-nm-thick undoped GaAs layer over the entire structure. To prevent a metallurgical reaction at the In/Au ohmic contacts, we kept the growth temperature fairly low of about 150℃. The surface morphology of the regrowth GaAs layer is smooth enough to proceed the following fabrication process. Contact windows were opened by removing the overgrown GaAs layer at ohmic contacts. Finally, Ti/Au was evaporated to form the gate electrode.

Figure 1 shows the SEM image of the QWT. We have fabricated multiple quantum wires of which nominal width was ranging from 100nm to 600nm. The effective 1D channel was somewhat reduced from the nominal width of QWs as shown by analyses of magneto-transport measurements.

## 1D ELECTRON TRANSPORT IN QUANTUM WIRES

Magnetic fields were applied perpendicular to the surface of multiple parallel QWs at 4.2K. Figure 2 shows magnetoresistance of the nominally 560nm wide QWs which shows average out magnetoresistance of 10 wires. Nevertheless, the well defined Shubnikov-de Haas oscillations have been observed as measured in the InAs single quantum wire.[7] This result indicates that each wire was uniformly made by the etching process using a phosphoric acid based etchant. The departure from the linearity of the Shubnikov-de Hass oscillations becomes more pronounced at low magnetic fields as the effective width of the QWs is reduced.

The sublevel index n is plotted as a function of the inverse magnetic field in Fig.3.

Closed circles refer to minima in the measured magnetoresistance shown in Fig.2. The dotted line is the theoretical result to fit the experimental data. By using the relation between sublevel index and inverse magnetic field fitted to the nonlinear experimental results,[8] we extracted 1D electron concentration $N_{1D}$ and the characteristic frequency $\omega_0$ defined to be the strength of the confinement. From this fitting, we found $N_{1D}=8\times10^6\mathrm{cm}^{-1}$ and $\omega_0=5\times10^{12}\mathrm{s}^{-1}$. It was also found that twelve 1D subbands are occupied without magnetic fields. With increasing magnetic fields, these subbands are depopulated. Each energy separation between 1D subband can be estimated to be 3.3meV, if the parabolic confined potential is assumed. The effective width W of 1D channel was determined by using $N_{1D}$ and $\omega_0$ to be 210nm, which is much smaller than the nominal width. The reduced effective channel width was found in the conductance measurements of the InAs single QW.[7]

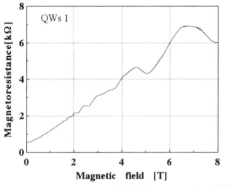

Fig.2 Magnetoresistance of QWs1 measured at 4.2K.

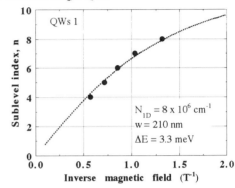

Fig.3 Sublevel index n vs inverse magnetic field $B^{-1}$.

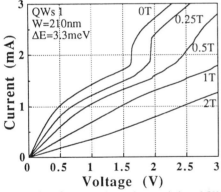

Fig.4 Current-voltage characteristics of QWs1.

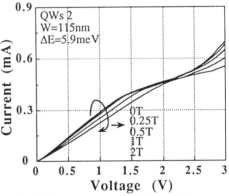

Fig.5 I-V characteristics of QWs2 at 4.2K.

In order to investigate 1D electron transport and hot electron effects, current-voltage characteristics have been measured at 4.2K. We compare the data of two QWs, QWs1 and QWs2, for the different effective width. Figure 4 shows the I-V characteristics of QWs1. We have measured another narrow QWs2 fabricated from the same InAs wafer. The effective channel width of QWs2 was estimated to be 115nm. In this narrow QWs, six subbands are occupied and the inter-subband energy separation is 5.9meV. Figure 5 shows the I-V characteristics of QWs2.

Applied magnetic fields can modulate the subband energy and scattering rates of 1D electrons.[9] The effects of magnetic fields have been clearly observed in Fig.4 and 5, especially in the wider QWs1. Since magnetic fields increase the energy separation between subbands, the threshold voltage of inter-subband scattering of hot electrons should be shifted under magnetic fields. If we argue the kink observed in I-V characteristics to be the threshold of inter-subband scattering of hot electrons, the current modulation by magnetic fields can be reasonably understood. In the narrow channel, the effect of magnetic field is small in comparison with that in wider QWs1. Effects of magnetoresistance and suppressing electron backscattering in 1D channel may be also responsible for the magnetic field-dependence observed in the present experiments.

Next, we discuss the I-V characteristics of InAs/AlGaSb QWTs measured at 77K together with those of conventional 2D Transistors. The gate length of the QWT was 2μm and that of the 2D transistor, 20μm. The channel width of the 2D transistor was 50μm. By optimizing conditions for the photolithography and the succeeding wet chemical etching, we successfully fabricated quantum wires with the width as narrow as 100nm. Figure 6 shows the I-Vcharacteristics of a QWT. By introducing the regrown GaAs barrier layer,

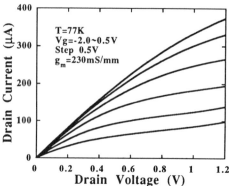

Fig.6 Device characteristics of the QWT

the gate leakage current was sufficiently suppressed for QWT structure.[10] The maximum transconductance, $g_m$ at $V_{DS}=1V$ is 230mS/mm for the QWT and 130mS/mm for the 2D transistor. The estimated drift mobilities in the QWT and the 2D transistor at $V_{DS}<1V$ were about $3 \times 10^3 cm^2/V \cdot s$ for both structures. Therefore, the difference in $g_m$ is possibly due to the difference in the device geometry. However, we did not observe any mobility decrease in QWTs caused by the surface roughness of the wires indicating the successful fabrication process and the usefulness of the QWTs.

In summary, we reported magneto-transport and high-field transport properties of InAs QWs and nonlinear characteristics have been discussed. Hot electron effects in InAs QWs should be quantitatively discussed in the near future.

## ACKNOWLEDGMENTS

The author wish to thank K.Koike and C.Dono for their technical assistance in MBE growth and Prof.M.Yano for helpful discussions. This work was party supported by Grant-in-Aid for Scientific Research on Priority Area "Quantum Coherent Electronics, Physics and Technology."

## REFERENCES

1. M.A. Read and W.P. Smith, Nanostructure Physics and Fabrication, Academic, Boston (1990).
2. H. Sakaki, Jpn. J. Appl. Phys. 19:L735 (1980).
3. J.P. Leburton, J. Appl. Phys. 56:2850 (1984).
4. K. Ismail, D.A. Autoniadis and H.I. Smith, Appl. Phys. Lett. 54:1130 (1989).
5. M. Inoue, K. Yoh and A. Nishida, Semicond. Sci. Technol. 9:L966 (1994).
6. K. Yoh, K. Kiyomi, A. Nishida and M. Inoue, Jpn. J. Appl. Phys. 31:4515 (1992).
7. K. Yoh, H. Taniguchi, K. Kiyomi, R. Sakamoto and M. Inoue, Semicond. Sci. Technol. 7:B295 (1992).
8. K.F. Berggrem, G. Roos and H.von Houten, Phys. Rev. B37:10118 (1988).
9. N. Telang and S. Bandyopadhyay, Phys. Rev. B 51:9728 (1995).
10. S. Osako, T. Sugihara, Y. Yamamoto, T. Maemoto, S. Sasa, M. Inoue and C. Hamaguchi, Micro Process Conference '95, Sendai (1995).

# NONEQUILIBRIUM TRANSPORT AND CURRENT INSTABILITIES IN QUANTUM POINT CONTACTS

S.M. Goodnick*, J.C. Smith[†], C. Berven[†], M.N. Wybourne[†]

*Department of Electrical and Computer Engineering
Oregon State University, Corvallis, OR 97331

[†]Department of Physics
University of Oregon, Eugene, OR 97403

## INTRODUCTION

The near equilibrium conductance properties of quantum point contacts has been an intense field of research over the past 7 years following the demonstration of conductance quantization in such structures.[1] However, the nonequilibrium transport properties of such systems are important for practical device applications involving nanoscale structures. Non-linear transport measurements in quantum point contacts have been reported in which oscillatory conductance with source-drain bias is observed associated with transport through successive one-dimensional subbands.[2] Such transport is 'ideal' in the sense that it may described in terms of the two-terminal Landauer-Büttiker model when extended to non-zero $V_{sd}$.

In the present paper, we report on non-ideal S-type negative differential conductance observed in quantum point contact structures.[3] Time dependent measurements are reported of random telegraph signals (RTS) in the measured current through a quantum constriction. This time dependent switching behavior is only observed in samples exhibiting SNDC in their current voltage characteristics, and is only found in the region of instability in the DC characteristics.

## EXPERIMENT

Electron-beam lithography was used to define split-gate quantum point contacts of physical width ($W$) 0.2 μm and lengths ranging from 0.0 to 0.5 μm, shown schematically in the inset to Fig. 1. The gate electrodes were made from 25 nm of Au thermally deposited on 7.5 nm of Ti and were fabricated on a molecular-beam epitaxy grown, uniform modulation doped GaAs/Al$_{0.27}$Ga$_{0.73}$As heterostructure. The GaAs cap and the doped AlGaAs layer both have a silicon donor concentration of $1\times10^{18}$ cm$^{-3}$. The point contact is formed by applying a negative

*Hot Carriers in Semiconductors*
Edited by K. Hess *et al.*, Plenum Press, New York, 1996

gate bias, $V_g$, to both gate electrodes. At $V_g = 0.0$ V and 4.2 K the material has an electron density and mobility of $3 \times 10^{11}$ cm$^{-2}$ and $1 \times 10^6$ cm$^2$/Vs, respectively. The corresponding Fermi wavevector and energy are 46 nm and 10.7 meV, respectively.

The devices were mounted in a pumped $^4$He cryostat fitted with an infrared light emitting diode ($\lambda = 830$ nm) for device illumination. Constant current measurements were used to study the $I$-$V$ characteristics at 1.2 K before and after infrared illumination. For small source-drain bias, the $I$-$V$ characteristics display well-defined conductance plateaus as a function of gate bias which are qualitatively unchanged after infrared illumination (Fig. 1). When cooled slowly the devices display ideal behavior at large source-drain bias and conductance oscillations consistent with the results of Martin and Moreno[2]. Infrared illumination or rapid quenching from room temperature produces one or more regions of current-controlled s-shaped differential conductance, as shown in Fig. 2. Details of the SNDC varied from sample to sample, and for different cool-downs of the same sample. Ideal behavior can be restored by annealing at or above 120 K for 12 hours. The annealing time and temperature are consistent with those required to deionize DX centers in Al$_{0.27}$Ga$_{0.73}$As [4].

The origin of the SNDC was investigated by applying a constant voltage across a load resistor in series with the device, and measuring the current through the point contact as a function of time. When the bias conditions are such that the load line crosses the SNDC, the current displays random telegraph switching between a high and low value. The two current values correspond to two distinct regions of constant differential conductance found on either side of the SNDC region. The transitions between the high and low states have an exponential behavior that can be explained by the transient response of the load resistor in series with the device, which is in parallel with an 830 pF capacitor. The capacitor represents device and parasitic capacitance in the experimental arrangement.

Figure 3 shows a typical device characteristic at a single gate voltage with the high and low states for various values of load resistor. Note that the load line need not intersect either region of constant differential conductance in order for the device to switch between the two states. The SNDC in the steady-state characteristic is a time average of the RTS. The relationship between the SNDC and illumination and thermal cycling of the device suggest that the RTS is due to the trapping/detrapping of electrons from shallow donor levels associated with DX centers in the AlGaAs. This trapping process appears to be current activated.

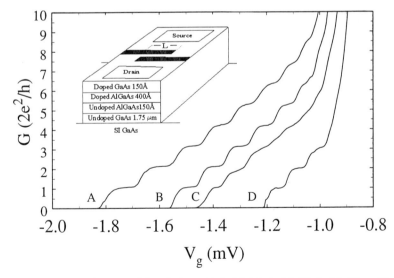

Figure 1. The equilibrium conductance of four devices with L=0.0 (A), L=0.1 (B), L=0.2 (C), and L=0.3μm (D). The inset is a schematic of the devices and the material.

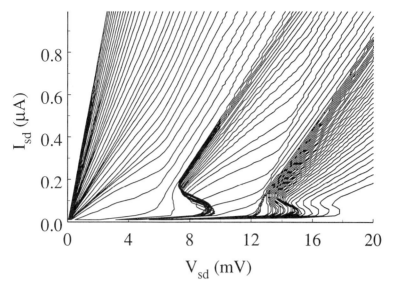

Figure 2. Current-voltage characteristic of a device that had two regions of SNDC along the source-drain bias axis.

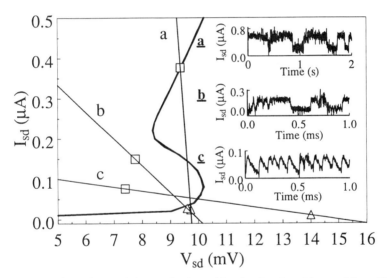

Figure 3. The current-voltage characteristic at a fixed gate bias showing three bias conditions. The switching occurs between a high current state (square) and a low current state (triangle). The inset shows the random telegraph signal associated with each load line.

## DISCUSSION

The SNDC shown in Fig.2 bears a resemblance to the behavior we reported earlier in measurements of nonlinear transport through split-gate quantum dot structures[5]. There we were able to explain SNDC observed in the *I-V* characteristics through an electron heating model in which hot electron runaway of the isolated carriers of the dot occurs due to the injection of energetic carriers from the drain[6]. The main difference between the present SNDC and that in quantum dots is first that the voltage scale for switching is a few meV in the single QPC case, whereas in quantum dots, switching occurred at hundreds of meV. Further, no RTS was observed in the dot case in contrast to the results shown in Fig. 3. The heating model cannot explain the time dependent switching of Fig. 3. In order to explain the phenomena of switching, we suggest that this behavior is due to the charging/discharging of shallow impurity states located in the vicinity of the point contact, in which the occupancy of the states is dependent on the current through the contact due to impact ionization. The dependence of the occurrence of SNDC on temperature and optical excitation gives a clue to its origin. When the sample is cooled slowly in the dark, the DX centers in the AlGaAs due to Si doping are completely filled, i.e. doubly occupied by electrons with an accompanying lattice distortion. The deep level state is not easily ionized by impact processes with free carriers. If some of these states are photoionized, the DX centers become shallow donors (with an accompanying lattice relaxation), releasing free carriers into the conduction band. These shallow donors are easily accessible in terms of the excess energy to electrons injected over a barrier for the voltage range shown in Fig. 2 (several meV). From the Si donor concentration we estimate there are ~800 DX centers in a volume $W^2d$, where $d$ is the thickness of the AlGaAs layer. If the switching is due to the action of individual uncorrelated traps, the number of photoionized DX centers in the vicinity of the point contact is probably too high to account for the well-defined, two-level switching we observe. This suggests that there is a collective fluctuation of an ensemble of interacting traps. Such a mechanism has been used to explain random telegraph signals observed in MOS tunnel diodes[7]. The SNDC occurs in this model if the change in occupancy of the traps is accompanied by a change of the barrier height of the potential at the saddle point of the point contact. The time dependent behavior of Fig. 3 just corresponds to charging and discharging of these states. When the current is large, the states are always unoccupied. For low current densities, the states are always occupied.

The above argument is a qualitative explanation of the behavior observed in Figs. 3 and 4. To make this argument slightly more quantitative, we model the non-linear transport through

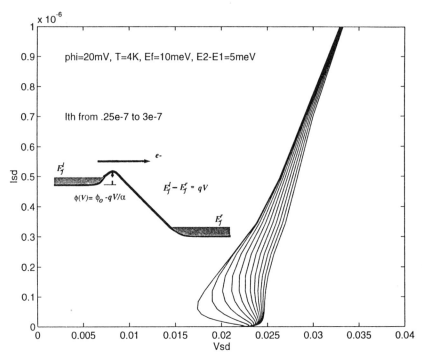

Figure 4. Calculated I-V for several values of $I_{th}$, with $\phi_0(0)$=20meV and $\alpha$=3. The inset shows a 1D model of the band profile under bias.

a point-contact using the band diagram shown in the inset of Fig. 4. The energy profile represents the 1-dimensional variation of the potential going from the source to the drain through the QPC. As the source-drain bias increases, the barrier height, $\phi$, decreases according to

$$\phi = \phi_0 - {}^{V_{sd}}\!/\!_\alpha$$

(1)

where $\phi_0$ is the zero bias barrier height due to the lateral depletion associated with the gate bias applied to the split gates. The parameter $\alpha$ is a barrier shape factor which determines the degree of barrier lowering with voltage drop. When (1) is used in the two terminal multi-channel Landauer-Büttiker formula, the oscillatory conductivity of Martin-Moreno et al[2] is obtained.

To explain the non-ideal switching behavior of Fig. 2, suppose that the ionization of a collection of traps causes the barrier height to change by an amount $\Delta\phi$. This trapping/detrapping depends critically on the current flowing in the channel due to an impact ionization process. A model for this is

$$\phi_0(I) = \phi_0 - \Delta\phi\left(1 - e^{-I/I_{th}}\right),$$

(2)

where $I_{th}$ is a threshold current for the impact ionization process. If we replace $\phi_0$ with $\phi_0(I)$ in Eq. 1, the resulting $I$-$V$ curves exhibit SNDC as shown in Fig. 4 for various values of the threshold current, $I_{th}$ (and a fixed occupied trap barrier height of 20 meV). The calculated SNDC becomes more and more pronounced as the threshold current for ionization becomes smaller and smaller.

## REFERENCES

1. C.W.J. Beenakker and H. van Houten, Quantum transport in semiconductor measurements, in "Solid State Physics," edited by H. Ehrenrich and Turnbull (Academic Press, New York, 1991), Vol. 44, and references therein.
2. L. Martin-Moreno, J.T. Nicholls, N.K. Patel, and M. Pepper, Non-linear conductance of a saddle-point constriction, J. Phys. Condens. Matter 4:1323 (1992).
3. C. Berven, M.N. Wybourne, A. Ecker, and S.M. Goodnick, Negative differential conductance in quantum waveguides, Phys. Rev. B 50:14639 (1994).
4. Z. Wilamowski, P.T. Suski, and W. Jantsch, DX puzzle: where are we now, Acta Physica Polonica A 82:561 (1992)
5. J.C. Wu, M.N. Wybourne, S.M. Goodnick, Negative differential conductance observed in a lateral double constriction device, Appl. Phys. Lett. 61:1 (1992)
6. S.M. Goodnick, J. Wu, C. Berven, M.N. Wybourne, and D.D. Smith, Hot-electron bistability in quantom dot structures, Phys. Rev. B, Rapid Communications 48:9150 (1993).
7. K.R. Farmer, C.T. Rogers, and R.A. Buhrman, Localized-State Interactions in Metal-Oxide-Semiconductor Tunnel Diodes, Phys. Rev. Lett. 58:2255 (1987).

# NONLINEAR COULOMB FRICTIONAL DRAG IN
# COUPLED QUANTUM WELLS AND WIRES

Ben Yu-Kuang Hu and Karsten Flensberg[*]

Mikroelektronik Centret
Danmarks Tekniske Universitet
DK-2800 Lyngby, Denmark

## INTRODUCTION

Coulomb interactions lead to many interesting phenomena in low-dimensional systems, one of which is the drag effect in coupled quantum wells.[1,2] These experiments involve placing two individually contacted quantum wells close together, driving a current through one layer. This drags the carriers in the second layer along, leading to measurable effects such as a voltage in the second layer. The magnitude of the response is directly related to the intralayer carrier-carrier interaction, and gives information among other things on the collective modes of the the coupled quantum well system.[2]

So far, studies have generally concentrated on the linear response regime. In this paper, we study the the effect of a large driving field on the system, which gives a nonlinear drag rate. We obtain analytic expressions for the leading order nonlinear terms for the drag rate, and discuss the viability of experimental observation of these effects. We show that drag in quantum wires is very sensitive to the relative densities of carriers in the wires.

## FORMALISM

The transresistivity is given by $\rho_{21} = J_1/E_2$ where $J_1$ is the current density in layer 1 (the driven layer) and $E_2$ is the electric field created in the second layer due to the drag effect when $J_2 = 0$. For isotropic parabolic bands, one can define a drag rate which is directly related to the momentum transfer per particle, $\tau_D^{-1} = n_1 e^2 \rho_{21}/m_1 = E_2/(E_1\tau_1)$, where $\tau_1$ is the intralayer momentum scattering.

Given that the distribution functions $f_i(\mathbf{k})$ in layers $i = 1, 2$ the momentum transfer rate from layer 1 to 2 is, to lowest nonvanishing order (i.e., second) in the screened interlayer interaction $W(q,\omega) = V_{12}(q)/\epsilon_{12}(q,\omega)$ (where $V_{12}$ and $\epsilon_{12}(q,\omega)$ are the bare interlayer Coulomb interaction and the dielectric function, respectively) is

$$n_2\left(\frac{\partial \mathbf{k}_2}{\partial t}\right)_{e-e} = 2\pi \int \frac{d\mathbf{q}}{(2\pi)^d} \mathbf{q} \int_{-\infty}^{\infty} d\omega |W(\mathbf{q},\omega)|^2 S_1(-\mathbf{q},-\omega)\, S_2(\mathbf{q},\omega) \qquad (1)$$

*Hot Carriers in Semiconductors*
Edited by K. Hess *et al.*, Plenum Press, New York, 1996

where

$$S_i(\mathbf{q}, \omega) = 2 \int \frac{d\mathbf{k}}{(2\pi)^d} \, f_i(\mathbf{k_i}) \left[1 - f_i(\mathbf{k_i})\right] \delta\left(\epsilon_{\mathbf{k_i+q}} - \epsilon_{\mathbf{k_i}} - \hbar\omega\right) \tag{2}$$

is the structure factor. This equation is general and can be used for arbitrary $f_i(\mathbf{k})$. In the case when the distribution functions are drifted Fermi Dirac functions, $f(\mathbf{k}) = \left[\exp(\hbar^2(\mathbf{k} - \mathbf{k_d})^2/2m) + 1\right]^{-1}$ one can use the fluctuation—dissipation theorem and properties of the Bose function $n_B(\hbar\omega) = \left[\exp(\hbar\omega/k_BT) - 1\right]^{-1}$ to write down the momentum transfer rate,[3] for $\mathbf{v_{d,1}} = \hbar\mathbf{k_{d,1}}/m_1 = \mathbf{v_d}$ and $\mathbf{v_{d,2}} = 0$,

$$n_2\left(\frac{\partial k_2}{\partial t}\right)_{e-e} = \frac{1}{2\pi} \int \frac{d\mathbf{q}}{(2\pi)^d} \, \mathbf{q} \int_{-\infty}^{\infty} d\omega |W(\mathbf{q},\omega)|^2 \chi_1''(q, -\omega + \mathbf{q}\cdot\mathbf{v_d})$$
$$\chi_2''(q,\omega) \, n_{B,1}\left(\hbar\omega - \mathbf{q}\cdot\mathbf{v_{d,1}}\right) n_{B,2}(-\hbar\omega) \tag{3}$$

where $\chi''$ is the imaginary part of the random phase approximation (RPA) susceptibility function. We use this form in the rest of this paper.

**Drag in quantum wires**

In quantum wires with only a single filled subband, the $\tau_D^{-1}$ to leading order in $T$ and $E$ is dominated by back-scattering, $\delta q \approx 2k_F$, because forward scattering, $\delta q \approx 0$, is not effective at removing momentum from the system. We find to leading order in $T$ and $E$, for identical wires with equal densities,

$$\tau_D^{-1} = \frac{|W(2k_F)|^2 m}{\hbar^4 k_F v_F} \left[\frac{m v_d v_F}{2\pi} + \frac{\pi k_B T}{36}\right]. \tag{4}$$

The $\delta q \approx 0$ scattering contributes only to higher order in $T$ and $E$. The region of validity of Eq. (4) depends on the size of $|W(2k_F)|^2$; this region is small for large wire separation. Within the RPA, the screened interaction vanishes logarithmically at $q = 2k_F$ as $W(2k_F + \delta q, \delta\omega) \sim \left[V(2k_F) |\log(\delta\omega + \delta q)|^2\right]^{-1}$. Thus, if one includes dynamical interlayer screening effects, the drag rate would have an additional $|\log(v_d)|^{-4}$ dependence.

When the densities of the wires are different there is a complete absence of any drag at $T = 0$ until $k_d = |k_{F,1} - k_{F,2}|$. This is because of the extremely restricted phase-space for scattering in one dimension. All the features described above can be seen in Fig. 1, which shows the $\tau_D^{-1}$ as functions of both $T$ and $v_d$ for $n_2/n_1 = 0.5$ and 1.

**Drag in quantum wells**

In the limit of not too large temperatures and electric fields, we can use the low $\omega$ and $q$ expansion $\chi'' = m^2\omega/2\pi\hbar^3 q k_F$. Furthermore, if we assume that $k_F d \ll 1$ the screened interaction can be approximated by $W(q) = \hbar^2\pi q/(2m q_{TF} \sinh qd)$. Substituting these into Eq. (3) gives

$$E_2 = \frac{8m^2}{4\pi e\hbar k_F^2 q_{TF}^2 n_2} \int \frac{d^2\mathbf{q}}{(2\pi)^2} \frac{q}{\sinh^2 qd} \int_{-\infty}^{\infty} d\omega \, \omega(-\omega + \mathbf{q}\cdot\mathbf{v_d}) n_B(-\hbar\omega) n_B(\hbar\omega - \hbar\mathbf{q}\cdot\mathbf{v_d})$$
$$\tag{5}$$

The integrals can be done analytically, yielding

$$\frac{1}{\tau_d} = \frac{\pi}{16}\zeta(3)\frac{1}{\hbar\epsilon_F(k_F d)^2(q_{TF}d)^2}\left[(k_B T)^2 + \alpha(\hbar v_d/2\pi d)^2\right] \tag{6}$$

where $\alpha = 15\zeta(5)/[4\zeta(3)] = 3.23485 \cdots$ and $\zeta$ is the Riemann zeta function.

Inserting numbers for the Gramila et al.'s experiment (Ref. 1), we get that at $T = 1$K the non-linear effect equals the linear effect for electric fields on the order of $E_1 \sim 45$V/m which should be achievable experimentally without causing significant heating of the electron gas.

Finally, we mention that it has been shown that collective modes can substantially affect the drag response.[2] An additional interesting feature of nonlinear drag is that at high drift velocities, the collective modes become unstable (the so-called two-stream instability).[4] At drift velocities just under the instability threshhold, the system becomes extremely polarizable, and the plasmon mediated drag noted in Ref. [2] should become very large, which should result in a significant rise in the drag rate.

To conclude, we have studied the nonlinear drag in quantum wires and quantum wells. For quantum wells and identical quantum wires, the drag rate goes as $aT^d + bE^d$ where $d$ is the dimensionality and $a$ and $b$ are system dependent quantities. The different power laws mirror the difference in density of states available for scattering in one and two dimensions. Finally, the drag in wires is extremely sensitive to the relative densities in the two wires.

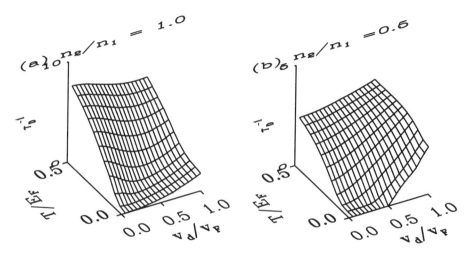

**Figure 1.** The drag rate as functions of relative drift velocity $v_d/v_F$ and temperature $T/E_F$, for quantum wires of widths and separation $k_F^{-1}$, $2e^2/\pi\epsilon_0 v_{F,1} = 1$, and $n_2/n_1 =$ (a) 1 and (b) 0.5. The unit used on the $z-$ axis is $10^{-3}me^4/(\pi\hbar^2\epsilon_0^2) \approx 5 \times 10^9$ s$^{-1}$ for GaAs. The static screening approximation is used. Note that in (b), when densities are mismatched, the drag rate suppressed for $v_d < 0.5v_F$.

## Acknowledgments

KF was supported by the Carlsberg Foundation.

## REFERENCES

* Present address: DFM, DTU, DK-2800 Lyngby, Denmark

[1] P. M. Solomon, et al., Phys. Rev. Lett. **63**, 2508 (1989); T. J. Gramila, et al., ibid. **66**, 1216 (1991); U. Sivan, et al., ibid. **68**, 1196 (1992).

[2] K. Flensberg and B. Y.-K. Hu, Phys. Rev. Lett. **73**, 3572 (1994), and references therein.

[3] H. L .Cui, et al., Superlatt. Microstruct. **13**, 221 (1993).

[4] B. Y.-K. Hu and J. W. Wilkins, Phys. Rev. B **43**, 14009 (1991), and references therein.

# HOT-ELECTRON SCATTERING FROM ELECTRON-HOLE
# PLASMA AND COUPLED PLASMON-LO-PHONON MODES
# IN GaAs QUANTUM WIRES

B. Tanatar[1], C. R. Bennett[2], N. C. Constantinou[2], and K. Güven[1]

[1]Department of Physics,
Bilkent University,
06533, Ankara, Turkey

[2]Department of Physics,
University of Essex,
Colchester, CO4 3SQ, England

ABSTRACT.- We present a fully dynamical and finite temperature study of the hot-electron momentum relaxation rate in a coupled system of electron-hole plasma and bulk LO-phonons in a quantum wire structure. Interactions of the scattered electron with neutral plasma components and phonons are treated on an equal footing within the random-phase approximation. Coupled mode effects substantially change the transport properties of the system at low temperatures.

## INTRODUCTION

Electronic devices making use of the hot-electron scattering phenomenon have been a subject of interest for some time.[1] In particular, hot-electron transistors with a base region made of high-mobility semiconducting material like GaAs offer a high-speed device. Theoretical work on the inelastic scattering rates of such systems for zero and finite temperature in various approximation schemes has been presented.[2,3] Relatively few studies[4] are devoted to the scattering and energy-loss rates of hot-electrons from neural plasmas as occur in photoexcited electron-hole systems. The purpose of this paper is to investigate the momentum relaxation rate of electrons in a quasi-one-dimensional (Q1D) structure. It has been noted and discussed in detail by Hu and Das Sarma[5] that the momentum relaxation rate is a more meaningful quantity to study the hot-electron scattering in quantum wires. We shall consider a quantum wire under intense photoexcitation, so that an equal number of electrons and holes are created, and scattering of electrons from the neutral plasma and from the LO-phonons will be taken into account. We have evaluated the momentum relaxation rate for the coupled plasmon-LO-phonon system; our conclusions are that coupled mode effects drastically change the transport properties of Q1D systems.[6]

The main assumptions that go into our calculations are as follows. (i) The electrons and holes are in their respective lowest subbands. We only consider the heavy holes, for simplicity. (ii) The electrons and holes are in a quasi-equilibrium state interacting with each other via the bare Coulomb interaction, $V_q$. Neutral plasma components interact with LO-phonons via the Fröhlich interaction. We neglect phonon confinement effects. (iii) Screening effects among the plasma components are treated within the random-phase approximation (RPA). We describe the hot-electron scattering within the Born approximation where the differential scattering rate is given by

$$P(q,\omega) = -2 V_q^\infty \operatorname{Im} \left[ \frac{1}{\varepsilon_T(q,\omega)} \right] \frac{1}{1 - e^{-\beta \omega}}, \tag{1}$$

in terms of the imaginary part of the total dielectric function of the neutral plasma-phonon system, $\varepsilon_T(q,\omega)$. Here $\beta = (k_B T)^{-1}$ is the inverse of the lattice temperature. The total scattering rate of a hot-electron with momentum $k$ is

$$\Gamma_k = \int \frac{dq}{2\pi} \frac{q}{k} \int d\omega \, P(q,\omega) \, \delta(\omega + kq/m - q^2/2m) \left[ 1 - f(k-q) \right], \tag{2}$$

where the $\delta$-function guarantees energy-momentum conservation in the scattering event, and the statistical factor describes the Pauli principle ($f(k)$ is the equilibrium Fermi distribution function). The total dielectric function for the coupled neutral plasma-phonon system is written as

$$\varepsilon_T(q,\omega) = 1 + \frac{\omega_{LO}^2 - \omega_{TO}^2}{\omega_{TO}^2 - \omega^2} - V_q^\infty \sum_{i=e,h} \Pi_i(q,\omega), \tag{3}$$

with $V_q^\infty = V_q/\epsilon_\infty$ is the bare Coulomb potential scaled by the high-frequency dielectric constant, and $\Pi_i(q,\omega)$ the Lindhard function for a Q1D system of electrons or holes.

## RESULTS AND DISCUSSION

The system we consider is a circular GaAs quantum wire of radius $R = 50\,\text{Å}$, and plasma density $n = 5 \times 10^5\,\text{cm}^{-1}$ such that only the lowest electron and hole subbands are occupied. The material parameters we use are tabulated by Adachi.[7] For the Q1D bare Coulomb interaction we employ the model developed by Gold and Ghazali.[8] The momentum relaxation rate $\Gamma_k$ for the electron-hole system (solid line) and for the coupled system of LO-phonons and neutral plasma (dotted line) at $T = 0$ is shown in Fig. 1a. We observe that in the absence of LO-phonons, scattering is predominantly from plasmons. As the dotted line indicates, the coupled mode system affects the relaxation rate significantly. There are two thresholds yielding a large $\Gamma_k$. The low energy threshold is due to the ordinary plasmons (collective excitations in the absence of holes and phonons), and the second one is due to the LO-phononlike mode (at a slightly higher value than $E_k = \omega_{LO}$ due to coupling). The acoustic plasmons as they exit the particle-hole band give rise to the low energy bump. In general, collective excitations of the coupled mode system give the dominant contribution to $\Gamma_k$ in Q1D structures.[6] The effect of finite temperature is to smooth the observed peaks in $\Gamma_k$ (Fig. 1b), and at room temperature results look similar to those obtained by assuming bare LO-phonons (not shown). The results obtained here indicate behavior different than the 2D and 3D structures.[4,9] As the growth technology and fabrication techniques for Q1D systems become more refined, our predictions could be experimentally tested.

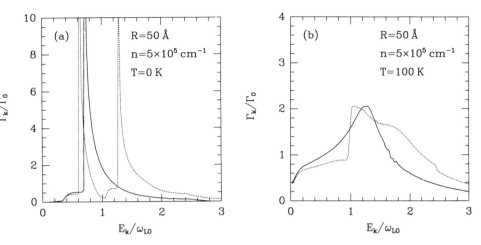

Fig. 1    (a) Momentum relaxation rate $\Gamma_k$ as a function of the hot-electron energy
at $T = 0$. (b) Same as (a) for $T = 100\,\mathrm{K}$. $\Gamma_0 = 8.7\,\mathrm{ps}^{-1}$.

This work is supported by the Academic Link Scheme of the British Council,
and TUBITAK under Grant No. TBAG-AY/77. Financial support from EPSRC is
gratefully acknowledged by C. R. B. and N. C. C.

## References

1.  A. F. J. Levi, J. R. Hayes, P. M. Platzman, and W. Weigmann, Phys. Rev. Lett.
    **55**, 2071 (1985); M. Heiblum, D. Galbi, and M. Weckwerth, *ibid.* **62**, 1057 (1989);
    for a review see also D. N. Mirlin and V. I. Perel, *Spectroscopy of Nonequilibrium
    Electrons and Phonons*, ed. C. V. Shank and B. P. Zakharchenya, (North-Holland,
    Amsterdam, 1992).

2.  J. M. Rorison and D. C. Herbert, J. Phys. C **19**, 6357 (1986); P. H. Beton, A. P.
    Long, and M. J. Kelly, Solid State Electron. **31**, 637 (1988); R. Jalabert and S.
    Das Sarma, Phys. Rev. B **41**, 3651 (1990).

3.  B. Y.-K. Hu and S. Das Sarma, Phys. Rev. B **44**, 8319 (1991); Semicond. Sci.
    Technol. **7**, B305 (1992).

4.  J. F. Young and P. J. Kelly, Phys. Rev. B **47**, 6316 (1993); J. F. Young, N. L.
    Henry, and P. J. Kelly, Solid State Electron. **32**, 1567 (1989); J. A. Kash, Phys.
    Rev. B **40**, 3455 (1989).

5.  B. Y.-K. Hu and S. Das Sarma, Appl. Phys. Lett. **61**, 1208 (1992); Phys. Rev. B
    **48**, 5469 (1993).

6.  C. R. Bennett, N. C. Constantinou, and B. Tanatar, submitted; C. R. Bennett,
    Ph. D. Thesis, Essex University, unpublished.

7.  S. Adachi, J. Appl. Phys. **58**, R1 (1985).

8.  A. Gold and A. Ghazali, Phys. Rev. B **41**, 7626 (1990).

9.  R. Jalabert and S. Das Sarma, Phys. Rev. B **40**, 9723 (1989).

# HELICAL QUANTUM WIRES

Alfred M. Kriman

Department of Electrical and Computer Engineering and
Center for Electrical and Electro-optic Materials
215 Bonner Hall
State University of New York at Buffalo
Buffalo, NY 14260

## INTRODUCTION

Bloch oscillations are associated with weakly scattered transport in a periodic band structure. Most periodic structures have band gaps at the reciprocal lattice boundaries. Mathematically, this is because the two states at the edge of the (1D) Brillouin zone can have equal energy only by means of an accidental degeneracy. Helical quantum-wire structures are an exception to the general rule: although they are periodic in one dimension, the additional glide symmetry makes the boundary states degenerate, so the band gap vanishes. Their periodic band structure resembles qualitatively that of a free electron represented in the repeated-zone scheme. The Hamiltonian of a helical quantum wire with parabolic potential cross section can be represented by means of creation and annihilation operators, and this formalism has been used in a numerical study of helical quantum wire band structure.

## MODEL

We consider a single-electron model with Hamiltonian

$$H = \frac{p^2}{2m} + \frac{1}{2}m\omega^2 \left[ (x-x_c(z))^2 + (y-y_c(z))^2 \right] . \tag{1}$$

Here $\mathbf{p}$ is the full three-dimensional momentum vector. The potential is parabolic in the $xy$ plane, with a center that depends on $z$. The center position, given by

$$(x_c,y_c) = R_c \left(\cos Qz , \sin Qz\right) , \tag{2}$$

twists at a fixed distance $R_c$ from the helix axis $z$, executing a $2\pi$ twist in a distance $\Lambda$ (reciprocal lattice vector $Q = 2\pi/\Lambda$). If $Q = 0$, this becomes an ordinary quantum wire parallel to the $z$ axis, with a lowest mode having spatial extent of order a length quantum

$$\ell \equiv (\hbar/m\omega)^{1/2}. \tag{3}$$

This Hamiltonian is invariant under neither infinitesimal translations not infinitesimal rotations about the $z$ axis. However, it is invariant under infinitesimal glide transformations, so that

$$k = p_z - Q\, L_z \tag{4}$$

is a good quantum number. Using a canonical transformation to separate out the glide motion allows the model to be solved by expressing $\Psi(x,y)$ as a sum of oscillator states. The Hamiltonian couples the amplitude in each state $(n,m)$ only to its "nearest neighbors" $(n\pm1,m)$ and $(n,m\pm1)$, so the eigenstates can then be found by ADI finite-difference methods.

## RESULTS

By appropriate scaling, it is possible to express results for this model in terms of universal or reduced plots. That is, although there are four model parameters — $m$, $\omega$, $R_c$, and $Q$ — we can allow two of these to set dimensional units and find a complete set of solutions by varying only two parameters independently. Specifically, if all lengths are scaled $\ell$, all actions and angular momenta by $\hbar$, all energies by $\hbar\omega$, and all other quantites by the derived unit of appropriate dimension, then the results depend only on the dimensionless model parameters $\ell Q$ and $R_c/\ell$. This has been done, and in the following all quantites are scaled as described.

In Figure 1, we display the ground miniband energy for $R_c/\ell = Q\ell = 1$. Three central Brillouin zones are shown. Qualitatively, the most prominent feature of this band structure is the absence of band gaps. This is not a case of small band gaps missed by the simulation or the plot, and the band gaps are missing in all simulations (*i.e.* all model parameters). The continuous glide symmetry here plays the same rôle as does pure translation symmetry in a constant potential: a constant potential is periodic with any period, but does not have band gaps because the finite translation symmetry does not exhaust the symmetry group of the Hamiltonian. Alternatively, a band gap may be viewed as an avoided level crossing. Level crossings are avoided only between levels that have identical symmetry quantum numbers—thus, an additional symmetry places Brillouin zone-edge states in different sectors, non-interacting by symmetry.

The energy of the ground state band shown in Figure 1 is approximately parabolic. Nonparabolicity is evident, however, from the fact that level crossings occur at energies that are in the ratio $1 : 3.69 :: 7.17 : 10.9$, rather than $1 : 4 : 9 : 16$.

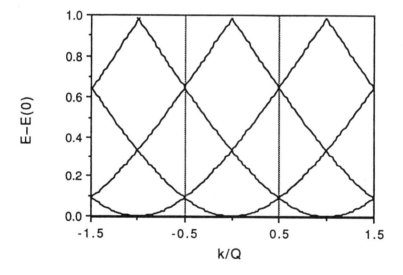

**Figure 1.** Ground minibands in the three central Brillouin zones. Helix radius is comparable to lowest miniband localization scale ($R_c = \ell$), and $Q\ell = 1$ (Helix period is $2\pi\ell$).

Nonzero $Q$ represents angular momentum that the electron must gain in order stay near the bottom of the potential while moving along the z direction. Figure 2 illustrates this effect. Orbital angular momentum is generally noninteger, confirming that it is not a good quantum number for this problem. If the electron were at a fixed distance $R_c$ from the axis, then $L_z$ would increase linearly with $k$. However, as $k$ increases the electron tends to move along the outside of the potential to provide necessary cetripetal force, so the effective radius increases, as does the slope of the curves in Figure 2.

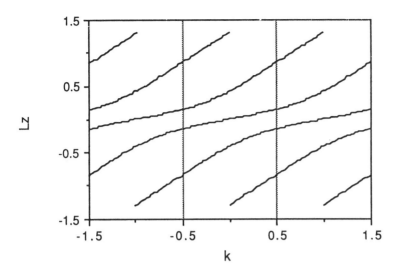

**Figure 2.** Average angular momentum of minibands shown in Figure 1.

In Figure 3, the average $L_z$ is shown along with its standard deviation, but for $R_c = 0.3162\ell$. Increasing standard deviation in $L_z$ also reflects radially broadened wavefunction. The increasing slope found in Figure 2 is again seen, but as $k$ approaches $2/\ell$, the slope decreases. This arises because tunneling between adjacent loops of the helix is beginning to compete with streaming motion along the helix.

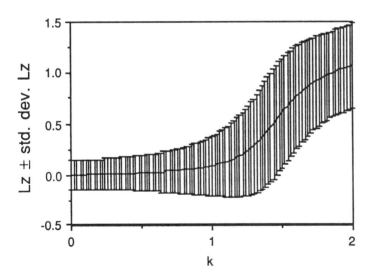

**Figure 3.** Angular momentum expectation value and standard deviation as functions of glide-translational momentum. Slope increases at intermediate $k$ as the electron rides up the outer side of the potential, and decreases at higher $k$ as tunneling begins to compete with gliding/orbital behavior within the wire.

# TRANSPORT SPECTROSCOPY
# OF SINGLE AND COUPLED QUANTUM DOTS
# IN SINGLE ELECTRON TUNNELING

Rolf J. Haug

Max-Planck-Institut für Festkörperforschung
Heisenbergstr. 1
D-70569 Stuttgart
Federal Republic of Germany

## INTRODUCTION

In recent years interest arose into the study of transport through quantum dots. The interest is driven by two reasons. First, small devices like quantum dots and charging effects [1] in such devices open the possibility for future applications in electronics as transistors, memory cells, sensors etc. Small devices are especially interesting in high frequency applications due to their small capacitances. Although different schemes for applications have been proposed, none of it is under development or even in production and it is not clear what will be of real use in the end. The second reason for the interest into these systems is a basic physics point-of-view. These systems are extremely interesting since single quantum dots can be seen as a sort of artificial atom. In this manner-of-speaking coupled quantum dots can be called artificial molecules. These quantum dots are similar to atoms or molecules since a certain number of electrons is confined in a potential. But, in contrast to real atoms, in these artificial systems the number of electrons, the size of the system and the strength of the confinement potential can easily be changed. Therefore, in these quantum dots parameters, such as the ratio between cyclotron frequency and confinement energy or coupling strength between two quantum dots etc, can be easily varied, which is not achievable in real atoms or molecules.

Zero-dimensional systems or quantum dots can be realized in semiconducting material. Although many different semiconductors have been used, one of the most widely used systems is the material system GaAs/AlGaAs grown by MBE. For the preparation of quantum dots one can distinguish two different basic structures: In so called vertical structures quantum wells are enclosed by barriers in the MBE growth and transport is studied in the growth direction, i.e. perpendicular to the surface of the wafer. The top and bottom of the wafer are doped and forming the emitter and collector for transport studies. Quantum dots can be generated here by etching fine pillars out of the material [2]. The second structure uses a two-dimensional electron gas in modulation doped AlGaAs/GaAs heterostructures. Quantum dots can be formed here by evaporating metallic gates on to the surface of the wafer creating a so-called split-gate device [3]. Applying negative gate voltages the two-dimensional electron gas underneath the gates is depleted, barriers can be formed and electrons are confined to regions in between different gates. By varying the gate voltages barrier heights and the number of confined electrons can be tuned. Emitter and collector are formed here by the two-dimensional electron gas and transport is studied in the two-dimensional electron gas parallel to the surface of the device.

## RESONANT TUNNELING
## THROUGH DOUBLE BARRIER STRUCTURES

To study transport through zero-dimensional systems enclosed by tunneling barriers, one of the simplest systems to consider is a double-barrier resonant tunneling structure.

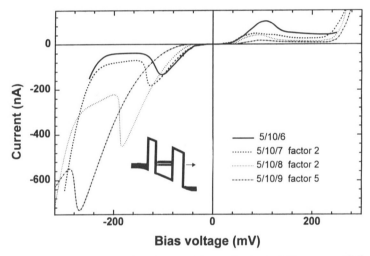

**Figure 1.** I/V characteristics of four pillars with typical diameters of about 200nm taken at 4.2K. The notation 5/10/6 characterizes a sample with an AlGaAs barrier of 5nm, a quantum well of 10nm and a second barrier of 6nm. For some of the devices the measured current was multiplied by a factor as noted in the figure.

Such a structure has the advantage, that the barriers are produced during the MBE growth process and that therefore their thickness is fixed. Here we have studied the dependence of transport properties on varying the ratio between the widths of the two barriers [4].

The double-barrier heterostructures used in this study were grown by molecular beam epitaxy on a $n^+$ GaAs substrate. The resonant tunneling structure is formed by a 10nm GaAs quantum well, a 5nm AlGaAs top barrier and a AlGaAs bottom barrier (with 30% Al) of varying thickness. Structures with bottom barriers of 6, 7, 8, 9nm have been grown. Undoped GaAs spacer layers with a thickness of 7nm separate the contact regions formed by 300nm thick GaAs layers doped with $4 \times 10^{23} m^{-3}$ Si.

Free standing pillars of varying diameter (varying from 200nm up to 10$\mu$m) have been produced from these four heterostructures. Typical I/V-characteristics of some of the smallest pillars (with typical diameters of 200nm) are shown in Fig. 1 for the four heterostructures used. Clear peaks and regions of negative differential conductance are seen for positive and for negative bias voltages similar to the behavior of large resonant-tunneling devices. For positive bias the peak position depends only slightly on the barrier asymmetry, since charge accumulation in the well is small and the electric field in the double-barrier region remains more or less constant. In contrast, for negative bias where the electrons tunnel first through the thinner barrier and leave the well through the thicker barrier (see schematical inset in Fig. 1) a strong shift of the peak to more negative bias voltages is observed for increasing barrier asymmetry due to charge accumulation in the quantum well. Calculations of the accumulated charge in the well show values up to $5.6 \times 10^{11} cm^{-2}$ (for the structure with 5 and 9nm thick barrriers [4]). The observed peak positions show no significant deviation from those of large area structures of both polarities. The peak-current density decreases exponentially with increasing barrier thickness, as expected. In comparing the peak curents obtained in the small pillars with the peak-current densities of the large devices from the same wafer one can obtain a measure for the conducting diameter of the pillars and the sidewall-depletion. Here, one assumes that the peak-current density is not changed in the small devices, which could be justified by the unchanged peak voltages. The sidewall-depletion values obtained for these structures were about 70 - 80nm. Interestingly, the conducting diameter appeared to be slightly larger for negative bias than for positive bias which could be explained by additional screening from the accumulated charge.

## SINGLE-ELECTRON TUNNELING IN RESONANT TUNNELING DEVICES

Whereas in the last section the peaks in the current expected also for large resonant-tunneling diodes were discussed, the onset of the current through the small pillars will be analyzed in this section. In general, the I/V characteristics at the onset of the current resembles a staircase. For negative bias (in charging direction) a clearer staircase is usually resolved for increasing asymmetry of barriers due to the increase of the accumulated charge in the well by single electrons. Because of the three-dimensional confinement of the quantum well in the small pillars one can speak in this current regime of tunneling through a quantum dot. The first onset of the current corresponds then to the tunneling through the ground state of the quantum dot for one electron. For the very asymmetric structures no additional structure is observed on the step [5, 6]. Here, one can conclude that the asymmetry of the two barriers is large enough to suppress tunneling via excited states of the quantum dot, i.e. the long escape time favors the occupation of the ground state. At the onset of the second step the quantum well is charged with two electrons. Whether the second electron is bound to the same quantum dot and tunneling occurs via the manybody ground state of the quantum dot for two electrons or whether the second electron is located in a second potential minimum can not really be extracted from the I/V characteristics taken at vanishing magnetic field. Different potential minima can originate from well-width fluctuations or from potential fluctuations due to residual dopings. To clarify the origin of the different steps and of the fluctuations occuring on top of the steps the study of the magnetic field dependence is quite essential. As an example, Fig. 2 shows I/V traces taken in magnetic fields applied parallel to the current. Data for the pillar with barrier thicknesses of 5 and 6nm are shown for positive bias, i.e. charging should be small. The first current step indicating the tunneling of the first electron shows an almost parabolic magnetic field dispersion as expected. Deviations from these behavior had been attributed to the jumps in the chemical potential in the emitter contact due to Landau quantization [5]. Whereas the steps observed in the I/V characteristics at zero magnetic field follow a magnetic-field dispersion which can be attributed to the tunneling via states in the same quantum dot, the oscillatory finestructure superimposed on the first (and to some extend also on the second and third step) shows a totally different magnetic field dispersion [7]. Similar to the influence of the chemical potential in the emitter on the position of the steps these fluctuations can be attributed to the local-density-of-states fluctuations in the emitter region due to the disorder caused by the doping (schematically depicted in the inset to Fig. 2) [4]. Therefore, in single-electron tunneling through these devices not only the magnetic field dispersion of ground and excited states of quantum dots can be studied, but also the emitter states can be investigated.

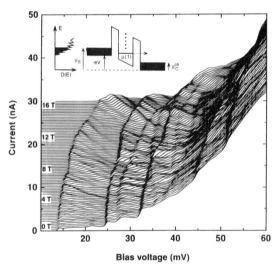

**Figure 2.** I/V characteristics of the pillar with barriers of 5nm and 6nm thickness for positive bias (non-charging direction) near the onset of the current. The magnetic fields varying from 0T up to 16T were applied in the direction of the current and the sample was immersed in the mixing chamber of a $^3$He/$^4$He-dilution refrigerator with a temperature of T=30mK.

## GATE CONTROLED SINGLE ELECTRON TRANSPORT

In contrast to the double-barrier structures discussed above, the use of the split-gate technology opens the possibility of varying the barriers in one structure. Due to the presence of the gates also the number of electrons accumulated in the quantum dot can be changed not only by increasing the bias voltage, but also by increasing the gate voltage. In vertical tunneling structures gates were only rarely used up to now [8]. In gate-controlled measurements excited many-body states of the quantum dots generated by an electron tunneling into the quantum dot can be distinguished from excited states generated by an electron leaving the quantum dot [9].

# EXCITATION SPECTROSCOPY OF COUPLED QUANTUM DOTS

Besides these investigations of single structures, double quantum dot systems are also studied now in detail. In transport experiments in the linear regime coupling effects between two dots have already been clearly identifed. Non only electrostatic coupling effects [10], but also quantum mechanical coupling of wavefunctions leading to the formation of molecular-like states were observed [6, 11]. Nonlinear transport experiments have not been performed in much detail. Figure 3 shows the differential conductance for a split gate geometry as presented on the left side of the figure. The two quantum dots formed underneath the surface of the device are different in size due to the slightly asymmetric gate structure. For a vanishing bias voltage $V_{DS}$ the conductance is almost periodic with large amplitude variations as expected for a double quantum dot system. Here the occurence of a conductance peak can be interpreted as the formation of a coherent state in the double dot. For a negative bias voltage the conductance peaks split similar to the observations in single dots [9, 12], but seem to sharpen up at the same time.

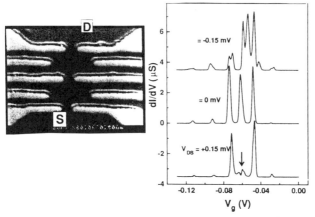

**Figure 3.** Differential conductance of a double-quantum dot system generated with the split-gate structure shown on the left side. Three traces for different bias voltage $V_{DS}$ are shown.

In a double quantum dot the state in one dot can be used as a spectrometer (without thermal smearing) for the second dot leading to this narrowing of the peaks [13]. For positive bias voltage no splitting of the conductance peaks is observed, but a drastic variation in the amplitude (shown by the arrow). The application of the bias voltage leads here to the destruction of some of the coherent states in the double dot. For this polarity a larger bias voltage is necessary to obtain a real splitting of the peaks. The strong difference on bias polarity is due to the asymmetric quantum dot structure. So, also for a double quantum dot system, nonlinear transport can be used in spectroscopy of the quantum dot states.

## ACKNOWLEDGEMENTS

Discussions with and contributions of T. Schmidt, J. Weis, R.H. Blick, M. Tewordt, D. Pfannkuche, V. Falko, K. v. Klitzing, K. Eberl, A. Förster, H. Lüth are gratefully acknowleged. Part of this work has been supported by the Bundesministerium für Bildung, Wissenschaft, Forschung und Technologie.

# References

[1] *Single Electron Tunneling and Mesoscopic Devices*, eds. H. Koch and H. Lübbig, Springer Series in Electronics and Photonics, Springer Berlin (1992); *Single Charge Tunneling*, eds. H. Grabert and M.H. Devoret, Plenum Press New York (1992).

[2] M. A. Reed, J.N. Randall, R.J. Aggarwal, R.J. Matyi, T.M. Moore and A.E. Wetsel, *Phys. Rev. Lett.* **60**, 535 (1988); M. Tewordt, V.J. Law, M.J. Kelly, R. Newbury, M. Pepper, D.C. Peacock, J.E.F. Frost, D.A. Ritchie and G.A.C. Jones, *J. Phys.: Condens. Matter* **2**, 8969 (1990); B. Su, V.J. Goldman, M. Santos, and M. Shayegan, *Appl. Phys. Lett.* **58**, 747 (1991); S. Tarucha, Y. Hirayama, T. Saku, and T. Kimura, *Phys. Rev. B.* **41**, 5459 (1990).

[3] U. Meirav, M.A. Kastner, and S.J. Wind, *Phys. Rev. Lett.* **63**, 1893 (1990); L.P. Kouwen-hoven, N.C. van der Vaart, A.T. Johnson, W. Kool, C.J.P.M. Harmans, J.G. Williamson, A.A.M. Staring, and C.T. Foxon, *Z. Phys. B* **85**, 367 (1991); R.J. Haug, K.Y. Lee, and J.M. Hong, in *Single-Electron Tunneling and Mesoscopic Devices*, Eds: H. Koch, H. Lübbig, Springer, Berlin (1992).

[4] T. Schmidt, M. Tewordt, R.J. Haug, K. v. Klitzing, A. Förster, and H. Lüth, *Solid State Electron.*; T. Schmidt, M. Tewordt, R.J. Haug, K. v. Klitzing, B. Schönherr, P. Grambow, A. Förster, and H. Lüth (to be published).

[5] T. Schmidt, M. Tewordt, R.H. Blick, R.J. Haug, D. Pfannkuche, K. v. Klitzing, A. Förster, and H. Lüth, *Phys. Rev. B* **51**, 5570 (1995).

[6] R.J. Haug, R.H. Blick, and T. Schmidt, *Physica B* **704** (1995).

[7] M.R. Deshpande, E.S. Hornbeck, P. Kozodoy, N.H. Dekker, J.W. Sleight, M.A. Reed, C.L. Fernando, and W.R. Frensley, *Semic. Sci. Techn.* **9**, 1919 (1994).

[8] P. Gueret, N. Blanc, R. Germann, and H. Rothuizen, *Phys. Rev. Lett.* **68**, 1896 (1992); M.W. Dellow, P.H. Beton, C.J.G.M. Langerak, T.J. Foster, P.C. Main, L. Eaves, M. Henini, S.P. Beaumont, and C.D.W. Wilkinson, *Phys. Rev. Lett.* **68**, 1754 (1992).

[9] J. Weis, R.J. Haug, K. v. Klitzing, and K. Ploog, *Phys. Rev. Lett.* **71**, 4019 (1993); R.J. Haug, J. Weis, K. v. Klitzing, and K. Ploog, *Physica B* **194-196**, 1253 (1994); R.J. Haug, *Advances in Solid State Physics* **34**, 219 (1994).

[10] F. Hofmann, T. Heinzel, D.A. Wharam, J.P. Kotthaus, G. Böhm, W. Klein, G. Tränkle, and G. Weimann, *Phys. Rev. B* **51**, 13872 (1995).

[11] R.H. Blick, R.J. Haug, J. Weis, D. Pfannkuche, K. v. Klitzing, and K. Eberl (to be published).

[12] A.T. Johnson, L.P. Kouwenhoven, W. de Jong, N.C. van der Vaart, C.J.P.M. Harmans, and C.T. Foxon, *Phys. Rev. Lett.* **69**, 1592 (1992); E.B. Foxman, P.L. McEuen, U. Meirav, N.S. Wingreen, Y. Meir, P.A. Belk, N.R. Belk, M.A. Kastner, and S.J. Wind, *Phys. Rev. B* **47**, 10020 (1993).

[13] N.C. van der Vaart, S.F. Godijn, Y.V. Nazarov, C.J.P.M. Harmans, J.E. Mooij, L.W. Molenkamp, and C.T. Foxon, *Phys. Rev. Lett.* **74**, 4702 (1995).

# BRAGG REFLECTION AND PHONON FOLDING EFFECTS ON HOPPING MAGNETOCONDUCTANCE IN SUPERLATTICES

Yuli Lyanda-Geller and Jean-Pierre Leburton

Beckman Institute for Advanced Science and Technology,
University of Illinois at Urbana-Champaign, Urbana, IL 61801

## I. INTRODUCTION

Transport in quantum microstructures (QMS) manifests such fundamental phenomena as quantum interference and electron localization. QMS can be designed to achieve arbitrary spectra of electronic states and are flexible to geometrical confinement providing new windows for technological innovation (see e.g.[1]). In the present paper we investigate the conductance mechanisms in Quantum Box Superlattices (QBSL), with discrete transverse spectrum, and demonstrate that a set of minima (antiresonances) occurs in the current-voltage characteristics. This effect may arise i) owing to the Bragg reflection (momentum selection rule for the interaction of acoustic phonons with the Wannier-Stark (WS) localized electrons) and ii) due to the gaps in folded phonon spectrum. We find that due to the electron localization in high electric field the longitudinal magnetoresistance of superlattices is negative over substantial range of magnetic field.

## ELECTRONIC MODEL

In 1D QBSL the electron motion is modulated by the periodic potential in the $z$-direction and is confined in the $xy$ plane. The Hamiltonian in the presence of the uniform electric field $\mathbf{F} \parallel z$ is $\mathcal{H}_0 = (E + \epsilon_n - eFd\nu)\,\delta_{\nu\nu'} + \Delta\,(\delta_{\nu,\nu'+1} + \delta_{\nu,\nu'-1})$, $\Delta$ is the tunneling matrix element (we consider only the tunneling between neighboring QB), $d$ is the period of QBSL , $\epsilon_n$ is the transverse energy and $n$ is the integer number. This Hamiltonian results in a bandspectrum with a width $4\Delta$ in zero electric field (for simplicity we consider the lowest londitudinal subband only). If the potential drop over the period of a structure, eFd, exceeds the collisional broadening of the electron levels $\hbar/\tau$, this subband splits into WS ladder of localized states with energies $E_{n\alpha} = \epsilon_n - eFd\alpha$, (see Fig.1a) and the wavefunctions $\Psi_{n\alpha} = \sum_\nu J_{\nu-\alpha}\psi_{n\nu}$, where $\psi_{n\nu}$ is the Wannier function describing transverse mode n in the separate quantum well $\nu$, $J_k(2\Delta/eFd)$ are the $k$-order Bessel functions, $\alpha$ is the quantum well index in the diagonal representation (see e.g.[2]). Therefore the band conduction breaks down and electrons move in the $z$-direction only by hopping between wells due to scattering; direct tunneling between the WS states manifests in the electron spectrum and in the wavefunctions.

There is an essential difference between conventional SL and QBSL. In SL the transverse spectrum is continuous and the energy conservation law can always be satisfied for any scattering process. In elastic scattering by impurities the energy conservation is due to an interchange betweeen longitudinal and transverse energies, while phonon assisted hopping is accompanied by a partial energy transfer to the lattice. In QBSL the transverse degree of freedom is characterized by a discrete spectrum $\epsilon_n$ and electron hopping exists only when the transverse energy spacing is equal to the separation between the WS levels or differs from the latter by the phonon energy. It follows then that elastic and optical phonon scattering should manifest in resonant peaks in the current. The only background mechanism for conduction in SL is scattering by acoustic phonons. However, we demonstrate (see also [2]) that acoustic-phonon-assisted hopping in QBSL is characterized by a set of antiresonant minima. These minima are related to the quasi 1D character of the phonon propagation in QBSL, which dramatically affects both the phonon spectrum and the electron-phonon interaction.

## ANTIRESONANCES AND RESONANCES IN HOPPING TRANSPORT

We describe the hopping conductance in QBSL by the following transparent formula

$$j_z = e \sum_{\alpha,\alpha',N,N'} (z_{\alpha,N} - z_{\alpha',N'})W^{\alpha,\alpha'}_{N,N'}, \tag{1}$$

where $z_{\alpha,N} - z_{\alpha',N'}$ is the electron displacement (the hopping length) upon the scattering $(\alpha, N \to \alpha', N')$, N stands for the set of indices describing the transverse states, $W^{\alpha,\alpha'}_{N,N'}$ is the scattering probability. For phonon-assisted hopping

$$W^{\alpha\alpha'}_{NN'} = \sum_{\mathbf{q}_\perp q_z} |C^{\alpha,\alpha'}_{N,N'}|^2 \delta(E_{\alpha N} - E_{\alpha'N'} + \hbar\omega_q)[(f_{\alpha N} - f_{\alpha'N'})N_q - f_{\alpha'N'}(1 - f_{\alpha N})], \tag{2}$$

$f_{\alpha N}$ is the electron distribution function, $C^{\alpha,\alpha'}_{N,N'}$ is the electron-phonon scattering amplitude, $\omega_q$ and $N_q$ are the phonon frequency and the occupation number. Critical features regarding this formula and hopping regime in microdevices are emphasized in [2].

It can be shown that $| C^{\alpha,\alpha'}_{N,N'} |^2$ is given by

$$| C^{\alpha,\alpha'}_{N,N'} |^2 = | V_{\mathbf{q}} |^2 | \langle N|e^{iq_\perp r_\perp}|N' \rangle |^2 J^2_{\alpha-\alpha'}\left(\frac{4\Delta}{eFd}\sin\frac{q_z d}{2}\right), \tag{3}$$

where $\mathbf{q}$ is the transferred momentum. We see that phonons with $q_z = \frac{2\pi n}{d}$ result in vanishing electron transition between different wells $\alpha$ ($J_{\alpha-\alpha'}(0) = 0$ at $\alpha \neq \alpha'$). Assume first that the dispersion of acoustical phonons is described by a constant speed of sound $s$ and that phonons propagate in the $z$−direction. Then the energy conservation law given by $\delta$-function in Eq.(2) determines $q_z$. At electric fields $F = nF_0$, n is the integer number, $F_0 = 2\pi\hbar s/ed^2$, phonons are ineffective. The scattering probability (3) and

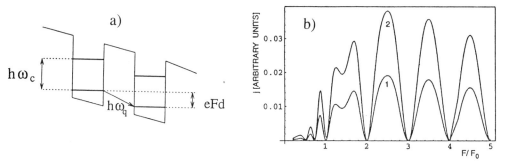

Figure 1: Hopping conductance in magnetically confined QBSL: a) Schematic representation of phonon-assisted transition in a QBSL with longitudinal WS localization and the first two Landau levels; b) Current in the conditions of partial occupation of the lowest Landau level. $F_0 = 2.17$ kV/cm. Curve 1: B=6 T; Curve 2: B=12 T.

the hopping current (1) vanish. This effect takes its origin in Bragg reflection: the transfer of phonon momentum equal to the momentum of reciprocal lattice does not change the longitudinal electron state. We have a specific momentum selection rule for the scattering of the WS localized electrons. The difference between the continuous transverse SL spectrum and QBSL with the discrete one is very important, and only the discrete spectrum results in a single value of $q_z$. If there are no other acoustic phonons in the structure this means the absence of the background current and the appearance of zero minima in the conductance, i.e. antiresonant effect. Let us consider the possibility of the observation of the momentum selection rule which is the result of the 1D propagation of phonons. It will be observable experimentally if either the density of states of 1D phonons is essential or the transverse scattering formfactor given by $F(\mathbf{q}_\perp) =| \langle N|e^{iq_\perp r_\perp}|N'\rangle |^2$ has a sharp maximum at $\mathbf{q}_\perp = 0$. For instance, such a maximum is realized in SL in magnetic field, when the transverse modes are Landau levels (LL). At the moment this configuration is the best way to simulate QBSL, although magnetic-field-confinement is specific. The 1D array of boxes with magnetic- and electric-field-controlled discrete spectrum is realized in strong electric field, due to WS quantization. If the energy difference between the initial and final WS levels satisfies the antiresonant condition, electron hopping occurs between partially filled LL with the same quantum number and the formfactor is $F(\mathbf{q}_\perp) \propto e^{-l_B^2 q_\perp^2}$, where $l_B$ is the magnetic length. Consequently, the energy conservation law takes the form $eFd \simeq \hbar s q_z$, if the inequality $eFd/\hbar s \gg q_\perp \sim l_B^{-1}$ or $(eFd)^2/ms^2 \gg \hbar \omega_c$, where $\omega_c$ is the cyclotron frequency, is satisfied. One sees that in high electric fields only the $q_z$ component of the phonon momentum is relevant and therefore hopping current reaches a minimum. We note that transitions to the next nearest QB states with the same number of LL are also in antiresonance, but transitions between different Landau levels result in finite current. Their contribution is much weaker than that of the transitions between nearest boxes, because phonon emission at low temperatures and $eFd < \hbar \omega_c$ requires a large hopping distance. Let us notice that one can also eliminate the transverse component of the phonon momentum by using an external source of phonons with a given direction of propagation. If the signal related to external phonons is extracted from the total current, the zero minima at the certain voltages can be found. On Fig.1b we present the current calculated taking into account the transitions between nearest QB for the case of magnetically confined QBSL.

The range of electric fields $eFd \gg \hbar/\tau$ turns out to be very interesting because of the occurrence of negative magnetoresistance. The physical origin of this effect in our case is that in the electric-field-induced localization regime the current is proportional to the scattering probability, in contrast to Drude current at small electric fields, which is proportional to the scattering time and inverse proportional to scattering probability. The latter in the case of transition between lowest LL is proportional to the "density of states": $\int d\mathbf{q}_\perp e^{-l_B^2 q_\perp^2} = 2\pi/l_B^2$ and increases with magnetic field. Correspondingly, the current in the WS localization regime also increases and we see (Fig.1b) that the bigger magnetic field the bigger the hopping current. We predict that this negative magnetoresistance will be observed as long as the probability $1/\tau$ is smaller than $eFd/\hbar$ with increasing magnetic field. In higher magnetic fields the WS localization is destroyed , and the magnetoresistance becomes positive.

Consider now the effect of folded acoustical phonons [3] on the QBSL conductance. The dispersion relation for folded phonons in periodic potential is of the same form as Kronig-Penney dispersion relation for electrons and their spectrum can be considered as linear only at small $q$, while at the Bragg plane the wave velocity is zero and a gap arises. Obviously, if the energy conservation law Eq.(2) requires that $\hbar\omega_q$ falls into the gap, the current vanishes, and an antiresonant plateau is to be observed. Note that if the growth direction of a structure is (001) the gaps in the phonon spectrum are only at the Brillouin zone edge $q_z = \pm\pi/d$ and its centre $q_z = 0$. Eq.(3) determines the conditions for current vanishing in the vicinity of the gaps. The current breakdown is not abrupt at $q_z = 0$ but at $q_z = \pm\pi$. This is another manifestation of the selection rule. Note that current peaks may arise due to the phonon "defects", which levels fall inside the folded phonon gap. These phonon modes come from the potential fluctuations [4].

## CONCLUSION

We have shown that the influence of quantum confinement on the electron and phonon states results in antiresonances in the hopping conductance of quantum box superlattices. The electron-phonon interaction momentum selection rule as well as the folded phonon gaps are to be observed in the magnetoconductance in the Wannier-Stark localization regime. We also predict negative magnetoresistance in high electric fields.

This work was supported by NSF under Grant No. ECS 91-08300 and by the US Joint Service Electronics Programm under contract No.N00014-90-J-1270.

# References

[1] F.Capasso et al. In: Resonant Tunneling in Semiconductors, ed by L.L.Chang et al. Plenum Press, NY, 1991.

[2] Yu.Lyanda-Geller and J.P.Leburton. Phys.Rev.B **52**, 2779 (1995).

[3] C.Colvard, R.Merlin, M.V.Klein and A.C.Gossard. Phys.Rev.Lett. **45**, 298 (1980).

[4] J.Hori in Spectral Properties of Disordered Chains and Lattices, Int.Ser. of Monographs on Natural Phylosophy, gen.ed. T der Haar, Pergamon, Oxford (1968).

# CHARGE GATING OF NANOMETER SCALE PILLAR ARRAYS

Z.A.K. Durrani[1], B.W. Alphenaar[2], H. Ahmed[1], and K. Köhler[3]

[1]University of Cambridge Microelectronics Research Centre, Cavendish Laboratory, Madingley Road, Cambridge CB3 0HE, UK

[2]Hitachi Cambridge Laboratory, Cavendish Laboratory, Madingley Road, Cambridge CB3 0HE, UK

[3]Fraunhofer-Institut für Angewandte Festkörperphysik, Tullastraße 72, D-79108 Freiburg, Germany

Electron transport through laterally confined resonant tunnelling diodes [1] (RTDs) can be influenced by both Coulomb charging [2,3] and low-dimensional quantum size effects [4]. In a laterally confined RTD, a quantum dot is formed by etching a double barrier heterostructure into a sub-micron diameter pillar. In principle, charging and quantum confinement energies can be increased by simply reducing the pillar dimensions. In practice, there has been a limit to the minimum device diameter due to the presence of surface states which deplete the pillar and block current flow [5].

We have characterised arrays of nanometer scale RTD pillars and observed behaviour which suggests that surface charge trapping can be used as a gate to control current flow. Measurements of more than $10^5$ RTD pillars in parallel reveal that the array resistance switches between two stable states with a current peak to valley ratio of 500:1. Scanning tunnelling microscope characterisation shows that the individual pillars traps charge which raises the conduction band and produces a potential barrier in the conduction path.

The nanoscale pillar arrays are fabricated using a novel `natural' lithographic technique [6]. First, we evaporate a thin gold film onto the semiconductor surface. It is well documented that for film thicknesses around 10 nm and less, gold forms a granular layer composed of disconnected islands [7]. The gold islands partially mask the underlying material so that plasma etching in a SiCl$_4$/Ar environment selectively removes the material between the islands and produces an array of semiconductor pillars. Fig.1 shows a scanning electron micrograph of a typical sample after etching. The pillars are densely packed and the diameters range from 20-50 nm.

The pillars are measured as follows [8]. An RTD pillar array is formed by etching a double barrier AlAs/GaAs heterostructure in the manner described above. In our structure, a 5.6 nm thick GaAs well is isolated between 2 nm thick AlAs barriers. The GaAs leads are Si doped n-type, with a concentration graded from $5 \times 10^{18}$ cm$^{-3}$ at the contacts to $2 \times 10^{17}$ cm$^{-3}$ near the barriers. Contact is made to the top of the pillars by evaporating a $30 \times 30$ micron gold pad. The gold pad does not fill the gaps between the pillars because the gold grain size is approximately the same as the pillar spacing. A Si$_3$N$_4$ insulating layer is used to support the wire bond. The back contact is AuGeNi annealed at 400° C for 10 s.

The result of a typical hysteresis measurement is shown in Fig.2. Starting at zero bias, the device is initially in a high resistance state. Moving to negative bias, the current jumps sharply to a low resistance state at –2.4 V. The device stays in the low resistance state until the bias crosses zero and increases to +1.4 V. Here, the device switches back to the high resistance state with a current peak to valley ratio of 500:1. Further measurements show that the high resistance state is completely stable at zero bias while the low resistance state is stable for at least one hour before eventually switching to the high resistance state. In comparison, measurements of a standard 10 micron diameter RTD made from the same material showed a room temperature peak to valley ratio of only 3:1 and no hysteretic switching behaviour [9]. Switching was observed in pillar arrays fabricated from two double barrier wafers which were obtained from separate growth facilities. However, no switching was observed in a similarly grown single barrier sample. This demonstrates that quantum well confinement is necessary for the switching effect.

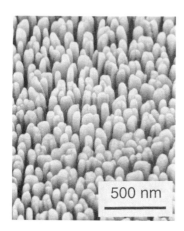

Figure 1. (a) Scanning electron micrograph of a typical pillar array sample. The Au mask is 5nm thick and the sample was etched for 3min. in 10 mTorr $SiCl_4$ :Ar at a ratio of 1:2 with a power density of 4 mWcm$^{-2}$

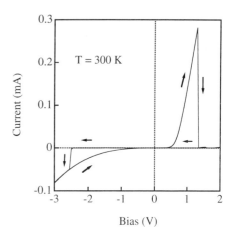

Figure 2. Current versus source-drain bias for a typical resonant tunnelling diode pillar array device. The peak to valley ratio is 500:1. The high and low resistance states are stable

Because the pillars are small and closely packed, it is impossible to contact a single device by conventional means. However, using a scanning tunnelling microscope (STM), we are able to locate and contact isolated pillars. A tungsten tip WA Technology STM operating at room temperature in air was used. We first locate an individual pillar using constant current scans. Once a pillar is located, the feedback is disabled and the gap is reduced bringing the tip in contact with the sample. A number of pillars were measured, with similar results.

Fig.3a shows the current voltage characteristics of a 50 nm diameter RTD pillar measured with an STM tip contact. The characteristics are strongly dependent on the bias history; stressing the device at –10 V for 60 s switches it into a low resistance 'on' state for the opposite bias. The 'on' state measurement shows a strong current peak and a negative differential resistance region (NDR). As the device is swept to a positive bias in the 'on' state, it turns 'off' and a second sweep without stressing results in a low current. Before a sweep, the device is held at zero bias to avoid transients. The measurements for 60 s and 600 s hold times demonstrate that the 'on' state is stable at zero bias. The device gradually turns 'off' at a high positive bias. Similar results can be obtained for either bias direction.

Both the on/off states and the sharp current peak and NDR of Fig.3a have been observed in the single barrier heterostructure sample as well as the double barrier heterostructure sample. This implies that these effects are not caused by the quantum well in the double barrier heterostructure.

The device is driven more strongly 'on' for larger stress bias. In Fig.3b, the device is 'off' if it is not stressed and little current flows. As a negative stress is applied, higher current is observed in the positive sweep and a clear current peak and negative differential resistance region (NDR) appears by the time the stress reaches –6V. This becomes higher and sharper with increasing stress. A shoulder also appears before this peak which changes into a second sharp peak by the time the stress increases to -11 V. This peak does not appear in the single barrier sample. It can be attributed to resonant tunnelling through the main quantum well level in the double barrier material.

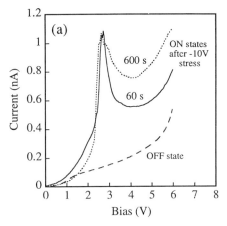 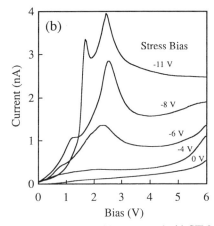

**Figure 3.** (a) Current versus voltage characteristics for a 50nm diameter RTD pillar measured with STM tip contact. The device was stressed at -10V tip bias for 60s before the solid and doted curves (ON state). Hold time at zero bias: 600s (doted) and 60s (solid). The dashed curve (OFF state) was measured with no pre-stress directly after the solid curve. (b) Dependence of the current versus voltage characteristics on the stress bias. The stress bias produces a gating effect.

To explain these results, we propose that trapped charge modifies the position of the conduction band in the pillars [5]. One possible mechanism is seen in Fig.4. On the surface of the pillar, a large density of localised states exist due to dangling bonds and strain. In equilibrium (Fig.4a), the Fermi energy $E_F$ is pinned within these surface states. This depletes the electron density in the centre of the pillar and forms a large barrier in the central lightly doped region. When a large negative stress bias is placed across the device (Fig.4b), electrons are forced into the surface states near the negative bias contact. The trapped charge acts as a gate and raises the conduction band edge, creating a potential barrier. Similarly, electrons are forced from the states into the positive bias contact and the resulting positively charged states lower the conduction band. Fig.4c shows the band diagram for the device at zero bias after charging. Because of the asymmetry of the charge gated barrier, the device is in the 'on' state for positive bias (as the bias increases, electrons pass over the barrier) and in the 'off' state for negative bias. In an 'on' sweep, NDR occurs as the device charges up and gradually turns 'off'. The resonant tunnelling peak is superimposed on the charge gating behaviour.

We now consider the multiple pillar device. This also shows 'on' and 'off' states similar to those in individual pillars. However, the characteristic is asymmetric and sharp transitions are observed between the states (Fig.2). Multiple pillar switching occurs only in double barrier material while the individual pillars in both single and double barrier material show charge gating. It is unclear why the positively charged state remains stable for so long. A more detailed analysis is needed which perhaps takes into account interactions among the pillars.

In conclusion, we have observed dramatic hysteretic switching behaviour in nanometer scale pillar arrays. We use a novel 'natural' lithography technique to fabricate arrays of closely packed pillars 20-50 nm in diameter from a GaAs/AlAs resonant tunnelling diode heterostructure. The array resistance switches between two stable states with a current peak to valley ratio of 500:1. STM measurements of individual pillars suggest that this effect is caused by the charging and de-charging of trap states which influence the energy of the confined levels. It is possible to use the charge trapping as a gate to control the device characteristics. We hope that further experiments will lead to a more complete theory to describe both single pillar and multiple pillar device behaviour.

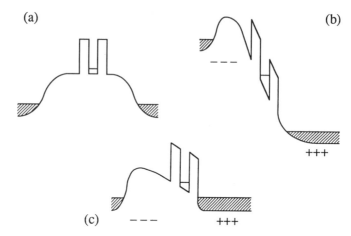

**Figure 4.** Schematic drawing of the charge gating mechanism. (a) At equilibrium, the surface states are pinned at the Fermi energy creating a depletion barrier in the lightly doped region. (b) Charging at a negative stress: The large bias forces electrons into states near the negative bias contact and forces electrons out of states near the positive bias contact. The negatively charged states raise the conduction band and produce a potential barrier. The positively charged states pull down the conduction band. (c) Device is returned to zero bias: It is ON for positive bias and OFF for negative bias.

## REFERENCES

1. For a recent review of resonant tunnelling diodes and their applications see F. Capasso, S. Sen and F. Beltram in "High-Speed Semiconductor Devices", edited by S.M. Sze, John Wiley & Sons, New York, (1990).
2. Atanas Groshev, Phys. Rev. B 42, 5895 (1990).
3. B. Su, V.J. Goldman, and J.E. Cunningham, Science 255, 313 (1992); Phys. Rev. B 46, 7644 (1992); M. Tewordt, L. Martín-Moreno, J.T. Nicholls, M. Pepper, M.J. Kelley, V.J. Law, D.A. Ritchie, J.E.F. Frost, and G.A.C. Jones, Phys. Rev . B 45 14407 (1992).
4. M.A. Reed, J.N. Randall, R.J. Aggarwal, R.J. Matyi, T.M. Moore, and A.E. Wetsel, Phys. Rev. Lett. 60, 535 (1988).
5. M.A. Reed, J.N. Randall, J.H. Luscombe, W.R. Frensley, R.J. Aggarwal, R.J. Matyi, T.M. Moore, and A.E. Wetsel, Festkörperprobleme 29, 267 (1989).
6. There have been many recent publications demonstrating similar natural lithographic techniques. See for example: K. Hiruma, T. Katsuyama, K. Ogawa, M. Koguchi, H. Kakaibayashi, and G.P. Morgan, Appl. Phys. Lett. 59, 431 (1991); C. Schönenberger, H. van Houten, and H.C. Doonkersloot, Europhys. Lett. 20, 249 (1992); M. Green, M. Garcia-Parajo, F. Khaleque and R. Murray, Appl. Phys. Lett. 66, 1234 (1995).
7. For a review describing granular metal films see B. Abeles, Appl. Sold. Sci. 6, 1 (1976).
8. B.W. Alphenaar, Z.A.K. Durrani, A.P. Heberle, and M. Wagner, Appl. Phys. Lett. 66, 1234 (1995).
9. C. J. Goodings, Ph.D. Thesis, University of Cambridge (1993).

# ENERGY RELAXATION IN QUANTUM DOTS:
## RECENT DEVELOPMENTS ON THE PHONON BOTTLENECK

Clivia M. Sotomayor Torres

Nanoelectronics Research Centre
Department of Electronics &Electrical Engineering
University of Glasgow
Glasgow G12 8QQ, Great Britain

## INTRODUCTION

The "phonon bottleneck" is a model proposed to explain the drop in the integrated emission intensity in semiconductor quantum dots and wires with decreasing lateral size L[1]. It is based on considerations of the energy spacing in these nanostructures which is typically less than the energy of an optical phonon thus leaving carrier relaxation to take place via electron-acoustic phonon scattering (see Fig.1). In turn the electron-acoustic phonon interaction had been calculated to be reduced by orders of magnitude as the dimensionality of the system decreases from 2 to 0 [2]. Thus, electrons in upper energy levels spend times longer than the nonradiative scattering times and they recombine nonradiatevely leading to a decrease of the luminescence from quantum dots. Very few electrons relax to the bottom of the conduction band with the right set of quantum numbers to recombine with thermalised holes.

Several theoretical and experimental approaches have been put forward addressing the phonon bottleneck. The main emphasis has been the search for mechanisms to bypass it, thereby enhancing interband transitions, in order to make use of novel optical properties associated with 0-dimensional semiconductors.

Some of these approaches include Auger processes,[3] electron-electron interaction[4], multiphonon processes[5] and the role of excited energy levels of the exciton[6] among others. These models have sought to include processes others than or in addition to the electron-acoustic phonon interactions. The role of statistical fluctuations of the dot size upon the phonon bottleneck and a more detailed theory of it has recently been published[7].

From the experimental point of view, suggestions involving highly demanding engineering specifications have been proposed, such as the design and realisation of quantum dot structures with energy level separation matching an integer number of longitudinal optical (LO) phonons[8]. Recently different semiconductor nanostructures have been investigated including self-organised quantum dots (to be discussed below), formed by island growth in the InAs-GaAs system, and deep etched Si-SiGe quantum dots [9].

## DEEP ETCHED QUANTUM DOTS

The bottleneck model[1] considers mainly nanostructure in the size regime where exciton-centre-of-mass quantisation is expected: lateral size L = 100 -200 nm in GaAs-based quantum dots and wires. Regular and symmetric nanostructures are realised using nanolithography with a typical size distribution of <5%. Several teams have shown

*Hot Carriers in Semiconductors*
Edited by K. Hess *et al.*, Plenum Press, New York, 1996

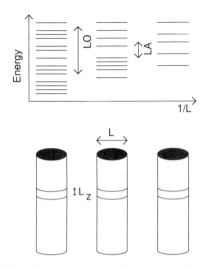

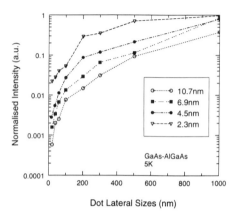

**Figure 1.** (top] Schematic of energy levels in quantum dots changing with lateral size L and compared to LO and LA phonon energies. [bottom] schematic of deep-etched dots showing L and $L_Z$.

**Figure 2.** Integrated emission intensity from quantum dots of different diameter $L^{13}$. The emission is strongest (weakest) for the 2.3 nm (10.7nm) quantum well agreeing with ref. 12.

indirectly that in the large size regime, surface effects dominated the emission intensity of quantum dots, which decreased dramatically with decreasing $L^{10}$. A good fit to the curve of the normalised emission intensity from dots plotted against dot diameters has been achieved by a combination of surface-dominated mechanisms, compounded in the concept of the surface recombination velocity for $L \geq 200$ nm, and the phonon bottleneck approach for $L < 200$ nm[11]. Moreover, based on the calculations of Bockelmann and Bastard[2] and Bockelmann[12] the effect of the electron-acoustic phonon scattering in dots of different diameters fabricated in wafers containing quantum wells of different width $L_Z$ was tested. It was found that although surface effects determine the emission strength to some degree, it is the electron-acoustic phonon scattering which has the dominant effect[13] (see Fig.2).

Most optical studies of dot emission have been carried out exciting hundreds and thousands of dots simultaneously. It is known from cathodoluminescence studies that not all dots emit light as a result of defects in the starting material and those introduced during fabrication[14]. In this respect the typical intensity from dots against L curves represents only the lower limit. Recent emission studies from a single quantum dot fabricated by focused laser interdiffusion in a single quantum well sample have revealed strong emission not only from the exciton ground state but also from its excited states suggestive of the phonon bottleneck being extended to the exciton-phonon interaction[15]. Calculations of scattering rates between exciton states and LA phonons have shown that when the phonon wavelength is smaller than the smallest dimension of the dot (usually $L_Z$), the exciton-phonon coupling becomes weak similar to the case of the electron-phonon interaction[6].

Emission studies of arrays of deep dry etched and of overgrown GaAs-GaAlAs quantum wires and dots with an intermediate step of wet etching prior to overgrowth showed that the emission strength is maintained for structures with L down to 6 nm[16]. In other words, no evidence of the phonon bottleneck was observed in these overgrown structures. We have shown previously that after overgrowth the emission strength from wires and dots is significantly enhanced in GaAs-AlAsAs[17]. Preliminary overgrowth studies showed that an initial polycrystalline growth around the dots may occur prior to planarisation[18]. Furthermore, even under improved overgrowth conditions, cylindrical pillars containing etched quantum dots experience a non-uniform strain field which shows up as a different splitting between the heavy- and the light-hole exciton from quantum wells with different $L_Z$ positioned along the cylinder[19]. The overgrowth process may result in strain fields and the presence of impurities that enhance carrier capture and therefore quench the phonon bottleneck, as will be discussed below.

# QUANTUM DOTS AND WIRES FORMED BY QUANTUM WELL INTERMIXING

Time-resolved studies have been carried out in intermixed quantum well forming dots and wires. The main attraction of this method is the absence of surfaces effects such as the optical "dead layer" which appears in the data analysis of many deep etched nanostructures. Measurements of decay times at various energies in the emission band from wires and dots revealed longer decay times in both wires and dots with decreasing L and higher electron temperatures for the smallest L, with the carrier temperature being significantly higher for dots than for wires, consistent with the phonon bottleneck model[20].

# STRAIN-INDUCED QUANTUM DOTS AND WIRES

This method proposed by Kash et al[21] also has the advantage of avoiding the "surface problem". In these wires and dots the conduction band is modulated in regions directly under a stressor (carbon, tungsten) resulting in a local lower bandgap region where electrons are confined. The emission spectrum is a combination of transitions from the strained (quantum wire) and unstrained (quantum well) areas with a typical double peak structure. From the temperature dependence of the ratio of these peaks and their behaviour under increasing laser power excitation, Zhang et al have recently concluded that there is no reduction in the emission efficiency at low temperatures[22]. Moreover, they showed that at any temperature up to about 70K the sum of the integrated intensities of the 2D and wire emissions amounts to the integrated intensity of the unpatterned quantum well. These author invoke the possibility of carrier tunnelling through the confining barrier, thermalization of 2D excitons into wire regions and thermal activation of 2D localised excitons, with the latter mechanism being least dominant.

A variation of this approach has been developed by patterning a pseudomorphic upper layer in a quantum well structure, resulting in a semiconductor stressor[23]. Decay time measurements under quasi-resonant conditions with the ground state showed that the decay times of the strain-defined wires are longer than that of the reference sample, being longer for decreasing L. Rather than postulating the phonon bottleneck these authors suggest the origin of the increased decay times to be the asymmetry of the conduction and valence bands in strain-induced wires and dots. New results on InGaAs strained dots have shown that the excited state luminescence from these dots is associated with the slower thermalization of both electrons and holes and therefore the assumption of a quickly thermalising hole made in the phonon-bottleneck model[1] is not fulfilled[24].

# QUANTUM DOTS FORMED BY ISLAND FORMATION

The Stranski-Krastanow growth mode has been shown to result in self-organised dots of, e. g., InAs on GaAs, when the InAs thickness exceeds the equivalent of 1.5-1.7 monolayers (ML) under a special range of growth conditions [25]. As with dots formed using stressors, the attraction here is the absence of surfaces with the additional benefits of higher dot densities (nearly $10^{11}$ cm$^{-2}$) and smaller lateral size (10 to 30 nm diameter depending on growth parameters). One disadvantage is the size spread (over 10%) which is much larger than for deep etched dots. These dot arrays emit very brightly up to room temperature and at low temperature the emission of individual dots has been observed with linewidth below 0.1 meV[26]. Recently, a diode structure based on self-organised quantum dots have shown lasing action[30].

Carrier lifetime measurements in InGaAs/GaAs self-organised dots have shown that the dot emission decay time is longer (880 ps) than that of a reference sample (330 ps)[28] which has been explained by the localised nature of excitons in these quantum dots at low temperatures. However, the authors point out that the dynamics between absorption and emission processes are not well understood and that the phonon bottleneck is not a main limiting mechanism despite the longer emission decay times of these dots, since their integrated emission intensity up to about 200 K is almost two orders of magnitude higher than that of the reference sample.

# SOME APPROACHES TO BYPASS THE PHONON BOTTLENECK

The careful design of the semiconductor structure has been suggested as one way to ensure electron-optical phonon interaction. For example, if the energy spacing between the extended and confined energy levels were $2\hbar\omega_{LO}$ then the phonon bottleneck would be minimised[8]. In practice, the stringent demands on fabrication would limit the value of this approach.

As mentioned above, the phonon bottleneck has been shown to apply also to the exciton-phonon interaction. Previous work by Bryant[4] considered the case of an electron correlated with a hole with the hole accommodating momentum not conserved during the electron-phonon energy relaxation. Bryant calculated scattering rates due to acoustic phonon emission in the case of a confined electron-hole pair in which the electron: (a) is in an excited stated, and (b) relaxes by phonon emission to the correlated electron-hole ground state. Nevertheless, it was found that correlation effects did not accommodate enough additional momentum to change substantially the electron-acoustic phonon scattering rates.

Second order effects in the electron-phonon interaction to consider multiphonon scattering revealed that if the separation between quantum dots states were $\hbar\omega_{LO} \pm \hbar\omega_{LA}$ then the relaxation rate would become comparable to radiative recombination times and thereby minimise the phonon bottleneck[5]. Quantum dot structures with states separated roughly by $\hbar\omega_{LO}$ would still be needed. Multi-phonon relaxation has been observed in GaAs quantum dots[29] although the origin of the relaxation mechanisms was not clear since the possibility of exciton localisation was probably the main reason for the multiphonon cascade.

We have seen that in stressed dots and wires and in self-organised nanostructures, strains plays a role not only in confining electronic wavefunctions but also introducing a local distortion in the lattice. It is possible that the local change in the crystal symmetry allows other phonons to be coupled to the electron/exciton. In particular our magneto-optical studies in submonolayer InAs in GaAs, the regime just prior to island formation, have shown that with increasing magnetic field, i.e., as the electronic wavefunction shrinks, anticrossing behaviour between Landau level transition involving longitudinal zone centre optical GaAs and InAs phonons as well as zone edge LA InAs phonons[30] (see Fig.3).

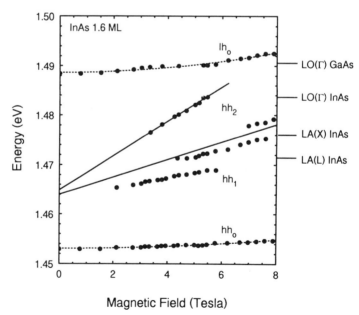

**Figure 3.** Transition energies as a function of magnetic field for 1.6 ML InAs embedded in GaAs obtained from photoluminescence excitation at 4K. Landau level anticrossing is clearly observed at energies comparable to the InAs LA(X) and LA(L) phonon energies. Similar results are obtained for a 1ML sample[30].

In a recent work[31], Sercel has considered the situation of a quantum dot formed by islanding and a nearby deep level. Following the deep level formalism of strong lattice coupling in deep centres, the situation is described in terms of an intraband nonradiative decay through multiphonon emission where the presence of a deep level could dramatically enhance the luminescence efficiency of such a quantum dot system. Furthermore, it is suggested that the formalism would apply to point defects and interface states. Comparing calculations to data from InGaAs/GaAs dots[28], scattering rates are obtained. It is shown how, depending on temperature and distance between the dot and the deep level, a situation may arise whereby relaxation times of 100 ps or less are possible thus bypassing the phonon bottleneck. Sercel's work is most stimulating because real situations are likely to have dots and impurities correlated in real space. More research is needed on the role of impurities in the relaxation process of wires and dots and this may benefit also from comparisons to scattering rates involving isoelectronic impurities.

It is likely that these two factors: strain and impurities are reponsible for the conflicting results obtained in dots and wires formed by stressors, which can also affect the self-organised dots as discussed by Sercel[31]. Self-organised structures are very promising for the study of relaxation since they offer the possibility of varying the strain and confinement conditions through a variety of material combinations.

## CONCLUSIONS

I have tried to review recent developments concerning the phonon bottleneck in quantum dots. As shown above direct evidence is not readily available as many of the dot and wire systems deviate from such a simple picture. Increased carrier temperature and decay times with decreasing lateral sizes in intermixed structures can be interpreted as evidence of the phonon bottleneck. Substantial work remains to be done to understand carrier dynamics and relaxation in deep etched and self-organised dots where strain and impurities may play the key role to bypass the phonon bottleneck. Infrared spectroscopy as well as deep level spectroscopy are yet to be carried out to probe systematically saturation in intersubband transitions and impurity distribution, respectively. The theoretical treatment of the role of isoelectronic impurities and strain in and around dots upon scattering rates remains to be done.

## ACKNOWLEDGEMENTS

This work has been based partly on work carried out in collaboration with P. D. Wang, Y. S. Tang and N. N. Ledentsov, H. Benisty and C. Weisbuch. Part of this work has been supported by a series of grants from UK Engineering and Physical Sciences Research Council (GR/H4474), the ESPRIT NANSDEV 3133 project, the NATO Linkage grant No. 921378 and the INTAS grant 94-481. The author wishes to thank F. Adler for interesting discussions.

## REFERENCES

1. H. Benisty, C. M. Sotomayor Torres, and C. Weisbuch, Intrinsic mechanism for the poor luminescence properties of quantum box systems, *Phys. Rev. B.* 44:10945 (1991)
2. U. Bockelmann and G. Bastard, Phonon scattering and energy relaxation rate in two-, one-,,, and zero-dimensional electron gases, *Phys. Rev. B,* 42:8947 (1990)
3. U. Bockelmann and T. Egeler, Electron relaxation in quantum dots by means of Auger processes,*Phys. Rev. B,* 46:15574 (1992)
4 G. Bryant, Excitons in zero dimensional nanostructures, *in*: "Optics of Excitons in Confined Systems", A. D'Andrea, R. Del Sole, R. Girlanda and A. Quattropani, Eds. Inst. of Physics, Bristol (1992)
5. T. Inoshita and H. Sakaki, Electron relaxation in a quantum dot: significancce of multiphonon processes, *Phys. Rev. B,* 46:7260 (1992)
6. U. Bockelmann, K. Brunner and G. Abstreiter, Relaxation and Radiative decay of excitons in GaAs quantum dots, *Solid State Electronics,* 37:1109 (1994).
7 H. Benisty, Reduced electron-phonon relaxation rates in quantum-box systems: theoretical analysis, *Phys. Rev. B.,* 51:13281 (1995).

8. Y. Arakawa, Fabrication of quantum wires and dots by MOCVD selective growth, *Solid State Electroncis*, 37:523 (1994) and references therien.

9. Y. S. Tang, W. X. Ni C. M. Sotomayor Torres and G. V. Hansson, Enhanced room temperature electroluminescence from Si-Si$_{0.7}$Ge$_{0.3}$ quantum dot diodes, to appear in *Electronic Lett .*, Nov 95

10. see, for example, A. Forchel, B. E. Maile, H. Leier, G. Mayer and R. German, Optical emission from quantum wires, *in*: "Science and Engineering of One- and Zero Dimensional Semiconductors", S. P. Beaumont and C. M. Sotomayor Torres, eds., Plenum, New York (1990) and C. M. Sotomayor Torres, Spectroscopy of semiconductor nanostructures, *in*: "Physics of Nanostructures", J. H. Davies and A. R. Long, eds., Institute of Physics, Bristol (1992) and references therein.

11. P. D. Wang, C. M. Sotomayor Torres, H. Benisty, C. Weisbuch and S. P. Beaumont, Radiative recombinaion in GaAs-Al$_x$Ga$_{1-x}$As quantum dots, *Appl. Phys. Lett.*, 61:946 (1992)

12. U. Bockelmann, Relaxation of hot carriers in semiconductor nanostructures, *in*: "Phonons in Semiconductor Nanostructures", J. P. Leburton, J. Pascual and C. M. Sotomayor Torres, eds., Kluwer, Dordrecht (1993)

13. P. D. Wang, C. M. Sotomayor Torres, H. McLelland, S. Thoms, M. C. Holland and C. R. Stanley, Photoluminescence intensity and multiple phonon Ramman scattering in quantum dots; evidence of the bottleneck effect, *Surf. Sci. 305:585 (1994)*

14. G. Williams, A. G. Cullis, C. M. Sotomayor Torres, S. Thoms, D. Lootens, P. Van Daele, S. P. Beaumont, C. R. Stanley and P. Demeester, Cathodoluminescence studies of GaAs-GaAlAs free standing dots, *in*: "Microscopy of Semiconductors", A. G. Cullis, ed., Institute of Physics, Bristol (1991).

15. K. Brunner, U. Bockelmann, G. Abstreiter, M. Walther, G. Böhm, G. Tränkle and G. Weimann, Photoluminescence from a single GaAs/AlGaAs quantum dot, *Phys. Rev. Lett.* 69:3216 (1992).

16. J. Y. Marzin, A. Izrael and L. Birotheau, Optical properties of etched quantum wires and dots, *Solid State Electronics*, 37:1091 (1994) and references therein.

17. H. E. G. Arnot, M. Watt, C. M. Sotomayor Torres, R. Glew, R. Cusco Cornet, J. Bates ans S. P. Beaumont, Photoluminescence of overgrown GaAs-GaAlAs quantum dots, *Superlattices and Microstructures*, 5:459 (1989).

18. R. Glew and H. E. G. Arnot, unpublished

19. G. Armelles and C. M. Sotomayor Torres, unpublished, periodic progress report No. 2, ESPRIT project 3133 NANSDEV, June 1991

20. F. Adler, M. Burkard, H. Schweizer, S. Benner, H. Haug, W. Klein, G. Tränkle and G. Weiman, Carrier cooling in intermixed GaAs/AlGaAs quantum dots and wires using high excitation and transient spectroscopy, *Phys. Stat. Sol.* (b) 188:241 (1995) and Adler et al in this volume.

21. see, for example, K. Kash, Optical properties of III-V semiconductor quantum wires and dots, *J. Lumin.* 46:69 (1990) and references therein.

22. Y. Zhang, M. D. Sturge, K. Kash, B. P. van der Gaag, A. S. Gozdz, L. T. Florez andd J. P. Harbison, Temperature dependence of luminescence efficiency, exciton transfer and exciton localization in GaAs/Al$_x$Ga$_{1-x}$As quantum wires and dots, *Phys. rev. B.*, 51:13303 (1995)

23. I. H. Tan, Y. L, Chang, R. Mirin, E. Hu, J. Merz, T. Yasuda and Y. Segawa, Observation of increased photoluminescence decay time in strain-induced quantum well dots, *Appl. Phys. Lett.* 62:1376 (1993).

24. H. Lipsanene, M. Sopanen and J. Ahopelto, Luminescence from excited states in strain-induced In$_x$Ga$_{1-x}$As quantum dots, *Phys. Rev. B*, 51:13868 (1995).

25. see, for example, D. Leonard, M. Krishnamurthy, C. M. Reaves, S. P. Denbaars and P. M. Petroff, Direct formation of quantum-sized dots from uniform coherent islands of InAs on GaAs surfaces, *Appl. Phys. Lett.*, 63:3203 (1993) and N. N. ledentsov, P. D. Wang, C. M. Sotomayor Torres, A. Yu. Egorov, M. V. Maximov, I. G. Tabatazde, V. M. Ustinov and P. S. Ko'pev, Optical studies on InAs transformation grown on (100) and (311) GaAs surfaces, *Phys. Rev. B* 50:12171 (1994) and references therein.

26. J. Y. Marzin, J. M. Gerard, A. Izrael, D. Barrier and G. Bastard, Photoluminescence of single InAs quantum dots by self organised growth on GaAs, Phys. Rev. Lett., 73:716 (1994)

27. N. Kirstädter, N. N. Ledentsov, M. Grundmann, D. Bimberg, U. Richter, S. S. Ruminov, P. Werner, J. Heyndenreich, V, M. Ustinov, M. V. Maximov, P. S. Ko'pev and Zh I. Alferov, Low threshold, large To injection laser emission from (InGa)As quantum dots, *Elec tronic Lett.*.30:1416 (1994).

28. G. Wang, S. Fafard, D. Leonard, J. E. Bowers, J. L. Merz and P. M. Petroff, Time-resolved optical characterisation of InGaAs/GaAs quantum dots. *Appl. Phys. Lett.*, 64:2815 (1994).

29. P. D. Wang and C. M. Sotomayor Torres, Multi-phonon relaxation in GaAs-AlGaAs quantum well dots, *J. Appl. Phys.*, 74:5047 (1993)

30. P. D. Wang, N. N. Ledentsov, C. M. Sotomayor Torres, I. N. Yassievich, A. Pakhommov, A. Yu. Egorov, P. S. Ko'pev and V. M. Ustinov, Magneto-optical properties in ultrathin InAs-GaAs quantum wells, *Phys. Rev. B* 50:1604 (1994).

31. P. C. Sercel, Multiphonon-assisted tunneling through deep levels: A rapid energy-relaxation mechanism in nonideal quantom-dot heterostructures, *Phys. Rev. B.* 51:14532 (1995)

# REDUCTION OF ENERGY RELAXATION RATE OF PHOTO-EXCITED HOT ELECTRONS IN QUASI-ONE AND ZERO DIMENSIONAL STRUCTURES

N. Sawaki, S. Niwa, M. Taya, T. Murakami, and T. Suzuki*

Dept. of Electronics, Nagoya University, Chikusa-ku,
Nagoya 464-01, Japan
*Present address: Nippondenso Co. Ltd., Kariya, 448, Japan

## INTRODUCTION

Energy relaxation rate and/or electron mobility in one (1D) or zero dimensional (0D) structures have been expected to be of different character from that in two (2D) or three dimensional (3D) one, because of the change in the scattering rate of phonons in a low dimensional structures.[1-3] Experimentally, however, the process induced defects have veiled the noble effects, and there have been few reports to find the evidence of the relaxation time enhancement In this report we will present an experimental evidence on the decrease of the electron-phonon scattering rate in samples made by wet chemical etching, which has been believed to introduce less defects during the fabrication processes.

## SAMPLES AND THE EXPERIMENTAL PROCEDURE

A 14nm nominally undoped quantum well (QW) was grown by MBE on a S.I.GaAs substrate, which was embedded between AlGaAs cladding layers, and covered by an n-type doped 40nm GaAs cap layer. Quasi- one (1D) and zero (0D) dimensional structures were developed by using laser holography and wet chemical etching.[4,5] The geometrical width is 280–600nm in 1D and 250–700nm in 0D samples. Because of the rather large sample size, the quantization energy is not large enough so that all electrons are in the lowest level. Thus, the samples are not of pure 1D or 0D, but of quasi-1D or quasi-0D.

To study the relaxation phenomena of hot electrons in 0D samples, we adopted an asymmetric double quantum well structure. We measured the tunneling escape time of electrons from the narrow well to the wide well, which is assisted by emission of the LO and/or LA phonons.[6] In case of 1D structure, the net cooling rate of photo-excited hot electrons is studied, which is modified by the heating effect due to the external electric field applied along the wire axis. The energy relaxation process are investigated using the photoluminescence intensity correlation method with a CPM dye laser (wave length 610nm, pulse width 150 fs) at 77K.[6] The temperature dependence of the electron mobility in our particular sample suggests that it is limited by the electron–LO phonon scattering at 77K($\mu$=48,000cm$^2$/Vs). Therefore, we might measure the relaxation process due to the electron–phonon scattering with less ambiguity, in the time scale on the order of picoseconds.

## ENERGY RELAXATION TIME IN 0D SAMPLES

First of all we will discuss on the emission rate of LA and LO phonons in a quantum well structure. As was discussed in detail by several authors,[6,7] if we utilize an asymmetric double quantum well structure, we can estimate the phonon emission rate via the tunneling escape time from the narrow well to the wide well. If the difference energy between the ground states, which is in the wide well, and the first excited states, which is in the narrow well, is larger than the LO phonon energy, 36meV in GaAs, the tunneling escape is governed by the emission of LO phonons. Otherwise, it is determined by the emission of LA phonons. We have prepared two kinds of samples. Sample A has a 10nm and a 7nm GaAs quantum well separated by a 4nm AlGaAs barrier layer, where the energy difference is $\Delta\varepsilon$=22meV. Sample B has a 10nm and a 5nm GaAs quantum well separated by 4nm AlGaAs barrier layer, where $\Delta\varepsilon$=45meV. Thus the tunneling escape time is determined by the LA phonon emission in sample A, while in sample B by LO phonon emission.

Figure 1 shows typical PL spectra for sample A. In case of 0D sample, the clear blue shift of the PL peak is observed. Experimentally, the amount of the blue shift is 2–5meV at 77K for a lateral dimension of the sample of 500–300nm. The amount of the blue shift in the wide well is slightly smaller than that in the narrow well. This is attributed to the fact that the lateral cinfinement effect (in x–y plane) is more enhanced in a narrower QW (z–axis confinement potential) sample. If the lateral size of the dot is larger than 500nm, on the other hand, we could not find the blue shift, but the red

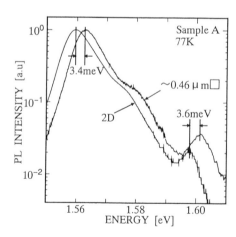
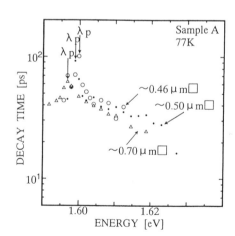

Fig.1 PL spectra of 0D sample       Fig.2 Decay time constant in 0D sample

shift has been observed, which is attributed to the indirect transition or the Stark shift of the PL energy due to the strong built in field at the edge of the 0D structure.[7]

Since the quantization energy due to the low dimensionality is estimated to be on the order of 2–5meV in the smallest sample, the additional quantization is not so important in the energy spectra at 77K, where the thermal energy is around 6meV. Nevertheless, the scattering matrix element itself should be affected because of the change in the basis states,[1] therefore we might expect a certain effect due to the low dimensionality at high temperatures.

Using the PL intensity correlation method with a CPM dye laser,[6] the decay time constant of the PL intensity associated with the narrow well has been measured, from which we estimate the tunneling time as well as the energy relaxation time of photoexcited hot electrons in the narrow well. In as grown 2D samples, the tunneling escape time was on the order of 1ns in sample A. In case of 0D samples, however, the decay time constant was substantially reduced, which is attributed to the reduction of the carrier lifetime due to the defects introduced by the fabrication processes as lithography. But, in small samples where we found the blue shift of the PL peak energy, the decay time constant becomes longer. Typical results are shown in Fig. 2. In case of sample A, the decay is dominated by the LA phonon process, so the decay time is longer than that in sample B (not shown in the figure), where the main process is due to LO phonon emission. In both cases, the smaller the lateral size is, the longer the decay time constant becomes, especially at higher energies. These results suggest that the both the LA and LO phonon emission rate are reduced by the additional quantization.

It should be remembered that the decay time constant at the PL peak energy corresponding to the wide well is on the order of 1 ns. That is, even in the small samples, the carrier life time due to the nonradiative decay process is much longer than those of the tunneling decay time from the narrow well into the wide well. This confirms us that the decay time constant shown in Fig. 3 is really due to emission of phonons in the quantum well structures.

## ENERGY RELAXATION IN 1D SAMPLES

In case of 1D samples, we measured the variation of the correlation curves at various electron energies under the electric field applied along the wire. We have found that the correlation curve is very sensitive to wave length measured, i.e., to the excess energy measured from the bottom of the quantum well. Because of the high carrier density in the QW $3.8 \times 10^{11} \text{cm}^{-2}$, we could not find clear blue shift in the PL spectra even at 4.2K,[4] but we could measure the transport properties in a wire as narrow as 300nm. In such narrow samples, electron temperature rise by an external electric field has been found to be more enhanced as compared to that in as grown 2D sample.[4,8] Figure 3 shows typical correlation signal in the wire structure with 280nm wide. By applying the external electric field along the wire axis, the curves exhibit more and more slower decay and finally it shows anti–correlation behaviour. It is notable that the field strength is as high as 37V/cm to get the anticorrelation Figure 4 shows the signal intensity at $\tau=0$ and E=37V/cm, as a function of the PL energy. Obviously, the negative sign of the correlation signal is obtained if the excess energy is nearly equal to 36meV, i.e., GaAs LO phonon energy. In the particular case as of Fig. 4, the bottom of QW is at 1.530eV and the correlation signal is negative around h$\nu$=1.563eV. These noble effect was observable only in 1D samples.

This result suggests that there has happened a temporal accumulation of carriers at this energy. Simple analyses of the decay properties show that the accumulation of carriers under the fields should be the result of that the cooling effect due to the phonon emission is overcome by the heating effect due to the external field. This again suggest that the phonon emission rate is reduced in the low dimensional structure.

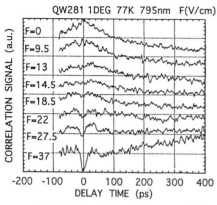

Fig.3 PL intensity correlation curves under electric fields

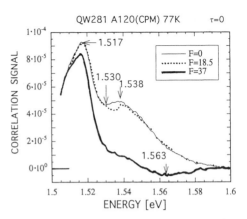

Fig.4 PL intensity correlation signal as a function of energy

## SUMMARY

The energy relaxation phenomena in low dimensional structures have been studied with the femtosecond PL intensity correlation method. It was found that the decay time constant becomes longer in smallest samples. The results suggest that the phonon emission rate is reduced in low dimensional structures. This work was partly supported by the Grant–in–Aid from the Ministry of Education, Science and Culture, and Mazda foundation.

## REFERENCES

1. U. Bockelmann and G. Bastard, Phys. Rev. B 42:8947(1990).
2. H. Benisty, C.M. S-Torres, and C. Weisbuch, Phys. Rev. B 44:10945(1991).
3. J. P. Leburton and D. Javanovic, Semicond. Sci. & Technol. 7:B202(1992).
4. S. Niwa et al., Jpn. J. Appl. Phys. 33:7180(1994).
5. G. Abstreiter, et al, "Optical Phenomena in Semiconductor Structures of Reduced Dimensions," D.J. Lockwood and A. Pinczuk, ed., Kluwer Academic Pub., 327(1993).
6. N. Sawaki et al., Appl. Phys. Lett. 55:1996(1989).
7. T. Matsusue et al., Phys. Rev. B 42:5719(1990).
8. S. Niwa, T. Suzuki, and N. Sawaki, Intern. Workshop on Mesoscopic Physics and Electronics, Tokyo,J-5:141(1995).

# COOLING OF HOT CARRIERS IN INTERMIXED GaAs/AlGaAs QUANTUM WIRES AND DOTS

F. Adler[1], M. Burkard[1], E. Binder[2], H. Schweizer[1],
S. Hallstein[3], W. W. Rühle[3], W. Klein[4], G. Tränkle[4], G. Weimann[4]

[1]4. Physikalisches Institut, Universität Stuttgart,
Pfaffenwaldring 57, 70550 Stuttgart, Germany
[2]Lehrstuhl für Theoretische Physik, BTU Cottbus,
Karl-Marx-Straße 17, 03013 Cottbus, Germany
[3]MPI für Festkörperforschung Stuttgart,
Heisenbergstraße 1, 70569 Stuttgart, Germany
[4]Walter Schottky Institut, Technische Universität München,
Am Coulombwall, 85748 Garching, Germany

## INTRODUCTION

In optoelectronic applications low dimensional structures are discussed as promising for devices with fast modulation response. Up to now the realized quantum wire and quantum dot lasers yield no improvement in modulation response. A possible reason that no improvement could be observed may be the changed carrier scattering and capture mechanisms in quasi 1D and 0D systems. Theoretical calculations predict bottleneck effects[1] which can be proved in low dimensional semiconductor systems with systematically varied sizes. The interest in these studies is twofold i) investigation of carrier scattering from regions of high to low dimensionality and ii) investigation of carrier phonon interaction in quasi 1D and 0D structures. We are able to distinguish these different processes on the one hand by variation of the laser excitation wavelength and on the other hand by the realization of quantum wires (QWWs) and quantum dots (QWDs) with variable size which allow the systematic observation of size dependent carrier capture processes from the barriers into the low dimensional wire/dot region.

## EXPERIMENTAL

For the fabrication of quantum wires and quantum dots we started with a high quality MBE-grown GaAs/Al$_{0.38}$Ga$_{0.68}$As structure with a 3.8 nm single quantum well (QW) 25 nm below the sample surface. Based on high resolution electron beam lithography we defined wire masks down to 75 nm and dot masks down to 45 nm lateral size to obtain buried nanostructures by masked implantation enhanced intermixing. This method has been described in detail in[2]. The photoluminescence (PL) studies were carried out at low temperature 2 K and 15 K. The measurement of transient spectra was performed either by time resolved single photon counting with a microchannel plate photo multiplier with a time resolution of 50 ps or a streak camera with a time resolution of 10 ps. The sample was excited by a mode-locked Nd:YAG laser followed by a synchronously pumped dye-laser (pulse width 4...7 ps) or by a titanium sapphire laser (pulse width 2 ps), respectively. Both detection principles enable the observation of transient spectra at various delay times after the laser pulse within a certain time window. We determined the time dependent carrier temperature by fitting a straight line to the high energy tail of the logarithmic plot of the dot or wire PL spectrum.

## RESULTS AND DISCUSSION

First we studied the hot carrier effect in QWWs and QWDs as a result of carrier capture from the higher dimensionl lateral barrier into the nanostructures. For this experiment the sample was excited nonresonantly (above the lateral barrier). The transient spectra for the narrowest QWD structures of about 45 nm diameter as well as the determined carrier temperatures of different QWDs with various sizes are shown in figure 1.

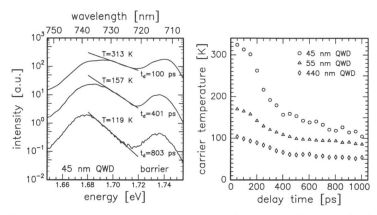

**Figure 1.** Transient spectra of 45 nm QWDs (1.68 eV) and the corresponding lateral barriers (1.73 eV) at various delay times after the laser pulse (left) and the evaluated carrier temperatures for different QWDs (right) vs. delay time.

The spectra (fig. 1 left side) show directly the processes of carrier capture and relaxation in systems with different dimensionality. The low dimensional quasi 0D system cools slower compared to the 2D barrier. Even for rather long delay times ~ 800 ps the QWDs (emission at 1.68 eV) appear overheated compared to the barrier (emission at 1.73 eV) indicated by the different slopes of the high energy tail in both spectra. This hot carrier effect scales with the size of the structures (fig. 1 right side). We find very high initial carrier temperatures in narrow QWDs compared to wider QWDs. And again in agreement with the qualitative finding taken from the direct view of the spectra we observe elevated carrier temperatures in smaller structures for long delay times > 800 ps. To obtain a separation of both the hot carrier effect caused by the carrier capture and the hot carrier effect caused by relaxation bottlenecks we performed an additional experiment. The sample excitation was carried out resonantly (below the lateral barrier) but with the same photon flux density. From the experimentally determined temperatures after resonant ($T_{res}$) and nonresonant ($T_{nonres}$) excitation we calculated the temperature difference $\Delta T = T_{nonres} - T_{res}$ for each structure as a function of time shown in fig. 2. Comparing wire (fig. 2 left side) and dot structures (fig. 2 right side) we observe as a striking feature a strong difference in $\Delta T$-values between the wire and the dot systems.

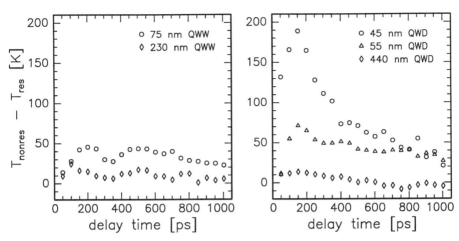

**Figure 2.** Transient temperature difference $\Delta T = T_{nonres} - T_{res}$ between nonresonantly and resonantly excited QWWs (left side) and QWDs (right side).

In detail we observe in large wire and dot structures similar temperature differences at early times followed by a decrease down to $\Delta T = 0$. The higher $\Delta T$-values at early times are caused by the high overshoot energy which has been transferred into the carrier system by nonresonant excitation and depends strongly on size (compare wires and dots in fig. 2). After approximately 250 ps this overshoot energy has been transferred to the phonon subsystem by carrier phonon scattering. The observed transfer time, however, is too large for pure electron phonon interaction and indicates that additional mechanisms at a longer time scale must be responsible for the slow decrease of the carrier temperature. During the first 150 ps we observe an increase of $\Delta T$ in the narrowest QWDs and QWWs. A well known effect is recombination heating[3,4] which is taken into account in our analysis, but will not discussed in detail here. More important effects contributing to an increase of $\Delta T$ can be attributed to carrier capture of barrier carriers into higher wire and dot levels in conjunction with reduced carrier relaxation inside the QWWs and QWDs. Carrier capture was also demonstrated by cw-measurements of the PL-intensities which were in good agreement with a capture length of about 40 nm [2].

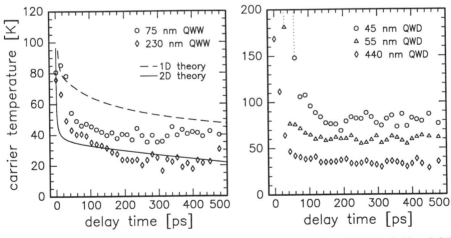

**Figure 3.** Transient carrier temperatures after resonant pulse excitation of QWWs (left) and QWDs (right) measured with a streak camera. Note the different temperature scales for QWWs and QWDs. Dashed and solid lines represent calculated cooling curves for 1D and 2D systems using a model described in the text.

The relaxation bottleneck effect with decreasing structure size can be analyzed directly from transient carrier temperatures after resonant excitation (cf. fig. 3). The narrow 45 nm and 55 nm QWDs show the highest initial carrier temperatures compared to the 2D like 440 nm QWDs (fig. 3 right side). We find a similar picture in the transient carrier temperature of QWWs (fig. 3 left side). As expected also for wires the cooling of the 75 nm wire starts at higher temperature compared to the 230 nm wire and again the temperatures of different systems with different size show smaller differences in wires than in dots. Similar experimental results have been obtained in the cooling behaviour of V-groove quantum wires[6]. We can explain the transient carrier cooling inside the low dimensional systems using a phenomenological model based on publications by Bimberg[6], Shah[7], Leo and Rühle[4]. By modification of the general formula for recombination heating [6] we obtain the differential equation (1) as a description for the time evolution of the carrier temperature T

$$\frac{dT}{dt} = \left(\frac{\partial E_{tn}}{\partial T}\right)_n^{-1} \left[\left(\left(\frac{\partial E_{tn}}{\partial n}\right)_T - \frac{E_{tn}}{n}\right) \cdot \frac{n}{\tau} + n \cdot \langle ELR \rangle\right] . \tag{1}$$

Here $E_{tn}$ denotes the total energy of the electron system, n is the carrier density and $\langle ELR \rangle$ the averaged energy loss rate per carrier which was taken from[8]. In eq. (1) we assumed a monoexponential recombination law and an energy independent recombination probability $\tau$. The result of this calculation for a 2D and 1D system involving only LO-Phonon scattering and recombination from the lowest subband of a 75 nm wire is given in fig. 3. The model is in principle in good agreement with the experimental findings. Two significant features have been found in the cooling behaviour in 1D and 0D systems compared to 2D and 3D systems. First the clear slowed carrier cooling during the first 250 ps (compare in fig. 3, 75 nm with 230 nm QWWs, left side and 45 nm with 440 nm QWDs, right side) second the higher carrier temperatures even after 500 ps in small systems indicating an impaired carrier relaxation with decreasing dimensionality. From the calculation this impaired carrier relaxation in low dimensional systems could be attributed to a weaker interaction between electrons and longitudinal optical phonons in low dimensions.

The authors would like to acknowledge the fruitful discussions with Prof. M. H. Pilkuhn, Dr. T. Kuhn and J. Schilp.

REFERENCES

1. H. Benisty, C. M. Sotomayor-Torrès, and C. Weisbuch, Intrinsic mechanism for the poor luminescence properties of quantum-box systems, *Phys. Rev. B* 44:10945 (1991).
2. M. Burkard, U. A. Griesinger, A. Menschig, H. Schweizer, H. Klein, G. Böhm, G. Tränkle, and G. Weimann, Progress in mask technology for ion implantation based nanofabrication, *J. Vac. Sci. Technol. B* 12:3677 (1994).
3. F. Adler, M. Burkard, H. Schweizer, S. Benner, H. Haug, W. Klein, G. Tränkle, and G. Weimann, Carrier cooling in intermixed GaAs/AlGaAs quantum dots and wires using high excitation and transient spectroscopy, *phys. stat. sol. (b)* 188:241 (1995).
4. K. Leo, W. W. Rühle and K. Ploog, Hot-carrier energy-loss rates in GaAs/AlGaAs quantum wells, *Phys. Rev. B* 38:1947 (1988).
5. A. C. Maciel, C. Kiener, L. Rota, and J. F. Ryan, U. Marti, D. Martin, F. Morier-Gemoud, and F. K. Reinhart, Hot carrier relaxation in GaAs V-groove quantum wires, *Appl. Phys. Lett.* 66:3039 (1995).
6. D. Bimberg, J. Mycielski, Recombination-induced heating of free carriers in a semiconductor, *Phys. Rev. B* 31:5490 (1985).
7. Jagdeep Shah, Hot carriers in quasi-2-D polar semiconductors, *IEEE J. of Quantum Electr.* 22:1728 (1986).
8. J. P. Leburton, Size effects on polar optical phonon scattering of 1-D and 2-D electron gas in synthetic semiconductors, *J. Appl. Phys.* 56:2850 (1984).

# IMPORTANCE OF ELECTRON-HOLE SCATTERING FOR HOT CARRIER RELAXATION IN LOW DIMENSIONAL QUANTUM STRUCTURES

I. Vurgaftman[†], K. Yeom, J. Hinckley and J. Singh

Department of Electrical Engineering and Computer Science,
The University of Michigan, Ann Arbor, MI 48109
[†]Presently with Naval Research Laboratory, Washington, DC 20375

Carrier thermalization in low dimensional structures such as quantum wires and quantum dots highlights the influence of the density of states (DOS) on the carrier-phonon and carrier-carrier scattering rates. In particular, the electron-phonon interaction in structures with singularities in unrenormalized electronic spectra is weakened since, once energy and momentum conservation requirements are satisfied, a final state may not necessarily be readily available for the scattering electron. The broadening of the electronic DOS due to the electron-phonon coupling and prominence of unusual energy loss channels can be thought of as consequences of this phenomenon, not present in higher dimensional semiconductor systems.

The problem can be summarized in the following form: can an electron in an excited state of a quantum dot attain the ground state before recombining nonradiatively if the phonon emission time is much longer than the nonradiative recombination time[1]? A possible energy loss mechanism is an Auger-type process by which the electron gives up the needed energy to electrons in the continuum outside of the quantum dot[2]. If a large density of holes is created in the dot along with electrons (e.g., by photoexcitation or injection in a laser diode structure), a distinct energy loss channel is brought into existence. The number of states in the valence band is much higher than in the conduction band for the dimensions of interest (see, e.g., for a structure intermediate between a quantum dot and wire, the DOS obtained by a finite-difference solution of the 1-band (conduction band) and 4-band (valence band) Hamiltonians in Fig. 1(a) and (b)). The hole-phonon scattering is thus a much more efficient process, and hole thermalization is relatively fast (of the order of a ps). Since electrons and holes are localized within the same small volume, the probability of energy exchange between them by Coulombic scattering is high, and the electrons are able to give up energy to holes which pass it on to the lattice allowing electron thermalization on a subnanosecond time scale.

In this paper, the results of a recently developed model[3,4] for Monte Carlo analysis of the electron thermalization process in small quantum dots ($< 8000$ nm$^3$) and narrow quantum wires ($< 400$ nm$^2$) investigated in connection with optical applications are summarized.

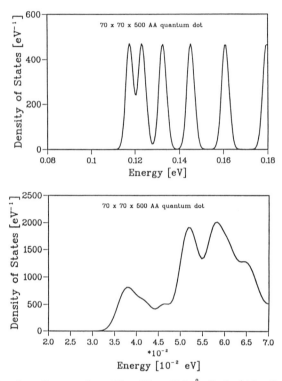

**Figure 1.** The density of states in a $70 \times 70 \times 500$ Å GaAs/Al$_{0.3}$Ga$_{0.7}$As structure in the conduction (a) and valence (b) bands.

Electron thermalization is simulated by including the polar optical and acoustic phonon as well as electron-hole scattering in the Born approximation based on the assumption of instantaneous hole thermalization for equal electron and hole densities of $9.6 \times 10^{18}$ cm$^{-3}$ (3 electrons and 3 holes in a single dot). Electrons are injected from the extended barrier states into the confined states in the quantum dot. The broadening of the energy levels is given by the imaginary part of the electron self-energy. Both optical and acoustic phonon scattering are found to be inefficient if any two successive states are separated by an energy significantly different from the optical phonon energy. The inclusion of electron-hole scattering allows the relaxation process to be speeded up by one to two orders of magnitude, as can be seen from Fig. 2. The relaxation time is of the order of several hundred ps and increases with the reduced size of the quantum dot.

If the structure is allowed to extend to a length much greater than the de Broglie wavelength in one dimension (the case of aquantum wire), phonon scattering is facilitated although resonances in the DOS remain. The inefficiency of electron-electron scattering in a structure with few subbands makes thermalization in the vicinity of the effective band edge slow if the band edge is off the scattering resonances with the higher subbands arising from peaks in the DOS, and relaxation occurs predominantly by acoustic phonon emission. If electron-hole scattering is included, however, the relaxation time can be reduced by as much as a factor of 2 for a carrier density of $10^{18}$

cm$^{-3}$ (see Fig. 3). These results demonstrate that electron-hole scattering in quantum dots and wires plays a role that is much more important than in higher dimensional structures.

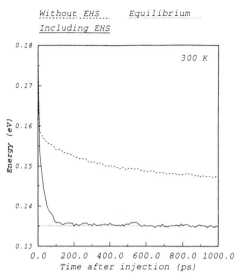

**Figure 2.** The mean electron energy as a function of time after injection for the $50 \times 250 \times 250$ Å GaAs/AlGaAs quantum dot at 300 K. The horizontal line represents the equilibrium mean energy.

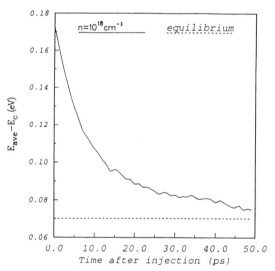

**Figure 3.** The time evolution of the mean electron energy in a $100 \times 100$ Å GaAs/AlGaAs quantum wire including the effects of electron-hole scattering.

## REFERENCES

1. H. Benisty, C. M. Sotomayor-Torres, and C. Weisbuch, *Phys. Rev. B* 44:10945 (1991).
2. U. Bockelmann and T. Egeler, *Phys. Rev. B* 46:15574 (1992).
3. I. Vurgaftman and J. Singh, *Appl. Phys. Lett.* 64:232 (1994).
4. I. Vurgaftman, Y. Lam and J. Singh, *Phys. Rev. B* 50:14309 (1994).

# HOT EXCITONS IN QUANTUM WELLS, WIRES, AND DOTS

J. Černe[1], H. Akiyama[2], M.S. Sherwin[1], S.J. Allen[1], T. Someya[2], S. Koshiba[2], H. Sakaki[2], Y. Arakawa[2], and Y. Nagamune[2]

[1]Physics Department and Center for Free-Electron Laser Studies, University of California, Santa Barbara
[2]Japanese Research Development Corporation and University of Tokyo

## INTRODUCTION

Advances in the growth of semiconductor quantum heterostructures have allowed the realization of 2-D, 1-D and 0-D quantum-confined systems. Much can be learned by investigating the equilibrium and non-equilibrium response of quantum-confined carriers to far-infrared (FIR) radiation. Investigations of photoexcited carriers are particularly interesting since neither doping nor contacts are required. We have previously observed that intense FIR radiation from UCSB's free-electron lasers (FEL) heats photoexcited carriers in quantum wells (QWs), thereby quenching excitonic photoluminescence (PL) at low lattice temperatures.[1] In this paper, we show preliminary results on the effects of FIR radiation on photoexcited carriers in quantum wires (QWIs) and dots (QDs).

## EXPERIMENT

Three undoped samples were investigated in this study. The first sample was grown by molecular beam epitaxy (MBE) to form fifty periods of 100 Å-wide GaAs quantum wells (QWs) with 150 Å-thick $Al_{0.3}Ga_{0.7}As$ barriers.[1] The second sample consisted of ridge quantum wires (QWIs) grown by MBE on a strip-line-patterned substrate.[2,3] The third sample was grown using selective metal-organic chemical vapor deposition (MOCVD) to form quantum dots (QDs) on a substrate that was masked by square $SiO_2$ windows.[4]

A schematic of the PL setup is shown in Figure 1. The PL measurement in the presence of intense FIR radiation was accomplished by photoexciting carriers and collecting PL from the front of the sample while the FIR radiation passed through the sample from the back. An argon ion laser was used to create electron-hole pairs in the undoped sample and the resulting PL was captured by the same optic-fiber which delivered the excitation radiation. This luminescence was reflected off a beam splitter into a .34 m monochromer and detected by a cooled GaAs photomultiplier tube. The $Ar^+$ all-line laser excitation intensity ranged from 10 to 50 W/cm$^2$ at the sample. The output of the argon ion laser was modulated acousto-optically to produce a 100 μs visible excitation pulse that coincided with the 5 μs FIR pulse at the sample. A boxcar integrator gated the PL signal during and after the FIR pulse (see right side of Figure 1). The unquenched, low temperature PL immediately after the FIR pulse was used to normalize the data.

The FIR transmission through the cryostat window and the sample substrate is included in the determination of the FIR intensity at the quantum heterostructure. The sample geometry is not accounted for in determining the magnitude of the FIR intensity for the QWIs and QDs. No FIR etalon effects are observed. The FIR intensity is estimated to an accuracy of ±20%.[1]

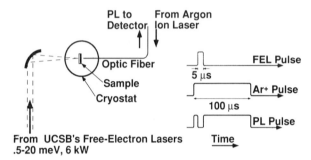

**Figure 1.** A schematic of the experimental setup. PL is detected during and after the FIR pulse has passed through the sample. The FIR radiation is polarized parallel to the plane of the quantum well. The polarization for the QWIs is either parallel or perpendicular to the wires. The timing of the laser pulses and PL is shown on the right side.

## RESULTS

In all three samples, the FIR radiation heats carriers without significantly affecting the lattice temperature. The heavy hole exciton PL amplitude is used as an indicator for the carrier temperature ($T_{carrier}$) between 10 and 100K.[1] The PL amplitude for a given FIR intensity at a low lattice temperature is compared to the PL amplitude at a higher lattice temperature (without FIR irradiation) to obtain an estimate of the carrier temperature. Since the FIR pulse length is much longer than the carrier energy relaxation time, the heating is steady state where the power absorbed by each carrier is equal to the power lost.

$$P_{abs} = I_{FIR}\,\sigma_{abs} = P_{lost} = k_B\,\Delta T/\,\tau_{energy} \qquad (1)$$
$$\text{where } \Delta T = T_{carrier} - T_{lattice}\,.$$

In Equation 1, $P_{abs}$, $I_{FIR}$, $\sigma_{abs}$, $P_{lost}$, $k_B$, $\Delta T$, and $\tau_{energy}$ are the power absorbed, FIR intensity, absorption cross-section, power lost, Boltzmann constant, temperature difference between the carrier and the lattice, and the energy relaxation time respectively. Grouping together all the measured quantities ($\Delta T$ and $I_{FIR}$) in Equation 1, allows the product of the absorption cross section and energy relaxation time to be isolated.

$$k_B\,\Delta T/\,I_{FIR} = \sigma_{abs}\tau_{energy} \qquad (2)$$

Assuming that $\tau_{energy}$ only depends on carrier temperature and is frequency independent, we plot $k_B\,\Delta T/\,I_{FIR}$ at a fixed carrier temperature to obtain the frequency dependence of $\sigma_{abs}$. The quantity $k_B\,\Delta T/\,I_{FIR}$ reflects the efficiency in carrier heating by the FIR.

In the QWs, the FEL polarization is parallel to the plane of the quantum wells. Figure 2 shows the frequency dependence of the carrier heating in the QWs. The efficiency $k_B \Delta T / I_{FIR}$ in heating carriers to 36.7K drops as $\omega^{-2}$. Despite strong excitonic correlations, this result suggests that the carrier heating is dominated by free-carrier, Drude absorption of the FIR radiation.[1]

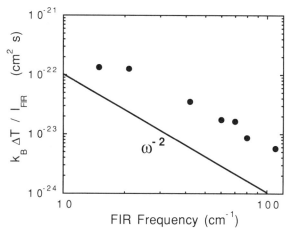

**Figure 2.** $k_B \Delta T / I_{FIR}$ vs. FIR frequency for the multiple quantum well sample. The lattice temperature was 9K and the carriers were heated to 37K. The efficiency in carrier heating decreases as $\omega^{-2}$, which is consistent with Drude free-carrier heating.

The QWIs can be excited with the FIR radiation polarized parallel or perpendicular to the wires. Figure 3 shows the polarization and frequency dependence in heating the carriers to 36.3K. Above 20 cm$^{-1}$ (2.5 meV), the frequency dependence of the absorption cross-section follows a Drude form for both polarizations, as in the QWs. Furthermore, in this frequency range the carrier heating is more efficient for the FIR polarized parallel to the wires. One may expect the parallel polarization to induce free-carrier absorption and efficiently heat the carriers since the motion along the wires is free whereas the motion perpendicular to the wires (i.e., in the confined direction) should be frozen and the expected intersubband transitions are at much higher frequencies. However, the Drude-like heating for the perpendicular polarization, and the cross-over below 20 cm$^{-1}$ (where the most efficient heating is for the perpendicular polarization) still require further explanation.

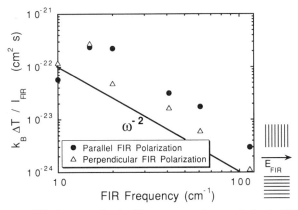

**Figure 3.** $k_B \Delta T / I_{FIR}$ vs. FIR frequency for the ridge quantum wire sample for the FIR polarized parallel or perpendicular to the wires. The lattice temperature was 23K and the carriers were heated to 36K.

In Figure 4, the frequency dependence of the heating efficiency $k_B \Delta T/I_{FIR}$ for the QDs is shown for carrier temperatures of 15, 30, and 50K. The heating shows a weak Drude trend, but at 6.3 meV the carrier heating is resonantly enhanced. The electronic intersubband spacing is estimated to be 6 meV in these QDs, which suggests that the resonant heating is due to FIR-induced intersubband transitions. Also note that the heating efficiency is smaller at higher carrier temperatures, which is consistent with a shorter energy relaxation time at these carrier temperatures. This effect is due to enhanced LO phonon emission and is seen in all three samples.

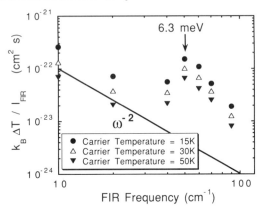

**Figure 4.** $k_B \Delta T/I_{FIR}$ vs. FIR frequency for the quantum dot sample at carrier temperatures of 15, 30 and 50K. The lattice temperature is 7K. For all three carrier temperatures, the heating efficiency reaches a maximum for FIR energy of 6.3 meV.

## CONCLUSIONS

Through this all-optical technique we are able to explore directly the absorption and relaxation mechanisms in lower dimensional quantum heterostructures. This method may be used to definitively demonstrate quantization in quantum dot structures, as indicated by the suggestive heating resonance. Further quantitative analysis must address the fundamental difficulty in separating purely dimensional effects from those due to sample and measurement artifacts. In our future investigations we hope to determine the strength of confinement and to test lower dimensional issues such as the phonon bottleneck.

## ACKNOWLEDGMENTS

This work has been supported by QUEST DMR 91-20007, ONR N00014-K-0692, and the JRDC Quantum Transition Project. The authors would also like to thank D.P. Enyeart at the Center for Free-Electron Studies for his technical support.

## REFERENCES

1. J. Černe et al., Phys. Rev. B **51**, 5253 (1995).
2. S. Koshiba et al., Solid-State Electron. **37**, 729 (1994).
3. H. Akiyama et al., Phys. Rev. Lett. **72,** 924 (1994).
4. Y. Nagamune et al., Appl. Phys. Lett. 64, 2495 (1994).

# ENERGY RELAXATION IN In$_{0.53}$Ga$_{0.47}$As/InP QUANTUM WIRES

F. Kieseling, W. Braun, P. Ils, K. H. Wang, and A. Forchel

Technische Physik, Universität Würzburg,
Am Hubland, D-97074 Würzburg, Germany

## 1. INTRODUCTION

In low-dimensional semiconductor structures significant modifications of the carrier scattering rates are expected due to the reduced number of final states that fulfill both energy and momentum conservation. On the one hand, the reduced rates for carrier-carrier-, carrier-phonon- and carrier-impurity-scattering make low-dimensional structures promising candidates for electronic devices with high carrier mobilities [1, 2]. On the other hand, the small scattering rates are necessarily connected with slow energy relaxation processes which may give rise to a reduced quantum efficiency and to a degraded high speed performance [3]. Another important process that may limit device efficiency and speed is the transfer of carriers into the active regions of low-dimensional semiconductor structures.

In the present work we study the transfer and relaxation of photoexcited carriers in deep etched In$_{0.53}$Ga$_{0.47}$As/InP quantum wires with widths between 34 and 247 nm by time-resolved photoluminescence (PL) spectroscopy. By varying the linear polarization of the nonresonant laser excitation we are able to create initial carrier distributions that are located mainly in the unpatterned InP barrier. The significant increase of the PL rise time with decreasing wire width observed under these excitation conditions gives evidence for a reduced carrier transfer into narrow wire structures. The temperature dependence of the PL rise time shows that this behaviour can be attributed to the formation of a potential barrier in the laterally confined InP wire region. For the investigation of the carrier relaxation processes transient high excitation spectra of different wire structures were recorded. A lineshape analysis of the spectra gives indication for a slower cooling of the electron-hole-plasma and a faster decrease of the plasma density in narrow wire structures.

## 2. PATTERNING TECHNOLOGY AND TIME-RESOLVED PL SPECTROSCOPY

For our investigations we have fabricated In$_{0.53}$Ga$_{0.47}$As/InP quantum wires from GSMBE grown single quantum well layers by high resolution electron beam lithography and deep wet chemical etching. The quantum well structure consists of a 200 nm thick InP buffer layer, a 5 nm wide In$_{0.53}$Ga$_{0.47}$As quantum well and an 8 nm thick InP cap layer. This fabrication technique allows a continuous variation of the structure sizes and ensures high

*Hot Carriers in Semiconductors*
Edited by K. Hess *et al.*, Plenum Press, New York, 1996

optical quality. A detailed description of the fabrication process is given in Ref. [4]. The dimensions of the resulting wire structures were determined from SEM micrographs. The height of the free-standing wires is approximately 40 nm. The wire width was varied between 247 and 34 nm.

The samples were held in a variable temperature cryostat at a temperature of 40 K and were excited at wavelengths of 850 and 930 nm, respectively, by 70 fs pulses of a mode-locked Ti:Sapphire laser operating at a repetition rate of 82 MHz. The photoluminescence emission of the sample and a delayed part of the laser pulse were focused onto a LiIO$_3$ crystal that generated the sum frequency of the two light waves [5]. The sum frequency signal was dispersed by a 0.25 m grating spectrometer and detected with a liquid nitrogen cooled CCD camera. The temporal evolution of the PL emission was recorded by varying the delay time of the laser pulse with respect to the exciting laser pulse and simultaneously detecting the intensity of the sum frequency signal. The overall time resolution of the system was about 300 fs.

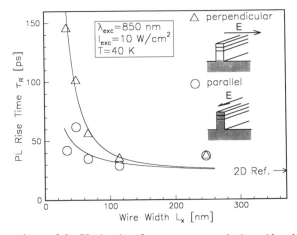

**Fig. 1.** Wire width dependence of the PL rise time for nonresonant excitation with polarization perpendicular (triangles) and parallel (circles) to the wire axis. The solid lines are a guide to the eye. The initial carrier distributions are depicted schematically by the shaded areas in the inset.

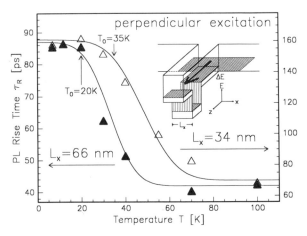

**Fig. 2.** Temperature dependence of the PL rise time for 66 (filled triangles) and 34 nm (open triangles) wide wires for nonresonant perpendicular excitation. The solid lines are a guide to the eye. The inset illustrates the carrier transfer across the conduction band potential barrier in the InP region of the wires.

## 3. CARRIER TRANSFER INTO DEEP ETCHED QUANTUM WIRES

In order to study the transfer of carriers into the wire structures we have investigated the onset of the PL emission after excitation at a wavelength of 850 nm, i. e. above the band gap of the InP barrier material. Calculations of the electric field distribution in deep etched semiconductor wire structures [6] have shown that an external electric field oriented perpendicular to the wire direction penetrates only little into the wire structure when the width of the structure becomes comparable to its height. In contrast, there is no such effect if the electric field is oriented parallel to the wire direction. Making use of this effect, we are able to create specific initial carrier distributions by varying the polarization of the exciting laser beam as shown schematically in the inset of Fig. 1: For excitation parallel to the wire axis carriers are generated both in the $In_{0.53}Ga_{0.47}As$ and in the InP regions of the wire structures. In contrast, for excitation perpendicular to the wire direction the absorption in the free-standing wire ridge is strongly suppressed and the initial carrier distribution is therefore mainly located in the unpatterned InP barrier material.

Fig. 1 shows the wire width dependence of the PL rise time $\tau_R$ as obtained by fitting the PL onset with the simple exponential expression $I(t)=I_0 \cdot [\exp(-t/\tau_R)-\exp(-t/\tau_L)]$, where $\tau_L$ is the carrier life time. For excitation perpendicular to the wire axis (triangles) we observe a pronounced increase of the PL rise time for wire widths below about 100 nm resulting in a rise time of about 145 ps for a wire width of 34 nm. The comparatively short rise times

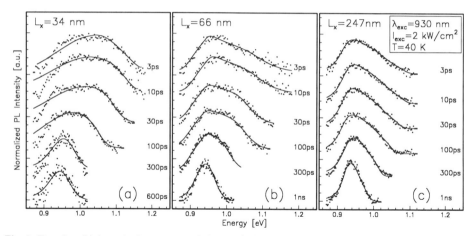

**Fig. 3.** Transient high excitation spectra of deep etched $In_{0.53}Ga_{0.47}As/InP$ wire structures with widths of (a) 34, (b) 66 and (c) 247 nm. The delay times are indicated at the different traces. The solid lines represent results of a model calculation (see text).

observed for parallel excitation (circles), which are almost independent of the wire width, show that the increase of the rise time is not due to a reduced carrier capture or carrier relaxation in the narrow wire structures. Consequently, the long PL rise times in the narrow wires must be attributed to a reduced efficiency of the carrier transfer from the unpatterned InP barrier into the $In_{0.53}Ga_{0.47}As$ wire regions.

The increase of the PL rise time with decreasing wire width observed for perpendicular excitation results from lateral confinement effects in the InP regions that lead to the formation of a potential barrier between the unpatterned InP material and the laterally structured InP wire region. This is illustrated schematically in the inset of Fig. 2. The carrier transfer process across this potential barrier can be thermally activated as is indicated by the temperature dependence of the PL rise time shown in Fig. 2 for wire widths of 34 (open triangles) and 66 nm (filled triangles). For both wire widths the rise time is almost constant up to a temperature $T_0$ that can be regarded as the activation temperature of the transfer process. For the 34 nm wide wires this temperature is about 35 K, for the 66 nm wide wires it is about 20 K. Above that temperature the PL rise time decreases drastically by almost a factor of two. The estimated activation energies $E_0 = k_B T_0$ of 3.0 and 1.7 meV for the 34 and 66 nm wide wire structures, respectively, compare fairly well with estimates of the lateral quantization energies of 4.5 and 1.2 meV in the free-standing InP wire region.

## 4. ENERGY RELAXATION IN HIGHLY EXCITED QUANTUM WIRES

For the investigation of the energy relaxation processes the wire structures were excited with high intensities of 2 kW/cm$^2$ at a wavelength of 930 nm, i. e. below the InP barrier band gap, in order to exclude the influence of carrier transfer processes as discussed in Sec. 2. The transient PL emission spectra of wire structures with widths of (a) 34, (b) 66 and (c) 247 nm are displayed in Fig. 3. Using a modified version of the model described in Ref. [7], we have performed a lineshape analysis of the transient spectra that yields the temporal evolution of the density and the temperature of the photogenerated electron-hole-plasma. The results of the model calculations are represented by the solid lines in Fig. 3.

Fig. 4 and 5 show the temporal evolution of the plasma density and the plasma temperature in the different wire structures. We observe an approximately wire width independent initial plasma density of about $10 \cdot 10^6$ cm$^{-1}$. However, in the narrow wires the plasma density decreases faster and the onset of the drop occurs at earlier delay times. This behaviour can be attributed to a reduction of the carrier lifetime in the narrow wire structures due to the increasing influence of surface recombination processes [8]. At short delay times the 34 and the 66 nm wide wire structures show plasma temperatures of more than 300 K that are significantly higher than the plasma temperature of only about 150 K in the 247 nm wide wires. For larger delay times the plasma temperatures of the narrow wires gradually approach the temperatures in the wide wires, but the highest plasma temperatures are still observed in the narrow wires. These results give evidence for a reduced energy relaxation in narrow wire structures that can be attributed to a reduction of the carrier scattering rates in one-dimensional structures.

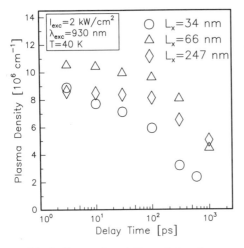

**Fig. 4.** Temporal evolution of the plasma density in deep etched In$_{0.53}$Ga$_{0.47}$As/InP wire structures with widths of 34 (circles), 66 (triangles) and 247 nm (diamonds).

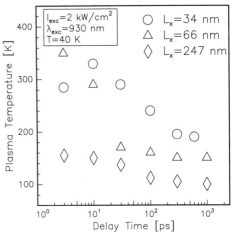

**Fig. 5.** Temporal evolution of the plasma temperature in deep etched In$_{0.53}$Ga$_{0.47}$As/InP wire structures with widths of 34 (circles), 66 (triangles) and 247 nm (diamonds).

## ACKNOWLEDGEMENT

We would like to thank L. Goldstein and Ph. Pagnod-Rossiaux, Alcatel-Alsthom Recherche, for supplying the high quality $In_{0.53}Ga_{0.47}As/InP$ quantum well structure. We are also grateful to S. Benner for supplying and modifying the program used for the lineshape analysis of the transient emission spectra. The financial support of this work by the Volkswagen Foundation, by the State of Bavaria and by the ESPRIT project NANOPT is gratefully acknowledged.

## REFERENCES

[1]  H. Sakaki, Jpn. J. Appl. Phys. **19**, L735 (1980)
[2]  H. L. Störmer, A. C. Gossard, W. Wiegmann, Appl. Phys. Lett. **39**, 912 (1981)
[3]  H. Benisty, C. M. Sotomayor-Torrès, C. Weisbuch, Phys. Rev. B **44**, 10945 (1991)
[4]  P. Ils, M. Michel, A. Forchel, I. Gyuro, M. Klenk, E. Zielinski, Appl. Phys. Lett. **64**, 496 (1994)
[5]  J. Shah, IEEE J. Quantum Electron. **24**, 276 (1988)
[6]  P. Ils, Ch. Gréus, A. Forchel, V. D. Kulakovskii, N. A. Gippius, S. G. Tikhodeev, Phys. Rev. B **51**, 4272 (1995)
[7]  S. Benner, H. Haug, Europhys. Lett. **16**, 579 (1991)
[8]  F. Kieseling, W. Braun, P. Ils, M. Michel, A. Forchel, I. Gyuro, M. Klenk, E. Zielinski, Phys. Rev. B **51**, 13809 (1995)

# RELAXATION PROCESSES IN GaAs and InGaAs V-SHAPED QUANTUM WIRES

R.Cingolani, R.Rinaldi, P.V.Giugno

Dipartimento Scienza dei Materiali, Universita' di Lecce, 73100 Lecce Italy

M.Lomascolo and M.DiDio

CNRSM-Pastis, SS 7 APPIA Km 713, 72100 Brindisi, Italy

U.Marti, D.Martin, F.K.Reinhart

Institute de Micro-Optoelectronique, EPFL, CH-15-Lausanne Switzerland

## Abstract

We investigated the relaxation of excitons and free-carriers in V-shaped GaAs and InGaAs quantum wires with distinct one-dimensional properties. Trapping of carrier in the wires and simultaneous recombination of higher index transitions are studied by spectrally and time resolved luminescence measurements performed at the relevant energies of the quantized states observed in photoluminescence excitation spectra. The effect of disorder in low-crystalline quality quantum wires (inferred by TEM studies) is also discussed.

V-shaped quantum wires have been demonstrated to be ideal systems for the study on two-dimensional quantum confinement and for subsequent optoelectronic applications [1,2]. In this work we concentrate on the relaxation dynamics of excitons and free - carriers confined in GaAs and InGaAs quantum wires of typical lateral width smaller than 20 nm. The samples were grown by Molecular Beam Epitaxy on holographically patterned GaAs substrates. A single thin quantum well deposited onto superlattice barriers ($[(GaAs)_m/(AlAs)_n]$, where m and n label the number of monolayers) realizes the lateral confinement due to the narrowing of the layer along the groove sidewalls. Systematic transmission electron microscopy analysis were performed in order to measure the actual profile of the quantum well. A solution of the full-two-dimensional Schroedinger equation incorporating the lateral potential obtained by the bent quantum well profile is used to evaluate the carrier confinement energies [3]. Time resolved photoluminescence spectra were measured by using a Ti:Sa laser delivering 2ps pulses at 82 MHz frequency in the range 730-850 nm and in the blue-green, by using the second harmonic generator. The detection system consisted of a synchro-scan streak camera with two-dimensional CCD acquisition, providing 5 ps resolution time. PLE experiments were performed by using a cw Ti:Sa laser and a double 0.85 m monochromator.

In Fig. 1 we plot the typical photoluminescence mapping on one of the investigated structures. In the patterned region the emission from quantum wire states (QWR) as well as from the surrounding superlattice barriers (BENT SL) can be identified. Excitation of the unpatterned region of the sample (reference part), clearly shows the emission from the planar quantum well (QW) and from the planar superlattice barrier (SL). In what follows we will show the different temporal evolution of these emission lines, and will draw some conclusion about the trapping efficiency of carriers in the wires.

In Fig.2a and 2b we show the PLE and PL spectra of a GaAs and InGaAs quantum wire structure at low temperature, respectively. In Fig.2a, a clear band filling is observed in the PL spectrum with increasing excitation intensity. The energy separation between the transitions amounts to about 20 meV, consistent with the calculated quantization energies of the quantum wire (16 nm lateral width). The PLE spectra exhibit distinct structures at energies corresponding to the main PL bands, allowing us to identify the $n=1, n=2$ and $n=3$ transitions. The spectral broadening and the Stokes shift are consistent with the average size fluctuations at the bottom of the groove observed by TEM [4]. Similar results are obtained for the InGaAs quantum wires, in which three transitions are well resolved (Fig.2b).

In all these samples the intersubband separation is smaller than the characteristic LO phonon energy. Therefore one expects that photogenerated carriers do not efficiently relax at the ground level transition, within their lifetime, and preferentially recombine from the higher index states. This is consistent with the observation of band filling luminescence even under very low continuous power excitation. This is actually shown in Fig.3, where we plot the temporal evolution of the $n=1, n=2$ and $n=3$ transitions for the GaAs quantum wires of Fig.2a. The $n=1$ transition exhibits a rather long decay time, of the order of 360 ps, with a long plateau region due to free- carrier formation coexisting with the exciton gas. The higher index transitions exhibit decreasing decay times (190 ps and 120 ps, respectively), indicating that relaxation occurs both through radiative recombination and non radiative decay via acoustic phonon-interaction or intercarrier scattering [5]. At low power excitation, in the excitonic regime, the three decay times are almost equal and there is no "plateau" region in the decay curve of the $n=1$ luminescence.

Similar results are obtained for the InGaAs quantum wires (Fig.4), though with decay times shorter by a factor two with respect to the GaAs case. It is interesting to note that the decay time shortens considerably with increasing temperature. This behavior is presently not fully understood. Qualitatively, the increase of the thermal energy in the relatively shallow quantum wire potential should lead to a progressive delocalization of the wavefunctions, with consequent decrease of the emission intensity and decay time. This is somewhat analogous the thermal escape of carriers from the quantum well induced by the temperature. However, the specific geometry of our confining potential and the competition of different recombination channels from the barrier and the wires, makes this interpretation more difficult than in the case of quantum wells. Experiments are in progress to clarify this issue.

We now turn to the carrier dynamics for excitation in the superlattice barrier. As reported in ref.[6,7], the existence of localized superlattice states can reduce the transfer efficiency into the quantum wires due to the reduced overlap of superlattice and quantum wire states. Localization in the barrier is usually monitored through the long decay time of the superlattice luminescence, longer that the quantum wire itself. In order to clarify this issue we have performed intensity dependent time resolved PL measurements, to compare the temporal evolution of the $n=1$ and superlattice barrier recombination. This is shown in Fig.5 for a 20 nm wide GaAs quantum wire. The $n=1$ emission exhibits a decay time of about 360 ps, independent of the pump power. On the contrary, the

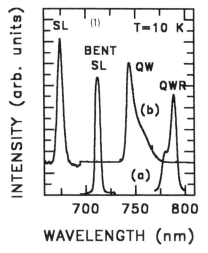

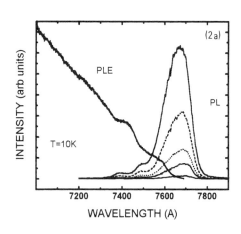

Fig.1 - Photoluminescence (PL) mapping of a 20nm GaAs quantum wires sample . (a) PL spectrum collected from the patterned region ; (b) PL spectrum collected from the unpatterned region.

Fig.2 - a) Photoluminescence excitation (PLE) and intensity dependent photoluminescence (PL) of a 16nm GaAs quantum wire sample . The excitation intensities for the PL spectra (starting from the bottom) are : $640mW/cm^2$, $6.4W/cm^2$, $64W/cm^2$, $127W/cm^2$ , $300W/cm^2$.

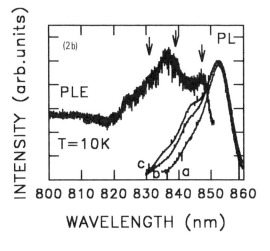

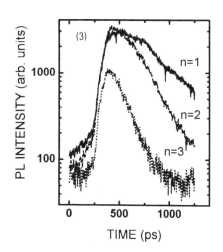

Fig.2b) PLE and intensity dependent PL spectra for a 16nm InGaAs quantum wire sample : (a) $1.5W/cm^2$, (b) $7.5W/cm^2$, (c)$15W/cm^2$. The arrows indicate calculated transition energies.

Fig.3 - Temporal evolution of n=1 ,n=2, and n=3 transitions of the GaAs sample of Fig.2a).

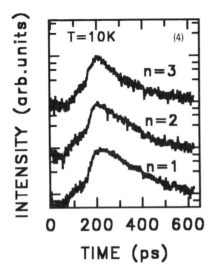
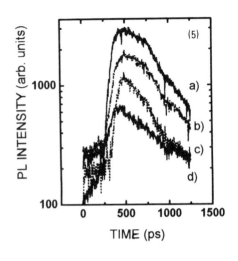

Fig.4 - Temporal evolution of n=1, n=2 and n=3 transitions of the InGaAs sample of Fig.2b). The exciting wavelength was set to 750 nm and the exciting power is $2.2kW/cm^2$.

Fig.5 - Temporal evolution of the n=1 and SL barrier recombinations for the GaAs wire of Fig.2a) at two different excitation intensities. a) n=1, $I=3.5W/cm^2$; b) n=1 , $I=175W/cm^2$; c) SL barrier , $I=3.5W/cm^2$ ; d) SL barrier , $I=175W/cm^2$.

superlattice barrier exhibits a shortening of both the decay and rise time with increasing power density, indicating the saturation of the localization process. Nevertheless, the superlattice decay time is found to be always faster than the quantum wire recombination, suggesting that the transfer efficiency is not bad. These results are apparently in contrast with those obtained by other authors [6,7]. However, this discrepancy is found to be strongly dependent on sample quality. In fact we have performed analogous experiments in samples with disordered superlattice barriers (non uniform well and barrier thickness) and eventually with evident stacking faults across the wire region. In this case we have found that the barrier decay is always longer or at best comparable to the quantum wire recombination, indicating the expected poor charge transfer, supporting the superlattice localization interpretation [7].

This work was partially supported by the HCM network of the European Community "Ultrafast" and by the Esprit project Nanopt.

### References

[1] R.Cingolani and R.Rinaldi, Rivista del Nuovo Cimento, **9**, 1-83 (1993)

[2] E.Kapon, Semiconductor and Semimetals, **40**, 259-335 (1994)

[3] R.Rinaldi et al. Phys. Rev. Lett. **73**, 2899 (1994)

[4] E.Molinari et al. Proc. Int.Conf. Physics Semiconductors, Vol.2, 1707 ( 1994), World Scientific. Publ. Ed. D.J.Lockwood

[5] A.Maciel et al. Semicond. Sci. Technol. **9**, 893 (1994)

[6] S.Haacke et al., Proc. 7th. Int.Conf. Modulated Semicond. Structures, Vol.1, 486 (1995)

[7] C.Kiener et al., Proc. 7th. Int.Conf. Modulated Semicond. Structures, Vol.1, 423 (1995)

# HOT CARRIER RELAXATION AND SPECTRAL HOLE BURNING IN QUANTUM WIRE AND QUANTUM WELL LASER STRUCTURES

I. Vurgaftman[†], Y. Lam[††], and J. Singh

Department of Electrical Engineering and Computer Science,
The University of Michigan, Ann Arbor, MI 48109
[†]Presently with Naval Research Laboratory, Washington, DC 20375
[††]Presently with Nanyang Technological University, Singapore 2263

The differential gain in quantum wires can be considerably greater than in quantum wells since the peak in the density of states at the band edge provides a much higher gain per injected carrier. In semiconductor lasers, however, the carriers are injected not at the band edge, but from the barriers at higher energies. If the carriers are continuously taken out by radiative recombination at the lasing wavelength, the rate at which they are replenished must balance the recombination rate. Otherwise, the occupation probability at the lasing wavelength is reduced with a concomitant drop in the gain. Also, when the laser is modulated, the carriers must be able to reach the band edge states in a time sufficient for an optical response characteristic of high injection. Therefore, carrier thermalization is of paramount importance in laser structures.

In recent years, it has become clear, however, that the description of the laser dynamics in terms of the intraband relaxation times is insufficiently accurate. The details of the thermalization process in a quantum structure are important as well. For example, if electron-electron scattering is quite strong, as in quantum wells, a spectral hole in the gain spectrum is unlikely to develop. Instead, the carrier temperature rises, and the entire gain spectrum saturates homogeneously. On the other hand, in situations in which carrier-carrier scattering is absent or weak, as in quantum wires, a spectral hole in the distribution function may develop depending on the bias level, injection conditions and the choice of the potential profile. In this paper, the effects of hot carrier relaxation on the operation of quantum wire and well lasers are examined and compared. The results summarized here have been obtained with Monte Carlo simulations tailored to quantum wells and quantum wires and have been presented in detail elsewhere[1,2,3].

The effect of carrier thermalization on the gain spectrum can be evaluated by a combination of macroscopic calculations based on the Fermi golden rule and electron-photon rate equations and a microscopic simulation of the carrier dynamics. When a photon density of $1.7 \times 10^{11}$ cm$^2$ is introduced into a 50ÅGaAs/AlGaAs quantum well, the zero-photon gain spectrum saturates noticeably yet no spectral hole develops neither in the distribution function nor in the gain profile (see Fig. 1). Curves A and B are used in calculating the nonlinear gain parameter by approximating the peak gain derivative with respect to the photon density[3]. Curve C demonstrates the difference between the gain spectrum obtained with the quasi-Fermi distribution producing the same peak gain (and, hence, the same optical response) and the real distribution function, which is primarily in the high-energy tail. The numerical value of the nonlinear gain parameter is $1.1 \times 10^{-17} \text{cm}^3$ leading to a relaxation-limited modulation bandwidth of $f_{max} = 78$ GHz for a cavity with a 1 ps photon lifetime.

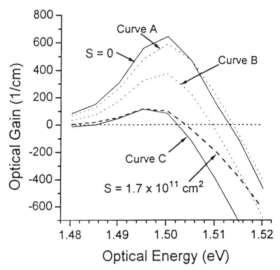

**Figure 1.** The optical gain spectra for the two steady-state solutions in the gain saturation simulation. Also shown are the optical gain spectra for the two steady-state solutions calculated using the quasi-Fermi function. Curves A and B are for the Fermi-based initial and final states respectively. Curve C, based on the quasi-Fermi distribution has the same peak gain as the dashed curve.

The gain spectrum saturation is qualitatively different in quantum wires. In structures with few subbands, electron-electron scattering is extremely inefficient. The only mechanisms capable of randomizing the electron ensemble are acoustic phonon scattering and electron-hole scattering. Even if acoustic phonon scattering is treated exactly with the possibility of inelastic events, the spectral hole cannot be filled up on the femtosecond time scale, as in quantum wells. For carriers injected at the barrier edges, the thermalization time for carriers increases considerably provided the band edge is off resonance with the respect to the conduction band offset. The latter condition is satisfied in quantum wires with a small cross section, which are particularly useful for laser applications. The average energy of the electron ensemble obtained by a Monte Carlo simulation is shown in Fig. 2. Considering the peculiarities of the carrier relaxaton process in quantum wires, the following model can be used to represent the nonlinear gain effect. A Boltzmann equation including optical phonon scattering only can be solved subject to the assumption of monoenergetic optical phonons. Next, the rest of scattering processes can be lumped together in a spectrally-varying broadening parameter. Once the broadening is performed, the distribution function to be used in determining the optical gain spectrum can be obtained. Spectral hole burning in quantum wire lasers can be exaggerated by adopting a carrier extraction time constant of 10 ps, corresponding to the far-above-threshold operation. The carrier distribution function for this case is shown in Fig. 3 along with the quasi-Fermi distribution function. The effect of slow carrier thermalization is seen to be twofold: i) a spectral hole is burned around the lasing wavelength, ii) the temperature of the electron gas is raised as made apparent from the high-energy tail of the function.

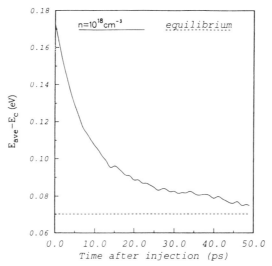

**Figure 2.** The time evolution of the mean electron energy in a $100 \times 100$ Å GaAs/AlGaAs quantum wire including the effects of electron-hole scattering.

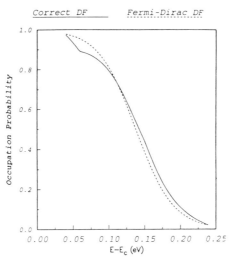

**Figure 3.** The distribution function in the presence of spectral hole burning as compared with the equilibrium quasi-Fermi distribution for an extraction time of 10 ps.

## REFERENCES

1. I. Vurgaftman, Y. Lam and J. Singh, *Phys. Rev. B* 50:14309 (1994).
2. I. Vurgaftman and J. Singh, *J. Appl. Phys.* 74:6451 (1993).
3. Y. Lam, and J. Singh, *IEEE J. Quant. Electron.* 30:2435 (1994).

# REDUCED CARRIER COOLING IN GAAS V-GROOVE QUANTUM WIRES DUE TO NON-EQUILIBRIUM PHONON POPULATION

J.M. Freyland,[1] K. Turner,[1] C. Kiener,[1] L. Rota,[1] A.C.Maciel[1]
J.F. Ryan,[1] U. Marti,[2] D. Martin,[2] F. Morier-Gemoud,[2] F.K. Reinhart[2]

[1]Clarendon Laboratory
Parks Road, Oxford OX1 3PU, U.K.

[2]Ecole Polytechnique Federale de Lausanne
CH-1015 Lausanne, Switzerland

## INTRODUCTION

Hot carrier dynamics in semiconductor quantum wires is the subject of much current interest, since unique characteristics of inter- and intra-subband carrier-carrier (CC) and carrier-phonon (CP) interactions in quasi-one-dimensional (1D) systems have been predicted [1]. Previous studies of carrier dynamics in etched rectangular GaAs/AlGaAs wires using time-resolved photoluminescence (PL) have reported that relaxation is slower than that measured in similar 2D systems [2,3]. Hot phonon reabsorption was suggested as a possible cause. In these experiments, however, both the confining barriers and the wires were strongly excited, and so the measured relaxation times are influenced by carrier capture into the wires. V-groove structures present an alternative method of realising quantum wires, and high optical quality has been obtained, as demonstrated by their application in efficient, low-threshold lasers [4]. Here we present picosecond time-resolved PL measurements of carrier relaxation in GaAs v-groove quantum wires which were directly photoexcited at relatively low density with respect to previous experiments. We compare our measurements with the results of a hybrid Monte Carlo simulation which allows us to evaluate the important energy relaxation processes.

## EXPERIMENTAL RESULTS

We have investigated samples grown by molecular beam epitaxy on non-planar GaAs substrates (see ref. [5] for a description of the sample). Time integrated luminescence[6] from our samples shows distinct 1D subbands with a splitting between the first two energy levels of $\sim 15$ meV. Photoluminescence was excited using subpicosecond pulses from a modelocked Ti:sapphire laser tuned to 1.630 eV. At this photon energy electrons are excited directly into the 1D states, well-below the confining potential barrier. The luminescence was analysed using a subtractive-dispersion double monochromator, and detected using a synchronous streak camera, with overall

*Hot Carriers in Semiconductors*
Edited by K. Hess *et al.*, Plenum Press, New York, 1996

temporal and spectral resolution of 20 ps and 1 nm respectively. The carrier density estimated from the experimental parameters was $5 \times 10^6$ cm$^{-1}$. Time-resolved measurements of the PL at different energies are shown in Fig. 1. At high energy, 55 meV above the $n = 1$ line (a), the risetime of the signal is within the instrumental resolution. At lower energies both the rise and decay times get systematically longer, consistent with hot carrier relaxation. At the $n = 1$ peak (1.57 eV) we observe a risetime on the order of 100 ps; this value is similar to that observed in 2D systems, and is a measure of the cooling of the electron-hole plasma.

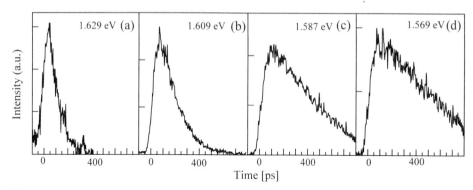

**Figure 1.** Time-resolved photoluminescence data from the GaAs V-groove quantum wire structure at four different energies

The time-dependent electron-hole plasma temperature has been obtained from the spectra by fitting the lineshapes using a multi-subband 1D DOS and a hot Fermi-Dirac distribution assuming a phenomenological broadening of 5 meV, which is consistent with the observed PL linewidth. The results are presented in Fig. 2(a). We find that at the earliest time resolvable in our experiment (20 ps) the carrier temperature $T_0$ is 110 K, and this subsequently drops to $\sim 60$ K after 300 ps. Fig. 2(b) shows the results of a similar experiment in another sample where the density is estimated to be $\sim 2.5 \times 10^6$ cm$^{-1}$. In this sample the inhomogeneous broadening is 8 meV, and a very similar relaxation is observed. These results appear to be similar to those obtained in measurements on 2D systems, though a direct comparison is not possible. We know from previous measurements of 2D systems under similar excitation conditions that hot phonon effects are important and reduce the energy relaxation rate.[7] This comparison suggests that hot phonons are important also in the 1D case; however, such a comparison is only qualitative, and a more detailed evaluation is required.

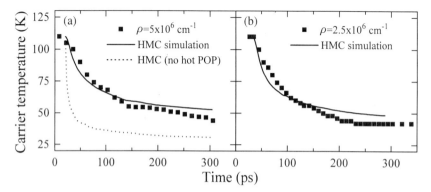

**Figure 2.** Time evolution of the carrier temperature for two different samples and carrier densities. The experimental data is represented by squares.

## MONTE CARLO SIMULATION AND DISCUSSION

A simple Monte Carlo (MC) approach was not suitable for the simulation of these experiments due to the extension to relatively long times. We have then adopted a hybrid Monte Carlo (HMC) method to calculate the time-dependent carrier temperature. Our approach is based on a multisubband electronic band structure for the V-groove quantum wire potential[6] and contains the usual MC algorithms for the treatment of the CP interaction including nonequilibrium phonons and degeneracy effects. As the first experimental point is available only at relatively late times ($\sim$ 20 ps) when the carrier distribution function is already thermalized to a hot Fermi function, we used a simplified approach for the treatment of CC scattering. This interaction is particularly important in these high density and low temperature conditions because it allows a relatively fast cooling of the carriers to the lattice temperature by pushing them above the phonon emission threshold. CC scattering is treated by imposing electron thermalisation to a Fermi distribution with the same total energy after a characteristic time $\tau_{cc}$. By running a full ensamble MC simulation in equivalent condition we found a thermalization time of $\sim$ 2 ps; this time is relatively long due both to the reduced efficiency of CC scattering in quantum wires and to the extremely degenerate conditions. The starting conditions for the HMC are then a hot Fermi distribution with temperature $T_0$ taken from the experimental investigation and a phonon distribution at the same temperature. Only bulk phonons are considered, which is expected to be reasonably accurate for wires with dimensions $\geq$ 10 nm.

The results of the HMC simulations are given by the solid curves in Fig. 2. For both samples and densities the agreement with the experimental values is excellent. The dotted curve in Fig. 2(a) corresponds to the case when phonon reabsorption is switched off; the cooling is much faster, in sharp contrast with the experiment. The difference between the two HMC curves shows the extent of hot phonon effects. These results give the best evidence to date that nonequilibrium phonon effects are significant in carrier dynamics in quantum wires.

We wish to acknowledge support from the European Commission under the "UL-TRAFAST", "NANOPT" and "HCM" programs as well as from the Austrian Science Foundation (project No P10048).

## REFERENCES

1. U. Bockelmann and G. Bastard, Phys. Rev. **B42** 8947 (1990).
2. G. Mayer, B.E. Maile, R. Germann, A. Forchel, P.Grambow and H.P. Meier, Appl. Phys. Lett. **56** 2016 (1990).
3. R. Cingolani, H. Lage, H. Kalt, L. Tapfer, D. Heitmann and K. Ploog, Phys. Rev. Lett. **67**, 891 (1991); Semicond. Sci. and Technol. **7** B287 (1992).
4. E. Kapon, D.M. Hwang and R. Bhat, Phys. Rev. Lett. **63** , 430 (1989).
5. A.C. Maciel, C. Kiener, L. Rota, J.F. Ryan, U.Marti, D. Martin, F.Mourier-Gemoud and F.K. Reinhart, Appl. Phys. Lett. **66**, 3039 (1995).
6. R. Rinaldi, R. Cingolani, F. Rossi, L. Rota, M. Ferrara, P. Lugli, E. Molinari, U. Marti, D. Martin, F. Morier-Gemoud and F.K. Reinhart, Proceedings of 20th Int. Symp. on GaAs and Related Compounds, Eds. H.S. Ruprecht and G. Weimann, IOP Conf. Series 136 233 (1994).
7. J.F. Ryan, M. Tatham D.J. Westland, C.T. Foxon, M.D. Scott and W.I. Wang, Proc. SPIE 942, 256 (1988).

# CARRIER TRAPPING INTO QUANTUM WIRES

K. Turner[1], J.M. Freyland[1], A.C. Maciel[1], C. Kiener[1], L. Rota[1],
J.F. Ryan[1], U. Marti[2], D. Martin[2], F. Morier-Gemoud[2], and
F.K. Reinhart[2]

[1]Clarendon Laboratory
Parks Road, Oxford OX1 3PU, U.K.

[2]Ecole Polytechnique Federale de Lausanne
CH-1015 Lausanne, Switzerland

## INTRODUCTION

Trapping into the active region is one of the most important design considerations in quantum well lasers. This process becomes crucial for the operation of
quantum wire devices since the very small active volume demands strong coupling
to the external region, otherwise trapping is greatly inhibited. We report the first
measurements of electron trapping from extended three-dimensional (3D) states into
one-dimensional (1D) states of a semiconductor quantum wire structure. Our joint
theoretical and experimental investigation shows that the trapping time is extremely
fast, ~10 ps.

We have investigated GaAs quantum wires which are formed by molecular beam
epitaxial growth in v-shaped channels etched on (001) GaAs substrates. An array
of grooves parallel to the ($1\bar{1}0$) crystallographic axis was produced by holographic
photolithography, each 250 nm wide and 110 nm deep. A 5nm GaAs quantum well
was embedded within $(GaAs)_8(AlAs)_4$ superlattice (SL) barriers, growing in a dis-
tinctive crescent shape[1] (see Fig. 1). The layer thickness at the bottom of the groove
was measured by transmission electron microscopy (TEM) to be 9.3nm; it decreases
rapidly with distance away from the centre of the groove to a value of 2.2nm where it
merges with the superlattice barriers. The combined effects of layer bending and narrowing give rise to a lateral confining potential which produces distinct 1D subbands[2]
with energy separations of ~15 meV. As there are no lateral or side-wall quantum
wells (c.f. 1), this structure has a relatively simple band structure which permits the
study of trapping directly from 3D extended states into 1D confined states.

## THEORETICAL MODEL

Since the localisation process depends critically on the shape of the potential and the resulting electron states, we first determined accurate energy levels

and wavefunctions for the quantum wire and the SL barrier. The full 2D cross-sectional potential profile was obtained from high-resolution TEM images of the v-groove structure. The 2D Schrödinger equation was then solved numerically using a plane-wave expansion[3]. Three distinct types of state are found: (i) 1D confined quantum wire (QWR) states, (ii) bulk-like extended superlattice (ESL) states, and (iii) "localised" superlattice (LSL) states. Fig. 1 shows the v-groove structure obtained from the TEM data, together with contour plots of electron probability density for states in each of these categories. The QWR state shown in Fig. 1(a), with quantum numbers $|n_z = 1, n_y = 7\rangle$, is strongly localised in the wire, which is typical of all states within the $|n_z = 1\rangle$ manifold. The ESL states, on the other hand, extend throughout the barrier region (Fig. 1(b)).

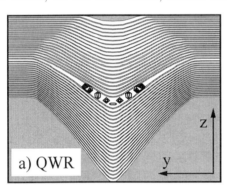

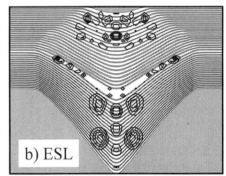

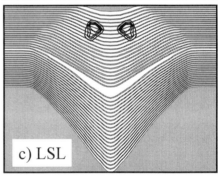

**Figure 1.** Potential profile of the GaAs v-groove structure with $(GaAs)_8 (AlAs)_4$ SL barriers obtained from TEM data (AlAs layers are shaded), together with contour plots of electronic probability density for: (a) the $|1,7\rangle$ QWR state, (b) a typical ESL state, (c) the lowest energy LSL state.

An important feature of the SL barriers, which is clearly evident in the TEM data, is that the layer thicknesses show systematic spatial variations: the layers below the wire are generally narrower than those above, and furthermore, there is a thickening of the GaAs layers near the centre of the groove, whereas the AlAs layers are more closely uniform. This leads to a lower average potential in the central region of the structure, which gives rise to "localised" states in the barrier which show up quite clearly in the band structure calculation. These states are "localised" in the $[1\bar{1}0]$ plane and spatially separated from the QWR region, but are extended in the $(1\bar{1}0)$ direction. Fig. 1(c) shows the lowest energy LSL state. The number of such states is dependent on the quality of the sample, and can be reduced by optimising the growth conditions.

A qualitative indication of the likely strength of the trapping process can be obtained from the spatial overlap of ESL and QWR states. We find that the overlap is relatively large for low-energy ESL states. However, the ESL states also overlap strongly with LSL states, which in turn have practically no overlap with the QWR states. Consequently, electrons which are localised into LSL states are unlikely to be trapped into the wires within their recombination time. Using these energy levels and wavefunctions we have computed scattering rates for longitudinal optical phonon emission assuming a bulk phonon approximation.

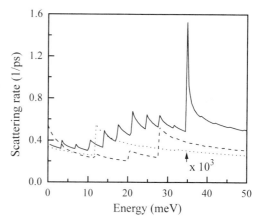

**Figure 2.** Calculated trapping rates for electron-LO phonon emission. The solid line represents capture from the ESL state into LSL states, and the dashed line capture into QWR states. The dotted line represents capture from the LSL state into QWR states.

Fig. 2 shows the rates as a function of initial state energy. Both ESL → QWR and ESL → LSL rates are on the order of 0.3ps$^{-1}$, whereas the LSL → QWR rate is three orders of magnitude lower. The sharp peaks arise from singularities in the ideal 1D density of states, but in reality they will be smoothed out by inhomogeneous broadening.

## EXPERIMENTAL RESULTS

Experimental evidence supporting both the band structure analysis, and the implications for carrier trapping, has been obtained using time-resolved photoluminescence (PL) spectroscopy. Dye laser pulses of 5 ps duration were used to photoexcite the v-groove structure, and time-resolved PL spectra were measured using a streak camera with ~30 ps resolution. Time-resolved PL data are presented in Fig. 3. The $|1,1\rangle$ QWR signal in Fig. 3(a) shows a rise time of ~150 ps, and a decay time of ~400 ps. The latter is similar to the lifetime measured in quantum wells, and underlines the high quality of the v-groove wires. The risetime is due to relaxation from higher-energy wire states[2]. The inset to Fig. 3(a) shows the QWR signal measured at much higher energy where intra-wire relaxation effects are less pronounced. The risetime is now within the resolution of the detection system, which directly confirms that trapping into the wires is rapid, occurring within 30 ps.

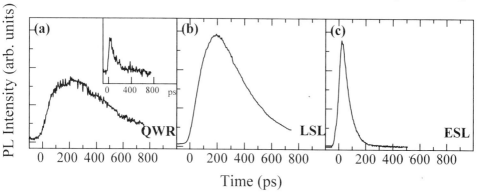

**Figure 3.** Time-resolved photoluminescence from: (a) the $|1,1\rangle$ QWR state, (b) LSL states, and (c) ESL states.

Fig. 3(b) shows the time-resolved LSL luminescence measured under the same experimental conditions as in Fig. 3(a). The risetime is ~150ps, which is similar to the $|1,1\rangle$ QWR behaviour, and is characteristic of energy relaxation within the LSL states. The decay time of ~400 ps is again typical of radiative recombination; however, it is significantly longer than the risetime of the QWR signal (Fig. 3(a) inset), indicating that trapping from the LSL into the QWR is not a significant effect. In contrast to this behaviour, Fig. 3(c) shows that the decay time of the ESL signal is extremely fast confirming that scattering out of the extended barrier states is very efficient. After deconvoluting the streak camera response we obtain a decay time of ~ 30ps. Although this value seems to be relatively larger than that obtained from the theory we must consider that the experiment was performed at high carrier density ($\sim 10^7 \text{cm}^{-1}$) where the effects of carrier-carrier scattering, phonon reabsorption, and final state filling reduce both the trapping and cooling rates[2].

Photoluminescence is sensitive to both electron and hole dynamics. Because of their larger effective mass, holes are likely to be trapped more rapidly than electrons, so that the decay of barrier photoluminescence is determined by hole trapping, whereas the risetime of wire luminescence is characteristic of electron trapping[4]. The data presented in Fig. 3 are unable to resolve any difference in these two times, but nevertheless they confirm the basic theoretical model described above.

It is important to compare these results with the trapping rates of electrons from bulk states into isolated quantum wells, i.e. 3D →2D trapping. Electron capture times in the range of 25ps were obtained in calculations for a separate confinement heterostructure with a 10 nm quantum well within a 100nm barrier[5]. The relatively long times arise from weak overlap of initial and final states. The use of barriers with an intentionally graded bandgap, which increases with distance from the quantum well, dramatically reduces the trapping time to ~1ps, due to the lowest energy states of the barrier being forced to overlap strongly with the confined states. Our results show that this mechanism is intrinsic to the v-groove quantum wire structure: thickening of both the GaAs quantum wire and SL layers near the centre of the v-groove forces the low energy ESL barrier states to overlap strongly in the lateral direction with the confined states of the wire. The trapping rate we obtain for the quantum wire, ~0.3 ps$^{-1}$, is close to that predicted for optimised graded-barrier quantum well lasers[5].

In conclusion, we have reported experimental and theoretical studies of carrier trapping in GaAs v-groove quantum wires. We find that trapping from extended superlattice states into the wires is extremely efficient, and is explained by the special nature of the self-organised growth process which results in strong overlap of initial and final states.

We wish to acknowledge financial support from EPSRC (UK), and the E.U. through the NANOPT and ULTRAFAST HCM programmes, and the Austrian Science Foundation.

# REFERENCES

1. E. Kapon, D.M. Hwang and R. Bhat, Phys. Rev. Lett. **63** , 430 (1989).
2. A.C. Maciel, C. Kiener, L. Rota, J.F. Ryan, U. Marti, D. Martin, F. Morier-Gemoud, F.K. Reinhart, Appl. Phys. Lett. **66**, 3039 (1995).
3. L. Rota, F. Rossi, P. Lugli and E. Molinari, Proc. Int. Workshop on Computational Electronics, University of Leeds, 236 (1993).
4. B. Deveaud,, J. Shah, T.C. Damen and W.T. Tsang, Appl. Phys. Lett. **52**, 1886 (1988).
5. P.W.M. Blom, C. Smit, J.E.M. Haverkort and J.H. Wolter, Phys. Rev. **B47**, 2027 (1992).

# ANOMALOUS BLUE-SHIFT IN PHOTOLUMINESCENCE FROM STRAINED InAs QUANTUM DOTS FABRICATED BY MBE

Kanji Yoh, Hayato Takeuchi* and Toshiya Saitoh

Research Center for Interface Quantum Electronics, Hokkaido University,
North 13, West 8, Sapporo 060, Japan

## INTRODUCTION

Recently, several fabrication techniques of quantum dots have been proposed such as selective growth technique[1,2], selective etching thechnique[3] and self-assembled formation technique[4-6]. However, effect of strain on quantum dots in lattice mismatched system has not so far been discussed, despite the importance of strained InAs heterostructure system for its potential high temperature operation of quantum effects [7]. This paper shows the photoluminescence (PL) and cathode luminescence(CL) measurement results of strained planar InAs quantum well grown by MBE and InAs quantum dots fabricated by wet chemical etching.

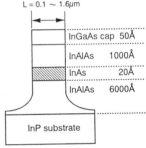

Figure 1. Schematic diagram of InAs quantum dot structure.

## PL RESULTS OF InAs QUANTUM DOTS

InAs dots based on InAs/ $In_{0.52}Al_{0.48}As$ hetero-structure was formed by wet chemical etching on InAs quantum well structures grown on InP by molecular beam epitaxy (MBE). Then, the column-shaped mesa structure are fabricated by the conventional optical lithography and the wet chemical etching as illustrated in Figure 1. Clear blue-shifts of photoluminescence of 22meV, 14meV, 3.5meV from InAs dots with the nominal lateral size of 2000Å, 4000Å and 1.6μm, respectively, can be seen in Figure 2. Energy shifts of these InAs dots with 20Å well together with 40Å well samples are plotted against lateral size in Figure 3. Measured energy shifts are always higher than the estimated value as drawn in the figure marked by 0%, which was calculated taking acount of lateral confinement effect only. Other solid curves with marks of 0.5%, 1.0%,..etc. indicate calculations of energy shifts assuming additional energy shift due to strain difference (maximum 3.2% in the present material system). This result indicate possible interplay of the inhomogeneous residual strain distribution in both planar and dot samples to cause the anomalous blue-shift of the PL peaks.

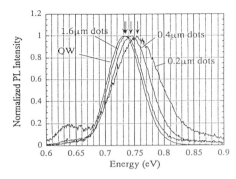

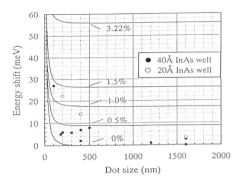

Figure 2. PL spectra of InAs quantum dots with the nominal lateral size of 2000, 4000 Å and 1.6μm. Clear blue-shifts of 22meV, 14meV, 3.5meV , respectively, can be seen.

Figure 3. PL peak energy shift versus dot size. Measured energy shifts are always higher than the estimated value drawn in solid curve and marked by 0%.

## AFM AND CL MEASUREMENT RESULTS ON InAs/InAlAs HETEROSTRUCTURE

In order to verify the inhomogeneous strain distribution nature of the strained InAs heterostructure surface morphology of the planar sample was investigated by AFM and the CL intensity distribution from the planar sample was measured. InAs well thickness of the planar samples before dot fabrication was 20Å - 40Å, since the critical thickness of the InAs in $In_{0.52}Al_{0.48}As/In_{0.53}Ga_{0.47}As$ system is calculated to be 45Å using Matthews-Blakeslee model[8].

Notion of uniform strain and uniform growth on lattice mismatched system, as Matthews and Blakeslee's model describes, up to the critical thickness have been widely accepted[9]. On the other hand, microscopically, inhomogeneous 'periodic' strain distribution was known and reported by Yao et al [10] on InGaAs/GaAs system for the first time by TEM analysis and associated 'periodic' nonuniform growth of the strained layer were reported on InGaAs[11] and other material systems[12].Figure 4 shows surface morphology by AFM(Fig.4(a) ) and CL image (Fig.4(b)) of the InAs/AlInAs planar quantum well with well thickness of 20Å. Even though the well thickness of 20Å is less than a half of the critical thickness (≈45Å), 'periodic' inhomogegeous growth is taken place during the overgrowth of 1000Å of AlInAs (Cf. Figure 1) resulting in 'periodic' inhomogeneous surface with an amplitude of ≈90Å. Note the similarity of the CL pattern image with the nonuniform thickness pattern. More intense luminescence is expected from less strained part of the InAs well where bandgap energy is the lowest.

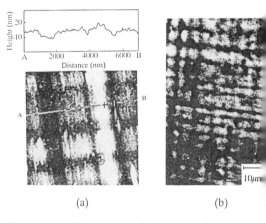

Figure 4. (a) AFM image and (b) CL image of InAs/$In_{0.52}Al_{0.48}$As heterostructure with 20Å quantum well

## CL MEASUREMENT RESULTS ON InAs/InAlAs QUANTUM DOTS

Figure 5 shows the CL image and CL spectra from each sigle dot of 1.6μm InAs dot. Regular array of CL image (Fig.5(a)) confirms that the excitons are spread over each dot, indicating no additional confinement exist. It can also be seen that the CL spectra from each single dot can be divided into two groups(Fig.5(b)). The peak energy and the line shape of the first group with lower peak energy are similar to that of PL and CL spectra of planar InAs/AlInAs sample. The second group, on the other hand, has wider energy range with higher peak energy, indicating the existance of InAs dots which retrain higher strain than average. These highly strained dots seem to contribute to increase the average peak energy of the PL intensity much higher than expected from quantum size effect. Figure 6 shows SEM picture and CL image of fixed photon energy from array of InAs dots with 30Å well. The InAs dots with the same luminescence energy form groups of stripes suggesting inhomogenious strain distribution of the quantum well.

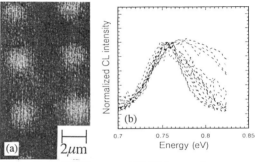

Figure 5. (a) AFM image and (b) CL image of InAs/In$_{0.52}$Al$_{0.48}$As heterostructure with 20Å quantum well.

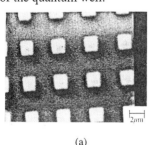 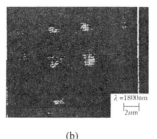 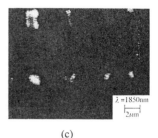

(a)                    (b)                    (c)

Fig.6 SEM picture and CL image of fixed photon energy from array of InAs dots with 30Å well. (a) SEM image (b) CL image of 1800nm wave length (c) CL image at 1850nm wave length

## CONCLUSION

Anomalous blue-shifts in photoluminescence (PL) was observed from patterned InAs dots based on InAs/InAlAs hetero-structure. Inhomogeneous 'periodic' strain in InAs/InAlAs hetero-structure were confirmed by Atomic Force Microscope(AFM) and cathode luminescence (CL) image. CL spectra from individual InAs dots suggested that the anomalous blue-shift originates from contributions of highly strained quantum dots which happened to be fabricated from highly strained area of inhomogeniously strained planar hetrostructure.

### Acknowledgement
The authors would like to thank Mr. A. Tanimura for technical assistance. This work was supported by a Grant-in-Aid for Scientific Research from the Ministry of Education, Science and Culture.

*Present address: SONY Corporation, Asahi-cho 4-14-1, Atsugi-shi, Kanagawa-ken 243 Japan

### References
1. T.Fukui and S.Ando, Superlattices and Microstructures, 2, 266 (1992).
2. Y.Arakawa, Solid-State Electron. 37, 523 (1994)
3. A.Schmidt et.al., Solid-State Electron. 37,1101 (1994)
4. J.M.Moison et al, Appl. Phys. Lett., 64, 196 (1993)
5. K. Nishi et.al., Proc. of 7th Int. Conf. of IPRM, 759 (1995)
6. D.Leonard et al, Appl.Phys.Lett., 63, 3203 (1993)
7. K.Yoh, M.Nishida and M.Inoue, Solid-State Electronics, Vol.37, pp.555-558 (1994)
8. J.W.Matthews and A.E.Blakeslee, J.Crystal Growth, 27, 118 (1974)
9. T.G.Andersson et al, Appl.Phys.Lett., 51, 752 (1987)
10. S.Guha et al, Appl.Phys.Lett., 57, 2110 (1990)
11. A.G.Cullis, J.Crystal Growth, 123, 333 (1992)

4. High Field Transport and Impact Ionization

# IMPACT IONIZATION IN SUBMICRON
# AND SUB-0.1 MICRON Si-MOSFETs

Nobuyuki Sano, Masaaki Tomizawa, and Akira Yoshii

NTT LSI Laboratories
3-1 Morinosato Wakamiya
Atsugi-shi, Kanagawa  243-01
Japan

## INTRODUCTION

Correct knowledge of impact ionization in semiconductors is an essential ingredient for precise analyses and predictions of device characteristics such as reliability and degradation. In particular, as the device size shrinks into deep-submicron regime, energy dependence of the impact ionization rate in low energy regions (below 3 eV) plays a much more important role than the cases in bulk.[1] This is because, in addition to an obvious reason that such ultra-small devices are to be operated at reduced applied voltages, the so-called nonlocal effects are so pronounced inside devices that many impact ionization events are actually induced by the electrons with energy below 3 eV (near threshold energy). Therefore, the ionization rate near threshold is of crucial importance for the investigations of hot carrier effects in ultra-small devices.

In spite of its importance, large discrepancy in the energy-dependence of the ionization rate in Si has been existed in the past.[2] This is mainly due to the fact that meaningful studies of impact ionization are possible only if the *correct* ionization rate is coupled with the *correct* transport analyses, *i.e*, when the ionization rate modeled empirically via, say, the Keldysh formula is incorporated in the Boltzmann transport equation, the realistic band structure of semiconductor has to be taken into account in both the ionization rate and the Boltzmann transport equation. The Full-Band Monte Carlo (FBMC) method currently provides the only way to solve this subtle problem with a satisfactory accuracy and is employed by various groups.[3-5] As a result, the discrepancy which existed in literature of the ionization rates among various theoretical results has now been greatly narrowed down. However, there still existed noticeable discrepancy among the first-principles calculations and the recent experimental results.[6] In fact, the discrepancy is even greater in low energy regions where its energy-dependence is of most importance for device analyses.

In the present paper, we give a possible explanation of the discrepancy between various ionization rates near threshold of Si. The present ionization rate is incorporated into a FBMC device simulator to investigate the impact ionization processes in submicron and sub-0.1 micron n-channel Si-MOSFET's. We shall point out some interesting features, which are not obvious from the arguments based on the conjectures from bulk properties, but rather peculiar under device structures.

*Hot Carriers in Semiconductors*
Edited by K. Hess *et al.*, Plenum Press, New York, 1996

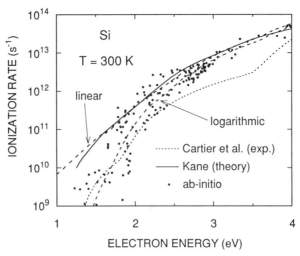

**Figure 1.** Wave-vector-dependent *ab-initio* ionization rates as a function of electron energy (solid circles). The dashed lines represent the energy-dependent ionization rates obtained by linear and logarithmic interpolation from the present *ab-initio* ionization rates, whereas the solid and dotted lines represent the ionization rates calculated by Kane and extracted from recent experiments, respectively.

## IMPACT IONIZATION RATE

The first-principles impact ionization rate (denoted, hereafter, as *ab-initio* ionization rate) is calculated under the Born approximation in which the electron-electron Coulomb interaction, along with the dielectric function for screening, is treated as a perturbation.[7] A main problem, except for a few quantum mechanical problems such as collisional broadening and intracollisional field effect, to evaluate *ab-initio* ionization rates, therefore, resides in how to reduce intensive numerical calculations required, rather than in its physical principle. We have fully employed the symmetry transformations for both the wave-functions and band structure of Si to reduce the memory and to enhance the calculation efficiency.[8]

*Ab-initio* impact ionization rates thus calculated are given by electron wave-vector (**k**-vecor) and shown in Fig. 1 as a function of electron energy. The intrinsic anisotropy of the ionization rate is manifested by large spreading of the ionization rates and this anisotropy is especially strong in low energy regions (below 3 eV), which are of most importance for device applications. Anisotropy is due to the strong restriction of both the energy and momentum conservation during ionization transitions. Above 3 eV, the number of final states available for the secondary and recoil electrons increases rapidly because other satellite valleys such as the L-valleys, in addition to the X-valleys, come to play. Hence, anisotropy diminishes as the primary electron energy increases. This finding also implies that impact ionization is actually dominated by the (joint-) density of states for the secondary and recoil carriers and is, in some sense, a *low-energy* process.[9] This is especially true under device structures, in which the quantization of electron gas at the drain edge forces to modify the ionization rate near threshold.[10-11]

For comparison, the previous *ab-initio* ionization rate by Kane[7], the ionization rate extracted from recent series of experiments[2], the ionization rates linearly and logarithmically interpolated from the present *ab-initio* **k**-dependent ionization rates are also shown in Fig. 1. Note that the ionization rate obtained from the linear interpolation is very close to that by Kane and both are even greater than any of the **k**-dependent *ab-initio* ionization rates at a given electron energy below 1.8 eV. This is because, when the interpolation is made with a finite mesh spacing, the energies at eight corners in a cubic mesh vary greatly, because of strong anisotropy, so that the ionization rate at some corner (**k**-point) could be extremely large. As a result, the linearly interpolated ionization rate overestimates its value in low energy regions, where intrinsic anisotropy is greatest.[6] Recall that the energy-dependent *ab-*

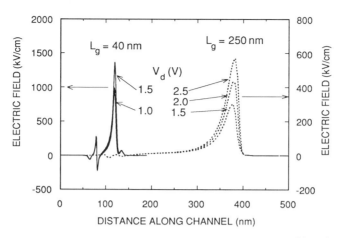

**Figure 2.** Electric field profiles as a function of distance along the channel of Si-MOSFET's with $L_g = 40$ nm (solid lines) and 250 nm (dotted lines) under various drain voltages. Notice that the scales of the electric field referred to two MOSFET's are different.

*initio* ionization rate by Kane was obtained by Monte Carlo sampling, which is equivalent to the present linear interpolation. On the other hand, the logarithmically interpolated ionization rate is much smaller near threshold and close to the rate extracted from experiments. Therefore, the discrepancy between the previous *ab-initio* and experimental ionization rates near threshold is mainly ascribed to numerical problems associated with intrinsic anisotropy of the ionization rates.

## IMPACT IONIZATION IN Si-MOSFET's

The FBMC method employed here is the standard one[3-6], and the coupling constants for the electron-phonon scatterings are chosen such that the magnitude of the scattering rates becomes similar to those from the first-principles.[5] Impact ionization is introduced by the present *ab-initio* ionization rates. We employ a typical MOSFET structure with two different gate lengths of $L_g = 250$ nm and 40 nm to represent, respectively, submicron and sub-0.1 micron devices. Figure 2 shows the electric field profiles obtained from the drift-diffusion (DD) simulator as a function of distance along the channel under various applied voltages. The gate voltage for each device is chosen such that the substrate current, estimated from the DD simulator, becomes maximum under the drain voltage of $V_d = 1.5$ V and fixed to this value for the other $V_d$. To save computation time, electron transport is simulated under the fixed electric field profiles. This is equivalent to ignoring the long-range electron-electron interaction, which along with the short-range electron-electron interaction actually affects the shape of the electron energy distributions in the drain regions, as discussed in depth by Fischetti *et al.*[10-11] The present simulations, therefore, provide only gross features of hot electron transport in devices rather than quantitative details.

### From Diffusive to Quasi-Ballistic Transport

Figure 3 shows the electron energy distributions integrated over the drain regions at 300 K and 77 K. There are quite a few features to be mentioned. The shapes of the energy distribution are very different for two devices; in 250 nm-gate MOSFET, electrons suffer several energy-dissipating scatterings before they reach the drain and, thus, the energy distribution in the regions between several hundreds meV and $qV_d$ ($q$: magnitude of electron charge) smoothly decreases. This *diffusive* transport property is also demonstrated by the fact that the number of electrons in these energy regions is much larger at 77 K because of

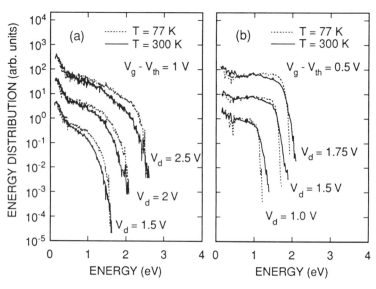

**Figure 3.** Electron energy distributions integrated over the drain regions of Si-MOSFET's with $L_g$ = (a) 250 nm and (b) 40 nm at 300 K (solid lines) and 77 K (dotted lines).

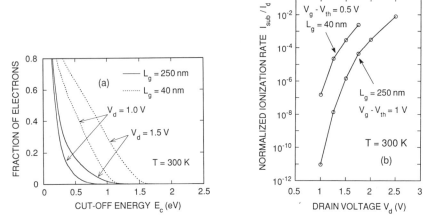

**Figure 4.** (a) Fraction of electrons with energy above the cut-off energy $E_c$ in the drain regions at 300 K for two MOSFET's with $L_g$ = 250 nm (solid lines) and 40 nm (dotted lines). (b) Normalized ionization rate $I_{sub}/I_d$ for two MOSFET's at 300 K as a function of drain voltage.

the reduced phonon scatterings. Above $qV_d$, the number of electrons rapidly decreases because the electrons gain only a limited amount of energy ($\sim qV_d$) during traveling the channel. On the other hand, in 40 nm-gate MOSFET, the energy distribution between several hundreds meV and $qV_d$ does not show significant energy-dissipation, and the contribution of ballistic electrons is much more pronounced, as shown in Fig. 3 (b). This is reasonable because the energy relaxation length is several tens of nm,[3] comparable to the present gate length. Above $qV_d$, the energy distribution is more clearly characterized by the thermal tail *approximately* equal to the lattice temperature.[12] In short, the electron energy distribution shows a clear transition from diffusive energy-dissipating transport regime in 250 nm-gate MOSFET to quasi-ballistic transport regime in 40 nm-gate MOSFET.

Such a transition of transport characteristics described above also implies that the fraction of electrons with respect to the whole electrons in the drain becomes larger as the device size decreases. Figure 4 (a) shows the number of electrons with energy above the cut-off energy $E_c$ (normalized by the total number of electrons). The number of electrons with en-

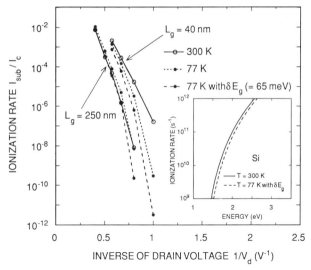

**Figure 5.** Normalized ionization rate $I_{sub}/I_d$ for two MOSFET's with $L_g$ = 250 nm and 40 nm at 300 K (solid lines) and 77 K (dotted lines). The dashed lines represent the normalized ionization rate at 77 K when the band-gap increase ($\delta E_g$ = 65 meV) is taken into account. The ionization rates per unit time at 300 K (solid line) and 77 K (dashed line) are shown as a function of electron energy to the inset.

ergy above $E_c$ decreases very rapidly in 250 nm-gate MOSFET, whereas it occupies quite a large portion in 40 nm-gate MOSFET. Therefore, as the device size shrinks, impact ionization tends to be more significant under the same $V_d$ (and under the maximum substrate current condition for $V_g$), as shown in Fig. 4 (b). In other words, even if the supply drain voltage is reduced, hot carrier effects could still exist in sub-0.1 micron devices and even more violent than in submicron devices. Finally, it should be pointed out that low-energy transport properties such as drift velocity, which are dominantly controlled by cold electrons in the channel in submicron or larger devices, could be also affected by the details of the band structure in sub-0.1 micron devices because a large portion of electrons can easily attain the energy up to $qV_d$ due to quasi-ballistic transport. Therefore, FBMC is actually indispensable even for the investigations of low-energy transport in sub-0.1 micron devices.

### Suppression of Impact Ionization at Low Temperature

Impact ionization is, in general, more significant when temperature is lowered. This is because phonon scatterings are suppressed as temperature decreases and, thus, the high energy tail of the electron energy distribution is more enhanced. According to the past experiments[13-15], however, this tendency is actually weak in devices and, in some cases (*i.e.*, under low $V_d$), even reversed (cross-over effect). We investigate this point via FBMC device simulations under the same device structures employed in the previous section. The normalized ionization rates $I_{sub}/I_d$ are calculated for two MOSFET's at 300 K and 77 K. The results are summarized in Fig. 5. The results with including the band-gap increase (*c.a.* 65 meV in Si) at 77K are also shown. Notice that impact ionization in devices is very sensitive to a slight change of the ionization rate near threshold. In bulk, such small changes of the band-gap at low temperature hardly affect ionization characteristics.

Suppression of impact ionization in devices at 77 K is clearly demonstrated in Fig. 5 and could be explained as follows. In 250 nm-gate MOSFET, the number of electrons below $qV_d$ responsible to ionizations is still larger at 77 K than that at 300 K because of diffusive transport (Fig. 3 a) and, therefore, suppression of ionizations at low $V_d$ is mainly ascribed to the band-gap increase. On the other hand, in 40 nm-gate MOSFET, the number of electrons below $qV_d$ is weakly dependent of temperature because of quasi-ballistic transport (Fig. 3 b).

Hence, the temperature dependence of $I_{sub}/I_d$ complies with that of the drift velocity in the drain because $I_{sub}/I_d$ is inversely proportional to drift velocity. Therefore, suppression of $I_{sub}/I_d$ at 77 K in 40 nm-gate MOSFET is ascribed to both quasi-ballistic transport and band-gap increase. The present FBMC results are fully consistent with experimental findings.[13-15]

## CONCLUSIONS

Large discrepancy existed in the ionization rates of Si has been greatly reduced in the past few years and the present study has further reduced the gap existed between *ab-initio* and *experimental* ionization rates near threshold. This is crucial for device analyses because the ionization process under device structures is very sensitive to the details of low energy behavior of the ionization rate, as demonstrated in this paper (and the paper by Fischetti *et al.*[11]). We have also demonstrated a clear transition of transport characteristics when the device size shrinks into sub-0.1 micron regime, which leads to various interesting effects unique under device structures such as suppression of impact ionization at low temperature.

## ACKNOWLEDGMENTS

The authors would like to thank Y. Imamura for his encouragement throughout the study. One of the authors (N. S.) would also like to thank M. V. Fischetti for discussions, especially, for his pointing out the effect of band-gap variation with respect to temperature, K. Taniguchi and M. Koyanagi for discussions about Monte Carlo simulations and experiments.

## REFERENCES

1. N. Sano, M. Tomizawa, and A. Yoshii, Nonlocality of ionization phenomena under nonuniform electric fields: a full-band monte carlo approach, *in*: "Computer Aided Innovation of New Materials II," M. Doyama, J. Kihara, M. Tanaka, and R. Yamamoto, ed., Elsevier, Amsterdam (1993).
2. See, for example, E. Cartier, M. V. Fischetti, E. A. Eklund, and F. R. McFeely, Impact ionization in silicon, *Appl. Phys. Lett.* 62:3339 (1993).
3. M. V. Fischetti and S. E. Laux, Monte carlo analysis of electron transport in small semiconductor devices including band-structure and space-charge effects, *Phys. Rev. B* 38:9721 (1988).
4. H. Shichijo, J. Y. Tang, J. Bude, and P. D. Yoder, Full band monte carlo program for electrons in silicon, *in*: "Monte Carlo Device Simulation: Full Band and Beyond," K. Hess, ed., Kluwer Academic, Boston (1991).
5. T. Kunikiyo, M. Takenaka, Y. Kamakura, M. Yamaji, H. Mizuno, M. Morifuji, K. Taniguchi, and C. Hamaguchi, A monte carlo simulation of anisotropic electron transport in silicon including full band structure and anisotropic impact-ionization model, *J. Appl. Phys.* 75:297 (1994).
6. N. Sano and A. Yoshii, Impact ionization rate near thresholds in Si, *J. Appl. Phys.* 75:5102 (1994).
7. E. O. Kane, Electron scattering by pair production in silicon, *Phys. Rev.* 159:624 (1967).
8. N. Sano and A. Yoshii, Impact ionization theory consistent with a realistic band structure of silicon, *Phys. Rev. B* 45:4171 (1992).
9. N. Sano and A. Yoshii, Impact-ionization model consistent with the band structure of semiconductors, *J. Appl. Phys.* 77:2020 (1995).
10. M. V. Fischetti, S. E. Laux, and E. Crabbe, Understanding hot-electron transport in silicon devices: is there a short-cut?, *J. Appl. Phys.* 78:1058 (1995).
11. M. V. Fischetti, S. E. Laux, and E. Crabbe, Monte carlo simulation of high-energy electron transport in silicon: is there a short-cut to happiness?, present volume.
12. B. Eitan, D. Frohman-Bentchkowsky, and J. Shappir, Impact ionization at very low voltages in silicon, *J. Appl. Phys.* 53:1244 (1982).
13. D. Lau, G. Gildenblat, C. G. Sodini, and D. E. Nelsen, Low temperature substrate current characterization of n-channel MOSFET's, *IEDM Tech. Dig.* p. 565 (1985).
14. A. K. Henning, N. N. Chan, and J. D. Plummer, Substrate current in n-channel and p-channel MOSFET's between 77 K and 300 K, *IEDM Tech. Dig.* p. 573 (1985).
15. M. Koyanagi, T. Matsumoto, M. Tsuno, T Shimatani, Y. Yoshida, and H. Watanabe, Impact ionization phenomena in 0.1 μm MOSFET at low temperature and low voltage, *IEDM Tech. Dig.* p. 341 (1993).

# UNIVERSAL RELATION BETWEEN AVALANCHE BREAKDOWN VOLTAGE AND BANDSTRUCTURE IN WIDE-GAP SEMICONDUCTORS

J. Allam[1] and J. P. R. David[2]

[1]Hitachi Cambridge Laboratory
Cavendish Laboratory
Cambridge, CB3 0HE, UK

[2]Dept. of Electronic and Electrical Engineering
University of Sheffield
Sheffield, S1 3JD, UK

## INTRODUCTION

Pair-production due to impact ionisation is an important process affecting the performance of semiconductor electronic devices. The secondary carriers lead to current multiplication in avalanche photodiodes and to microwave generation in IMPATT diodes. On the other hand, avalanche breakdown imposes a fundamental limit on the applied field for all semiconductor junctions. Hence from a technological point of view it is very important to have a simple means of predicting the breakdown voltage ($V_b$) in new semiconductor structures.

Impact ionisation is also of interest from a physical point of view, as it represents a rigorous test of our understanding of carrier transport at high energies. Recently, considerable progress has been made in the first-principles calculation of ionisation coefficients, incorporating the effects of the full bandstructure on the carrier kinematics,[1] scattering dynamics[2] and impact ionisation cross-section[3]. The calculation has been extended to incorporate non-classical aspects of the transport.[4] The stage is set for a full numerical-theoretical understanding of impact ionisation. However, relatively few materials have been studied to date. Due to the numerical complexity, no simple relation between the ionisation coefficients and easily-measurable properties of the energy bands (e.g. bandgaps, effective masses) has been expected.

Early experimental studies of the variation of ionisation coefficients in GaAs with orientation, alloy composition and temperature suggested a strong dependence of the impact ionisation coefficients on the bandstructure.[5] However, these results were not reproduced in later experimental[6] or theoretical[1] work. We have previously studied impact ionisation in a variety of materials (Si, Ge, GaAs, $Si_xGe_{1-x}$, $Al_xGa_{1-x}As$, $(Al_xGa_{1-x})_{0.52}In_{0.48}P$, $Al_xGa_{1-x}As/GaAs$ superlattices and multiple quantum wells, and $GaAs/In_xGa_{1-x}As$ strained layer superlattices) using bandstructure variation by the application of hydrostatic pressure, compositional variation and layer width variation.[7-11] The variation of $V_b$ with bandstructure variation was found to be rather weak; however we have found that the results point to a universal relation between the avalanche breakdown voltage and the bandstructure in wide band-gap semiconductors.

*Hot Carriers in Semiconductors*
Edited by K. Hess *et al.*, Plenum Press, New York, 1996

# APPROXIMATE RELATION BETWEEN $V_b$ AND THE BANDGAP

In 1966, Sze and Gibbons[12] reported approximate "universal" formulae for the breakdown voltage, determined from the experimentally-measured ionisation coefficients of Ge, Si, GaAs and GaP. The formula for abrupt pn junctions is

$$V_b \approx 60 \cdot (E_g/1.1)^{3/2} (N_B/10^{16} cm^{-3})^{-3/4} \qquad (1)$$

where $E_g$ is the bandgap and $N_B$ is the background doping density. From the same data, we have calculated a corresponding formula for the breakdown voltage of a p-i-n diode with a 1 μm thick i-layer

$$V_b \approx 30 \cdot (E_g/1.1). \qquad (2)$$

It is not known whether the apparent linear dependence of $V_b$ on $E_g$ holds for other materials, or how accurate the expression is. Nevertheless, these formulae have been widely used to predict $V_b$. We shall investigate the validity of Equation (2) for a wide range of materials, and will demonstrate a more accurate relation for wide bandgap semiconductors which depends not on $E_g$ but on an average energy gap <E> as described below.

# EXPERIMENTAL MEASUREMENT OF $V_b$ IN $Al_xGa_{1-x}As$ AND $(Al_xGa_{1-x})_{0.52}In_{0.48}P$ ALLOYS: $V_b$ SCALING WITH <E>

In attempting to investigate the scaling of $V_b$ with bandstructure for a range of materials, we encounter two problems. Firstly, ionisation coefficients for many materials are not well determined, with large discrepencies amongst the published values. Secondly, parameters of the bandstructure used in determining <E> are not well known in all cases. We have therefore directly measured $V_b$ in two alloy systems: $Al_xGa_{1-x}As$ and $(Al_xGa_{1-x})_{0.52}In_{0.48}P$. Both are lattice-matched to GaAs, and exhibit a direct-indirect bandgap transition within the range of composition studied. The bandgaps of AlGaAs are well known.[13] Recently, Prins et al.[14] have measured the bandgaps in $(Al_xGa_{1-x})_{0.52}In_{0.48}P$.

The breakdown voltage in $Al_xGa_{1-x}As$ and $(Al_xGa_{1-x})_{0.52}In_{0.48}P$ increases linearly with the composition x.[11] The absence of deviation from linearity at the direct-indirect transition indicates that the band-gap is not the important parameter in determining $V_b$. In figure 1, we show $V_b$ for 1μm p-i-n diodes as a function of the direct band-gap at the Γ point ($E_\Gamma$). $V_b$ is normalised to that of GaAs, calculated for a device with the same background doping and depletion width). Clearly the variation of $V_b$ is not well described by a single relation $V_b \propto E_g$.

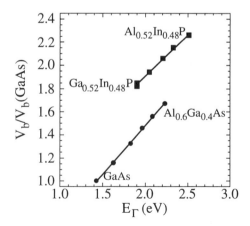

**Figure 1.** Breakdown voltage (normalised to that of GaAs) as a function of direct bandgap $E_\Gamma$ for $Al_xGa_{1-x}As$ and $(Al_xGa_{1-x})_{0.52}In_{0.48}P$.

Our previous hydrostatic pressure measurements[9] on breakdown in GaAs showed that the effective ionisation threshold energy scaled with an average conduction band energy, approximated by

$$<E> = (1/8) (E_\Gamma + 3E_X + 4E_L) \qquad (3)$$

where $E_X$ and $E_L$ are the conduction band minima at the X and L points of the Brillouin zone and are measured relative to the valence band maximum at $\Gamma$. We therefore plot experimentally-measured values of $V_b$ for the AlGaAs and AlGaInP alloys against $<E>$ (figure 2) and obtain the following linear relationship

$$\frac{V_b}{V_b(GaAs)} = \frac{\langle E \rangle}{0.50 eV} - 2.50 \qquad (4).$$

where $V_b(GaAs)$ is the breakdown voltage of a $1\mu m$ GaAs p-i-n diode ($\approx 33V$). Deviations from linearity are well within the uncertainty associated with determination of $<E>$.

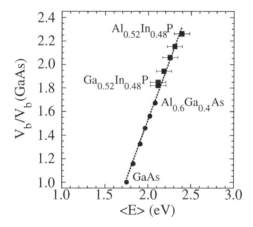

**Figure 2.** Normalised breakdown voltage as a function of $<E>$ for $Al_xGa_{1-x}As$ and $(Al_xGa_{1-x})_{0.52}In_{0.48}P$.

## $V_b(<E>)$ DETERMINED FROM IONISATION COEFFICIENTS

We have investigated the application of Equation (4) to other materials. Figure 3(a) shows (as a function of bandgap $E_g$) the value of $V_b$ measured directly for the AlGaAs and AlGaInP alloys, or calculated from published values of the ionisation coefficients for the materials InP, $Al_{0.48}In_{0.52}As$, GaP, Si, Ge, AlGaAsSb and $In_{0.53}Ga_{0.47}As$. Although a general trend of $V_b$ increasing with $E_g$ is followed, deviations of up to 30% are observed compared to the Sze formula.

The same data are plotted in figure 3(b) as a function of $<E>$. All the wide-gap semiconductors ($E_g > 1eV$, $<E> > 1.5eV$) with the exception of GaP obey the linear relationship described by equation (4). The agreement is remarkable given the uncertainty in $V_b$ and $<E>$. The apparent discrepency of GaP may relate to an experimental artefact. As an example, we compare the breakdown voltage in GaAs and InP. Despite having a smaller bandgap, the breakdown voltage in InP is $\approx 30\%$ higher than in GaAs, as a result of the larger value of $<E>$ (i.e. larger intervalley separation).

We interpret the existence of this relation as evidence for the distribution between conduction-band minima of final electron states of the pair-production process. If the intervalley separations are small compared to the bandgap, the distribution of final electron states is determined primarily by the density of states. Due to the high scattering rates experienced by carriers which initiate impact ionisation, their momentum is randomised and

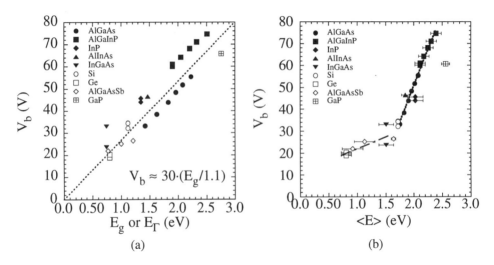

**Figure 3.** Breakdown voltage determined from ionisation coefficients published in the literature, plotted as a function of (a) $E_g$ or $E_\Gamma$ and (b) $<E>$.

hence initial states corresponding to pair-production with final electron states in the different conduction-band minima are widely available. The energy barrier for pair-production is largely determined by $<E>$. A full discussion of the interpretation of Equation (4) and the implications for simple theories will be reported elsewhere.[15]

## CONCLUSIONS

In conclusion, we have found a new simple relation between the avalanche breakdown voltage and an average conduction-band energy. The relation applies to wide band-gap tetrahedally-coordinated semiconductors, and is a result of the distribution in the Brillouin zone of final electron states from pair-production. The existence of such a simple relation allows us to select materials and engineer structures with optimised breakdown characteristics.

## REFERENCES

[1]H. Shichijo and K. Hess, Phys. Rev. B **23**, 4197 (1981).

[2]M. V. Fischetti and S. E. Laux, Phys. Rev. B**38**, 9721 (1988).

[3]N. Sano and A. Yoshii, J. Appl. Phys. **77**, 2020 (1995).

[4]J. Bude, K. Hess and G. J. Iafrate, Phys. Rev. B**45**, 10958 (1992).

[5]T.P. Pearsall, F. Capasso, R.E. Nahory, M.A. Pollack & J.R. Chelikowsky, Solid-State Electronics, **21**, 297 (1978).

[6]V.M. Robbins, S.C. Smith and G.E. Stillman, Appl. Phys. Lett. **52**, 296 (1988).

[7]J. Allam, High Pressure Research **9**, 231 (1992).

[8]I. K. Czajkowski, J. Allam, A.R. Adams and M.A. Gell, J. Appl. Phys. **71**, 3821 (1992).

[9]J. Allam, A.R. Adams, M.A. Pate and J.S. Roberts, *GaAs and Related Compounds* (Inst. Phys. Conf. Ser. No. 112), 375 (1990).

[10]J. P. R. David, J. Allam, A. R. Adams, J. S. Roberts, R. Grey, G. Rees and P. N. Robson, Appl. Phys. Lett. **66**, 2876 (1995).

[11]J.P.R. David, M. Hopkinson and M.A. Pate, Electron. Lett. **30**, 909 (1994).

[12]S. M. Sze and G. Gibbons, Appl. Phys. Lett. **8**, 111 (1966).

[13]S. Adachi, EMIS Data Review 7, (INSPEC, 1993)

[14]A.D. Prins, J.L. Sly, A.T. Meney, D.J. Dunstan, E.P. O'Reilly and A.R. Adams, Proc. 22$^{nd}$ Int. Conf. Physics Semicond., 727 (1995).

[15]J. Allam (to be published).

# SPATIO–TEMPORAL DYNAMICS OF FILAMENT FORMATION INDUCED BY IMPURITY IMPACT IONIZATION IN GaAs

M. Gaa, R. E. Kunz, E. Schöll

Institut für Theoretische Physik, Technische Universität Berlin,
Hardenbergstraße 36, 10623 Berlin, Germany

We present a two-dimensional simulation showing the spatio-temporal dynamics of the formation of a current filament. The model we use is based on the classical rate equations with microscopic transport parameters which were gained from a Monte Carlo simulation for homogeneously doped material.

## INTRODUCTION

Due to the inherent nonlinearities of the underlying generation-recombination (GR) processes in semiconductors, impact ionization of trapped carriers in high electric fields – often associated with S-shaped negative differential conductivity (SNDC) of the sample – constitutes a physical mechanism that gives rise to a menagerie of spatio-temporal bifurcation scenarios[1-3]. Self-organized transitions of a spatially homogeneous steady state to a variety of filamentary structures are prominent examples for possible bifurcations. These current filaments have been predicted theoretically[1,4-9] and found experimentally[10-12] in a variety of semiconductor materials, e.g., p-Ge or n-GaAs, at liquid Helium temperatures. Recently the evolution of such structures starting from a high resistivity state has received increasing interest[13,14].

Since the experimental bistability in the transition regime can be explained in terms of standard GR kinetics (including impact ionization) only if at least two impurity levels are taken into account[4,15,16], we model the infinite hydrogenlike energy spectrum of the shallow donors in n-GaAs by the ground state and an "effective" excited state close to the band edge. In this case, the average state of a given system can be characterized by the spatial distributions of number densities of the carriers in the impurity ground state, in the excited state, and in the conduction band, given by $n_1(\underline{x}, t)$, $n_2(\underline{x}, t)$ and $n(\underline{x}, t)$, respectively, where $\underline{x}$ is the spatial coordinate and $t$ denotes time.

For an n-type semiconductor the temporal evolution of $n$, $n_1$ and $n_2$ is then governed by the rate equations:

$$\dot{n} - \frac{1}{e} \nabla \cdot j = \phi \equiv X_1^s n_2 - T_1^s (N_D - n_1 - n_2) n + X_1 n n_1 + X_1^* n n_2, \tag{1}$$

$$\dot{n}_1 = \phi_1 \equiv T^* n_2 - X^* n_1 - X_1 n n_1, \tag{2}$$

$$\dot{n}_2 = \phi_2 \equiv -\phi - \phi_1, \tag{3}$$

where the dot denotes the partial derivative with respect to time, $j \equiv e(\mu n \underline{\mathcal{E}} + D \nabla n)$ is the current density, $N_D$ is the total density of donors, $\underline{\mathcal{E}}$ the electric field and $\mu$ the mobility of electrons. We

*Hot Carriers in Semiconductors*
Edited by K. Hess *et al.*, Plenum Press, New York, 1996

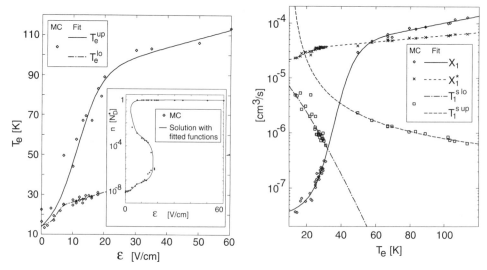

**Figure 1.** (left) Electron temperature $T_e$ as a function of the electric field $\mathcal{E}$: MC data and fitted analytic representation. Material parameters: $N_D = 7 \times 10^{15}\text{cm}^{-3}$, $N_A = 2 \times 10^{15}\text{cm}^{-3}$, $\mu = 10^5\text{cm}^2/\text{Vs}$, $T_L = 4.2$K. Inset: stationary $n(\mathcal{E})$ characteristic with GR parameters given in the caption of Fig. 2.

**Figure 2.** (right) GR coefficients as a function of the electron temperature $T_e$: MC data and fitted analytic representations. $X_1^s = 1.17 \times 10^6\text{s}^{-1}$, $X^* = 3.36 \times 10^3\text{s}^{-1}$, $T^* = 4.10 \times 10^7\text{s}^{-1}$, $\tau_M = 7.83 \times 10^{-14}$s (dielectric relaxation time).

assume the validity of the Einstein relation for the diffusion coefficient $D = \mu k_B T_L/e$, where $T_L$ is the lattice temperature and $e$ the electron charge. $X_1^s$ is the thermal ionization coefficient of the excited level, $T_1^s$ is its electron capture coefficient, $X_1$, $X_1^*$ are the electron impact ionization coefficients from the ground and excited level, respectively, $X^*$, $T^*$ denote the transition coefficients from the ground level to the excited level and vice versa, respectively.

The electric field $\underline{\mathcal{E}}$ must satisfy Poisson's equation

$$\epsilon \nabla \cdot \underline{\mathcal{E}} = e(N_D^* - n - n_1 - n_2), \tag{4}$$

where $\epsilon$ is the dielectric constant, $N_D^* = N_D - N_A$ holds and $N_A$ is the density of compensating acceptors.

## MONTE CARLO SIMULATION

The dependence of the GR coefficients on the dynamic variables was obtained by a spatially homogeneous single particle Monte Carlo (MC) simulation as described in[17]. As band-impurity processes thermal generation from the excited donor level, recombination into the excited level (Lax-Abakumov) and impact ionization from both the ground and the excited donor level were included. The relevant intraband scattering processes were elastic ionized impurity scattering (Conwell-Weisskopf approximation) and inelastic acoustic deformation potential scattering. (Optical phonon scattering is neglected because of the low lattice temperature, although optical phonon emission becomes relevant for energies above 36meV[18]. However those states are not frequently populated except at highest fields[19].) As a result of the MC simulation the impact ionization coefficients $X_1$ and $X_1^*$ as well as the capture coefficient $T_1^s$ depend not only on the electric field $\mathcal{E}$, but also on the electron concentration $n$. This dependence reflects a higher electron temperature $T_e^{up}$ in the upper branch of the $n(\mathcal{E})$ characteristics as compared to the values $T_e^{lo}$ on the lower and the middle branch (Fig. 1). Instead of $\mathcal{E}$, we use the electron temperature $T_e$ in addition to $n$ to parametrize $X_1$, $X_1^*$ and $T_1^s$ (Fig. 2). By fitting the MC data (dots in Figs. 1,2) using smooth analytic functions (lines in Figs. 1,2), the functional dependencies of $T_e$ on $\mathcal{E}$ as well as $X_1$, $X_1^*$ and $T_1^s$ on $T_e$ could be obtained (cf. inset of

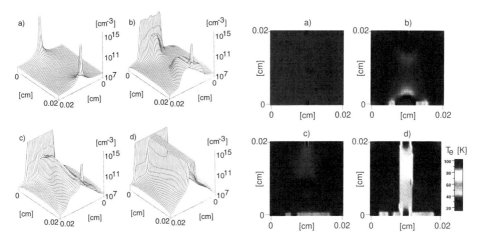

**Figure 3.** (left) Temporal evolution of the electron density $n$: a) $t = 0$ns, $U = 0$V, b) $t = 0.5$ns, $U = 0.48$V, c) $t = 1.0$ns, $U = 0.48$V, d) $t = 2.5$ns, $U = 0.48$V. The applied voltage corresponds to $\mathcal{E} = 24$V/cm in the homogeneous $n(\mathcal{E})$ characteristic of Fig. 1.

**Figure 4.** (right) Electron temperature distribution $T_e(\underline{x}, t)$: a) $t = 0$ns, $U = 0$V, b) $t = 0.5$ns, $U = 0.48$V, c) $t = 1.0$ns, $U = 0.48$V, d) $t = 2.5$ns, $U = 0.48$V.

Fig. 1). In this manner the stationary $n(\mathcal{E})$ characteristic, obtained from the MC simulation (dots in inset of Fig. 1), could be reproduced from equations (1-3) using the analytic representations (lines in inset of Fig. 1).

## FILAMENT FORMATION

To study the nascence of current filaments we solve equations (1-4) numerically as described in[14]. We choose a square sample geometry with side length $L_x = L_y = 0.02$cm. We model point contacts by applying Dirichlet boundary conditions to two opposite regions of length $L_c = 8 \times 10^{-4}$cm at the centers of the sample edges. At the contacts $n$ is fixed to a value $n_D = 5 \times 10^{15}$cm$^{-3}$ to model Ohmic contacts. All other boundaries are treated as insulating where the components of the current density $\underline{j}$ and the electric field $\mathcal{E}$ perpendicular to the boundaries vanish. We start our simulation at time $t_0 \equiv 0$ with the sample in thermal and chemical equilibrium, i.e. vanishing currents and fluxes. Due to the Ohmic nature of the contacts we find small regions in the vicinity of the contacts in which the electron concentration is largely enhanced compared to the bulk (Fig. 3 a).

Within 1ps the voltage is linearly increased from $U = 0$V to $U = 0.48$V. Due to the assumed voltage control the electric field $\mathcal{E}$ reacts quasi-instantaneously giving a dipole-like electric field distribution and enlarged areas of increased electron density at both the cathode and the anode. Subsequently impact ionization multiplies the electron concentration at the cathode and establishes a front that moves towards the anode (Fig. 3 b). The propagation of the front is accompanied by a high field domain associated with an increased electron temperature $T_e$ (Fig. 4 b). Although the electric field behind the front is smaller than in front of it for reasons of current conservation the increased electron density in regions passed by the front is almost conserved because recombination is much slower than generation. Hence impact ionization downstream is encouraged, whereas further generation upstream is inhibited. When the front meets the region of increased carrier density around the anode, a rudimentary filament is formed (Fig. 3 c). Finally impact ionization leads to a uniform increase of electron density until the filament reaches its mature state (Fig. 3 d). During this process the total current increases rapidly (Fig. 5). When the rudimentary filament is established the growth rate shows a local minimum. Afterwards the transition to the mature filament occurs rapidly. When

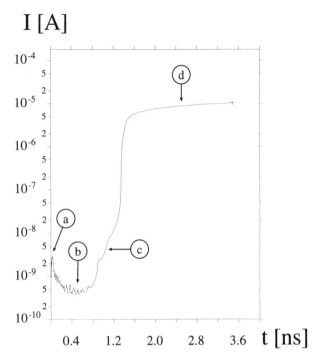

**Figure 5.** Total current $I(t)$ during the formation of a filament. The letters correspond to the electron density profiles in Fig. 3.

the filament reaches its mature state the electron temperature within the filament is increased much more strongly than outside (Fig. 4 d).

Hence from our simulation we find three stages of impact ionization breakdown: a stage of front creation and propagation (stage I) followed by a stage of stagnation after the front has reached the anode (stage II) and one in which the rudimentary filament grows to a mature filament (stage III). Whereas there are indications from 1D longitudinal simulations that the stagnation of the breakdown process in the second stage results from a seesaw-like mode, in which a high field domain moves back and forth between the anode and the cathode[13], the mechanism for the slowing down of the increase in $I(t)$ in the 2D case is not clear.

In this paper we have presented theoretical results showing the dynamics of the nascence of a current filament. Using GR parameters gained from MC simulations and elaborate numerical techniques the basic mechanisms responsible for the formation of current filaments in semiconductors have been worked out. An extension of the MC method to inhomogeneously ($\delta$-) doped GaAs and comparison with experiment has been presented elsewhere[17,19].

## REFERENCES

1. E. Schöll, *Nonequilibrium Phase Transitions in Semiconductors* (Springer, Berlin, 1987).
2. E. Schöll, in *Handbook on Semiconductors*, 2nd ed., edited by P. T. Landsberg (North Holland, Amsterdam, 1992), Vol. 1.
3. E. Schöll and A. Wacker, in *Nonlinear Dynamics and Pattern Formation in Semiconductors and Devices*, edited by F. J. Niedernostheide (Springer, Berlin, 1995), pp. 21–45.
4. E. Schöll, Z. Phys. B **48**, 153 (1982).
5. E. Schöll and D. Drasdo, Z. Phys. B **81**, 183 (1990).
6. R. Kunz and E. Schöll, Z. Phys. B **89**, 289 (1992).
7. G. Hüpper, K. Pyragas, and E. Schöll, Phys. Rev. B **47**, 15515 (1993).
8. G. Hüpper, K. Pyragas, and E. Schöll, Phys. Rev. B **48**, 17633 (1993).
9. A. Wacker and E. Schöll, Z. Phys. B **93**, 431 (1994).

10. F.-J. Niedernostheide, B. S. Kerner, and H.-G. Purwins, Phys. Rev. B **46**, 7559 (1992).
11. J. Spangler, U. Margull, and W. Prettl, Phys. Rev. B **45**, 12137 (1992).
12. J. Peinke, J. Parisi, O. Rössler, and R. Stoop, *Encounter with Chaos* (Springer, Berlin, Heidelberg, 1992).
13. R. E. Kunz and E. Schöll, Z. Phys. B (1995), (in press).
14. R. E. Kunz, Doctoral Thesis, TU Berlin, 1995.
15. E. Schöll, Z. Phys. B **46**, 23 (1982).
16. A. A. Kastalsky, phys. stat. sol. **15**, 599 (1973).
17. B. Kehrer, W. Quade, and E. Schöll, Phys. Rev. B **51**, 7725 (1995).
18. H. Kostial, Th. Ihn, P.Kleinert, R. Hey, M. Asche, and F. Koch, Phys. Rev. B **47**, 4485 (1993).
19. H. Kostial, M. Asche, R. Hey, K. Ploog, B. Kehrer, W. Quade, and E. Schöll, Semicond. Sci. Technol. **10**, 775 (1995).

# INTRA-CENTER MID-INFRARED ELECTROLUMINESCENCE FROM IMPACT IONIZATION OF Fe-DEEP LEVELS IN InP BASED STRUCTURES

Gaetano Scamarcio, Federico Capasso, Albert L. Hutchinson, Tawee Tanbun-Ek, Deborah Sivco, and Alfred Y. Cho

AT&T, Bell Laboratories,
Murray Hill, New Jersey 07974

## ABSTRACT

Narrow-band mid-infrared recombination radiation from impact ionization of transition-metal ions in semiconductors is reported. A large and abrupt increase of electroluminescence at 3.5 μm, enhanced by the presence of high electric field domains, is observed at cryogenic temperatures above the onset for the transferred-electron effect in n-type InP doped with Fe. The spectrum shows four sharp lines corresponding to the symmetry allowed $d$-shell transitions of $Fe^{2+}$ ions.

## INTRODUCTION

Recombination radiation from impact ionization of *shallow* impurities is a well known phenomenon[1-3]. During an impact event, free electrons accelerated by low electric fields can transfer enough energy to promote a trapped electron into a conduction band state. Radiative transitions from the conduction band to the impurity ground state give rise to *broad bands* in the far-infrared range 70-2000 μm. An efficient method of creating impact ionization is to generate high-field domains in n-type materials. In this way, *broad* band-to-band luminescence and even coherent radiation associated with recombination of electrons and holes generated by band-to-band impact ionization have been previously observed in GaAs [4].

In this paper we report the first observation of *narrow-band* mid-infrared emission at 3.5μm from impact ionization of $Fe^{3+/2+}$ acceptor *deep* levels in InP, enhanced by the presence of high electric field domains.

## EXPERIMENT

Two kinds of planar structures were investigated. The first one (sample A) consists of an InP channel layer doped with Si and Fe to $[Si] \approx [Fe] \approx 10^{17} cm^{-3}$, grown by

MOCVD on semi-insulating InP substrate. The second one (sample B) is a GaInAs/AlInAs multiple heterostructure grown by MBE lattice matched on InP substrate doped to $[Fe] \approx 5 \times 10^{16}$. Here a 2D-electron-gas is created by modulation doping and electrons can transfer in the substrate by thermionic emission. The samples were processed into $150 \times 150 \mu m^2$ mesas and standard Ni/Au/Ge/Ag or Ni/Au/Sn/Ti ohmic contacts were deposited. Voltage pulses of duration in the range 50 ns - 50 μs were applied along the 40 μm thick channel with a repetition rate in the range 10 Hz - 50 kHz while keeping the sample in a He flow cryostat. Optical power measurements were performed with a 77K-cooled InSb detector. The detector-preamplifier overall time constant is 1μs. The use of Si lenses gives a short wavelength cut-off of $\approx 1.1$ μm. The spectrum of the emitted radiation was measured with a FTIR spectrometer in the step-scan operation mode with lock-in or boxcar amplification techniques.

## RESULTS AND DISCUSSION

When a voltage $\geq 12V$ (electric field $F \geq 3$ kV/cm) is applied across the sample, emission of mid-infrared radiation is detected. The electroluminescence spectrum of Fig. 1 unambiguously demonstrate that the emission of radiation is associated to intra-center transitions within $^5D$ states of $Fe^{2+}$ ions substitutional to In [5]. The sharp zero-phonon lines at 2843.6 cm$^{-1}$, 2830.1 cm$^{-1}$, 2819.1 cm$^{-1}$, and 2801.8 cm$^{-1}$ are ascribed to the four symmetry-allowed transitions among the $^5T_2$ and $^5E$ levels (inset (b)).

Figure 2 shows a representative current and optical power versus voltage curve measured at 5 K for a devices of sample A. The current reaches a maximum at 36.5 V (~9.0 kV/cm). Further increase of the bias voltage drives the system into a negative differential resistance (NDR) regime, as clearly shown by the negative slope of the I-V characteristics. This is a well known consequence of electric-field induced $\Gamma \rightarrow L$ transfer.

The electrical instability associated with NDR leads to the formation of high-field domains with $F \geq 10^5$ V/cm. Free electrons accelerated by the domain field can gain enough energy for impact ionization of $Fe^{2+}$ ions (threshold energy ~ 0.63 eV). By using ZnSe optics, transparent both in the near- and mid-infrared portion of the spectrum, we have verified that band-gap luminescence around 0.9 μm, due to band-to-band impact ionization, is always weaker than the Fe-related luminescence and often below the detection limit. This is consistent with the small value of the band-to-band impact ionization coefficient in InP, which is about one order of magnitude smaller than that in GaAs for electric fields ~ $10^5$ V/cm.

The break in the I-V characteristics of Fig. 2 is due to the sudden increase of the free-electron density related to impact ionization phenomena. A kink, corresponding to an abrupt transition to a higher current level is directly observable also in the oscilloscope traces.

In the range of electric fields before the break in the I-V characteristics the optical power grows non linearly with F, reaching measured values of the order of 5-6 nW. In this range radiative recombination predominantly arises from direct impact excitation of $Fe^{2+}$ $^5T_2$ levels. The striking feature is that together with the appearance of the electrical instabilities in the oscilloscope traces and the breaks in the I-V characteristics the optical power abruptly increases reaching values as high as 40 nW.

The large and sudden increase of the mid-infrared emission with well defined spectral features at 3.5 μm, concomitant with the abrupt increase of the current, strongly suggests that impact ionization of Fe, enhanced by the formation of high-field domains in the transferred-electron effect regime, is the cause of the mid-infrared electroluminescence and supports the following microscopic picture (see inset of Fig. 1).

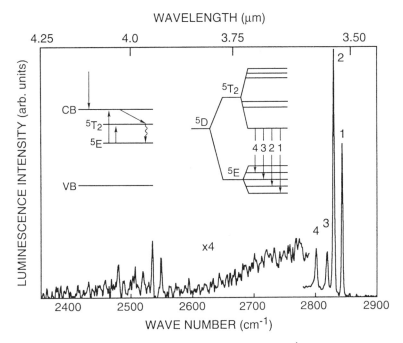

Fig. 1 - Electroluminescence spectra recorded at 5 K with 40V (~ $10^4$V/cm). The four sharp zero-phonon lines at 5 K are labeled according to inset. The insets show a sketch of impact excitation and impact ionization processes, and the level scheme of $Fe^{2+}$ in InP, considering crystal-field and spin-orbit interactions. The broader band extending in the range 2600-2800 cm$^{-1}$ and the peaks in the range 2470-2550 cm$^{-1}$ are phonon sidebands.

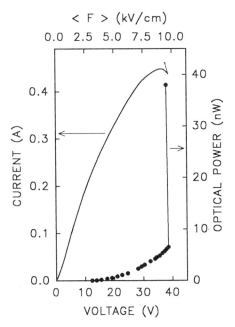

Fig. 2 - Typical current and optical power vs bias voltage at 5 K The collection efficiency of the apparatus is $\eta_c \approx 3.5 \times 10^{-3}$.

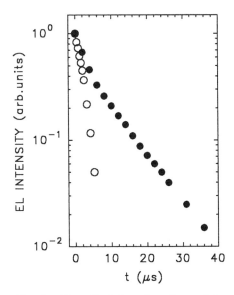

Fig. 3 - Normalized time dependence of the $Fe^{2+}$ electroluminescence decay after the end of the pulse below (hollow circles) and above (filled circles) the onset for high-field domains formation.

355

Hot electrons accelerated in high-field domains impact ionize a large number of $Fe^{2+}$ ions which are left in their neutral state. Radiation is emitted in the second step of a cascade process in which first a free electron is captured into an excited state of $Fe^{2+}$, and then recombines to the ground state.

Additional evidence for the existence of two different regimes comes from the time dependence of the electroluminescence decay after the end of the voltage pulse. Below the threshold for high-field domains formation, a decay time $\leq 3$ µs is measured. This results from both the excited state lifetime and the finite time-constant of our detection system. In contrast, in the high-field domain regime, a slower decay, characterized by a time constant $\approx 10$µs follows a relatively fast initial decrease. The fast decay is ascribed to electrons occupying the $^5T_2$ excited states at the end of the pulse and is determined by the lifetime $\tau$ of these levels. The slow decay is mainly determined by the capture time $\tau_c$ necessary for excess electrons in the conduction band to be re-captured into a $^5T_2$ state.

In steady-state, the optical output power can be written as $P = \eta_c \, \eta_r \, \hbar\omega \, N^* V_a / \tau$, where $\eta_r$ is the radiative efficiency, $\hbar\omega$ is the photon energy, $N^*$ is the density of $^5T_2$ excited states, and $V_a$ is the active volume. An estimation of the maximum optical power, assuming an ionization rate large enough to have all $Fe^{2+}$ ionized gives $P_{max} \approx 5$ nW, which is smaller by a factor $\approx 8$ than the maximum measured value. This large discrepancy can be explained considering a field induced diffusion of hot carriers deeply into the semi-insulating InP substrate where additional $Fe^{2+}$ is present. Evidence of the latter phenomenon comes from electroluminescence experiments carried-out on sample B, where Fe is present only in the semi-insulating substrate. The peaks in the electroluminescence spectra of sample B coincide with those of sample A. Since the threshold field for $\Gamma \rightarrow L$ transfer in GaInAs is $\approx 1.5$ kV/cm, much lower than that in InP, high-field domains are easier to form. Hence, hot electrons can overcome the conduction band discontinuity ($\Delta E_c \approx 0.2$ eV) at the GaInAs/InP substrate interface and impact ionize $Fe^{2+}$ in InP.

The fabrication of optimized structures including optical confinement and feedback would open the possibility of a new *electrically pumped* solid-state laser operating in the mid-infrared. It is worth to cite that laser action at 3.5 µm in InP:Fe has already been demonstrated by *optical pumping* [6]. As a final remark we note that narrow-band intracenter electroluminescence associated with impact ionization of deep impurities, strongly enhanced by the presence of high-field domains, should be observable also in other semiconductor/transition-metal systems, due to the fact that the transferred-electron effect is a general property of many III-V direct gap semiconductors and that transition-metal impurities typically introduce deep levels in the band gap.

## ACKNOWLEDGMENTS

This work is partly supported by Cons. Ind. Technobiochip, Marciana, Italy and CNR-NATO.

## REFERENCES

1. S. H. Koenig, and R. D. Brown, III, Phys. Rev. Lett. 4, 170 (1960)
2. E. Gornik, Phys. Rev. Lett. 29, 595 (1972)
3. I.Melngailis, G.E.Stillman, J.O.Dimmock, C.M.Wolfe, Phys.Rev.Lett. 23, 1111(1969)
4. P.D.Southgate, J. Appl. Phys. 38, 4589 (1967), IEEE, J. Quant. El. QE-4, 179 (1968)
5. K. Pressel, K. Thonke, A, Dornen, G. Pensl, Phys. Rev. B43, 2239 (1991)
6. P. B. Klein, J. E. Furneaux, and R. L. Henry, Appl. Phys. Lett. 42, 638 (1983)

# NOVEL IMPACT IONIZATION MODEL USING SECOND- AND FOURTH-ORDER MOMENTS OF DISTRIBUTION FUNCTION FOR GENERALIZED MOMENT CONSERVATION EQUATIONS

Ken-ichiro Sonoda, Mitsuru Yamaji, Kenji Taniguchi, and Chihiro Hamaguchi

Department of Electronic Engineering,
Osaka University, Suita, Osaka, 565 Japan

## INTRODUCTION

Device degradation caused by hot carriers has been main concern from the reliability point of view. Because secondary-generated carriers created by impact ionization (I.I.) have great influence on the degradation of gate oxide, accurate modeling of I.I. is necessary.

We propose an I.I. model which is formulated using second- and fourth-order moments of distribution function for precise description of I.I. in inhomogeneous electric field. A set of moment conservation equations for carrier transport is also presented to perform practical device simulation with the I.I. model.

## IMPACT IONIZATION MODEL

To investigate the I.I. phenomena in inhomogeneous field, we use the Monte Carlo (MC) simulation program with analytical multi-valley band structure, in which phonon scattering[1] and the impact ionization[2] are implemented as a function of electron energy.

Figure 1 shows calculated average energy, $\langle \varepsilon \rangle$, normalized average square energy, $\xi \equiv \sqrt{(3/5)\langle \varepsilon^2 \rangle}/\langle \varepsilon \rangle$, and normalized impact ionization generation rate, $G_{ii}/n$, in inhomogeneous electric field. ($\langle A \rangle$ means $\int A f d\mathbf{k} / \int f d\mathbf{k}$ hereafter.) Note that $\xi$ equals one when the distribution function is Maxwellian. The generation rate due to impact-ionization depends not only on average energy but also on the value, $\xi$. The figure also shows that $\xi$ increases in the decreasing field, meaning that there exists the high energy tail in the distribution function in spite of the sharp decrease of the average energy.

We evaluate correct impact ionization rate by using both the average energy, $\langle \varepsilon \rangle$, and the normalized fourth-order moment, $\xi$, of the distribution function. Figure 2 shows e-h pair generation rate calculated in the exponential increasing fields as shown in Fig. 1 for different peak electric fields. In the figure, the I.I. rate are plotted as a function of the inverse of the average energy with several $\xi$'s as a parameter. The I.I.

*Hot Carriers in Semiconductors*
Edited by K. Hess *et al.*, Plenum Press, New York, 1996

generation rate is empirically expressed as

$$G_{ii} = ng_{ii0} \exp\left(-\frac{\varepsilon_{c0} \exp\left(-\gamma\xi\right)}{\langle\varepsilon\rangle}\right),$$ (1)

where $g_{ii0} = 1.39 \times 10^{14} \text{s}^{-1}$, $\varepsilon_{c0} = 35.6\text{eV}$, and $\gamma = 1.59$.

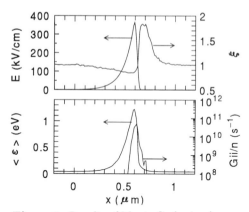

**Figure 1.** Results of Monte Carlo simulation in inhomogeneous electric field. The field increases exponentially and decrease linearly.

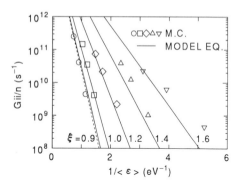

**Figure 2.** Normalized I.I. generation rate as a function of inverse average energy for different $\xi$ values. Solid lines are obtained from the model equation (1). Dashed line indicates the generation rate in homogeneous electric field.

## MOMENT CONSERVATION EQUATIONS

The use of the new I.I. model (1) in device simulation requires both second-order moment, $\langle\varepsilon\rangle$, and fourth-order one, $\langle\varepsilon^2\rangle$. The fourth-order moment is numerically calculated from the conservation equation for $\langle\varepsilon^2\rangle$ incorporated in the hydrodynamic model[3]. The conservation equations derived from the Boltzmann transport equation (BTE) are

$$\nabla \cdot (n\langle\boldsymbol{u}\varepsilon^2\rangle) = -2q\boldsymbol{E} \cdot \boldsymbol{S} - n\frac{\langle\varepsilon^2\rangle - \langle\varepsilon^2\rangle_0}{\tau_{\langle\varepsilon^2\rangle}} - U_{\langle\varepsilon^2\rangle}$$ (2)

$$n\langle\boldsymbol{u}\varepsilon^2\rangle = \frac{\tau_{\langle\boldsymbol{u}\varepsilon^2\rangle}}{\tau_{\langle\boldsymbol{u}\rangle}}\frac{7}{3}\left(\langle\varepsilon^2\rangle\frac{\boldsymbol{J}}{-q} - \frac{kT_n}{q}n\mu\nabla\langle\varepsilon^2\rangle\right),$$ (3)

where $\boldsymbol{u}$ is the group velocity of an electron, $\boldsymbol{E}$ is the electric field, $\boldsymbol{J} \equiv -qn\langle\boldsymbol{u}\rangle$ is the electric current density, $\boldsymbol{S} \equiv n\langle\boldsymbol{u}\varepsilon\rangle$ is the energy flux, $\langle\varepsilon^2\rangle_0$ is the fourth-order moment at thermal equilibrium, $U_{\langle\varepsilon^2\rangle}$ is the net loss rate of $\langle\varepsilon^2\rangle$ due to generation-recombination process, $\tau_{\langle A\rangle}$ is the relaxation time of $\langle A\rangle$, $\mu$ is the electron mobility, and $T_n$ is the electron temperature defined by $3kT_n/2 \equiv \langle\varepsilon\rangle$. The relaxation time, $\tau_{\langle A\rangle}$, is derived from the Monte Carlo simulation.

## RESULTS AND DISCUSSIONS

Figure 3 shows the calculated I.I. generation rate, $G_{ii}$, in an $n^+nn^+$ structure with the use of different I.I. models together with MC results. Parameters used in the I.I. models are calibrated to provide the same I.I. rate obtained from MC simulation in

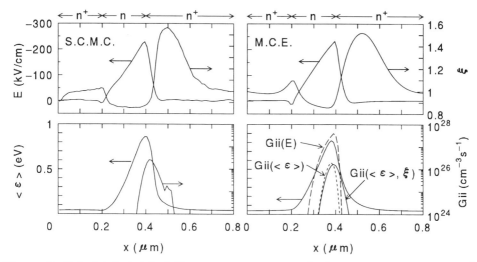

**Figure 3.** Calculated electric field, average energy, normalized average square energy, and impact ionization rate calculated by self-consistent Monte Carlo (left row) and moment conservation equations (right row) in a 1-D $n^+nn^+$ device ($n^+ = 5\times10^{17}\text{cm}^{-3}$, $n = 5\times10^{15}\text{cm}^{-3}$) at V= 3V.

homogeneous electric field. In the MC simulation, the BTE and Poisson equation are solved self-consistently.

It should be noted that the local field model, $G_{ii}(E)$, overestimates maximum I.I. rate nearly one order of magnitude while it underestimates the generation rate in the decreasing field region. The maximum I.I. rate is significantly improved with the use of $G_{ii}(\langle\varepsilon\rangle)$ which is expressed as a function of average energy. This model, however, still underestimates $G_{ii}$ in the decreasing electric field region. In contrast with the two models described above, the new model ($G_{ii}(\langle\varepsilon\rangle,\xi)$), provides the high I.I. rate in field decreasing region, which agrees well with the MC result.

## CONCLUSION

We proposed a new I.I. model including second- and fourth-order moments of the distribution function, which is applicable for spatially varying electric field. The validity of the new model was verified through the comparison between the numerical calculation based on the generalized moment conservation equations and MC simulation in the $n^+nn^+$ structure.

## REFERENCES

1. T. Kunikiyo, M. Takenaka, Y. Kamakura, M. Yamaji, H. Mizuno, M. Morifuji, K. Taniguchi, and C. Hamaguchi, "Monte Carlo Simulation of Anisotropic Electron Transport in Silicon Including Full Band Structure and Anisotropic Impact-Ionization Model," *J. Appl. Phys.,* vol. 75, p. 297 (1994).
2. Y. Kamakura, H. Mizuno, M. Yamaji, M. Morifuji, K. Taniguchi, C. Hamaguchi, T. Kunikiyo and M. Takenaka, "Impact Ionization Model for Full Band Monte Carlo Simulation," *J. Appl. Phys.,* vol. 75, p. 3500 (1994).
3. R. Thoma, A. Emunds, B. Meinerzhagen, H. Peifer and W. L. Engl, "A Generalized Hydrodynamic Model Capable of Incorporating Monte Carlo Results," *IEDM Tech. Dig.,* p. 139 (1989).

# SCREENING INDUCED IMPACT IONIZATION IN AlGaAs PIN PHOTODETECTORS UNDER HIGH OPTICAL ILLUMINATION

R.A.Dudley, A.J.Vickers

Department Of Physics
University Of Essex
Wivenhoe Park
Colchester
Essex
C04 3SQ
United Kingdom

## Abstract

We present a study of $Al_{0.3}Ga_{0.7}As$ 200μm diameter PIN photodiodes excited with high power, pico-second laser pulses. All devices were found to exhibit a reduced bandwidth with increasing optical power, confirming recent experimental[1] and numerical results[2,3] on electric field screening within photo-detecting devices. Monte Carlo and drift diffusion simulations show the fields time evolution within our devices by solving Poisson's equation.[2,3] Not only do these results show a reduction in the field across most of the device, but that compensating regions of high field are generated near the p-i and i-n junctions, on time scales less than 1ns. Experimental results show proof of impact ionisation occurring in our devices at fields of $\approx 150$kV/cm, lower than typically observed using c.w measurements, $>250$kV/cm.[4] We suggest that the regions of high field generated by carrier screening, become sufficient to induce carrier multiplication, giving rise to impact ionisation at an apparently lower field than suggested by the applied bias and 'i' region width.

## Introduction

PIN photodiodes are currently the primary choice for high-speed photo-detection, particularly in telecommunications, due to their stable, linear behaviour. However, with the demands of a growing network, higher capacities require higher bit rates. Increased device bandwidth generally translates into smaller devices to overcome capacitance and transit time effects. Reduced dimensions produce higher charge densities within the device, and hence stronger influences from mechanisms such as carrier screening. Under these regimes the PIN diode is no longer a linear well behaved device, but in fact shows significant degradation at high carrier densities. In this paper we investigate some of the properties of PIN photo-diodes under high illumination, and present the first, to our knowledge, experimental evidence of induced impact ionisation by carrier screening.

## Experimental

In our experiments, bulk $Al_{0.3}Ga_{0.7}As$ PIN diodes were fabricated by wet chemical etching down to an n+ GaAs substrate, producing 200μm diameter mesas incorporating an annular top contact, (p+ and n+ of $10^{18}$ cm$^{-3}$, intrinsic $10^{15}$ cm$^{-3}$, all regions $\approx 1$μm thick). Finished mesas were cleaved into 250μm squares and bonded, using conductive epoxy, onto 50Ω semi-Insulating GaAs coplanar transmission lines, terminated with SMA connectors. All diodes exhibited reverse breakdowns in excess of 30 volts, and leakage currents of a few micro-amps.

Optical excitation of the diodes was provided by a 5ps Nd:YAG pumped R6G dual dye jet laser operating at 585nm, producing pulses with a 76MHz repetition rate. Focusing onto the p+ top contact was achieved through a x10 microscope objective and CCD camera viewing system. Resulting temporal traces were recorded using a 25ps sampling oscilloscope, and logged by an interfaced computer. Measured device speeds of 600-700ps FWHM are somewhat larger than the predicted RC limitation of 150ps, suggesting an influence from minority carrier diffusion through the p+ region into the intrinsic layer.

Effects of increased bias and incident optical power produced two distinct regimes. At powers below 1mW the diodes response reveals full width half maxima that initially decrease, stabilising above 3-5 volts, figure 1. The initial speed increase arises from a greater diffusion efficiency of minority carriers across the p+ layer and into the intrinsic region. Above zero bias, carriers are swept away more rapidly from the p-i contact, creating a larger diffusion gradient across the p+ region and more rapid transport. Eventually, the rate of carrier escape from the p-i contact will saturate, at which stage no further speed increase can occur. The enhanced device speed produces a complementary variation in the temporal respons peek height, creating an initial rise followed by a region of saturation, figure 2. On the time scales of even the slowest responses here, carrier recombination processes are not occurring in significant amounts. Thus, all photo-generated carriers produced within the device can be assumed to escape.

At optical powers above 2 milli-Watts the devices response becomes significantly non-linear. Two main effects were observed, firstly with increasing power the FWHM of the devices are seen to increase, figure 1, secondly the peak signal height increases rapidly when the applied field is above 150kV cm[-1] and the optical power is in excess of 3mW, figure 2. Further, the threshold field at which the peak height rapidly increases is inversely proportional to the incident laser power, figure 2 insert. Two process are thought to produce this non-linear behaviour, namely electric field screening and carrier multiplication.

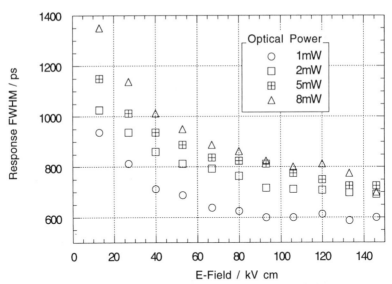

**Figure 1**: Diodes FWHM dependence on field and optical power.

Field screening results from the interaction of mobile charge carriers with the field of an illuminated semiconductor region. As carriers drift across the i region of a externally biased pin, the magnitude of the field in the intrinsic layer is altered according Poisson's equation. It is well known that under high illumination, the field is reduced across the majority of the active region, producing compensating high fields at each contact.[5] Kaul et al's[6] simple analysis shows that the carrier densities generated within our devices are sufficient to generate field screening. Further, Viallet et al shows deviations from the non-illuminated field pattern occurring at carrier generations equivalent to <2mW shining on a 100μm diode. The consequent slowing of carriers in the low field region creates a device with an overall reduced bandwidth, as reported by Kaul and Denton et al.[7]

Increasing peak heights with high bias and illumination suggest a carrier multiplication mechanism. However, observed fields of 150kV cm[-1] are somewhat lower than those typically seen in impact ionisation studies[5], >300 kV cm[-1] are more typical. Such differences in threshold are thought to result from assuming the diodes field is simply a ratio of applied volts and i region width. At low optical powers this assumption is valid, but as Poisson's equation shows, the field cannot be regarded as a simple ratio under high illumination. Calculated profiles such as Viallet's[7] show regions of high field near the contacts, produced by screening mechanisms. While suggested by Viallet, impact ionisation has not been thought to occur in these regions due to limited transit distances and field strengths. However, recent suggestions by Dunn[2] indicate that fields can become sufficient to stimulate multiplication, but have not been confirmed with any published experimental evidence. Further, Wilson[8] suggests that a multiplication of two or three times could be generated across a region of only a few hundred nanometers when the field is above the ionisation threshold.

## Conclusion

In conclusion we have shown that under certain bias and illumination strengths, impact ionisation can be self induced within a device. In some circumstances this will be undesirable and operating conditions will need to be changed to avoid it.

## Acknowledgements

Both authors would like to thank EPSRC. R.Dudley is grateful to Essex University Physics Department for support.

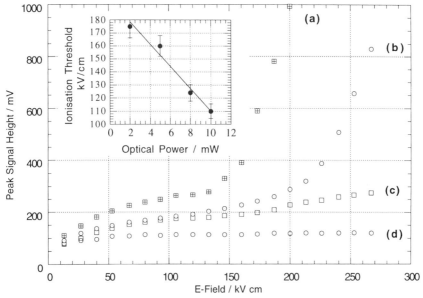

**Figure 2:** Peak response height with increasing field and optical intensity, (a)8mW (b)5mW (c)2mW (d)1mW. Insert shows the threshold field for multiplication with increased optical power.

1. K.Williams R.Esman M.Dagenais, "Effects of high space charge fields on the response of microwave photodetectors", IEEE P.Tech.Lett, Vol 6 No 5, p639, May 1994.
2. G.Dunn G.Rees J.David, "Monte carlo simulation of impact ionisation and current multiplication in pin diodes", TBP.
3. S.Wilson S.Brand R.Abram, "Spatial impact ionisation transients in GaAs: their origin and characteristic", Semiconductor Sci Tech, 9, p1174, 1994.
4. V.Robbins S.Smith G.Stillman, "Impact ionisation in AlGaAs for x=0.1-0.4", Appl.Phys.Lett, 52, p296, (1988).
5. J.Viallet, E.Mottet, L.Le Huerou, C.Boisrobert, "Photodiode for coherent detection: Modeling and experimental results", Journal De Physique, 49, C4-321, 1988.
6. D.Kuhl, F.Hieronymi, E.Holger, Bottcher, T.Wolf, "Influence of space charges on the impulse response of InGaAs MSM photodetectors", IEEE J.Light.Tech, Vol10 No6, p753, 1992.
7. M.Dentan, B.DeCremoux, "Numerical simulation of the non-linear response of a pin photodiode under high illumination", IEEE J.Light.Tech, vol 8, no 8, p1137, 1990.
8. S.Wilson S.Brand R.Abram, "Spatio-Temporal impact ionisation transients: a luck drift model study in GaAs", Solid State Electronics, Vol 38 No2, Feb 95.

# INTERVALLEY SCATTERING AS A FUNCTION OF TEMPERATURE IN GaAs USING TIME-RESOLVED VISIBLE PUMP – IR PROBE ABSORPTION SPECTROSCOPY

Michael A. Cavicchia, Wubao Wang, and R. R. Alfano

Institute for Ultrafast Spectroscopy and Lasers, New York State Center for Advanced Technology for Ultrafast Photonic Materials and Applications, Physics Department, The City College and The Graduate School of the City University of New York, New York, New York 10031

Understanding the detailed physics underlying intervalley scattering in multi-valley semiconductors is crucial for the efficient design of high-speed electronic and microwave devices and has been the subject of intense study over the past decade. Several experimental and theoretical studies based on ultrafast spectroscopy have been performed to obtain the scattering times for hot electrons going between the central, $\Gamma_6$, valley and the satellite, $X_6$ and $L_6$, conduction band valleys in GaAs under various conditions.[1,2] The techniques previously used involved investigations of the hot electron distribution only in the central, i.e., k=0, conduction band valley.

In this report, we have performed time-resolved visible pump - IR probe absorption measurements to isolate the dynamics of hot electrons in the $X_6$ satellite valley, at $k \neq 0$, of GaAs as a function of lattice temperature.

A 500-fs visible (585 nm) pump pulse is used to photoexcite electrons from the valence bands to the central, $\Gamma_6$, valley. Electrons excited from the heavy-hole and light-hole bands obtain sufficient kinetic energy to undergo intervalley scattering to the satellite, $X_6$ and $L_6$, valleys (see Fig. 1 (a)). A 500-fs IR probe pulse (3.3 $\mu$m) is used to monitor the induced absorption. The induced absorption is due to free carrier absorption (FCA), inter-valence band absorption (IVA), and $X_6 \rightarrow X_7$ inter-conduction band absorption (ICA).

The results for the temporal behavior of the induced IR absorption at temperatures of 4°K, 77°K, and 290°K are shown in Fig. 2. The three curves in Fig. 2 are characterized by resolution-limited rise times and a two component decay. The initial decay time varies from 11-ps to 33-ps as the lattice temperature is decreased from 290°K to 4°K. The initial decay is followed by much slower decay that is essentially flat on the time scale of the measurement.

The total induced IR absorption at 3.3-$\mu$m is due to free carrier absorption (FCA), hh → split-off and lh → split-off inter-valence band absorption (IVA), and $X_6 \rightarrow X_7$ inter-conduction band absorption (ICA).[3] The induced FCA is expected to have a resolution-limited rise and then decay slowly, on the time scale of the electron-hole recombination or the plasma diffusion, i.e., hundreds of picoseconds for this carrier density and temperature. Thus, the temporal decay of the induced FCA will be essentially flat on the time scale of the

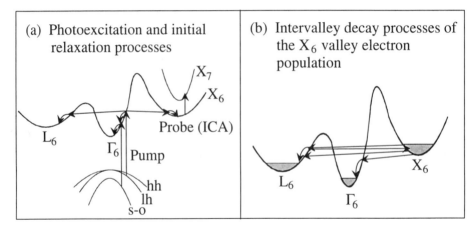

**Fig. 1** Schematic diagram of the pump, probe, and relaxation in the band structure of GaAs, where the $\Gamma_6$, $L_6$, and $X_6$ valleys of the conduction band are indicated. (a) The photoexcitation (pump) transitions, initial relaxation processes, and ICA (probe) transition are shown by arrows. (b) The intervalley and intravalley scattering processes which contribute to the intervalley decay time of the $X_6$ valley electron population are indicated by arrows.

present experiment. This expected behavior was observed to occur for the same pumping condition but with a probe wavelength of 4.7-μm, which is below the threshold for ICA or IVA (see inset of Fig. 2).

The induced IVA should have a temporal behavior similar to that of the FCA. Direct measurements[4] have shown that the photoexcited holes thermalize and scatter to the heavy-hole band within the first few hundred femtoseconds after photoexcitation. Thus, after a resolution-limited rise only a slow, flat decay component for the induced IVA is expected.

The $X_6 \rightarrow X_7$ ICA is directly proportional to the $X_6$ valley electron population. After reaching a maximum, the decay process of the electron population in the $X_6$ valley will consist of multiple scatterings back and forth between the $X_6$ and the $L_6$ and $\Gamma_6$ valleys before the average kinetic energy of the electrons falls below the minimum energy of the $X_6$ valley. Eventually all of the electrons will transfer to the lower energy $L_6$ and $\Gamma_6$ valleys and the $X_6 \rightarrow X_7$ ICA will decay to zero. Therefore, we attribute the initial decay component of the total induced IR absorption to the decay of the ICA. The decay time of the ICA is the effective intervalley decay time of the $X_6$ valley electron population.

Stanton and Bailey[5] (SB) have shown that the intervalley scattering times and population decay times are not simply related to each other. For high-energy electrons, the intervalley scattering time, $\tau_{inter}$, is faster than the intravalley scattering time, $\tau_{intra}$. By modeling time-resolved experiments using a rate equation analysis SB have shown that when $\tau_{inter} < \tau_{intra}$, the decay time of an electron population in an upper valley is limited by the slower intravalley relaxation of the electron energy in the lower valley. Therefore, the decay time of the electron population in the $X_6$ valley depends not only on the $X_6 \rightarrow \Gamma_6$, $L_6$ intervalley scattering rate, *but also depends strongly on* the intravalley relaxation rate in the $L_6$ and $\Gamma_6$ valleys. This decay process is depicted schematically in Fig. 1(b).

The temperature dependence of the intravalley relaxation rate by emission of polar-LO phonons can be estimated using the standard formula given by Conwell.[6] The intravalley relaxation rate can be shown to vary less than one percent over the temperature range of 4°K to 290°K for the electron energies of this experiment. Therefore, the intravalley polar-LO phonon scattering has no appreciable influence on the observed temperature dependence of the decay time of the $X_6$ valley electron population. *The temperature dependence of the effective intervalley decay time is due solely to the temperature dependence of the $X_6 \rightarrow \Gamma_6$, $L_6$ intervalley*

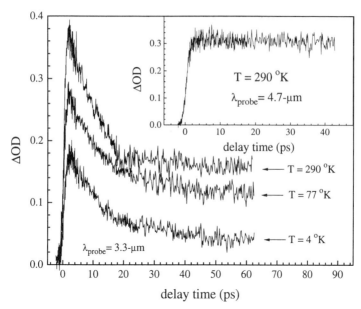

**Fig. 2** Time-resolved IR absorption at 3.3-μm, for a 585-nm pump pulse, measured at crystal temperatures of 4°K, 77°K, and 290°K. The inset shows the induced absorption at 4.7-μm for identical pumping conditions at a crystal temperature of 290°K.

*scattering time.* Thus, the $X_6 \rightarrow \Gamma_6$, $L_6$ intervalley scattering time decreases by a factor of ~3 as the temperature is reduced from 290°K to 4°K. This result is in excellent agreement with the "rigid-pseudoion" model calculations of Zollner, et. al.[7]

To obtain a complete picture of the temperature dependence of the $X_6 \rightarrow \Gamma_6$, $L_6$ intervalley scattering time to compare with theory, the time-resolved measurements of the induced IR absorption were extended to temperature intervals of ~20°K between 4°K and 290°K. For each temperature the effective intervalley decay time of the $X_6$ valley electron population was determined. The results are plotted in Fig. 3. The temperature dependence of the intervalley scattering time depends on which phonon modes are involved in the scattering. The deformation potentials for the interaction are essentially temperature independent. The main temperature dependence comes from the temperature dependence of the phonon occupation numbers according to the Bose-Einstein function.

Zollner, et. al.,[7] have performed a calculation for the average intervalley scattering time, $\langle \tau \rangle$, for an electron at a given **k** point. In their formalism, they define an intervalley phonon spectral function which corresponds to the density of phonon states weighted with the relevant intervalley deformation potential and the final electron density-of-states. When the intervalley phonon spectral function is taken to be a δ-function with an "effective" deformation potential, $D_{eff}$, at an "effective" phonon energy, $\Omega_{eff}$, then their expression for the intervalley scattering time, $\langle \tau \rangle$, reduces to the Conwell approximation:[6]

$$\frac{1}{\langle \tau(T_L) \rangle} = \frac{2\pi\hbar}{\rho\Omega_{eff}} D_{eff}^2 g(\mathcal{E}_k) \left[ N_{eff}(T_L, \Omega_{eff}) + \frac{1}{2} \right], \qquad (1)$$

where $\rho$ is the crystal density; $g(\mathcal{E}_k) = N_V m^* [2m^* \mathcal{E}_k]^{1/2}/(2\pi^2\hbar^3)$ is the final electron density-of-states, where $N_V$ is the degeneracy of the final valley and $m^*$ is the electron effective mass in the final valley; and $N_{eff}$ is the Bose-Einstein occupation factor evaluated at $\Omega_{eff}$. By fitting the data points in Fig. 3 with equation (1), an accurate determination of $\Omega_{eff}$ is obtained. The solid line in Fig. 3 is a model calculation using the Conwell approximation for the

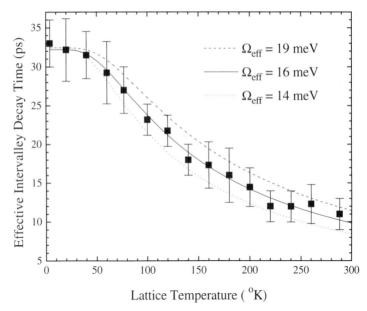

**Fig. 3** Temperature dependence of the effective intervalley decay time of the $X_6$ valley electron population obtained from the $X_6 \rightarrow X_7$ ICA. The solid line is the best theoretical fit to the data for $X_6 \rightarrow \Gamma_6$, $L_6$ intervalley scattering with $\Omega_{eff} = 16$-meV. The upper (dashed) and the lower (dotted) curves are obtained with values of $\Omega_{eff} = 19$-meV and $\Omega_{eff} = 14$-meV, respectively, and normalized at 4°K.

temperature dependence of the intervalley scattering time, multiplied by a temperature independent normalization constant, with an effective phonon energy of $\Omega_{eff} = (16 \pm 3)$ meV.

The value of $\Omega_{eff}$ obtained from the intervalley phonon spectral function of Zollner, et. al., for electrons in the $X_6$ valley is about 25-meV. The much lower measured value for $\Omega_{eff}$ indicates a stronger participation of the transverse acoustic phonon modes in intervalley scattering from the $X_6$ valley than expected from the theory.

This research was supported by the New York State Technology Foundation and NASA.

## References

1. D. W. Bailey, C. J. Stanton, K. Hess, M. J. LaGasse, R. W. Schoenlein, and J. G. Fujimoto, Femtosecond studies of intervalley scattering in GaAs and $Al_xGa_{1-x}As$, *Solid State Electron.* **32**:1491 (1989).
2. M. J. Kann, A. M. Kriman and D. K. Ferry, Effect of electron-electron scattering on intervalley transition rates of photoexcited carriers in GaAs, *Phys. Rev. B* **41**:12 659 (1990).
3. W. B. Wang, N. Ockman, M. Yan, and R. R. Alfano, Determination of $X_6$ valley hot electron dynamics and the intervalley $X_6 \rightarrow \Gamma_6$ scattering time in GaAs, *J. Lumin* **50**:347 (1992).
4. X. Q. Zhou, K. Leo, and H. Kurz, Ultrafast relaxation of photoexcited holes in n-doped III-V compounds studied by femtosecond luminescence, *Phys. Rev. B* **45**:3886 (1992).
5. C. J. Stanton and D. W. Bailey, Rate equations for the study of femtosecond intervalley scattering in compound semiconductors, *Phys. Rev. B* **45**:8369 (1992).
6. E. Conwell, High field transport in semiconductors, *in:* "Solid State Physics, Advances in Research and Applications", Supp. 9, F. Seitz, D. Turnbull, and H. Ehrenreich, eds., Academic Press, New York (1967).
7. S. Zollner, S. Gopalan, and M. Cardona, Microscopic theory of intervalley scattering in GaAs: k dependence of deformation potentials and scattering rates, *J. Appl. Phys.* **68**:1682 (1990).

# REAL TIME IMAGING OF PROPAGATING HIGH FIELD DOMAINS IN SEMI-INSULATING GaAs

F.Piazza, P.C.M.Christianen and J.C.Maan

High Field Magnet Laboratory and Research Institute for Materials,
University of Nijmegen,
Toernooiveld, 6525 ED Nijmegen,
The Netherlands

## INTRODUCTION

Semi-Insulating (SI) GaAs shows spontaneous current oscillations when a DC bias voltage in the order of 1 kV/cm is applied, related to the formation of high field domains. This self organized structure is commonly explained by a negative differential resistance due to field enhanced capture of electrons, presumably in EL2 traps. Although a model has been elaborated since 1970[1], many input parameters and their dependence on electric field, which are generally poorly known, must be introduced. Measuring only current and voltage at the electrodes does not provide for sufficient information for any quantitative description. Therefore we present results obtained with a time resolved, non-invasive technique that allows a quantitative measurement of the voltage profile for a domain traveling in the sample. From those measurements, we obtain all the necessary information relevant for the description of the domain structure, and have direct access to the electric field dependence of the trapping coefficient, which is the origin of the domain formation.

## EXPERIMENT

The experimental set-up is based upon the longitudinal electro-optic effect of a $Bi_{12}SiO_{20}$ (BSO) crystal to transform a voltage distribution in a phase shift of light. The crystal has a transparent electrode on the front and a dielectric mirror on the back side and is put with the back side on top of the sample. An expanded polarized pulsed laser beam enters the crystal through the transparent electrode and is reflected back by the mirror through an analyzer. Grounding the top electrode of the BSO, the phase shift induced by the voltage difference between the two sides of the crystal can be recorded as a light intensity distribution on a CCD camera. The quantitative relation between the voltage and the phase shift is obtained by an in-situ calibration. The laser pulses and the readout of the camera are synchronized with the current oscillations to resolve the measurement in time[2]. The sample is a <100> LEC grown single crystal GaAs-wafer: the mobility is 0.66 $m^2$/Vs and the electron density $1.9 \ 10^{13} \ m^{-3}$, with a thickness of 0.5 mm and having the NiAuGe contacts 8 mm apart.

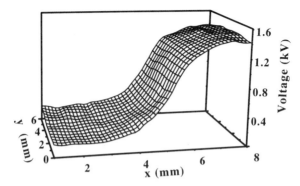

**Figure 1.** Voltage distribution as function of the position on the sample. The anode and the cathode are 8 mm apart in x direction. The applied voltage is 1.6 kV, the delay with respect to the current pulse is 50 ms and the exposure time 0.2 ms .

For bias voltages below 0.6 kV, the sample behaves ohmic, while for higher voltages the current oscillates between two levels. When the current level is low a high resistivity, high electric field region exists, and the voltage drops by 80% in a region of about 2 mm (Fig 1). This region is formed at the cathode and moves through the sample toward the anode with a constant velocity and without changing shape. Upon arrival of a domain at the anode the current rapidly increases by a factor of ten and remains large during the whole process of annihilation of one domain at the anode and the simultaneous creation of the next one at the cathode, to drop only when the completely formed domain leaves the cathode. Therefore it is clear that the peak in the current is not simply caused by the absence of a high resistivity region in the sample, as usually stated, but it is related to the formation-annihilation mechanism in a more complex way.

The experiment shows the presence of a non-linear voltage profile and the electric field and the charge distribution can be obtained by differentiating the measured voltage along the axis of the sample as presented in Fig.2 a) for a domain propagating in the bulk under an external bias of 1.4 kV. The domain consists of a 4 mm wide almost gaussian peak of the electric field sustained by a symmetric charge dipole. The electric field has a peak value of $6 \cdot 10^5$ V/m, larger then the $3 \cdot 10^5$ V/m that are necessary to activate intervalley scattering in GaAs. To obtain the measured space charge density a carrier density of $2 \cdot 10^{17}$ m$^{-3}$ is required, that is four orders of magnitude larger than the density of free electrons. Thus, the slow domains, in contrast with the Gunn domains, are not formed by a modulation in space of the density of electrons in the conduction band but must involve electrons trapped in impurities. This detailed quantitative information has not been available before and forms the basis for our description of the domain.

## DISCUSSION

The basic equations underlying the behavior of the slow domains are Poisson equation and charge conservation, including current drift and rate equations for free and trapped electrons as formulated originally by Sacks and Milnes[1]. Because of the importance of EL2 as trap the new available information about EL2 defects[3,4] is included. The concentration of EL2 defects in GaAs is about $2 \cdot 10^{22}$ m$^{-3}$ and, at equilibrium and room temperature about $3 \cdot 10^{21}$ m$^{-3}$ of them are ionized in a positive charge state to compensate the different densities of fully ionized shallow donors and acceptors. The important quantities for the formation of the domains are therefore $n$, the free electron density, and $n_t$, the fluctuation in density of the filled EL2 states; depending on the sign of the fluctuation, excess positive charge is originated by donors and excess negative charge by acceptors. The ratio between $n$ and the density of trapped electrons is determined by the trapping and emission coefficients ($C_n$ and $X_n$ respectively), which appear in the rate equation. The profile of $n$ is determined by the current via the continuity equation and $n+n_t$ is related to the electric field through the Poisson equation. Using our experimental data and including the measured constant domain velocity, we have calculated $n(x)$ and $n_t(x)$ (Fig.2 b) ) directly from the experiment. The high resistivity of the domain is found to be due to the 80% depletion of free electrons in the domain and the charge dipole that is necessary to increase the field is almost entirely produced by a fluctuation in $n_t$. It is important to note that, although $n_t$ is four orders of magnitude larger than $n$, it is still five orders of magnitude smaller than the density of filled states at equilibrium. Therefore the variation in the ratio between free and trapped electrons is completely determined by $n$ only and the reduction of $n$ in the region of high electric field is a direct evidence for the field enhanced trapping. The possibility to build up the observed charge dipole without any appreciable contribution from $n$ and with a small effect on the density of filled states is a necessary condition for the existence of the slow domain.

Unfortunately the knowledge of $n$ and $n_t$ is not sufficient to calculate independently $C_n$ and $X_n$, which eventually are microscopically responsible for this phenomenon. Since the velocity of the domain is so slow compared to the trapping-emission process, the free and trapped electrons are locally, in quasi equilibrium and only the ratio between the two coefficients can be determined directly. In literature, $X_n$ is commonly considered constant up to $10^7$ V/m [5], when the Poole-Frenkel effect becomes important. The behavior of $C_n$ (Fig.3), evaluated under this approximation, is, for low electric field, in good agreement with the theoretical model of Sacks and Milnes, calculated supposing an electrostatic barrier around EL2 defects; after a very small plateau $C_n$ raises almost linearly by a factor of ten up to a value of $33 \cdot 10^{-16}$ m$^3$/s in correspondence with an electric field of $2 \cdot 10^5$ V/cm. For higher electric fields $C_n$ decreases: this effect is not explained in any model and can be due to a real reduction of the trapping coefficient or to an enhancement of the emission coefficient. More experimental work is necessary to discriminate between the two possibilities.

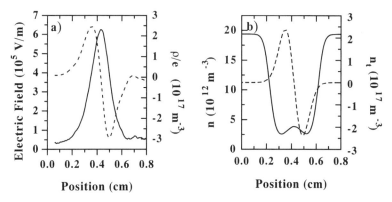

**Figure 2.** a) Electric field (solid line) and charge distribution (dotted line) in the domain. The bias voltage is 1.4 kV and the time delay with respect to the current peak is 150 ms. The electric field and the charge density divided by electron charge are obtained via differentiat ion of the directly measured voltage profile. b) calculated $n$, free electron density (solid line) and $n_t$, the fluctuation of trapped electrons in EL2 defects around the equilibrium value (dotted line).

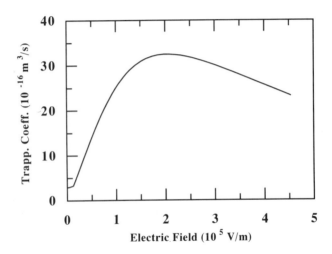

**Figure 3.** Trapping coefficient $C_n$ as function of the electric field obtained from the data assuming the emission coefficient $X_n$ constant.

In summary we have presented experimental data of slow domains in SI-GaAs in which all relevant quantities are obtained locally and time resolved. From this quantitative experimental description the underlying non-linearities related to trapping and detrapping of electrons are evaluated and the domains are explained.

1. H.K.Sacks and A.G.Milnes, *Int. J. Electronics* 28:565 (1970).
2. B.Willing and J.C.Maan, *Phys. Rev. B* 49:13995 (1994).
3. D.C.Look, *Semicond. and Semimetal* 38: 91 (1993).
4. M.Kaminska and E.R.Weber, *Semicond. and Semimetal* 39:91 (1993).
5. S.Makram-Ebeid and M.Lannoo, *Phys. Rev. B* 25: 6406 (1982).

# HOT ELECTRON EXCITATION OF LUMINESCENT IMPURITIES IN AC THIN-FILM ELECTROLUMINESCENT DEVICES

S. Pennathur, I. Lee, M. Reigrotzki‡, R. Redmer‡, K. Streicher, T. K. Plant, J. F. Wager, P. Vogl†, and S. M. Goodnick

Department of Electrical and Computer Engineering, Oregon State University, Corvallis, OR 97331-3211, USA

† Walter Schottky Institute, Technical University Munich, D-85748 Garching, Germany

‡ Fachbereich Physik, Universitaet Rostock, Germany

## INTRODUCTION

AC thin-film electroluminescent (ACTFEL) devices are an essential technology for future flat-panel display applications. The basic structure of a typical ACTFEL device consists of a phosphor layer sandwiched between two insulating layers and a pair of electrodes. The phosphor consists of a wide bandgap semiconductor such as ZnS ($E_g$=3.7 eV) heavily doped ($\sim 1\%$) with a luminescent impurity such as $Mn^{+2}$. Light emission is achieved by the application of large AC voltages to the layer structure such that electrons are injected into the high-field phosphor layer from interface states at the insulator-semiconductor interface either due to tunneling or due to field emission. For sufficiently high-fields (typically 1-2 MV/cm), electrons in the phosphor layer may gain sufficient energy to impact excite electrons in the luminescent impurities from the ground to excited states, which subsequently undergo radiative decay emitting photons[1]. The luminescent and power conversion efficiency of such devices critically depend upon the high-field distribution of electrons in the conduction bands of the phosphor, and the impact excitation rate for exciting luminescent impurities. Thus, an understanding of high-field carrier transport in the phosphor layer and of the physics of different threshold processes such as band-to-band impact ionization and impact excitation of luminescent impurities is essential for ACTFEL device design.

## FULL BAND MONTE CARLO MODEL FOR ZnS

The band structure of ZnS has been calculated using the nonlocal empirical pseudopotential method[2] (EPM), in which the calculated reflectivity spectrum is fit to recent uv spectroscopic ellipsometric data[3] for this material. For the purpose of full-band Monte Carlo simulations, the dispersion relation and density of states for the first four

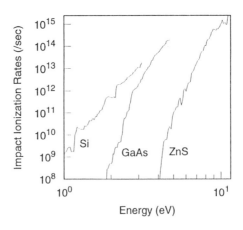

**Figure 1.** Calculated average impact ionization rates for Si, GaAs and ZnS.

conduction bands have been used to describe the carrier dynamics. The nonlocal correction to the pseudopotential was included, along with a corresponding correction for calculating velocities as well as effective masses[4]. The calculated effective masses and intervalley separation energies corresponding to the nonlocal EPM as well as the local band structure calculation used by Brennan[5] in earlier Monte Carlo work on ZnS are listed in Table 1. The main difference is that the intervalley separation in the nonlocal case is considerably smaller than the earlier local calculation.

The energy values and energy derivatives (i.e. velocities) corresponding to an irreducible wedge in the Brillouin zone are computed and stored so that the carrier energies and velocities during the course of the Monte Carlo simulation can be determined (interpolated, when necessary) from the tabulated full-band dispersion relation. The scattering rates computed using a nonparabolic band model are normalized at high energies using the ratio of the full band to nonparabolic density of states. In addition to all the relevant phonon and impurity scattering including polar optical, acoustic, and nonpolar optical phonon scattering as well as elastic scattering due to ionized and neutral impurities[6], the full-band Monte Carlo model also includes both band-to-band impact ionization and impact excitation of luminescent impurity atoms. The impact excitation rate for $Mn^{2+}$ is calculated based on the exchange reaction proposed by Bringuier[7], which is simplified by using a parabolic band approximation for a short range scattering potential normalized by the nonlocal EPM density of states. An averaged energy-dependent band-to-band impact ionization rate[8] computed using an EPM derived full band dispersion is incorporated into the Monte Carlo model. A $q$-dependent dielectric function derived by Levine and Louie[9] is used to model the interaction between conduction and valence electrons in the calculation of the ionization rates.

## RESULTS AND DISCUSSION

Owing to the lack of information about the deformation potentials of the nonpolar intervalley scattering mechanism in the second and higher conduction bands, the values used in the simulations are obtained by tuning them to fit the measured electric field variation of impact ionization coefficient[10] in ZnS as shown in Fig. 2. The simulated

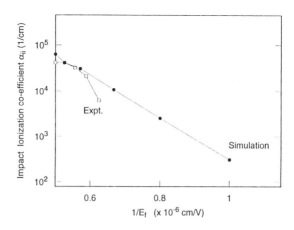

**Figure 2.** Measured and simulated impact ionization coefficient, $\alpha_{ii}$, as a function of inverse electric field.

electron energy distribution for three different phosphor electric field strengths is shown in Fig. 3a, compared with the calculated impact excitation cross-section of $Mn^{2+}$. The distribution becomes hotter (i.e. increasing average energy) with increasing phosphor fields. For Mn centers (used for yellow luminescence) with an excitation threshold energy of about 2.1 eV, it is seen that a considerable number of the electrons in the ensemble are energetic enough to cause impact excitation. From the distributions shown, we estimate that about 2.1%, 15.3 %, and 34.4% of the electrons possess energy above 2.1 eV at fields of 1 MV/cm, 1.5 MV/cm and 2.0 MV/cm respectively.

An estimate of the quantum yield (a measure of the per-electron contribution to the luminescence) is obtained by tracking the number of impact excitation events per electron across a 0.5 $\mu$m phosphor layer. Figure 3b shows a plot of this impact excitation quantum yield as a function of the phosphor field for Mn luminescent centers. It shows a threshold of approximately 0.8 MV/cm, increasing monotonically for increasing fields. This monotonic increase is consistent with experimentally observed trends in the brightness-voltage curves[12] of ACTFEL devices. Saturation of the quantum yield at high electric fields is concomitant with the domination of band-to-band impact ionization at these fields. It is noted that collision broadening has not been treated in this work despite the fairly high scattering rates in ZnS.

**Table 1.** Calculated effective masses and intervalley separation energies.

|  | Valley | Nonlocal EPM | CB local EPM[11] |
|---|---|---|---|
| effective | $\Gamma$ | 0.18 | 0.20 |
| mass | L | 0.28 | 0.30 |
| $(m_0)$ | X | 0.46 | 0.46 |
| intervalley | $\Gamma$-L | 1.16 eV | 1.48 eV |
| separation | $\Gamma$-X | 0.94 eV | 1.48 eV |

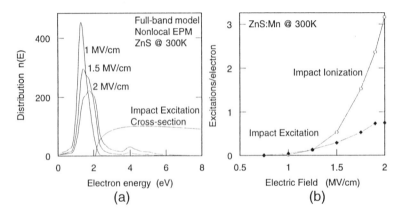

**Figure 3.** (a) Calculated energy distribution of carriers for various electric field strengths compared with the calculated impact excitation cross-section of $Mn^{2+}$. (b) Excitations/electron for impact ionization and excitation processes.

## CONCLUSIONS

A full-band Monte Carlo simulation of high-field transport in the ZnS phosphor of an ACTFEL device is presented, using the band structure of ZnS based on a nonlocal empirical pseudopotential calculation. The steady-state electron energy distribution for typical phosphor fields in the range of 1 MV/cm to 2 MV/cm reveal a substantial fraction of electrons with energies in excess of the excitation threshold energies of typical luminescent centers such as $Mn^{2+}$. While band-to-band impact ionization plays a crucial role in stabilizing the electron energy distributions at the highest fields, the impact excitation yield is found to be strongly affected by the presence of impact ionization and the associated threshold fields for this mechanism.

## REFERENCES

[1] D. C. Krupka, *J. Appl. Phys.* 43:476 (1972).

[2] J. R. Chelikovsky and M. L. Cohen, *Phys. Rev. B* 14:556 (1976).

[3] J. Barth J, R. L. Johnson and M. Cardona, *Handbook of Optical Constants of Solids II* (Boston: Academic) pp. 213 (1991).

[4] M. M. Rieger and P. Vogl, *Phys. Rev. B* 48:276 (1993).

[5] K. Brennan, *J. Appl. Phys.* 64:4024 (1988).

[6] K. Bhattacharyya, S. M. Goodnick and J. F. Wager, *J. Appl. Phys.* 73:3390 (1993).

[7] E. Bringuier and K. Bhattacharyya, To appear in *Phys. Rev. B.*

[8] M. Reigrotzki, M. Stobbe, R. Redmer and W. Schattke, To appear in *Phys. Rev. B.*

[9] Z. H. Levine and S. G. Louie, *Phys. Rev. B* 25:6310 (1982).

[10] N. E. Rigby, T. D. Thompson and J. W. Allen, *J. Phys. C: Solid State Phys.* 21:3295 (1988).

[11] M. L. Cohen and T. K. Bergstresser, *Phys. Rev.* 166:789 (1966).

[12] D. H. Smith, *J. Lumin.* 23:209 (1981).

# SPATIALLY MODULATED HOT CARRIER TRANSPORT AND LIGHT DIFFRACTION IN NON-UNIFORM MICROWAVE ELECTRIC FIELDS

L.Subačius[1], V.Gružinskis[1], E.Starikov[1], P.Shiktorov[1], and K.Jarašiūnas[2]

[1]Semiconductor Physics Institute, A Goštauto 11, LT2600 Vilnius, Lithuania
[2]Vilnius University, Saulėtekio ave. 9-3, LT2054 Vilnius, Lithuania

## INTRODUCTION

Temporal and spatial modulation of light-induced free carrier plasma and its heating by applied electric field lead to novel effects in transport phenomena, which may be used for various purposes such as contactless techniques for measurements of kinetic coefficients and nonequilibrium carrier dynamics, high-speed optoelectronics, holography, etc. Combining light diffraction on light-induced free carrier (FC) transient grating (TG) with an electron gas heating by uniform microwave (MW) field in GaAs, we estimated hot carrier diffusion coefficients and their field dependencies[1]. Hot carrier transport and non-linear optical phenomena which arise in TG due to non-uniform carrier heating are not sufficiently studied.

In this paper, we present a modeling results of spatially modulated nonequilibrium carrier transport and of non-linear optical phenomena for the case of non-uniform electron gas heating. A comparison with the experimental data on light self-diffraction efficiency in GaAs and Si crystals in external MW field is carried out.

## TRANSIENT GRATING TECHNIQUE

To account for the main physical processes inherent in a light-induced free carrier transient grating the basic continuity equations were written in the form[2]:

$$\frac{\partial N}{\partial t} = \alpha I(x,t) - \gamma N - \frac{\partial}{\partial x}\left[ N\mu_n(E)E(x,t) - D_n(E)\frac{\partial N}{\partial x} - N\tau_v(E)\frac{\partial \phi(E)}{\partial x} \right], \qquad (1)$$

$$\frac{\partial P}{\partial t} = \alpha I(x,t) - \gamma P - \frac{\partial}{\partial x}\left[ P\mu_p E(x,t) - D_p \frac{\partial P}{\partial x} \right], \qquad (2)$$

where $N(x,t)$ and $P(x,t)$ are the electron and hole concentrations, $\alpha$ is the absorption coefficient, $I(x,t)=I_0 f(t)(1+cos(2\pi x/\lambda)$ is the incident light interference pattern, $\lambda$ is the spatial period of the grating, *f(t)* describes the temporal shape of the laser pulse, $\gamma$ is the recombination rate, $\mu_n(E)$, $D_n(E)$, $\tau_v(E)$, $\phi(E)$ are the Monte Carlo calculated parameters of the extended drift-diffusion model[2], $\mu_p$ and $D_p$ are the hole mobility and diffusion coefficient taken at the thermal equilibrium, $E(x,t)$ is the electric field obtained from a simultaneous solution of the Poisson equation.

The experiments have been performed in high resistivity bulk GaAs and Si crystals by using interference pattern of 10 ns duration of YAG laser ($\lambda$=1.06 $\mu$m) and synchronously applying the MW field in 10 GHz frequency range. The grating period $\Lambda$=15...50 $\mu$m, photoexited carrier concentration $\Delta N \approx \Delta P = 10^{16}...10^{17}$ cm$^{-3}$ and MW electric field amplitude $E_m$=0...7 kV/cm were monitored. The time integrated self-diffracted beam intensity on transient grating was measured under carrier heating by non-uniform MW field $E_m \parallel grad$N.

## RESULTS AND DISCUSSION

Instantaneous spatial profiles of internal electric field $E_{int}$ and of modified electron concentration N/m* are given in Figure1 for GaAs crystals at various $E_m$. The calculations revealed very strong local internal field (up to ~80 kV/cm at $E_m$=8 kV/cm, $\Lambda$=25 $\mu$m) and formation of Gunn-domain in the minima of light-interference pattern. The amplitude of $E_{int}$ follows the oscillations of MW field, and the value of $E_{int}$ depends on $E_m$, FC concentration and its spatial profile. The localization of Gunn-domain in narrow region (in few $\mu$m) leads to many spatial Fourier harmonics of $E_{int}$. Peculiarities of electron heating effects are seen also in spatial dependencies of modified electron concentration N(E)/m*(E), which value in grating minima decrease with increasing $E_m$ (see Figure1b); here electron concentration decreases due to nonuniform carrier heating and subsequent carrier thermo-diffusion (the last term in eq.(1)), while the increase of m* in GaAs is governed by hot electron inter-valley repopulation. In Si crystals, the spatial profiles of $E_{int}$ and N(E)/m*(E) are rather close to sinusoidal shape; here the field-domain, localized in grating minima, determines up to 3-fold increase of $E_m$, while increase of m* due to valley non-parabolicity is negligible.

For the further analysis of nonuniform MW field effect we calculated the corresponding changes of refractive indices by free carrier $\Delta n_{FC}$ and electro-optical nonlinearities $\Delta n_{EO}$, taking into account two Fourier harmonics. Calculations revealed significant influence of internal field formation on refractive indices modulation in TG. The field-caused peculiarities of $\Delta n_{FC}$ exist at the very beginning of laser pulse and after its action, i.e. when nonequilibrium carrier concentration is low enough. Meanwhile, field-domain induced electro-optical effect dominates during the laser pulse action.

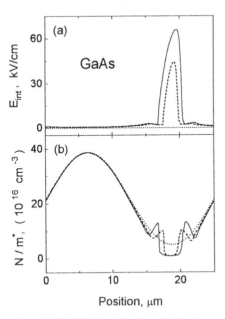

**Figure1.** Spatial dependencies of instantaneous values of internal electric field (a) and modified electron concentration N/m* (b) at the peak intensities of MW and laser pulses (t=12 ns) for GaAs. $E_m$, kV/cm: dot - 0, dash - 4, solid - 8.

Therefore, the expected variations of refractive index modulation seen in time-integrated regime of light self-diffraction on transient gratings are determined mainly by field induced nonlinear electro-optical mechanism. The field induced changes of diffraction intensity on hot carrier grating are negligible in given time-domain.

As for the electro-optic effects, the anisotropic feature of photorefractive nonlinearity predicts zero linear effects for crystals of cubic symmetry (as Si) and for used (001)-oriented GaAs samples. Thus, we analyzed the quadratic electro-optic effect $\Delta n_{EO} \sim E_{int}^2$, which may take place in GaAs, and the electrorefractive photorefractivity $\Delta n_{PR} \sim E_{int}^2$ due to Franz-Keldysh effect in Si for used laser frequency.

The self-diffraction intensity $I_1^* \sim \Delta n_{EO}^2$ into direction of the first order takes into account the overlapping of diffraction from the first and the second spatial Fourier harmonics of $E_{int}$, and integrates the instantaneous values of diffraction intensities. The calculated field dependency $I_1^*(E_m)$ and measured field-induced changes of normalized diffraction efficiencies $\eta = [I_1^*(E_m)-I_1^*(0)]/I_1^*(0)$ are shown in Figure 2. The comparison of calculated field dependencies of self-diffraction efficiencies with the experimental data leads to conclusion that the electro-optical mechanisms of enhancement of diffraction intensity are dominant in time-domain of ns-pulse both for GaAs and Si crystals.

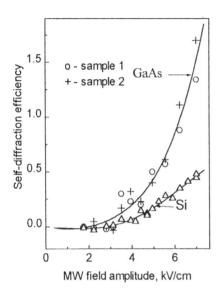

**Figure 2.** Field dependencies of normalized self-diffraction efficiency at grating period $\Lambda=25$ μm for GaAs and Si crystals. Points - experimental, lines - calculation results.

The experimental results obtained by varying TG period $\Lambda$ exhibited the stronger effect of external MW field on normalized diffraction efficiencies at smaller transient grating periods. This peculiarity also follows from our calculations, because the diffusion-governed free carrier TG decay is very sensitive to grating period, while the value of $I_{EO}^*$ is not so sensitive to variation of $\Lambda$, both in GaAs and Si crystals.

## ACKNOWLEDGMENTS

The research described in this publication was made possible in part by Grant LA9000 from the International Science Foundation, and Grant LI1100 from the Joint Program of the Government of Lithuania and the International Science Foundation.

## REFERENCES

1. L.Subačius, and K.Jarašiūnas, Experimental studies of light - microwave field interaction and nonequilibrium carrier transport in GaAs, in: Proceedings of the European Gallium Arsenide and Related III - V Compounds Applications Symposium, Torino, Italy (1994).
2. V.Gružinskis, E.Starikov, and P.Shiktorov, Conservation equations for hot carriers - I. Transport models, Solid- State Electronics, **36** : 1055 (1993).

# CARRIER ACCUMULATION IN MOMENTUM
# SPACE UNDER INTENSE MICROWAVE FIELDS

Tatsumi Kurosawa and Norihisa Ishida

Department of Physics
Chuo University
Bunkyo-ku, Tokyo 112, Japan

## ABSTRACT

Hot carrier dynamics under intense microwave fields is investigated theoretically for the case that the dominant scattering process is optical phonon emission. If the microwave intensity is appropriate, an accumulated distribution of carriers in momentum space appears. This is a new type of dynamic population inversion and is found to cause various peculiar dc field responses under realistic physical conditions.

## Accumulated Distribution of Carriers

In many semiconductors, the carrier strongly interacts with the optical phonon. In pure crystals at low temperatures, the collision time is long when the energy is less than the optical phonon energy $\hbar\omega_{op}$, while it is very short if the energy is larger than $\hbar\omega_{op}$. It has long been known that the system of such carriers exhibits characteristic behaviours in strong electric fields.[1] As described below, the system shows peculiar properties also under intense microwave fields.[2]

Under microwave fields, carriers make sinusoidal motion in momentum space. If the amplitude is appropriately large, the following situation occurs: Carriers on most trajectories are accelerated above $\hbar\omega_{op}$ and scattered into low energy region within a period of the microwave field, while some carriers belonging to the region as shown by the hatching in Fig.1 do not arrive at $\hbar\omega_{op}$ and continue the free motion for their long collision time. As a consequence, the carriers are accumulated in this region: Figure 2 shows the distribution function calculated with one-dimensional model. In these figures, $p_{op}$ denotes the momentum corresponding to $\hbar\omega_{op}$. The width of the accumulation region $p_w$ is given by $2(p_{op} - eE_1/\omega)$, in which $E_1$ is the microwave amplitude, $\omega$ the angular frequency. The degree of accumulation is roughly proportional to $\omega\tau_a$, where $\tau_a$ is the average collision time of the accumulated carriers. The accumulated distribution is a new type of dynamic population inversion and causes various peculiar effects.

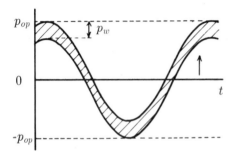

**Figure 1.** Formation of accumulated distribution (shown by hatching).

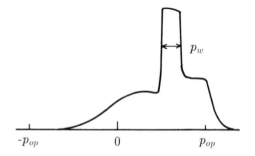

**Figure 2.** Distribution function at the time shown by the arrow in Fig.1.

## DC Responses

Among others the response to dc electric fields ($E_d$) parallel to the microwave field exhibits many remarkable features. Figures 3 and 4 show examples of the response (drift velocity $v_d$ ) to $E_d$, where $v_{op}$ is the carrier velocity corresponding to $\hbar\omega_{op}$ and $E_d$ is expressed in dimensionless forms. Figure 3 is calculated with one-dimensional model, which does not mean the quantum wire case but corresponds to the 1D-projection of the motion in 3D momentum space. The assumed collision frequency is nearly the same with that in Fig.4. Figure 4 shows a more realistic 3D case, heavy holes in Ge at low temperatures with $2 \times 10^{12}$ cm$^{-3}$ charged centers. The effects of the band warping on the carrier motion are fully taken into account. Two figures show very similar features. The 1D model is treated easily with an iterative method but the results represent farely well the main features of the corresponding 3D case in general, thus it is useful as a guide for complicated 3D problems.

The first significant feature is the strong nonlinearity found at very small $E_d$. This low field nonlinearity is due to the decrease in the accumulated carriers that have long collision time and large mobility. The fractional decrease is roughly given by $eE_d\tau_a/p_w$, which determines the degree of nonlinearity. The second feature is the complicated structures in the $v_d$-$E_d$ relation often accompanying NDC. The arrows in Figs.3 and 4 denote the dc field strength satisfying the relation, $eE_d^{(k)}(k/2)T_p = p_w$, where $T_p$ is the period of the microwave and $k$ is an odd integer (indicated together with the arrows). The field $E_d^{(k)}$ represents the critical field where a carrier group of long collision time (high mobility) stepwise disappears: Above $E_d^{(k)}$ all the carriers are accelerated above $\hbar\omega_{op}$ within the time $(k/2+1)T_p$.

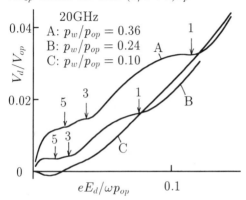

**Figure 3.** The $v_d$-$E_d$ relation in 1D model; similar parameters to Fig.4.

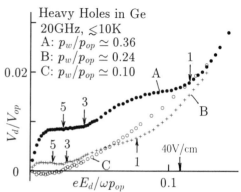

**Figure 4.** The $v_d$-$E_d$ relation for heavy holes in Ge. On the arrows, see the text.

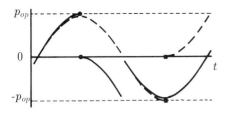

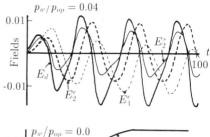

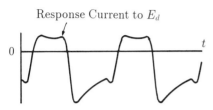

Response Current to $E_d$

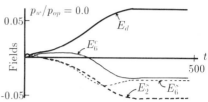

**Figure 5.** Explanation for the negative mobility (upper trace) and response current to $E_d$ calculated by 1D model.

**Figure 6.** Development of the field components caused by the negative mobility (1D model).

## Negative Mobility and Related Instabilities

As $p_w$ approaches 0 the structures vanish, instead the low field response becomes negative. In general the low field mobility has a sharp dip as a function of $p_w$ around $p_w = 0$ and becomes negative when $\omega\tau_a$ is suitably large, say $> 5$. Figure 5 depicts the origin of the small mobility, showing typical trajectories of carriers causing the negative response (upper trace); broken curves represent the trajectories in the absence of $E_d$ and solid ones are for $E_d \neq 0$. The lower trace shows the waveform of the response current to $E_d$ calculated with the one-dimensional model, which is in accord with the upper trace and gives a negative average.

The response current is found to be accompanied by strong higher harmonic components, especially of $2\omega$. Because of the negative mobility the system is unstable against the growth of the dc field, but the higher harmonic components may grow even more rapidly. The carriers respond nonlinearly to the latter too. The motion of this coupled system may differ depending on the physical conditions. For instance if the higher harmonic waves rapidly run away, the instability will be similar to the usual negative resistance case. Another extreme is the case that the higher harmonic waves are perfectly confined in the system and the medium is lossless except for the loss due to the carriers. Figure 6 shows two typical examples of the development of the main field components in the latter case. Here for the applied microwave field $E_1 \sin \omega t$, the generated field is taken to be $E_d + \sqrt{2} \sum_n (E_n^c \cos n\omega t + E_n^s \sin n\omega t)$. Both the ordinate and abscissa are represented in dimensionless forms; i.e. the unit of the electric fields is $p_{op}\omega/e$ and that of the time is $\kappa m^*\omega/4\pi N e^2$, where $N$ and $\kappa$ are the carrier density and the dielectric constant of the matter. In the upper figure the system makes a chaotic oscillation, while in the lower case it converges to a stationary state.

## REFERENCES

1. For example, S.Komiyama, T.Kurosawa, and T.Masumi, Chapter 6, *in:* "Hot Electron Transport in Semiconductors," L.Reggiani, ed., Springer-Verlag, Berlin (1985).
2. A more detailed presentation for a part of this article will appear in: N.Ishida and T.Kurosawa, *J. Phys. Soc. Jpn.* 64: 2996 (1995).

# CURENT INSTABILITIES IN BIPOLAR HETEROSTRUCTURES

Viktoras Gružinskis[1], Jevgenij Starikov[1], Pavel Shiktorov[1],
Luca Varani[2], and Lino Reggiani[3]

[1] Semiconductor Physics Institute, Goštauto 11, 2600 Vilnius, Lithuania
[2] Centre d' Electronique de Montpellier Universitè Montpellier II,
34095 Montpellier Cedex 5, France
[3] Dipartimento di Scienza dei Materiali ed Istituto Nazionale di Fisica della
Materia, Università di Lecce, Via Arnesano, 73100 Lecce, Italy

## INTRODUCTION

Solid-state devices with S-type negative differential conductance (NDC) have a wide application in microelectronics as fast switchers and microwave power generators. Recently, significant attention has been devoted to the heterostructure hot-electron diode (HHED)[1] where S-type NDC and high-frequency oscillations up to 100 $GHz$ have been obtained[1-4]. To improve the high-frequency performance of the HHED we propose here to replace the anode $n^+$ contact with a $p^+$ layer thus introducing the possibility for an additional hole current and a second switching of the device to a higher conductance state.

## THEORETICAL MODEL AND RESULTS

We consider the vertical transport in a layered heterostructure which consists of an active region placed between heavily doped GaAs $n^+$ and $p^+$ contacts. The active region contains three layers A, B and C. The $n$-GaAs layer A is placed at the $n^+$ contact and serves as a drift region for electrons and holes. The switching time and generation frequency depends mainly on the length of this layer. The $n - Al_xGa_{1-x}As$ layer B creates a barrier of about 0.53 eV for holes with $x = 1$. Because of the high resistivity of $Al_xGa_{1-x}As$ as compared with that of GaAs, the B layer must be as thin as possible. The undoped GaAs layer C (spacer) is introduced between the B layer and the $p^+$ contact to minimize the dispersion of momentum distribution of holes entering the respective barrier. The best device performance can be achieved when the C layer thickness is of about $100 \div 300$ Å and the thickness of A-B and B-C heterojunctions are $20 \div 100$ Å wide. All the results presented below are obtained for a structure with an active region length of 800 Å. The length of the layer A is of 600 Å with doping concentration of $n = 10^{17} cm^{-3}$. The layers B and C are 100 Å thick each. The layers B doping concentration is $n = 10^{17} cm^{-3}$ and C is undoped. The thickness of the A-B and B-C heterojunctions is assumed to be the same and equal to 40 Å. The doping concentration of contacts is $n^+ = p^+ = 10^{19}\ cm^{-3}$. Carrier transport throughout the structure is investigated by a direct solution of the coupled Poisson and Boltzmann equations performed with a Monte Carlo Particle technique. Since carrier transport times thorugh the active region are considerably shorter than the electron-hole recombination time, this last process is neglected. A three-valley conduction band and a single heavy-hole valence band is considered by using spherically symmetric nonparabolic dispersion laws in all bands. The peculiarity of the structure is that the A and C layers serve as heating regions for electrons and holes, respectively, and the intermediate B layer is the place where potential barriers prevent electrons and holes from penetrating into the diode. At low applied voltages, both barriers act as closed gates and the system is in the lowest conductance-state. At increasing voltages a first high conductance-state is achieved when the electron barrier acts as an open gate owing to thermionic emission of electrons via upper valleys. By further increasing the voltage, the onset of a second higher conductance-state, which corresponds to the situation where also the hole barrier acts as an open gate, takes place. The above three states (labelled respectively as I, II, and III) correspond to a conductance which is controlled, respectively, by a cold electron, a hot electron, and a hot electron-hole current. A switching from one state to another should be responsible for a large variation of the total current at practically the same applied voltage which, in turn, should lead to the appearance of S-type regions in the current-voltage characteristic. The simulation of the current transport throughout the given structure shows that the transition between states I and II did not lead to S-type NDC while the transition between states II and III shows a strong S-type NDC. Figure 1 presents the current voltage characteristic of the structure at $T = 10$ and $300\ K$. Figure 2 reports the results of a simulation performed by connecting the diode in series with a load resistance $R = 2.5 \times 10^{-12}\ \Omega m^2$ and in parallel with a capacitance $C = 20 \times C_d$, where $C_d$ is the geometrical capacitance of the structure. Here we evaluate the time variation of the voltage drop on $R$, $U_R(t)$, when the voltage applied to the whole circuit is sweeping as $U_a(t) = at$, with $a = 10^{11}\ V/s$. Three different temperatures $T = 10\ ,150$ and $300\ K$ are considered. The fast switching of $U_R(t)$ (switching time $t_{sw} < 3\ ps$) is independent from $U_a(t)$. To clarify the switching mechanism, the evolution of the potential drop on the structure, $U_d(t)$, and the average concentrations of $X$-valley electrons $n_X(t)$ and holes $p(t)$ in the A layer are presented in Fig. 3 at $T = 150\ K$. Both $n_X(t)$ and $p(t)$ are normalized to the average concentration of electrons in A layer $n(t)$. The fast growth of $n_X(t)$ for $U_d(t) > 1\ V$, indicates the onset of thermionic emission of electrons over the barrier, that is the I-II transition. The further growth of $U_d(t)$ is followed by a slower increase of the electron current, which is typical of short $n^+nn^+$ structures. When the current becomes large enough to produce a voltage drop on the C region sufficient for holes to overcome their barrier, the II-III sets in, i.e. $p(t)$ increases from 0 to 0.9. The growth time of $p(t)$ corresponds approximately to $t_{sw}$. When R is sufficiently high, the structure can operate as a microwave generator. This is shown in Fig. 4 where current oscillations through $R$ are clearly seen under the application of a constant voltage $U_0$ on the whole circuit. Here the following values for the parameters are used: $U_0 = 11\ V$, $R = 10^{-10}\ \Omega m^2$, $T = 80\ K$, and an Al fraction in B layer of $x = 0.7$. Accordingly, we find an oscillation frequency $f = 1.2\ THz$ and a microwave power conversion efficiency of 0.26%. Due to the high current flowing in the circuit, the diameter of the structure, $d$, must be small. For $d = 5\ \mu m$, the structure generates a 45 $mW$ power in a load resistance of 5 $\Omega$. The generation frequency can be tuned down to very low frequencies by taking appropriate values of $C$. It must be noted that, when the Al fraction $x$ in the B layer increases, the generation frequency decreases while the efficiency and power increase. For example, with $x = 1$, $U_0 = 20\ V$, $R = 10^{-10}\ \Omega m^2$, $T = 80\ K$ the generation frequency is $f = 0.95\ THz$, the efficiency of 0.5%, and the generation power for $d = 5\ \mu m$ of about 300 $mW$. The same structure at $T = 300\ K$ generates with $f = 0.7\ THz$, an efficiency of 0.08%, and the generation power of 35 $mW$.

## ACKNOWLEDGMENTS

This work is performed within the European Laboratory for Electronic Noise (ELEN) supported

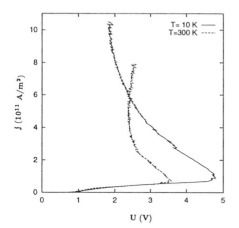

Fig. 1 - Current voltage characteristic of the structure for an Al fraction in the B layer of $x = 1$.

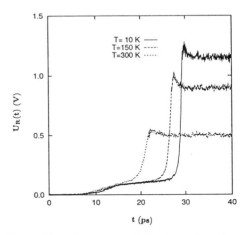

Fig. 2 - Time dependence of the voltage drop between $R$ terminals under the influence of a linear increase of the applied voltage for an Al fraction in the B layer of $x = 0.7$.

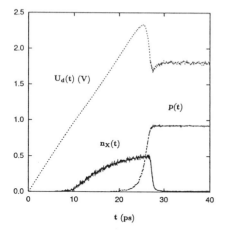

Fig. 3 - Time dependence of the voltage drop between the diode terminals, $U_d(t)$, and of the relative concentration of $X$-valley electrons, $n_X(t)$, and holes in the layer A, $p(t)$, within the same conditions of Fig. 2.

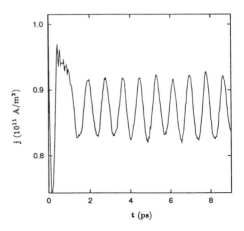

Fig. 4 - Current oscillations in the load resistance when a constant voltage $U_0 = 11\ V$ is applied to the whole circuit.

through the contract ERBCHRXCT920047. Partial support from CNRS PECO/CEI/2003 bilateral cooperation project is gratefully acknowledged.

## REFERENCES

1. K. Hess, T. K. Higman, M. A. Emanuel, and J. J. Coleman, J. Appl. Phys., **60**, 3775 (1986).
2. A. M. Belyantsev, A. A. Ignatov, V. I. Piskarev, M. A. Sinitsyn, V. I. Shashkin, B. S. Yavich, and M. L. Yakovlev, JETP Lett., **43**, 437 (1986).
3. A. Reklaitis and G. Mykolaitis, Sol. State Electron., **37**, 147 (1993).
4. A. Wacker and E. Scholl, Appl. Phys. Lett., **59**, 1702 (1991).

# DRIFTED HOT ELECTRON DISTRIBUTION FUNCTION AND MEAN FREE PATH INVESTIGATED BY FIR-EMISSION FROM GaAs/AlGaAs HETEROSTRUCTURES

C. Wirner[1], C. Kiener[2], W. Boxleitner[2], E. Gornik[2], G. Böhm[3], and G. Weimann[3]

[1]Osaka University, Dept. of Electronic Engineering, Suita-City, 565 Osaka, Japan
[2]Institut für Festkörperelektronik, TU Wien, Floragasse, A-1040 Wien, Austria
[3]Walter-Schotty-Institut, TU München, Am Coulombwall, D-85448 Garching, Germany

## ABSTRACT

We have investigated FIR-Emission from a periodically modulated high mobility GaAs/AlGaAs heterostructure. Smith-Purcell type radiation directly correlated to the drifted distribution function of the carriers is detected at 200nm modulation period. The onset of LO-phonon emission is clearly observed. The mean free path of the hot carrier distribution is derived by measuring the emission spectrum at different modulation period lengths. A transition from Smith-Purcell type radiation to collective plasmon emission is found by increasing the height of the modulation potential.

## 1. INTRODUCTION

The transport characteristics of hot electrons in AlGaAs/GaAs heterostructures are of great interest due to the very high electron mobilities and very low electron densities in these structures[1]. Deviations of the drifted hot electron distribution function from a drifted hot Fermi distribution have been predicted if the drift velocity exceeds the Fermi velocity and the electron-electron interaction is weak compared to LO-phonon emission[2].

It has been demonstrated experimentally that hot electron transport in low density heterostructures is indeed dominated by LO Phonon scattering reducing the mean free path even in high mobility samples to values shorter than 1μm and resulting in a saturation of the drift velocity to $2 \cdot 10^7 cm/s$[3-5].

In this paper we present FIR-emission from a periodically modulated drifting hot electron gas as an experimental tool to study directly the shape of the drifted hot electron distribution function and to derive its mean free path in high mobility GaAs/AlGaAs heterostructures.

## 2. THE HOT CARRIER DISTRIBUTION FUNCTION

In 1953 Smith and Purcell discovered that electrons moving close to a periodic grating potential with its grating vector parallel to the direction of motion interact with the potential resulting in an energy loss via photon emission[6]. The analog effect in a high mobility heterostructure is realised by creating a periodic potential on the top surface of the sample acting on the carriers driven by an external electric field[7].

Theoretically the emission spectrum is calculated by second order perturbation theory. Assuming very low electron density and sufficiently high electric fields the emission power $P(\omega)$ in the energy interval $[\hbar\omega, \hbar\omega + d(\hbar\omega)]$ is

$$P(\omega) \propto |V_0|^2 \sum_{\pm q} \left\langle |q| \cdot \int dk_y f[k_x(q,\omega), k_y] \right\rangle \qquad (1)$$

where $V_0$ is the strength of the modulation potential, $q = 2\pi/a$ the grating momentum and $f(k_x, k_y)$ the distribution function drifted in x-direction. As is evident from equation (1) the emission spectrum $P(\omega)$ directly reflects the drifted distribution function $f(\mathbf{k})$.

In our experiment a sample is used having an electron density of $n_{2d} = 6 \cdot 10^{10}$ cm$^{-2}$ and a mobility of $\mu = 8 \cdot 10^5$ cm$^2$/Vs. The periodic potential is fabricated by nanostructuring the surface and wet chemical etching. FIR-emission is detected by a magnetic field tunable InSb detector.

Figure 1(a) (solid lines) shows the electric field dependence of the detected radiation in the case current flow parallel to the grating vector representing the intersection of the distribution function driven by the external electric field. Emission energies shift to higher values with increasing electric field. In addition the slope of the high energy decay becomes steeper at 75 V/cm. A discussion of the undrifted case represented by the geometry with current flow perpendicular to the grating vector is omitted here and can be found elsewhere[8]

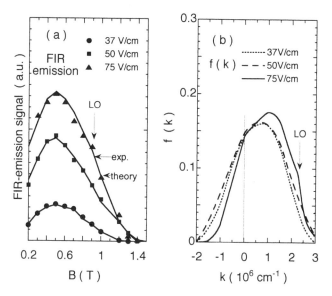

**Figure 1.** (a) Solid lines: FIR-Emission spectrum recorded from a high mobility GaAs/AlGaAs heterostructure with a potential modulation period of a=200nm at three different external electric fields. Current flow is parallel to the grating vector. Symbols: Theoretical emission spectrum from a calculated distribution function shown in Figure 1b. (b) Theoretical distribution function in direction of the external electric field calculated using a Boltzmann integration numerical technique.

In order to relate the results to the drifted hot carrier distribution function first a theoretical distribution is determined by transforming the Boltzmann-equation into an integral equation which was solved numerically taking all scattering events into account. The theoretical ansatz is explained in detail elsewhere[2]. The result is displayed in Figure 1(b) for three different electric fields. Then the emission power of this theoretical distribution is calculated according to equation (1).

The theoretical emission signals (Figure 1, symbols) are in good agreement with the experimental data. The steep slope arising at 75 V/cm in the experimental data (Figure 1a) is clearly related to the onset of LO-phonon emission found also in the theoretical distribution function (Figure 1b). All features of the distribution function are reflected in the experimental spectra. The FIR-emission experiment is thus suitable to directly record the heated distribution function. Note that in our low density structure electron-electron interaction is weak compared to LO-phonon emission resulting in deviations of the distribution from a heated Fermi-Dirac distribution.

## 3. THE MEAN FREE PATH

Smith-Purcell type FIR-emission is present only if the modulation period length is longer than the mean free path of the carriers. Thus studying the changes of the FIR-emission

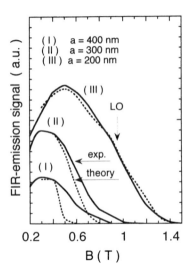

**Figure 2.** Solid lines: FIR-Emission spectrum from a high mobility GaAs/AlGaAs heterostructure having different grating period lengths of a=200nm, a=300nm, and a=400nm at an electric field of 75 V/cm. Current flow is parallel to the grating vector. Dashed lines: Corresponding calculated emission signals from a theoretical distribution function.

signal as a function of the modulation period length will provide valuable information about the mean free path of the hot carriers.

Figure 2 (solid lines) displays FIR-Emission spectra recorded from samples having different grating period lengths of a=200nm, a=300nm and a=400nm at an electric field of 75 V/cm. Current flow is parallel to the grating vector. With increasing the modulation length an intensity loss and a shift of the maximum emission intensity to smaller energies is found. The high energy tail becomes significantly smoother with increasing grating period and all spectra extend to almost the same energy level.

The corresponding theoretical emission intensity emitted from a distribution calculated using numerical techniques is shown as dashed lines in Figure 2. At modulation periods longer than a=200nm a clear broadening of the experimental spectra compared to the theoretical results is found. The experimental signal extends to significantly higher energies than predicted by theory. The discrepancy increases with increasing modulation period.

In order to relate the broadening of the spectra to the limited mean free path $\lambda$ of the hot carriers we assume that carriers having a mean free path $\lambda$ will interact with a potential of the form

$$V_q(x,\lambda) = V_0 \cos (qx) \exp (-x/\lambda) \qquad (2)$$

where q is the grating vector and $\lambda$ is the average interaction length of the hot electrons to be determined[9]. Performing a Fourier analysis of the interaction potential (2) the emission power for an arbitrary grating vector Q is calculated according to equation (1). The total emission power is then determined by performing the integral over all contributions from arbitrary momentum transfers Q of the electron-grating interaction:

$$P(\omega,\lambda) = \int dQ \, P_Q(\omega,\lambda) \qquad (3)$$

Figure 3 shows a comparison of the experimental data recorded at the grating period a=400nm (Current flow parallel to the grating vector) with calculations assuming different mean free paths. While the emission signal extends to remarkably higher energies in the limit of the mean free path much longer than the grating period length ( $\lambda \gg a$ ) a good agreement is found when the interaction length is limited to $\lambda = 300$nm. This result is confirmed for the samples with a=300nm and a=200nm. In the latter case no broadening effect is evident (cf. Figure 2) indicating that only at grating periods a $\leq 200$ nm the emission spectrum directly reflects the hot carrier distribution function.

In order to confirm the ansatz made in equation (2) the mean free path $\lambda^{th}$ is calculated directly from the collision-free flight time of the driven carriers taking into account all scattering mechanisms.

The results (in direction of the applied electric field) are displayed in Figure 4. LO-phonon emission results in a remarkable reduction of $\lambda^{th}$ at high wave vectors. $\lambda^{th}$ are

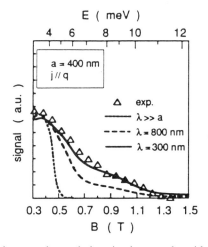

**Figure 3.** Comparison of the experimental data in the sample with a grating period a=400nm (Current flow parallel to the grating vector) with calculations assuming a mean free path $\lambda \gg a$, $\lambda = 800$nm, $\lambda = 300$nm.

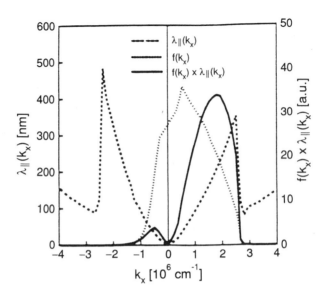

**Figure 4.** Intersection lines of the theoretically calculated two-dimensional mean free path $\lambda$th (broken curve), the distribution function $f(k_x, k_y)$ (dotted curve) and the weighted MFP $\lambda$th $\cdot f(k_x, k_y)$ (full curve) in electric field direction.

mean free paths which electrons can reach according to their initial momentum k. For comparison of $\lambda^{th}$ with the value obtained by the Fourier analysis one has to take into account the occupation number of the electron wave vector k according to the distribution function of the carriers. The full curve in Figure 4 represents $\lambda^{th}$ weighted by the distribution function. It is evident that mainly MFP values from 200nm to 300nm contribute to the emission process. These values are in good agreement with $\lambda=300$nm determined by the Fourier analysis confirming that the Fourier analysis model is a good approximation and the interpretation of the average interaction length in equation (2) as ensemble value of the mean free path of the electric field driven carriers is correct.

## 4. PLASMON EMISSION

In this paragraph we present FIR-emission signals recorded at different heights of the periodic potential. The height of the periodic potential is changed by etching the sample stepwise and recording the FIR emission signal at the different etch steps. Etching not only increases the potential height but also decreases the electron mobility and the drift velocity. Important sample data at the different etch steps are recorded in Table 1.

Figure 5 shows FIR-Emission spectra at different strengths of the periodic confinement potential in the geometry current flow parallel to the grating vector recorded at 75 V/cm.

**Table 1**

| Etchstep | Electr. density ( $10^{11}$ cm$^{-2}$ ) | Mobility ( $10^5$ cm$^2$ / Vs ) | Drift at 75 V/cm ( $10^7$ cm / s ) | Potential height ( meV ) |
|---|---|---|---|---|
| 0 (2DEG) | 0.8 | 9 | 1.46 | 0 |
| 1 | 0.6 | 8 | 1.42 | 0.6 |
| 2 | 0.6 | 4.8 | 1.2 | 0.8 |
| 3 | 0.55 | 3 | 1 | 1.2 |
| 4 | 0.55 | 1 | 0.63 | 1.5 |

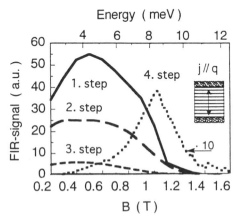

**Figure 5.** FIR-Emission spectrum from a GaAs/AlGaAs heterostructure having different potential strengths (indicated by etch steps). Current flow is parallel to the grating vector.

An intensity loss is observed with increasing the etch depth and hence the potential modulation. In addition the steep decay of the emission signal at high energies is only observed in the first etch step and missing already in the second etch step. At the fourth etch step the emission signal completely changes to a narrowband signal having its intensity maximum at 8.5 meV. Similar signals are also observed in the case current flow perpendicular to the grating vector.

The effects of the increasing potential on the emission signal at the first three etch steps are similar to what is observed with increasing the grating period length. Therefore we attribute the changes to the reduction of mobility, drift velocity and hence mean free path of the carriers induced by increasing the potential strength. At the last etch step the mean free path is too short to see effects related to the drifted distribution function. However electrons can couple collectively to the strong periodic potential emitting plasmon radiation. The emission energy is in reasonable agreement with theoretical predictions, shifted to somewhat higher energies as was also observed and explained by Hoepfel et al[10].

The experimental results clearly show that Smith-Purcell-type radiation of quasifree electrons can only be observed in samples with mean free paths sufficiently longer than the modulation period while collective plasmon emission is present at short mean free paths in a sufficiently stong periodic interaction potential.


## 5. CONCLUSION

In conclusion we have studied FIR-Emission from a periodically modulated GaAs/AlGaAs heterostructure. In samples where the mean free path of the carriers is significantly longer than the modulation period length the emission spectrum directly reflects the drifted hot carrier distribution function. A non equilibrium distribution dominated by LO-phonon emission is observed. The mean free path of the distribution is derived from the broadening of the emission spectrum at grating period lengths comparable with the mean free path of the carriers. By increasing the modulation potential strength and at the same time decreasing the mean free path a transition from a broadband Smith-Purcell-type FIR-emission to a collective narrowband plasmon emission is observed.


## REFERENCES

1   K. Hirakawa and H. Sakaki, Phys.Rev. **B 33**, 8291 (1986)
2   C. Kiener and E. Vass, J.Appl.Phys. **64**, 6365 (1988)
3   G. Zandler, C. Kiener, W. Boxleitner, E. Vass, C.Wirner, E. Gornik and G. Weimann, J. Appl.Phys. **70** (11), 6842 (1991)
4   W.T. Masselink, N. Braslau, D. LaTulipe, W.I. Wang, and S.L. Wright, Solid State Electron. **31**, 337 (1988)
5   A. Palevski, M. Heiblum, C.P. Umbach, C.M. Knoedler, A.N. Broers, and R.H. Koch, Phys.Rev.Lett. **62**, 1776 (1989)
6   S.J. Smith and E.M. Purcell, Phys.Rev. **92**, 1069 (1953)
7   E. Gornik, R. Christanell, R. Lassnig, W. Beinstingl, K. Berthold and G. Weimann, Solid State Electron. **31** (3/4), 751 (1988)
8   C. Wirner, C. Kiener, W. Boxleitner, M. Witzany, E. Gornik, P. Vogl, G. Böhm, and G. Weimann, Phys.Rev.Lett. **70** (17), 2609 (1993)
9   C. Kiener, C. Wirner, W. Boxleitner, E. Gornik, G. Böhm, and G. Weimann, Semicond.Sci.Technol. **9**, 193 (1994)
10  R.A. Höpfel, and G. Weimann, Appl.Phys.Lett. **46**, 291 (1985)

# HOT-ELECTRON DIFFUSION AND VELOCITY FLUCTUATIONS
# IN QUANTUM WELL STRUCTURES

V. Aninkevičius[1], V. Bareikis[1], R. Katilius[1], P.M. Koenraad[2], P.S. Kop'ev[3],
J. Liberis[1], I. Matulionienė[1], A. Matulionis[1], V.M. Ustinov[3],
W.C. van der Vleuten[2], and J.H. Wolter[2]

[1]Semiconductor Physics Institute, Vilnius 2600, Lithuania
[2]University of Technology, Eindhoven 5600, The Netherlands
[3]A.F. Ioffe Institute, St. Petersburg 194021, Russia

## INTRODUCTION

Sensitivity of hot-electron fluctuations to electron scattering at elevated energies has been exploited to investigate real-space transfer[1,2] and other kinetic processes[1] in selectively-doped quantum-well (QW) structures. In this technique, transverse freedom of originally two-dimensional electrons is controlled by their kinetic energy acquired in the external electric field applied in the plane of the QW. Lattice temperature, interelectrodal distance, barrier height, QW width are varied to excite and resolve intrasubband, intersubband, real-space-transfer and other fluctuations[3]. The technique is complementary to the optical ones developed to study kinetic process in QW structures (see, e.g.[4-6]).

The field-controlled interplay between planar and transverse kinetic processes helps observing sources of hot-electron velocity fluctuations absent in bulk semiconductors. Our goal is to resolve longitudinal hot-electron velocity fluctuations caused by transitions from confined to partially extended states in δ-doped GaAs and in δ-doped GaAs/AlGaAs heterostructure containing a thin barrier of AlAs inside the QW. The electron transverse freedom being increased with heating, the planar mobility increases in the δ-doped GaAs, but it decreases in the selectively doped GaAs/AlGaAs heterostructure. In both cases the excess longitudinal velocity fluctuations and diffusion demonstrate maxima at electric field below the threshold for the intervalley transfer. The time constants of the related kinetic processes are estimated. The contribution of transverse tunneling to the longitudinal diffusion of hot electrons is observed for the first time.

## DEFINITIONS

We shall deal with longitudinal hot-electron velocity fluctuations measured at 10 GHz frequency in the direction of applied electric field. Transmission-line-model structures with alloyed Ni-Ge-Au ohmic contacts were prepared and investigated by microsecond-time-domain pulsed technique used to avoid Joule overheat of the lattice at high fields.

The hot-electron equivalent noise temperature $T_n$ measured in the direction of electric field $E$ is obtained from the spectral density of noise power $P_n$ emitted by the biased sample into the matched waveguide in the frequency band $\Delta f$

$$P_n(E) = kT_n(E)\Delta f. \tag{1}$$

The longitudinal hot-electron diffusion coefficient $D$ is determined from the experimental data on $T_n$ and differential mobility $\mu'$:

$$D(E) = (k/e)T_n(E)\mu'(E). \tag{2}$$

The contribution to the longitudinal diffusion resulting from back-and-forth transfer between two weakly-coupled subbands can be accounted for by:

$$\Delta D(E) = \frac{n_1 n_2}{(n_1 + n_2)^2}(v_1 - v_2)^2\tau, \tag{3}$$

where $v_1$, $n_1$ and $v_2$, $n_2$ are the electron drift velocities and densities in the subbands, and $\tau$ is the intersubband relaxation time constant. The contribution is caused by the difference of longitudinal drift velocities in the subbands. According to Eq.(3), a steep dependence of the ratio $n_1/n_2$ on electric field forms maxima of $n_1 n_2$ and $\Delta D(E)$ at around $n_1 = n_2$; a longer time constant $\tau$ favours a higher maximum of $\Delta D(E)$.

## INTERSUBBAND TRANSITIONS IN δ-DOPED GaAs

Planar-doped GaAs was grown by MBE technique at low temperatures (480°C) onto (100) oriented undoped GaAs substrates. The 2 nm Si-doped layer provided $1.43 \cdot 10^{12} \text{cm}^{-2}$ two-dimensional electron gas for the V-shaped δ-doped QW; the low-field electron Hall mobility was 3920 cm$^2$/Vs at 4.2K.

The electric field applied in the plane of the well heats the electrons, and they become in position to occupy partially extended states at elevated energies. After the transition the electron finds itself farther away from the donor plane, and the rate of electron scattering by the donors is reduced.

Figure 1 illustrates the effects of hot-electron transition from confined to partially extended states. The increase of electron mobility is evidenced at electric fields up to 3 kV/cm. The transfer is also accompanied by the steady-state hot-electron fluctuations demonstrating a well expressed maximum of noise temperature. This is a strong evidence that we observe fluctuations due to back-and-forth transitions between the low-mobility and the high-mobility subbands in the symmetric V-shaped δ-doped QW.

The decrease of mobility in the field range E > 3 kV/cm (Figure 1) results from the intervalley transfer of the high-energy

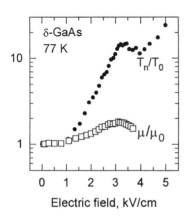

**Figure 1.** Field-dependent electron mobility and steady-state equivalent noise temperature at 77K for δ-doped GaAs ($\mu_0 = 4300$ cm$^2$/Vs).

high-mobility electrons. Eventually, at fields over 4 kV/cm the intervalley transfer fluctuations prevail over those caused by the confined-to-extended-state transitions.

Data of Figure 1 and Eq. (2) lead to around 400 $cm^2/s$ value for the maximum excess diffusion coefficient resulting from the back-and-forth intersubband transitions in the δ-doped QW at fields around 3 kV/cm. By assuming $10^7 cm/s$ for the velocity difference $v_2-v_1$, we obtain a rough estimate for the intersubband transfer time: $\tau = 15$ ps. Such value of the time constant favours manifestation of the intersubband maximum of the field-dependent longitudinal noise temperature at 10 GHz.

## TRANSVERSE TUNNELING

MBE-grown AlGaAs/GaAs/AlAs/GaAs heterostructure consisted of a δ-doped GaAs and undoped GaAs layers separated by a thin 2.5 nm spacer of pure AlAs and covered with an undoped AlGaAs cap layer. The plane of Si donors was centered inside the 1 nm GaAs layer and provided $1.3 \cdot 10^{12} cm^{-2}$ two-dimensional electron gas for the undoped GaAs layer. The low-field electron mobility was 35000 $cm^2/Vs$ at 77K.

Figure 2 illustrates the energy diagram obtained from a Schrödinger-Poisson model. At equilibrium the mobile electrons occupy the right-hand QW in the undoped GaAs. The δ-doped narrow QW is empty: its lowest subband is located well above the occupied subband in the wide QW. The barrier of AlAs is high. It is higher as compared to the intervalley separation gap in GaAs, and electrons should exercise the intervalley transfer in GaAs before jumping *over* the barrier from the occupied QW into the empty one.

Experimental data on longitudinal hot-electron diffusion for the QW structure and a high-mobility GaAs are compared in Figure 3. The source of excess diffusion specific to the QW structure is observed at the electric fields too low for the intervalley transfer to appear in the high-mobility GaAs. Therefore, one is forced to think that the maximum of diffusion coefficient cannot be ascribed to the electron jumps over the AlAs barrier.

We think that transverse tunneling of hot electrons is responsible for the observed maximum of longitudinal diffusion at fields below the threshold for the intervalley transfer

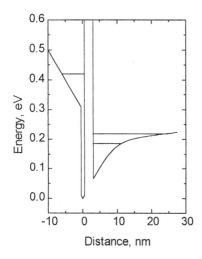

**Figure 2.** Energy diagram for the heterostructure consisting (from right to left) of pure GaAs, pure AlAs, δ-doped GaAs and undoped AlGaAs layers.

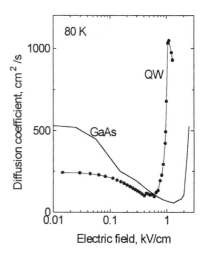

**Figure 3.** Longitudinal hot-electron diffusion for AlGaAs/GaAs/AlAs/GaAs QW structure (circles, L = 18 μm) and lightly doped GaAs (solid line[7]).

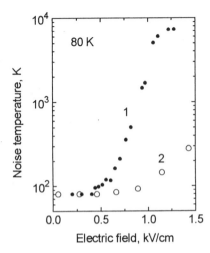

**Figure 4.** Field-dependent noise temperature for AlGaAs / GaAs / AlAs / GaAs heterostructures. Interelectrodal distance: 1 - 18 μm, 2 - 3 μm.

(see Figure 3). In the framework of Eq.(3) under assumption $v_1$-$v_2 \sim 2 \cdot 10^7$ cm/s we obtain an estimation of the tunneling time constant: $\tau \sim 10$ ps.

An alternative way to estimate the time constant is based on suppression of longitudinal hot-electron fluctuations in short samples[7]. Figure 4 presents our experimental data on longitudinal noise temperature for 18 μm and 3 μm samples of the AlGaAs/GaAs/AlAs/GaAs structures. The 3 μm sample seems to be too short for manifestation of the noise source observed in the long sample at fields over 500 V/cm. This source is caused, as discussed, by the longitudinal velocity fluctuations related to transverse tunneling. Its partial suppression in the short sample means that the electron transit time $t_{tr}$ along the QW from one electrode to another is comparable to the transverse-tunneling time constant $\tau$. By assuming $v = 2 \cdot 10^7$ cm/s for the hot-electron drift velocity, we obtain $t_{tr} = 15$ ps.

The values for the transverse-tunneling time constant estimated in two independent ways are in a reasonably good agreement. A similar value of 11 ps has been reported for electron *resonant* tunneling across 6 nm $Al_{.6}Ga_{.4}As$ barriers[6].

## ACKNOWLEDGMENTS

The research described in this publication was made possible in part by Grant LHP 100 from the Joint Program of the Government of Lithuania and the International Science Foundation, by Grant P5/94 from the Lithuanian State Science and Studies Foundation, PECO Grants ERBCIPDCT 94007 and 940020, and COPERNICUS Grant CP941180.

## REFERENCES

1. V.Aninkevičius, V.Bareikis, R.Katilius, P.S.Kop'ev, M.R.Leys, J.Liberis, and A.Matulionis, Hot-electron noise and diffusion in AlGaAs/GaAs, *Semicond. Sci. Technol.* 9:576 (1994).
2. Z.S.Gribnikov, K.Hess, and G.A.Kosinovsky, Nonlocal and nonlinear transport in semiconductors: real-space transfer effects, *J. Appl. Phys.* 77:1337 (1995).
3. V.Bareikis, R.Katilius, and A.Matulionis, High-frequency noise in heterostructures, *in:* "Noise in Physical Systems and 1/f Fluctuations", Proceedings of the 13th International Conference, Palanga, Lithuania, edited by V.Bareikis and R.Katilius (World Scientific, Singapore 1995), p.14.
4. J.Shah, Ultrafast luminescence spectroscopy of semiconductors: carrier relaxation, transport and tunneling, *in:* "Spectroscopy of Nonequilibrium Electrons and Phonons", ed. by S.V.Shank and B.P.Zakharchenya (North-Holland, Amsterdam 1992) p.57.
5. M.C.Tatham and J.F.Ryan, Inter- and intra-subband relaxation of hot carriers in quantum wells probed by time-resolved Raman spectroscopy, *Semicond. Sci. Technol.* 7:B102 (1992).
6. A.P.Heberle, X.Q.Zhou, A.Takeuchi, W.W.Rühle, and K.Köhler, Dependence of resonant electron and hole tunneling times between quantum wells on barrier thickness, *Semicond. Sci. Technol.* 9:519 (1994).
7. V.Bareikis, J.Liberis, I.Matulionienė A.Matulionis, and P.Sakalas, Experiments on hot electron noise in semiconductor materials for high-speed devices, *IEEE Trans. Electron Devices,* **ED-41:**2050 (1994).

# MONTE CARLO ANALYSIS OF HIGH-FIELD TRANSPORT
# UNDER CLASSICAL SIZE-EFFECT CONDITIONS

Oleg M. Bulashenko,[1] Luca Varani,[2] Pascal Piva,[2] and Lino Reggiani[3]

[1]Universidad Carlos III de Madrid, Butarque 15, 28911 Leganés, Spain
[2]Centre d'Electronique de Montpellier, Université Montpellier II,
 34095 Montpellier Cedex 5, France
[3]Dipartimento di Scienza dei Materiali, Istituto Nazionale di Fisica della
 Materia, Università di Lecce, via Arnesano, 73100 Lecce, Italy

## INTRODUCTION

When reducing the dimensions of a device, carrier transport goes from diffusive to ballistic regime and both current-voltage characteristics and Johnson-Nyquist noise caused by carrier scattering processes acquire new features. In diffusive regime the characteristic scattering processes are spatially isotropic, the current autocorrelation functions exhibit a standard exponential behavior and, therefore, the spectral densities of current fluctuations are Lorentzians. In ballistic regime the transport is mostly determined by transit times, which are either between the contacts or between the boundaries, and the associated scattering processes are spatially anisotropic depending on the sample geometry. In this regime, the current autocorrelation functions become non-exponential giving rise to: a redistribution of the noise at higher frequencies (blue-shift), a deviation of the noise spectrum from the Lorentzian shape and, in particular cases, a series of geometrical resonances in the spectrum.[1-5]

In this contribution we present the results obtained from a Monte Carlo simulation for carrier mobility and non-equilibrium current fluctuations under size-effect conditions. This situation can occur in thin films, multilayers, wide ballistic wells, etc., where the electron scattering at the boundary is essentially of diffuse type.

## MODEL DESCRIPTION

We consider a thin semiconductor slab, with the thickness $d$ being smaller than, or comparable to, the electron mean free path $\lambda$ (classical size-effect). The electric field $E$ applied along the surfaces of the slab is assumed to be uniform (flat band model). The carrier concentration is considered under nondegenerate conditions and carrier-carrier interaction is neglected. For the sake of simplicity we take a single spherical and parabolic band and treat the electron transport within an energy independent relaxation-time model,[6] implying maxwellization of the velocity components to the lattice temperature $T$ after any bulk scattering event. This approach implies simple formulas to generate the stochastic time between collisions and the carrier velocity after scattering.[7] We consider entirely diffuse electron scattering at the slab surfaces, but with two different types of reflection: (i) nonelastic scattering (NS) with energy dissipation on the surface (relaxation to the lattice temperature); (ii) elastic scattering (ES) with energy conservation on the surface (reflection on the surface roughness).

To summarize, our problem can be formulated in terms of two dimensionless parameters: (i) the ballistic parameter $\gamma = \lambda/d$ ; (ii) the electric field strength $\delta = E/E_0$ , where $E_0 = (2m \, k_B T)^{1/2}/e\tau$ is a characteristic electric field, $m$ the effective mass, $k_B$ the Boltzmann constant, $\tau$ the relaxation time, $e$ the electron charge. In the limit of unbounded sample ($\gamma \to 0$), the model admits analytical expressions[6] which we used to verify the Monte Carlo algorithm.


## RESULTS AND DISCUSSION

Within our microscopic model the system is found to exhibit a linear velocity-field characteristics with a mobility dependent on the slab thickness (Fig.1). Here the mobility is a decreasing function of $\gamma$ due to an increasing importance of the surface scattering for $\gamma \gg 1$, and for the NS case the mobility is seen to be larger.

The spectral densities of the current fluctuations $S_I(\omega)$ are calculated through the velocity autocorrelation functions.[5,7] For weak fields $\delta \ll 1$ large deviations from a Lorentzian spectrum are observed for $\gamma \gg 1$, thereby indicating the size-effect. For high fields $\delta \geq 1$ the signal-to-noise ratio defined as $I(\omega)/(S_I(\omega)\Delta\omega)^{1/2}$ is found to be quite different for the NS and ES cases. One of the striking result is the presence of a maximum at $\gamma \sim 1$ only for the NS case (Fig.2), while for the ES case this ratio has a maximum as a function of field for the fixed thickness of the slab (Fig.3). As a consequence, the noise characteristic of the sample could be improved by an appropriate choice of the applied field and of the sample geometry, e.g., scaling down its size (until the transverse thickness becomes comparable with the carri-

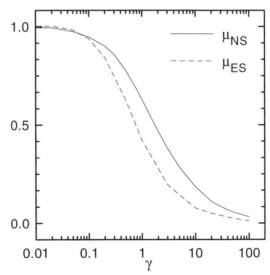

**Figure 1.** Mobility of carriers $\mu/\mu_0$ for nonelastic scattering (solid line) and elastic scattering (dashed line) at the surfaces of the slab vs ballistic dimensionless length $\gamma$. Here $\mu_0 = e\tau/m$.

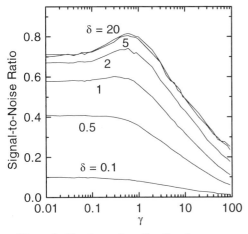

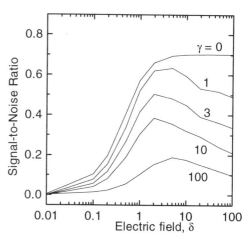

**Figure 2.** Signal-to-noise ratio at low frequency (nonelastic case) versus ballistic length γ for different values of field strength δ.

**Figure 3.** Signal-to-noise ratio at low frequency (elastic case) versus electric field strength δ for different values of ballistic length γ.

er mean free path) and by adopting carrier cooling at the boundaries or increasing the electric field and by adopting energy conservation at the boundaries. This result, which we believe to be of interest from an applied point of view, comes from the joint action of the electric field and the surface scattering, both processes being spatially anisotropic. This effect is reminiscent of that found in bulk samples[8] with the size effect acting on the much larger length-scale of energy relaxation.

In conclusion, in contrast to bulk systems we have shown that small ballistic systems exhibit current fluctuations which are spatially correlated. As direct consequences, anisotropic effects are brought into the system, non-Lorentzian spectra show up even at equilibrium, and strong electric fields can improve the signal-to-noise ratio.

Support by Ministerio de Educación y Ciencia (Spain) and Ministero dell' Università e della Ricerca Scientifica e Tecnologica (Italy) is acknowledged.

## REFERENCES

1. T. Kuhn, L. Reggiani, and L.Varani, Correlation functions and quantized noise in mesoscopic systems, *Superlattices and Microstructures* 11:205 (1992).
2. T. Kuhn, L. Reggiani, and L.Varani, Non-equilibrium noise in mesoscopic systems: effects of transit times, *Semicond. Sci. Technol.* 7:B495 (1992).
3. O.M. Bulashenko, O.V. Kochelap, and V.A. Kochelap, Size effect on current fluctuations in thin metal films: Monte Carlo approach, *Phys.Rev.B* 45:14308 (1992).
4. O.M. Bulashenko and V.A. Kochelap, Johnson-Nyquist noise for a 2D electron gas in a narrow channel, *J.Phys.Cond.Matter* 5:L469 (1993).
5. O.M. Bulashenko and V.A. Kochelap, Geometrical resonances in the Nyquist noise spectrum for ballistic conductors, *Preprint IC/93/240*, ICTP, Trieste, Italy (1993).
6. C.J. Stanton and J.W. Wilkins, Nonequilibrium current fluctuations in semiconductors: A Boltzmann-equation-Green-function approach, *Phys.Rev.B* 35:9722 (1987).
7. O.M. Bulashenko, L. Varani, and L. Reggiani, Monte Carlo analysis of non-equilibrium current fluctuations under classical size-effect conditions, in: "Noise," V. Bareikis and R. Katilius, eds., World Scientific, Singapore, (1995) p.197.
8. V.A. Kochelap, V.N. Sokolov, and N.A. Zakhleniuk, Limitation and suppression of hot-electron fluctuations in submicrometer semiconductor structures, *Phys.Rev.B* 48:2304 (1993).

# MONTE CARLO CALCULATION OF HOT-CARRIER THERMAL CONDUCTIVITY IN SEMICONDUCTORS

Paola Golinelli[1], Rossella Brunetti[1], Luca Varani[2],
Lino Reggiani[3], and Massimo Rudan[4]

[1]Istituto Nazionale di Fisica della Materia,
Dipartimento di Fisica, Università di Modena, Italy
[2]Centre d'Electronique de Montpellier, Universitè Montpellier II, France
[3]Istituto Nazionale di Fisica della Materia,
Dipartimento di Scienza dei Materiali, Università di Lecce, Italy
[4]DEIS, Università di Bologna, Italy

## INTRODUCTION

A study of the thermal conductivity of charge carriers in semiconductor materials and devices subject to very high electric fields is recognized to be necessary in order to have an accurate description of the transport process. To this purpose a generalization of the Wiedemann-Franz law under high electric fields is a crucial step for any hydrodynamic modeling of semiconductors.[1,2] This work presents two different theoretical approaches which account for hot-carrier effects in the thermal-conduction process. Numerical results for the thermal conductivity and Lorentz number obtained from a Monte Carlo simulation of electrons in Si are compared and discussed.

## THEORY AND RESULTS

The first theoretical approach that can be used to estimate the thermal conductivity $\kappa(E,\omega)$ and the differential electrical conductivity $\sigma_d(E,\omega)$ of the carriers is based on a set of correlation functions describing the microscopic fluxes around the steady-state:[3]

$$\kappa(E,\omega) = \frac{n}{KT_n^2(E,\omega)} \frac{I_{11}(E,\omega)I_{22}(E,\omega) - I_{12}(E,\omega)I_{21}(E,\omega)}{I_{11}(E,\omega)} \tag{1}$$

$$\sigma_d(E,\omega) = \frac{e^2 n}{KT_n(E,\omega)} I_{11}(E,\omega) \tag{2}$$

$$I_{\nu\mu}(E,\omega) = \int_0^\infty dt\, \overline{\delta j_\mu(t')\, \delta j_\nu(t'+t)}\, \exp(i\omega t), \qquad \nu,\mu \equiv 1,2. \tag{3}$$

Here $n$ is the carrier concentration, $K$ the Boltzmann constant, $e$ the electron charge, $j_1 = v(\mathbf{k})$, $j_2 = v(\mathbf{k})\epsilon(\mathbf{k})$, $\epsilon(\mathbf{k})$ and $v(\mathbf{k})$ being the carrier energy and the velocity along the direction of the electric field $E$, $T_n(E,\omega)$ the carrier noise temperature, and $\delta j_\mu(t)$ the fluctuation of $j_\mu(t)$

*Hot Carriers in Semiconductors*
Edited by K. Hess *et al.*, Plenum Press, New York, 1996

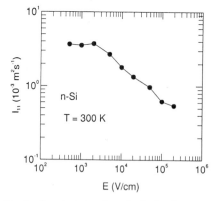

Fig. 1: Spectral density of velocity fluctuations at $\omega = 0$ for electrons in Si at 300 K as a function of the electric field strength.

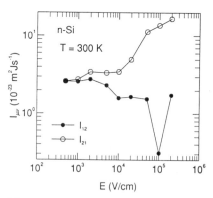

Fig. 2: Spectral density of the cross-correlations between fluctuations in velocity and energy flux at $\omega = 0$ for electrons in Si at 300 K as a function of the electric field strength.

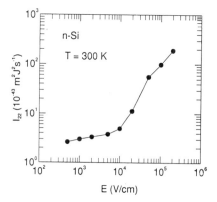

Fig. 3: Spectral density of the energy flux fluctuations at $\omega = 0$ for electrons in Si at 300 K as a function of the electric field strength.

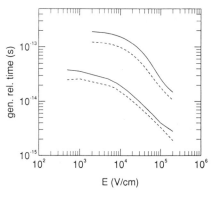

Fig. 4: Generalized relaxation times $\tau_q$ (dashed) and $\tau_p$ (continuous), computed from (4) (lower pair of curves) and with the spherical-harmonics expansion[2] (upper curves). In both cases $\tau_p$ is longer than $\tau_q$.

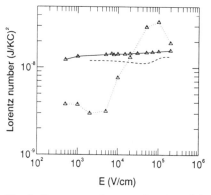

Fig. 5: Lorentz number as a function of the electric field strength. The dotted line refers to the calculation based on (1), the continuous one to that based on (5). The dashed line refers to the result of the spherical-harmonics expansion[2] which are reported for comparison.

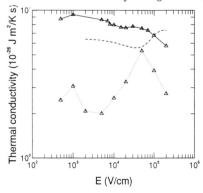

Fig. 6: Thermal conductivity per unit concentration as a function of the electric field strength. The symbols and lines have the same meaning as those in Fig. 5.

around its average value. In (3), the bar denotes ensemble average over $t'$, and $\omega$ the angular frequency.

The second theoretical approach[2] is based on the velocity and energy-flux generalized relaxation-times $\tau_p(E)$ and $\tau_q(E)$ defined as:

$$\tau_p(E) \int j_1 \left(\frac{\partial f}{\partial t}\right)_c dk = \int j_1 f(\mathbf{k}) \, d\mathbf{k}, \qquad \tau_q(E) \int j_2 \left(\frac{\partial f}{\partial t}\right)_c dk = \int j_2 f(\mathbf{k}) \, d\mathbf{k}, \qquad (4)$$

where $f(\mathbf{k})$ and $\partial f/\partial t)_c$ are the carrier distribution function and the collision integral, respectively. These two quantities are related to the thermal conductivity by[2]

$$\kappa(E) = \frac{5}{2} \left(\frac{K}{e}\right)^2 \frac{\tau_q(E)}{\tau_p(E)} \sigma(E) T_e(E), \qquad (5)$$

where $\sigma(E)$ is the electrical conductivity and $T_e(E)$ the temperature associated to the carrier average energy. Calculations of different quantities depending on the electric field are performed for the case of electrons in Si at 300 K. Details of the physical model used in the Monte Carlo simulations can be found elsewhere.[4] We remark that the $\omega = 0$ condition is considered here, and that $T_n(E)$ is replaced with the electron temperature $T_e(E)$, which represents a reasonable approximation in the present case. Figs. 1 to 3 report the spectral density of the fluctuating quantities in (3) at $\omega = 0$, as a function of the electric field strength. In our case the behavior of $I_{11}$ is well understood since $I_{11} = D_l$, $D_l$ being the longitudinal diffusion coefficient. The cross-correlation terms coincide at low fields according to the Onsager relations, but differ significantly at increasing fields. Their behavior reflects the different role played by energy and momentum relaxations of the respective correlation functions, both relaxations being coupled because of hot-carrier conditions. In particular, the prevalence of carrier heating with respect to energy-momentum coupling is responsible for the steep increase of $I_{12}$ for fields above $10^5$ kV/cm. The behavior of $I_{22}$ reflects the dominant role played by the increase of the carrier average energy with electric field strength. The generalized relaxation times of (4) are shown in Fig. 4, where the same quantities calculated by a solution of the Boltzmann equation performed with a sperical-harmonic expansion[2] are also shown for comparison. Despite the quantitative difference, to be ascribed to the more accurate description of the scattering mechanisms implemented in the Monte Carlo procedure used here,[4] the behavior is qualitatively similar, as reflected by the Lorentz "number" $L = \kappa/(\sigma T_e)$. The values of the latter are shown in Fig. 5 along with those calculated with (1-2). Finally, the thermal conductivity is shown in Fig. 6, where the result obtained by the spherical-harmonics expansion is also reported.[2] In conclusion, two different theoretical approaches presented in this paper attempt to generalize the definition of the carrier thermal conductivity to high-field transport in semiconductors. Preliminary results differ by a factor 5 at most. Further theoretical and numerical investigations on the models are currently being carried out. Since at present no experimental data is available in this respect, the implementation of the two methods in a device simulator can give indirect indications on the validity of the two approaches.

## ACKNOWLEDGMENTS

The authors gratefully thank Carlo Jacoboni for fruitful discussions, and M. Cristina Vecchi for kindly providing computer results. This work is funded in part by ADEQUAT (JESSI BT11) ESPRIT-8002.

## REFERENCES

1. A. M. Anile and S. Pennisi, Phys. Rev. B46, 13186 (1992).
2. M. Rudan, M. C. Vecchi, and D. Ventura in "The Pitman Research Notes in Mathematics", Longman (1995), to be published.
3. R. Thoma, K. P. Westerholz, H. J. Peifer, and W. L. Engl, Semicond. Sci. Technol. 7, B328 (1992).
4. R. Brunetti, C. Jacoboni, F. Nava, L. Reggiani, G. Bosman, and R. J. J. Zijlstra, J. Appl. Phys. 52(11), 6713 (1981).

# A MULTIPLICATION SCHEME WITH VARIABLE WEIGHTS FOR ENSEMBLE MONTE CARLO SIMULATION OF HOT-ELECTRON TAILS

Andrea Pacelli,[1,2] Amanda W. Duncan,[2] and Umberto Ravaioli[2]

[1] Politecnico di Milano, Dipartimento di Elettronica e Informazione
Piazza Leonardo Da Vinci 32, 20133 Milano, Italy
[2] Beckman Institute, University of Illinois at Urbana-Champaign
405 N. Mathews Ave., Urbana, IL 61801

## INTRODUCTION

The use of the Monte Carlo (MC) technique for simulation of semiconductor devices is very expensive in terms of computational resources. This is mainly due to the high statistical error in the modeling of rare events. Several variance-reduction schemes have been proposed in the literature, based on the concept of splitting or repetition of trajectories, and averaging of the results.[1−6] Many of the methods are only applicable to single-particle simulations, because they assume steady state and do not account for time-dependent Coulomb interaction between particles.[1, 2, 3] Schemes suitable for ensemble simulation are basically extensions of the one-particle procedure, and require some tuning to the particular device structure being simulated.[4] In this work we present an extension of the splitting/gathering scheme of Ref. 5. A flexible procedure is derived that allows the simulation of particles of different weights to obtain a balanced sampling of the phase space in a time-dependent, self-consistent framework. Results are presented for bulk, one-dimensional, and two-dimensional simulations.

## VARIANCE-REDUCTION TECHNIQUE

Variance-reduction schemes for ensemble simulations are based on the use of variable weight factors for the MC particles.[4, 5] We assign to the $i$-th particle in the simulation a weight $w_i$, and assume that each MC particle represents a number of electrons or holes (with associated mass and charge) proportional to its weight. In a conventional simulation, all particles have the same weight. The variance-reduction technique consist of manipulating the weights to equalize the sampling of phase space, using a larger number of particles with smaller weights in sparsely populated regions. For example, a large particle can be split into many identical particles, each with a fraction of the weight of the original one. Since the number of particles in the simulation cannot increase indefinitely, in the long run some gatherings will have to be performed to reduce the total number of MC particles. In our context, gathering means reducing a set of $N$ particles to only one particle without affecting the result of the simulation. This needs to be done with care. If the states of the $N$ particles are averaged into one larger particle, unphysical effects may arise. Consider for example two particles with opposite wave vectors in a simple parabolic conduction band. A particle with the average momentum would lie at the bottom of the valley, thus violating conservation of energy.

A more correct way of performing the gathering would be to select one particle at random from the "gathering set," removing the remaining ones from the simulation. Since conservation of the total charge is desirable, we assign all the weight of the removed particles to the one that is retained. The gathering can be performed in such a way that, while momentum and energy are not conserved for any particular random choice, the expected values of the observables remain unchanged. We call $p_i$ the probability that the $i$-th particle be selected. The expected

*Hot Carriers in Semiconductors*
Edited by K. Hess *et al.*, Plenum Press, New York, 1996

change of weight for the $i$-th particle is

$$\Delta w_i = p_i \sum_{j=1, j \neq i}^{N} w_j - (1 - p_i) w_i. \tag{1}$$

Since the ensemble averages are computed by weighted sums over all particles, we simply require the expected change of weight to be zero. This is obtained if

$$p_i = \frac{w_i}{\sum_{j=1}^{N} w_j}. \tag{2}$$

We note that no assumption has yet been made on the gathering set. In the following we will assume for convenience that the phase space of the simulated physical system is divided into discrete sub-domains or "bins." The bins will usually be energy intervals and/or device regions, however, the partitioning is essentially arbitrary.

The possible gathering strategies can be distinguished between *local* and *non-local*. A local gathering set includes particles from only one bin in phase space, whereas a non-local set includes particles from many bins. The simplest local strategy is based on the number of particles: the gathering is applied to particles in the bins that contain many particles and are supposedly over-sampled. This algorithm is not optimal because it only uses the information given by the number of particles, not their weight. The next logical step is to locate the bins with particles that do not contribute significantly to the estimators, because much larger particles are present in the same bin. This corresponds to performing the gathering on bins that have a large ratio between the largest and smallest particle. Once the bin has been selected, the gathering is performed on the smallest particles, thus creating one larger particle. This weight-equalizing scheme can fail for bins that have a large extension in phase space, spanning many states with widely different occupations. Smaller particles will exist in the more sparsely populated part of the bin. If the smallest particles are chosen for the gathering, it is possible that the program will systematically deplete the least occupied part of the bin, thus increasing the variance of the distribution function close to the boundaries between bins.

A more flexible scheme is obtained by applying the gathering procedure in its most general form, using a non-local gathering set. One possible algorithm consists of keeping a linked list of small particles lying in bins that contain much larger particles. We can reasonably assume that these particles are redundant in the simulation, i.e., do not contribute significantly to the estimators. This approach, straightforward to implement, has several advantages. First of all, it does not suffer from the problems of the weight-equalizing scheme. In fact, the selection criterion for redundant particles provides an additional degree of freedom that can be exploited to preserve the small, but significant, particles. Moreover, the bins are immediately cleared of redundant particles, which are not counted when re-equalizing the occupation of bins, improving the accuracy of the sampling. Finally, the search for small particles can be performed at the beginning of the population-balancing step, speeding up the procedure considerably. The local sets can be used as a "back-up" of the non-local strategy, if additional equalization is needed for higher accuracy.

## SIMULATION RESULTS

The variance-reduction algorithm has first been applied to electron transport in bulk silicon. A model with two conduction bands has been used for electron kinematics, where the band structure of silicon has been computed with the local empirical pseudopotential method.[7] The scattering processes considered are inelastic acoustic and optical phonons, and impact ionization.[8, 9] Figure 1 shows the simulation results for a uniform electric field of 100 kV/cm. The energy range has been partitioned into energy bins, each 50 meV wide. The figure also reports the result from a simulation where a simple local scheme has been used. In this case the statistical error is very high, resulting in an underestimation of the high-energy tail. As mentioned before, the local scheme is applied also in the non-local approach, but it performs more efficiently after the redundant particles have been removed.

In the context of self-consistent simulations, two aspects of the variance-reduction technique can introduce problems: the nonuniform weight of the particles and the random changes of weight. The use of variable-weight particles in a self-consistent simulation can lead to an increase in the numerical error when the weights have a large dynamic range. In this case, the charge appearing in the Poisson equation may be dominated by a few large particles, and the resulting potential profile will be exceedingly noisy. In this work, we solve the problem by requiring a minimum number of "Poisson particles" to take care of electrostatic balance. This is automatically obtained by forcing the weight of the particles to be no larger than the weight used in the conventional simulation. The random changes in particle weights cause unphysical disturbances to the Poisson equation. Our scheme always keeps a constant charge, and thus ensures global electrostatic neutrality of the device. Therefore one must be concerned about rearrangement, rather than about creation or destruction of charge. If the bins in real space are small enough, local gathering strategies will only have a minor impact on the results. Non-local gathering is performed for very small particles, so that the effect on the overall electrostatic balance is negligible.

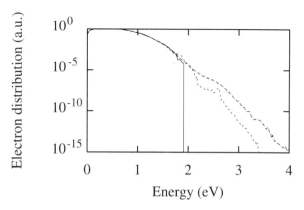

Figure 1: Energy distribution for electrons in silicon with a uniform field of 100 kV/cm. Comparison of the conventional MC simulation (solid line) with a non-local (dashed) and local (dotted) schemes.

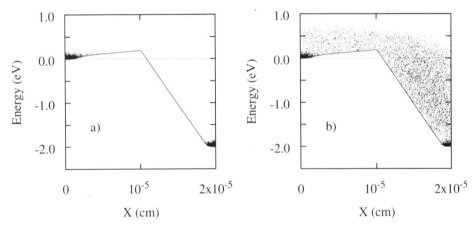

Figure 2: Energy profile and distribution of particles in for the triangular-barrier diode. a) conventional simulation, b) with variance reduction.

As a second example we consider a one-dimensional triangular-barrier diode, with the same physical model of the bulk simulation of Fig. 1. Figure 2 shows the potential profile and distribution of particles from a conventional self-consistent MC simulation, and with application of the variance-reduction technique. In the case of Fig. 2a, only a very small fraction of the electrons diffuse through the retarding potential of the barrier, to be swept by the the high field towards the exit contact. Conventional MC simulations suffer thus from lack of samples in the high-field region. In the case of Fig 2b, the number of particles has been doubled, and the variance-reduction algorithm has been applied so that the maximum weight of particles is the same as for the simulation in Fig. 2a. The position-energy space is now densely occupied along the entire device, with a large number of samples available in the high-field region.

Finally, we consider a simple n-MOSFET test structure, with uniform source, drain and substrate doping. The oxide thickness is 100 Å. The metallurgical channel length is 0.5 $\mu$m, source and drain dopings are $5 \times 10^{19}$ cm$^{-3}$, the substrate p-doping is $5 \times 10^{16}$ cm$^{-3}$. Bias voltages of 4 V and 3 V have been applied to the drain and gate, respectively. We have computed the oxide-injection current using the variance-reduction technique to amplify the weak high-energy tails along the entire device. A nonlinear two-dimensional Poisson solver has been employed, including holes in the simulation self-consistently under the assumption of a constant hole quasi-Fermi level. Poisson's equation is solved every 2 fs of simulated time, in order to resolve plasma oscillations in the high-doping regions.[4] Figure 3a shows the distribution of electrons as a function of position along the channel and of kinetic energy. In this simulation, a dynamic range of $10^{20}$ has been allowed for the weights, resolving the Maxwellian tail in the initial part of the channel well above 1 eV. Figure 3b shows the position dependence of the oxide injection current, separating the thermionic and tunneling components (here treated in

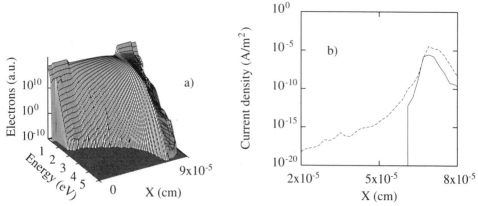

Figure 3: MC simulation of a 0.5 $\mu$m n-MOSFET. a) Position/energy distribution of electrons along the device, b) Oxide current density. Solid line: thermionic current. Dashed line: tunneling current.

the WKB approximation). The abrupt step in the thermionic current around $X = 600$ nm is due to the finite dynamic range of weights. The soft increase of the tunneling component reflects the smooth dependence of transmission probability on electron energy. Thermionic emission only depends on conservation of energy and crystal momentum parallel to the interface, so that a sharp increase occurs as the energy supplied by the field approaches the barrier height.

In all cases presented, the use of the variance-reduction procedure only takes a very small fraction of the total computer time. This is due to the simplicity of the scheme and to the efficiency of the non-local sweep of the ensemble of particles. The method is thus applicable with no significant penalty even when the number of particles and phase-space bins is very high (e.g., 100,000 particles and 10,000 bins).

## ACKNOWLEDGEMENT

This work has been supported by the Volta-Badoni Fellowship of the Associazione Elettrotecnica ed Elettronica Italiana (A. P.), by a Computational Science Graduate Fellowship of the US Department of Energy (A. D.) and by the Joint Services Electronics Program, grant N00014-90-J-1270 (U. R.).

1. A. Phillips, Jr. and P. J. Price, Monte Carlo calculations on hot electron energy tails, *Appl. Phys. Lett.* 30(10):528 (1977).

2. E. Sangiorgi, B. Riccò, and F. Venturi, MOS$^2$: An efficient *MO*nte Carlo *S*imulator for *MOS* devices, *IEEE Trans. Comput.-Aided Design Integrated Circuits* CAD-7(2):259 (1988).

3. U. A. Ranawake, C. Huster, P. M. Lester, and S. M. Goodnick, PMC-3D: A parallel three-dimensional Monte Carlo semiconductor device simulator, *IEEE Trans. Comput.-Aided Design Integrated Circuits* 13(6):712 (1994).

4. M. V. Fischetti and S. E. Laux, Monte Carlo analysis of electron transport in small semiconductor devices including band-structure and space-charge effects, *Phys. Rev. B* 38(14):9721 (1988).

5. F. Venturi, R. K. Smith, E. Sangiorgi, M. R. Pinto, and B. Riccò, A general purpose device simulator coupling Poisson and Monte Carlo transport with applications to deep submicron MOSFET's, *IEEE Trans. Comput.-Aided Design Integrated Circuits* CAD-8(4):360 (1989).

6. D. Liebig, P. Lugli, P. Vogl, M. Claassen, and W. Harth, Tunneling and ionization phenomena in GaAs PIN diodes, *Microelectronic Engineering* 19:127 (1992).

7. M. L. Cohen and T. K. Bergstresser, Band structures and pseudopotential form factors for fourteen semiconductors of the diamond and zinc-blende structures, *Phys. Rev.* 141(2):789 (1966).

8. J. Y. Tang and K. Hess, Impact ionization of electrons in silicon (steady state), *J. Appl. Phys.* 54(9):5139 (1983).

9. J. Bude, K. Hess, and G. J. Iafrate, Impact ionization in semiconductors: Effects of high electric fields and high scattering rates, *Phys. Rev. B* 45(19):10958 (1992).

# HOT-CARRIER NOISE UNDER DEGENERATE CONDITIONS

Patrick Tadyszak[1], Alain Cappy[1], Francois Danneville[1], Lino Reggiani[2],
Luca Varani[3] and Lucio Rota[4]

[1]Institut d'Electronique et de Microelectronique du Nord U.M.R. CNRS
n. 9929, Departement Hyperfrequences et Semiconducteurs,
Domain Scientifique et Universitaire de Villeneuve D' Ascq,
BP 69, 59652 Villeneuve D'Ascq Cedex, France
[2]Istituto Nazionale di Fisica della Materia, Dipartimento di Scienza
dei Materiali, Università di Lecce, Via Arnesano, 73100 Lecce, Italy
[3]Centre d'Electronique de Montpellier, U.R.A. CNRS n. 391
Université Montpellier II, 34095 Montpellier Cedex 5, France
[4]Clarendon Laboratory, Department of Physics, University of Oxford,
Parks Road, Oxford, OX1 3PU, UK

## INTRODUCTION

Degeneracy plays an important role in most electronic devices when regions with high
carrier concentration exist and/or low temperatures are considered. After the seminal paper
of Bosi and Jacoboni[1], several authors[2-4] have investigated degeneracy effects by using the
Monte Carlo (MC) technique. While most of the interest has been addressed to study average
quantities, to our knowledge electronic noise has never been considered. Therefore, it is the
aim of this paper to analyse some interesting effects associated with the many-body character
of degeneracy on noise in the presence of an external electric field of arbitrary strength. As
interesting feature, we remind that because of degeneracy no simple relation (e.g. Price's
relationship[5]) exists between noise and diffusion. Thus, modelling of noise under degenerate
conditions cannot relate on the knowledge of diffusion, and its determination rests on the
calculation of the correspondent current (or voltage) correlation function.

## THEORY AND NUMERICAL RESULTS

As a model for the material, we consider $n$-type Si at $T = 300\ K$ with one equivalent
band spherical and non-parabolic (or parabolic) with lattice scattering only. This model well
reproduces first order kinetic coefficients and correlation functions when the field is applied
along the $< 111 >$ crystallographic direction in the absence of degeneracy effects[6]. Degeneracy
is included in a MC code following previous works[1-4]. In the simulation we use $N = 1 \times
10^4 \div 3 \times 10^4$ electrons and a time step of $1\ fs$ to calculate time series. To clearly evidence
degeneracy effects, we take $n = 10^{21}\ cm^{-3}$ which corresponds to a Fermi energy of 0.85 and 1.2
$eV$ for nonparabolic and parabolic band, respectively, and ionized impurity scattering has been
neglected. The dependences with electric field strength of the average quantities: drift velocity
$v_d$, mean energy $< \epsilon >$ and variance of the drift-velocity fluctuations $C_{vd}(0)$ are reported
in Figs. 1 to 3. Each figure compares the results with (full dots) and without (open dots)

*Hot Carriers in Semiconductors*
Edited by K. Hess *et al.*, Plenum Press, New York, 1996

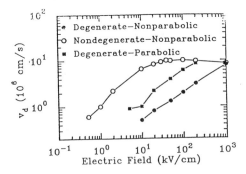

Fig. 1 - Drift velocity as a function of electric field for n-Si at $T = 300\ K$ with $n = 10^{21}\ cm^{-3}$. Full and open dots includes and neglects exclusion principle, respectively. Calculations are performed with a parabolic and nonparabolic band model.

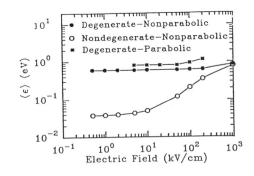

Fig. 2 - Carrier average energy as a function of electric field for the same conditions as Fig. 1.

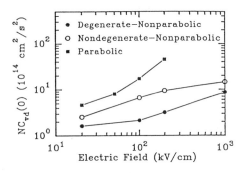

Fig. 3 - Variance of drift-velocity fluctuations as a function of electric field for the same conditions as Fig. 1. Full squares refer to a parabolic band for which case inclusion and neglect of exclusion principle give the same results.

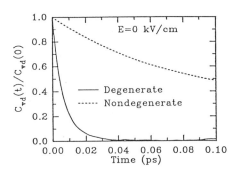

Fig. 4 - Normalized correlation functions of drift-velocity fluctuations at thermodynamic equilibrium ($E = 0$) for the same conditions as Fig. 1.

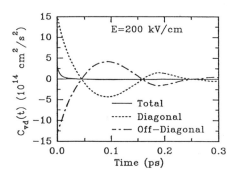

Fig 5. - Decomposition of the correlation function of drift-velocity fluctuations at $E = 200\ kV/cm$ for the same conditions as Fig. 1.

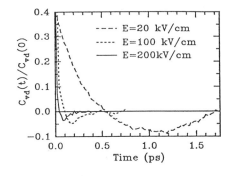

Fig 6. - Normalized correlation functions of drift velocity fluctuations at different electric fields. To emphasize the negative minimum results at long times are reported.

degeneracy effects. For the parabolic case the net effect of degeneracy is found: (i) to lower the low-field mobility because of a shorter scattering time associated with carriers interacting at the Fermi energy. As a consequence, the non-Ohmic behavior is less pronounced (see Fig. 1). (ii) To pin the carrier average energy at the Fermi value until carrier heating yields the average energy comparable with the Fermi value (see Fig. 2). (iii) To be irrelevant for the variance of drift-velocity fluctuations (see Fig. 3) which well fits an electron temperature model: $NC_{vd}(0) = K_B T_e/m$ with the electron temperature $T_e$ calculated from the average energy obtained by the MC simulation without degeneracy as $\frac{3}{2}K_B T_e = < \epsilon >$, $K_B$ being the Boltzmann constant and $m$ the carrier effective mass. For the nonparabolic case the net effect of degeneracy is found: (i) to decrease the drift velocity and to increase the average energy at the same field. For this reason, calculations have been extended up to 1 $MV/cm$. (ii) To reduce substantially the variance of drift-velocity fluctuations. To analyze the effect of degeneracy on electronic noise we have evaluated the correlation functions of drift-velocity fluctuations as well as of single-particle velocity fluctuations[6]. We consider the nonparabolic case only. Figures 4 and 5 shows the correlation function of drift-velocity fluctuations at thermodynamic equilibrium and for an applied field of 200 $kV/cm$, respectively. At equilibrium, see Fig. 4, an exponential decay on the momentum time-scale is clearly exhibited. Here the faster decay for degenerate conditions is due to the shorter relaxation time. In the presence of an electric field (see Fig. 5) the time decay strongly deviates from the exponential behavior and a minimum with a negative part of the correlation shows up. Such a minimum shifts at shorter times linearly with increasing fields. We associate this behavior with carriers crossing ballistically the degenerate Fermi sphere. Due to the very high carrier density, the Fermi sphere is almost completely full even when an external electric field is applied. Carriers that are moving with a negative velocity, i.e. against the field are slowed down and enter the Fermi sphere, at this point they cannot scatter any more because there are no final states available. Carriers continue their free flight until they reach the boundary of the Fermi sphere with a positive velocity, at this point they can again scatter. As the scattering with phonons is isotropic a large fraction of carriers is then scattered back at negative velocity and the process is repeated. This periodic motion becomes more evident when considering the diagonal and off-diagonal part of $C_{vd}(t)$ which exhibits an oscillatory behavior[7]. We remark that the separation into a diagonal and off-diagonal part should be handled carefully when treating degeneracy, since, in contrast to what happens in a semiclassical MC simulation, in a fully quantum mechanical treatment particles should be considered indistinguishable. For completeness, Fig. 6 reports the long-time behavior of the normalized correlation functions at several values of the applied field. Here the minimum of the correlation functions shifts to shorter times and becomes less pronounced at increasing values of the field.

## ACKNOWLEDGMENTS

This work is performed within the European Laboratory for Electronic Noise (ELEN) supported through the contract ERBCHRXCT920047.

## REFERENCES

1. S. Bosi and C. Jacoboni, J. Phys. C: Solid-State Phys. **9**, 315 (1976).
2. P. Lugli and D.K. Ferry, IEEE Trans. Electron Dev., **ED-32**, 2431 (1985).
3. M.V. Fischetti and S.E. Laux, Phys. Rev **B38**, 9721 (1988).
4. N.S. Mansour, K. Diff and K.F. Brennan, J. Appl. Phys. **70**, 6854 (1991).
5. P.J. Price: in "Fluctuation Phenomena in Solids", Ed. by R.E. Burgess (Academic Press, New York, 1965) p. 355.
6. L. Varani and L. Reggiani, Rivista Nuovo Cimento **17**, 1 (1994).
7. L. Varani, Proc. 13th Int. Conf. on Noise in Physical Systems and 1/f Fluctuations, Eds. V. Bareikis and R. Katilius, World Scientific (Singapore, 1995) p. 203.

# QUANTUM TRANSPORT WITH ELECTRON-PHONON INTERACTION
# IN THE WIGNER-FUNCTION FORMALISM

Rossella Brunetti[1], Carlo Jacoboni[1], and Mihail Nedjalkov[2]

[1] Istituto Nazionale di Struttura della Materia,
  Dipartimento di Fisica, Università di Modena,  Italy
[2] Center for Informatics and Computer Technology,
  Bulgarian Academy of Sciences, Sofia, Bulgaria

## INTRODUCTION

The interest of electron transport in mesoscopic systems is well known both for electronic applications and for investigations of basic quantum physics[1]. The use of the Wigner function has been found particularly appropriate for the theoretical analysis of such a problem[2] since it combines the rigorous approach of quantum mechanics with the more familiar representation of the phase space. However, while ballistic coherent transport has been deeply investigated in the recent literature, dissipative scattering by phonons has been included in the theory only by means of the relaxation-time approximation[3], heavily deteriorating the quality of the rigorous quantum approach.

In this communication we present a theory of quantum electron transport based of an extension of the Wigner function that includes electron-phonon coupling. An interaction scheme is used, where the electron potential profile (band engineering plus external applied voltage) is already accounted for in the unperturbed dynamics. This paper is mainly focused on the presentation of the basic aspects of the theory, and for the moment practical examples are limited to few simple cases.

## THEORY

A three-dimensional system of independent electrons interacting with phonons is considered here with translational invariance along two directions (x-y). The unperturbed Hamiltonian $H_0$ of the system contains the electron Hamiltonian, including the potential profile and the free-phonon term:

$$H_0 = H_e + H_p = \frac{p^2}{2m} + V(z) + \sum_q a_q^+ a_q \, \hbar\omega_q \tag{1}$$

where $p$ and $m$ are the electron momentum and mass, respectively, $V(z)$ is the electron potential profile (including the applied voltage), $a_q$ and $a_q^+$ are the annihilation and creation operators of the phonon mode $q$ with frequency $\omega_q$. $H_{e\text{-}p}$ is the electron-phonon interaction:

$$\mathbf{H_{e\text{-}p}} = \sum_q i\hbar F(q)[\mathbf{a}_q e^{i\mathbf{qr}} - \mathbf{a}_q^+ e^{-i\mathbf{qr}}] \qquad (2)$$

where F(q) is a function that depends on the type of electron-phonon interaction. In our case polar optical phonons have been considered.

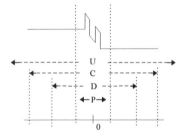

Several characteristic space regions are of interest for this problem, as indicated in Fig. 1. P is the "potential region", inside which the potential profile can vary; D is the "device region", where the Wigner function is to be evaluated; C is the "correlation region", within which any correlation of the electron wavefunctions from inside the device is exhausted; U is a much larger region, used to normalize the wavefunctions.

Fig. 1 - Characteristic lenghts considered in this problem. A potential profile is shown above as an example.

The present theoretical approach starts from the definition of the Wigner function, generalized to include phonons in the system, together with one electron (or, equivalently, many independent electrons):

$$f(\mathbf{r},\mathbf{p},n_q,n_q') = \frac{1}{h^3}\int d\mathbf{r}' e^{-i\mathbf{pr}'/h}\rho(\mathbf{r}+\mathbf{r}'/2,n_q;\mathbf{r}-\mathbf{r}'/2,n_q') \qquad (3)$$

where $\rho$ is the density matrix of the electron-phonon system.

For any given basis $\{|\varphi_l\rangle\}$ for the electron states, it is possible to move from the Wigner function to the density matrix representation and vice versa by means of the coefficients

$$f_{lm}(\mathbf{r},\mathbf{p}) = \frac{1}{h^3}\int d\mathbf{r}'\, e^{-i\mathbf{pr}'/h}\langle \mathbf{r}+\mathbf{r}'/2\,|\varphi_l\rangle\langle\varphi_m|\,\mathbf{r}-\mathbf{r}'/2\rangle. \qquad (4)$$

If we consider the density-matrix operator in the interaction picture with respect to the unperturbed hamiltonian $\mathbf{H}_o$, its corresponding Wigner function $\tilde{f}$ satisfies the equation of motion

$$\frac{\partial}{\partial t}\tilde{f}(\mathbf{r},\mathbf{p},n_q,n_q',t) = \frac{1}{h^3}\int d\mathbf{r}'\, e^{-i\mathbf{pr}'/h}\langle \mathbf{r}+\mathbf{r}'/2,n_q\,|[\tilde{\mathscr{H}}'(t),\tilde{\rho}(t)]|\,\mathbf{r}-\mathbf{r}'/2,n_q'\rangle \qquad (5)$$

where $\tilde{\mathscr{H}}' = \tilde{H}_{e\text{-}p}/i\hbar$. After formal integration, we obtain

$$\tilde{f}(\mathbf{r},\mathbf{p},n_q,n_q',t) = \tilde{f}(\mathbf{r},\mathbf{p},n_q,n_q',0) + \int_0^t dt' \sum_{nn'} f_{nn'}(\mathbf{r},\mathbf{p}) \times$$

$$\times \sum_{\substack{mm\\q}} h^3\int d\mathbf{r}'\int d\mathbf{p}'\{\tilde{\mathscr{H}}'(nn_q,mm_q,t')f^*_{mn'}(\mathbf{r}'\,\mathbf{p}')\,\tilde{f}(\mathbf{r}',\mathbf{p}',m_q,n'_q,t')$$

$$-f^*_{nm}(\mathbf{r}',\mathbf{p}')\tilde{f}(\mathbf{r}',\mathbf{p}',n_q,m_q,t')\tilde{\mathscr{H}}'(m,m_q,n',n'_q\,t')\} \qquad (6)$$

In order to return to the Schrödinger picture we make use again of the coefficients in Eq.(4) to obtain the density matrix, then transform the density matrix to the Schrödinger picture, and finally we return to the Wigner function by means of the same coefficients.

Eq.(6) can be substituted into itself to obtain the Neumann series for our problem. The zero order term, after transformation to the Schroedinger picture, yields the ballistic evolution of the Wigner function. Terms of first order in the interaction hamiltonian correspond to non-diagonal terms in the phonon state and are of no interest to our present problem. Terms of second order yield contributions corresponding to one phonon scattering. For a real emission process, for example, the correction of the Wigner function is given by

$$\Delta f^{(2,r,E)}(r,p,t) = 2\mathcal{R}e\sum_q F^2(q)\langle n_q+1\rangle\sum_{ll'} f_{ll'}(r,p)\sum_{ms}\mathcal{C}^*(m,q,l)\mathcal{C}(s,q,l') \times$$

$$\times \mathcal{T}^{(Er)}(t,l,q,l',s,m)h^3\int dr'\int dp' f^*_{ms}(r'p')f(r',p',0) \tag{7}$$

where $\mathcal{T}^{(Er)}(t,l,q,l',s,m)$ is

$$\frac{e^{-i(\omega_m-\omega_s)t}-e^{-i(\omega_l-\omega_{l'})t}}{-(\omega_{l'}-\omega_s+\omega_q)(\omega_m-\omega_l+\omega_{l'}-\omega_s)}+\frac{e^{-i(\omega_m-\omega_{l'}-\omega_q)t}-e^{-i(\omega_l-\omega_{l'})t}}{(\omega_{l'}-\omega_s+\omega_q)(\omega_m-\omega_l-\omega_q)} \tag{8}$$

Here $\hbar\omega_n$ is the energy associated to the electron state $n$ and $\mathcal{C}(m,q,l)$ is the matrix element of $e^{iqr}$ between the states $m$ and $l$.

The Schrödinger equation has been solved numerically inside the region P. Outside this region the potential profile is constant, and the solution can be continued analically. For each particular energy above the bias voltage two degenerate states are found and their combinations have been obtained corresponding to the scattering states[4]. The density of states of such eigenfunctions, as function of the incoming wavevector **k**, can be found as limit of the equivalent density inside a box, when the dimension of the box is let to increase indefinitely. It has a peculiar form, with a divergence at a value of **k** corresponding to an electron coming from the low-voltage side with a kinetic energy equal to the bias potential energy.

The terms in the series expansion (as that in Eq.(7)) relate the Wigner function at any given (**r**, **p**) at time t to its initial value at all phase-space points (**r'**, **p'**). In our physical problem, however, the Wigner function is assumed to be known at the initial time t = 0 inside the device and at all times at the two boundaries of the region D (see Fig. 1) for values of **p** corresponding to entering momenta. It is possible to show that if the boundaries are so far away from the potential region P that the wave functions of the entering electrons at the boundaries are uncorrelated to the region P, the integral in Eq. (7) can be substituted by three terms, as simbolically indicated in Fig. 2. One term is identical to the original expression in Eq. (7), with the integral over the initial condition limited to the points internal to the D region. The other two terms are evaluated at the two boundaries of the D region, divided by the entering velocity of the electron, and integrated over the possible incoming momenta.

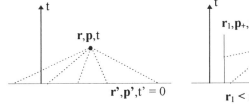

Fig. 2 - The initial condition in the integral in Eq. (7) can be substituted by the boundary conditions of entering electrons

For simplicity we assumed for the entering Wigner function the form of a Maxwell distribution.

## NUMERICAL RESULTS

In principle, the Neumann series obtained above can be evaluated with a Monte Carlo technique. A number of problems arise on this respect related to the convergence of the series and to a possible divergence of the variance in its evaluation. Such problems can in principle be solved, as will be discussed elsewhere. Formidable numerical difficulties remain, however, related to interference of large quantum oscillations. Suitable numerical techniques are being developed to overcome them.

In Fig. 3 the unperturbed (ballistic) Wigner function is shown, as obtained from the scattering states entering the device with a Maxwellian distribution. The potential profile is a ramp of uniform electric field (E = 5 kV/cm) inside a region of width L = 20 nm. At low momentum values oscillations can be seen due to interference effects related to quantum reflections. In the region of positive positions a peak is present that represents electrons accelerated by the electric field.

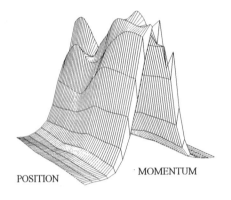

POSITION    MOMENTUM

Fig. 3 - Unperturbed Wigner function obtained with scattering states entering with a Maxwellian distribution a region with a potential ramp

The integral in Eq. (7) has been evaluated with full numerical integration over *m, s, l,* and *l'*, and a Monte Carlo selection of **r'**, **p'** and **q**. A typical contribution of such sampling is shown in Fig. 4 where, for a given **q**, the correction to the Wigner function due to real and virtual emissions is shown. As in semiclassical theory, the scattering process decreases *f* at the values of **r** and **p** of the initial state and increases it at **r** and **p** of the final state of the collision.

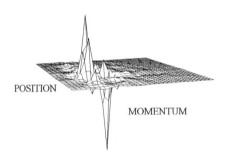

POSITION

MOMENTUM

Fig. 4 - Sample of second-order perturbative correction corresponding to a real and virtual phonon emission process.

### Acknowledgements

This work has been made possible by the partial support of the European Research Office and, through it, by the A.R.O. and the O.N.R. The Authors gratefully acknowledge Dr. A. Abramo for scientific discussions and Marco Pascoli for his assistance in the preparation of the paper.

### REFERENCES

[1] See, for example, Proc. NATO ASI on *Quantum Transport in Ultrasmall Devices*, Plenum, in press, and *Mesoscopic Phenomena in Solids*, Modern Problems in Condensed Matter Sciences, Vol. 30 (1991).
[2] W. R. Frensley, Rev. Mod. Phys. **62**(3), 745 (1990).
[3] See, for example, N.C. Kluksdahl, A. M. Kriman, D. K. Ferry, and C. Ringhofer, Phys. Rev. B **39**(11), 7720 (1989).
[4] A. M. Kriman, N.C. Kluksdahl, and D. K. Ferry, Phys. Rev. B **36**(11), 5953 (1987).

# SELF-CONSISTENT TIME DEPENDENT SOLUTIONS TO THE QUANTUM LIOUVILLE EQUATION IN THE COORDINATE REPRESENTATION: APPLICATION TO BARRIER STRUCTURES[*]

H. L. Grubin[1], T. R. Govindan[1], and D. K. Ferry[2]

[1]Scientific Research Associates, Inc.; Glastonbury, Connecticut
[2]Arizona State University; Tempe, AZ

## INTRODUCTION

We report on *transient accurate* self-consistent solutions of the quantum Liouville equation in the coordinate representation:

$$i\hbar\frac{\partial\rho(x,x',t)}{\partial t} = -\frac{\hbar^2}{2m}\left(\frac{\partial^2}{\partial x^2} - \frac{\partial^2}{\partial x'^2}\right)\rho(x,x',t) + \left[(V(x)-V(x')) - (E_F(x) - E_F(x'))\right]\rho(x,x',t)$$

where $E_F(x)$ and $E_F(x')$ represent quasi-Fermi levels subject to the constraint that the kinetic energy of entering and exiting carriers are equal [Grubin (1995)]. Dissipation is included and current, $J_{Total}$, contains both *displacement and conduction* contributions. Under time independent conditions the conduction current, $j$, for a scattering rate $\Gamma$ satisfies the condition that: $E_F(L) - E_F(0) = -j\int_0^L dxm\Gamma(x)/\rho(x)$. The boundary conditions incorporate current via the density matrix equivalent of a displaced Fermi-Dirac distribution [ Grubin *et al.* (1993)].

For this discussion, *V(x=0)=0*, and for a given downstream potential energy *V(L)*, the system is brought to steady state. A step change to a new constant value is introduced for *V(L)*, and the resulting time dependent behavior is computed at fixed increments in time.

## THE RESULTS

Three symmetric 120 nm length structures were studied: (i) $N^+NN^+$ [$N^+=10^{24}/m^3$ (30 nm long), $N^-=10^{22}/m^3$ (30 nm long)], three 200 meV barriers [4 nm barriers, 4 nm wells]; (ii) $N^+NN^+$, two 300 meV barriers [5nm barriers, 6 nm well] and (iii) as (ii) but uniformly doped with $N=10^{24}/m^3$. We assumed Fermi statistics, parameters appropriate to GaAs, except for the barriers, and a constant effective mass. While the length of the structure is too small for any real device studies, the choice achieved small time steps. A step change in voltage to 150 meV from a steady state of 100 meV, was applied.

*The triple barrier structure*: At 100 meV the steady state current was 1.477 x 10⁹ amps/m². At 150 meV steady state yielded a current value of 2.657 x 10⁹ amps/m². The time dependent space charge distribution following the voltage change was followed through 420 fs. The time step was 70fs, and the transient current at the end of each time increment is shown in figure 1. 70 fs after application of the step change in voltage the charge is generally increased everywhere within the structure. (This is true for the double barrier structure as well.) *For the regions surrounding the barriers* we find little time variation in density although the cladding region voltage undergoes changes, suggesting that the dominant time dependent behavior in the cladding region is due to displacement current contributions. *For the barrier region,* matters are different; the time dependent changes in voltage are reduced. At 70 fs after application of the voltage step the charge within the barriers exceeds that of the lower bias steady state; see figure 2, which displays denstiy, within the central 20 nm region. At 140 fs there is a decrease in charge within the three barriers and a corresponding increase in charge between the barriers. These results are consistent with an interpretation that a fraction of the charge present at 70 fs has tunneled through the barriers at 140 fs. Excess charge in the wells would subsequently tunnel back toward the respective barriers at a later time, etc., until a steady state is reached.

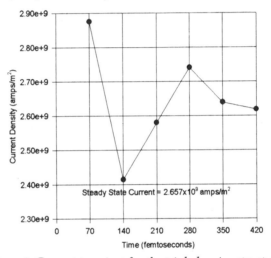

*Figure1. Current transient for the triple barrier structure.*

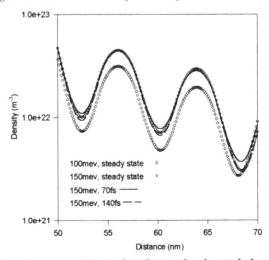

*Figure 2. Transient space charge distribution for the triple barrier struture.*

*The Double Barrier Structures*: The situation for the $N^+N^-N^+$ double barrier structure is similar. Here at a bias of 100 meV, the steady state current is 6.102 x $10^8$ amps/m², at 150 meV the current is 1.047 x $10^9$ amps/m². *For the regions surrounding the barriers,* the transient density distribution shows considerably more structure in the cladding regions than for the triple barrier structure and suggests that conduction current contributions as well as displacement current contributions are significant. *For the barrier region,* figure 3, charge appears to tunnel to the center well, where the peak density exceeds its steady state value. Tunneling out of this region increases the charge in the barrier. This tunneling into and out of the quantum well may be driving the time dependence of the device.

The situation for the double barrier structure embedded in a uniformly doped $10^{24}$/m³ region also shows time dependent behavior, but the initial and final state disributions of charge are not significantly different, and the steady state distibution of charge appears to be reached in the early time stages. Dielectric relaxation may be playing a significant role.

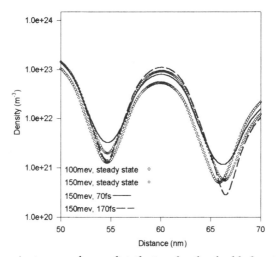

*Figure3. Transient space charge distribution for the double barrier structure.*

## SUMMARY AND REMARKS

The calculations discussed here demonstrate the richness of the transient phenomena. There is tunneling through a barrier and into the quantum well, coupled to displacement current effects, and contributions arising due to the magnitude of the density within the structure. We have not attempted to extract a tunneling time from these studies; in actual device studies, the tunneling times would be dressed by dielectric contributions and the variability of the self consistent field. In most of the calculations the current reaches its steady state value at approximately 500 fs, but weak oscillations beyond this are expected. We have not carried this further. The details of the result will be depend upon the scattering parameters.

## ACKNOWLEDGEMENTS

* Supported by ONR

## REFERENCES

Grubin, H. L., Govindan, T. R, Kreskovsky, J..P. and Stroscio, M.A., 1993, *Sol. State Electron.* 36:1697
Grubin, H. L. 1995 in "Quantum Transport in Ultrasmall Devices" Eds. D K. Ferry, H. L. Grubin, C. Jacaboni and A. Jauho, Plenum, New York.

# FREE–CARRIER SCREENING IN COUPLED ELECTRON–PHONON SYSTEMS OUT OF EQUILIBRIUM

W. Pötz[1] and U. Hohenester[2]

[1] Physics Department
University of Illinois at Chicago
Chicago, IL 60607

[2] Theoretische Physik
Universität Graz
A–8010 Graz, Austria

## INTRODUCTION

Recent experimental studies of non–equilibrium carrier dynamics on sub–picosecond time scales have verified the presence of (partially) coherent carrier motion in semiconductor microstructures. Loss of phase coherence is generally believed to be caused by the Coulomb and electron–phonon interaction and, for macroscopic phenomena, by structural imperfections of the sample. A theoretical approach to study coherent carrier motion, including carrier generation, and its loss due to interaction processes must generally go beyond the single–band Boltzmann equation.[1-4] We have recently presented a derivation of generalized Boltzmann–Bloch (BB) equations within non–equilibrium Green's function techniques.[5] Just like the original Dyson equation, these BB equations contain the self–energy of the various interactions. When selecting relevant diagrams to approximate self–energies, such as the Fock contribution,[5] proper screening of the interactions is important to arrive at realistic dephasing rates. (It is beyond the scope of this contribution to review the work on screening in many–body systems. Following a suggestion of the referee, we just give one example and refer to work cited therein.[6] ) Each interaction, such as the electron–electron Coulomb ($c$) and electron–phonon ($ep$) interaction, contributes to screening. This, in general, causes a hybridization of all (two–particle) interactions into one net interaction.[5] This is illustrated in Fig. 1, at the example of screening in

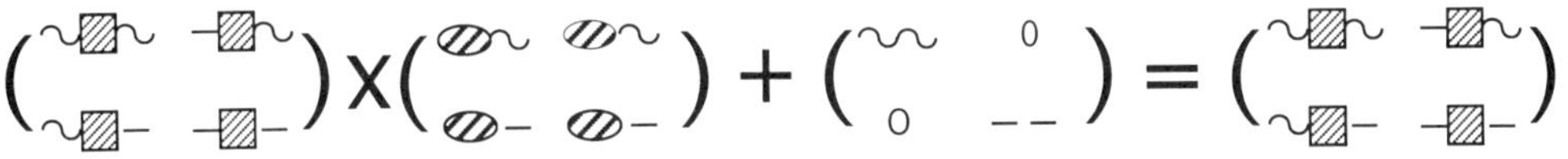

$$\begin{pmatrix} & \\ & \end{pmatrix} \times \begin{pmatrix} & \\ & \end{pmatrix} + \begin{pmatrix} 0 & \\ & 0 \end{pmatrix} = \begin{pmatrix} & \\ & \end{pmatrix}$$

**Fig. 1.** Relation between screened (squares) and unscreened $c$ (dashed line) and $ep$ vertices. Shaded bubble represents the polarization.

the presence of the $c$ and $ep$ interaction. This Dyson–type equation relates the screened interactions $V_{ij}$ to the bare interactions $v_j$, $i,j = c, ep$. In every electron self–energy diagram which contains a two–particle vertex a summation over all elements $V_{ij}$ must be performed. This leads to a net screened interaction

$$\{\underline{V}\} = \{1 - v\Pi\}^{-1} \otimes \{\underline{v}\} = \{\underline{v}\} \otimes \{1 - \Pi v\}^{-1} \quad ,$$

where $v = v_c + v_{ep}$.[7] Thus, the net screened interaction is the self–screened version of the sum of all bare interactions. Physically, screening is provided by the rearrangement of conduction electrons, as well as lattice ions. This net interaction may now be used in self–energy diagrams which enter into the carrier BB equations.

## SCREENING IN THE PRESENCE OF PHONON NON–EQUILIBRIUM

In order to account for significant build–up of phonons ("hot phonons") within BB equations, phonon BB equations must be solved in parallel to electron BB equations. In this case, lattice deformations are already accounted for by the phonon Green's function and must not be incorporated again in form of screening. This situation will now be considered in more detail.

The $(c)$ and $(ep)$ interaction, respectively, give rise to the interaction vertices

$$(v_c)_{\alpha\beta\gamma\delta}(\mathbf{q})\underline{1}\delta(t-t')$$

and

$$\underline{v}_{ep}{}^{(\sigma)}_{\alpha\beta\gamma\delta}(t,t',\mathbf{q}) = \sum_{mm'} M^{(\sigma m)}_{\alpha\gamma}(\mathbf{q})M^{(\sigma m')}_{\beta\delta}(-\mathbf{q})\left[\underline{g}_{o\sigma}(t,t',\mathbf{q}) + \underline{g}_{o\sigma}(t',t,-\mathbf{q})\right] \quad . \tag{1}$$

Here, $g_o(t,t',\mathbf{q})$ is the free phonon Green's function (GF), $M^{(\sigma m)}_{\alpha\gamma}(\mathbf{q})$ is the bare electron–phonon matrix element for phonon type $\sigma$ and coupling mechanism $m$. (1) represents an effective electron–electron interaction mediated by the exchange of a single phonon. It has two contributions, both of which provide net propagation of wave vector $\mathbf{q}$ from t' to t.

The BB equations for the interacting phonon GF $g_\sigma(t,t',\mathbf{q})$ arises from the Dyson equation represented graphically in Fig. 2. Using the electron–hole–pair approximation

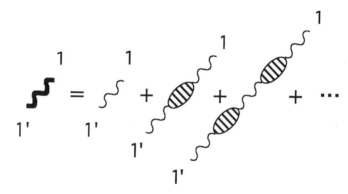

**Fig. 2.** Dyson equation for the phonon Green's function (thick wavy line) in terms of the free–phonon Green's function (thin wavy line) and the polarization (shaded bubble).

for the self–energy and ignoring coupling among different phonon modes, as well as interference between emission and absorption processes, leads to the standard phonon Boltzmann equation in the *unscreened ep* matrix elements.

Using the phonon GF to construct the vertex $V_p$ for the self–screened $ep$ interaction one obtains

$$(\underline{V}_{ep})^{(\sigma)}_{\alpha\beta\gamma\delta}(t,t',\mathbf{q}) = \sum_{mm'} M^{(\sigma m)}_{\alpha\beta}(\mathbf{q})M^{(\sigma m')}_{\alpha'\beta'}(-\mathbf{q})\left[\underline{g}_\sigma(t,t',\mathbf{q}) + \underline{g}_\sigma(t',t,-\mathbf{q})\right] \quad .$$

However, comparison between Fig. 1 and Fig. 2 shows that this vertex already accounts for self–screening contributions from the lattice. Therefore, the relation between bare and screened interactions in Fig. 1 must be modified to

$$\begin{pmatrix} \{\underline{V}_{cc}\} & \{\underline{V}_{cp}\} \\ \{\underline{V}_{pc}\} & \{\underline{V}_{pp}\} \end{pmatrix} = \begin{pmatrix} \{\underline{v}_c\} & 0 \\ 0 & \{\underline{V}_{ep}\} \end{pmatrix} + \begin{pmatrix} \{\underline{v}_c\} & \{\underline{v}_c\} \\ \{\underline{V}_{ep}\} & 0 \end{pmatrix} \otimes \{\underline{\Pi}\} \otimes \begin{pmatrix} \{\underline{V}_{cc}\} & \{\underline{V}_{cp}\} \\ \{\underline{V}_{pc}\} & \{\underline{V}_{pp}\} \end{pmatrix} \quad ,$$

which is depicted in Fig. 3.

The resulting net interaction is again obtained by summation over all types of vertices and is

$$\{\underline{V}\} = \sum_{i,j=c,p} \{V_{ij}\} = \left\{(1 + V_{ep}\Pi)^{-1} - v_c\Pi\right\}^{-1} \otimes \{\underline{v_c}\} + \left\{1 - (1 + V_{ep}\Pi)v_c\Pi\right\}^{-1} \otimes \{V_{ep}\} \ .$$

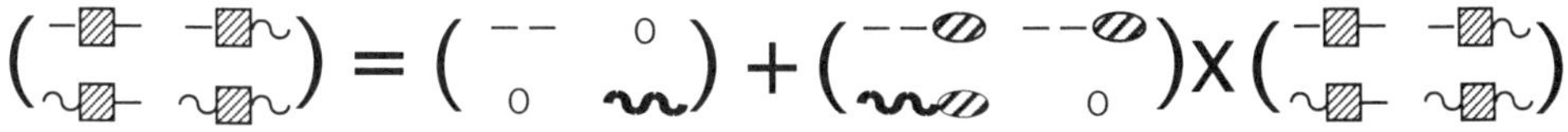

**Fig. 3.** Relation between screened (squares) vertices and unscreened $c$ (dashed line) and non–equilibrium $ep$ (thick wavy line) vertices. Shaded bubble represents the polarization.

Neglecting terms of order $(V_{ep})^2$ and higher one arrives at the simpler expression

$$\{\underline{V}\} = \left\{1 - (v_c + V_{ep})\Pi\right\}^{-1} \otimes \{\underline{v_c}\} + \left\{1 - v_c\Pi\right\}^{-1} \otimes \{V_{ep}\} \ .$$

This leads to the physical intuitive result that the net screened interaction for electrons in an interacting $ep$ system, in which phonon non–equilibrium is treated dynamically, is the sum of two interactions: the electron–electron Coulomb interaction screened by both the electron–electron Coulomb and the electron–phonon interaction, and the self–screened electron–phonon interaction screened by the electron–electron Coulomb interaction.

## SUMMARY AND CONCLUSIONS

We have discussed what we consider the simplest version of free–carrier screening in coupled out–of–equilibrium electron–phonon systems within Boltzmann–Bloch (BB) equations. If the phonon non–equilibrium is of minor importance, the net carrier–carrier interaction which is used in the electron Fock self–energy diagram is the self–screened hybrid electron–electron–phonon interaction. If phonons are driven out of equilibrium to a considerable degree phonon BB equations must be solved in parallel to the electron BB equations. In lowest order, the $ep$ matrix elements which enter the *phonon* BB equations should not be screened, the $c$ interaction in the electron BB equation should be screened both by the $c$ and $ep$ interaction, and the (self–screened) $ep$ interaction in the electron BB equation should be screened only by the $c$ interaction. A more detailed account of this investigation which accounts for mixing of different phonon modes, as well as "non–energy–conserving" diagrams with the emission (absorption) of two phonons will be given elsewhere.

## REFERENCES

1. A. V. Kuznetsov, Phys. Rev. B **44**:8721 (1991).
2. T. Kuhn and F. Rossi, Phys. Rev. B **46**:7496 (1992).
3. F. J. Adler, P. Kocevar, J. Schilp, T. Kuhn, and F. Rossi, Semiconductor Science and Technology **9**:446 (1994).
4. W. Pötz, M. Žiger, and P. Kocevar, Phys. Rev. B **52**:1959 (1995); M. Žiger, W. Pötz, and P. Kocevar, SPIE Proceedings Vol. **2142**:87 (1994); M. Žiger and W. Pötz, in *Proceedings of the 3$^{rd}$ International Workshop on Computational Electronics*, Portland Oregon, May 18–20, 1994.
5. W. Pötz and U. Hohenester, unpublished; see also, U. Hohenester and W. Pötz, these proceedings.
6. H. C. Tso and N. J. Horing, Phys. Rev. B **44**, 8886 (1991).
7. Notation as in J. Rammer and H. Smith, Rev. Mod. Phys. **58**:323 (1986); and Ref. 4.

# IMPACT OF DYNAMIC SCREENING AND CARRIER-CARRIER SCATTERING ON NON-EQUILIBRIUM DUAL CARRIER PLASMAS

J. E. Bair and J. P. Krusius

Cornell University, Schools of Electrical Engineering
and Applied Engineering Physics
Ithaca, New York

## ABSTRACT

The effect of free carrier screening model on the physical behavior of non-equilibrium electron-hole plasmas is explored. Simplified approaches are found to fail badly for both the description of ultrafast carrier relaxation and high field transport. The existence of significant transient structure is found to make it unlikely any simple model could provide an adequate description.

## INTRODUCTION

The importance of the carrier-carrier interaction in determining the observed behavior of non-equilibrium electron-hole plasmas has been appreciated for some time. However due to the complexity of this many-body process, practical methods of including it, that are both accurate and computationally efficient, in standard treatments of carrier dynamics have remained illusive. Early efforts almost exclusively focused on the simple static screening model. While the inadequacies of this approach for the treatment ultrafast carrier relaxation have been widely reported[1,2], it still continues to be used in the treatment of carrier transport[3] due to its simplicity. More recently the related "multi-static" or "quasi-dynamic" approach has been used with considerable success in the treatment of ultrafast carrier relaxation[2,4].

In this work we examine the consequences of these simplified models for the treatment of non-equilibrium electron-hole plasmas. For this purpose, we compare Monte Carlo results obtained within the static and quasi-dynamic for ultrafast carrier relaxation and high field transport to results obtained with a unique dynamic screening formulation that allows true RPA screening[5,6,7]. The accuracy of the dynamically screened Monte Carlo simulation has been extensively verified for the treatment of ultrafast carrier relaxation[6,7].

## RESULTS

Recent reports have found that the quasi-dynamic or multi-static approach to free carrier screening produces excellent correlations with experiment[2,4], and is essentially equivalent to the more sophisticated molecular dynamics approach[2]. The equivalence of the results obtained from quasi-dynamic and true dynamic screening for the observed time dependence of the probe transmission in optical pulse probe experiments has been confirmed by our simulations. However when the microscopic details underlying these results are examined, the apparent success of the pseudo-dynamic approach disappears. As shown in Fig. 1a, the electron-electron scattering rate for the dynamic and pseudo-dynamic screening methods are quite comparable, but there are very large differences in the rates of energy exchange. When the hole-hole and electron-hole scattering rates are examined, the errors for the pseudo-dynamic model are found to be as large, or larger, than those for static screening. Similar difficulties are found in the energy exchange rates.

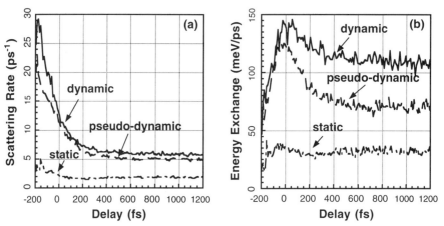

**Figure 1.**    Comparison of electron-electron scattering rates (a), and energy transfer rates (b), for dynamic, static and pseudo-dynamic screening models. Simulations were performed for excitation 57 meV above the band edge in $In_{0.53}Ga_{0.47}As$, with an excited carrier density of 2.5 x $10^{17}$ cm$^{-3}$.

The signs of the errors associated with the various processes are such that a significant cancellation of errors occurs when they are combined, such as in the observed probe transmission. Further analysis shows that the good results obtained by simulations using this screening model are the result of a fortunate cancellation of errors. Thus the good correlation with experiment cannot be taken as confirmation of the microscopic physics that underlies them.

The simplified treatment of screening does not fare any better in the study of high field transport in dense electron-hole plasmas, as is demonstrated in Fig. 2. Here we see the static screening approach seriously underestimates the effect of the electron-hole interaction on high field transport in GaAs. Consistent with the results noted above, the primary shortcoming of the static approach is that the transfer of energy from the electrons to the holes is greatly underestimated.

The difficulties experienced in developing a simple approach to free carrier screening are only part of a larger problem. While the general form of the equilibrium free carrier dielectric function is well known, its true complexity in non-equilibrium is not fully appreciated. The non-equilibrium free carrier distribution function often has considerable unexpected structure that can not be anticipated from equilibrium. For example in the first few 100 fs of the relaxation of optically excited carriers, the free carrier dielectric function has several ridges not present in equilibrium[7]. These new features can have large effects on the carrier-carrier scattering and energy exchange rates. These considerations make it unlikely that any simple approach to free carrier screening could fully account for the behavior of the non-equilibrium electron-hole plasma.

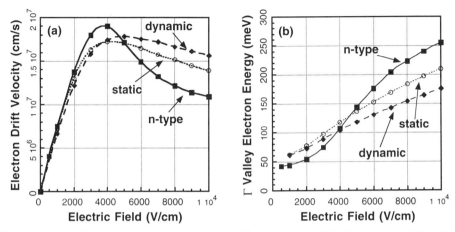

**Figure 2**.  Ensemble averaged steady state electron drift velocity (a), and $\Gamma$ electron energy (b), for high field transport in GaAs. Curves showing results for a $1 \times 10^{18}$ cm$^{-3}$ electron hole plasma using static and dynamic screening are compared to those for a high purity $1 \times 10^{14}$ cm$^{-3}$ n-type sample.

## CONCLUSION

It was found that none of the currently available simple models of free carrier screening produce satisfactory results for the treatment of carrier-carrier scattering in non-equilibrium electron-hole plasmas. Further, due to the complex transient structure of the non-equilibrium free carrier dielectric function it is unlikely that any simple model could provide an adequate description under all conditions.

## ACKNOWLEDGMENT

This work was supported by the Joint Service Electronics Program (Contract Number F49620-93-C-0016).

## REFERENCES

1.    J. F. Young, N. L. Henry, and P. J. Kelley, Solid State Elec. **32**, 1567 (1989).
2.    L. Rota, P. Lugli, T. Elsaesser, and J. Shah, in <u>Ultrafast Lasers Probe Phenomena in Semiconductors and Superconductors,</u> Robert R. Alfano, Editor, Proc. SPIE 1677, p. 146 (1992).
3.    R. P. Joshi, A. N. Dharamso, and J. McAdoo, Appl. Phys. Lett. **64**, 3611 (1994).
4.    M. Ulman, D. W. Bailey, L. H. Acioli, F. G. Vallee, C. J. Stanton, E. P. Ippen, and J. G. Fujimoto, Phys. Rev. B **47**, 10267 (1993).
5.    J. E. Bair and J. P. Krusius, submitted to Journal of Applied Physics.
6.    J. E. Bair, D. Cohen, J. P. Krusius, and C. R. Pollock, Phys. Rev. B **50**, 4355 (1994).
7.    J. E. Bair, Ph.D. Thesis, Cornell Univ, Ithaca NY, August (1995).

# COLLISION DURATION FOR POLAR OPTICAL AND INTERVALLEY PHONON SCATTERING

Paolo Bordone, Dragica Vasileska, and David K. Ferry

Center for Solid State Electronics Research
Arizona State University
Tempe, Arizona 85287-6206

## INTRODUCTION

The use of femtosecond laser pulses to excite plasmas in semiconductors has become a major method of studying fast processes.[1] The transition times from the $\Gamma$ valley to the satellite X and L valleys are comparable to the reciprocal of the frequency of the phonons involved, bringing into question the use of standard perturbation-theory approaches. Our aim is to evaluate the time required to emit a phonon, either the intravalley LO or the intervalley, by a nearly-free electron in semiconductors. The leading idea of our work is that the so-called "collision duration" is related to the time required to build up correlation between the initial and final state, and then to destroy this correlation as the collision is completed. The calculations are developed using a non-equilibrium Green's function formalism, which allows us to evaluate explicitly the effects of the correlations in time.

## THEORETICAL APPROACH

Our approach follows the pioneering ideas developed by Kuhn and Rossi[2] and Haug[3] to study ultrafast processes in photoexcited semiconductors. The electron interacts with the optical phonon (both polar and intervalley) through the polarization field of the phonon, first building up a correlation between the initial and final states, and then breaking up this correlation as the collision is completed. Therefore, we will define the collision duration time as the time in which the initial and final states are correlated.

Our starting point are the equations of motion for the non-equilibrium Green's function in the matrix notation,[4-6] where for the Green's function and self-energy matices we use the ones written in terms of the less-than, greater-than, time-ordered and anti-time-ordered functions. The self-energy $\Sigma$ is usually a two-point function. *Here, however, we make a crucial deviation from the normal approach.* We follow the approach of Kuhn-Rossi[2] and Haug[3] (who work with the electromagnetic field of the photon) by introducing only the polarization field of the longitudinal polar optical phonon.[7,8] This choice of the self-energy corresponds to the electron interacting with the dipole *field* of the phonon and makes the self-energy matrix diagonal, as the self-energy is a single-point function (function of a single time and position). Thus, we can separate the equations for the less-than functions from the matrix formulation.

## RESULTS AND DISCUSSION

Developing the calculations, whose details are given in Ref. [7], we derive an analytical expression for the time derivative of $G^<(\mathbf{k},t)$, which represents the probability for a carrier to end up in a final state $\mathbf{k}$ as a consequence of the emission of an optical phonon (either polar or intervalley) as a function of time. For the polar optical phonon case we find for the probality of scattering into the final state [which is the derivative of $G^<(\mathbf{k},t)$]

$$
P(k,t) = \left( \frac{e^2}{\hbar \eta \omega_0} \right) \sqrt{\frac{2m}{\hbar}} \frac{1}{(2\pi)^2} \frac{(n_0 + 1)}{\sqrt{\omega_k}}
$$
$$
\times \left\{ \frac{\pi}{2} \ln \left( \frac{\left( \sqrt{\omega_k + \omega_0 + i\gamma} + \sqrt{\omega_k} \right) \left( \sqrt{\omega_k + \omega_0 - i\gamma} + \sqrt{\omega_k} \right)}{\left( \sqrt{\omega_k + \omega_0 + i\gamma} - \sqrt{\omega_k} \right) \left( \sqrt{\omega_k + \omega_0 - i\gamma} - \sqrt{\omega_k} \right)} \right) \right.
$$
$$
\left. -2e^{-\gamma t} \int_{\omega_k + \omega_0}^{\infty} d\alpha \left[ \frac{\gamma}{\gamma^2 + \alpha^2} \sin(\alpha t) + \frac{\alpha}{\gamma^2 + \alpha^2} \cos(\alpha t) \right] \mathrm{arc\,cot} \left( \frac{\sqrt{\alpha - (\omega_k + \omega_0)}}{\sqrt{\omega_k}} \right) \right\}
$$

(1)

while the result for the intervalley phonon is

$$
P(k,t) = \frac{1}{(2\pi)^2} \left( \frac{D^2}{\hbar \rho \omega_0} \right) \left( \frac{2m}{\hbar} \right)^{3/2} (n_0 + 1) \left\{ \frac{\pi}{2} \left( \sqrt{\omega_k + \omega_0 + i\gamma} + \sqrt{\omega_k + \omega_0 - i\gamma} \right) \right.
$$
$$
\left. -e^{-\gamma t} \int_{\omega_k + \omega_0}^{\infty} d\alpha \left[ \frac{\gamma}{\gamma^2 + \alpha^2} \sin(\alpha t) + \frac{\alpha}{\gamma^2 + \alpha^2} \cos(\alpha t) \right] \sqrt{\alpha - (\omega_k + \omega_0)} \right\}.
$$

(2)

where $\omega_k$ is the final state frequency, $\gamma$ is the lifetime of the initial state, and $n_0$ and $\omega_0$ are the equilibrium phonon population and frequency respectively, considered to be independent of the phonon wave-vector. In both cases the first term on the RHS is the collisioned-broadened "Fermi golden rule" (FGR) while the second represents the time dependent buildup of the scattering process. In the polar-optical phonon case, the integral over the initial states frequency cancels the contribution of the FGR at $t=0$, thus inhibiting the scattering process. On the other hand, this integral is divergent for $\alpha \to \infty$ in the intervalley phonon case. This means that, for the latter process, the electron can't access the entire spectrum of initial state frequencies. To avoid this problem we use a cut-off for the integral. To choose such a cut-off we impose the condition of zero scattering rate at $t=0$. This way of proceeding finds its justification in two main arguments: 1) the cut-off frequency turns out to be very high, the range of kinetic energies available for the carriers in the satellite valley has a width going from 0.6 eV up to 3 eV. This means that, in any case, we cover the range of the important frequencies for excited carriers in a typical semiconductor, 2) the results that we obtain are in perfect agreement with the ones found for the polar-optical phonon scattering[7] where no cut-off restriction is imposed (in practice, the cutoff is at a sufficiently high energy that further contributions to the integral are in the noise; this will be discussed elsewhere).

For the purpose of the simulation we consider a finite collision duration only for the intervalley optical phonon emission. Since it is a much faster process than the polar optical emission, it is expected to show a stronger effect. To treat the finite collision duration within what is normally a semiclassical ensemble Monte Carlo approach, we introduce a secondary self-scattering process.[8,9] We use a rejection technique in conjunction with the probability function given in (2) and shown in Fig. 1 for 3 different values of the final state energy. Collisions are considered as completed (the long-time limit), but the rejection process is introduced to discard the collision event if the time elapsed from the previous scattering event is too short relative to a collision duration.[9]

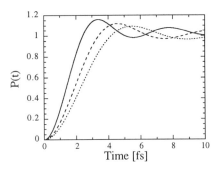

**Figure 1.** Scattering rate (normalized to the "Fermi golden rule") for three different values of the final state frequency for $\gamma=10^{14}$ s$^{-1}$ and $\omega_0=5.5\times10^{13}$ s$^{-1}$. $\omega_k=0.5\omega_0$ dotted line, $\omega_k=\omega_0$ dashed line and $\omega_k=2\omega_0$ solid line.

We use a 3-valley ensemble Monte Carlo simulation for electrons excited in GaAs at 300 K by a 2.0 eV photon. It is assumed in the simulation that the electrons are created by a 20 fs laser pulse, and the response of the electrons is followed for a subsequent time. Electrons are excited from all 3 valence bands. In Fig. 2, the occupations of the L and X valleys are plotted as a function of time after the onset of the laser pulse. In the earlier times after the laser pulse the main effect of the finite collision duration is to reduce the number of electrons transferred to the satellite valleys of about 20%. This is due to the fact that the finite collision duration decreases the number of the intervalley processes allowed in the short time limit. This can be noticed even in the different slope of the curves with and without finite collision duration. The time derivative of the electron population is related to the scattering probability, which is lowered in the short time limit by the collision duration effect. At later times, the oscillations of the scattering rate shown in Fig. 1 can induce an increase of the scattering with respect to the FGR result, leading to a change in the slope of the population curve, and eventually even to an increase in the electron popolation in the long time limit.

## CONCLUSIONS

We have derived an analytical expression for the probability for a carrier to end up in a final state $\omega_k$ as a consequence of the emission of a polar-optical or of an intervalley phonon as a function of time. The probability rises rapidly to the FGR result, and oscillations in the probability are damped by the decay of the initial state. Our calculations show that the collision duration time can be defined as the time over which the polarization of the initial state, induced by the phonon potential, decays. We then analyze the effect of the finite collision duration on Monte Carlo simulations of femtosecond laser excitation for the case of the intervalley process, and find that such a finite collision duration mainly reduces the number of electrons transferred to the satellite valleys of about 20%.

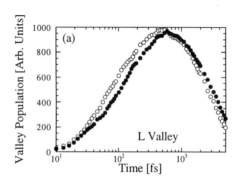

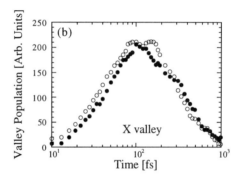

**Figure 2.** Relative population of the L (a) and X (b) valleys as a function of time after the excitation pulses. The solid symbols are for results which include collision retardation, while the open symbols neglect it.

## Acknowledgments

This work was supported by the U.S. Office of Naval Research.

## REFERENCES

1. See, e.g., J. Shah, *Sol.-State Electron.* **32**:1051 (1989), and references contained therein.
2. T. Kuhn and F. Rossi, *Phys. Rev. Lett.* **69**:977 (1992); *Phys. Rev. B* **46**:7496 (1992).
3. H. Haug, *Phys. Stat. Sol. (b)* **173**:139 (1992).
4. G. D. Mahan, *Phys. Repts.* **145**:251 (1987).
5. J. Rammer and H. Smith, *Rev. Mod. Phys.* **58**:323 (1986).
6. D. K. Ferry and H. L. Grubin. "Solid State Physics," H. Ehrenreich, ed., Academic Press, New York, in press.
7. P. Bordone, D. Vasileska, and D. K. Ferry, submitted for publication.
8. J. R. Barker and D. K. Ferry, *Phys. Rev. Lett.* **42**:1779 (1979).
9. D. K. Ferry, A. M. Kriman, H. Hida, and S. Yamaguchi, *Phys. Rev. Lett.* **67**: 633 (1991).

**Acoustic Phonon Assisted Energy Relaxation of
2D Electron Gases**

A.J.Vickers*, N.Balkan*, M.Cankurtaran**, H.Çelik**

\* University of Essex, Department of Physics, Colchester, UK
\** Hacettepe University, Department of Physics, Ankara, Turkey.

**Introduction**

The subject of acoustic phonon assisted energy relaxation in 2D systems has had a considerable amount of attention over the past fifteen years[1-4]. A number of groups have worked on heterojunctions and have found that they can only fit their models to the experimental results by assuming a value around 15eV for the deformation potential[1,4]. In all cases were experimental results have been compared to models the following assumptions have been made.
1. The acoustic phonons are bulk phonons in GaAs/AlGaAs,
2. The 2D system is taken as an ideal quantum well and
3. The approximation that $q_z >> q_\parallel$ is normally adopted to overcome the problem of obtaining an analytical solution.

In a previous paper one of us[5] reported on the results of a numerical calculation of the problem which removed the necessity of the third approximation above. In that work we showed that it was possible to achieve close fits to the experimental results whilst using the bulk GaAs value for the deformation potential (7eV). In this paper we study the problem further by looking at the well width dependence of the warm electron energy relaxation rates in GaAs/Ga$_{1-x}$Al$_x$As quantum wells. We compare the results of the numerical model with an analytical model[6] and also compare both models to a set of experimental results on a range of quantum well samples with varying well widths.

**Theory**

A complete outline of the theoretical basis for the numerical model is given in reference 5 and will not be repeated here. The important points are that no approximations regarding simplification of the calculation are made other than points 1 and 2 given above and that 7eV is used for the deformation potential. Screening is not included in the calculations which is likely to be a poor approximation especially for the piezoelectric interaction. Including screening would not be a trivial matter as it is clear that static screening is not applicable in 2D systems and hence dynamic screening would have to be used. The numerical results are compared to analytical expressions derived by Ridley[6] using approximations 1-3 above. In the high temperature (equipartition) regime the total power loss per carrier is given by

$$p = F(T_E, T_L)(C_{np} + C_p)(k_B T_E - k_B T_L)$$

where

$$C_{np} = \frac{3\Xi^2 m^{*2} n_{2D}}{2\rho \hbar^3 L_z} \quad \text{and} \quad C_p = \frac{3e^2 \kappa_{av}^2 m^{*2} v_s^2}{4\pi \varepsilon \hbar^3 L_z}$$

are the magnitudes of the polar and non-polar interactions, $T_E$ and $T_L$ are the electron and lattice temperatures respectively. $\Xi, \kappa_{av}, and, L_z,$ are the deformation potential, the average electromagnetic coupling coefficient, and the well width. The F factor is a correction factor which corrects for the fact that

*Hot Carriers in Semiconductors*
Edited by K. Hess *et al.*, Plenum Press, New York, 1996

most experiments, including those presented here, are undertaken at electron temperatures between the zero-point and equipartition regimes[2].

## Experimental Details

The parameters of the samples used in the experiments are given in Table 1.
All the samples were grown using MBE, and then fabricated into Hall bars. Ohmic contacts were formed by alloying Au/Ge/Ni.

| Sample | No.of wells | %Al.Conc. | $L_z$ (Å) | $n_{2D}$ (m$^{-2}$) | $T_L$ (K) |
|--------|-------------|-----------|-----------|---------------------|-----------|
| A | 10 | 32 | 51 | $9.0 \times 10^{15}$ | 1.55 |
| B | 10 | 33 | 75 | $1.1 \times 10^{16}$ | 1.70 |
| C | 10 | 32 | 78 | $9.9 \times 10^{15}$ | 1.64 |
| D | 10 | 32 | 106 | $9.1 \times 10^{15}$ | 1.70 |
| E | 10 | 32 | 145 | $1.08 \times 10^{16}$ | 1.57 |

**Table 1: Sample parameters (GaAs/Ga$_{1-x}$Al$_x$As)**

Electron temperatures were determined using the well known Shubnikov de Haas (SdH) technique in which the amplitude of the observed oscillations in the magnetoresistance is used as an electron temperature thermometer. The power loss per carrier is determined experimentally from the steady state value of the power input which must equal the steady state power loss ( $P = e\mu(E)^2$ ).

## Results and Discussion

Figure 1 shows comparisons of the power loss per carrier as a function of well width between the numerical results and the analytical results at two different electron temperatures. The lattice temperature and carrier density are fixed at 2.0K and $1\times10^{16}$m$^{-2}$ respectively.

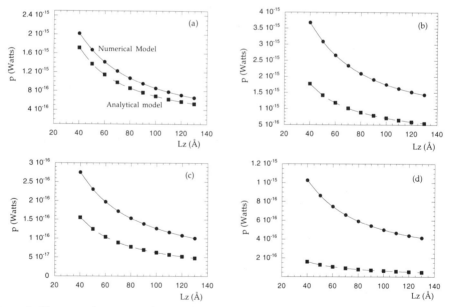

Figure 1. The power loss per carrier as a function of well width: (a) and (c) Deformation potential contributions at $T_E$=4.0K and 2.5K. (b) and (d) Piezoelectric contributions at $T_E$=4.0K and 2.5K.

The results show clearly that both calculations predict a 1/L well width dependence. At an electron temperature of 4.0K the results are extremely close especially for the deformation potential calculations (Fig 1a). At 2.5K the results of the two calculations are not as close but the trend against well width is the same.

Figure 2 shows the results of the experiments. The power loss per carrier is of similar size to the total power loss determined by the models although the models both predict a slightly higher power loss. The experimental results have a well width dependence similar to the predicted dependence in the range 50-100Å. The experimental trend above 100Å appears to indicate a subsequent increase in the power loss per carrier. Both the numerical and the analytical models only take into account the first subband which becomes an increasingly poor approximation as the well width is increased. The experimentally observed increase in the power loss per carrier above 100Å could be indicative of the increasing importance of higher subbands.

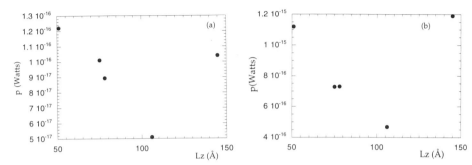

Figure 2. The experimental results of power loss as a function of well width at (a)$T_E$=2.5K and (b) $T_E$=4.0K.

The inclusion of higher subbands would intuitively lead to an increase in the power loss per carrier. The prediction of higher power loss per carrier by both models could be indicative of the need to include screening in the calculations. Screening has a larger effect on the polar interaction and we note that if we were to disregard the polar contribution, assuming that screening would make its contribution very small, then the deformation potential contribution alone would match very closely to the experimental results.

## Conclusion

We have shown that two models for the acoustic phonon power loss in 2D systems match very closely to experimental results obtained on a range of well width samples. In particular the well width dependence is shown clearly to be 1/L for well widths below 100Å. Above 100Å we believe that higher sub-band contributions play a role tending to make the power loss increase. This was not included in the models. Also we believe that our results show the need to include dynamic screening in the models.

## Acknowledgements

We are grateful (HC & MC to TUBITAK, and AJV &NB to EPSRC) for support and to Prof. B.K.Ridley and Drs. N.Constantinou, R.Gupta, and N.A.Zakhleniuk for many useful discussions.

## References

1. K.Hirakawa, H.Sakaki, Energy relaxation of two-dimensional electrons and the deformation potential constant in selectively doped AlGaAs/GaAs heterojunctions, *Appl. Phys. Lett.*, 49:889(1986)
2. M.E.Daniels, B.K.Ridley, and E.Emeny, Hot electron energy relaxation via acoustic phonon emission in GaAs/AlGaAs single and multiple quantum wells, *Solid State Electr.*, 32:1207(1989)
3. A. Straw, A.J.Vickers, and J.S.Roberts, Energy relaxation in GaInAs/AlInAs heterojunctions and GaAs/AlGaAs multiple quantum wells, *Solid State Electr.*, 32:1539(1989)
4.S.J.Manion, M.Artaki, M.A.Emanuel,J.J.Coleman, and K.Hess, Electron energy loss rates in AlGaAs/GaAs heterostructures at low temperatures, *Phys.Rev*, B35:9203(1987)
5. A.J.Vickers, Electron power loss in AlGaAs/GaAs quantum wells at intermediate electron temperatures, *Phys. Rev.*, B46:13315(1992)
6. B.K.Ridley, Hot Electrons in low-dimensional structures, *Rep.Prog.Phys.*,54:169(1991)

# MULTIMODE REGIME OF HOT ACOUSTIC PHONON PROPAGATION IN TWO DIMENSIONAL LAYERS

N. A. Bannov, V. V. Mitin, and F. T. Vasko*

Department of Electrical and Computer Engineering,
Wayne State University Detroit, MI 48202

## INTRODUCTION

At present there is a considerable interest toward new type of nanostructures: free-standing quantum wells (FSQWs) and free-standing quantum wires (FSQWIs). An important peculiarity of free-standing structures is the quantization of acoustic phonons. The acoustic phonon quantization has been observed both in optical and electrical experiments.[1,2] The acoustic phonon modes in free-standing structures and their interaction with electrons have been studied in a number of papers (see e.g. Refs.[3-5] and references therein). Due to complicated spectrum of acoustic phonons, their propagation in FSQWs and FSQWIs differs significantly from propagation of bulk acoustic phonons. This is important for heat pulses propagation in such structures and for heat removal from free-standing structures.

In this report we present results of our investigation of the hot acoustic phonon propagation in FSQWs. We have derived the kinetic equation for confined acoustic phonons interacting with 2D electron gas and analyzed it for the case of phonon transport in the above mentioned structures. We have obtained the renormalization of the acoustic phonon frequencies and relaxation rates due to phonon interactions with electrons through the deformation potential.

## PHONON MODES AND INTERACTION WITH ELECTRONS

There are three different types of acoustic modes in FSQWs: shear waves, dilatational waves and flexural waves (see e.g. Ref.[5] and references therein). Shear waves are similar to transverse waves in bulk material. Dilatational and flexural waves represent hybrids of logitudinal and transverse waves and differ by their symmetry: dilatational waves are symmetric and flexural waves are antisymmetric in respect to mid-plane of FSQW. We will denote the frequencies of acoustic modes as $\omega_m^{(\alpha)}(\mathbf{q}_{\parallel})$, where $\alpha$ is the type of mode (e.g., dilatational), $m = 0, 1, 2, \ldots$ is the mode number, $\mathbf{q}_{\parallel}$ is a phonon in-plane wave vector.

The Hamiltonian for phonon interactions with electrons through the deformation potential, $H_{e-p}$, is expressed in terms of electron operators, $a_k$, and phonon operators, $b_q$, in the standard form $H_{e-p} = \sum_{k,k',q}[w(k,k',q)a_{k'}^{\dagger}a_k b_q + H.C.]$, where $q$ denotes the set of phonon variables, $q = (m, \alpha, \mathbf{q}_{\parallel})$, $k$ denotes the set of electron variables, $k = (n, \mathbf{k}_{\parallel})$ ( $n$ is an electron subband number, $\mathbf{k}_{\parallel}$ is an electron in-plane wave vector), $w(k,k',q) = \delta_{\mathbf{k}'_{\parallel},\mathbf{k}_{\parallel}-\mathbf{q}_{\parallel}} < n|\Gamma_m^{(\alpha)}|n' >$. Here $< n|\Gamma_m^{(\alpha)}|n' >$ is the matrix

*Hot Carriers in Semiconductors*
Edited by K. Hess *et al.*, Plenum Press, New York, 1996

element for the transition, the function $\Gamma_m^{(\alpha)}$ is given by the formulae

$$\Gamma_m^{(\alpha)}(z) = \sqrt{\frac{E_a^2}{2\mathcal{A}\rho\omega_m^{(\alpha)}(\mathbf{q}_\parallel)}} \left( i\mathbf{q}_\parallel \mathbf{w}_m^{(\alpha)}(\mathbf{q}_\parallel, z) + \frac{\partial \mathrm{w}_{z,m}^{(\alpha)}(\mathbf{q}_\parallel, z)}{\partial z} \right) ,$$

$\mathbf{w}_m^{(\alpha)}(\mathbf{q}_\parallel, z)$ is the normalized vector of relative dispalcement in a phonon mode ( $m, \alpha$ ), which is given in Ref.[5], axis $z$ is perpendicular to FSQW, $E_a$ is the deformation potential constant, $\mathcal{A}$ is the normalization area; we use such units, that $\hbar = 1$.

## KINETIC EQUARTION FOR PHONONS

The kinetic equation for one particle phonon density matrix, $\sigma_{q,q'}$ , may be derived by averaging the Liouville equation over all electron variables and all, but one, phonon variables, and retaining quadratic in $H_{e-p}$ terms (see e.g. Ref.[6]). It has the following form

$$\frac{\partial \sigma_{q,q'}}{\partial t} + i(\omega_q - \omega_{q'})\sigma_{q,q'} = -i \sum_{k,k',Q} \frac{w(k, k', q')w^*(k, k', Q)}{\varepsilon_k - \varepsilon_{k'} + \omega_q - i\lambda}[(1 - f_k)f_{k'}\delta_{q,Q} + (f_{k'} - f_k)\sigma_{q,Q}]$$

$$+ i \sum_{k,k',Q} \frac{w(k, k', Q)w^*(k, k', q)}{\varepsilon_k - \varepsilon_{k'} + \omega_{q'} + i\lambda}[(1 - f_k)f_{k'}\delta_{Q,q'} + (f_{k'} - f_k)\sigma_{Q,q'}] \qquad (1)$$

It is assumed, that the electron density matrix is diagonal and reduced to the distribution function $f_k$. We will also assume, that phonon decay rate and phonon frequency renormalization are relatively small in respect to differences between frequencies of different modes. Then, we may neglect the nondiagonal in respect to discrete variables $m, \alpha$ terms in $\sigma_{q,q'}$ and introduce Wigner function

$$N_m^{(\alpha)}(\mathbf{q}_\parallel, \mathbf{r}_\parallel) = \sum_{\Delta \mathbf{q}_\parallel} \sigma_{\mathbf{q}_\parallel + \frac{\Delta \mathbf{q}_\parallel}{2}; \mathbf{q}_\parallel - \frac{\Delta \mathbf{q}_\parallel}{2}}^{(m,\alpha)} \exp\left(i\Delta \mathbf{q}_\parallel \mathbf{r}_\parallel\right) .$$

If the spacial variations of the Wigner function are small, Eq. (1) takes the following form

$$\frac{\partial N_m^{(\alpha)}(\mathbf{q}_\parallel, \mathbf{r}_\parallel)}{\partial t} + \mathbf{s}_m^{(\alpha)}(\mathbf{q}_\parallel)\frac{\partial N_m^{(\alpha)}(\mathbf{q}_\parallel, \mathbf{r}_\parallel)}{\partial \mathbf{r}_\parallel} = I_{sp} - \nu_m^{(\alpha)}(\mathbf{q}_\parallel)N_m^{(\alpha)}(\mathbf{q}_\parallel, \mathbf{r}_\parallel) , \qquad (2)$$

which which is similar to the classical kinetic equation for phonons.

## RELAXATION RATES AND RENORMALIZED FREQUENCIES

The renormalized acoustic phonon velocity in Eq. (2) is equal to

$$\mathbf{s}_m^{(\alpha)}(\mathbf{q}_\parallel) = \frac{\partial \omega_m^{(\alpha)}(\mathbf{q}_\parallel)}{\partial \mathbf{q}_\parallel} + \left( \frac{\partial \Delta\Omega(\mathbf{q}_{1\parallel}, \mathbf{q}_{2\parallel})}{\partial \mathbf{q}_{1\parallel}} - \frac{\partial \Delta\Omega(\mathbf{q}_{1\parallel}, \mathbf{q}_{2\parallel})}{\partial \mathbf{q}_{2\parallel}} \right)_{\mathbf{q}_{1\parallel}=\mathbf{q}_{2\parallel}=\mathbf{q}_\parallel} ,$$

where correction to phonon frequencies due to electron-phonon interaction is given by the formulae ($\mathcal{P}$ stands for the principal value symbol)

$$\Delta\Omega(\mathbf{q}_{1\parallel}, \mathbf{q}_{2\parallel}) = 2 \sum_{n,n',k_\parallel} \mathcal{P} \left[ \frac{|<n|\Gamma_m^{(\alpha)}|n'>|^2}{\varepsilon_{n,k_\parallel} - \varepsilon_{n',k_\parallel+q_{1\parallel}} + \omega_m^{(\alpha)}(\mathbf{q}_{2\parallel})} \right] (f_{n,k_\parallel} - f_{n',k_\parallel+q_{1\parallel}}) ,$$

the phonon relaxation rate is determined by the formulae

$$\nu_m^{(\alpha)}(\mathbf{q}_\parallel) = 4\pi \sum_{n,n',k_\parallel} |<n|\Gamma_m^{(\alpha)}|n'>|^2 \, \delta(\varepsilon_{n,k_\parallel} - \varepsilon_{n',k_\parallel+q_\parallel} + \omega_m^{(\alpha)}(\mathbf{q}_\parallel)) \, (f_{n,k_\parallel} - f_{n',k_\parallel+q_\parallel}) ,$$

and the rate of the spontaneous phonon emission is given by the formulae

$$I_{sp} = 4\pi \sum_{n,n',\mathbf{k}_\parallel} |<n|\Gamma_m^{(\alpha)}|n'>|^2 \, \delta(\varepsilon_{n,\mathbf{k}_\parallel} - \varepsilon_{n',\mathbf{k}_\parallel+\mathbf{q}_\parallel} + \omega_m^{(\alpha)}(\mathbf{q}_\parallel)) \, (1 - f_{n,\mathbf{k}_\parallel}) f_{n',\mathbf{k}_\parallel+\mathbf{q}_\parallel} \, .$$

We have calculated phonon transport parameters for several the lowest modes and results for $\nu_m^{(\alpha)}(q_\parallel)$ are presented on Figs. 1 and 2. The cut-off at in-plane phonon wave vector $3.1 \times 10^6 cm^{-1}$ in Fig. 1 is due to high degeneracy of the electron gas for $T = 4.2K$; this characteristic wave vector is equal to the doubled Fermi wave vector. The minimum in the relaxation rate for zeroth mode at in-plane phonon wave vector $3.2 \times 10^6 cm^{-1}$ in Fig. 2 is associated with a peculiarity of zeroth mode discussed in Ref.[5] It is interesting to compare the obtained relaxation rates with the acoustic phonon decay rate due to phonon-phonon interactions. The estimate for spontaneous decay rate of LO acoustic phonons in bulk GaAs at T=4.2 K is given by the formulae:[7] $\nu_{sp} = 12 \, (q/10^6 cm^{-1})^5 \, s^{-1}$. Therefore, the phonon relaxation due to the electron scattering dominates in the phonon wave vector range from 0 to $3 \times 10^6 cm^{-1}$.

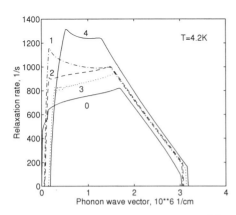

**Figure 1.** Phonon relaxation rates, $\nu_m^{(\alpha)}(q_\parallel)$, for five the lowest dilatational phonon modes, $m = 0, 1, 2, 3, 4$. GaAs FSQW of width $100\mathring{A}$, electron Fermi energy $10meV$, lattice temperature $T=4.2K$.

**Figure 2.** Phonon relaxation rates, $\nu_m^{(\alpha)}(q_\parallel)$, for five the lowest dilatational phonon modes, $m = 0, 1, 2, 3, 4$. GaAs FSQW of width $100\mathring{A}$, electron Fermi energy $10meV$, lattice temperature $T=30K$.

*Acknowledgment* - This work was supported by ARO.

* *Permanent address*- Institute of Semiconductor Physics, Kiev 252650, Ukraine.

[1] B. Bhadra, M. Grimsditch, I. Schuller, F. Nizzoli, Brillouin scattering from unsupported Al film, *Phys. Rev. B* 39:12456 (1989).
[2] J. Seyler and M. N. Wybourne, Acoustic waveguide modes observed in electrically heated metal wires, *Phys. Rev. Lett.* 69:1427 (1992).
[3] M. A. Stroscio, and K. W. Kim, Piezoelectric scattering of carriers from confined acoustic modes in sylindrical quantum wires, *Phys. Rev. B* 48:1936 (1993).
[4] N. Nishiguchi, Guided acoustic phonons in quantum wires: theory of phonon fiber, *Jap. J. Appl. Phys.* 33:2852 (1994).
[5] N. Bannov, V. Aristov, V. Mitin, and M. A. Stroscio, Electron relaxation times due to deformation potential interactions of electrons with confined acoustic phonons in a free-standing quantum well, *Phys.Rev.B* 51:9930 (1995).
[6] F. T. Vasko, Emission of acoustic phonons by two-dimensional electrons, *Sov. Phys. Sol. State* 30:1207 (1988).
[7] S. Tamura, Spontaneous decay rates of LA phonons in quasi-isotropic solids, *Phys.Rev.B.* 31:2574 (1985).

# HIGH FIELD ELECTRON TRANSPORT IN MODULATION DOPED Si:SiGe QUANTUM WELLS

T J Thornton[1], A Matsumura[1,2] and J. Fernández[2]

Imperial College of Science, Technology & Medicine
[1] Department of Electrical & Electronic Engineering
   Exhibition Road, London SW7 2BT, UK
[2] IRC for Semiconductor Materials
   Prince Consort Road, London SW7 2BZ, UK

## INTRODUCTION

The transport properties of modulation doped Si:SiGe quantum wells make them ideally suited for high performance FET applications primarily because of their enhanced electron and hole mobilities. At liquid helium temperatures electron mobilities in excess of $10^5$ cm$^2$/Vs are now routinely achieved[1,2]. At room temperature, electron mobilities larger than those normally found in MOSFETs or doped silicon have been reported [3,4] and similar improvements are expected for p-channel devices [5]. However, the advantages of high mobility will only be realised if the low field behaviour translates into high velocity at the large electric fields normally encountered during device operation. For this reason there is considerable interest in the high field transport properties of Si:SiGe quantum wells. Recent theoretical work has predicted enhanced electron velocities in strained silicon quantum wells [6,7,8]. In this paper we present velocity-field measurements of n-channel modulation doped Si:SiGe quantum wells made at room temperature and 77K. Preliminary results show that the electron velocity at high field is significantly larger than that in silicon MOSFETs.

## LAYER STRUCTURE AND MATERIAL CHARACTERISATION

We have grown n-channel Si:SiGe quantum wells using gas source molecular beam epitaxy (GS-MBE) [4]. The layers are modulation doped with arsenic as the donor source and a typical layer structure is shown in Figure 1. By varying the quantum well thickness, spacer layer thickness and doping concentration we have achieved electron mobilities in the range 60 - 95,000 cm$^2$/Vs at 4.2 K and 850 - 1400 cm$^2$/Vs at room temperature. The magnetoresistance and Hall resistance of a sample with $\mu = 95,000$ cm$^2$/Vs at 4.2 K is shown in Figure 2. The well defined Shubnikov - de Haas oscillations, onset of spin splitting and quantum Hall effect all attest to the high quality of these layers.

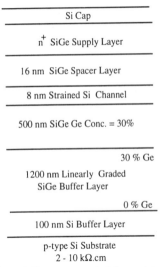

| |
|---|
| Si Cap |
| n+ SiGe Supply Layer |
| 16 nm SiGe Spacer Layer |
| 8 nm Strained Si Channel |
| 500 nm SiGe Ge Conc. = 30% |
| 30 % Ge |
| 1200 nm Linearly Graded SiGe Buffer Layer |
| 0 % Ge |
| 100 nm Si Buffer Layer |
| p-type Si Substrate 2 - 10 kΩ.cm |

Figure 1: Layer structure of the modulation doped quantum wells used in this work

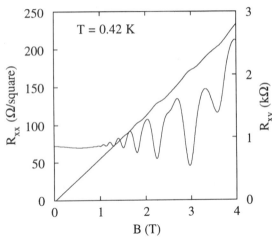

Figure 2: Magnetoresistance of a sample with $\mu$=95,000 cm$^2$/Vs and n=8.6x10$^{11}$ cm$^{-2}$

The Hall density and mobility in the temperature range 77-300K are shown in Figure 3. The room temperature mobility measured at extremely low fields is 1300 cm$^2$/Vs and the electron concentration is 10$^{12}$ cm$^{-2}$. On cooling, the electron concentration initially begins to fall but settles at a value of 5 x 10$^{11}$ cm$^{-2}$ for temperatures less than 200K. At 77K the Hall mobility is typically in the range 8 - 12,000 cm$^2$/Vs.

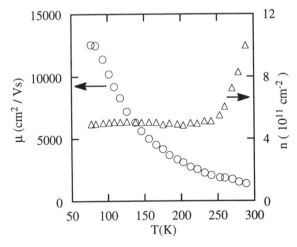

Figure 3: The variation of $\mu$ and n as a function of temperature

## PULSED VELOCITY-FIELD MEASUREMENTS

We have measured the velocity-field characteristics of two similar samples at 300 K and the data is shown in Figure 4. The samples are processed into Hall bars of width 80 $\mu$m and length 1.4 mm. The Hall bars are defined by a deep mesa etch which penetrates the p-type silicon substrate. A deep etch is important to prevent current flowing in the SiGe buffer layers. Voltage pulses of duration 200 ns derived from a charged transmission line are applied to the Hall bars at a repetition rate of 100 ms to avoid sample heating. The current pulse and electric field along the Hall bar are measured using a storage oscilloscope and from the electron concentration at room temperature we can extract the electron velocity. Up to fields of 1000 V/cm the v-E relationship is quite linear for both Hall bars corresponding to mobilities of 1200 and 850 cm²/Vs, very similar to the values measured using small d.c. currents. However, for fields above 2kV/cm a sub-linear behaviour is apparent and the effective mobility begins to decrease. We believe this is the beginning of velocity saturation ie the 2DEG is entering the warm-electron regime. At fields above 3.5 kV/cm a distortion in the current pulse develops for reasons we have not yet identified.

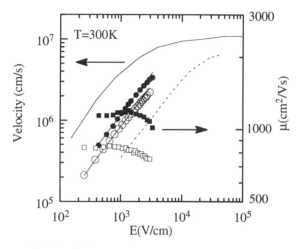

Figure 4: Velocity-field curves taken at 300K for two samples with different low field mobility (open and closed symbols). The effective mobility is also plotted

The dashed line in Figure 4 is the v-E curve for a MOSFET taken from Fang and Fowler [9]. By comparison, the electron velocity in the Si:SiGe quantum well is significantly larger but still not as high as predicted by the Monte Carlo simulations of Miyata et al [6] reproduced as a solid line in Figure 4. We suggest the following reasons why our measured v-E curves are smaller than the theoretical predictions. The background doping in these samples is ~$10^{16}$ cm$^{-3}$ and ionised impurity scattering may still be limiting the electron velocity in the regime before velocity saturation occurs. The background doping may also be responsible for parallel conduction in the buffer layers as well as in the heavily doped supply layer. However, the parallel conduction is insufficient to explain the difference between our measurements and the calculations. One source of error arises from the 2-terminal nature of the measurements. A voltage drop will occur across each of the contacts and the E-field that we measure is therefore an upper bound. The total contact resistance at 300K is typically ~10 kΩ compared to a total channel resistance of ~150kΩ and the error is too small to explain the smaller velocity that we observe.

At 77K the parallel conduction is strongly suppressed because of carrier freeze-out in the buffer and supply layers. The v-E data taken at 77K are shown in Figure 5 along with the results predicted for this temperature by Miyata et al [6]. At 77K the current distortion at high fields is not observed and the measurments were extended to the highest values we could achieve with the voltage supply used to charge the transmission line. The velocity at a given field is again smaller than the predicted results.

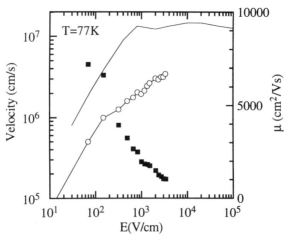

Figure 5: Velocity-field curve (open circles) and effective mobilty (solid squares) measured at 77K. The solid line is reproduced from Miyata et al [6].

## CONCLUSIONS

By means of pulsed field measurements we have been able to extract the v-E curves for modulation doped Si:SiGe quantum wells. Although smaller than the theoretical predictions the velocity we measure at a given field is larger than that observed in silicon MOSFETs suggesting that these structures may have useful FET applications.

It is a pleasure to acknowledge discussions with D K Ferry regarding the pulsed field measurements.

## REFERENCES

1. K Ismail et al "Extremely HIgh Electron Mobility in Si/SiGe Modulation Doped Heterostrcutures" *Appl. Phys. Letts.* 66: 1077 (1995)
2. D Tobben, F Schaffler, A Zrenner and G Abstreiter "Magnetotransport in High Quality Si/SiGe Heterostructures" *Phys. Rev. B* 46: 4344 (1992)
3. S F Nelson, K Ismail, J O Chu and B S Meyerson "Room Temperature Mobility in Strained Si/SiGe Heterostructures" *Appl. Phys. Letts.* 63: 367 (1993)
4. A Matsumura et al "Characterisation of N-Channel Si/SiGe Modulation Doped Structures" *Semicond. Sci and Technology* to be published
5. Y H Xie et al "Very High Mobility 2DHG in Si/GeSi/Ge Structures Grown by MBE" *Appl. Phys. Letts.* 63: 2263 (1993)
6. H Miyata, T Yamada and D K Ferry " Electron Transport Properties of a Strained Layer on a Relaxed SiGe Substrate by MC Simulation" *Appl. Phys. Letts.* 62: 2661 (1993)
7. Th Vogelsang and K R Hofmann " Electron Transport in Strained Si Layers on SiGe Substrates" *Appl. Phys. Letts.* 63: 186 (1993)
8. K Miyatsuji et al "High Field Transport of Hot Electrons in Strained Si/SiGe Heterostructures" *Semicond. Sci. Technol* 9: 772 (1994)
9. F F Fang and A B Fowler "Hot Electron Effects and Saturation Velocities in Silicon Inversion Layers" *J. Appl. Phys.* 41: 1825 (1970)

# HOT CARRIER TRANSPORT IN SiGe/Si TWO-DIMENSIONAL HOLE GASES

G. Brunthaler[1], G. Bauer[1], G. Braithwaite[2], N.L. Mattey[2], P. Phillips[2], E.H.C. Parker[2] and T.E. Whall[2]

[1] Institut für Halbleiterphysik, Universität Linz, A-4040 Linz, Austria
[2] Department of Physics, University of Warwick, Coventry, CV4 7AL, England.

**Abstract:** The hot carrier energy loss rate in a two dimensional hole gas in compressively strained SiGe quantum wells has been studied for samples with a Ge content (x= 0.2) and carrier concentrations ranging from $3 \times 10^{11}$ to $7 \times 10^{11}$ cm$^{-2}$. The energy loss in this highly non-parabolic system is dominated by acoustic phonon deformation potential scattering, whereas the piezoelectric interaction is negligible.

## INTRODUCTION

Recently Si/SiGe heterojunctions have attracted a lot of interest since Si/SiGe/Si n-p-n heterobipolar transistors with high transit frequencies have been realised by several groups. In this context, the characterization of SiGe p-channel devices is of considerable technological relevance. Recently, some of the present authors have determined the effective heavy hole masses in remote doped pseudomorphic Si/SiGe quantum well structures with $0.05 < x < 0.3$.[1-3] In this paper we describe hot carrier Shubnikov-de Haas (SdH) and B=0 resistivity measurements and their theoretical analysis in order to determine which mechanisms dominate the power loss of two dimensional (2D) heavy holes in coherently strained, remotely doped Si/ SiGe structures at lattice temperatures of T = 0.35 and 2 K. The Si/SiGe samples were grown by solid source MBE on (001) Si substrates, consisting of a 300 nm undoped Si buffer followed by a 30 nm Si$_{1-x}$Ge$_x$ (x = 0.2) alloy layer, an undoped Si spacer layer and a 50 nm boron doped cap layer. A self consistent model was used to design the structures to give hole sheet densities in the range of $3.0 \times 10^{11}$ to $7.7 \times 10^{11}$ cm$^{-2}$. Double crystal x-ray diffraction measurements confirmed that the SiGe layers were fully strained. Magnetotransport measurements were carried out on Hall bars fabricated by photolithography/wet etching.

## RESULTS AND DISCUSSION

In order to evaluate the energy loss of hot holes, information on the heavy hole masses as a function of Ge content and carrier concentration is required. Such data were obtained from the analysis of ohmic Shubnikov-de Haas (SdH) measurements, evaluating the temperature dependence of the SdH oscillations. For this analysis effects of temperature dependent screening, weak localisation and hole-hole interactions were taken into account as described in Refs. 1-3. In this analysis, the mass is an adjustable parameter to obtain unity gradient in a plot of $\ln[\Delta\rho_m(T)/\rho_0(T)]$ versus $[\ln(\xi/\sinh\xi) - \pi/\omega_c\tau_q(T)]$ where $\Delta\rho_m$ is the peak value of the longitudinal resistivity extracted using a cubic spline interpolation, $\rho_0$ is the Boltzmann resistance, $\xi=2\pi^2 kT/\hbar\omega_c$, $\omega_c=eB/m^*$, and $\tau_q$ is the quantum lifetime (which was assumed to be equal to the transport lifetime). In all samples investigated just the first hole subband was populated. The SiGe layers are under biaxial compressive strain, which has important consequences for the values of heavy hole masses. It was found, that the hole effective mass increases with hole sheet density as well as with magnetic field, an effect which is quite substantial for Ge-contents of about 0.2. The non-parabolicity may be described by $\hbar^2 k^2/2m^*_{(k=0)} = \varepsilon[1+c\varepsilon/\Delta]$ with $\hbar k$ being the carrier momentum, $\varepsilon$ their energy, c being the nonparabolicity factor (= 0.4) and $\Delta$ is the heavy-light hole splitting. The hole masses used for the further analysis can be approximated by $m^* = (0.20 + 0.113 p_s/(10^{16}\ cm^{-2}))m_0$.

For the determination of the hot carrier power loss as a function of the carrier temperature, SdH oscillation amplitudes were used as the most sensitive physical parameter for degenerate carrier statistics for this purpose.[4-8] The damping of the SdH oscillations as a function of applied electric field was measured at lattice temperatures of T=0.35K and T= 2K for samples with hole concentrations ranging from 2.6 to 7.7 x$10^{11}$cm$^{-2}$ and Ge contents from 13 to 20%. As an example, data are shown in Fig.1 where the lattice temperature dependence from T=0.35K to 1.5K is compared with the electric field dependence in the range from about 2mV/cm to 300mV/cm keeping the lattice temperature fixed at T= 0.35 K.

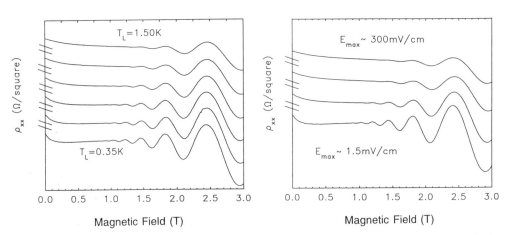

**Figure 1.** Longitudinal magnetoresistance for a) different lattice temperatures from 0.35 to 1.50 K and b) for different electric fields from 1.5 to 300 mV/cm at $T_l = 0.35$ K. The damping of Shubnikov de Haas oscillations and the change of the zero-magnetic field resistivity with increasing carrier temperature is visible.

The resulting energy loss rate as a function of hot heavy hole temperature is shown in Figs 2a and 2b for lattice temperatures of 2 and 0.35 K, respectively for the samples with 20% Ge content in the SiGe channels. These power losses were compared with numerical calculations for deformation potential and piezoelectric scattering, with and without screening. The average energy loss per carrier was estimated by calculating the energy gained by the phonons from the holes and deviding by the number of carriers which participate as outlined in Refs. 4 and 5. For the deformation potential scattering the spherically averaged elastic coefficients $C_l =(3C_{11}+2C_{12} +4C_{44})/5$ and $C_t = (C_{11}-C_{12}+3C_{44})/5$, interpolating linearly between their values for Si and Ge, were used from which also the sound velocities were obtained. Furthermore, for the deformation potential constant which is obtained

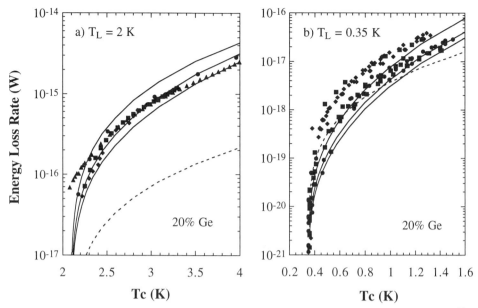

**Figure 2.** Hot carrier energy loss rate in $Si_{0.8}Ge_{0.2}$ channels with hole concentrations from 3.0 to $7.7x10^{11}$ cm$^{-2}$ at $T_l = 2$ K and b) $T_l = 0.35$ K. The experimentally determined energy loss rate (symbols) and the calculated one for deformation potential (solid lines for $p_s = 3$, 5 and $7x10^{11}$ cm$^{-2}$ and $D = 4.5$ eV) and piezoelectric coupling (dashed line for $p_s = 3x10^{11}$ cm$^{-2}$ and $e_{14} = 1.6x10^{-2}$ Cm$^{-2}$) are compard.

from the fit of the data, represents an effective deformation potential, which is related to the valence band deformation parameters a, b, and d and to $C_l$ and $C_t$ according the expressions given by Wiley.[9]

It turns out that with increasing carrier concentration the calculated energy loss rate per carrier is somewhat decreased as shown in Fig. 2a and 2b. Experimentally such a dependence cannot be found to the same extent. Furthermore, we would like to emphasize, that for the measurements taken at T=0.35 K, there is some scatter in the data, i.e. a non-systematic variation with carrier concentration, which does not allow to extract a definitive trend in this respect for the comparison with theory. However, at rather small carrier temperatures, it seems that the power loss due to deformation potential scattering, does not account completely for the observed experimental power loss. We used an effective deformation potential constant of 4.5 eV, which is in good agreement with the values quoted by Wiley[9] if interpolated between bulk Si and Ge.

In this context, the question of the screening of the deformation potential interaction for the hot hole-acoustic phonon interaction arises. Conventionally, for the calculation of the energy loss of hot electrons at low temperatures, it has been customary to include at least static screening. However, as outlined by Fischetti and Laux,[10] treating screening in the simple static approximation may result in an underestimation of the deformation potential interaction. At low temperatures dynamic screening matters most of course, but it is quite difficult to consider it correctly due to the necessity for degeneracy corrections and for the screening effects of intersubband plasmons etc. With static screening, a value for the deformation potential constant of about 15 eV would result, a factor of three higher than the effective deformation potentials for holes in Si or Ge.[9]

Furthermore the energy loss due to the interaction of hot holes with acoustic phonon modes via the piezoelectric interaction was considered as well. There are previous assertions on the importance of piezoelectric scattering for the energy loss in the $Si_{0.8}Ge_{0.2}/Si$ 2D system by Xie et al[11]. The physical origin of the piezoelectric coupling was attributed to quite large strain-induced changes in the ionicity or of scattering of acoustic phonons from ordered domains via the piezoelectric coupling. A quite high value for $e_{14}$ was deduced, only a factor of three smaller than the corresponding value for InAs ($4.5x10^{-2}$ Cm$^{-2}$)[12]. This high value for the piezoecoupling constant $e_{14} = 1.6x10^{-2}$ Cm$^{-2}$ ($h_{14} = e_{14}/\varepsilon_s\varepsilon_0 = 1.4x10^8$ V/m) is quite astonishing considering the fact that the main origin for piezoelectricity in SiGe alloys is the built-in strain. Any observation of a piezoelectric effect in strained SiGe would be quite interesting because of the implications of a certain *ionic* character in these IV-IV alloys. On the other hand, it is well known that any attempts to find a transverse optic(TO)-longitudinal optic(LO) mode splitting in SiGe alloys have so far been unsuccessful. This lack of evidence for TO-LO mode splitting finds its solution in pseudopotential calculations of the transverse effective charge $e_T^*$ for a hypothetical completely ordered $Si_{0.5}Ge_{0.5}$ alloy performed by Olego et al[13]. The corresponding value for $e_T^*$ turns out to be 0.60, which is much smaller than that of InAs (2.53) (Ref. 13, p 220). The transverse effective charge $e_T^*$ is related to the piezoelectric charge by $e_p^* = Z^*/\zeta - e_T^* (1-\zeta)/\zeta = -0.10$ (for $Z^* = 0.11$, $\zeta = 0.70$) which in turn determines the piezoelectric constant $e_{14} = \zeta e_p^* e/2\pi a^2 = 3.2 \times 10^{-2}$ $C/m^{-2}$.[13,14] For an ordered arrangement of the Ge atoms in $Si_{0.8}Ge_{0.2}$ the piezoelectric constant should at least be rescaled by the Ge content ratio 0.2/0.5 to $1.24 \times 10^{-2}$ $Cm^{-2}$ and furthermore even smaller in a disordered alloy.

Piezoelectric scattering, if it exists at all, should become remarkable in the total energy loss to the acoustic phonon modes just for the lowest lattice temperatures. Thus, despite the arguments against piezoelectric scattering, we have included in the calculated energy loss this contribution (see Fig. 2b) Even at T=0.35 K and assuming for $e_{14} = 1.6 \times 10^{-2}$ $Cm^{-2}$ the comparison with experiment shows that this contribution is indeed negligible. Consequently, the energy loss data at low lattice temperatures do not support the idea that the acoustic phonon scattering includes a piezoelectric component.

In conclusion, we obtain a good fit for hot hole power loss data for deformation potential coupling (coupling constant: 4.5 eV for $Si_{0.8}Ge_{0.2}$) of acoustic phonons at low carrier temperatures ($T_1 = 2K$). At T= 0.35 K the experimentally observed energy loss is somewhat higher than the calculated one, the origin of this discrepancy is not yet known. Furthermore we have shown that the value of the deformation potential coupling constant, which follows from our calculations is in good agreement with previously determined values based on temperature dependent measurements of the hole mobility in Si and Ge, if properly weighted for the SiGe alloys. From the hot carrier data there is *no evidence for piezoelectric acoustic phonon scattering* in SiGe alloys with a Ge content of x = 0.2. Calculations of the piezoelectric coupling constant $e_p^*$ for completely ordered $Si_{0.8}Ge_{0.2}$ show that its value is about 30 % of that for InAs. Since in real SiGe layers a high degree of disorder is expected, the absence of piezoelectric scattering in hot hole transport is supported by the theoretical considerations.

We thank P.Vogl for helpful discussions, this work was supported by FWF and GME, Vienna and the UK-SERC.

## REFERENCES

1. T.E.Whall, N.L.Mattey, A.D.Plews, P.J.Philips, O.A.Mironov, R.J.Nicholas, M.J.Kearney, Appl.Phys.Lett. **64**, 357 (1994).
2. T.E.Whall, A.D.Plews, N.L.Mattey, E.H.C.Parker, Appl.Phys.Lett. **65**, 3362 (1994).
3. T.E.Whall, A.D.Plews, N.L.Mattey, P.J.Philips, U.Ekenberg, Appl.Phys.Lett. **66**, 2724 (1995).
4. H.Kahlert, G.Bauer, Phys.Rev.B7, 2670 (1973).
5. S.J.Manion, M.Artaki, M.A.Emanuel, J.J.Goldman, K.Hess, Phys.Rev. B**35**, 9203 (1987).
6. B.K.Ridley, Rep. Prog. Phys. **54**, 169 (1991).
7. G. Stöger, G. Brunthaler, G. Bauer, K. Ismail, B.S. Meyerson, J. Lutz and F. Kuchar, Phys. Rev. B**49**, 10417 (1994).
8. K.Ismail, M.Arafa,K.L.Saenger, J.O.Chu, B.S.Meyerson, Appl.Phys.Lett. **66**, 1077 (1995).
9. J.D.Wiley, in Semicondcutors and Semimetals, eds. R.K.Willardson and A.C.Beer (Academic Press, New York, 1975) Vol.10, p.91.
10. M.V. Fischetti and S.E.Laux, Phys.Rev. B**48**, 2244 (1993).
11. Y. Xie, R. People, J.C. Bean and K.W. Vecht, Appl. Phys. Lett. **49**, 283 (1986); J. Vac. Sci. Techn. 135, 744 (1987).
12. Th. Grave, K. Hübner, in Physics of Group IV Elements and III-V Compounds, Ed. O. Madelung, Landolt-Börnstein, New Series, Group III, Vol. 17a, (Springer Verlag, Berlin 1982) p 302.
13. D. Olego, M. Cardona, P. Vogl, Phys. Rev. B**25**, 3878 (1982).
14. W.A.Harrison, Electronic Structure and the Properties of Solids, (Freeman and Co, San Francisco, 1980), pp.224.

# HOT HOLE EFFECTS IN STRAINED MQW
# HETEROSTRUCTURES Ge/Ge$_{1-x}$Si$_x$

V.Ya.Aleshkin, A.A.Andronov, N.A.Bekin, I.V.Erofeeva, V.I.Gavrilenko,
Z.F.Krasil'nik, O.A.Kuznetsov, M.D.Moldavskaya and V.V.Nikonorov

Institute for Physics of Microstructures of Russian Academy of Sciences
46, Uljanov St., 603600, Nizhny Novgorod, Russia

The paper deals with the first investigations of the 2D hot hole effects in multilayer heterostructures Ge/Ge$_{1-x}$Si$_x$ aimed at the realization of dynamical heating and intraband population inversion of carriers in strong electric fields. Ge/Ge$_{1-x}$Si$_x$ heterostructures (x $\approx$ 0.1, d$_{Ge}$ $\approx$ 200 A) were grown on Ge(111) substrates by CVD technique. In these heterostructures the quantum wells in the valence band are realized in thin germanium layers which are uniformly compressed in the plane ofthe structure because of the mismatch of lattice periods of Ge and GeSi. The deformation may be represented as a result of hydrostatic compression and equivalent uniaxial tension of Ge layers; the latter is responsible for the valence band splitting. Earlier we have observed in undoped heterostructures the CR line of photoexcited 2D holes ($m_c \approx 0.07m_0$) in 2-mm wavelength range at liquid helium temperature[1]. The halfwidth of CR line corresponds to the high value of the hole mobility $\mu$ $\approx 10^5$ cm$^2$/V·s thus testifying the quality of the heterostructures.

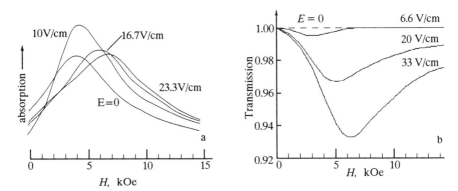

**Figure 1.** CR spectra in undoped sample #306 in d.c. electric fields ($\lambda$ = 2.1 mm, T = 4.2 K); a - CR absorption by photoexcited holes, b - absolute measurements of changes in sample transmission due to the impact ionization of residual shallow acceptors.

Hot hole CR in undoped MQW heterostructures $Ge/Ge_{1-x}Si_x$ was investigated for the first time in d.c. and pulsed electric fields. The electric voltage was applied to the sample via alloyed ohmic contacts deposited on the surface of the heterostructure. Application of d.c. fields about few V/cm results in the remarkable shift of CR line to higher magnetic fields. This effect is clearly seen for both the photoexcited holes (Figure 1a) and the holes produced by impact ionization of residual shallow acceptors (Figure 1b). The shift is shown to be due to the carrier heating and to the strong nonparabolicity of the energy-momentum law (Figure 2). In higher electric fields the CR investigations were carried out using the pulsed technique (Figure 3a,b). In the latter case the holes were excited at the impact ionization of residual shallow acceptors in the heterostructure thus modulating the mm-wave absorption in the sample[2].

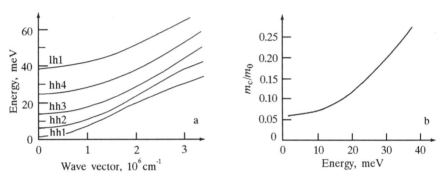

**Figure 2.** Calculated 2D hole energy spectra in $Ge/Ge_{1-x}Si_x$ heterostructure #306 (a) and energy dependence of cyclotron resonance mass in the lowest subband hh1 (b) ($d_{QW}$= 200 A, x = 0.12, **k** ‖ [112], $P_{equiv.}$= 4 kbar).

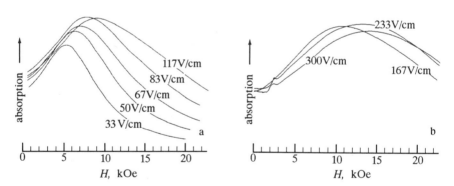

**Figure 3.** Hole CR absorption spectra in undoped sample #306 in pulsed electric fields ($\lambda$=2.3mm, T=4.2K).

The obtained absolute values of CR transmission (Figure 1b) together with the relative increase of the integral intensity of CR line with the rise of pulsed electric fields up to its saturation at E ≥ 200 V/cm (Figure 3a,b) corresponding to the complete ionization of shallow acceptors allow to estimate the total concentration of residual impurities in the heterostructure: $n_S \approx 2 \cdot 10^{11}$ cm$^{-2}$ (per 162 layers of the structure) that roughly corresponds to the volume concentration $n \approx 3 \cdot 10^{14}$ cm$^{-3}$. The technique used makes it possible to estimate the binding energy $\mathcal{E}_B$ of acceptor centers in quantum-well heterostructures

Ge/Ge$_{1-x}$Si$_x$. Because of adiabatic heating of the sample (together with the substrate) during the strong electric field pulse its temperature increases significantly over 4.2K that results in the thermoionization of shallow acceptors behind the trailing edge of the pulse[2]. Measuring the intensity of CR absorption line just after the electric field pulse as a function of the dissipated Joule energy and taking into account the temperature dependence of the specific heat of the germanium substrate we obtain the temperature dependence of the concentration and estimate $\mathcal{E}_B \approx$ 2-4 meV that is sufficiently smaller than that in bulk Ge because of the strain induced valence band splitting.

In strong electric fields an enormous CR line shift to higher effective mass region up to $m_c \approx 0.3m_0$ at $E \approx$ 300 V/cm (Figure 3a,b) was revealed that corresponds to the carrier heating up to $T_e \geq$ 200 K. The latter proves the realization of the streaming motion of the carriers and indicates the possibility of the population inversions of 2D holes in $\mathbf{E} \perp \mathbf{H}$ fields just similar to those in bulk p-Ge[3].

The increase of the hole effective masses at the carrier heating results in the Gunn effect at the lateral charge transport that was revealed in selectively doped ($n_s =2 \cdot 10^{11}$ cm$^{-2}$) MQW heterostructure Ge/Ge$_{1-x}$Si$_x$ #125 (T = 4.2 K, $d_{QW}$= 175 A, x = 0.08, $P_{equiv.}$= 4 kbar)[4]. This current instability seems to be quite similar to that observed in uniaxially compressed bulk p-Ge at $\mathbf{E} \| \mathbf{P}$[5,6].

Hot hole far IR emission from selectively doped sample #125 was detected in strong electric fields up to 500 V/cm. The radiation from both the plane and the edge of the sample was observed. The preliminary investigations of the emission intensity dependences on electric and magnetic fields as well as spectral studies of the radiation using the tunable by magnetic field filter n-InSb show that the radiation results predominantly from the indirect intraband transitions of hot carriers.

The research described in this publication was maid possible in part by Grant #94-02-05445 from Russian Foundation for Fundamental Investigations, Grant #1-028 from Russian Scientific Program "Physics of Solid State Nanostructures", Grant #R8H000 from International Science Foundation, Grant #R8H300 from International Science Foundation and Russian Government and Grant #94-842 from INTAS.

## References

1. V.I.Gavrilenko, I.N.Kozlov, O.A.Kuznetsov, M.D.Moldavskaya, V.V.Nikonorov, L.K.Orlov and A.L.Chernov, Cyclotron resonance of charge carriers in strained Ge/Ge$_{1-x}$Si$_x$ heterostructures, *JETP Lett.* 59: 348 (1994).
2. V.I.Gavrilenko, A.L.Korotkov, Z.F.Krasil'nik and V.V.Nikonorov, Far IR cyclotron resonance and luminescence of hot holes in p-Ge, *Opt. Quantum Electronics* 23: S163 (1991).
3. *Opt. Quantum Electronics* 23, N2, Special Issue on Far-infrared Semiconductor Lasers (1991).
4. V.I.Gavrilenko, I.N.Kozlov, O.A.Kuznetsov, M.D.Moldavskaya and V.V.Nikonorov, Hot carriers in multi-quantum-well heterostructures Ge/Ge$_{1-x}$Si$_x$, *in* "Extended Abstracts of the 1994 International Conference on Solid State Devices and Materials", Japan Society of Applied Physics, Yokohama (1994). , p.503
5. J.E.Smith, J.C.McGroddy, M.I.Nathan, Bulk current instabilities in uniaxially strained germanium, *Phys.Rev.* 186: 727 (1969).
6. V.I.Gavrilenko, A.V.Galyagin, V.V.Nikonorov and P.N.Tsereteli, On the nature of the "stimulated" FIR emission of hot carriers in uniaxially stressed p-Ge in the absence of the magnetic field", *Lithuanian Journal of Physics (Suppl.)* 32: 161 (1992).

# HOLE TRANSPORT IN A STRAINED SI LAYER GROWN ON A RELAXED $\langle\,001\,\rangle$-$Si_{(1-x)}Ge_x$ SUBSTRATE

J.E. Dijkstra, W.Th. Wenckebach

Faculty of Applied Physics, Delft University of Technology
P.O. Box 5046, 2600 GA Delft, The Netherlands

Transport of holes in strained silicon layers grown on a $(001)Si_{(1-x)}Ge_x$ substrate is gaining interest because of the possibility to fabricate high performance PMOS devices with this material. Nayak[1] reported the observation of a high mobility p-channel metal-oxide-semiconductor field-effect transistor on strained Si. This high mobility is attributed to the strain of the Si in the p-channel. Later, in a theoretical article Nayak[2] calculated the in-plane mobility in strained Si using $\vec{k}\cdot\vec{p}$ theory and a relaxation time method for vanishing small electric fields. The results of his calculations show an increase of the mobility with a factor 6 for strained Si grown on a $Si_{0.8}Ge_{0.2}$ substrate as compared to unstrained Si.
In real MOS devices the electric field in the p-channel can be quite high, so one needs to extend Nayaks work to strong electric fields. Therefore in this article we use the Monte Carlo method to calculate the hole drift velocity as a function of strain and electric field up to 5 kV/cm. Our calculations show that the gain in mobility or drift velocity in strained Si compared to unstrained silicon decreases at higher electric fields.

The valence band model used in the Monte Carlo simulation is a $\vec{k}\cdot\vec{p}$ model including the spin-orbit, light and heavy hole band (see Hinckley[3]). This results in a $6\times 6$ hamiltonian matrix. The exact solution of this hamiltonian automatically includes the anisotropy and strong non-parabolicity of the valence band. The influence of strain on the energy spectrum of the holes is implemented using deformation potential theory.
The scatter processes used in the simulations are acoustical and optical phonon scattering. The optical and acoustical phonon scatter processes are implemented using the same deformation potential model as used to implement strain. For all these scatter processes the exact solutions of the $6\times 6$ $\vec{k}\cdot\vec{p}$ hamiltonian are used in calculating the matrix elements of the scatterprocess. The acoustical phonon scatter process is modeled elasticly because the energy of an acoustical phonon is much smaller than the average anergy of the hole. In modelling the optical phonon process the optical phonon energy is chosen to be constant and independent of the wavevector. So the only way in which the hole can loose energy is by emitting an optical phonon.
Since in the Monte Carlo simulation we have to calculate very often the energy of the hole as a function of $\vec{k}$ and the eigenvectors of the hole it is necessary to be able to calculate them

*Hot Carriers in Semiconductors*
Edited by K. Hess *et al.*, Plenum Press, New York, 1996

very quickly. By deriving an exact analytical expression for the energies and the eigenvectors of the hole in the $\vec{k} \cdot \vec{p}$ model, the calculation of these quantities was speeded up by two orders of magnitude.

For the simulations we use the usual Monte Carlo scheme described by Jacoboni[4]. In this M.C. scheme an artificial process called selfscattering is needed to make the total scatter rate constant i.e. independent of the wavevector of the hole. Using this technique it is very simple to generate randomly the time between two consecutive colissions, the time of free flight. We use the same technique for calculating the total scatter rate of each scatter process separately. In this case we overestimate the differential scatter rate from state $\vec{k}$ in band $M$ to state $\vec{k}'$ in band $N$ by a constant independent of the final state of the hole. The integration over all final states now becomes very simple and can be performed very fast with aid of a table of the density of states in the final band. At the end of a free flight it must be decided which scatter process will scatter the hole. This is done in the usual way using the overestimated total scatter rates of the separate processes. As a result the band and the energy of the hole after scattering are decided. Finally the wave vector of the hole after scattering must be chosen. The direction of the wavevector after scattering is chosen randomly. Then the overestimated differential scatter rate is compared to the correct scatter rate and with aid of the standard rejection technique it is decided whether this wave vector is accepted or rejected. If it is rejected a self scattering occurs and the next free flight starts. The extra selfscattering introduced in choosing the wavevector after scattering corrects for overestimation of the differential scatter rate.

We calculated the drift mobility $\mu_E = v_d(E)/E$ with $v_d$ the drift velocity, in strained Si for several strain levels as a function of the electric field. The strain is indicated by the germanium content of the relaxed $Si_{(1-x)}Ge_x$ $\langle 001 \rangle$ substrate on which the strained Si layer is grown. The in plane direction of the electric field is $\langle 100 \rangle$, the out of plane direction is $\langle 001 \rangle$. In figure 1 the driftmobilities in the in plane and out of plane direction of the electric field are

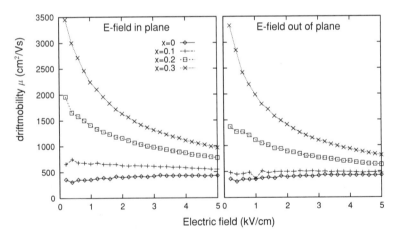

**Figure 1:** The driftmobility in strained pure Si at T=300 K. The direction of the electric field is either in plane (left) or out of plane (right).

shown. The error in the calculation varies from 20 % at 200 V/cm to less than 5 % at the highest electric fields. The parameters we used for the simulation in Si are given in Szmulowicz[5]. The value we found for the mobility in unstrained pure Si is $\sim 450$ cm²/Vs. This value can be compared to 505 cm²/Vs found by Szmulowicz using a relaxation time method with the same parameters.

From figure 1 it can be seen that the mobility in unstrained Si is almost independent of the electric field. The strong inter- and intraband scattering in Si which causes the poor low field mobility also causes the linear transport region to extend up to 5 kV/cm.

As the strain in the silicon increases also the drift mobility increases. This is caused by reduced scattering of the hole in strained Si due to strain induced changes of the valence band structure. The mean collision times for strained and unstrained Si (extracted from the simulations) showing this reduction are depicted in figure 2.

The increase of the drift mobility depends strongly on the electric field. At low fields the increase reaches the values stated by Nayak. At higher fields the increase is much smaller. The reduction of the increase of the drift mobility at higher electric fields is caused by onset of the saturation of the drift velocity. Due to the reduced scattering in strained Si the drift velocity saturates at lower electric fields.

The drift mobility in strained Si at high field is close to the mobility for unstrained Si because strain affects the bandstructure more in the vicinity of $\vec{k}=0$ than at bigger $\vec{k}$ values and at higher fields the hole will be on the average further away from $\vec{k}=0$ feeling a (not so altered) bandstructure.

We can conclude that the increase in low field drift mobility with a factor 6 in strained Si with x=0.2 compared to unstrained Si as shown by Nayak can be found also from our calculations. In addition we show that at an electric field of 5 kV/cm the drift mobility in strained Si drops to 2 times the drift mobility in unstrained Si at the same electric field.

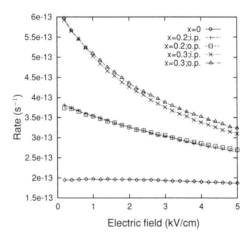

**Figure 2:** The mean collision time in strained pure Si at T=300 K. The electric field is in the in plane (i.p.) or out of plane (o.p.) direction.

### REFERENCES

1. D.K. Nayak, J.C.S. Woo, J.S. Park, K.L. Wang, and K.P. MacWilliams. High-mobility p-channel metal-oxide-semiconductor field-effect transistor on strained Si. *Appl. Phys. Lett.*, 62(22): 2853–2855, May 1993.
2. D.K. Nayak and S.K. Chun. Low-field mobility of strained Si on (100) $Si_{(1-x)}Ge_x$ substrate. *Appl. Phys. Lett.*, 64(19):2514–2516, May 1994.
3. J.M. Hinckley and J. Singh. Hole transport theory in pseudomorphic $Si_{(1-x)}Ge_x$ alloys grown on Si(001) substrates. *Phys. Rev. B.*, 41(5):2912–2926, February 1990.
4. C. Jacoboni and L. Reggiani. The Monte Carlo method for the solution of charge transport in semiconductors with applications to covalent materials. *Rev. Mod. Phys.*, 55(3):645–705, July 1983.
5. F. Szmulowicz. Calculation of optical- and acoustic-phonon-limited conductivity and hall mobilities for p-type silicon and germanium. *Phys. Rev. B.*, 28(10):5943–5962, November 1983.

# THE STATIC MANY-SOLITON DISTRIBUTION OF BIPOLAR PLASMA IN MANY-VALLEY SEMICONDUCTOR Ge

O. G. Sarbey, M. N. Vinoslavskiy, and D. A. Khizhnyak

Institute of Physics, Ukrainian Academy of Sciences,
Prospect Nauki 46, Kiev-22, 252022, Ukraine

In recent times detailed experimental investigations of spontaneous current instabilities, associated with creation of driving space-charge domains in monopolar carrier system due to impurity breakdown at liquid-helium temperature under dc electric field bias in a bulk of pure p-Ge have been carried out, see [1,2] and ref. there. The theoretical model of solitary space-charge waves was elaborated in [3] and proved to be in good agreement with these experimental results.

In previous paper [4] we reported on the investigation of solitary static and moving thermodiffusional autosolitons (AS's) in quasi-neutral electron-hole plasma (EHP), heated by dc electric field in n-Ge samples at T=77K. These AS's were predicted by theory, e.g. [5]. The AS's are spontaneously created, when the plasma concentration n and dc electric field E have reached their threshold values (n>1*$10^{14}$cm$^{-3}$, Edc>100V/cm). The AS's exhibit themselves as oriented perpendicular to a current direction narrow strata with high field strength (up to Eas=4000 V/cm) in its center, high carrier temperature (up to Te=1500K) and decreased plasma concentration. The AS formation mechanism was associated in [5] with nonlinear dependence of energy relaxation time $\tau_\varepsilon$ on carrier temperature Te near the Debye temperature Td: at Te<Td=438K (for Ge) $\tau_\varepsilon$ decreases, and at Te>Td $\tau_\varepsilon$ - increases.

Usually the AS's appear either in high field exclusion region near the positive contact, or in the vicinity of high inhomogeneities of field E in a bulk of specimen. Due to initial carrier heating near the Debye temperature at one of these places the rate of energy dissipation from carriers to lattice decreases, the positive feed-back for increasing of carrier temperature arises and the AS appears. The stationary state of such AS is provided, when a thermodiffusional flux of hot carriers outside from AS center is counterbalanced by a contrary directed diffusional flux of cold carriers to AS center. If AS appears in the middle part of a sample, anisotropy of the electron kinetic parameters causes the AS motion in the direction of more mobile carriers drift: that is in the direction of minority carriers (holes) drift - to negative contact in the samples with the field, applied along <111> axes, or in the direction of majority carriers (electrons) drift - to positive contact for <100> oriented E. The unmoving state of an AS may take place if it appears near the positive contact in the <100> oriented samples.

In present paper we demonstrate the formation of truly static many-soliton distribution of EHP in Ge samples with <110> field orientation at T=77K. The samples were cut from intrinsic n-Ge with a donor concentration Nd=1*$10^{13}$ cm$^{-3}$ with the dimensions of 0.05*0.1*0.8cm$^3$. The EHP with concentration of n=1*$10^{14}$- 5*$10^{16}$cm$^{-3}$ was photogenerated uniformly along a sample by white light illumination pulse I with time duration of $t_I$=1ms. Simultaneously, the EHP was heated by dc electric field of values

Edc<300V/cm (voltage/sample length) under bias of rectangular pulse voltage. A redistribution of an electric field along the specimen was registered by a set of 26 contact potential probes with spacing between two neighboring ones of about 0.15-0.4mm.

When applied pulsed voltage is much higher than the threshold value, necessary for AS formation [4], and the concentration of EHP, increasing in time (as light illumination increases) reaches its first threshold value $n_1$, the AS with electric field of about 2600V/cm appears near the positive contact similar to that observed in <100> and <111>- specimens (Fig.1, curve $t_1$). The current J, increasing with growth of light illumination I at the beginning of the pulse I, decreases a little at some value $n_1$, when the AS have appeared (see insertion in Fig.1). With further growth of n the AS is destroyed at some value $n_2>n_1$, and then up to four static high field AS's appear in the middle part of a sample at $n_3>n_2$ (Fig.1, curve $t_2$). These latter form some grating of carrier concentration n and carrier temperature Te with period d of about 1mm.

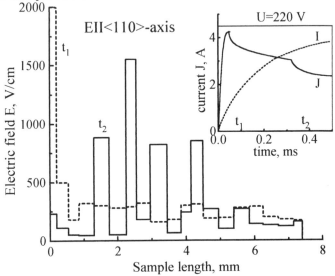

Figure 1. The distribution of field strength along the <110>-samples: $t_1$-solitary static AS at $n_1$ (time moment t1 in insertion); $t_2$- four static AS's at $n_3$ (time moment t2 in insertion).

The existing theory [5] predicts the formation of an array of the static AS's in a semiconductor symmetrical bipolar plasma (with equal kinetic parameters of electrons The existing theory [5] predicts the formation of an array of the static AS's in a semiconductor symmetrical bipolar plasma (with equal kinetic parameters of electrons and holes) and an array of the driving to one of the contacts AS's in nonsymmetrical plasma (with different EHP concentration: n-p=Nd>0 and different values of effective mass and mobilities of electrons and holes). In order to understand the causes of appearance of the several static AS's in a middle part of <110>-sample in nonsymmetrical EHP and the absence of AS motion to one of the contacts (as there is in <100> and <111>-samples) the electron and hole temperatures, mobilities, momentum and energy relaxation times have been calculated from energy balance equation with taking into account the carrier scattering mechanisms on the acoustic and optic phonons together with intervalley and electron-hole scattering of two groups of electrons in different valleys and of two groups of heavy and light holes.

It is well known that the electron mobilities in <110>specimens in high field strength have an intermediate values between ones in <100> and <111>-samples. The field E dependence of a ratio of electrons and holes mobilities for typical experimental values of local plasma concentrations $n=5*10^{14}-5*10^{15}cm^{-3}$ in high field region of AS center and $n=1*10^{15}- 5*10^{16}cm^{-3}$ in low field region outside the AS (Eh=50-100V/cm) is shown in Fig. 2. A comparison of these data with qualitative analysis of bipolar drift equation for EHP exhibits that the static state of AS in <110>-samples at sufficiently high values of n may realize due to competition of two contrary directed motions of bipolar plasma: i) the plasma motion to negative contact-in hole drift direction at decreased EHP concentration $n<1*10^{15}cm^{-3}$ and field $E< 800V/cm$ (because $b=\mu_e/\mu_h<1$, see Fig. 2) and ii) the plasma motion to positive contact-in electron drift direction at high values $n>1*10^{15}cm^{-3}$ and $E>800V/cm$, for which $b>1$ (Fig. 2).

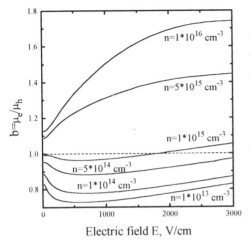
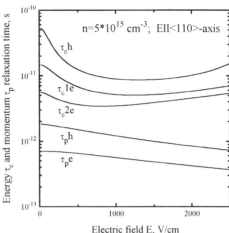

Figure 2. Calculated field E dependence of ratio of electron and hole mobilities for different values n.

Figure 3. Calculated field E dependence of energy and momentum relaxation times for $n=1*10^{15}cm^{-3}$.

The theoretical model [5] require the following condition for AS creation in quasi-neutral EHP: i) $\tau_p<<\tau_{ee}<<\tau_\varepsilon$ ($\tau_p\sim T^\alpha$, $\tau_\varepsilon\sim T^s$ and $\tau_{ee}$ are the times of momentum and energy relaxation and inter-electron collisions, respectively), ii) $l_\varepsilon/L<<1$ (L and $l_\varepsilon$ are the diffusion length and that of the energy relaxation of carriers) and iii) $\alpha+s>-1$ ($\alpha=d(\ln \tau_p)/d(\ln T)$, $s=d(\ln \tau_\varepsilon)/d(\ln T)$). These conditions were satisfied in our experiments as it is looking from the data of Fig. 3, calculated for n-Ge <110>-samples at T=77K: i)$\tau_p\sim1*10^{-12}sec<<\tau_\varepsilon\sim1*10^{-11}sec.$; ii) $L\sim1*10^{-2}cm$, $l_\varepsilon\sim1*10^{-5}cm$; iii) $\alpha=-1/2$ and s>-1/2 at $E>800V/cm$ (s=0 at Debye temperature Td). The high electric power, dissipated in the static AS (of about 4000-6000(W/cm)), leads to local nonhomogeneous heating of the crystal lattice (up to $\delta T=20-40(K)$) near the positive contact and to cooling of carriers, and in its turn disturbs the iii)-condition. The first AS near the positive contact disappears. Now these conditions may be realized at the field inhomogeneities in a middle part of a sample. The field strength in all four AS is smaller than that of in the first solitary AS and the field strength outside AS's is although smaller, than that of beyond the solitary AS, so the both AS-types could exist at the same voltage on the sample.

1. S.W. Teitsworth, R.M. Westervelt, and E. E. Haller, Nonlinearoscillations and chaos in electrical breakdown in Ge, Phys.Rev. Lett. 51:825 (1983).
2. A.M. Kahn, D.J. Mar, and R.M. Westervelt, Spatial measurements near the instability threshold in ultrapure Ge, Phys. Rev. B, 45:8342 (1991).
3. I.R. Cantalapiedra, L.L. Bonilla, M.J. Bergman, and S.W.Teitsworth, Solitary wave dynamics in extrinsic semiconductors under dc voltage bias, Phys.Rev.B,48:12278 (1994).
4. O.G. Sarbey, and M.N. Vinoslavskiy, Evolution of thermodiffusional autosolitons in non-equilibrium electron-hole plasma in Ge, Semicond. Sci. Technol.9:573 (1994).
5. B.S. Kerner, V.V. Osipov, Autosolitons, Sov.Phys.-Usp.,32:101(1989).

# HOT-HOLE STREAMING IN UNIAXIALLY COMPRESSED GERMANIUM

I.V.Altukhov, M.S.Kagan, K.A.Korolev,
and V.P.Sinis

Institute of Radioengineering and Electronics
of Russian Academy of Sciences
11, Mokhovaya
Moscow, 103907
RUSSIA

The streaming motion of charge carriers occurs in sufficiently strong electric field provided that the threshold scattering mechanism (e.g. optical-phonon emission) exists. It is the cyclic process consisted of scattering-free acceleration of holes up to the optical-phonon energy, emission of an optical phonon and the return to the region of small energies (so called "a source") after which the process repeats. In this case, the distribution function in momentum space is stretched out along the field direction. The uniaxial stress removes the degeneracy of the valence band of Ge at $k = 0$ and splits it into two subbands separated by the energy gap $\Delta$ proportional to the stress P. In this work we studied spontaneous far-infrared radiation from stressed p-Ge. It will be shown that the limiting energy for ballistic acceleration of holes in the higher-energy subband is the energy corresponding to the hole transition to the lower-energy band via optical-phonon emission and the dynamics of hole heating in stressed p-Ge is strongly influenced by inter-subband optical-phonon scattering.

Fig.1 represents the electric field dependence of spontaneous far-IR radiation (Ge<Ga> detector, $h\nu \approx 10$ meV) at different pressures [1]. For zero stress, the maximum in radiation at $E = E_c$ has been explained [2] by the onset of a streaming motion of light holes. The condition for it is $eE_c\tau = p_0$ ($p_0 = (2m_l\varepsilon_0)^{1/2}$ is the momentum value corresponding to the optical phonon energy, $\varepsilon_0$, $m_l$ is the light-hole effective mass, $\tau$ is mean free time). The band structure of uniaxially compressed p-Ge is shown in Fig.2. It is clear that the momentum value corresponding to the optical-phonon scattering of holes within the upper subband is $p_{01} = [2m_l(\varepsilon_0+\Delta)]^{1/2}$, ($\Delta$ is the energy gap between strain-split valence subbands) and has to increase with increasing pressure. So the field of a maximum of radiation, $E_c$, has to increase. Fig.1 shows that this field does not change with stress. On the other hand,

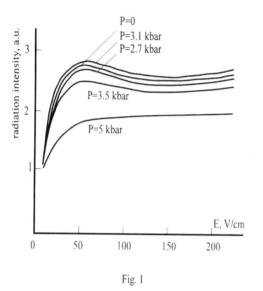

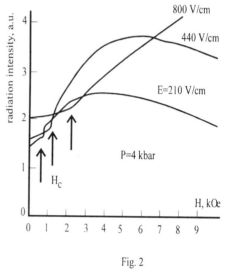

Fig. 1

Fig. 2

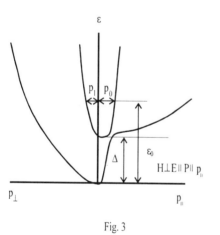

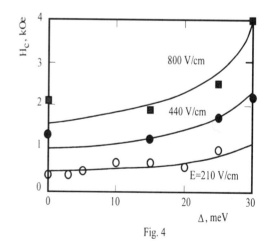

Fig. 3

Fig. 4

the holes in the upper subband with the energy $\varepsilon_0$-$\Delta$ can get into the lower band emitting the optical phonon. The corresponding momentum value, $p_{02}$, and hence $E_c$ are independent of pressure. Thus, the "ceiling" for ballistic motion of holes in the upper band is the energy $\varepsilon_0$-$\Delta$ at which intensive inter-subband hole transitions due to optical-phonon emission occur.

This limiting energy for the ballistic acceleration in the upper band has also to manifest itself when a transverse magnetic field is applied. The dependences of spontaneous radiation on magnetic field H for different electric fields E are shown in Fig.3. The steep rise of radiation beginning at H = $H_c$ indicates that a magnetic trap in the upper hole subband appears [2]. One can deduce the equation for the ballistic trajectory of a hole moving in crossed fields that is $d\varepsilon = dp_\perp cE/H$ where $p_\perp$ is the momentum component in the direction perpendicular to E and H, c is the light velocity. For $\mathbf{E} \| \mathbf{P}$ and $\varepsilon_0$-$\Delta$ as a limiting energy for ballistic acceleration of holes, we obtain the threshold magnetic field for magnetic trap formation: $H_c = cE[m_\perp/2(\varepsilon_0-\Delta)]^{1/2}$ ($m_\perp$ is the effective mass in the $p_\perp$ direction). Fig.4 shows that the stress dependences of $H_c$ from this expression and the experiment are in a good agreement. Note that the streaming motion within the upper band only gives $H_c$ independent of pressure as $p_\perp$ does not depend on P in this case.

The intensive inter-subband optical-phonon scattering beginning at the upper-subband hole energy $\varepsilon_0$-$\Delta$ is a cause of peculiar hole dynamics in stressed Ge differing from that in unstressed material. The streaming motion of holes in one subband only has to include the hole return to the initial region of small energies (a source) in the same band after emission of an optical phonon. The existence of such a source is a reason for the distribution function to be stretched out along the electric field direction. In stressed Ge, the streaming (as a cyclic process) in the higher-energy subband turns out to be broken as holes reaching the energy $\varepsilon_0$-$\Delta$ depart from the upper band into the lower one. The source in the upper band can not appear in this case since the hole return to the bottom of the same band via optical-phonon emission is absent. This is evident, in particular, under streaming conditions for the lower band when the holes are uniformly distributed along the field direction. Although the upper-band distribution function is not stretched out in this case, the hole motion is ballistic at sufficiently strong electric fields, but it is not the streaming in usual sense. When magnetic field is applied closed trajectories (the magnetic trap) in the passive energy region, $\varepsilon < \varepsilon_0$-$\Delta$, appear, and, as a result, the inter-subband optical-phonon scattering weakens. This is the cause of increasing radiation intensity with magnetic field , as against a strain-free situation.

The research described in this publication was made possible in part by Grant No.N7H300 from the International Science Foundation and Russian Government.

References

1. I.V.Altukhov, M.S.Kagan, K.A.Korolev, V.P.Sinis, and F.A.Smirnov.
      Sov.Phys.-JETP, V.74 (2), P.404 (1992)
2. A.A.Andronov. "Infrared and Millimetre Waves", edited by K.J.Button,
      V.16 (Academic Press, Orlando, 1986) p.146

# APPEARANCE OF A LARGE 'HALL' CURRENT COMPONENT
# PARALLEL TO B IN P-GE IN STRONG E ⊥ B FIELDS

R.C. Strijbos, S.I. Schets, and W.Th. Wenckebach

Faculty of Applied Physics, Delft University of Technology,
P.O. Box 5046, 2600 GA Delft, The Netherlands

We have investigated the transport properties of light and heavy holes in lightly doped $p$-type germanium in strong crossed $\mathbf{E} = (E_x, 0, 0)$ and $\mathbf{B} = (0, 0, B_z)$ fields at low temperatures using a Monte Carlo simulation program that accounts for the anisotropy of the band structure. Since optical phonon emission is the strongly dominant scatter process, it is customary to discriminate between streaming holes, which are repeatedly accelerated by the electric field and scattered due to optical phonon emission, and accumulated holes, moving along closed cyclotron orbits below the optical phonon energy $\varepsilon_{op}$. The latter can only be scattered by a-coustic phonons and ionized impurities and, thus, have a much longer relaxation time.

In a spherical band model, the hole system in strong $\mathbf{E} \perp \mathbf{B}$ fields can be characterized by the parameter $\zeta^n = v_{op}^n / v_{dr}$,[1,2] where $v_{dr} = E/B$ is the average drift velocity in the $\mathbf{E} \times \mathbf{B}$ ($-y$) direction and $v_{op}^n = \sqrt{2\varepsilon_{op}/m^n}$ is the velocity of holes with effective mass $m^n$ and energy $\varepsilon_{op}$. Three situations can be discerned: total streaming ($\zeta^n < 1$), partial streaming and partial accumulation ($1 < \zeta^n < 2$), and total accumulation ($\zeta^n > 2$). It is clear that in the isotropic picture, due to the mirror symmetry in the principal $k_x k_y$-plane, there will be no current component along the magnetic field direction in any of these three situations.

The magnitude of these current components in an anisotropic model has been investigated by one of us (S.I.S) using a Monte Carlo program, simulating light and heavy hole trajectories in crossed electric and magnetic fields. The anisotropy was accounted for by using the parabolic 'warped' energy dispersion relation and by including the so-called overlapfactors, derived from the analytic eigenfunctions of the parabolic four-band model of Bir and Pikus,[3] in the expressions of the differential scatter rates. A detailed description of the Monte Carlo program is given elsewhere.[4]

The calculations show that both the mean wavevector and the drift velocity can indeed have large components along the magnetic field direction if there is no mirror symmetry with respect to the principal plane perpendicular to $\mathbf{B}$.[4] In our calculations $\mathbf{B}$ ($z$-axis) is chosen along the $[\bar{1}\bar{1}1]$ direction. The result is shown in Fig. 1: both $|k_z|$ and $|v_z|$ increase substantially when $\zeta^n > 1$ and reach a maximum at $\zeta^n \approx 2$. For heavy holes, the mean wavevector is max-

*Hot Carriers in Semiconductors*
Edited by K. Hess *et al.*, Plenum Press, New York, 1996

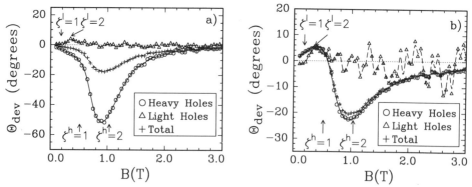

**Figure 1:** Angle of deviation towards **B** direction for a) mean wavevector and b) mean drift velocity as a function of magnetic field strength for heavy holes, light holes, and for light and heavy holes together; $\mathbf{E} = (E_x, 0, 0) \parallel [1\bar{1}0] = 1\,\mathrm{kVcm}^{-1}$, $\mathbf{B} = (0, 0, B_z) \parallel [\bar{1}\bar{1}1]$, $T = 10\mathrm{K}$, $n_l = p_0 = 1.5 \cdot 10^{14}\mathrm{cm}^{-3}$. The arrows denote the magnetic field strengths, where — in the isotropic model — holes start to accumulate ($\zeta^{l,h} = 1$) and all holes are accumulated ($\zeta^{l,h} = 2$).

imally about 52 degrees out of the $\langle \bar{1}\bar{1}1 \rangle$ plane (Fig. 1a), while the drift velocity deviates up to 22 degrees (Fig. 1b). For light holes, the effects are smaller and in opposite direction.

For $\zeta < 1$ all holes are in streaming motion. The elongated hole distribution in **k** space moves away from the electric field direction due to the Lorentz force. This causes effects similar to the Sasaki-Shibuya effect (SSE) for streaming holes in a strong electric field,[5,6] but now not only affecting the current in the $xy$-plane $\perp$ **B**, but also yielding a current along the $z$-axis $\parallel$ **B** due to the absence of a mirror plane in the band structure for **B** $\parallel [\bar{1}\bar{1}1]$. Two mechanisms contribute to this effect, as illustrated in Fig. 2a for heavy holes. First, since the velocity $\mathbf{v}^n(\mathbf{k}) = \frac{1}{\hbar}\nabla_{\mathbf{k}}\varepsilon^n(\mathbf{k})$ is directed perpendicular to the warped equi-energy surface, the direction of the average velocity deviates from the direction of the average wavevector in a direction corresponding to a lower effective mass. In addition to this mechanism based on the anisotropy of band structure, there is also a contribution from the anisotropy of the relaxation time.[6] Since the average time between two optical phonon emissions $T_{op}$ is larger for holes below the streaming axis with a larger effective mass $m^n$, these holes have a relatively larger contribution to the average current and wavevector than the holes above this axis. As a consequence, the mean wavevector deviates in the direction of higher effective mass. Both effects are observed in the simulation results of Fig.1 for $\zeta < 1$.

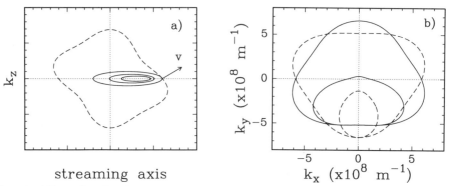

**Figure 2:** a) Illustration of the Sasaki-Shibuya effect: a contour plot of the distribution function is drawn in a $\langle \bar{1}0\bar{1} \rangle$ plane, assuming the streaming direction deviates $30^0$ from the electric field. The dashed line is the equi-energy line $\varepsilon = \varepsilon_{op}$; **E** $\parallel [1\bar{1}0]$, **B**, $\mathbf{k}_z \parallel [\bar{1}\bar{1}1]$. b) Illustration of the lack of mirror symmetry of band structure with respect to a $< \bar{1}\bar{1}1 >$ plane. Cross-sections of the accumulation region and the encompassing equi-energy surface $\varepsilon = \varepsilon_{op}$ are drawn for $k_z = -$cst (solid line) and $k_z =$ cst (dashed line); **E** $\parallel [1\bar{1}0]$; **B**, $\mathbf{k}_z \parallel [\bar{1}\bar{1}1]$.

In order to explain the much larger effect for accumulated holes ($\zeta > 1$), that we call the Parallel Hall Effect (PHE), we refer to Fig. 2b, where cross sections of the accumulation region and the equi-energy surface $\varepsilon^n(\mathbf{k}) = \varepsilon_{op}$ in the planes $k_z = \pm$cst are drawn for the case that the magnetic field $\mathbf{B}$ is directed along a $[\bar{1}\bar{1}1]$ direction. The band structure transforms as the point group $D_{3d}$ and has no mirror plane perpendicular to the $[\bar{1}\bar{1}1]$ axis. Contrary to the isotropic case, now the accumulation region is extremely asymmetric. Most notably, for $k_z > 0$ it is much larger than for $k_z < 0$. If we assume that the probabilities for scattering into both parts of the accumulation region are approximately equal, averaging of both $k_z$ and $v_z$ over the asymmetric hole distribution leads to an 'effective' component parallel to the magnetic field, as observed in Fig. 1 for $\zeta > 1$. As more and more holes become accumulated between $\zeta = 1$ and $\zeta = 2$, both $v_z$ and $k_z$ become larger. However, at still larger values of $\zeta$, the 'centre of gravity' of the accumulation region (in the isotropic model the point $C = (0, m^n E/\hbar B)$) moves closer to the $\Gamma$ point and the accumulation region eventually becomes bounded by the equi-energy surface $\varepsilon^n(\mathbf{k}) = \varepsilon_{op}$. Thus, the accumulation region recovers inversion symmetry with respect to the $\Gamma$ point (see Fig. 2b), and $v_z$ and $k_z$ return to zero.

We finally show that the PHE can have large consequences for the operation of $p$-Ge hot-hole lasers, especially when operated in Voigt configuration. These far-infrared lasers are based on a population inversion between accumulated light holes and (partially) streaming heavy holes and normally operate in the electric and magnetic field region where $\zeta^h \approx 1.4$.[7] Then, for a crystal cut along the directions of Fig. 1, the PHE tries to deflect the average drift velocity up to 10 degrees in the direction of $-\mathbf{B}$. As in the Hall effect, charge is built up at the sample sides to compensate for this deviation of the drift direction. The electric field component $E_{PHE} \parallel \mathbf{B}$ due to these charged sample sides accelerates the accumulated light holes along the magnetic field direction so they can be scattered out of the accumulation region by optical phonon emission, resulting in a decrease of the small signal gain. If the sample is tilted, $E_{PHE}$ can be compensated by the $z$-component of the applied electric field, resulting in larger population inversion and a stronger laser action. This anomalous behaviour has been observed by Gavrilenko et al.[8] in the luminescence of the light-heavy hole transition when tilting the sample electrodes with respect to $\mathbf{B} \parallel [111]$. In their case the maximum of luminescence was shifted by 1 or 2 degrees only and was incorrectly attributed to the SSE.[8] A more prominent demonstration of the PHE is found when monitoring the laser emission of a $p$-Ge crystal with $\mathbf{E} \parallel [1\bar{1}0]$, $\mathbf{B} \parallel [\bar{1}\bar{1}1]$ (as in Fig. 1) that is mounted in Voigt configuration. Preliminary experiments indicate that the crystal has to be tilted approximately 10 degrees in order to obtain maximum output power. Further results will be presented in a forthcoming paper.

## REFERENCES

1. I.I. Vosilyus and I.B. Levinson, Galvanomagnetic effects in strong electric fields during nonelastic electron scattering, *Sov. Phys.-JETP* 25:673 (1967).
2. S. Komiyama and T. Masumi, Streaming motion and population inversion of hot electrons in silver halides at crossed electric and magnetic fields, *Phys. Rev. B* 20:5192 (1979).
3. G. L. Bir and G. E. Pikus, Symmetry and Strain-induced Effects in Semiconductors, John Wiley & Sons, New York (1974).
4. S. I. Schets, Monte Carlo simulation of hole transport in p-type germanium, Master's thesis, Delft University of Technology (1994).
5. W. Sasaki, M. Shibuya, K. Mizuguchi, and G.M. Hatoyama, Anisotropy of hot electrons in germanium, *J. Phys. Chem. Solids* 8:250 (1959).
6. W.E.K. Gibbs, Anisotropy in the conductivity of hot holes in germanium, *J. Appl. Phys.* 33:3369 (1962).
7. *Opt. Quantum Electron.* 23, special issue far-infrared semiconductor lasers (1991).
8. V.I. Gavrilenko, A.L. Korotkov, Z.F. Krasil'nik, and V.V. Nikonorov, Far IR cyclotron resonance and luminescence of hot holes in $p$-Ge, *Opt. Quantum Electron.* 23:S163 (1991).

5. Hot Electrons in Devices

# MONTE CARLO SIMULATION OF HIGH-ENERGY ELECTRON TRANSPORT IN SILICON: IS THERE A SHORT-CUT TO HAPPINESS?

M. V. Fischetti, S. E. Laux, and E. Crabbé

IBM Research Division
T.J. Watson Research Center
P.O Box 218, Yorktown Heights, NY 10598

## INTRODUCTION

Impact ionization and injection into the $SiO_2$ gate insulator, the most commonly studied phenomena in the context of hot-electron transport in Si devices, have been investigated using a large variety of physical models, from full-band Monte Carlo simulations[1,2], to parabolic-band approximations coupled to Monte Carlo[3] or simplified[4,5] solution to the transport equation, to empirical fits to the electron energy distribution[6-8], to analytic models based on Maxwellian distribution functions and the lucky-electron concept[9]. We have argued in the past[10] that most of these attempts were premature, since we were still missing some basic knowledge of high-energy ($> 1$ eV) transport in Si. A look at the top frames of Fig. 1 clarifies the reasons of our past pessimism: Even restricting ourselves to publications dealing with 'physically based' Monte Carlo simulations, we see that a huge uncertainty existed, as recently as a few years ago, on our knowledge of the rates for the most important scattering processes for electrons in Si (with phonons[1,11,12] and impact ionization[12-17]). While these models were tailored to reproduce a limited amount of experimental data under homogeneous conditions, it is clear that there was little hope of accurately predicting hot-electron transport in the most general situation.

In this paper we reconsider our earlier skepticism: In the past couple of years several groups have reached consensus if not on all of the details of a transport model, at least on the (obvious) necessity of using a full-band structure to handle transport, and on the overall magnitude of the electron-phonon scattering rates, whether theoretically computed[19,20] or determined empirically[12,18,21]. Similarly, triggered by additional experimental results[22], revised calculations of the pair-production rate[23-25] have resulted in a consistent picture, which enables us to predict pair production in submicron devices[26,27]. The bottom frames of Fig. 1 illustrate clearly how the uncertainty has been reduced to an acceptable level.

These recent developments allow us to handle hot-electron effects in Si devices with more confidence. Indeed, satisfactory agreement between experiments and simulations can be obtained using this 'standard model'[27]. Yet, difficulties remain: A correct analysis of both impact-ionization and of gate currents requires not only a correct kinematics (band structure) and 'correct' electron-phonon and impact-ionization scattering rates, but also the inclusion of computationally expensive elements. Space limitations force us to refer the readers to Refs. 12 and 27 for details of the full-band model we have employed and for a more extensive bibliography. Here we shall only present a few selected results and emphasize some salient characteristics of hot-electron transport: The main role played by interparticle Coulomb interactions, which are crucial in determining the shape of the high-energy tails of the electron distribution; the necessity of handling dielectric screening within a dynamic model; exchange and correlation effects, very important in the heavily-doped drain, where most ionization events occur; the strong nonlocality of the ionization processes; the well-known - but often underemphasized - non-Maxwellian form of the distribution function in regions where the high-energy component matters. In addition, the analysis of gate currents is also affected by controversial issues, such as conservation of parallel momentum, inclusion of image-force corrections, and scattering in the oxide.

*Hot Carriers in Semiconductors*
Edited by K. Hess *et al.*, Plenum Press, New York, 1996

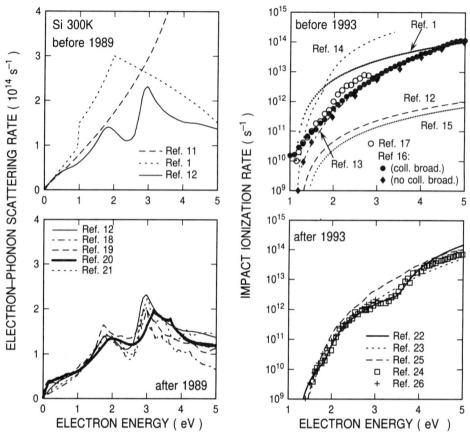

**Figure 1.** Electron-phonon (left) and impact-ionization (right) scattering rates in Si versus electron energy at 300 K, averaged over equienergy-surfaces, taken from the past (top) and more recent (bottom) literature.

Accounting for all of these effects is at the limit of our computational ability. While we readily admit that these simulations cannot be performed routinely, yet, there seems to be no 'short-cut': As hot-electron effects often consist of part-per-million (or less!) fluctuations from macroscopic averages, small errors and/or approximations to the physics are likely to yield dramatically incorrect results.

## IMPACT IONIZATION IN MOSFETS

In the high-field region around the drain-substrate junction of n-channel MOSFETs, electrons accelerated by the longitudinal source-to-drain electric field can reach energies in excess of the threshold-energy for pair production[28]. Most of the holes so generated will eventually drift away and be collected at the substrate contact, giving rise to a substrate current. This current has been studied in great detail for several reasons: First, it is a monitor of the hot-electron damage to the gate insulator, a simple proportionality between the substrate current, $I_{sub}$, and the gate current, $I_{gate}$, having been suggested as a general rule[9]. Secondly, it affects directly the operation of the device, being an obvious measure of the breakdown behavior, of the 'snap-back'[29] effect, or of the 'kink effect' in silicon-on-insulator (SOI) MOSFETs[30].

Let's consider first the $0.25\mu m$ gate-length n-channel MOSFET described in Ref. 27 and see how the calculated generation rate is affected by Coulomb interaction. In Fig. 2 (a) we show the calculated substrate current compared to experimental data. Keeping in mind the 'usual' uncertainty about the doping profile[31], the agreement is satisfactory. Yet, we could only obtain such an agreement when properly accounting for the dynamically-screened electron-electron interaction: When all Coulomb interactions are suppressed (the long-range component by running in a 'frozen field' configuration, having fixed the field to its time-averaged value, and explictly suppressing the short-range component,

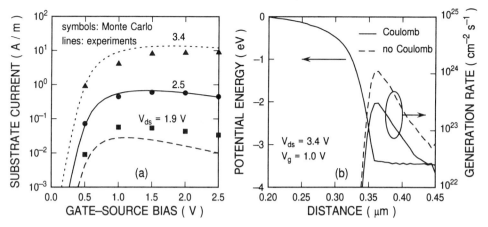

**Figure 2.** (a) Measured and calculated substrate current as a function of gate-source bias at three values of the source-to-drain bias, $V_{ds}$, in a 0.15 $\mu$m effective channel-length MOSFET. (b) Potential energy at the Si-SiO$_2$ interface and pair generation rate along the channel.

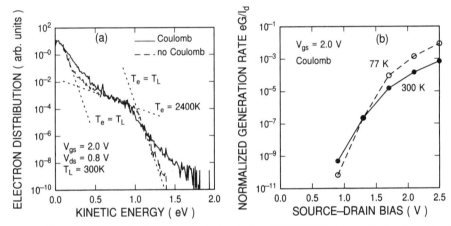

**Figure 3.** (a) Electron energy distribution at the drain-end of an SOI MOSFET with an effective channel-length of about 0.18 $\mu$m, calculated with and without the inclusion of short- and long-range interparticle Coulomb interactions. (b) Generation rate, $G$, normalized to the drain current as a function of source-drain bias at two temperatures.

treated as another Monte Carlo scattering process[12,27]), the generation rate (and, therefore, the substrate current) is significantly overestimated, as illustrated in Fig. 2 (b). Indeed, most of the ionization events occur beyond the high-field region, just inside the metallurgical channel-drain junction. It is in this region that the hot channel electrons reach their maximum kinetic energy. This intrinsically non-local property is particularly important in shorter devices and cannot be handled by any local model, such as models based on the first few moments of the transport equation. Another interesting effect of the Coulomb interaction is shown in Fig. 3. In this case we have considered an SOI MOSFET of similar dimensions (0.18 $\mu$m effective channel-length) biased *below* the expected ionization threshold, at a source-drain bias, $V_{ds}$, of only 0.8 V. We have considered the electron energy distribution at a position along the channel where the average energy reaches its maximum value. The distribution obtained without the inclusion of Coulomb interactions exhibits some intuitively understood features (dashed line, Fig. 3(a)): At low energy we see the thermal contribution of the electrons in the drain. The middle portion of the distribution (between about 0.5 and 0.9 eV) approaches a Maxwellian tail at an effective electron temperature, $T_e$, of about 2400 K, corresponding roughly to the maximum average energy reached by the electrons at the end of the channel. At about 0.9 eV, the distribution is abruptly 'truncated': This energy represents the maximum kinetic energy a ballistic electron launched from the source (whose Fermi level is about 0.1 eV, because of the high doping) can reach. Energies in excess of this value can only be obtained by electrons 'lucky' enough to absorb phonons. Above this cut-off energy, the distribution drops exponentially with an effective temperature roughly equal to the lattice temperature[32,33]. As one could expect naively, there is an almost negligible amount of

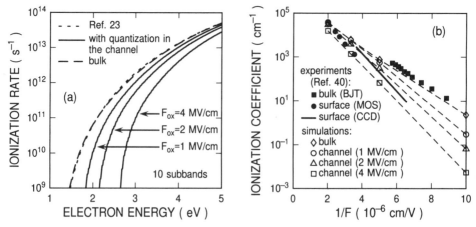

**Figure 4.** (a) Impact ionization rate calculated using the random-k approximation accounting for quantization in a (triangular well) channel at various values of the (normal) field in the insulator. The ionization rate in the bulk is compared to the calculation reported in Ref. 23. (b) Ionization coefficient in inversion layers as a function of inverse longitudinal field compared to the experimental data reported in Ref. 40. No values of the oxide field are reported in that reference, so that only a qualitative agreement can be inferred.

impact ionization in the device at such low bias: We must rely on those very rare carriers which are both 'lucky' not to emit phonons *and* lucky enough to absorb a few. On the contrary, the inclusion of Coulomb interactions results, as expected, in a 'smoother' distribution: The thermalizing effect of the interparticle interactions is very evident, as the distribution moves towards a Maxwellian, although the interparticle interactions have not had the 'time' to fully thermalize the distribution. Yet, a significant population is observable above the ionization threshold, as the fast channel electrons interact with the thermal carriers in the drain. The large amount of energy and momentum transferred in this (exchange) collision makes for very weak screening. For this reason dynamic exchange effects are crucial. The net effect, shown in Fig. 3(b), is the occurence of ionization even at such low bias[34,35]. As we move to a lower lattice temperature, enhanced screening, a cooler initial distribution in the drain and, most important, a larger gap, result in a lower ionization at a sub-bandgap bias, as indeed observed experimentally[36−38] and in recent simulations of sub-0.1 μm devices[39].

A final effect which influences the generation rate at positive gate-to-source bias is caused by the quantization effects in the inversion layer: The ionization threshold is effectively increased, since no (final) states are available to the recoil and secondary electrons at energies below the bottom of the first subband. As shown in Fig. 4(a), the ionization rate, computed using the random-k approximation[13], but accounting for the two-dimensional density of states in the channel[27], is a strong function of the 'normal' (or oxide) field. Indeed, a reduction of the ionization coefficient moving from the bulk to inversion layers had been reported in the past[40]. Figure 4(b) shows the ionization coefficient in an ideal 'triangular well' approximation, compared to the experimental data reported by Slotboom *et al.*[40].

## INJECTION INTO SILICON DIOXIDE

The classical experiment by Ning, Osborn, and Yu[41] (in which electrons optically generated in the negatively biased substrate of an n-channel MOSFET are accelerated towards the channel - grounded via the source and drain contacts - and the resulting probability of injection into the SiO$_2$ insulator is measured as the ratio of the gate-to-channel currents) has become an essential calibration tool against which many models have been tested[1,12,27,42,43]. Besides the obvious concerns about the band-structure model, the scattering processes, and the numerical noise, important questions remain either unanswered or controversial: Should we account for conservation of (crystal) momentum as an electron attempts to enter the amorphous oxide[27], or should we assume that the disorder somehow provides the missing momentum[44]? Is the process of injecting an electron into the oxide so fast that the surface charges do not have the time to redistribute themselves and set-up the image force[45], so that image-force corrections should be ignored[42]? Or, instead, is the characteristic injection time large enough to allow a full redistribution of the surface charges so that the classical, static image potential (corrected for many-body effects[46]) should be used[47,48]? In Fig.5 we show results of simulations we have performed of the device labeled 15-12-8

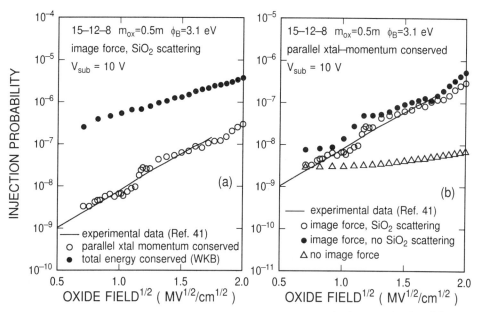

**Figure 5.** (a) Probability of electron injection into SiO$_2$ in the device 12-15-8 of Ref. 41 as a function of the square root of the oxide field. Results obtained by accounting for conservation of total energy only (solid symbols) or also conservation of parallel momentum (open symbols) are compared to the experimental data (line). (b) As in (a), but now the effects of the inclusion or exclusion of image force effects and scattering in the image-forced-rounded barrier are shown. The Monte Carlo model of Ref. 52 has been used to treat transport in SiO$_2$.

in Ref. 41. Apparently, 'conventional' wisdom results in a satisfactory agreement: parallel momentum is indeed conserved at least for the <100>-oriented interface of interest here, possibly because of the short-range order exhibited by the interfacial region of the SiO$_2$ film[49,50]; image-force corrections do indeed apply, since both tunneling[51] and transmission over the Si-SiO$_2$ barrier[52] occur in a time-scale much longer that the inverse frequency of the surface (valence) plasmons.

## CONCLUSIONS

While the success achieved in our simulations is indeed promising, we should recall that in Ref. 27 we have expressed some concerns: Despite the high computational cost of our simulations, it is perhaps depressing to think of the many approximations we have been forced to employ, either because of sheer ignorance of the correct physics, or because of computational constraints. For example, dynamic screening, of crucial importance, had to be treated assuming that only the thermal electrons in the drain (as far as substrate current calculations are concerned) or in the inversion layer (in the case of the simulation of electron injection into the gate insulator) contribute to screening as in the case of a gas at equilibrium and within the random-phase-approximation. The details of the dynamic response of a two-dimensional electron gas have also been neglected. Additional concerns remain about how to model properly the transport of electrons across the Si-SiO$_2$ interface, about correlation effects[12], and, last but not least, the correct implementation of scattering processes, such as the correct contribution of each phonon mode to the total scattering rate, or possible anisotropic effects, possibly significant at energies near threshold, of the ionization processes.

We may conclude by saying that the recent work on the electron-phonon interaction and ionization processes has lifted most of the skepticism we had expressed in Ref. 10: We can indeed come close to predicting qualitatively (and often quantitatively) high-energy transport in Si devices. However, the price we must pay is a hefty one: Extremely large computation times are required. This renders our simulations unsuitable to routine use. Yet, alternative simplified models seem even less useful: In Ref. 27 we have shown how the effects of Coulomb interactions, the nonlocality of the pair production processes, and the form of the distribution functions (as in Fig. 3(a), for example), not amenable to simple 'fitted' analytic expressions, render 'brute-force' simulations a necessity. We must conclude with the dull remark that, at least for the time being, we see no short-cuts to a happy situation in which hot-electrons effects in Si devices may be accurately predicted in a less painful way.

## AKNOWLEDGMENTS

We are grateful to J. Bude, C. Hamaguchi, and G. Jin for having made available to us their work before publication, and to K. Hess, J.M. Higman, and P.D. Yoder for fruitful discussions. We are also pleased to aknowledge the National Center for Computational Electronics (NCCE) at the Universtity of Illinois, Urbana-Champaign, for ideas which are the source of the 'standard model' presented in the introductory section. One of us (MVF) would like to thank in particular N. Sano for many discussions on impact ionization.

## REFERENCES

1. J.Y. Tang and K. Hess, *J. Appl. Phys.* 54:5139 (1983).
2. J.J. Ellis-Monaghan, K.W. Kim, and M.A. Littlejohn, *J. Appl. Phys.* 75:5087 (1994).
3. B. Riccó, E. Sangiorgi, F. Venturi, and P. Lugli, *IEDM Tech. Dig.* 559 (1989).
4. S.-L. Wang, N. Goldsman, Q. Lin, and J. Frey, *Solid-State Electron.* 36:833 (1993).
5. C. Fiegna, F. Venturi, E. Sangiorgi, and B. Riccó, *IEDM Tech. Dig.* 451 (1990).
6. C. Fiegna, F. Venturi, M. Melanotte, E. Sangiorgi, and B. Riccó, *IEEE Trans. Electron. Dev.* 38:603 (1991).
7. K. Rahamat, J. White, and D.A. Antoniadis, *IEEE Trans. CAD* 12:817 (1993).
8. J.G. Rollins, V. Axelrad, and S. Motzny, *Microelectron. Eng.* 19:265 (1992).
9. S. Tam, P.-K. Ko, and C. Hu, *IEEE Trans. Electron. Dev.* ED-31:1116 (1984).
10. M.V. Fischetti, S.E. Laux, and D.J. DiMaria, *Appl. Surf. Sci.* 39:578 (1989).
11. C. Jacoboni and L. Reggiani, *Rev. Mod. Phys.* 55:645 (1983).
12. M.V. Fischetti and S.E. Laux, *Phys. Rev. B* 38:9721 (1988).
13. E.O. Kane, *Phts. Rev.* 159:624 (1967).
14. R. Thoma, H.J. Pfeifer, W.L. Engl, W. Quade, R. Brunetti, and C. Jacoboni, *J. Appl. Phys.* 69:2300 (1991).
15. Th. Vogelsang and W. Hänsch, *J. Appl. Phys.* 70:1493 (1991).
16. J. Bude, K. Hess, and G.J. Iafrate, *Phys. Rev. B* 45:10958 (1992).
17. N. Sano and A. Yoshii, *Phys. Rev. B* 45:4171 (1992).
18. H. Shichijo, J.Y. Tang, J. Bude, and P.D. Yoder, *in* "Monte Carlo Device Simulation: Full Band and Beyond", Karl Hess ed., Kluwer Academic, Boston (1991), p. 285.
19. M.V. Fischetti and J.M. Higman, *in* "Monte Carlo Device Simulation: Full Band and Beyond", Karl Hess ed., Kluwer Academic, Boston (1991), p. 123.
20. P. D. Yoder, Ph.D. Thesis, University of Illinois at Urbana-Champaign, 1994.
21. T. Kunikiyo, M. Takenaka, Y. Kamakura, M. Yamaji, H. Mizuno, M. Morifuji, K. Taniguchi and C. Hamaguchi *J. Appl. Phys.* 75:297 (1994).
22. E. Cartier, M.V. Fischetti, E.A. Eklund, and F.R. McFeely, *Appl. Phys. Lett.* 62:3339 (1993).
23. Y. Kamakura, H. Mizuno, M. Yamaji, M. Morifuji, K. Taniguchi, C. Hamaguchi, T. Kunikiyo, and M. Takenaka, *J. Appl. Phys.* 75:3500 (1994).
24. J. Kolnik, Y. Wang, I. H. Oguzman, and K.F. Brennan, *J. Appl. Phys.* 76:3542 (1994).
25. N. Sano and A. Yoshii, *J. Appl. Phys.* 75:5102 (1994).
26. J. Bude and M. Mastrapasqua, *IEEE Trans. Electron. Dev.* (to be published, 1995).
27. M.V. Fischetti, S.E. Laux, and E. Crabbé, *J. Appl. Phys.* 78:1058 (1995).
28. T.Kamata, K. Tanabashi, and K. Kobayashi, *Japn. J. Appl. Phys.*, 15:1127 (1975).
29. S. M. Sze, "Physics of Semiconductor Devices", Wiley, New York (1981).
30. B. Dierickx, L. Warmerdam, E. Simoen, J. Vermeiren, and C. Claeys, *IEEE Trans. Electron. Dev.* 35:1120 (1988).
31. V.M. Agostinelli Jr., K. Hasnat, T.J. Bordelon, D.B. Lemersal, A.F. Tasch, and C.M. Maziar, *Solid-State Electron.* 37:1627 (1994).
32. A. Lacaita, *Semicond. Sci. Technol.*, 7:B590 (1992).
33. C.C.C. Leung and P.A. Child, *Appl. Phys. Lett.*, 66:162 (1995).
34. S. Tam, F.-C. Hsu, C. Hu, R.S. Muller, and P.-K. Ko, *IEEE Electron Device Lett.*, EDL-4:249 (1983).
35. L. Machanda, R.H. Storz, R.H. Yan, K.F. Lee, and E.H. Westerwick, *IEDM Tech. Dig.*, p. 994 (1992).
36. B. Eitan, D. Frohman-Bentchowsky, and J. Shappir, *J. Appl. Phys.*, 53:1244 (1982).
37. A.K. Henning, N.N. Chan, J.T. Watt, and J.D. Plummer, *IEEE Trans. Electron. Dev.*, ED-34:64 (1987).
38. D. Esseni, L. Selmi, R. Bez, E. Sangiorgi, and B. Riccó, *IEDM Tech. Dig.*, p. 307 (1994).
39. N. Sano, M. Tomizawa, and A. Yoshii, Impact ionization in submicron and sub-0.1 micron Si-MOSFETs, this volume.
40. J. Slotboom, G. Streutker, G.J.T. Davids, and P.B. Hartog, *IEDM Tech. Digest* p. 494 (1987).
41. T.H. Ning, C.M. Osburn, and H.N. Yu, *J. Appl. Phys.* 48:286 (1977).
42. C. Fiegna, E. Sangiorgi, and L. Selmi, *IEEE Trans. Electron. Dev.*, 40:2018 (1993).
43. G. Jin and R.W. Dutton, *J. Appl. Phys.* (to be published, Sept. 1 1995).
44. Z.A. Weinberg, *Solid-State Electron*, 20:11 (1977).
45. Z.A. Weinberg and A. Hartstein, *Solid-State Commun.*, 20:179 (1976).
46. P.A. Serena, J.M. Soler, and N. Garcia, *Phys. Rev. B*, 34:6767 (1986).
47. G. Binning, N. Garcia, H. Rohrer, J.M. Soler, and F. Flores, *Phys. Rev. B*, 30:4816 (1984).
48. K. Puri and W.L. Schaich, *Phys. Rev. B*, 28:1781 (1983).
49. K. Hübner, *in* "The Physics of SiO₂ and its Interfaces", S.T. Pantelides ed., Pergamon, New York, (1978), p. 111.
50. E.J. Grunthaner, P.J. Grunthaner, M.H. Hecht, and D. Lawson, *in* "Insulating Films on Semiconductors", J.J. Simonne and J. Buxo ed., North-Holland, Amsterdam (1985), p. 1.
51. M. Büttiker and R. Landauer, *Phys. Rev. Lett.*, 49:1739 (1982).
52. M.V. Fischetti, D.J. DiMaria, S.D. Brorson, T.N. Theis, and J.R. Kirtley, *Phys. Rev. B* 31:8124 (1985).

# NOISE-TEMPERATURE SPECTRUM OF HOT ELECTRONS IN $n^+nn^+$ DIODES

Viktoras Gružinskis[1], Jevgenij Starikov[1], Pavel Shiktorov[1],
Lino Reggiani[2] and Luca Varani[3]

[1] Semiconductor Physics Institute, Goštauto 11, 2600 Vilnius, Lithuania
[2] Dipartimento di Scienza dei Materiali ed Istituto Nazionale di Fisica della Materia, Università di Lecce, Via Arnesano, 73100 Lecce, Italy
[3] Centre d' Electronique de Montpellier, Universitè Montpellier II, 34095 Montpellier Cedex 5, France

## INTRODUCTION

One of the most used methods to investigate electronic-noise in semiconductor two-terminal structures at high frequencies ($f \geq 0.5\ GHz$) is the measurement of the noise temperature, $T_n(f)$. It consists in the determination of the "black-body" equivalent-temperature under the condition that, within a given frequency bandwidth $\Delta f$ centered on $f$, the thermal noise-power generated by the "black-body" coincides with the maximum noise-power, $P_n$, generated by the structure and extracted in a matched output-circuit[1]. Then, by definition it is:

$$T_n(f) = \frac{P_n(f)}{k_B \Delta f} \tag{1}$$

where $k_B$ is the Boltzmann constant. The aim of this work is to calculate the frequency and voltage dependence of the noise temperature in a $n^+nn^+$ GaAs diode at a kinetic level, thus allowing for a direct comparison with experimental data[2].

## THEORETICAL MODEL AND RESULTS

The noise temperature spectrum $T_n(f, U)$ is calculated from the relation:

$$T_n(f, U) = \frac{S_I(f, U)}{4k_B Re[Y(f, U)]} \tag{2}$$

where $S_I(f, U)$ is the spectral density of total current fluctuations, $U$ the constant voltage applied between the diode terminals, and $Re[Y(f, U)]$ the real part of the diode small-signal admittance. The validity of Eq. (2) in the whole frequency range is subjected to the constraint $Re[Y(f)] > 0$. As a rule, the above situation is realized for applied voltages slightly below the threshold for the Gunn-effect to occur. The two quantities defining $T_n$ in Eq. (2) are calculated for a GaAs $n^+nn^+$ diode at the lattice temperature $T = 300\ K$, being this an interesting prototype of more complex structures, using an ensemble Monte Carlo simulator for $S_I(f)$ [3] and a closed hydrodynamic approach for $Y(f)$ [4]. This mixed scheme combines the advantages of consistency and numerical precision for the determination of different quantities, which are obtained within an accuracy of 10 % for $S_I(f)$ and of 1 % for $Y(f)$. Results are reported in Figs. 1 to 4 for the case of a diode, sufficiently long to prevent overshoot phenomena, with the following parameters: length of the $n$-region 7.5 $\mu m$, cathode and anode lengths of 0.5 $\mu m$ each, doping concentration in the $n$- and $n^+$-regions of $10^{15}\ cm^{-3}$ and $2 \times 10^{16}\ cm^{-3}$, respectively. Since the diode becomes unstable at $U > 3.1\ V$, the following analysis is limited to voltages below this threshold value. The spectra of $Re[Y(f)]$ and $S_I(f)$, calculated for several values of the applied voltages, are shown in Figs. 1 and 2, respectively. At $U < 1.5\ V$, both $Re[Y(f)]$ and $S_I(f)$ remain nearly flat up to frequencies of about 100 $GHz$ and then begin to decrease towards a $1/f^2$ slope. The appearance of a resonant peak at $f = 1.2\ THz$ reflects the existence of an additional noise source related to the plasma frequency of the $n$ and $n^+$ regions. A similar peak is also observed in the frequency dependence of $Re[Y(f)]$. Since electron heating at the homojunctions is negligible, the frequency behavior of $Re[Y(f)]$ and $S_I(f)$ at the resonant peak is practically independent from the applied voltage. When $U$ approaches the threshold value, an additional resonant behavior of $Re[Y(f)]$ and $S_I(f)$ appears in the intermediate frequency range $f = 20 \div 30\ GHz$. At $U = 2.8\ V$ (i.e. slightly below the threshold value) $Re[Y(f)]$ exhibits negative values in a narrow frequency range where a pronounced peak of $S_I(f)$ is found (see Fig. 2, the curves for $U \geq 2.8\ V$). This second peak is associated with spontaneous oscillations of the current at the transit-time frequency, which are caused by the spontaneous formation of Gunn-domains and their subsequent drift through the $n$-region. The oscillations are also responsible for a considerable increase of $S_I$ in the low-frequency range of the spectrum (see Fig. 2). Figure 3 shows the $T_n(f)$ spectrum calculated from Eq. (2) using the results presented in Figs. 1 and 2. Figure 4 reports the $T_n(f)$ results of Fig. 3 at 10 $GHz$ together with available experimental data[2]. We have found an excellent agreement between theory and experiments that proves the soundness of the theory here developed and the reliability of experiments. At $U < 0.4\ V$, $T_n(f)$ is practically independent of frequency (see Fig. 3, solid curve) and equals the lattice temperature thus fulfilling the Nyquist theorem (see Fig. 4). Then, a systematic increase of $T_n$ with $U$ is observed. In the intermediate region $0.4 < U < 1.5\ V$ the increase of $T_n$ is smooth and mostly associated with the decrease of $Re[Y(U)]$. At $U \geq 1.5\ V$, heating is sufficient for carrier transfer to upper valleys. Accordingly, the increase of $T_n$ with $U$ becomes steeper. As it follows from Fig. 3, such an increase takes place mostly in the low-frequency region of the $T_n$ spectrum at frequencies below the transit-time value $f = 30\ GHz$. At higher frequencies, $T_n$ decreases reaching values comparable with the electron temperature. By comparing Figs. 1-4, we conclude that the steeper increase of $T_n$ at low-frequency is associated with both a decrease of $Re[Y(U)]$ and an increase of $S_I(U)$. For $U \leq 2.6\ V$, the increase of $S_I(U)$ plays a minor role when compared with the decrease of $Re[Y(U)]$, while above 2.6 $V$ also the increase of $S_I(U)$ becomes quite significant, thus contributing to a further increase in the slope of $T_n$ with $U$. Accordingly, due to a resonance behavior of both $Re[Y(f)]$ and $S_I(f)$ near the transit-time frequency $T_n(f)$ exhibits a spike which quickly goes to infinity so that $T_n(f)$ becomes undefined inside the amplification band $f = 20 \div 30\ GHz$. Near $U = 3.1\ V$ the slope of $T_n(U)$ becomes practically vertical (see Fig. 4). This corresponds to the onset of microwave power self-generation at the frequency of about 25 $GHz$. Above this threshold voltage $T_n(f)$ becomes undefined in the whole frequency range.

## ACKNOWLEDGMENTS

This work is performed within the European Laboratory for Electronic Noise (ELEN) supported through the contract ERBCHRXCT920047. Partial support from CNRS PECO/CEI/2003 bilateral cooperation project is gratefully acknowledged.

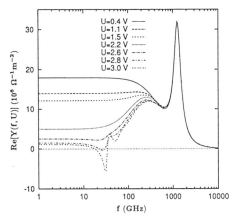

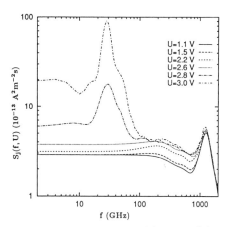

Fig. 1 - Frequency dependence of the real part of the small-signal admittance for a $n^+nn^+$ GaAs diode at $T = 300\ K$ with $n$- and $n^+$-region doping of $10^{15}\ cm^{-3}$ and $2 \times 10^{16}\ cm^{-3}$, respectively; the lengths of the cathode, $n$-region and anode are of 0.5, 7.5 and 0.5 $\mu m$.

Fig. 2 - Frequency dependence of the spectral density of current fluctuations for the same diode of Fig. 1.

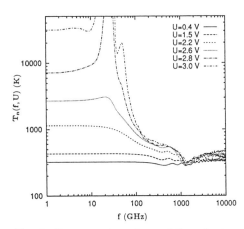

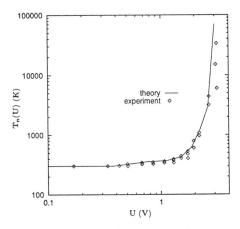

Fig. 3 - Frequency dependence of the noise temperature calculated using Eq. (2) from the admittance and noise spectra presented in Figs. 1 and 2, respectively.

Fig. 4 - Voltage dependence of the noise temperature at $f = 10\ GHz$ in the same diode of Fig. 1. Symbols refer to experiments and curve to theoretical calculations performed within a Monte Carlo and hydrodynamic scheme.

# REFERENCES

1. J.P. Nougier, in III-V Microelectronics, Ed. by J.P. Nougier (Elsevier Science Publ. B.V. 1991) p. 183.
2. V. Bareikis, J. Liberis, I. Matulioniene, A. Matulionis, P. Sakalas, IEEE Trans. Electron. Devices, ED-41, 2050 (1994).
3. V. Mitin, V. Gružinskis, E. Starikov, P. Shiktorov, J. Appl. Phys., 75, 935 (1994).
4. V. Gružinskis, E. Starikov, P. Shiktorov, L. Reggiani, M. Saraniti, L. Varani, Lith. J. Phys., 32 Suppl., 169 (1992).

# 2-DIMENSIONAL MOSFET ANALYSIS INCLUDING IMPACT IONIZATION BY SELF-CONSISTENT SOLUTION OF THE BOLTZMANN TRANSPORT AND POISSON EQUATIONS USING A GENERALIZED SPHERICAL HARMONIC EXPANSION METHOD

W-C Liang, Y-J. Wu, K. Hennacy, S. Singh,
N. Goldsman, and I. Mayergoyz

Department of Electrical Engineering
University of Maryland, College Park, MD 20742

## INTRODUCTION

We present a new 2-D MOSFET simulation tool which employs a spherical harmonic expansion to deterministically solve the Boltzmann transport equation (BTE), and thereby provide the distribution function for the entire device. The unique aspects of the approach are: (i) The spherical harmonic formulation of the BTE is performed to arbitrarily high order; (ii) Self-Consistent deterministic solution of the BTE and Poisson equations is achieved for the 2-D MOSFET structure; (iii) Impact Ionization is included to provide values for substrate current which are based on the distribution function. (iv) Self-consistently determining the distribution function and potential requires approx. 30 minutes on a Spark5. (v) Solving the 2-D Boltzmann equation alone (as a post-processor) requires approx. one minute on a Spark5, whereas similar Monte Carlo calculations are significantly longer.

The device model consists of the BTE for electrons, the current-continuity equation for holes and the Poisson equation:

$$\nabla_{\mathbf{r}}^2 \phi(\mathbf{r}) = \frac{e}{\epsilon_s} \left[ \frac{1}{4\pi^3} \int f(\mathbf{k}, \mathbf{r}) d\mathbf{k} - p(\mathbf{r}) + N_A(\mathbf{r}) - N_D(\mathbf{r}) \right] \tag{1}$$

$$\frac{1}{\hbar} \nabla_{\mathbf{k}} \varepsilon \cdot \nabla_{\mathbf{r}} f(\mathbf{k}, \mathbf{r}) + \frac{e}{\hbar} \nabla_{\mathbf{r}} \phi(\mathbf{r}) \cdot \nabla_{\mathbf{k}} f(\mathbf{k}, \mathbf{r}) = \left[ \frac{\partial f(\mathbf{k}, \mathbf{r})}{\partial t} \right]_{phon} + \left[ \frac{\partial f(\mathbf{k}, \mathbf{r})}{\partial t} \right]_{imp-ion} \tag{2}$$

$$\nabla_{\mathbf{r}} \cdot [\mu_p p(\mathbf{r}) \nabla_{\mathbf{r}} \phi(\mathbf{r}) + \mu_p V_t \nabla_{\mathbf{r}} p(\mathbf{r})] = R(\phi, n, p) \tag{3}$$

The unknowns in the above system are distribution function $f(\mathbf{k}, \mathbf{r})$, the potential $\phi(\mathbf{r})$, and the hole concentration $p(\mathbf{r})$. We use the band-structure of and transport model of

[1,2]. The first collision term accounts for acoustic, optical and intervalley phonon scattering. The impact ionization rate is calculated using the random-k approximation[3].

## 2-D FORMULATION TO ARBITRARILY
## HIGH-ORDER AND NUMERICAL APPROACH

For steady-state 2-D MOSFET simulation, the BTE is a 5-dimensional integro-differential equation, and is therefore extremely difficult to solve. Using the SH expansion method, the BTE is reduced into a 3-dimensional system of differential-difference equations which is tractable for MOSFET simulation. In contrast to other recent works, which were based on a 1st order SH expansion[4,5], we have generalized the expansion approach and formulated the BTE to arbitrarily high-order SH accuracy. With this approach, the momentum distribution function is expressed in terms of an infinite series of spherical harmonics: $f(\vec{r}, \vec{k}) = \sum_{l=0}^{\infty} \sum_{m=-l}^{l} f_l^m(\vec{r}, \varepsilon) Y_l^m(\theta, \phi)$. Here, $f_l^m(\vec{r}, \varepsilon)$ represent the unknown expansion coefficients; and the spherical harmonics basis functions $Y_l^m(\theta, \phi)$ provide the angular dependence of the distribution function. By substituting the spherical harmonics expansion of the distribution function into the BTE[6], taking advantage of the recurrence and orthogonal relationships between spherical harmonics, and changing the variable energy $\varepsilon$ to Hamiltonian $H$[5], we derive the following expression for the unknown coefficient $f_l^m(\vec{r}, H)$. The powerful aspect of the technique is that the equation for each $f_l^m(\vec{r}, H)$ has the same form, and can therefore be automatically generated to arbitrarily high order.

$$\sum_{i=1}^{3} \sqrt{\frac{2}{m}} \frac{\sqrt{\gamma}}{\gamma'} \left\{ \left[ \frac{\partial}{\partial x_i} - qE_i \frac{\gamma'}{\gamma} \frac{l-1}{2} \right] \hat{a}_i^- + \left[ \frac{\partial}{\partial x_i} + qE_i \frac{\gamma'}{\gamma} \frac{l+2}{2} \right] \hat{a}_i^+ \right\} f_l^m = \left[ \frac{\partial f_l^m}{\partial t} \right]_{coll} \quad (4)$$

where $\gamma$ is the dispersion relation; $E_i$ is the electric field in the $i$-axis direction; $\hat{a}_i^-$ and $\hat{a}_i^+$ are the lowering and raising operators which account for the coupling between the equations for the other coefficients.

To solve the SH-expanded BTE, we transform the BTE into a self-adjoint form, and then perform a Scharfetter-Gummel-type discretization. We then overcome problems typically associated with 3-dimensional calculations by using a fixed point SOR iterative solution technique. This method avoids direct solution of large matrix equations (and is easily parallelized). The 2-D Poisson and Hole Continuity equations are solved with standard methods. The Boltzmann-Poisson-Hole Continuity system is solved self-consistently using a decoupled Gummel-type iterative process.

## SIMULATION RESULTS FOR A 2-D MOSFET

We performed example calculations on a $0.5\mu m$ channel length nMOSFET. We show the results for an applied bias of $V_{ds} = 2V$ and $V_{gs} = 3V$. While we have not made quantitative comparisons, our results for substrate current and distribution averages are in qualitative agreement with typical experimental and hydrodynamic simulations[7], respectively. Fig. 1, shows the electron concentration within the device, which is obtained by numerical integration of the distribution function. Fig. 2 shows the behavior of the energy distribution function along the MOSFET channel. As would be expected, the distribution function is non-Maxwellian in the high-field drain region. Fig. 3 shows the distribution function along a plane in the substrate. Since the solution in the low-current substrate region must be Maxwellian, the Maxwellian form obtained provides a check as to the accuracy of the calculation. Fig. 4 shows the electron

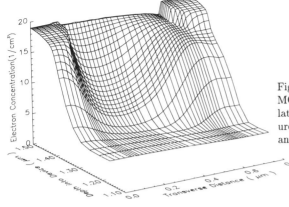

Fig. 1: Electron concentration in a $0.5\mu m$ MOSFET with $V_{gs} = 3V, V_{ds} = 2V$ calculated with the BTE device simulator. (Figures 2 through 5 refer to the same device and bias.)

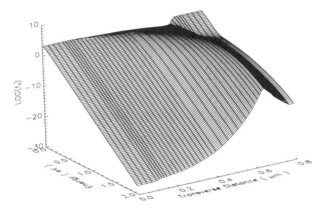

Fig. 2: Calculated distribution function along MOSFET channel. The non-Maxwellian shape quanitfies hot-electron phenomena near drain.

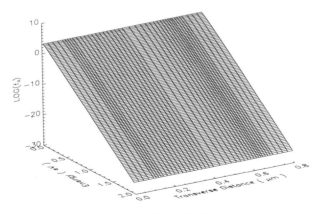

Fig. 3: Calculated distribution function in MOSFET substrate. The Maxwellian shape indicates near equilibrium operation in this region.

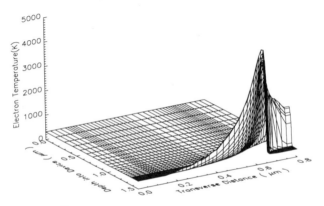

Fig. 4: Electron temperature calculated by directly integrating the variance of the distribution function.

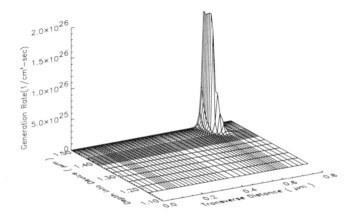

Fig. 5: Carrier generation rate due to impact ionization calculated by integration of collision integral over momentum space.

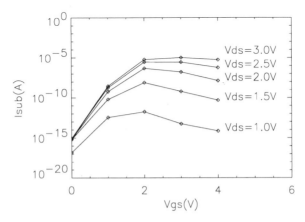

Fig. 6: MOSFET total substrate current due to impact ionization, over a wide range of biases.

temperature which was obtained through an integration of the distribution function. Fig. 5 shows the electron-hole pair generation rate from impact ionization. Fig. 6 shows the simulated substrate current resulting from impact ionization.

[1] R. Brunetti, C. Jacoboni, *Sol.-State. Elec.*, v. 32, no. 12, pp. 1663–1667, 1989.

[2] C. Fiegna, E. Sangiorgi, *IEEE Trans. Elec. Devs.*, v. 40, no. 3, pp. 619–627, 1993.

[3] E. O. Kane, *Physical Review*, v. 159, p. 624, 1967.

[4] H. Lin, N. Goldsman, I. Mayergoyz, *Sol.-State Elec.*, v. 35, no. 6, pp. 769, 1992.

[5] D. Venture, A. Gnudi, G. Baccarani, F. Odeh, *Sol.-State Elec.*, v. 36, no. 4, pp. 575, 1993.

[6] K. Hennacy, PhD. Thesis, University of Maryland, 1993.

[7] Q. Lin, N. Goldsman, G. Tai, *Sol.-State. Elec.*, v. 37, no. 2, pp. 359, 1994.

# 3D Discrete Dopant Effects on Small Semiconductor Device Physics

J-R. Zhou[*] and D. K. Ferry
Department of Electrical Engineering

Arizona State University, Tempe, AZ 85287-6206

## INTRODUCTION

The requirements of high speed and performance and cost effectiveness demand of VLSI chips a continuing push in miniaturization. As a consequence, the design rule (or the effective gate length) has been reduced from several microns down to < 0.1 μm envisioned within a few years. While the expected 256 Mb chip requires the use of quarter micron design rules, it is expected that we will need 0.1-0.15 μm rules for gigabit chips, and the extrapolation of the down scaling trends will extend the design rules to sub-0.1 μm for a terabit memory chip in the near future. With this scenario for VLSI technology development, the deep understanding of the device physics governing device operation is the key to successful device design, as different physical mechanism impose different level of effects on different device scales.

In general, semiconductor device operation depends on the use of potential barriers (such as the gate) to control carrier transport through the device. Although a successful device design is quite complicated and involves many aspects, the device engineering mostly addresses optimal device structure and manipulates impurity profiles to obtain best control of the carrier flow though the device. This becomes increasingly difficult as the device scale becomes smaller. New problems keep occurring which hinder the high performance requirement. Well known problems include hot carrier effects, short channel effects, etc. Here, we discuss a potential problem caused by impurity fluctuations which can not be perfectly controlled as devices become too small.

Impurities in semiconductor devices are randomly distributed during the processing. Although electron transport in the devices always experiences the effect of a random distribution of the impurities, the statistical contribution of these effects to the electronic performance of large devices is negligible, due to ensemble averaging, and a simplified uniform background impurity distribution (the average of the impurity charges in space) is adequate to describe the effect of the fixed charge in the devices. Only devices with a small active volume are susceptible to large fluctuations of the local impurity

---

[*] Present address: VLSI Technology, San Antonio, Texas

concentration, and these will have noticeable conductance fluctuations as a result. As device scaling continues to the deep submicron regime, the active device region will contain so few dopant atoms that the statistical fluctuation of the dopant, either in total number or in spatial distribution in the device, will cause non-negligible effects on device performance. The anticipated effects include: 1) quantum mechanical effects due to the conservation of phase interference of the electron waves propagating through the semiconductor such as universal conductance fluctuation, weak localization,[1] etc.; 2) classical statistical fluctuations, such as current level shift and threshold voltage shift due to the total dopant number fluctuation and/or distribution. A full description of the quantum mechanical effects is beyond the scope of this paper. Only a few attempts have been devoted to study the effect of random atomistic impurities on device performance, and the initial research[2] us es a drift-diffusion model to simulate the random impurity effect on sub-0.1 μm MOSFET devices. We investigate the classical effects from the simulations of 3-dimensional device structure of HEMTs by using hydrodynamic equations,[3] with the discrete 3-dimensional random impurity distribution and fluctuation included.[4]

## DEVICE SIMULATION

The discrete impurity region is only defined in the highly-doped layer, which in the present case is a δ-doped layer in the GaAlAs. The charge in a discrete cell is set to be either one or zero following a distribution scheme. By doing this, it is presumed that the ionized charge is a point charge. For the grid size used (3 nm), the cell size is much larger than the atomic size, thus the point charge model is reasonable. However, the accuracy in the placement of the impurities is of the order of the grid size, since a realistic distribution may have impurities fall between grid points, regardless what grid scheme is used. To integrate Poisson's equation numerically with the point charge distribution, some approximation is certainly required when approaching the point charge, and we do not expect to numerically integrate the delta function impurity location exactly. The corresponding uniform doping in the highly-doped layer corresponds to $1.5 \times 10^{18}$ cm$^{-3}$. The total number of dopant in the discrete region is determined by taking the total charge in the region divided by a single ion charge. The distribution of the discrete charge is performed by using a series of random numbers. When the random number for a cell is greater than $1-r$, where $r$ is the ratio of the total number of dopants in the region to the total discrete cells in the region, the cell is assigned an ion charge. This scheme satisfies that the requirement that the mean of the total discrete charge equal the uniform doping. For an ensemble of 5000 different initializations, of a device with 42 nm gate length, the number of impurities under the gate has a mean value of 36 (approximately one tenth of the mean value for the total discrete impurities, since the volume of the gate region is one tenth of the volume of the total discrete region), with a standard deviation of 5.99. It is clear that there will be a corresponding variation in the properties of the devices themselves. The device simulation method is the same as that we used previously,[3] which solves a set of hydrodynamic equations that describes the conservation of particle, momentum and energy, in conjunction with Poisson's equation.

The simulated device structure is a domain of 0.36 μm (L) × 0.1 μm (H) × W, with W in the range of 0.042 μm to 0.162 μm. The thickness of the GaAlAs layer is 40 nm, and results for two different gate lengths are discussed here. Figures 1-3 illustrate the results for a δ-doped HEMT. The structure is for a width of 42 nm with a gate length of 24 nm. Here, as in the remainder of this paper, the bias potentials are $V_D$=1.5 V and $V_G$=0.3 V. In Fig. 1, the density in the channel is plotted for conditions near pinch-off, as may be seen from the very small amount of density under the gate. Note the very inhomogeneous nature of the channel electron density, a direct result of the random nature of the doping. In Fig. 2, the velocity in the channel is illustrated, while Fig. 3 depicts the electron temperature. The inhomogeneities of the density are clearly seen to carry through to the transport parameters.

## 3D HEMT: The 2D channel electron density showing random impurity effect

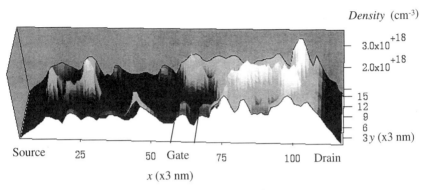

**Figure 1.** Density in the channel for a δ-doped HEMT.

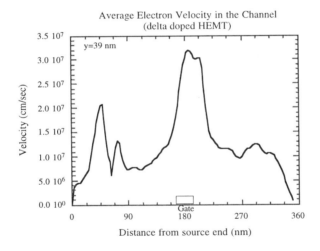

**Figure 2.** Electron velocity in the channel for the device of Fig. 1.

## EXCHANGE AND CORRELATION POTENTIALS

Inclusion of a full quantum mechanical description in the device simulation level is generally impractical. However, the density functional approach[5,6] gives a simple way of including many-body exchange and correlation effects in the calculation of electronic structure, and has been found to be successful in many electronic structure applications. In general, the so-called exchange-correlation potential engergy introduced appears in the potential energy as an unknown functional of the electron density. In reality, however, the simplest approximation to the exchange-correlation potential, the local-density-functional

(LDA) approximation, works quite well. There are a number of different forms of the LDA potential, which give similar results. We implement here one introduced by Stern and Das Sarma.[6] The exchange potential energy per volume is given as

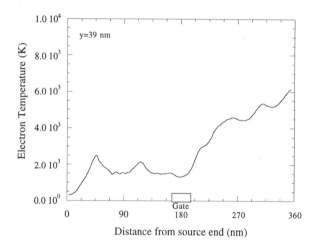

**Figure 3.** Electron temperature in the channel for the device of Fig. 1.

$$V_{ex}(\mathbf{r}) = - C_X \, n(\mathbf{r})^{4/3} \tag{1}$$

where $n(\mathbf{r})$ is the local density, $\mathbf{r}$ is the spatial coordinate, and

$$C_X = \frac{3e^2}{16\pi\varepsilon}\left(\frac{3}{\pi}\right)^{1/3}$$

is an estimated prefactor for best fit to known energy levels. Then, the average force on a single electron generated by the exchange potential energy change at a particular position $\mathbf{r}$ is given by

$$\mathbf{F}(\mathbf{r}) = -\frac{1}{n(\mathbf{r})} \, \nabla_r V_{ex}(\mathbf{r}) = C_X \, \nabla_r n(\mathbf{r})^{4/3}. \tag{2}$$

By including this force into the hydrodynamic equations, the expected exchange-correlation effects are oberved for our 3D device simulation. Figure 4 shows the change in the density distribution of a 24 nm gate AlGaAs/GaAs HEMT device, by comparing the density along the channel for the case with and without the exchange potential. This effect is rather small, as the exchange energy works mainly to increase the local self-energy (deepen the potential wells) and this is "tuned out" in some sense by the self-consistent potential solutions. However, in Fig. 5, the velocity along the channel shows a large effect as a result of the introduction of the exchange potential, again because local fields are heavily modified by the existence of this term.

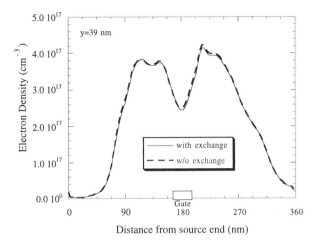

**Figure 4**. Carrier density in the channel along one longitudinal cut, illustrating the effect on the density of the exchange and correlation corrections.

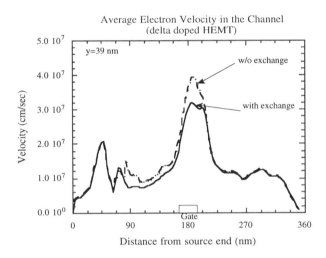

**Figure 5**. Electron velocity in the channel along one longitudinal cut, illustrating the effect on the velocity of the exchange and correlation corrections.

## CONCLUSION

We have investigated the effect of random impurity fluctuations and distribution on small device operations. For the device structure simulated here, the results suggest that the effects of random impurity fluctuation and distribution can cause current fluctuation as large as 50 per cent for small devices if the total gate area is very small. As expected, the random impurity effect on the device performance is reduced with an increase in the gate length or gate width. But the effect can be effective even for devices with gate lengths close to 0.1 μm. A full 3-D simulation of the device including the random impurity and the role of the local exchange and correlation potentials is important in ultra-submicron device modeling. We feel that a combination of limited 3-D simulation plus statistical method might be useful in providing an applicable method to estimate the random impurity effect on device performance as device down scaling to 0.1 μm range.

## Acknowledgments

The authors would like to thank E. Zaremba (Queen's University, Ontario), H. L. Grubin, and D. Vasileska for many helpful and stimulating discussions. This work is supported by the Office of Naval Research.

## REFERENCES

1.  D. K. Ferry, Y. Takagaki and J. R. Zhou, "Future ULSI: Transport Physics in Semiconductor Nanostructures," *Jpn. J. Appl. Phys*. 33:873 (1994).
2   H.-S. Wong and Y. Taur, "Three-Dimensional 'Atomistic' Simulation of Discrete Random Dopant Distribution Effect in Sub-0.1μm MOSFET's," *IEDM*, p. 705, 1993.
3   J.-R. Zhou and D. K. Ferry, "Simulation of ultra-small GaAs MESFET using quantum moment equations," *IEEE Trans. Electron Devices* 39:473 (1992).
4.  J.-R. Zhou and D. K. Ferry, "3D Simulation of Deep-Submicron Devices: How Impurity Atoms Affect Conductance," *Comp. Sci. Engr*. 2(2):30 (1995).
5.  E. Zaremba and H. C. Tso, "Thomas-Fermi-Dirac-von Weizsacher hydrodynamics in parabolic wells," ***Phys. Rev. B***. 49:8147 (1994).
6.  F. Stern and S. Das Sarma, "Electron energy levels in GaAs-Ga$_{1-x}$Al$_x$As heterojunctions," ***Phys. Rev. B***. 30:840 (1994).

# MICROSCOPIC ANALYSIS OF NOISE BEHAVIOR IN SEMICONDUCTOR DEVICES BY THE CELLULAR AUTOMATON METHOD

A. Rein, G. Zandler, M. Saraniti, P. Lugli and P. Vogl

Physik Department and Walter Schottky Institut
TU München, D-85747 Garching, FRG

## INTRODUCTION

Noise is one of the crucial features in modern semiconductor devices. Nevertheless, only a few microscopic investigations of intrinsic noise behavior of semiconductor devices in the GHz regime have been performed in the past [1]. Although the dependency of intrinsic noise on the doping concentration, applied voltages and the influence of hot carrier effects is an important issue, systematic investigations are hampered by the fact that very long time sequences have to be simulated for the calculation of the relevant correlation functions. For semiconductor devices, such simulations are usually performed with the Monte Carlo method, self-consistently coupled to a Poisson equation. In [2], we have shown a highly efficient cellular automaton (CA) approach to be physically equivalent to the Monte Carlo technique for macroscopic transport quantities. However, our initial implementation of the CA method showed discrete lattice effects that led to a slightly enhanced carrier diffusion and hampered an accurate calculation of correlation functions. We have recently developed a new scheme for the CA that eliminates this problem.

The basic features of this new CA approach are sketched briefly in the following section. Then we discuss the general behavior of current autocorrelation functions in highly doped devices. In particular, results for the high frequency noise behavior in Si-MOSFET's are presented.

## NEW DEVELOPMENTS IN THE CELLULAR AUTOMATON APPROACH

Recently, the full Boltzmann equation (BE) has been transformed into a CA, where the kinetic terms of the BE are replaced by hopping probabilities in such a way that the equations of motion are fulfilled on the average for an ensemble of quasi particles [2]. In an explicit procedure, the drift term of the BE has been transformed into probabilistic field scattering rates. This corresponds to a substitution of the free flight by a random walk. For high electric fields, this procedure leads to artificial diffusion effects on the k-space lattice. Associated with this diffusion is an enhancement of kinetic energy, entropy and longitudinal real space diffusion. In principle, this error can be reduced by a sufficiently small lattice constant $\Delta k$. Nevertheless, to study fluctuations accurately, the memory requirements become unacceptable for a three dimensional momentum space.

Fortunately, it is possible to transform the drift term of the BE into a set of *deterministic* scattering rules of the CA. This procedure completely suppresses this statistical error. The main point is to replace the probabilistic scattering rate by a discrete free flight. We derive these scattering rules by calculating the number of time steps $N$ that a particle needs to change its momentum by an amount equal to the lattice constant $\Delta k$ in k-space. This procedure confines the statistical error to one k-cell. Therefore, only of the order of $10^3$ 3-D k-cells are required for a nonparabolic band structure up to 2 eV. The lattice we have chosen is a hexagonal closed-packed structure, where each cell has twelve nearest neighbors. A rigorous proof of this new approach will be published elsewhere. Importantly, we found that the new implementation of the CA does not require more computer time per iteration than our earlier two-dimensional implementation [2]. For bulk silicon at room temperature, the speed-up in computer time, compared to a Monte Carlo method that includes the momentum and real space dynamics, is approximately 20 + 2.5F, where F is the electric field in units of 100 kV/cm.

In order to study the noise behavior of a stochastic variable $A = \langle A \rangle + \delta A(t)$, the quantity of main interest is the spectral density $S_A(\omega) = \int_{-\infty}^{+\infty} C_A(t) e^{i\omega t} dt$, where $C_A(t) = \langle \delta A(t) \delta A(0) \rangle$ denotes the corresponding autocorrelation function. To test the new CA approach in determining correlation functions, a comparison with Monte Carlo (MC) results has been carried out. In Fig. 1(a), we depict the calculated normalized autocorrelation function of the drift velocity fluctuations as a function of time. Excellent agreement between both techniques is found. In particular, the negative part in the correlation function that reflects a pronounced hot carrier effect is accurately reproduced by the CA method.

## NOISE IN HIGHLY DOPED SEMICONDUCTOR DEVICES

### General aspects

It is well established [3] [4], that noise correlation functions can exhibit damped oscillations already at equilibrium. This behavior originates in plasma oscillations in highly doped regions. In contrast to the voltage autocorrelation function, the (vectorial) fluctuations in the current can only couple to the (scalar) density fluctuations if the system is spatially inhomogeneous and contains, for example, a depletion region. However, the detailed conditions for the formation of these microscopic current oscillations have not been investigated so far.

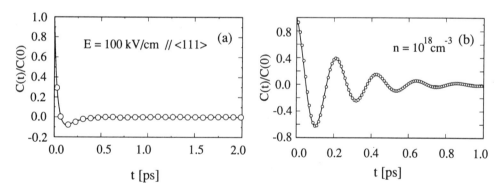

**Fig 1.** (a) Comparison between MC and CA results for the autocorrelation function of the velocity, calculated as a function of time. The material is Silicon at room temperature. The electric field strength equals 100 kV/cm and is applied along the <111>-direction. (b) Comparison between CA simulation and an analytical model for the current correlation function in a n-i junction. The doping concentration in the n-region is $10^{18}$cm$^{-3}$.

The simplest structure containing all basic features of this effect is a n-i junction at equilibrium. Employing the CA method, we have calculated the normalized current correlation function depicted in Fig. 1(a). Hereby, Dirichlet and von Neumann conditions have been applied at the n- and i-side, respectively. The behavior of this correlation function can be understood in terms of a simple driven and damped harmonic oscillator model (cf. [3]). In this model, the n-region oscillates with the plasma frequency $\omega_P \sim \sqrt{n}$ and is damped according to the inverse momentum relaxation time $\tau_m$ that has been obtained from the CA calculations. The random driving force is assumed to have a white noise spectrum and to represent the fluctuating space charge at the n-i junction. Within this model, the current autocorrelation function $C_J(t)$ can be calculated analytically and agrees quantitatively with the numerical results from the CA calculations, as shown in Fig. 1(b). This analytical model yields a minimal carrier concentration for the appearance of these oscillations, namely $\omega_p > 1/(2\tau_m)$. We were able to verify numerically for an n-i-n structure that these oscillations scale as $< (\delta J/J)^2 > \sim 1/N$ where $J$ is the current and $N$ the number of particles. Consequently, these oscillations vanish in the thermodynamic limit. Moreover, we have checked that they also vanish in homogeneously doped devices. In this case, the random force tends to zero.

## Intrinsic current noise in MOSFET's

In this work we are interested in the frequency range around 100 GHz, where velocity and number fluctuations (shot noise) are dominant. In such a frequency range, the noise spectral density of the current in the drain region is independent of frequency and will be denoted by S(0). We have studied the dependence of S(0) on applied voltages for a 0.1 $\mu$m MOSFET. It possesses a doping concentration of $10^{19}$cm$^{-3}$ and $10^{17}$cm$^{-3}$ for the n$^{++}$ region and the p-buffer, respectively, and a gate length of 0.1 $\mu$m and an oxide thickness of 5 nm.

Utilizing the numerically efficient CA method for the carrier dynamics of $10^5$ particles in a MOSFET, the frequent solution of the Poisson equation (each fs) can be a bottleneck of the entire simulation. We have overcome this difficulty by implementing a fast multigrid Poisson solver [5]. Source, drain and gate currents have been monitored by employing the method proposed in [6]. In order to calculate reliable autocorrelation functions for these terminal currents, time sequences of 300 ps were necessary.

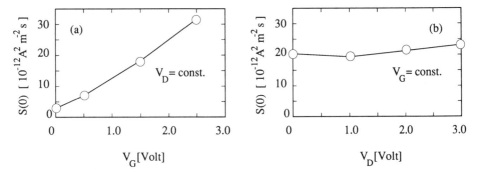

**Fig 2.** Low frequency part of the spectral density of the drain current, S(0), in a 0.1 $\mu$m MOSFET as a function of the applied voltage. (a) S(0) increases linearly with the gate voltage for constant drain voltage. (b) S(0) remains approximately constant as a function of the applied drain voltage for fixed gate voltage $V_G$=1.5 V.

The influence of the gate and drain voltage on the spectral density S(0) is depicted in Fig. 2. We predict a strong enhancement of S(0) with increasing gate voltage but almost no change with drain voltage, in agreement with earlier results for planar GaAs MESFET's [4]. These results can be understood qualitatively as follows. By applying a positive gate voltage, the number of carriers in the channel increases while the mean energy of the carriers remains approximately constant. Therefore, it appears plausible that S(0) increases with increasing gate voltage. In Fig. 3(a), we depict the computed $C_J(t)/C_J(0)$ for different gate voltages, where all current fluctuations due to density and velocity fluctuations and the displacement current have been included in the calculations. The increase in S(0) is seen to stem from a diminishing of the plasma oscillations in the n$^{++}$drain contact region, leading to a less negative $C_J(t)$ in the integral $S(0) = \int_0^\infty C(t)dt$ for a positive gate voltage. By increasing the drain voltage for the same device, (see Fig. 3) only small changes in the current autocorrelation function are detected. To understand the qualitative physics of this finding, we note that the energy of the carriers near the end of the channel increases significantly with drain bias. Within a simple relaxation time model for the velocity autocorrelation function, one obtains $S(0) \propto\, <(\delta v)^2> \tau_m$. With increasing drain voltage, the kinetic energy of the carriers increases whereas the momentum relaxation time decreases. These effects tend to compensate each other and give a small net change of S(0). We note that the diffusion constant in bulk Si actually decreases with increasing field.

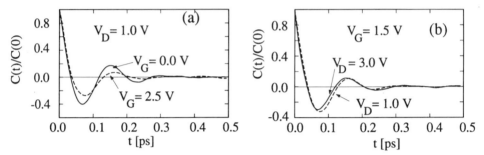

**Fig 3.** Autocorrelation functions of the drain current in a 0.1 $\mu$m MOSFET. (a) The normalized current correlation function for different gate voltages. (b) The same quantity, but for different drain voltages.

## CONCLUSION

In conclusion, we have found the new CA method to be a computationally efficient yet accurate tool to investigate higher order correlation functions that determine the intrinsic carrier noise in semiconductor devices. Oscillations in the current autocorrelation functions of highly doped devices have been explained in terms of plasma oscillations that are driven by the fluctuations of the space charge in the depletion zone. The strong dependence of the drain current noise on gate bias in short channel MOSFET's has been analyzed. It can be explained by the increasing number of carriers with gate bias. We have found no significant influence of hot carrier effects onto the drain current noise. This originates in a compensation between the increase in the carrier energy and the decrease in the momentum relaxation time.

This work was supported by Siemens AG, project SFE II and by the Deutsche Forschungsgemeinschaft (SFB 348).

## REFERENCES

1.  L. Varani and L. Reggiani, "Microscopic theory of electronic noise in semiconductor unipolar structures," *La Rivista del Nuovo Cimento*, vol. 17, no. 7, p. 1, 1994 and references therein.
2.  K. Kometer, G. Zandler, K. Kometer, and P. Vogl, "Lattice-gas cellular automaton method for semiclassical transport in semiconductors," *Phys. Rev. B*, vol. 46, p. 1238, 1992.
3.  J. Zimmermann and E. Constant, "To hot carrier diffusion noise calculation in unipolar semiconducting components," *Solid state electronics*, vol. 23, no. 9, p. 915, 1980.
4.  T. Gonzalez, D. Pardi, L. Varani, and L. Reggiani, "Monte Carlo simulation of electronic noise in MESFET's," in *GAAS 94 proceedings of the european gallium arsenide and related compounds*, p. 385, Politecnico di Torino, Apr. 1994 and references therein.
5.  M. Saraniti, A. Rein, G. Zandler, P. Vogl, and P. Lugli, "An efficient multigrid poisson solver for device simulations," *sub. to IEEE transactions on computer aided design of intergated circuits and systems*, 1995.
6.  V. Gruzinskis, S. Kersulis, and A. Reklatis, "An effcient Monte Carlo particle technique for two-dimensional transistor modelling," *Semicond. Sci. Technol.*, vol. 6, pp. 602–606, 1991.

# SUBPICOSECOND RAMAN STUDIES OF ELECTRON VELOCITY
# OVERSHOOT IN A GaAs-BASED NANOSTRUCTURE SEMICONDUCTOR

E.D. Grann[1], K.T. Tsen[1], D.K. Ferry[2], A. Salvador[3], A. Botcharev[3],
and H. Morkoc[3]

[1]Department of Physics and Astronomy
[2]Department of Electrical Engineering
Arizona State University, Tempe, AZ 85287
[3]Coordinated Science Laboratory
University of Illinois, Urbana, IL 61801

## INTRODUCTION

The recent development of laser sources with picosecond and subpicosecond pulse widths has led to optical methods which can directly probe electron transport phenomena on very short time scales.  In these short time regimes, the transport properties of electrons are known to be quite different from those observed under steady state conditions.[1-4] Understanding these nonequilibrium phenomena is the key to taking advantage of these novel transient transport properties.

A fundamental parameter of interest in a semiconductor is the electron distribution function.  On very short time scales, electrons which are photoexcited in a polar semiconductor such as GaAs exhibit extremely nonequilibrium distributions, far from either a Maxwell-Boltzmann or Fermi-Dirac distribution.  The application of a high external electric field enhances the nonequilibrium nature of the electron distribution.

Raman scattering from single particle excitations has long been known as an effective probe of the velocity distribution of electrons moving under the influence of an electric field in semiconductors.  Mooradian and McWhorter[5] first demonstrated this technique using a Q-switched Nd:YAG laser to probe the velocity distribution of electrons in n-GaAs (n $\approx 10^{15}$ cm$^{-3}$) at liquid helium temperatures and for applied fields up to 2 kV/cm. Ralph and Wolga[6] used single particle scattering (SPS) difference spectra to monitor changes in the electron distribution function in a GaAs-based n$^+$ -n$^-$ -n$^+$ device for fields up to 1 kV/cm.  Grann et al.[7] used transient picosecond Raman spectroscopy to determine electron drift velocities as well as electron distribution functions in undoped, bulk GaAs for a variety of applied electric field intensities, photon energies, and photoexcited electron densities at T $\approx 80$ K.

In this paper we report on the direct determination of the electron distribution functions  in undoped GaAs-based p-i-n nanostructure semiconductor in the presence of high electric fields using transient subpicosecond Raman spectroscopy.  The electron distribution functions and drift velocities have been studied as a function of electric field intensity. All of our experimental results are compared with ensemble Monte-Carlo simulations.

## SAMPLE AND EXPERIMENTAL TECHNIQUE

The GaAs-based p-i-n nanostructure sample was grown by molecular beam epitaxy on a (100)-oriented GaAs substrate. The details of the sample have been described elsewhere.[7]

The laser pulses used in this experiment had an energy of 1.951 eV and a pulse width of $\approx$ 600 fs. They were generated by a double-jet DCM dye laser which was pumped by the second harmonic of a mode-locked Nd:YAG laser operating at 76 MHz.

All of the SPS experiments were conducted at T $\approx$ 80 K in the $Z(X,Y)\overline{Z}$ scattering geometry, where Z=(001), X=(100), and Y=(010). This orientation scatters light from only single particle excitations associated with spin density fluctuations.[8,9] The SPS cross section is inversely proportional to the square of the effective mass; therefore our experiment predominantly probes electron transport in the $\Gamma$ valley even though holes are simultaneously present. In our SPS experiments, photons from the same pulse were used to both excite and probe electrons, hence, the results represent an average over the laser pulse width. The backward scattered light was collected and analyzed by a double monochromator and a CCD detector.

## EXPERIMENTAL RESULTS AND ANALYSIS

First of all, we note that the measured SPS spectra can be easily converted into electron distribution functions by using Eq. (24) of Ref. 10 with the assumption that the effect of electron collisions and the momentum dependence of the matrix elements can be neglected.

Figs. 1 (a) and (b) show the measured electron distribution functions for an electron density of n $\approx 10^{17}$ cm$^{-3}$ and for electric field intensities of 20 and 25 kV/cm, respectively. The results of EMC simulations are also shown. Qualitatively, the fits are reasonably good except for the regions around -1 x $10^8$ cm/sec and 1.2 x $10^8$ cm/sec. The deviation close to 1.2 x $10^8$ cm/sec is most probably due to the specific details of the hyperbolic band assumed in the EMC simulations. The large number of electrons in the negative velocity region (between -0.5 x $10^8$ cm/sec and -1.0 x $10^8$ cm/sec), which are not seen in the Raman experiment, is very likely due to the manner in which electron scattering at the interface is handled in the EMC simulations. In our EMC calculations, we assume that all the electrons that reach the AlAs-GaAs interface on the p-type region of the sample suffer diffusive scattering. It is obvious that if some of these electrons were allowed to suffer backscattering, the fit of the distribution function in the spectral range (from -0.5 to -1.0 x $10^8$ cm/sec) would improve.

The deduced electron drift velocities are shown in Table 1. Also listed are the drift velocities deduced from SPS spectra taken under exactly the same experimental conditions except that the laser pulse width was 3 ps. We notice that the drift velocities deduced from our current subpicosecond Raman experiments are much larger than those from the picosecond Raman measurements. Since the picosecond Raman experiments essentially probed the steady-state electron transport, we conclude that the electron transport under our experimental conditions is in the velocity overshoot regime.

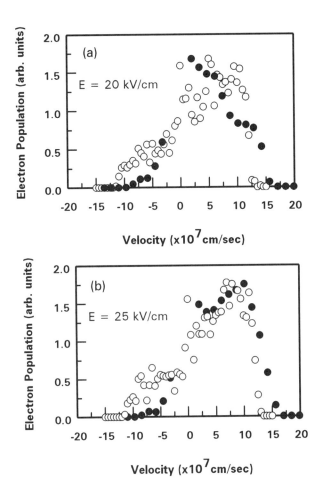

**Figure 1.** Measured electron velocity distributions (closed circles) along the direction of applied electric field are compared with those of ensemble Monte Carlo simulations (open circles) for electric field intensities (a) E= 20 kV/cm; and (b) E= 25 kV/cm, respectively. Qualitatively, the fit is good except for regions around $-1.0 \times 10^8$ and $1.2 \times 10^8$ cm/sec.

**Table 1.** The electron drift velocities deduced from SPS spectra as well as calculated from EMC simulations for different electric field intensities and different laser pulse widths.

| $V_d$ <br> E | Drift Velocity (cm/sec) (600fs) | Drift Velocity (cm/sec) (EMC, 600fs) | Drift Velocity (cm/sec) (3ps) | Drift Velocity (cm/sec) (EMC, 3ps) |
|---|---|---|---|---|
| 20 kV/cm | $(4.5\pm0.8) \times 10^7$ | $(3.1\pm0.3) \times 10^7$ | $(2.5\pm0.5) \times 10^7$ | $(2.2\pm0.2) \times 10^7$ |
| 25 kV/cm | $(5.8\pm0.9) \times 10^7$ | $(3.5\pm0.4) \times 10^7$ | $(2.5\pm0.5) \times 10^7$ | $(2.2\pm0.2) \times 10^7$ |

## CONCLUSION

In conclusion, electron velocity overshoot in a GaAs-based p-i-n nanostructure semiconductor has been observed by transient sub-picosecond Raman spectroscopy. Extremely non-equilibrium electron velocity distributions as well as very high drift velocities associated with such transient transport phenomenon were directly measured. Our experimental results are shown to be in agreement with ensemble Monte Carlo simulations.

## ACKNOWLEDGMENTS

This work is supported in part by the National Science Foundation under Grant No. DMR-9301100 and by the Office of Naval Research.

## REFERENCES

1. E. Constant, Non-steady-state carrier transport in semiconductors in perspective with submicrometer devices, in : "Hot Electron Transport in Semiconductors", L. Reggiani, ed., Berlin, Springer, (1985).
2. J. Shah, and R.F. Leheny, Hot carriers in semiconductors probed by picosecond techniques, in: "Semiconductors Probed by Ultrafast Laser Spectroscopy, Vol.1" R.R. Alfano, ed., Academic Press, New York (1984),
3. J.G. Ruch, Electron dynamics in short channel field-effect transistors, *IEEE Tran. Electro. Devices*, vol. ED-14, pp. 652-654, 1972.
4. T.J. Maloney and J. Frey, Transient and steady-state electron transport properties of GaAs and InP, *J. Appl. Phys.*, vol. 48, pp. 781-787, 1977.
5. A. Mooradian and A.L. McWhorter, Light scattering from hot electrons in semiconductors, in: "Proceedings of the 10th International Conference on the Physics of Semiconductors, S.P. Keller, J.C. Hansel, and F. Stern, eds., U.S. Atomic Energy Commision, Oak Ridge, TN, (1970).
6. S.E. Ralph and G.J. Wolga, Field-induced nonequilibrium carrier distributions in GaAs probed by electronic Raman scattering, *Phys. Rev. B* 42:11353, (1990).
7. E.D. Grann, S.J. Sheih, K.T. Tsen, O.F. Sankey, S.E. Günçer, D.K. Ferry, A. Salvador, A. Botcharev, and H. Morkoc, Transient picosecond Raman studies of high-field electron transport in GaAs-based p-i-n nanostructure semiconductors, *Phys. Rev. B* 51:1631 (1995).
8. M.V. Klein, Electronic Raman scattering, in: "Light Scattering in Solids I", M. Cardona and G. Guntherodt, eds., Berlin, Springer, (1983).
9. G. Abstreiter, M. Cardona and A. Pinczuk, "Light Scattering by Free Carrier Excitations in Semiconductors," in Light Scattering in Solids IV, M. Cardona and G. Guntherodt, eds., Berlin, Springer, (1983).
10. C. Chia, O.F. Sankey, and K.T. Tsen, Theoretical studies of transient Raman scattering of non-equilibrium carriers in semiconductors - effects of carrier collisions, *Mod. Phys. Lett. B* 7:331 (1993).
11. D.K. Ferry, M.J. Kann, A.M. Kriman, and R.P. Joshi, Molecular dynamics extensions of Monte Carlo simulation in semiconductor device modeling, *Comp. Phys. Commun.* 67:119 (1991).

# NUCLEATION OF SPACE-CHARGE WAVES IN AN EXTRINSIC SEMICONDUCTOR WITH NONUNIFORM IMPURITY PROFILE

Michael J. Bergmann,[1] Stephen W. Teitsworth,[1] and Luis L. Bonilla[2]

[1]Department of Physics
Duke University, Durham, NC 27708-0305
[2]Universidad Carlos III de Madrid
Escuela Politécnica Superior 28913 Leganés, Madrid, Spain

## INTRODUCTION

Moving space charge waves are observed experimentally in ultrapure bulk p-type Ge under voltage bias.[1] In many models of bulk semiconductor systems, it is assumed that the impurity profile is spatially uniform.[2] We investigate here the effect that a small notch variation in the impurity profile (e.g., see Figure 1) has on the transition from stationary to periodic behavior in a model of closely-compensated p-type Ge under voltage bias. Closely-compensated samples have a compensation ratio $\alpha$ (ratio of acceptor concentration to donor concentration) $\sim 1$.

We use a standard drift-diffusion description of the hole current, electric-field-dependent generation and recombination of free holes, and Poisson's law to model electrical conduction.[3-5] We formulate the problem as a "reduced equation" in terms of a single spatially-dependent field variable, the dimensionless electric field $E(x,\tau)$, and the dimensionless, spatially-homogeneous current density $J(\tau)$.[4,5]

## BIFURCATION FROM STATIONARY TO PERIODIC BEHAVIOR

At small voltage biases and for a spatially-uniform impurity profile, the electric field is stationary, and outside of a small region near the injecting contact it is uniform.[5] In closely-compensated samples ($1.1 \lesssim \alpha \lesssim 1.8$), the stationary state becomes spatially dependent.[4,5] As the spatially-averaged electric field (voltage divided by sample length) approaches the value for impact-ionization-mediated impurity breakdown, a step-like electric-field profile develops.[5] The larger resistivity of the injecting contact causes a high electric-field region to develop at the injecting contact. At increased voltage,

*Hot Carriers in Semiconductors*
Edited by K. Hess *et al.*, Plenum Press, New York, 1996

the extent of the high-field region grows until it extends beyond a critical length and becomes unstable to periodic pulses.[4,5] The pulses nucleate from the high-field region and decay rapidly as they enter the low-field region of the sample.[5]

From the stationary limit of the reduced equation (see Ref. 4), we construct the stationary-state of the electric field $E_{ss}(x)$,

$$\frac{dE_{ss}(x)}{dx} = \frac{J - J_{sh}(\alpha, E)}{V(E)}, \tag{1}$$

where $J_{sh}(\alpha, E)$ is the stationary, spatially-homogeneous current density and $V(E)$ is the field-dependent drift velocity.[4,5] We solve Eq. (1) subject to an Ohmic boundary condition, $E_{ss}(x = 0) = \rho_0 J$, at the injecting contact and a voltage-bias constraint, $\int_0^L E_{ss}(x)\,dx = \phi$; where, $\rho_0$, $L$, and $\phi$ are the dimensionless contact resistance, sample length, and voltage bias, respectively. In Figure 1(b), $J_{sh}(\alpha, E)$ is shown for $\alpha = 1.21$ and $\alpha = 1.2085$, along with the contact current density $J_{con}(E) = E/\rho_0$. Equation (1) and the boundary conditions specify a unique electric-field profile for each voltage bias.

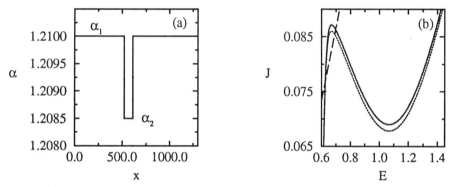

**Figure 1.** (a) A notched compensation-ratio profile, $\alpha(x)$, with $\alpha_1 = 1.21$, $\alpha_2 = 1.2085$, and $W = 90$. (b) The $J_{sh}(\alpha, E)$ curves for $\alpha = 1.21$ (solid line), $\alpha = 1.2085$ (dotted line), and $J_{con}(E) = E/\rho_0$ with $\rho_0 = 7.5$ (dashed line). The unit of length corresponds to $\sim 0.011$ mm, current density $\sim 20.5$ mA, resistivity $\sim 78\ \Omega$ cm, and electric field $\sim 10.0$ V/cm.

For the spatially-uniform compensation profile, the high-field region becomes unstable as the current density approaches $J_c$, where $J_c$ is defined by the intersection of $J_{con}(E)$ with the region of $J_{sh}(\alpha, E)$ with negative differential conductivity.[4,5] The high-field region then extends into the sample at approximately $E_2(J_c)$. The constant curve $J = J_c$ intersects $J_{sh}(\alpha, E)$ three times; $E_i(J_c)$ corresponds to the electric-field value at an intersection, indexed in ascending order of magnitude. The low-field value of the step-like profile is at $E_1(J_c)$.

For a notched compensation profile, a reduction in $\alpha$ produces a region of increased resistivity. Depending on the difference $(\alpha_1 - \alpha_2)$ and the width of the notch region $W$ (which extends from $x_1$ to $x_2$), the electric field at $x_2$ can reach $E_2(J, \alpha_1)$ for $J < J_c$. A region of $E_2(J, \alpha_1)$ then develops that extends from $x_2$ towards the receiving contact. As in the case of a spatially-uniform compensation profile, the extent of the $E_2(J, \alpha_1)$ region increases with voltage until it becomes unstable to periodic pulses, as shown in Figure 2(a). We consider the case where $\alpha_2 < \alpha_1$ and the local maximum of $J_{sh}(\alpha_2, E) < J_c(\alpha_1)$. If these conditions are not satisfied, the electric field at $x_2$ never reaches $E_2(J, \alpha_1)$ for $J < J_c$, and, therefore, the stationary state always loses stability near the injecting contact, rather than in the region following the notch.

Integrating Eq. (1) determines the critical notch width,

$$\int_{E_1(\alpha_1, J_c)}^{E_2(\alpha_1, J_c)} \frac{V(E)\, dE}{J_c - J_{\text{sh}}(\alpha_2, E)} = W_c. \tag{2}$$

For a given $\alpha_1$ and $\alpha_2$ the stationary state will lose stability to periodic pulses near the injecting contact for $W < W_c$ and in the region after the notch for $W > W_c$. In Figure 2(b), we show the critical notch width $W_c$ as a smooth curve, determined from Eq. (2), as a function of $\alpha_2$ for $\alpha_1 = 1.21$. We also performed numerical simulations with the full time-dependent reduced equation to determine the well width at which nucleation changes from contact to notch. The data points in Figure 2(b) indicate the crossover as determined from the simulations; the uncertainty is on the order of the size of the points.

This work is supported by the National Science Foundation grant DMR-9157539 and the NATO travel grant CRG-900284.

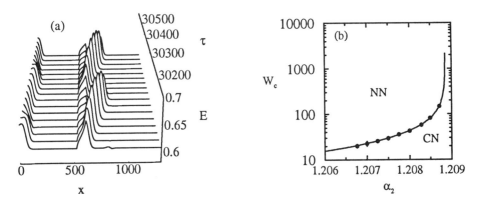

Figure 2. (a) $E(x, \tau)$ for a voltage bias slightly above the onset to periodic behavior. The $\alpha(x)$ profile is as shown in Figure 1(a) ($W > W_c$). The unit of time corresponds to $\sim 2.1$ μs. (b) The critical notch width $W_c(\alpha_2)$ for $\alpha_1 = 1.21$. NN $\equiv$ notch nucleation, CN $\equiv$ contact nucleation.

REFERENCES

1. A. M. Kahn, D. J. Mar, and R. M. Westervelt, Spatial measurements of moving space-charge domains in p-type ultrapure germanium, *Phys. Rev. B.* 43:9740 (1991).
2. M. P. Shaw, V. V. Mitin, E. Schöll, and H. L. Grubin. "The Physics of Instabilities in Solid State Electron Devices," Plenum Press, New York (1992).
3. S. W. Teitsworth, The Physics of Space Charge Instabilities and Temporal Chaos in Extrinsic Photoconductors, *Appl. Phys. A.* 48:127 (1989).
4. S. W. Teitsworth, M. J. Bergmann, and L. L. Bonilla, Space charge instabilities and nonlinear waves in extrinsic semiconductors, *in:* "Nonlinear Dynamics and Pattern Formation in Semiconductors and Devices", Vol. 79 of *Springer Proceedings in Physics*, F.-J. Niedernostheide, ed., Springer-Verlag, Berlin-Heidelberg (1995).
5. M. J. Bergmann, S. W. Teitsworth, L. L. Bonilla, and I. R. Cantalapiedra, *Phys. Rev. B.* (submitted).

# TRANSMISSION PROPERTIES OF RESONANT CAVITIES
# AND ROUGH QUANTUM WELLS

A. Abramo[1], P. Casarini[1], and C. Jacoboni[1]

[1]Istituto Nazionale di Fisica della Materia,
Dipartimento di Fisica, Modena, Italy

## INTRODUCTION

The progress of semiconductor technologies has recently led to the possibility of tailoring so called *mesoscopic* solid state structures.[1] Since in such devices partial or total confinement of carriers takes places, charge transport can no longer neglect the quantum behavior of particles. Also in ULSI devices, quickly approaching these limits, the study of coherent and non-coherent quantum phenomena is gaining importance.

In particular, the analysis of the transmitting properties of resonant cavities and rough quantum wells can be significant for the comprehension of more complex physical situations, such as electron mobility in MOSFET inversion layers.[2,3]

In this work we present the study of the transmission properties of resonant cavities in presence of surface roughness, localized Coulomb impurities and two-dimensional (2D) potential profiles. Preliminary results concerning the transit time of the coherent scattering states through such devices are also presented.

## THE THEORETICAL APPROACH

We have solved the time independent envelope-function equation in a generic 2D domain. It consists of an *open boundary system*, i.e. a potential region surrounded by adducing leads, as sketched in Fig.1. The solution is computed only inside the potential region, imposing

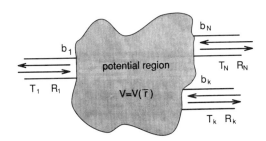

**Figure 1.** Schematic representation of a generic *open boundary* problem.

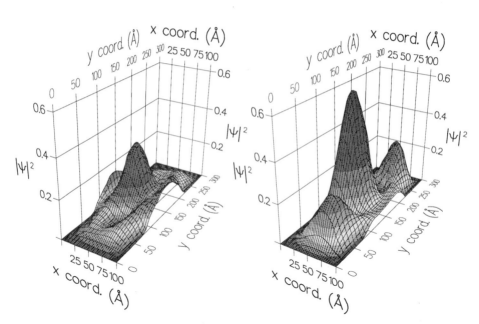

**Figure 2.** $|\Psi|^2$ of the scattering state through
a resonant cavity with (right) and without (left)
the presence of a Coulomb impurity in its central
position.

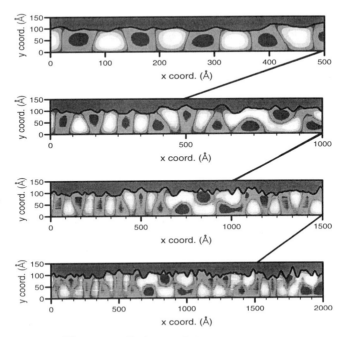

**Figure 3.** Real part of the scattering state
through rough channels of different lengths but
same roughness profile.

zero wave function at the border of the region and a superposition of incoming and reflected plane waves and evanescent modes at each lead boundary.[4] The solution is computed using the finite elements scheme over a triangular mesh.

## RESULTS

We have applied the simulator to the study of the transmitting properties of a resonant cavity with a localized screened Coulomb potential, simulating the presence of a dopant impurity, and of rough channels of different lengths.

Fig.2 shows the $|\Psi|^2$ of the scattering state through a resonant cavity with and without a Coulomb impurity in its central position. A part from a small tunneling through the shallow Coulomb potential, the impurity strongly opposes to the wave propagation through the cavity.

We also investigated the transmission behavior of quantum wells with non planar boundaries, as it happens at the surface of FETs. [5,6] We used a random roughness profile with Gaussian correlation. The parameter used are $\Delta = 1$ nm and $\Lambda = 2.04$ nm.

Fig.3 shows the real part of the scattering state through rough channels of different length but with exactly the same roughness profile. As the length increases, the coherent state deviates from the *plane wave* behavior, showing also transversal mode mixing phenomena. This is due to interference phenomena that add up as the length increases.

Fig.4 shows the transmission coefficients for the same channels of Fig.3. The interference phenomena add up reducing the transmissivity of the channels. This can also seen looking at the conductance of the channels, shown in Fig.5 for the 50 nm and 200 nm length channels, where the degradation of the channel conductance is observed for increasing lengths.

We have generalized to the 2D case the calculation of the *phase time*[7] to compute the transmission delay of a coherent scattering state through a device.[8] Fig.6 shows the transit time and the transmission coefficient of the 50 nm rough channel. The transit time increases in correspondence of the resonances of the transmission coefficient.[9]

Finally, we applied the simulator to the study of an electro-optical device[10,11] used as focusing lens. Fig.7 shows the $|\Psi|^2$ and $\mathbf{J}_\Psi$, together with the contour of the potential lens. The bi-concave shaped potential barrier (0.25 eV) focuses the electrons towards the center of the channel, possibly reducing the effect of surface roughness on the transport properties.

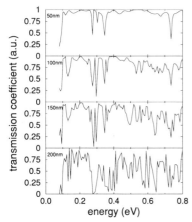

**Figure 4.** Transmission coefficients for the rough channels of Fig.3.

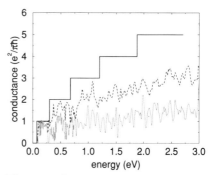

**Figure 5.** Conductance of rough channels of different lengths: dashed line: 50 nm device; dotted line: 200 nm device; solid line: conductance of an ideal channel of the same width.

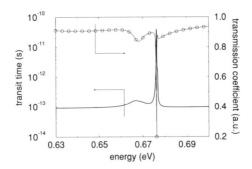

**Figure 6.** Transmission coefficient (dashed line) and transient time for the 50 nm rough channel of Fig.3.

Focusing electronic lens

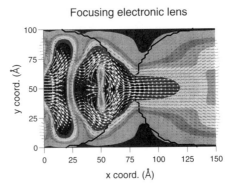

**Figure 7.** $|\Psi|^2$ (grey shades), $\mathbf{J}_\Psi$ (arrows) and contour of the focusing lens potential (solid line) for the electronic focusing device.

## Acknowledgments

This work has been partially supported by the A.R.O and O.N.R. through the E.R.O.

## REFERENCES

1. V.M. Agranovich and A.A. Maradudin *Mesoscopic phenomena in solids.* Amsterdam: North-Holland, 1991.
2. M.R. Pinto, E. Sangiorgi, and J. Bude *IEEE Electron Device Lett.*, vol. 14, p. 375, 1993.
3. J.A. Cooper and D.F. Nelson *J. Appl. Phys.*, vol. 54, p. 1445, 1983.
4. C. Lent and D.J. Kirkner *J. Appl. Phys.*, vol. 67, p. 6353, 1990.
5. S.M. Goodnick, D.K. Ferry, C.W. Wilmsen, Z. Lilental, D. Fathy, and O.L. Krivanek *Phys. Rev. B*, vol. 32, p. 8171, 1985.
6. A. Abramo, J. Bude, and F. Venturi and.R. Pinto in *IEDM Tech. Dig.*, p. 731, 1994.
7. E.H. Hauge, J.P. Falck, and T.A. Fjeldly *Phys. Rev. B*, vol. 36, p. 4203, 1987.
8. A. Abramo, P. Casarini, and C. Jacoboni *to be published.*
9. P.J. Price *Appl. Phys. Lett.*, vol. 62, p. 289, 1993.
10. J. Spector, H.L. Stormer, K.W. Baldwin, L.N. Pfeiffer, and K.W. Wast *Appl. Phys. Lett.*, vol. 56, p. 1290, 1990.
11. U. Sivan, M. Heiblum, C.P. Umbach, and H. Shtrikman *Phys. Rev. B*, vol. 41, p. 7937, 1990.

# TUNNELING SPECTROSCOPY AS A PROBE OF HOT ELECTRONS IN THE UPPER LANDAU LEVEL OF A 2DEG

BRA Neves, L Eaves, N Mori, M Henini and OH Hughes

Dept. of Physics, University of Nottingham, Nottingham, NG7 2RD, United Kingdom

We have employed magnetotunnelling spectroscopy to study AlAs/GaAs double barrier resonant tunnelling diodes (DBRTD). These DBRTD show new additional peaks at voltages below the main resonant peak for B>2T due to tunneling transitions from the 2D emitter to quantum well in which the Landau level index p is not conserved. By monitoring the intensities of these new peaks, we are able to measure directly the relative electron population of the Landau levels in the emitter 2DEG. Analysis of the data gives an electron distribution in the 2DEG which is far from thermal equilibrium and a relaxation time between the two lowest Landau levels $\tau_r > 100$ ns.

In the past 10 years there has been great interest in magnetotunnelling studies of double barrier resonant tunnelling diodes (DBRTD) due to their potential applications as high speed devices and also because they are an attractive system for the study of quantum transport. When a magnetic field is applied parallel to the current flow, the emitter and quantum well states are quantised into Landau levels. Resonant tunnelling with transition between Landau level indices in the emitter ($p_e$) and the well ($p_w$) has been observed by several authors when $p_w > p_e$ (positive $\Delta p$)[1,2] . Another effect of the magnetic field applied perpendicular to a 2DEG is the enormous decrease in the acoustic phonon relaxation rate[3]. Due to the constraints of momentum and energy conservation, the electron relaxation time for acoustic phonon processes is much longer than in the zero field case. Therefore, a non-equilibrium electronic energy distribution can be achieved in the 2DEG in the presence of a quantising magnetic field. In this paper, we investigate additional new tunneling peaks in a DBRTD which correspond to a tunneling process with negative $\Delta p$ Landau level transition. The intensity of these peaks is directly proportional to the relative occupation of the Landau levels in the emitter 2DEG. Therefore, by monitoring and modelling the peak intensity as a function of magnetic field, we probe the presence of hot electrons in the upper Landau level of the emitter 2DEG and find a relaxation time greater than 100ns.

The experiment was carried out on devices fabricated from a AlAs/GaAs double-barrier heterostructure grown on a semi-insulating GaAs (001)-substrate using molecular beam epitaxy. The DBRTD comprises a 59Å-GaAs well sandwiched between two 47Å-AlAs barriers. A 200Å-GaAs spacer layer was grown adjacent to each barrier to prevent diffusion/segregation of Si donors to the active region from the doped region, which consisted of 2500Å-GaAs with doping graded from $2x10^{16}$ cm$^{-3}$ to $2x10^{18}$ cm$^{-3}$. The wafer was processed into (3 x 50) $\mu$m cross-shaped devices using a photolithographic process with selective wet etching[4].

The I(V) characteristics of the devices were measured at low temperature (T = 4.2K) with magnetic field B, applied parallel to the current and typical results are shown in figure 1. At low fields (B < 2T), we observe only the main resonance at $\approx 0.69$V and the LO-phonon assisted tunneling peaks at $\approx 0.82$V and $\approx 0.93$V indicated by * in fig. 1[1,2]. However, for higher field values (B > 2T), two additional new peaks at voltages below the resonant peak are observed (see inset of figure 1). These two peaks move to lower voltages with increasing field and their origin extrapolates back to the same voltage at B=0. At voltages beyond the resonant peak, two series of current peaks develop from the LO-phonon assisted tunneling peaks and are due to tunneling involving changes in the Landau level indices ($\Delta p \neq 0$)[1,2].

We now focus our attention on the new peaks represented by lines $P_{-1}$ and $P_{-2}$ in figure 1a. We identify them as $\Delta p \neq 0$ elastic transitions without the mediation of a LO-phonon. In particular, $P_{-2}$ represents a $\Delta p = p_w - p_e = -2$ transition and $P_{-1}$ a $\Delta p = -1$ transition. Therefore $P_{-2}$ ($P_{-1}$) represents tunneling from the emitter Landau level $p_e = 2$ ($p_e = 1$) to the well Landau level $p_w = 0$. In order to confirm this interpretation, we devised an experiment to enhance their intensity. Following the work of Leadbeater *et al.*[5] we made I(V) measurements with tilted magnetic fields, i. e., we kept the parallel component of the field $B_{//}$ constant and varied the perpendicular component $B_{\perp}$.

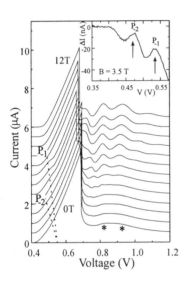 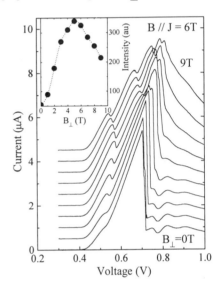

**Figure 1 - a)** The I(V) characteristics at 4.2K and for magnetic fields at 1T intervals between 0 (bottom) and 12 T (top). The dotted lines indicate the negative $\Delta p$ transitions. The inset shows the region of the threshold voltage in more detail and plots the difference $\Delta I = I(V, B = 3.5T) - I(V, B = 0)$ in the I(V) curves. The arrows indicate both $P_{-1}$ and $P_{-2}$ peaks. **b)** I(V) curves for $B_{//}=6T$ and $0 \leq B_{\perp} \leq 9T$. The curves are shifted for clarity. The inset of the figure shows the intensity dependence of the $P_{-1}$ peak as a function of $B_{\perp}$.

We carried out tilted field experiments for various values of the $B_{//}$ component and in fig. 1b we show the I(V) curves for $B_{//} = 6T$ and $0 \leq B_{\perp} \leq 9T$. For this value of the $B_{//}$ component, we observe only the $P_{-1}$ peak at around $\approx 0.52V$. We can clearly see an overall enhancement of the $P_{-1}$ intensity as $B_{\perp}$ increases. The inset of figure 1 shows the $P_{-1}$ intensity variation with $B_{\perp}$. We can observe a maximum intensity at $B_{\perp}=5T$. Using a simple model for the $\Delta p \neq 0$ transitions in tilted magnetic fields[5] we can explain this behaviour as follows: within a perturbation theory approximation, the overlap integral $I_{pp'}$ for the $\Delta p \neq 0$ transitions can be written as[5]:

$$I_{pp'} = \int \phi_p^*(y - Y)e^{i\Delta k_0(y-Y)}\phi_{p'}(y - Y)dy , \qquad (1)$$

where $\phi_p(y - Y)$ is a standard simple harmonic oscillator centered on y=Y, where $Y = -\hbar k_z / eB_{//}$ and $\hbar k_0 = eB_{\perp}\Delta s$ is the momentum change in the y direction due to the action of the Lorentz force over a distance $\Delta s$, equal to the average separation of the electrons in the emitter and in the quantum well. Considering, for this case, $p = p_w = 0$ and $p' = p_e = 1$, the overlap integral may be simply evaluated, giving $2|I_{01}|^2 = \alpha^2 e^{-\alpha^2/2}$, where $\alpha = \Delta k_0 \ell_B$ and $\ell_B = (\hbar / eB_{//})^{1/2}$. Taking the mean distance $x_0$ of an electron in the emitter from the barrier to be $\approx 10nm$, emitter barrier width b=4.7nm, well width w=6nm gives $\Delta s = x_0 + b + w/2 = 17.7\,nm$. The maximum amplitude of the $P_{-1}$ peak is thus attained when $d|I_{01}|^2/d\alpha = 0$, which gives $\alpha = \sqrt{2}$. For $B_{//}=6T$, $\ell_B = 10.4\,nm$, hence $\alpha = 0.28 B_\perp$. Therefore, we obtain a maximum for the $P_{-1}$ peak at $B_\perp = 5.01T$, in excellent agreement with our experimental observations.

Returning to the B//J case, we now consider how the intensity of both $P_{-1}$ and $P_{-2}$ peaks vary as a function of B. Applying the same method used to create the inset of Figure 1a (subtract I(V, B=0) from each I(V) curve) we determine directly the intensity (current) of both $P_{-1}$ and $P_{-2}$ peaks. The result is plotted in figure 2. We can see that $P_{-2}$ is visible for B > 2T, has a maximum intensity at B = 2.5T and disappears at B=4T, while $P_{-1}$ intensity grows linearly for 3T < B < 7T, reaching a maximum at B = 7T and then is rapidly quenched and completely disappears for B > 8T. This behaviour is explained as follows: as the magnetic field increases, the degeneracy of the Landau levels increases. Since the 2DEG density is constant (at a constant bias), as the magnetic field increases, the upper Landau levels are depopulated as the bottom Landau levels can accommodate more electrons, and eventually the upper Landau levels are completely emptied. This explains the disappearance of both $P_{-2}$ and $P_{-1}$ peaks at specific values of magnetic field. We can thus estimate the emitter 2DEG density $n_e$ using the cut-off values of both $P_{-2}$ and $P_{-1}$ peaks; and we found $n_e = 3.8 \times 10^{11}$ cm$^{-2}$. From magneto-oscillations in the I(B) curves of this DBRTD[6], we obtained $n_e = 3.6 \times 10^{11}$ cm$^{-2}$ at V=0.55V, in excellent agreement with the previous value. Therefore, for 4T < B < 8T, only the Landau levels $p_e = 0,1$ in the emitter are occupied. However, the sharp drop of the $P_{-1}$ peak is inconsistent with a thermal electron distribution in the emitter 2DEG.

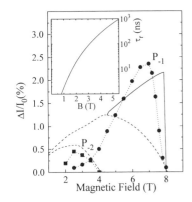

**Figure 2** - Plot of the relative amplitudes of the non-Landau quantum number conserving transitions in I(V). The dotted lines are a guide for the eye. The solid curve is derived from our model, assuming equal populations in the P=0 and P=1 Landau levels. The dashed curves assume a Fermi-Dirac distribution for the Landau levels at 4.2K. The inset shows the calculation of the phonon relaxation time as a function of the applied magnetic field.

We can use the amplitude of the $P_{-1}$ peak as a means of measuring the relative electron distribution in the two occupied Landau levels ($p_e = 0,1$). In order to do so, we have developed a transfer Hamiltonian model to describe the tunneling process from the emitter 2DEG into the quantum well[7]. The tunnel current is proportional to the number of electrons in the well if we assume an energy-independent tunneling rate from the well to the collector, i.e., $I \propto \sum_j n_j^W$, where $n_j^W$ is the number of electrons in the j-th Landau level in the well and is calculated using the following expression:

$$n_j^W = \frac{\sum_i S_{ij}\Delta(\varepsilon_i^e + eV_1 - \varepsilon_j^W)n_i^e}{\sum_i S_{ij}\Delta(\varepsilon_i^e + eV_1 - \varepsilon_j^W) + \gamma_c} \quad (2)$$

where $n_i^e$ is the number of electrons in the i-th Landau level in the emitter 2DEG, $\varepsilon_i^e$ ($\varepsilon_j^W$) is the subband level in the emitter (well), $\Delta(\varepsilon) = \Gamma^2/(\varepsilon^2 + \Gamma^2)$ characterizes the energy conservation, $V_1$ is the effective applied voltage between emitter and well, $\gamma_c$ is the dimensionless parameter characterizing charge build-up in the well and $S_{ij}$ is the tunneling probability between the i-th level in the emitter and the j-th level in the well (which accounts for both direct and elastic tunneling processes). Assuming a slowly varying interface roughness at the interfaces of the emitter barrier, we have

$$S_{ij} = (1 + 2\kappa^2\alpha^2)\delta_{ij} + 2\sum_q K(q)J_{ij}^2(q) \quad (3)$$

where $1/\kappa$ is the penetration depth of the electron wave function in the emitter barrier, $\alpha$ is the mean height of the interface roughness, q is the two-dimensional wave vector parallel to the interface, $K(q)$ represents the correlation of the interface roughness which is assumed to be Gaussian and $J_{ij}(q)$ is the matrix element between the simple harmonic oscillator states[8]. Using equations (2) and (3), we can calculate the tunneling current and the results are plotted in figure 2. The dashed curves assume a thermalized (4.2K) Fermi-Dirac electron distribution for the emitter 2DEG. The $P_{-2}$ peak is very well fitted by the curve. However, the agreement of the thermalized distribution with the $P_{-1}$ peak is very poor: it predicts a maximum amplitude for $P_{-1}$ at around 4T and then a gradual fall-off up to 8T. This gradual fall-off reflects the gradual depopulation of the Landau level $p_e=1$ as the $p_e=0$ Landau level population increases due to its increasing degeneracy. But, as clearly seen, this is *not* what is happening in the emitter 2DEG. The experimental behaviour of $P_{-1}$ indicates a much higher electron population at $p_e=1$ than the expected thermalized population at high magnetic fields. A further calculation was carried assuming equal populations for both Landau levels in the range $4.5 < B < 8T$. The result is plotted as a full line in figure 2 and shows a much closer agreement with the experimental data. This remarkable result indicates a high density of hot electrons in the upper Landau level and therefore a very slow energy relaxation time between the two Landau levels in the emitter, i.e., for the hot electrons to be present, the inter-Landau level transition time is, at least, as long as the tunneling time $\tau_t$. Using the relation $J = n_e e / \tau_t$, where J is the current density, we find $\tau_t = 108$ ns at $V = 500$ mV. Thus, the relaxation time $\tau_r$ from the upper Landau level to the lower is comparable, or longer than $\tau_t$.

In the absence of a magnetic field, the relaxation time by acoustic phonon emission for electrons in a 2DEG is approximately 1 ns. However several experimental and theoretical papers have recently demonstrated that, in the presence of a quantising magnetic field, this scattering process is much slower. Considering energy and momentum conservation in single phonon emission processes, one finds a strong suppression of the emission rate when the emitted phonon wavenumber required for the energy conservation exceeds the inverse confinement length of the 2DEG along the confining direction[3]. Using a modified version[7] of the method described in ref. 3, we estimated the acoustic phonon relaxation time for 1-phonon process for the conditions of our experiment. The results, shown in the inset of figure 2, clearly indicate a relaxation time longer than 100 ns for $B > 3.5T$, in excellent agreement with our experimental evidence. It is likely that under these conditions, emission of THz radiation (far infra-red)[9] and/or 2-phonon (with opposite momentum) processes[10] lead the relaxation of the hot electrons in the upper Landau level of the emitter 2DEG.

In conclusion, we have observed new additional peaks in the I(V) curves of a DBRTD undergoing magnetotunnelling. These peaks enable us to monitor the relative populations of the Landau levels in the emitter 2DEG. We have probed the presence of hot electrons in the upper Landau level and found that the electron relaxation time is in excess of 100 ns.

This work is supported by EPSERC. BRAN, NM and LE acknowledge CNPq (Brazil), the British Council and the EPSRC, respectively for financial support.

**REFERENCES:**
1) M. L. Leadbeater, E. S. Alves, L. Eaves, M. Henini, O.H. Hughes, A. C. Celeste, J. C. Portal, G. Hill and M.A. Pate; Phys. Rev B **39**, 3438 (1989)
2) V. J. Goldman, D.C. Tsui and J. E. Cunningham; Phys. Rev B **36**, 7635 (1987)
3) G.A. Toombs, F.W. Sheard, D. Neilson and L.J. Challis; Solid Satate Commun. **64**, 577 (1987)
4) J. Wang, P.H. Beton, N. Mori, H. Buhmann, L. Mansouri, L. Eaves, P.C. Main, T.J. Foster and M. Henini; Appl. Phys. Lett. **65**, 1124 (1994)
5) M.L. Leadbeater, F.W. Sheard and L.Eaves, Semicond. Sci. Technol. **6**, 1021, (1991)
6) M.L. Leadbeater, E.S. Alves, F.W. Sheard, L.Eaves, M. Henini, O.H. Hughes and G.A. Toombs; J. Phys.:Condens. Matter **1**, 10605 (1989)
7) N. Mori *et al.*, unpublished
8) R. Kubo, S.J. Miyake and N. Hashitume, Solid State Physics, Vol.**17**, 279 (1965)
9) W. Heiss, B.N. Murdin, C.J.G.M. Langerak, G.M.H. Knippels, I. Maran, K. Unterrainer, E. Gornik, C.R. Pidgeon, N.J. Hovenir, W.T. Wenckenbach and G. Weinmann, Semicond. Sci. Technol. **9**, 1554 (1994)
10) V.I. Fal'ko and L.J. Challis, J. Phys.: Condens. Matter **5**, 3945 (1993)

# IDENTIFICATION OF TUNNELLING MECHANISMS THROUGH GaAs/AlAs/GaAs SINGLE BARRIER STRUCTURES

J. J. Finley[1], R. J. Teissier[2], M. S. Skolnick[1], J. W. Cockburn[1], R. Grey[3], G. Hill[3], and M. A. Pate[3]

[1]Department of Physics, University of Sheffield, Sheffield, S3 7RH, United Kingdom
[2]Laboratoire de Microstructures et Microelectronique, 196 av. Henri Ravera, 92225 Bagneux Cedex, France
[3]Department of Electronic and Electrical Engineering, EPSRC Central Facility, University of Sheffield, Sheffield S1 3JD, United Kingdom

## Abstract

We have identified the $\Gamma$-X and X-$\Gamma$ intervalley tunnelling mechanisms in GaAs/AlAs/GaAs single barrier p-i-n structures by employing electroluminescence and transport techniques. We show that the $\Gamma$-X-$\Gamma$ tunnelling process proceeds by either elastic transfer into $X_z$ states or momentum conserving phonon assisted transfer involving $X_{xy}$ states. The tunnelling process is found to be strongly sequential, carriers relaxing in the barrier before tunnelling out occurs. The $\Gamma$–$\Gamma$–$\Gamma$ component of the tunnel current is observed to be more than two orders of magnitude weaker than transfer via X states.

In recent years there has been considerable interest in both the fundamental physics and applications of electron tunnelling in semiconductor heterostructures. Transport through single barrier GaAs/AlAs/GaAs structures is particularly interesting since the lowest conduction band electron states in GaAs and AlAs arise from different points of the Brillouin zone (BZ)[1-3]. The lowest bulk conduction band state in AlAs is along the [100] direction, close to the X point of the BZ, whilst in GaAs it is at the zone centre ($\Gamma$ point). Resonant tunnelling in such structures arises from tunnelling into quasi bound states in the AlAs barrier derived from the X point, at which the AlAs behaves as a quantum well and the GaAs as a barrier[1-3] (Fig 1a). The X point in AlAs is highly anisotropic with the effective mass in the growth (z) direction ($m^*_{xz}{\sim}1.1m_0$)[4] being more than four times heavier than in the transverse (x,y) directions ($m^*_{xxy}{\sim}0.26m_0$)[5]. Quantum confinement lifts the threefold degeneracy of the X states, with the expectation that the lowest quasi bound state will be the high effective mass $X_z$ state. In addition to the effects of quantum confinement, the small lattice mismatch between AlAs and GaAs results in the AlAs layer being under a state of bi-axial compression[6]. This has the effect that the $X_{xy}$ states are reduced in energy relative to the $X_z$ states. Thus for wide AlAs layers, in which the effects of quantum confinement are small, the lowest confined X state will have transverse (xy) character at flat band condition ($\sim$1.51V). At higher bias the nature of the lowest confined state in the barrier may change from $X_{xy}$ to $X_z$ (Fig 1b) due to the greater Stark shift of the latter by virtue of its larger effective mass.

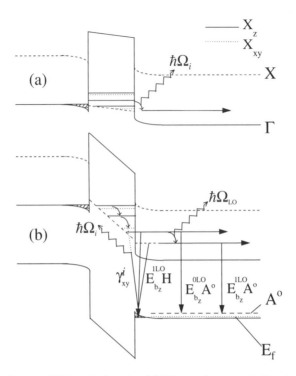

**Figure 1.** Band edge diagram: (a) $X_{xy}$ state lowest and (b) $X_z$ state lowest, including observed EL transitions.

In the present work electroluminescence (EL) and transport measurements are employed to determine the $\Gamma$-X and X-$\Gamma$ transport mechanisms through a series of single barrier p-i-n GaAs/AlAs/GaAs heterostructures of increasing AlAs layer width. When the structures are forward biased beyond the flat band condition electrons (holes) accumulate in a 2D layer adjacent to the n-type (p-type) GaAs/AlAs interface. As the bias increases, the energy separation between the emitter Fermi energy ($E_f$) and the $i^{th}$ confined X state in the barrier ($X_i$) decreases until $E_f$ becomes aligned with $X_i + \hbar^2 k_f^2/2m_x^*$ and resonant tunnelling may proceed via $X_i$ (fig 1a).

Electrons confined in the X potential of the AlAs layer may generate EL, via type-II recombination with the holes confined in the 2D accumulation layer close to the barrier (Fig 1b). Populations of hot electrons are also produced by tunnelling out of the barrier into the p-type collector, with the characteristic production of EL arising from their recombination with holes bound to neutral acceptors, denoted by e,$A^o$ [7,8].

The single barrier structures studied comprised the following layers: 0.5µm n+ (n=2x10^18 cm^-3) GaAs buffer layer, 50nm n=1x10^17 cm^-3 GaAs emitter, 50Å undoped GaAs spacer, 60,80 or 100Å undoped AlAs barrier, 50Å undoped GaAs spacer , 0.5µm p=1x10^17 cm^-3 collector and 0.5nm GaAs p+ top contact. Further experimental details of the EL experiments can be found in ref 8.

Differential conductance ($\sigma_d(V)$) characteristics, as a function of bias, are presented in fig 2 for the three samples. At forward bias in excess of 1.70V, marked features are observed in the $\sigma_d(V)$ characteristics, which correspond to $\Gamma$-X resonant tunnelling of emitter electrons into both $X_z$ and $X_{xy}$ barrier states. Magneto transport measurements[9] (not presented here) allow the emitter electron density and hence the electric field in the barrier to be determined independently at a given forward bias. Based on these results, a self consistent calculation of the relative energy separation of emitter ($\Gamma$) and barrier (X) states was carried out. This

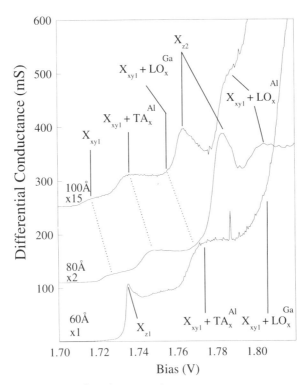

**Figure 2.** $\sigma_d(V)$ characteristics of 60Å, 80Å, and 100Å barrier samples, including assignment of resonances with X states.

calculation allows all resonances observed in $\sigma_d(V)$ to be attributed to elastic transfer into $X_z$ states (labelled $X_{z_n}$ n=1,2.. in fig 2) and momentum conserving (MC) phonon assisted transfer into $X_{xy}$ states (labelled $X_{xy_n}+phonon$) involving X point AlAs TA ($TA_x^{Al}$) and LO ($LO_x^{Al}$) and GaAs LO ($LO_x^{Ga}$) vibrational modes. Details and results of the calculation will be presented elsewhere. The $X_z$ features occur without phonon assistance, due to the relatively strong $\Gamma$-X intermixing by the potential discontinuity across the GaAs/AlAs interface in the z direction. The $X_{xy}$-$\Gamma$ intermixing by contrast is much weaker, arising from in-plane potential fluctuations[10], with the result that MC phonon participation is required for the $\Gamma$-X transfer into the barrier. The above interpretation of the resonances observed in $\sigma_d(V)$ contrasts with that in earlier studies where tunnelling only into $X_z$ states was considered and the participation of MC phonons was not identified[1-3].

Figure 3 shows EL spectra obtained from the 80 and 100Å samples at a forward bias of 1.760V. At around 1730meV a sharp peak, labelled $X_{xy}$, is observed which exhibits three satellite peaks to lower energy labelled $\gamma_{xy}^i$ (i=1,2, and 3). The energy separations $X_{xy}$-$\gamma_{xy}^i$ are 12±2meV, 30±2 and 48±2meV for i=1,2 and 3 respectively. From comparison with PL studies of type-II superlattices[10-12] we identify the relatively weak line $X_{xy}$ as arising from the zero phonon (ZP) excitonic recombination, involving electrons in the lowest lying $X_{xy}$ state in the barrier with holes bound in the hole accumulation layer (fig 1b). The energy spacing of the lines labelled $\gamma_{xy}^i$ allow us to identify these transitions as MC phonon replicas of $X_{xy}$ involving $TA_x^{Al}$, $LO_x^{Ga}$ and $LO_x^{Al}$ phonons respectively[11]. At around 1670meV a strong line labelled $E_{b_{xy}}{}^{0LO}A^o$ is observed. This feature exhibits a number of zone centre GaAs LO ($\hbar\omega$=36meV) phonon satellites to lower energy, only the first of which $E_{b_{xy}}{}^{1LO}A^o$ is shown in fig 3. The observation of a $LO_\Gamma^{Ga}$ phonon cascade is a clear signature of the relaxation of hot electrons[7]. We therefore identify the series $E_{b_{xy}}{}^{iLO}A^o$(i=1,2,..) as being due to e,A$^o$

519

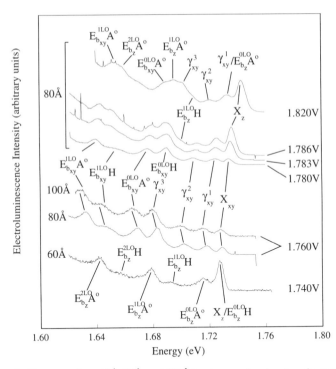

**Figure 3.** EL spectra from 60Å, 80Å, and 100Å barrier samples at various forward bias.

recombination of both ballistic electrons (i=0) and electrons which have relaxed their energy by emission of $iLO_\Gamma^{Ga}$ phonons in the collector[7]. A much weaker $LO_\Gamma^{Ga}$ phonon cascade is observed to originate from the $\gamma_{xy}^3$ phonon satellite, and is attributed to the recombination of ballistic and quasi ballistic electrons close to the barrier with the hole accumulation layer (fig 1b). We conclude that the ballistic population, leading to both the $E_{b_{xy}}^{iLO}H$ and $E_{b_{xy}}^{iLO}A^{\circ}$ series is injected into the collector with MC $LO_x^{Al}$ and $LO_x^{Ga}$ phonon assistance from the lowest $X_{xy}$ state of the barrier. The $E_{b_{xy}}^{iLO}A^{\circ}$ series is at ~12meV lower energy than the $E_{b_{xy}}^{iLO}H$, corresponding to the energy separation between $A^{\circ}$ and the hole Fermi energy in the collector.

The EL spectra obtained from the 60Å sample differ considerably from those obtained from the wider barrier samples. A single representative spectrum, taken at 1.740V, is presented in fig 3. Following the previous discussion we attribute the series labelled $E_{b_z}^{iLO}A^{\circ}$ as being due to the e,$A^{\circ}$ recombination of ballistic electrons in the collector. At 12meV to higher energy than $E_{b_z}^{iLO}A^{\circ}$ is a very intense line labelled $X_z/E_{b_z}^{0LO}H$. This line arises from the type II ZP recombination of electrons confined in the lowest $X_z$ state in the barrier with holes bound in the hole accumulation layer. Due to the stronger $X_z$-$\Gamma$ mixing the ZP line is expected to be several orders of magnitude stronger than any MC phonon satellites, which are consequently not observed. We conclude that the ballistic population in the collector is produced by direct injection from the lowest $X_z$ state, without MC phonon assistance. The $E_{b_z}^{iLO}H$ (i≠0) series arises from recombination of small populations of hot carriers close to the barrier with holes bound in the hole accumulation layer. The EL spectra obtained from this sample are qualitatively similar to those obtained from single barrier $Al_xGa_{(1-x)}As$ (x=0.38) structures[8], but in this case the ballistic population is injected from the lowest $X_z$ state of the barrier and not from the emitter directly.

At forward bias in the range 1.783 to 1.786V the EL spectra obtained from the 80Å sample exhibit a sudden change in character (fig 3). The ZP line becomes more intense relative to the $\gamma^j_{xy}$ satellites. In addition a very weak feature labelled $E_{b_z}^{1LO}H$ is observed at 36±2meV to lower energy than the ZP line. We attribute the change in the intensity of the ZP line and the appearance of a $LO_\Gamma^{Ga}$ satellite as being due to a change in nature of the lowest confined state in the barrier from $X_{xy}$ to $X_z$. For V>1.786V the EL spectra obtained from the 80 and 100Å samples are complicated by the observation of features arising from both $X_{xy}$ and $X_z$ states simultaneously, as indicated in fig 3 at 1.820V for the 80Å barrier.

Despite the fact that at higher bias, tunnelling occurs into excited X states, type II EL is only observed from the lowest X state in the barrier. This indicates that the inter-subband relaxation time between the X valleys is much more rapid than the tunnelling out time from the barrier and clearly demonstrates that the $\Gamma$-X-$\Gamma$ tunnelling process is strongly sequential in nature.

The EL energy position of the ZP line for the 60,80 and 100Å samples as a function of electric field in the barrier is shown in fig 4. The position of the ZP line of the 60Å sample exhibits a distinct change in slope at electric fields in excess of 98kV/cm. We identify this feature as corresponding to the onset of resonant $\Gamma-X-\Gamma$ tunnelling through the barrier. This attribution is in excellent agreement with $\sigma_d(V)$ measurements (fig 2) in which the first resonance in the 60Å sample was observed to be at 1.735V≡98kV/cm. Thus at electric fields below 98kV/cm the EL arises solely from the e,A° recombination of hot carriers injected non-resonantly ($\Gamma$-$\Gamma$-$\Gamma$), directly from the emitter. The relative intensity of EL spectra taken before and after the onset of resonant tunnelling via X states indicates that the non resonant ($\Gamma$-$\Gamma$-$\Gamma$) component of the current through the barrier is more than two orders of magnitude weaker than the resonant ($\Gamma$-X-$\Gamma$) component, at the first resonance.

In summary transport and EL studies have permitted very clear identifications of the tunnelling mechanisms through AlAs single barrier structures. We have clearly shown that both $\Gamma$-X and X-$\Gamma$ transfer proceeds elastically via $X_z$ states and with MC phonon assistance via $X_{xy}$ states. The tunnelling process has been shown to be strongly sequential in nature, relaxation taking place in the X valley before tunnelling out of the barrier occurs.

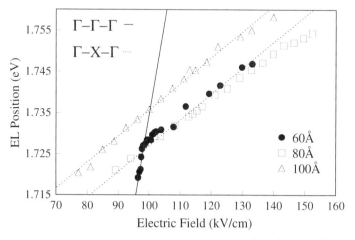

**Figure 4.** EL position of zero phonon type II transition for 60Å, 80Å, and 100Å samples.

## References

1 E.E.Mendez, L.L.Chang. (1990) Surface Science **229** 173-176
2 D.Landheer, H.C.Liu, M.Buchanan,R.Stoner. (1989) Appl. Phys. Lett. 54 **18**

3       R.Beresford, L.F.Luo, W.I.Wang. (1989).Appl. Phys. Lett. **55** 15

4       S.Adachi. (1985). J Appl. Phys. **58** 3

5       J.J.Finley, R.J.Teissier, M.S.Skolnick. (unpublished)

6       H.W. Van Kesteren,E.C. Cosman. (1989). Phys. Rev. B **39** 18

7       S.A.Lyon, C.L.Petersen. (1992).Semicond. Sci. Technol. **7**

8       R.J.Teissier, J.J.Finley, J.W.Cockburn,M.S.Skolnick . (1995).Phys Rev B **15** 4

9       L.Eaves, G.A.Toombs,F.W.Sheard,C.A.Payling,M.Leadbeater. (1988).Appl. Phys. Lett. **52** pg 212

10     P.Dawson,T.Foxon,H.W.Van Kesteren. (1990).  Semicond. Sci. Technol. **5**

11     E.Finkman, M.D.Sturge, M.-H.Meynadier, R.E.Nahory, M.C.Tamargo, D.M.Hwang and C.C.Chang. (1987) Journal of Luminescence **39** 57-54.

12     W.R.Tribe, S.G.Lyapin, P.C.Klipstein, G.W.Smith, R.Grey. (1994).Superlattices and Microstructures Vol **15 page 293**

# HOT ELECTRONS IN THE TUNNEL-COUPLED QUANTUM WELLS UNDER INTERSUBBAND EXCITATION

Yu. N. Soldatenko and F. T. Vasko

Institute of Semiconductor Physics, NAS of the Ukraine
Prospekt Nauki 45, Kiev-28, 252650, Ukraine

Nonequilibrium electron distribution in the tunnel-coupled double quantum wells (DQWs) may be achieved both under usual excitation methods (passing of current or interband photoexcitation), and due to intersubband transitions. The latter case is realized in recent experiments,[1,2] where both infrared and submillimeter (SM) pumping were used. Under resonance intersubband excitation, except of nonlinearity mechanism caused by the change of the electron temperature $T_e$, an additional mechanism caused by the redistribution of electrons between these states is important. Such a redistribution causes change of the transverse self-consistent electric field which changes the intersubband energy and, consequently, abruptly increases or decreases (depending on DQWs parameters and excitation frequency) the efficiency of pumping. In this paper we consider conditions for the transformation between linear and nonlinear excitation regimes, and also for saturation with increase of the pumping intensity. Moreover, the conditions for realization of the *bistability* in DQWs under resonance SM pumping are considered. The self-consistent (Hartree approximation) description of the energy states in DQWs and balance equation for concentration are used for calculation of rectification voltage (i.e. transverse voltage induced by the pumping) and nonlinear absorption. These results are compared with experimental data for SM excitation.[2]

The populations $n_\pm$ of the tunnel-coupled ground states $|\pm\rangle$ are determined by the system of balance equations

$$\nu_R(n_- - n_+) - (\partial n_+/\partial t)_T = 0, \qquad n_- + n_+ = n_{2D} \tag{1}$$

Assuming that the energy splitting $\Delta_T$ is small in comparison with LO-phonon energy $\hbar\omega_{LO}$ and

$$T_e \ll \hbar\omega_{LO} - \Delta_T, \tag{2}$$

we can neglect intersubband transitions with LO-phonon emission and take into account the quasielastic scattering only. So we can use approximation $(\partial n_+/\partial t)_T \simeq n_+/\tau_T$ for tunnel relaxation rate and suppose that the tunnel relaxation time $\tau_T$ is constant. The photoexcitation rate is given by

$$\nu_R = (2\pi/\hbar)(eE_\perp Z/\hbar\omega)^2 T^2 f_\Gamma(\Delta_T - \hbar\omega), \tag{3}$$

where $E_\perp$ is the transverse component of electric field, $\omega$ is the frequency, $T$ is the tunneling matrix element, $Z$ is the distance between the centers of QWs, and $f_\Gamma(E)$ describes the resonance absorption peak with halfwidth $\Gamma$.

The ground state splitting $\Delta_T = \sqrt{\Delta^2 + (2T)^2}$ is calculated in the self-consistent approach using the "rigid" orbitals $\varphi_{jz}$ (the effect of the self-consistent field on these orbitals is neglected). The wave functions for $|\pm\rangle$ states are presented as $\psi_l^\pm \varphi_{lz} + \psi_r^\pm \varphi_{rz}$. The columns $\psi_j^\pm$ ($j = l, r$) are determined from the matrix Hamiltonian,[3]

$$\frac{p^2}{2m} + \begin{vmatrix} \Delta/2 & T \\ T & -\Delta/2 \end{vmatrix}, \quad \Delta = \Delta_0 + U_l - U_r, \tag{4}$$

where $\Delta_0$ is the level splitting without tunneling and redistribution of the electrons between $|\pm\rangle$ states due to pumping.

The diagonal matrix elements (nondiagonal contributions are small in comparison with $T$) of the self-consistent electrons potential $U_H(z)$, determined from Poisson equation, are

$$U_j = \int_{-\infty}^{+\infty} dz \varphi_{jz}^2 U_H(z) \quad , U_H(z) = \varepsilon_c \int_{-\infty}^z dz'(z - z')n_e(z')/n_{2D}, \tag{5}$$

where we introduce $\varepsilon_c = 2\pi e^2 n_{2D}/\kappa Z$ ($\kappa$ is the uniform dielectric constant). The electron distribution $n_e(z)$ may be expressed by through $n_j$ ($j = \pm$)

$$n_e(z) = \sum_{j=\pm} n_j \left[ \left|\psi_l^j\right|^2 \varphi_{lz}^2 + \left|\psi_r^j\right|^2 \varphi_{rz}^2 \right]. \tag{6}$$

Substituting Eqs.(5),(6) into Eq.(4) we obtain nonlinear equation for level splitting $\Delta$

$$\Delta = \Delta_0 - \alpha \varepsilon_c (\Delta/\Delta_T)(n_- - n_+)/n_{2D} \tag{7}$$

$$\alpha = \int_{-\infty}^{+\infty} dz \left( \varphi_{rz}^2 - \varphi_{lz}^2 \right) \int_{-\infty}^z dz' \left( \varphi_{lz'}^2 - \varphi_{rz'}^2 \right)(z - z')/Z$$

where concentrations $n_\pm$ is obtained from Eqs.(1) and resonantly depend on $\Delta_T$.

We consider the case of small change of the level splitting $\delta = \Delta_T - \bar{\Delta}_T \ll \bar{\Delta}_T$ caused by population of $|+\rangle$ ($\bar{\Delta}_T$ is the level splitting without excitation), and use condition $\nu_R \tau_T \ll 1$, when linear redistribution is realized, $n_+ \simeq \nu_R \tau_T n_{2D}$. Under this assumptions, the equation for $\delta$ obtained from Eq.(7) is

$$\delta/E_c = f_\Gamma(\delta - \varepsilon_\omega)/f_\Gamma(0), \quad E_c = \varepsilon_c \alpha \nu_R^{max} \tau_T / \left\{ 1 + \alpha \varepsilon_c/\bar{\Delta}_T \right\} \tag{8}$$

where $\varepsilon_\omega = \hbar\omega - \bar{\Delta}_T$ is detuning energy of excitation.

Eq.(8) has one or three solutions, depending on parameters $\varepsilon_\omega/\Gamma$ and $E_c/\Gamma$. Therefore, under resonance SM pumping the bistability may be realized, see Fig.1. This effect is caused by the charge transfer and a subsequent, self-consistent shift of the energy levels which feeds back to absorption rate. Strong tunnel coupling of the ground states in DQWs consideraby decrease the bistability threshold in comparison with the case of pumping in the excited state.[4]

Rectification voltage in DQWs caused by the resonance SM pumping is determined by the change of $\Delta_T$ and is equal to $(\bar{\Delta}_T/\bar{\Delta})\delta/\alpha$. As the upper ground population is low, the relative absorption is proportional to $f_\Gamma(\delta - \varepsilon_\omega)$ or $\delta/E_c$. In Fig.2 we present comparison of this model with experimental data.[2] In the calculations we use experimental value for $(\partial \Delta_T/\partial V_{bias})$ from Ref.[5]. Another parameters ($T = 5.4meV$, $\Gamma = 0.7meV$, $n = 1.9 \cdot 10^{11}$ $cm^{-2}$ and $\tau_T = 300ps$) are in agreement with Refs.[2,5]. Field dependencies of the rectification voltage and the induced change of energy splitting, shown in Figs.3, is calculated for the same DQWs parameters.

In conclusion, we have demonstrated three nonlinear regimes under intersubband excitation due to: i) intrasubband heating [if (2) is broken], ii) upper ground state occupation (if $\nu_R \tau_T \gg 1$), and iii) self-consistent renormalization of the ground state splitting.

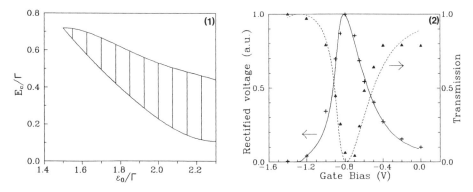

**Figure 1.** Region of parameters (shaded) when bistability is realized.
**Figure 2.** Rectification voltage and transmission for DQWs with parameters presented in Ref.2. Experimental data is shown by symbols.

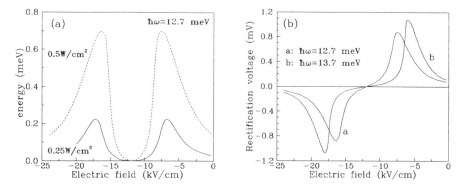

**Figure 3.** The pumping induced change of $\delta$ (a) and rectification voltage (b) in the DQWs from Ref.2.

**Acknowledgement:**This work has been supported in part by Grant N U65200 from the Joint Fund of the Government of Ukraine and International Science Foundation.

1. J. Faist, C. Sirtori, F. Capasso, L. Pfeiffer, and K.W. West, Phonon limited intersubband lifetimes and linewidths in two-dimensional electron gas, *Appl. Phys. Lett.* 64, 872 (1994).
2. J.N. Heyman, K.Unterrainer, K. Craig, B. Galdrikian, M.S. Sherwin, K. Campman, P.F. Hopkins, and A.C. Gossard, Temperature and intensity dependence of intersubband relaxation rates from photovoltage and absorption, *Phys.Rev.Lett.* 74:2682 (1995).
3. F.T. Vasko, and O.E. Raichev, Longitudinal transport of electrons in tunnel-coupled quantum wells with nonsymmetric scattering, *JETP*,80, (1995).
4. M. Seto, and M. Melm, Charge-transfer-induced optical bistability in asymmetric quantum-well structure, *Appl. Phys. Lett.* 60, 859 (1992).
5. J.N. Heyman, K. Craig, B. Galdrikian, M.S. Sherwin, K. Campman, P.F. Hopkins, S. Fafard, and A.C. Gossard, Resonant harmonic generation and dynamic screening in a double quantum well, *Phys. Rev. Lett.* 72, 2183 (1994).

# SELF-INDUCED PERSISTENT PHOTOCONDUCTIVITY
# IN RESONANT TUNNELING DEVICES

B.R.A. Neves[1], E.S. Alves[1], J.F. Sampaio[1], A.G. de Oliveira[1],
M.V.B. Moreira[1], and L. Eaves[2]

[1]Departamento de Física, ICEx, Universidade Federal de Minas Gerais, Caixa Postal
702, CEP 30161-970, Belo Horizonte, MG, Brazil
[2]Physics Department, University of Nottingham, Nottingham NG7 2RD, UK

Persistent photoconductivity (PPC)[1,2] in modulation-doped semiconductor heterostructures is a well known effect which is due to electron emission from DX centers in the (AlGa)As layer when the device is illuminated at low temperatures. Double barrier resonant tunneling devices have their electronic properties strongly affected by small changes in the conduction band profile[3,4]. The change in the charge state of the DX centers modifies the potential profile of the heterostructure and here we consider a device designed to be sensitive to this change. Persistent photoconductivity has been observed[5] in resonant tunneling structures, arising from Si DX centers in the (AlGa)As barriers which have diffused from the contact layers into the barrier regions.

In this work we report on the observation of a self-induced persistent photoconductivity (SIPPC) effect in a specially designed resonant tunneling device (RTD). The SIPPC produces a permanent shift of the resonant peak to lower voltages and is caused by light generated in the RTD itself by recombination of electron-hole pairs created by impact ionization caused by hot electrons in the depletion layer. This process is controlled by the applied voltage to the device.

The resonant tunneling devices were grown by molecular beam epitaxy (MBE) and consisted of the following layers, in order of growth, from the $n^+$-Si doped (100) oriented GaAs substrate: (i) 1μm of GaAs, Si doped to $2x10^{18}$ cm$^{-3}$; (ii) 500 Å of GaAs, doped to $2x10^{17}$ cm$^{-3}$; (iii) 500 Å of GaAs, doped to $2x10^{16}$ cm$^{-3}$; (iv) 35 Å undoped GaAs; (v) 36 Å of undoped AlAs barrier; (vi) 96 Å of undoped $Al_{0.20}Ga_{0.80}As$ well; (vii) 36 Å of undoped AlAs barrier; (viii) 35 Å undoped GaAs; (ix) 500 Å of GaAs, doped to $2x10^{16}$ cm$^{-3}$; (x) 500 Å of GaAs, doped to $2x10^{17}$ cm$^{-3}$; (xi) 1μm of GaAs, Si doped to $2x10^{18}$ cm$^{-3}$. The samples have been etched into 100 μm-mesas using standard photolithographic techniques.

The device was cooled to 4 K in the dark and a series of $I(V)$ curves were taken after several exposures to light provided by an infrared LED, with radiation energy below the GaAs band gap. The inset in Fig. 1 shows the $I(V)$ characteristics of the device at 4.2K in the dark and after illumination. In the dark, a well defined resonant peak is observed at ~1.82 V and it is attributed to electrons tunneling from the emitter into the first quasi-bound well state. After each light exposure, the resonant peak gradually moved to lower voltages until it reached ~1.73 V; further illumination produced no change in the peak position. The position of the resonant peak $V_p$ versus exposure (current across the LED times the time that it remains on) is shown in Fig. 1. The peak is permanently shifted after each exposure only returning to its dark value after the device being warmed up to temperatures above ~170 K. The dependence of the resonant peak position with temperature has also been investigated and the results will be published elsewhere.

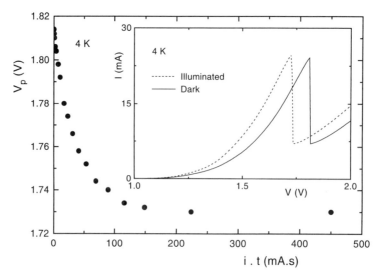

**Figure 1**. Plot of the resonant peak position versus exposure for the resonant tunneling device at 4K. Inset: The current-voltage characteristics of the device after being cooled down to 4K in the dark and after illumination.

The effects observed in Fig. 1 can be explained by a simple model[5]. The narrow emitter spacer layers together with a high growth temperature (630 °C) enhances the diffusion and segregation of Si donors from the highly doped contacts into the barrier regions, where DX centers[6,7] are then expected to be created. When the device is cooled down in the dark, electrons are trapped in the DX levels and the quasi-bound well states have their energies determined primarily by the conduction band profile of the heterostructure. By shining light on the device at low temperature the DX centers are ionized leaving a positive space charge in the barriers, which modifies the heterostructure potential lowering the energy of the quasi-bound well state relative to the emitter states. Therefore, a smaller applied voltage will be required to bring the well states on resonance with the emitter states, producing the observed shift of the resonant peak to lower voltages. The shift is persistent because the recapture of the electrons by the ionized donors is impeded by a potential barrier[6]. The effects shown in Fig. 1 have also been observed in a similar device with a GaAs quantum well.

A novel and interesting effect is observed when the device is biased above a certain threshold after being cooled down in the dark. The experiment was done by applying a voltage $V_a$ through the device for 30s after which a whole $I(V)$ characteristics is taken in the voltage range $0 < V < V_a$. The peak position is recorded for each applied voltage $V_a$ and the result is shown in Fig. 2. Notice that for $V_a > 2.8$V, in each $I(V)$ curve the peak is gradually shifted to lower voltages, reaching its minimum position at $V_a \cong 3.4$V; for higher applied voltages no change is observed in the peak position. The shift of the resonant peak is approximately the same for both forward and reverse applied bias directions. The observed shift is persistent: the peak only returns to its original position at $V_a=0$, after the device being thermally recycled to temperatures above ~170K. The observed self-induced persistent photoconductivity is caused by light generated in the RTD itself by recombination of electron-hole pairs created by impact ionization of hot electrons in the depletion layer of the device. When the device is biased above $V_a$~2.8V, hot electrons, produced by the high electric field across the device, create electron-hole pairs by impact ionization. The photons generated by recombination of the electron-hole pairs, photo-ionize the DX centers in the device. As described before, the change in the charge state of the DX centers modifies the potential profile for the electrons and a permanent shift to lower voltages is observed in the resonant peak position. The magnitude of this shift depends on the illumination intensity which is controlled by the applied bias to the device. This model explains the data shown in Fig. 2 and it is confirmed by the electroluminescence measurements carried out on the same structure as discussed below.

Electroluminescence measurements under different applied biases have been carried out on the RTD after cooling it down in the dark and the results are shown in the inset of Fig. 2. In order to allow to more light to come out from the device, the structures were processed into annular mesas. An electroluminescence peak, corresponding to GaAs band-gap radiation energy, appears for $V_a > 2.8$V and its intensity rapidly increases with the applied bias. We attribute the observed electroluminescence as due to recombination of electron-hole pairs which are created by impact-ionization of hot electrons accelerated by the high electric field in the collector depletion layer. One should notice that the onset of the electroluminescence peak is at the same voltage where we observe the resonant peak starting to shift, as shown in Fig. 2. This is a clear evidence that both effects are produced by light generated in the RTD itself.

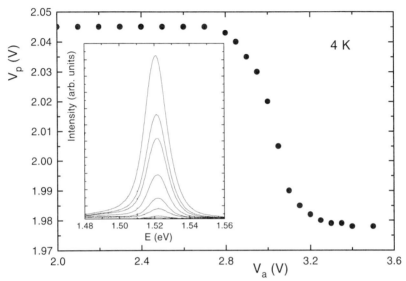

**Figure 2.** The dependence of the resonant peak position $V_p$ on the applied bias $V_a$ across the resonant tunneling device. The inset shows the electroluminescence spectra of the double barrier device under applied biases from 2.9, to 3.6V in 0.1V steps (bottom to top). No electroluminescence is observed for $V_a < 2.8$V.

In summary, we have demonstrated persistent photoconductivity in resonant tunneling structures due to DX centers created by diffusion of Si donors from the contact layers in the barrier regions. A novel self-induced persistent photoconductivity has also been verified, caused by light generated in the resonant tunneling device itself. The photons are created by recombination of electron-hole pairs produced by impact ionization in the collector depletion layer of hot electrons due to a high applied electric field.

This work was partialy supported by Fapemig, Finep and CNPq (Brazil).

## REFERENCES

[1] H.L.Stormer, R. Dingle, A.C. Gossard, W. Wiegmann, *Solid State Commun.* **29**, 705 (1979).

[2] H.L. Stormer, A.C. Gossard, W. Wiegmann, K. Baldwin, Appl. Phys. Lett. 39, 912 (1981)

[3] E.S. Alves, M.L. Leadbeater, L. Eaves, M. Henini, O.H. Hughes, *Solid State Electronics* **32**, 1627 (1989).

[4] B.R.A. Neves, E.S. Alves, J.F. de Sampaio, A.G. de Oliveira, E.A. Meneses, *Braz. J. Phys.* **24**,203 (1994).

[5] T.C.L.G. Sollner, H.Q. Le, C.A. Correa, and W.D. Goodhue, *Appl. Phys. Lett.* **47**, 36 (1985).

[6] P.M. Mooney, *J. Appl. Phys.* **67**, R1 (1990).

[7] J.F.Sampaio, A.S.Chaves, G.M.Ribeiro, P.S.S.Guimaraes, R.P. de Carvalho, and A.G. de Oliveira, *Phys. Rev. B* **44**, 10933 (1991).

# COULOMB INTERACTION EFFECT ON THE ELECTRON TUNNELING IN DOUBLE QUANTUM WELLS

O. E. Raichev and F. T. Vasko

Institute of Semiconductor Physics, NAS of the Ukraine
Prospekt Nauki 45, Kiev-28, 252650, Ukraine

Coulomb interaction of the photoexcited carriers in the double quantum wells (DQW's) modifies evolution of the electron distribution in these systems. First of all, this interaction renormalizes the splitting energy $\Delta$ of the tunnel-coupled electron levels. This effect has influence on the tunneling relaxation rate of the photoexcited electrons in DQW's with large electron densities[1-4]. Apart from the discussed effect, Coulomb interaction leads to the scattering between the electrons, which enhances the scattering-assisted nonresonant tunneling relaxation rate in DQW's. This mechanism, similar to Auger transitions between the size quantization subbands,[5] is discussed in this paper. We calculate the tunneling relaxation rate in DQW's due to electron-electron ($ee$) and electron-hole ($eh$) scattering and compare it with the partial tunneling relaxation rates due to the other scattering mechanisms.

Different cases of photoexcitation are shown in Fig.1 (a): interband[1,2] excitation, when the electrons and holes are created in the conduction and valence bands of DQW's; and intersubband excitation by infrared[3] or submillimeter[4] radiation, when only non-equilibrium electrons are present. After the excitation, the carriers relax to the equili-

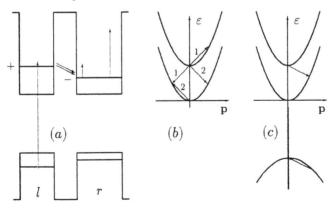

**Figure 1.** (a) Band diagram and photoexcitation schemes for the DQW's; (b) two kinds of Auger-like transitions under $ee$ scattering; (c) Auger-like transition under $eh$ scattering.

*Hot Carriers in Semiconductors*
Edited by K. Hess *et al.*, Plenum Press, New York, 1996

brium distributions and localize near extrema of the lowest subbands in each well (we assume that the characteristic electron energy is small in comparison with $\Delta$). Then, slow tunneling relaxation of the electrons between the tunnel-coupled + (upper) and − (lower) states takes place; coupling of the valence-band states and recombination are neglected. Possible channels of the Auger-like transitions are shown in Fig.1 (b,c).

The concentration balance equation for the electrons in + state may be written in the following way

$$\frac{dn_t^+}{dt} = -\left[\sum_k \nu_k(\Delta_t) + \nu_{ee}(\Delta_t, n_t^+, n_t^-) + \nu_{eh}(\Delta_t, p^l, p^r)\right] n_t^+, \tag{1}$$

where $n_t^\pm$ are the electron concentration in $\pm$ subbands, $p^l$ and $p^r$ are the hole concentrations in the lowest left- and right-well valence-band states. In Eq.(1) we introduce tunneling relaxation rates due to $ee$ and $eh$ scattering ($\nu_{ee}$ and $\nu_{eh}$, respectively) and relaxation rates $\nu_k$ due to the other scattering processes: impurities, interface roughnesses, acoustic and optical phonons (index $k$ specifies proper mechanism). Time dependence of $\nu_k$ is connected with the Coulomb renormalization of the splitting energy, roughly estimated in the Hartree approximation as $\Delta_t = \Delta + (4\pi e^2 Z/\epsilon)[n_t^+ + p^r - n_t^- - p^l]$, where $e$ is the electron charge, $\epsilon$ is the dielectric constant, and $Z$ is the interwell distance.

In order to calculate $\nu_{ee}$, we consider the kinetic equation for the interacting electrons. In the second order of Coulomb interaction we obtain (see Ref.6, for example)

$$\frac{\partial f_\alpha}{\partial t} = \frac{2\pi}{\hbar} \sum_{\beta\gamma\delta} \sum_{QQ_1} \left\{ \left(\alpha \left|e^{-iQr}\right|\gamma\right) \left(\gamma \left|e^{iQ_1r}\right|\alpha\right) \left(\beta \left|e^{iQr}\right|\delta\right) \left(\delta \left|e^{-iQ_1r}\right|\beta\right) \right.$$
$$\left. -Re\left[\left(\alpha \left|e^{-iQr}\right|\gamma\right) \left(\gamma \left|e^{iQ_1r}\right|\beta\right) \left(\beta \left|e^{iQr}\right|\delta\right) \left(\delta \left|e^{-iQ_1r}\right|\alpha\right)\right]\right\} v_Q v_{Q_1}$$
$$\times \delta(\varepsilon_\alpha + \varepsilon_\beta - \varepsilon_\gamma - \varepsilon_\delta)[f_\alpha f_\beta(1-f_\gamma)(1-f_\delta) - f_\gamma f_\delta(1-f_\delta)(1-f_\alpha)], \tag{2}$$

where $v_Q$ is the Fourier component of the Coulomb potential and $Q$ is the three-dimensional wave vector. The first and second terms in the braces correspond to the direct Coulomb and exchange contribution; $\varepsilon_\alpha$ and $f_\alpha$ are the energy spectrum and distribution functions of the electrons in the state $|\alpha) = |jsp) = F_j(z)|s) \exp(ipx/\hbar)$, where $x$ and $p$ are the in-plane coordinate and momentum, $F_j(z)$ is the envelope function of $j$−th subband, and $|s)$ is the spin function. Functions $F_j(z)$ are expressed through the linear combinations of the right- and left-well orbitals according to $F_+(z) = F_l(z) + F_r(z)T/\Delta$, $F_-(z) = F_r(z) - F_l(z)T/\Delta$, where $T$ is the tunneling matrix element, which describes tunnel coupling. Equation (2) is valid under conditions $\hbar\nu_{ee} \ll \Delta_t$.

Taking into account $(dn_t^+/dt) = 2\sum_p(\partial f_{+p}/\partial t)$, we perform an analytical calculation of $\nu_{ee}$, which is possible under assumption that the characteristic electron energy (temperature or Fermi energy) is small in comparison with both $\Delta$ and $\hbar^2/(mZ^2)$ ($m$ is the effective mass of the electron). We obtain

$$\nu_{ee}(\Delta, n^+, n^-) = \frac{\pi^2 e^4 T^2}{\hbar\epsilon^2\Delta^3}\left[n^+ w_l^2(q_c) + n^- w_r^2(q_c)\right], \tag{3}$$

where the dimensionless factors $w_l(q_c)$ and $w_r(q_c)$ are given by

$$w_i(q) = \int dz \int dz' \exp(-q|z-z'|)|F_i(z)|^2 \left(|F_r(z')|^2 - |F_l(z')|^2\right), \tag{4}$$

and $\hbar q_c \simeq \sqrt{m\Delta}$ is the momentum transferred in the transition. The time dependence of $\nu_{ee}$ exists not only due to the renormalization of $\Delta_t$, but also due to the time dependence of the electron concentrations $n_t^+$ and $n_t^-$.

Equation (2) can be applied to the calculation of the $eh$ relaxation rate $\nu_{eh}$, if we include the valence-band states in the sum in the right-hand part and ignore the exchange interaction term. Approximating the valence-band spectrum in the left and right wells by parabolas, we can obtain formulas similar to (3) and (4).

Below, in Fig.2, we present comparison of the Coulomb-assisted tunneling relaxation rate given by Eqs.(3),(4) with the tunneling relaxation rates $\nu_k$ due to the other scattering mechanisms. We consider the $GaAs/Ga_{0.65}Al_{0.35}As$ DQW's with barrier width 4.5 nm, narrow well width 5 nm and wide well width 7 nm. The solid lines show $ee$ relaxation rate $\nu_{ee}$ ($n^- = 10^{11}$ $cm^{-2}$ and $n^+ = 0$) and total relaxation rate $\nu_{ee} + \nu_{eh}$ in the case when the holes are also present ($n^- = p^r = 10^{11}$ $cm^{-2}$ and $n^+ = p^l = 0$). The dashed lines describe PO phonon emission and scattering on the interface roughnesses with one monolayer height and 2 nm autocorrelation length. As the electron concentration exceeds $10^{11}$ $cm^{-2}$, the Auger-like tunneling relaxation may dominate over the elastic scattering-assisted mechanisms.

In conclusion, we have demonstrated two mechanisms of *nonexponential* tunneling relaxation, which exists due to time dependence of the effective relaxation rate in Eq.(1). The first is the renormalization of level splitting energy, and the second is $ee-$ or $eh-$scattering. Both these mechanisms are important at high electron concentration.

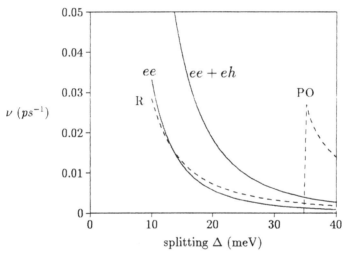

**Figure 2.** Nonresonant tunneling relaxation rate caused by different scattering mechanisms. Solid lines: Auger-like rate under $ee$ scattering, and combined $ee$ and $eh$ scattering. Dashes: polar optical phonon scattering (PO) and interface roughness scattering (R).

Acknowledgement: This work has been supported by the International Science Foundation Grant No U65000.

1. M.G.W.Alexander, M. Nido, W.W.Rühle, R.Sauer, K.Ploog, K.Köhler, and W.T.Tsang, Tunneling between two wells: $In_{0.53}Ga_{0.47}As/InP$ versus $GaAs/Al_{0.35}Ga_{0.65}As$, *Solid-State Electron.* 32:1621 (1989).
2. D.H.Levi, D.R.Wake, M.V.Klein, S.Kumar, and H.Morkoc, Density dependence of nonresonant tunneling in asymmetric coupled quantum wells, *Phys.Rev.B* 45:4274 (1992).
3. J.Faist, C.Sirtori, F.Capasso, L.Pfeiffer, and K.W.West, Phonon limited intersubband lifetimes and linewidths in a two-dimensional electron gas, *Appl.Phys.Lett.* 64:872 (1994).
4. J.N.Heyman, K.Unterrainer, K.Craig, B.Galdrikian, M.S.Sherwin, K.Campman, P.F.Hopkins, A.C.Gossard, Temperature and intensity dependence of intersubband relaxation rates from photovoltage and absorption, *Phys.Rev.Lett.* 74:2682 (1995).
5. S.Borenstain and J.Katz, Intersubband Auger recombination and population inversion in quantum-well subbands, *Phys.Rev.B* 39:10852 (1989).
6. E.V.Bakhanova, M.V.Strikha, and F.T.Vasko, Nonequilibrium carriers recombination channels in uniaxially stressed gapless semiconductors, *Phys.Stat.Solidi (b)* 164:157 (1991).

# PLASMON SATELLITES OF RESONANTLY TUNNELLING HOLES

B.R.A. Neves,[1] T.J. Foster,[1] L. Eaves,[1] P.C. Main,[1] M. Henini,[1] D.J. Fisher,[2] M.L.F. Lerch,[2] A.D. Martin[2] and C. Zhang[2]

[1]Dept. of Physics, University of Nottingham, Nottingham, NG7 2RD, UK
[2]Dept. of Physics, University of Wollongong, Wollongong, NSW 2522, Australia

Hole resonant tunnelling is investigated in an asymmetric p-type RTD which exhibits intrinsic multistability on the first light-hole (LH1) resonant peak. This device exhibits two previously unseen satellite peaks just beyond the LH1 resonant peak. Both satellites are strongly dependent on temperature and quench above 18K. We have modeled the tunneling transitions in our devices and show that these features are due to plasmon-assisted resonant tunneling of holes.

Recently, resonant tunneling diodes (RTDs) have proved to be a novel means of investigating the interaction between hot electrons and plasmon excitations of a two-dimensional electron gas confined in a quantum well. Using a negative output resistance (NOR) circuit, Lerch et al.[1] discovered the presence of a plasmon-assisted tunneling satellite in the I(V) characteristics of an RTD. The satellite was observed in the voltage overhang region of a device especially designed to produce strong space-charge buildup in the quantum well resonance. This overhang gives rise to the well-known bistability effect when the RTD is measured with conventional voltage sweep circuit[2,3]. The plasmon-assisted resonant tunneling process was considered theoretically by Zhang[4] and it was shown that the plasmon is an excitation of the degenerate electron gas in the quantum well which arises from an interaction with a hot electron injected from the emitter accumulation layer.

In this paper, we investigate hole tunneling in a p-type resonant tunneling diode, and observe similar satellites corresponding to hole plasmons. The special interest in studying this type of device is that the band structure of the hole states in the quantum well are much more complex than those for electrons. In particular, plasmons associated with inter- and intra-hole subband transitions are expected. We investigate the effect of temperature on the hole-plasmon features.

A schematic band diagram of the device used in this experiment is shown in Fig. 1a. It consists of a 4.5-nm emitter AlAs-barrier and a 5.7-nm collector AlAs-barrier enclosing a 4.2-nm GaAs quantum well. The asymmetric nature of the device gives rise to hole space charge buildup in the well for resonances in one bias direction. Undoped GaAs spacer layers of 5.1-nm separate the barriers from graded p-type (Be doped) GaAs contact layers. Full details of the composition are given in ref. 5. The current-voltage characteristics of the device, I(V), reveal several resonant features corresponding to tunneling into heavy hole (HHn) and light hole (LHn) subbands on the quantum well. Here, n is the quantum number of the envelope function. Figure 1b shows the I(V) characteristics of the RTD biased on the LH1 resonance at 4.2K. This is the only resonance which shows an intrinsic bistability effect. The full I(V) curve is shown in ref. 5.

The NOR circuit allows us to probe the voltage overhang region, which is inaccessible with a conventional voltage sweep circuit. Two satellite features can be observed beyond the main resonance. These are indicated by arrows in Fig. 1.b. Although the voltage position of the first feature is before the resonant peak, its energy is higher than that of the resonance. This effect is caused by the hole-charge feedback mechanism in the quantum well[1,6]. In the region of current falloff, a shoulder-like structure is present in the I(V) around the current value I≈60μA. This is due to oscillations in the NOR circuit, which, due to its negative resistance, is not stable in the entire voltage range[6]. However, the circuit is stable elsewhere, particularly on the regions of interest, i.e., the resonant peak and the valley region after it.

We have also measured the temperature dependence of the I(V) curves. In the inset of Fig.1b, it is plotted the behaviour of the second additional peak (V~0.81V) for temperatures up to 20K. The peak is completely quenched at temperatures above 20K. The other secondary peak has the same behaviour with temperature (not shown in the figure)

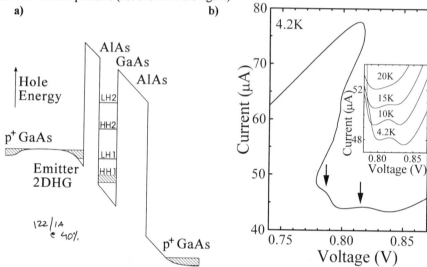

**Figure 1 - a)** Schematic band diagram of the investigated p-type RTD biased into the LH1 resonance, which presents intrinsic multistability. **b)** I(V) characteristics of the p-type RTD measured with the NOR circuit in the bias region of the LH1 resonance. The measurements were carried out at 4.2K. The arrows indicate the plasmon-assisted tunneling peaks. The inset of fig.1.b shows the temperature dependence for the second plasmon-related peak.

The behaviour of both secondary peaks with temperature and their relative voltage position to the main resonant peak (energy) are completely *inconsistent* with LO-phonon assisted tunneling treated in previous works[2,3]. Therefore we interpreted the secondary peaks as plasmon-related peaks. In order to confirm such interpretation, we have used a theoretical model to calculate the hole tunneling transitions in our RTD.

The tunneling rate for an electron coupled to electronic (or lattice) excitations while it is in the well can be written as[4]

$$T(\varepsilon_p) = \frac{\Gamma_L\Gamma_R}{\Gamma}\int_{-\infty}^{\infty}d\sigma\exp\left[\frac{-\Gamma|\sigma|}{2}+i(\varepsilon_{k_z}-\varepsilon_c)\sigma-\sum_q\frac{\upsilon_q}{\pi}\int_{-\infty}^{\infty}d\omega(1+N_q(\omega))\Im m\left(\frac{1}{\in(q,\omega)}\right)\left(\frac{i\sigma}{\omega-\Delta}+\frac{e^{-i\sigma(\omega-\Delta)}-1}{(\omega-\Delta)^2}\right)\right]$$

(1)

In eq.(1), $\Gamma_L$ ($\Gamma_R$) is the tunneling bandwidth through the left (right) barrier; $\varepsilon_{k_z}$ ($\varepsilon_c$) is the quasi level in the emitter (well), $\Delta$ is the energy transfer of the tunneling particle on the x-y plane, $N_q(\omega)$ is the boson distribution function and $\in(q,\omega)$ is the dielectric function of the system. We now apply the above result to hole tunneling. We denote the ground state energy difference between the heavy and light holes in the resonant well by $\omega_0 = E_{lh} - E_{hh}$ and assume that only the lowest level (heavy hole subband HH1) has a significant charge buildup due to rapid scattering of holes from the LH1 level (see Fig. 1a). We use the data in ref. 5 which gives the value of the sheet density in the quantum well as a function of bias. When the system is under such bias that the energy of the incoming holes is slightly higher than that of the light hole subband in the well, the following plasmon assisted tunneling can take place: (i) a hole tunnels into the well, emitting an intra-subband heavy hole plasmon and matches its energy with the light hole subband, $E_e = E_{lh} + \omega_{hh}$; and (ii) a hole tunnels into the well, emitting an inter-subband heavy hole plasmon, matching its energy with the light hole subband, $E_e = E_{lh} + \omega_{hh}^*$, where $\omega_{hh}^* = \left[\omega_{hh}^2 + \omega_0^2\right]^{1/2}$ is the inter-subband energy. The light hole plasmon excitation is not considered due to much lower population of the light hole subband. To obtain a qualitative and quantitative picture of the peak separation and relative strength of processes (i) and (ii), we shall use a simplified model in which both light holes and heavy holes are treated as free two-dimensional hole gases. Both band mixing and finite wavefunction spreading will be neglected. Assuming intra and inter-subband modes are well separated, the imaginary part of the inverse dielectric function including both intra and inter-subband plasmons can be written as,

$$\Im m \frac{1}{\epsilon(q,\omega)} = \frac{\pi}{2}\left\{\omega_{hh}\left[\delta(\omega - \omega_{hh}) - \delta(\omega + \omega_{hh})\right] + \frac{\omega_{hh}^2}{(\omega_0^2 + \omega_{hh}^2)^{1/2}}\left[\delta(\omega - \omega_{hh}^*) - \delta(\omega + \omega_{hh}^*)\right]\right\} \qquad (2)$$

The tunneling current I can be calculated with eqs. (1,2) and

$$I = e\sum_k T(k)\left[f_e(\varepsilon) - f_c(\omega + eV)\right]$$

(3) where $f_e$ ($f_c$) is the Fermi energy distribution of the emitter (collector) and V is the applied bias. The calculated current is plotted in figure 2 as a function of the hole energy. It shows two extra resonant peaks due to intra- and inter-subband plasmon-assisted tunneling. The first peak (closer to the resonance) represents tunneling of holes with the excitation of intra-subband plasmons in the HH1 state in the well. The excitation of more energetic inter-subband plasmons is responsible for the second additional peak in the calculated curve. In the calculation, the quantum-well charge feedback mechanism is neglected and therefore, the resonant peak in the theoretical simulation is not tilted as it is in the experimental results. The temperature dependence of the hole-plasmon peak is very similar to the electron-plasmon peak observed in asymmetric n-type RTDs[1], although the hole-plasmon peak is quenched at somewhat smaller temperatures.

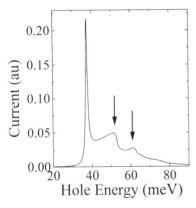

Figure 2 - Calculated tunneling probability for this p-RTD. We can see two peaks associated with both intra- and inter-subband plasmons (arrows).

In summary, we have measured the tunneling process in an asymmetric p-type RTD which shows significant charge build up in the quantum well in resonance. Using a NOR circuit, we were able to probe the region of voltage overhang in the LH1 resonance, which shows bistability when measured with a normal circuit. We observed two additional peaks beyond the LH1 resonance, which quench at temperatures > 20K. Using a theoretical model we assign them as plasmon-assisted tunneling of holes involving intra- and inter-subband plamons.

This work was supported by EPSERC. BRAN and LE acknowledge CNPq (Brazil) and EPSRC for financial support, respectively.

**References:**
[1] M.L.F. Lerch et al. Solid State Electron. **37**, 961 (1994)
[2] E.S. Alves et al. Electron. Lett **24**, 1190, (1988)
[3] V.J. Goldman, D.C. Tsui and J.E. Cunningham, Phys .Rev. Lett. **58**, 1256 (1987)
[4] C. Zhang, M.L.F. Lerch, A.D. Martin, P.E. Simonds and L. Eaves, Phys. Rev. Lett. **72**, 3397 (1994)
[5] R.K. Hayden, L. Eaves, M. Henini, D.K. Maude and J.C. Portal, Phys. Rev. B **49**, 10745, (1994)
[6] A.D. Martin, M.L.F. Lerch, P.E. Simmonds and L. Eaves, Appl. Phys. Lett. **64**, 1248 (1994).

# CYCLOTRON RESONANCE OBSERVATION OF HEAVY AND LIGHT SPIN SUBBANDS IN AN ULTRA-HIGH MOBILITY 2D HOLE GAS IN GaAs/(Al,Ga)As: EVIDENCE FOR COUPLED MAGNETOPLASMONS AND MANY BODY EFFECTS

B.E.Cole[1], S.O.Hill[2], A. Polisskii[2], Y. Imanaka[3],
Y. Shimamoto[3], J. Singleton[2], J.M. Chamberlain[1], N.Miura[3],
M. Henini[1], T. Cheng[1] and P. Goy[4]

[1] Department of Physics, University of Nottingham,
    Nottingham, NG7 2RD, UK
[2] Department of Physics, University of Oxford, The Clarendon
    Laboratory, Parks Road, Oxford, OX1 3PU, UK
[3] Institute for Solid State Phyics, University of Tokyo, Minato-Ku,
    Tokyo 106, Japan
[4] Laboratoire de Physique de l'École Normale Supérieure,
    24 rue Lhomond, 75231 Paris Cédex 05, France

## INTRODUCTION

Many-body effects may be conveniently studied in the two dimensional hole system (2DHS) formed in a GaAs/(Al,Ga)As heterostructure. This 2DHS exhibits extreme valence subband non-parabolicity and a large zero-field splitting of the dispersion relationships of the two spin projections[1]. In consequence, a complex cyclotron resonance (CR) spectrum might be expected[2]. However, the experimental study of CR in this Communication shows that such a complex spectrum is in fact absent in most cases. This absence is discussed in terms of a collective cyclotron motion due to Coulomb interactions between holes. Additional evidence for such a collective motion effect is provided by the observation of a very strong temperature dependence of the apparent CR position.

## EXPERIMENTAL

We have used CR absorption spectroscopy over more than two orders of magnitude of energy and at magnetic fields of up to 40T to study a variety of 2DHSs with areal hole densities ranging from 0.38 to 3.3 x $10^{11}$cm$^{-2}$. Samples were grown by Molecular Beam Epitaxy with layer structure: semi-insulating substrate, $1\mu$m undoped GaAs, 50-period superlattice, $0.5\mu$m undoped GaAs, $Al_{0.33}Ga_{0.67}As$ spacer layer, 40nm doped $Al_{0.33}Ga_{0.67}As$ and 1.7nm undoped GaAs capping layer. The details of samples (respectively: identification, orientation, hole concentration [x$10^{15}$m$^{-2}$], low-temperature mobility [$m^2V^{-1}s^{-1}$] and dopant species) are as follows: #1 (NU938), (311)A, 0.6, 40, Si; #2 (NU942), (311)A, 0.8, 90, Si; #3 (NU939), (311)A, 3.3, 7.8, Be; #4 (NU877), (100), 2.8, 14, Be. CR absorption measurements were carried out at magnetic fields of up to 40T and sample temperatures from 4K-20K, using an optically pumped molecular gas laser (163$\mu$m wavelength)and a long-pulse magnet[3]. Millimetre wave measurements between 25 and 220 GHz were made using a 'Millimetre Wave Vector Analyser' (MVNA)[4] at fields of up to 8T for sample temperatures between 1.4K and 4.2K.

## RESULTS AND DISCUSSION

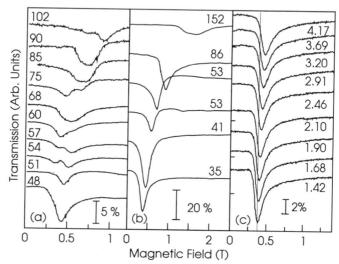

**Figure 1.** (a) Millimetre-wave transmission of #1 at 1.4K, as a function of magnetic field, for various frequencies (indicated above each trace, in GHz). (b) Millimetre-wave transmission of #2 at various frequencies, at 1.4K. (c) Temperature dependence of CR for #2 at 32 GHz. (sample temperatures indicated in K above each trace).

Millimetre wave data for #1 and #2 are shown in Figure 1(a) and 1(b); CRs are observed as dips in the transmission. Note that at some frequencies, the absorptions are split; this indicates that more than one CR transition is taking place. Consideration of a single-particle picture of Landau states suggests that at these low fields two CRs might be expected, because the two different spin subbands (split by the asymmetric confining potential) are both populated. Furthermore, the Landau splitting ~kT, so that more than one CR transition will be possible within each subband. The most surprising aspect of Figures 1(a) and 1(b), therefore, is not that two CR absorptions are occasionally seen but that in many cases only one CR is noted under conditions when multiple transitions are possible. The explanation for this, it is suggested, is that a collective CR mode, analogous to that proposed for two dimensional electron systems (2DES)[5], may occur due to strong Coulomb coupling between the holes. This coupling constrains the cyclotron motion to a single frequency at a "weighted average" of all of the possible CR frequencies.

Further evidence for the coupled nature of cyclotron motion in these systems comes from the temperature dependence of the resonant field, in those cases when only a single resonance is observed. Figure 1(c) shows the pronounced upshift in resonant field for #2 as the sample is warmed from 1.4 to 4.2K. A similar shift is found over a range of frequencies from 32 to 220 GHz[7], and hole densities from 0.38 to 1.28x10$^{11}$ cm$^{-2}$. The shift is apparent in (100) as well as (311) samples.

The large temperature dependence of the CR position can hardly be a single-particle effect because, in a non-interacting system, the hole states are determined by the Hamiltonian and the temperature only defines the distribution of the holes between the states; in an interacting system, however, the hole excitations may, in principle, be temperature dependent.

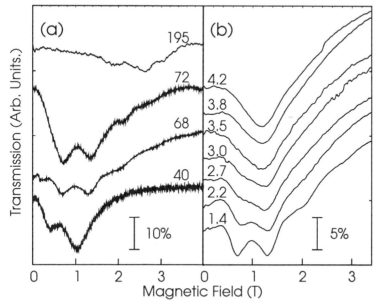

**Figure 2.** (a) Transmission as a function of magnetic field for #3, at various frequencies (indicated above traces, in GHz). (b) Temperature dependance of transmission for #3 at 68 GHz.

Absorption spectra over a range of energies for #3 (a sample of high hole density and comparatively low mobility) are shown in figure 2(a). Two or more resonant absorptions and a strongly energy-dependent lineshape are evident. In the limit of very low magnetic fields the measured CR masses should correspond to the 'classical' effective masses at the Fermi energy, for the two spin branches. The measurement at 40GHz gives values of $m^*=0.8m_e$ and $0.3m_e$, in agreement with calculations[1]. At intermediate field the transitions cannot reliably be assigned in view of the absence of Landau level calculations for (311) orientation samples. Qualitatively, the effect of the magnetic field is to reduce the mass difference between spin subbands; however, strong level mixing means the concept of independent spin states becomes meaningless at higher fields. The temperature dependance at 68GHz for this sample is shown in Figure 2(b). Several discrete transitions may be observed; the effect of increasing temperature is merely to broaden the resonances so that they overlap, rather than causing a shift in their positions.

In Figure 3 we compare the temperature dependence of the CR of (311) (#3) and (100) (#4) oriented samples at 7.6 meV energy. The shift in absorption peak fields with temperature is characteristic of collective mode behaviour, and is ascribed to the movement of the "weighted average" resonant field; this shift results from changes with temperature of the population of the next excited state. Differences in the amount of the field shift observed for the two samples illustrated are attributed to changes in the ordering and separation of the relevent levels in these samples; such changes are a consequence of different anti-crossing[6] behaviour for these particular sample growth directions.

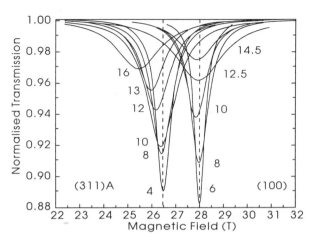

**Figure 3.** Temperature dependence of a (311) (#3) and (100) (#4) sample measured at 7.6meV; temperatures are marked in K.

## CONCLUSION

Observations of CR at millimetre wave and FIR frequencies over a large range of energy and magnetic field demonstrate that collective effects are significant in 2DHS, but that the criteria for the observation of such behaviour are complex. It is generally concluded that collective modes may be more readily observed in samples with a higher mobility; this is apparent in the present work (compare #1 and #2, Figure 1) and in early studies of low mobility 2DHS[8]. The filling factor also appears to be relevent in this context, and a lower filling factor is an indicator that collective mode behaviour might be seen.

## ACKNOWLEDGEMENTS

We acknowledge support from: EPSRC in the United Kingdom, the European Community, The British Council and Monbusho (Japan).

## REFERENCES

1. U. Ekenberg and M. Altarelli, Subbands and Landau levels in the 2D Hole gas at the GaAs-AlGaAs interface, *Phys. Rev. B* 32(6):3712 (1985).
2. J.G. Michels, S. Hill, R. J. Warburton, G.M. Summers, P. Gee, J. Singleton, R.J. Nicholas, C.T. Foxon and J.J. Harris, Cyclotron resonance to 100mK of a GaAs heterojunction in the ultra quantum limit, *Surface Sci.* 305:33 (1994).
3. N. Miura, H. Nojiri and Y. Imanaka, High magnetic fields in semiconductor physics, *in* "Proc. 22nd Int. Conf.on Physics of Semiconductors", D. Lockwood, ed., World Scientific, Singapore (1995).
4. French Patent CNRS-ENS 1989; US Patent nr 5 119 035 (June 2nd 1992).
5. N.R. Cooper and J.T. Chalker, Theory of spin-split cyclotron resonance in the extreme quantum limit, *Phys. Rev. Lett.* 72(13):2057 (1994).
6. U. Ekenburg, *priv. comm.*(1992); V. Nakov, *priv. comm.*
7. S. Hill, B. Cole, J. Singleton, J.M. Chamberlain, P.J. Rodgers, T.J.B.M. Janssen, P.A.Pattenden, B.L. Gallagher, G. Hill, M. Henini, High magnetic field millimetre and submillimetre spectroscopy of ultra high mobility 2D hole systems, *Physica B* 211:440 (1995).
8. H.L.Stormer, Z Schlesinger, A. Chang, D.C.Tsui, A.C. Gossard and W.Wiegmann, Energy structure and quantized Hall effect of two dimensional holes, *Phys.Rev.Lett.* 51:126 (1983).

# QUENCHING OF BISTABILITY BY PHOTOASSISTED TUNNELING THROUGH A SEMICONDUCTOR DOUBLE BARRIER

Jesus Iñarrea and Gloria Platero

*Instituto de Ciencia de Materiales (CSIC) and Departamento de Fisica de la Materia*
*Condensada C-III, Universidad Autonoma, Cantoblanco, 28049 Madrid, Spain.*

## I. INTRODUCTION

In this work we have analyzed the effect of an homogeneus photon field on the intrinsic bistability region of the sequential tunneling current through a double barrier (DB). Before switching on the light we have followed Goldman[1] considering the Coulomb interaction within the framework of the mean field approximation. There are three regions spatially separated: emitter, well and collector, where the charge is accumulated. The potential profile through the whole heterostructure is not abrupt and accumulation and depletion layers in the emitter and collector respectively are built up. Following ref. 1 , the effective charge distribution in the emitter is : $\Sigma_1 = eN(E_F)\Delta_1\delta_1$, where $N(E_F)$ is the three dimensional density of states in the emitter and $\Delta_1$ and $\delta_1$ are the voltage drop and the width in the emitter respectively. We have considered then that the charge in the emitter, well and collector is distributed as two dimensional sheets of charge in the middle of the three different regions. The electrostatic fields and potentials due to these two dimensional charge distributions are obtained by means of the Poisson equation and imposing charge neutrality in the system. Now we have a set of equations, Poisson and Schrodinger which can be solved following an iteraction procedure up to reach convergence. In order to obtain the charge density into the well once the external bias is applied, we follow our previous work[2,3] , where the sequential current through a DB is obtained imposing current conservation through the whole structure. This condition, which takes into account the relaxation proceses into the well and the finite width of the resonant state due to its coupling with the continuum[3], determines the Fermi energy into the well,i.e., the charge density in this region. The modification of the electrostatic potential profile due to the charge distribution through the structure as well as the sequential current is obtained consistently by iteraction. The application of a laser to the heterostructure modifies the tunneling current: the light assists the electron tunneling through photon absorption and emission processes[2]. We have considered linearly polarized light in the FIR regime in the growth direction and we have evaluated the photoassisted sequential tunneling current through the DB[2,3]. One of the main effects produced by illuminating the sample is to change the amount of electronic charge stored in the well: The effect of this charge on the electrostatic potential through the heterostructure is then included in the iteration procedure mentioned above in order to reach consistence in the current. Finally the tunneling current in increasing and decreasing bias is calculated in a mean field approximation and in the presence of light. We have evaluated this effect for different sample config-

*Hot Carriers in Semiconductors*
Edited by K. Hess *et al.*, Plenum Press, New York, 1996

urations, photon frequencies and electromagnetic field intensities: for all the cases we observe a reduction of the main bistability region and the appearence , in some of them, of new bistability regions which can be controlled modifying the parameters mentioned above.

## II. BISTABILITY WITH LIGHT

In fig. 1.a the current density through a $100\overset{\circ}{A} - 50\overset{\circ}{A} - 100\overset{\circ}{A}$ $GaAs - Al_{.3}Ga_{.7}As$ DB is calculated consistently in the presence of a laser (dotted line) for both increasing and decreasing bias. The continuous line corresponds to the current with no light present. We have considered an electromagnetic field of intensity F=8.10$^6$ V/m and $\hbar w$ =13 meV. We observe that the bistability region is reduced due to the light and that two new ones appear in the presence of light. These two new regions of bias where the current is bistable appear in this case at both sides of the central one and can be understood looking at the behaviour of the well state $E_r$ as a function of V (fig.1.b). Let us analyze firstable the case of increasing bias: As the well state crosses $E_F$, the well becomes charged and the induced electrostatic fields produce a non linear dependence of the energy of $E_r$ with V for both cases with and without light. As V increases, the light increases the charge into the well and $E_r$ deviates more from the linearity than in the case without an external laser applied, and it corresponds to the first peaked structure in J (fig.1.a). For higher bias $E_r$ is one photon over $E_c$ (bottom of the emitter conduction band). At this energy the electrons have a finite probability of emitting a photon below $E_c$ and then the light acts discharging the well and $E_r$ drops abruptly as V increases. This effect only occurs in the presence of light and it explains the first peak in the current density and the first bistability region as we will discuss below. The further increasing of V in the region which corresponds to the resonant state over $E_c$ and below one photon energy gives a contribution to J smaller than the values obtained without light. The reason is again the decrease of the charge density into the well due to the light which induces photon emission for $0 < E_r < \hbar w$. Due to this effect, the strong non linear effect of the charge on the electrostatic potentials brings, for a certain bias, $E_r$ below $E_c$ and J decreases abruptly for lower bias than for the case without light (fig. 1.a). Then, in the presence of light and at a certain bias $E_r$ drops abruptly into the range of energy: -$\hbar w < E_r < 0$ . At energies for the resonant state below zero and without illuminating the sample, the current should drop to zero by energy and momentum conservation considerations. However if the light is present, the current also flows due to the fact that at these energies the electrons in the emitter have a finite probability of emitting a photon and tunnel resonantly. Therefore, if we follow the dotted curve in fig. 1.a the current density at bias where $E_r$ is below $E_c$ is different from zero and presents a small peak. As V slightly increases $E_r$ becomes $-\hbar w$ and J drops to zero and the well becomes discharged and if V is decreased the current begins to flow at a different bias than the corresponding to the current cut off in increasing bias. This is the origin of a new bistability region in the current (J). Going further in decreasing V there is a finite current in the main bistability region which is zero in the decreasing bias direction for the case without light and which corresponds to $E_r$ below $E_c$ .The effect of the light charging the well is small in this bias region in decreasing direction and $E_r$ increases abruptly its energy practically at the same bias for both cases, with and without light, corresponding to the well state crossing $E_c$. Therefore the main bistability is not reduced in the left hand side but in the right one due to the light, as explained above. Going further in decreasing bias, $E_r$ increases abruptly its

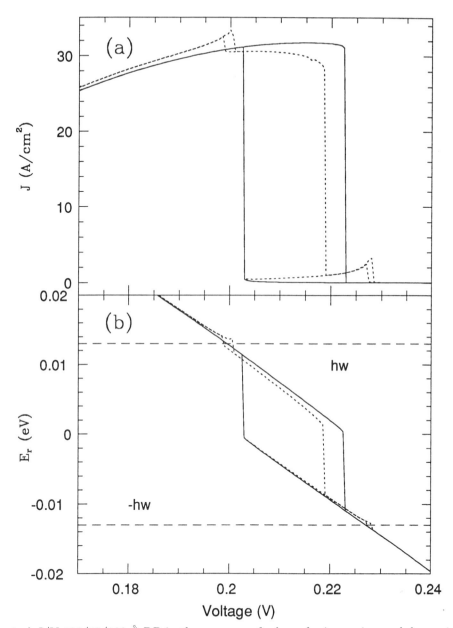

1.a) J/V 100/50/100 Å DB in the presence of a laser for increasing and decreasing bias, with (dotted line) and without (continuous line) light. 1.b) $E_r$ as a function of V with (dotted line) and without (continuous line) light. $F=8.10^6 V/m$, $\hbar w=13$ meV.

energy up to a value below 13 meV ($0 < E_r < \hbar w$), therefore J is smaller than in the case without light due to the discharging effect of the light in this range of energy (as discussed above in increasing bias direction).Decreasing further the bias $E_r$ becomes larger than one photon energy and the photon emission below $E_c$ has zero probability to occur, then the current increases further and the second satellite bistable structure in J is obtained.

In summary, the light assists the tunneling through absorption and emission processes and modifies significantly the charge occupation into the well modifying the current bistability. This effect increases for low photon frequencies, high intense fields, and asymmetric barriers. We observe new satellite bistability regions and negative differential resistance regions associated with them in the current density. We consider a Hartree model and we did not include exchange in the calculation. It is known that the exchange reduces the bistability region which in our model is overestimated[4], however we expect the same qualitatively features due to the light as we have described here. We expect also that these new regions of bistability give new features in the dynamical current.

This work has been supported in part by the Comision Interministerial de Ciencia y Tecnologia of Spain under contract MAT 94-0982-c02-02 and by the Acción Integrada Hispano-Alemana HA84.

REFERENCES
1. V.J.Goldman, D.C.Tsui and J.E.Cunningham,Resonant tunneling in magnetic fields: Evidence for space-charge buildup, Phys. Rev. Lett. **58** 1256 (1987).
2. J.Iñarrea, G.Platero and C.Tejedor,Coherent and sequential photoassisted tunneling through a semiconductor double-barrier structure , Phys. Rev. B **50** 4581 (1994).
3. Jesus Iñarrea, Gloria Platero,Magnetotunneling through a semiconductor double barrier structure assisted by light, Phys. Rev. B, **51** 5244 (1995).
4. N.Zou,M.Willander,I.Linnerud,U.Hanke,K.A.Chao and Y.M.Galperin, Suppression of intrinsic bistability by the exchange-correlation effect in resonant-tunneling structures, Phys Rev B,**49** 2193 (1994).

# POSITION DEPENDENT AC POTENTIAL VERSUS HOMOGENEOUS IRRADIATION APPLIED TO RESONANT HETEROSTRUCTURES

Ramón Aguado, Jesús Iñarrea and Gloria Platero

*Instituto de Ciencia de Materiales (CSIC) and Departamento de Fisica de la Materia Condensada C-III, Universidad Autonoma, Cantoblanco, 28049 Madrid, Spain.*

## I. INTRODUCTION

In the last years several works have been devoted to the analysis of the effect of AC fields on the transport properties of resonant heterostructures, however there is not yet a systematic discussion of the different experimental situations corresponding to an AC potential applied through the sample and oscillating with a position-dependent dephasing and the case where the whole sample is homogeneously illuminated[1]. In spite of the increasing interest in this field, most of the theoretical work has been done considering the first configuration with the approximation that the interaction does not cause transitions between electronic states. We have analyzed both configurations including, in the case of homogeneous illumination the coupling of electronic states due to the interaction with the electromagnetic field and we have shown that additional tunneling channels appear due to this interaction which are reflected in the transmission and current density.

## II. RESULTS FOR THE TWO DIFFERENT CONFIGURATIONS

We have studied the two configurations discussed in the previous section. In order to analyze coherent resonant tunneling we have extended the GTH formalism (Generalized Transfer Hamiltonian)[2] to include the effect of an AC potential. The stationary transmission probability can be written as:

$$P_{RL} = \frac{2\pi}{\hbar} \sum_{n,m=-\infty}^{\infty} J_n^2 \left(\frac{V_{AC1}}{\hbar\omega_0}\right) J_m^2 \left(\frac{V_{AC2}}{\hbar\omega_0}\right) \delta(\epsilon_{p_R} - \epsilon_{k_L} - n\hbar\omega_0 - m\hbar\omega_0)$$

$$\mid \langle p_R | V_L + V_R G^+(\epsilon_{k_L} + m\hbar\omega_0) V_L | k_L \rangle \mid^2 \tag{1}$$

Where the Bessel functions $J_n$ and $J_m$ account for the photoside bands associated to the left and right states respectively, and which contribute as additional channels to the transmission[3]. In this case we consider the amplitude of the AC field constant within each spatial region, therefore the field does not couple different electronic states (off-diagonal terms) within each spatial region. We have obtained as well the transmission probability of tunneling through a homogeneously illuminated heterostructure, considering that the electron-photon coupling contains the electronic matrix elements of the momentum including both diagonal and off-diagonal terms. The first ones give the photoside bands weighted by the Bessel functions whose arguments contain the momentum matrix elements which are very small : $\beta_k = \frac{eF\langle k|P_z|k\rangle}{m^*\hbar\omega^2}$. Therefore just the lowest order ones give a contribution to the transmission. This was not the situation in the previous case where many photoside bands contribute for arbitrary amplitude of the AC potential. The off-diagonal terms are in the case of homogeneous illumination, the ones which behave as additional tunneling channels compared with the case without the external field and the only ones which modify the transmission with respect to the non illuminated case. We have plotted the transmission coefficient T for a GaAs-AlGaAs double barrier for both configurations (fig.1) and for $\frac{eFd}{\hbar\omega_0} = 0.77$. The main difference between them is that the momentum matrix elements are very small, therefore, in the case of homogeneous light the Bessel functions of higher order than zero are negligible and the intensity of the corresponding photoside band can be neglected and to consider just the main peak (m=0) makes sense. For the case of an AC field, where we consider that the barriers oscillate dephased in $\pi$, the argument of the Bessel functions does not contain those elements and Bessel functions up to forth order give a non negligible contribution to J. This fact is reflected in the four satellites which appear at both sides of the main central peak (Bessel function of zero-order) (fig. 1.a). However we observe in fig. 1.b the presence of two satellites in T. These side-peaks have another origin than the photoside bands [1]. They come from the mixing of electronic states due to the light and which show up in the hamiltonian as the off-diagonal matrix elements of the electronic momentum in the coupling term. In this case just the main proccesses, involving absorption and emission of one photon are considered[1]. Therefore, the tunneling channels for the two configurations are different: in the case of an AC potential the off- diagonal terms cancel if the time dependent field is considered constant within each spatial region (emitter, well and collector) and the main tunneling channels (the only ones within this approximation) are the photoside bands: as the energy of the emitter photoside bands concides with the energy of the photoside bands in the well additional contributions to T and J occur. This contribution can be important even for high order photoside bands if the ratio $\frac{V_{AC}}{\hbar\omega_0}$ is of the order or higher than one. In the case of homogeneous light the off-diagonal terms are those which modify the current density. Those channels, involving different electronic states contribute in principle also with all their photoside bands, however, as the argument of the Bessel functions ( i.e., the intensity of the m photoside band ) remains very small, just the zero index band gives a contribution to J: the three peaks in T come from the main bands (index zero) correspond to three electronic states which differ in one photon energy and which are coupled by the field. We have also plotted J for different AC fields (fig.2): As main features we observe that the threshold current moves to lower bias. Also a step-like behaviour is observed in the current. We can explain these features in terms of the photoside bands: there are photoside bands associated to electronic states close to the Fermi energy $E_F$ which contribute to the resonant tunneling even when the resonant state $E_r$ is higher in energy than $E_F$. If $E_r$ is higher than $E_F$ in several photon energies the photoside bands which produce the current have a low spectral weight and the contribution to the current is small. As far as $E_r$ becomes closer to $E_F$, increasing the bias, the lower indexes photoside bands, i.e., those which are more intense in the spectral function become aligned with $E_r$ and therefore their contribution to the current increases. Once $E_r$ crosses $E_F$ J increases but remains smaller than the current with no AC applied. That is because the spectral function has finite weight in all the photoside bands and not only in the main one whose weight is smaller than one. In this case only a small number of them tunnels resonantly and the effect of the field is to reduce the current. For homogeneous illumination the effect of the light is much smaller and it

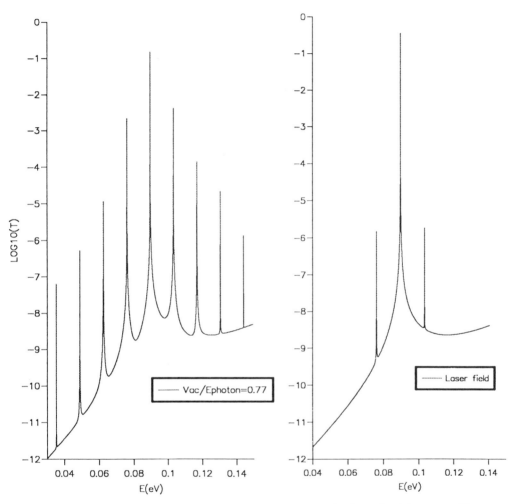

1. $Log_{10}$ T(E) for a double barrier GaAs-AlGaAs $(100/50/100\ \mathring{A}\ )$ (a)AC field:$\frac{V_{AC}}{\hbar\omega_0} = 0.77$. (b)Homogeneous light : $F = 4.10^5 \frac{Volt}{m}, \hbar\omega_0 = 13.6 meV$

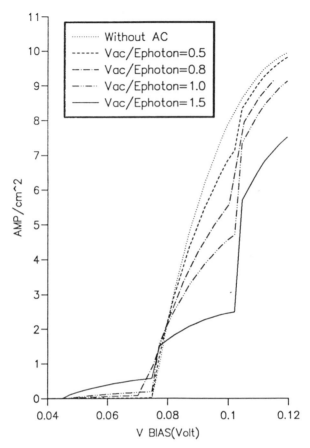

2. J(V) for the same configuration as in fig. 1(a) for different ratios of intensity and energy of the AC field ( $\hbar\omega_0 = 13.6meV$ ).

can be observed in the current difference as a function of V [1].

This work has been supported by the CICYT under contract MAT 94-0982 -c02-02 and by the Acción Integrada Hispano-Alemana HA84.

REFERENCES
1. J.Iñarrea, G.Platero and C.Tejedor, "Coherent and sequential photoassisted tunneling through a semiconductor double barrier structure", Phys. Rev. B **50** 4581 (1994)
2. L.Brey G.Platero and C.Tejedor, "Generalized transfer Hamiltonian for the study of resonant tunneling", Phys. Rev. B **38** 10507 (1988)
3. R.Aguado, G.Platero and J.Iñarrea," Coherent resonant tunneling in time dependent fields", to be published.

# PATTERNS IN BISTABLE RESONANT TUNNELING SYSTEMS

V. A. Kochelap[1], B.A.Glavin[1], V.V.Mitin[2]

[1]Institute of Semiconductor Physics, National Academy of Sciences,
252650 Pr. Nauka 45, Kiev, Ukraine
[2]Department of Electrical and Computer Engineering Wayne State University,
Detroit, MI 48202

## INTRODUCTION

In this work we present a previously unreported phenomenon of formation of nonuniform transverse structures in resonant tunneling systems demonstrating bistable (Z-like) current-voltage characteristics (CVC)[1]. The bistability is a result of electron charge accumulation in quantum well layer that, in turn, leads to significant change in potential profile and probability of tunneling. Such an electrostatic feedback is also a reason of spontaneous formation of transverse (with respect to the tunneling current) patterns (nonuniform structures). These structures are nonuniform distributions of the built-in charge and associated with them nonuniform tunneling current. We have developed a self-consistent theory of these patterns and calculated spatially nonuniform stationary structures of built-in charge and current filaments. We have done some conclusions about the nonstationary patterns.

## MODEL

We consider semiconductor double barrier heterostructure as a system consisting of three parts, emitter, quantum well and collector, separated by barriers. The quantum well holds a resonant level and is weakly coupled with other parts by tunneling of the electrons between the resonant level and the electrodes. The coupling is assumed weak so that the width of the resonant level is much less than the Fermi energy of the electrons in the emitter. Analysis of such a model shows that characteristic scale of the patterns, $l$ is about or larger than $v_F\tau$, where $v_F$ is the Fermi velocity, $\tau$ is characteristic time of the electrons in the quantum well with respect to tunneling, scattering, etc. Since $l$ is much greater than both, electron de-Broglie wavelength and thickness of the heterostructures,

*Hot Carriers in Semiconductors*
Edited by K. Hess *et al.*, Plenum Press, New York, 1996

one can use classical Boltzmann equation for two-dimensional electrons

$$\frac{\partial f}{\partial t} + \mathbf{F}\frac{\partial f}{\partial \mathbf{p}} + \frac{\mathbf{p}}{m}\frac{\partial f}{\partial \mathbf{r}} = G(\mathbf{r}, \mathbf{p}, t) - \frac{f(\mathbf{r}, \mathbf{p}, t)}{\tau} \tag{1}$$

Here $f$ – electron distribution function, $\mathbf{r} \equiv \{x, y\}$ is the two-dimensional vector of position, $\mathbf{p}$ is the electron momentum, $\mathbf{F}$ is the electrostatic force. $G(\mathbf{r}, \mathbf{p}, t)$ is the generation function, i.e. the tunneling rate of the electrons from the electrodes to the quantum well. $G$ depends on the electrostatic potential in the well, $\Phi$, because this potential determines position of the resonant level with respect to the bottom of the conduction band of the electrodes. In general case $\tau$ is also function of $\Phi$. For finite lateral dimensions of the heterostructure, $L_x, L_y$ one should impose boundary conditions at the sides of the heterostructure. In this case the patterns are strongly dependent on these conditions.

For simplicity we suppose that the quantum well is much narrower than the thickness of the double barrier structure, $d$. Then in Hartry approximation one can write the Poisson equation for $\Phi$:

$$\Delta\Phi = \frac{4\pi e}{\epsilon}\delta(z - a)\int d\mathbf{p}f(\mathbf{r}, \mathbf{p}, t), \tag{2}$$

where $\epsilon$ is dielectric constant, $a$ is the thickness of the emitter barrier. The boundary conditions for $\Phi$ are $\Phi(z = 0) = 0$, $\Phi(z = d) = V$, where $V$ is an external voltage bias applied to the heterostructure. For calculation of the function $G$ one can use both approaches applicable for resonant tunneling processes: coherent and sequential tunneling models. Both approaches lead to almost the same results. We assume also the Fermi distributions for the electrons in the electrodes. Thus, if the function $G$ is found the system of the equations (1), (2) describes self-consistently nonlinear electron transport through the double barrier heterostructure.

## RESULTS

First we have calculated the built-in charge, the voltage $\Phi(z)$ and current for the case of absence of the transverse patterns (uniform problem). We have assumed the coherent tunneling mechanism and set the following parameters: the resonant energy $E_r = 0.1\ eV$, the Fermi energy $E_F = 0.05\ eV$, the effective mass $m = 0.67\ m_0$, $a = 1.5\ nm$ and $a/d = 1/3$. In Fig.1 CVC with the bistable region $(V_l, V_h)$ is shown for such a case.

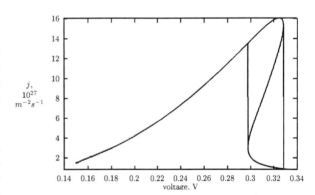

**Figure 1.** Calculated current-voltage characteristic of resonant tunneling diode in uniform case.

Analysing the nonuniform problem we have found that for the external voltage within the edges of the bistable region $(V - V_l \ll V_l,\ V_h - V \ll V_h)$ the spatial scales of the patterns are much greater than $l = v_F\tau$. Thus Eq. (1) can be reduced to drift-diffusion type of equation.

The equation contains nonlinear "generation- recombination" terms related to the injection of the electrons in the well and their escape. Since in the uniform case "generation-recombination" balance permits three stationary solutions in the bistable region, the drift-diffusion equation gives infinite number of the patterns. For finite in-plane dimensions of the heterostructure one should impose boundary conditions on the sides of the heterostructure. They

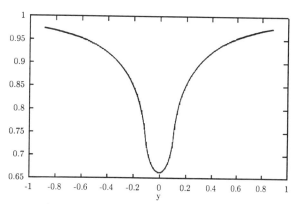

**Figure 2.** Density of built-in charge for the filament-like state.

select one or several patterns, which can be realized. For large dimensions $L_x, L_y$ we have found three different stationary patterns: 1) filament of tunneling current and peak of built-in charge, 2) dip of these values, 3) kink-like transition between two stable almost uniform solutions (exists only at certain bias $V_c$).

The same patterns have been investigated for the voltage inside the bistable region. In this case the spatial scale is of the order of $l$. In Fig.2 the density of the built-in charge is shown as function of $y$ at the voltage $V = V_l + (V_h - V_l)/4$ ($y$ is in units $l$, density is in units of the value which corresponds to the high-current uniform state). The kink-like transition is shown in Fig.3 for $V = V_c \equiv V_l + (V_h - V_l)/2$.

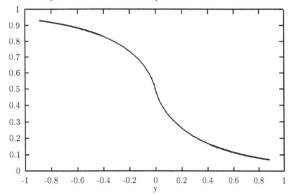

**Figure 3.** Density of built-in charge for the kink-like state.

For infinite ideal systems position of the patterns 1 and 2 is indefinite, but in fact they may be pinned by some inhomogeneity (impurities, nonuniform illumination, etc). The kink-like pattern and corresponding $V = V_c$ may be stabilized in result of interaction of the diode with external electric circuit.

The study has showed that there are critical perturbations, which inspire propagation of the switching waves in the quantum well layer. Velocity of these waves is about $v_F$. This result allows one to determine the role of the boundary conditions on the sides of the heterostructure. For example, if the boundaries reflect the electrons, CVC remains unchanged. If the boundaries inject or absorb additionally the electrons, the switching waves are inspired and some portions of CVC becomes unstable. These boundary conditions may be controlled by additional electrode (the pilot) at common quantum well (the base). We have considered the heterostructure with such a pilot and showed that by means of the pilot it is possible to control the patterns and CVC.

1. V. J. Goldman, D. C. Tsui and J. E. Cunninham, Phys.Rev.Lett., 58:1256 (1987).

# VALLEY CURRENT REDUCTION IN LATERAL RESONANT TUNNELING

Zhi-an Shao, Wolfgang Porod, and Craig S. Lent

Department of Electrical Engineering
University of Notre Dame
Notre Dame, IN 46556

## INTRODUCTION

In resonant tunneling devices it is advantageous to minimize the valley current to enable clear digital switching and reduce off-state power dissipation. High current peak-to-valley ratio also gives high power output and better dc-to-ac conversion efficiency.[1] In a double-barrier resonant tunneling (DBRT) device, the current peak-to-valley ratio is limited even in the ballistic transport regime,[2] since the transmission minimum has a finite value.

Recent studies show that the wave nature of the electron can give rise to new devices in the nanometer scale. Sols *et al.*[3] and Datta[4] independently proposed the concept of stub tuners. In this device, the transmission through a narrow channel is modified by changing the length of a side branch. Transistor action can be achieved by adjusting the applied gate voltage on the side branch. Experimental evidence of quantum interference effect in this structure has been presented by Miller *et al.*[5] and Aihara *et al.*[6] Weisshaar and co-workers investigated the transmission characteristics of split-gate devices.[7] A negative differential resistance region is obtained in the *I-V* characteristics at low temperature.

In previous work, we have investigated the transmission amplitude for quantum waveguides with resonantly-coupled cavities. We find that each quasi-bound state in the resonant cavity leads to a zero-pole pair in the complex-energy plane.[8] Here, we apply this idea to explore the possibility of engineering the placement of these zero-pole pairs in the complex-energy plane in an effort to minimize the valley current in 2D lateral DBRT structures.

## TRANSMISSION ZERO ENGINEERING

Combining the transmission features of the double-barrier resonant tunneling devices and the quantum waveguides with resonantly-coupled cavities, we propose to insert a resonant cavity (stub here) in the quantum well region of a DBRT structure (see fig. 1). We expect that the valley current can be reduced in this proposed device due to the zero transmission feature of the stub.

For a lateral DBRT device, the first transmission peak, $E_{pdbrt}$, is determined by the energy of the first propagating mode of the channel, $E_{Wch} = \hbar^2\pi^2/(2m^*W_{ch}^2)$, and the first quasi-bound state energy of the corresponding 1D DBRT structure, $E_1 = \hbar^2\pi^2/(2m^*W_s^2)$. Due to the inserted stub, in our proposed device, the first transmission peak, $E_{pstub}$, is determined by the bound state energy of the stub in the y-direction, $E_L = \hbar^2\pi^2/(2m^*L^2)$, and $E_1$ approximately. Since $L$ is always larger than $W_{ch}$ (see fig. 1), the position of the first transmission peak for the proposed device is lower than that of the corresponding lateral DBRT device, i.e., $E_{pstub} < E_{pdbrt}$ (see fig. 2).

In order to utilize the resonant nature of the first transmission peak and transmission zero in the proposed device, the peak energy must be higher than the first propagating mode energy of the channel, i.e., $E_{pstub} > E_{Wch}$. We can achieve this condition by engineering the device parameters appropriately.

In resonant tunneling devices, the peak and valley currents are related to the transmission maximum and minimum, respectively. Since the transmission minimum is zero in the proposed device (see fig. 2), its valley current is drastically reduced compared to the lateral DBRT devices. Further, note that since $E_{pstub} < E_{pdbrt}$ and the position of $E_{pstub}$ can be engineered, a controllable low peak voltage can be obtained in the proposed device, where peak voltage refers to the voltage at which the maximum current is reached.

We numerically modeled our proposed device. We first solve the 2D effective-mass Schrödinger eq. by the finite element method to find the transmission probability $T(E,V)$ as a function of energy at different biases, and then we calculate the $I$-$V$ characteristics by,

$$I(V) = \frac{2e}{h}\int_0^\infty [f(E) - f(E + eV)] T(E, V)\, dE$$

where $f(E)$ is the carrier distribution function. We choose a Fermi energy $of$ $E_f$=5 meV in our calculation, which corresponds to a carrier density of $5.96\times10^5$ (1/cm).

## EXAMPLE

Based on the above analysis, we choose a set of parameters for the structure shown in fig. 1: $L$=20.0 nm, $W_{ch}$=10.0 nm, $L_b$=1.5 nm, $W_s$=8.0 nm, and $V_h$=0.3 eV. In fig. 1, the shaded areas represent the tunneling barriers, and the device potential profile with an applied bias is also shown. In fig. 2, we show the transmission probabilities of this structure (solid line) and the corresponding lateral DBRT structure (dashed line) at zero bias. We have

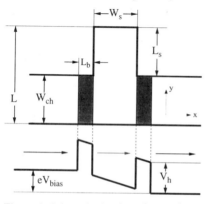

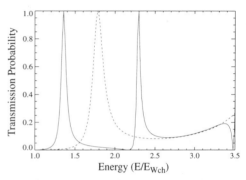

**Figure 1**. Schematic drawing of a quasi-1D DBRT structure with an inserted stub.

**Figure 2**. Transmission probabilities of a quasi-1D structure shown in fig. 1; see text.

used the energy of the first propagating mode of the channel, $E_{Wch}$, as the unit of the energy. Note that transmission zeros exist in the proposed device structure.

We show the *I-V* characteristics of the proposed device at zero temperature (solid line) in fig. 3. The result of the corresponding lateral DBRT device is also shown (dashed line). It can be seen that the current peak-to-valley ratio of the proposed device (1457) is drastically improved over the corresponding lateral DBRT device (18.56). Obviously, a low peak voltage is also obtained in the proposed device, which is desirable in device applications.

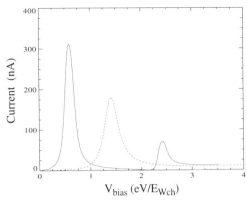

**Figure 3**. *I-V* characteristics of a DBRT structure with an inserted stub (solid line) and the corresponding lateral DBRT structure (dashed line).

In DBRT devices, a lower peak voltage is obtained when the well width is large ($E_1=\hbar^2\pi^2/(2m^*W_s^2)$). But, this also leads to a lower current peak-to-valley ratio since $E_1$ and $E_2$ are closer in this case. Therefore, there is a conflict between the two requirements in the device design. Here, by inserting a resonant cavity in the quantum well region of a DBRT structure, both higher current peak-to-valley ratio and lower peak voltage are achieved.

## SUMMARY

We have proposed a DBRT device with resonantly-coupled cavities. Through engineering the zero transmission feature, we find that the valley current in the proposed device is drastically reduced compared to the lateral DBRT device. As a result, the current peak-to-valley ratio is increased. A controllable lower peak voltage is also achieved in the proposed device.

**Acknowledgments:** This work was supported in part by ARPA/ONR and AFOSR.

1. S. M. Sze, Microwave Diodes, in: "High Speed Semiconductor Devices," S. M. Sze, ed., John Wiley & Sons, New York, (1990).
2. K. Hess and G. J. Iafrate, Theory and Applications of Near Ballistic Transport in Semiconductors, *Proceedings of the IEEE* **76**, 519:532 (1988).
3. F. Sols, M. Macucci, U. Ravaioli, and K. Hess, Theory for a Quantum Modulator Transistor, *J. Appl. Phys.* **66**, 3892:3906 (1989).
4. S. Datta, Quantum Devices, *Superlattices and Microstructures* **6**, 83:93 (1989).
5. D. C. Miller, R. K. Lake, S. Datta, M. S. Lundstrom, M. R. Melloch, and R. Reifenberger, Quantum Devices and Transistors, in: "Nanostructure Physics and Fabrication," M. A. Reed and W. P. Kirk, ed., Academic, Boston, (1989).
6. K. Aihara, M. Yamamoto, and T. Mizutani, Three-Terminal Conductance Modulation of a Quantum Interference Device Using a Quantum Wire with a Stub Structure, *Appl. Phys. Lett.* **63**, 3595:3597 (1993).
7. A. Weisshaar, J. Lary, S. M. Goodnick, and V. K. Tripathi, Negative Differential Resistance in a Resonant Quantum Wire Structure, *IEEE Elect. Dev. Lett.* **12**, 2:4 (1991); Analysis and Modeling of Quantum Waveguide Structures and Devices, *J. Appl. Phys.* **70**, 355:366 (1991).
8. W. Porod, Z. Shao, and C.S. Lent, Transmission Resonances and Zeros in Quantum Waveguides with Resonantly-Coupled Cavities, *Appl. Phys. Lett.* **61**, 1350:1352 (1992) and Resonance-Antiresonance Line Shape for Transmission in Quantum Waveguides with Resonantly-Coupled Cavities, *Phys. Rev. B* **48**, 8495:8498 (1993); Z. Shao, W. Porod, and C.S. Lent, Transmission Resonances and Zeros in Quantum Waveguide Systems with Attached Resonators, *Phys. Rev. B* **49**, 7453:7465 (1994).

# MAGNETOTUNNELING SPECTROSCOPY OF A QUANTUM WELL WITH A MIXED STABLE/CHAOTIC CLASSICAL PHASE SPACE

T.M. Fromhold[1], T.J. Foster[1], P.B. Wilkinson[1], L. Eaves[1], F.W. Sheard[1], M. Henini[1], N. Miura[2], T. Takamasu[2], D.K. Maude[3], and J.C. Portal[3]

[1]Department of Physics, University of Nottingham,
Nottingham NG7 2RD, UK
[2]Institute for Solid State Physics, University of Tokyo,
Roppongi, Minato-ku, Tokyo 106, Japan
[3]SNCI-CNRS, 38042 Grenoble, France

## INTRODUCTION

The classical phase space for hot electrons resonantly injected into a wide quantum well (QW) exhibits a transition from stable regular motion to strong classical chaos when an applied magnetic field $B$ is tilted away from the normal to the well walls[1]. The corresponding quantised energy level spectra and eigenfunctions of this experimentally accessible system have been investigated in the regime of strong classical chaos using resonant tunneling spectroscopy of very wide QWs ($\sim 100$ nm)[1,2]. In this domain, distinct series of periodic resonant peaks can be identified and related to periodic fluctuations[3] in the density of levels in the QW and periodic patterning of the wavefunctions[4].

In this paper we report new magnetotunneling studies of a 22 nm wide GaAs/(AlGa)As QW which is sufficiently narrow that the classical phase space for injected electrons contains islands of stable orbits *and* a sea of strongly chaotic trajectories for a range of tilt angles $\theta$ relative to the tunneling direction. In contrast to previous studies of QWs with a mixed classical phase space[2], the tunneling electrons can only access unstable orbits in the QW. We show that resonant features in the current-voltage characteristic $I(V)$ of the resonant tunneling diode (RTD) are controlled by a single unstable periodic orbit in the QW. The effects of very high tilted magnetic fields (up to 37 T) on the quantum properties of a 60 nm wide QW are investigated experimentally.

## CLASSICAL AND QUANTUM PROPERTIES OF THE 22 nm WIDE QW

Apart from the well width $w$, the composition of the RTD is identical to that given in Ref. 5. Under a bias voltage $V$, electrons tunnel into the QW from a two-dimensional electron gas (2DEG) in the emitter contact[1,2]. The $\theta$-dependence of the classical phase space for electrons injected into the QW is shown by the Poincaré sections[3] in Fig. 1.

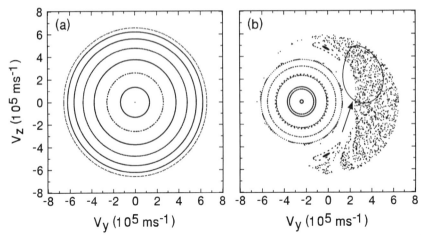

**Figure 1.** Poincaré sections for the 22 nm wide QW showing $(v_y, v_z)$ for 3400 collisions with the LH barrier generated from 17 different starting velocities consistent with the electron injection energy at $B$ = 17 T, $V$ = 488 mV and $\theta$ = 0° (a), and 40° (b). Solid curve (arrowed) in (b): range of lateral velocity components for electrons entering the QW from the 2DEG.

The scattered points show the velocity components $(v_y, v_z)$ in the plane of the QW each time the electron hits the left-hand (LH) barrier interface. The in-plane component of $B$ is parallel to the z-axis[4]. At $\theta$ = 0°, the electrons execute cyclotron motion about $B$, and so $(v_y, v_z)$ lie on concentric circles (Fig. 1(a)). When $\theta$ = 40° (Fig. 1(b)), islands of stable orbits persist at the LH side of the Poincaré section. However, there is also a sea of chaos within which the classical orbits are highly unstable. The size of the chaotic sea increases with increasing $B$ until it fills all of the phase space when $B \approx$ 40 T. This regime of strong classical chaos will be described elsewhere[6]. Here we consider the magnetic field value $B$ = 17 T for which the classical phase space has a mixed stable/chaotic character.

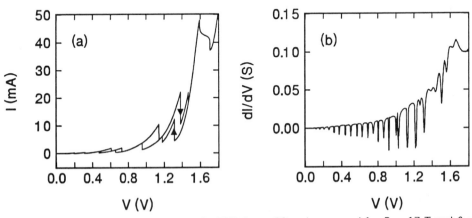

**Figure 2.** Tunneling characteristics for the RTD ($w$ = 22 nm) measured for $B$ = 17 T and $\theta$ = 0° (a), and 40° (b). Arrows in (a) indicate upsweep and downsweep of $V$.

To demonstrate how the onset of chaotic motion influences the tunneling process, Fig. 2 shows tunneling characteristics measured for $\theta$ = 0° (a) and 40° (b). When $\theta$ = 0°, resonant tunneling into two-dimensional subbands in the QW generates a series of resonant peaks in $I(V)$[5]. The peak positions are different for upsweep and downsweep of $V$ due to charge buildup in the QW[7]. Tilting $B$ at 40° to the tunneling direction (Fig. 2(b)) suppresses this bistability by reducing the charge buildup in the QW[6]. At $\theta = 40°$, a series of periodic resonant peaks is observed in $I(V)$ with a voltage spacing $\Delta V \approx 60$ mV.

To investigate the origin of these resonant peaks, we have calculated the energy level spectrum of the QW as described in Ref. 4. Conduction band nonparabolicity is included by using a mean effective mass $m^* = 0.067\ m_e(1 + \alpha K(V))$, where $K(V)$ is the kinetic energy of electrons at the centre of the QW under bias $V$, and $\alpha = 2$ eV$^{-1}$.

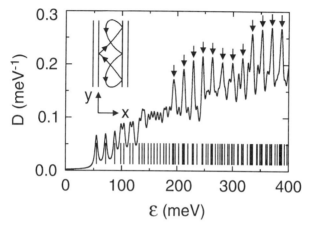

**Figure 3.** Energy levels (vertical lines) and density of levels $D$ versus electron energy $\epsilon$ measured from conduction band edge at RH side of the 22 nm wide QW. Arrowed maxima in $D$ originate from periodic clustering of energy levels associated with the unstable periodic orbit inset projected onto the $x$-$y$ plane. $B = 17$ T, $\theta = 40°$, $V = 488$ mV, $a = 2$ eV$^{-1}$.

Fig. 3 shows energy levels calculated for the QW, together with the density of levels $D$ obtained by broadening each energy level to a width $\Gamma = 6$ meV consistent with the electron scattering rate[1]. Because $\Gamma$ greatly exceeds the mean level spacing, individual energy levels are not usually resolved. However, periodic fluctuations in the *density* of levels (maxima arrowed in Fig. 3) can be identified and related to an unstable but periodic classical orbit in the QW in accordance with the Gutzwiller trace formula[3]. The energy period of the oscillations $\Delta \epsilon = 17.5$ meV is consistent with the value $h/T_P = 17.2$ meV obtained from the period $T_P = 0.24$ ps of the orbit shown inset in Fig. 3. Electrons in this orbit make three successive collisions on the right-hand (RH) barrier per period. The starting velocity for the orbit at the LH barrier is consistent with the range of velocity components of electrons entering the QW (solid curve in Fig. 1(b)). Since this periodic orbit is accessible to the tunneling electrons, the associated Gutzwiller oscillations in $D(\epsilon)$ are expected to generate periodic resonant peaks in $I(V)$. We estimate the voltage spacing of these peaks to be $\Delta V = fh/eT_P \approx 70$ mV, where $f \approx 4$ is determined from the potential distribution across the device and the mean position of an electron in the QW[1]. The good agreement between this calculated voltage spacing and the experimental value of $\approx 60$ mV shows that the observed oscillatory structure could originate from the Gutzwiller fluctuations in $D(\epsilon)$. However, our recent calculations of eigenstates for a 120 nm wide QW have shown that the wavefunctions corresponding to regular subsets of *individual* energy levels reveal regions of high probability density or 'scars' along the paths of particular unstable but periodic classical orbits in the QW[4]. This scarring of individual wavefunctions occurs with the same energy interval as the periodic clustering of *groups* of energy eigenvalues, and might also be expected to generate periodic resonant peaks in $I(V)$. Consequently, calculations of the wavefunctions of the 22 nm wide QW, and their influence on the tunneling characteristics of the RTD, are required to determine whether the observed resonant structure originates from Gutzwiller fluctuations in $D(\epsilon)$ or from periodic scarring of individual eigenfunctions, and are in progress.

## HIGH MAGNETIC FIELD STUDIES OF A 60 nm WIDE QW

We have investigated experimentally how the $I(V)$ characteristics of a RTD containing a 60 nm wide QW[2] evolves with increasing $B$. Tilted magnetic fields up to 37 T were used to generate strongly chaotic orbits in the QW over wide ranges of $V$ and $\theta$.

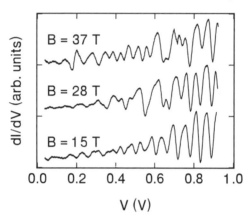

**Figure 4.** Tunneling characteristics for a RTD containing a 60 nm wide QW with $\theta = 40°$ and magnetic field values indicated.

Fig. 4 shows $dI/dV$ versus $V$ plots for several $B$ values at $\theta = 40°$. When $B = 15$ T, the observed series of periodic resonances can be related to periodic orbits similar to that shown in Fig. 3[6]. Orbits of this type exist for a wide range of $B$, and so the associated resonant peaks are seen in each $dI/dV$ plot in Fig. 4. The question of whether the resonant peaks reported here originate from clusters of energy levels in the QW or from individual scarred states can, in principle, be resolved by measuring $I(V)$ in tilted magnetic fields which are large enough to permit the resolution of individual energy levels in the QW.

TMF and LE are supported by EPSRC Advanced and Senior Fellowships.

## REFERENCES

1. T.M. Fromhold, L. Eaves, F.W. Sheard, M.L. Leadbeater, T.J. Foster, and P.C. Main, *Phys. Rev. Lett.* 72:2608 (1994).
2. T.M. Fromhold, A. Fogarty, L. Eaves, F.W. Sheard, M. Henini, T.J. Foster, P.C. Main, and G. Hill, *Phys. Rev. B* 51:18029 (1995).
3. M.C. Gutzwiller, "Chaos in Classical and Quantum Mechanics", Springer-Verlag, New York (1990).
4. T.M. Fromhold, P.B. Wilkinson, F.W. Sheard, and L. Eaves, to be published in *Phys. Rev. Lett.*.
5. M. Henini, M.L. Leadbeater, E.S. Alves, L. Eaves, and O.H. Hughes, *J. Phys.: Condens. Matter* 1:3025 (1989).
6. P.B. Wilkinson, T.M. Fromhold, L. Eaves, F.W. Sheard, N. Miura, and T. Takamasu, to be published.
7. F.W. Sheard and G.A. Toombs, *Appl. Phys. Lett.* 52:1228 (1988).

# SPECTRAL HOLE BURNING AND CARRIER-CARRIER INTERACTION IN SEMICONDUCTOR QUANTUM WELL LASERS: A MONTE CARLO INVESTIGATION

L. Rota,[1,2] M. Grupen,[2] and K. Hess[2]

[1]Clarendon Laboratory, University of Oxford
Parks Road, Oxford OX1 3PU, U.K.

[2]Beckman Institute, University of Illinois
Urbana, Illinois 61801

## INTRODUCTION

Carrier trapping in semiconductor quantum wells is a process of fundamental importance for optoelectronic device applications. After several years of both theoretical and experimental research a clear picture of this process is not yet available and many contradictory results can be found in the literature. The first quantum approach[1] indicated a quite slow trapping rate and a strong well width dependence, but was followed by several experimental investigations[2−4] which showed a fast trapping rate and no evidence of well width dependence. A more recent and systematic report[5] was able to partially reconcile theory and experiment indicating at least a weak dependence of the trapping process on well width. Moreover most of the research carried out, both on the experimental and theoretical side, is valid only for low carrier densities while quantum well lasers work at very high density. In this paper we investigate the effect of non-linear phenomena on the carrier capture and related laser function. We show that carrier-carrier (CC) scattering must always be included in the analysis of quantum well lasers because under lasing condition it is responsible for a large fraction of the capture process. Finally we show that, even though the distribution function inside quantum well lasers is always very close to the Fermi shape, a very small difference in the lasing region (spectral *hole* burning) can have significant effect on the laser performance.

## THEORETICAL MODEL

Many discrepancies in previous works are due to an incorrect consideration of the trapping rate. The usual definition is simply the probability for a bulk-like particle to be scattered into the quantum well. Although this definition is formally correct, it is of limited use for the assessment of laser operation. Due to the normalization of the bulk-like wavefunctions, systems with different sizes have different rates but can fill the quantum well in the same time. It is this time that is important for all device applications. Instead of using the traditional method to study lifetimes based on rate equations we decided then to use a more powerfull Monte Carlo (MC) simulation and

*Hot Carriers in Semiconductors*
Edited by K. Hess *et al.*, Plenum Press, New York, 1996

to define the trapping rate as the number of particles trapped inside the well per unit time. This rate does not depend on the size of the bulk-like device region (not considering small changes in the overlap form factors). We consider a single GaAs quantum well laser with a well width of 80 Å and an undoped separate confinement heterostructure (SCH) of 3000 Å. The percentage of aluminum in the SCH and the doped cladding was 30 % and 60 %, respectively.

Our MC simulator is based on a full wavevector representation, i.e. there is no diffusion in real space and the effect of diffusion is automatically and (in the coherent limit[6]) correctly considered using the full orthonormal set of wavefunctions for both quantum well and bulk-like states. We include three quantum well bands: light and heavy holes, and electrons. The L valley lies energetically above the quantum well barrier but can contribute to carrier trapping through a two step process. Carriers are first scattered from the bulk-like state into the confined L valley, subsequently scattering down into the $\Gamma$ quantum well valley. The same bands are included for bulk-like particles which for the MC simulation process are considered fully 3D (i.e. with continuous values of $k_z$, where $z$ is the growth direction of the well). With this approximation all carrier-phonon and carrier-carrier scattering processes restricted to the bulk-like region are simple, fast, and require the use of little computer memory. This approximation is quite accurate in our case due to the large extension of the SCH (3000 Å) and a typical spacing of the energy levels on the order of one meV. Nevertheless, as the trapping mechanisms are of fundamental importance to the operation of the device, both carrier-phonon and carrier-carrier trapping rates were computed using the correct confined wavefunctions. Scattering mechanisms considered in the simulation included LO phonons for electrons and holes, equivalent intervalley scattering for the L valley electrons, along with intraband and interband TO phonons for holes. LO and TO phonons were also included for the capture and escape processes, which in our MC model simply appear as a different type of interband process. Furthermore, due to the significant number of phonon emissions that each carrier must undergo from the high energy injection point to the bottom of the quantum well, where lasing takes place, we included the effect produced by the non-equilibrium phonon distribution. We have considered the simple bulk phonon model because it has been shown that there is a sum rule valid in a large range of quantum wells that states that the total contribution of confined and interface phonons is very similar to that produced by the bulk phonon model.[7] All forms of CC scattering were considered including the Auger like processes contributing to the carrier capture and escape from the well. These processes can be schematically divided into three groups:

$$3D + 2D \longleftrightarrow 2D + 2D \quad ; \quad 3D + 3D \longleftrightarrow 2D + 3D \quad ; \quad 3D + 3D \longleftrightarrow 2D + 2D$$

where 3D and 2D indicate, respectively, a particle in a bulk-like state or in the quantum well. The capture processes must be read from left to right while from right to left we have the corresponding escape processes. During the device operation the density inside the quantum well is much higher than the bulk density, and the largest effect arises mainly from the first of these terms. When dealing with CC scattering one of the most delicate points is the screening model. We have used a static screening model that is almost the only feasible approach within a MC simulation. At high density it is well known that static screening results in an overestimate of the screening effect, so we should consider our results as the lower limit for CC interaction. Finally, since we assume stationary conditions, we have included stimulated recombination as a scattering mechanism that eliminates an electron-hole pair at the lasing energy near the band gap of the quantum well, and we reintroduce the carriers at the top of the separate confinement region (under stationary condition the number of pairs entering the device must be equal to the number of photons generated, if dark recombination is ignored).

# RESULTS

As the turn on delay for a semiconductor laser is of the order of the spontaneous recombination, i.e. approximately 1 ns, it is computationally unfeasible to simulate the entire process by a MC simulation. We therefore used as a preprocessor a drift diffusion quantum well laser simulator (the MINILASE-II[8] code developed at the Beckman Institute) to obtain the photon occupation number and the various carrier densities in both the quantum well and the bulk-like regions. For our structure these values have already been shown to be in extremely good agreement with experimental results.[9] We then insert these values in the MC program and let it reach a new stationary condition. This process gives valuable information on the validity of the drift diffusion model and the MINILASE-II simulator which includes a phenomenological trapping time. If we assume that this time gives an accurate description of the physical process then the MC simulation should arrive at the same stationary state. However, our MC model contains a more accurate microscopic description of all scattering mechanisms that contribute to the carrier capture. If therefore the simulation reaches a new steady state we can feed back this information into the drift diffusion model in order to obtain a more accurate simulator. Figure 1 shows the carrier densities for electrons and holes as a function of time when the MC simulation is started for both carriers in the well (solid lines, right hand scale) and in the bulk region (dashed lines, left hand scale). As we can see there is a significant drop of the bulk electron density that corresponds to only a small increase of the quantum well density (left and right hand scale respectively). This is due to the fact that the 3D density is much smaller than the 2D, and a drop of 35% only results in a small percentage increase of charge carriers in the quantum well. Our example shows that the trapping rate used in the drift diffusion simulator was on the low side. On the other hand we can see that there is almost no change in the hole density. This in part is a proof that the trapping time for holes is more accurate in MINILASE-II, but it also reflects the fact that the hole system is closer to equilibrium, i.e. the quasi-Fermi level for the 2D and 3D carriers are almost identical.

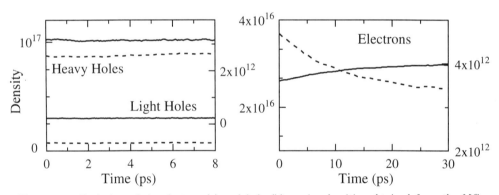

**Figure 1.** Evolution of the electron (a) and hole (b) carrier densities obtained from the MC simulation for both the quantum well particles (solid lines) and the bulk-like particles (dashed lines).

After the system has reached the new stationary condition, we can deduce from the MC simulation the effective capture probability distribution, i.e. the number of carriers entering per unit time at each energy point inside the quantum well. This is shown in Figure 2 for (a) light holes, (b) heavy holes, and (c) electrons. The solid line represents the total capture distribution while the dashed line corresponds to trapping due to phonons only; the area between the solid and the dashed line is then representative of all capture processes due to CC scattering. Three particularly important points can be extracted from this result: (*i*) the trapping due to CC scattering is as significant as that due to phonons, (*ii*) carriers trapped through the CC scattering can be found at relatively low energy and are then less likely to escape from the well than those trapped by phonon emission, (*iii*) a large fraction of carriers trapped in the quantum well have total energies exceeding the barrier energies (indicated by the vertical dotted lines). These particles should be considered only virtually trapped as the probability to scatter back into the bulk-like region is relatively high. Nevertheless, the probability to emit an intrasubband phonon is also high, and thus they effectively contribute to the overall trapping process.

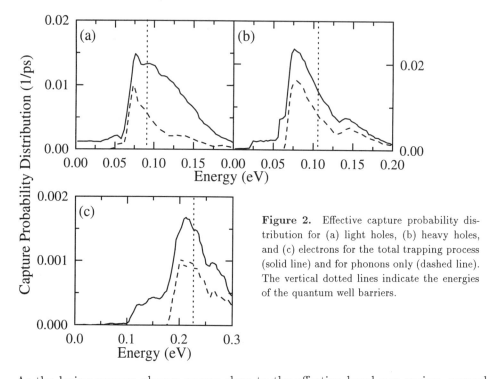

**Figure 2.** Effective capture probability distribution for (a) light holes, (b) heavy holes, and (c) electrons for the total trapping process (solid line) and for phonons only (dashed line). The vertical dotted lines indicate the energies of the quantum well barriers.

As the lasing process always occurs close to the effective band gap region, several phonons are required to take the carriers from the top of the bulk region to the bottom of the quantum well for each generated photon. At least for electrons the cooling requires the emission of LO-phonons in a relatively small region of the phase space and this creates a non-equilibrium phonon distribution. In turn, the hot phonon distribution modifies all carrier-phonon scattering rates enhancing phonon reabsorption (but emission too), and this often results in a slower overall cooling and trapping. The non-equilibrium LO distribution decays partially due to reabsorption but mainly through the decay of a small wave-vector LO-phonon into two acoustic phonons. This decay time can be computed from first principle lattice calculations or can be extracted from experimental Raman measurements of the hot-phonon lifetime (in GaAs at room temperature a typical value is 5 ps). It is relatively easy to include the hot-phonon effect in the MC simulation using an internal self scattering procedure and a simple exponential decay of the population. The hot-phonon population resulting under stationary lasing conditions is shown in Figure 3 (solid line) and compared with the equilibrium distribution at room temperature (dotted line). As we can see, in a significant range of wave-vectors the distribution is heated and shows, at small wave-vectors, the typical peak characteristic of the LO-phonon emission process. By performing two different simulations with and without hot-phonons we could verify that even though the phonon population is out of equilibrium it does not significantly influence the carrier temperature. This differs from a recent report by Alam and Lundstrom[10]. The difference can, in our opinion, be attributed to several factors.

First, we have considered a barrier with a 30% concentration of Al which corresponds to a confinement energy of 250 meV compared to the 400 to 600 meV confinement potential considered in [10]. The higher confinement energy necessarily results in a higher number of phonon emissions. A possibly larger contribution to the discrepancy comes from the fact that reference [10] considered a system with electrons only. This, at first, seems reasonable because the holes do not contribute significantly to the generation of hot-phonons due to the lower injection energy and the larger phonon wave-vectors (due to the larger effective mass). Therefore the density of states for phonons in that region is much larger and the population much less modified by an equivalent number of emissions. Also, the cooling for holes mainly occurs through the emission of TO-phonons which cannot easily become hot due to the isotropic scattering and the high density of available final states.

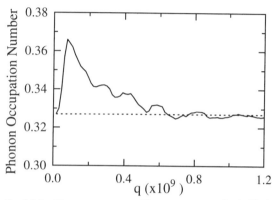

**Figure 3.** Hot-phonon (solid line) and equilibrium phonon (dotted line) distributions.

For these reasons holes are always "cold." However, even if one can neglect their contribution to the hot-phonon population, one cannot neglect at all the effect of holes on the carrier dynamics. In particular, CC interactions are strong enough to maintain always an equal temperature of electrons and holes, and this does not permit the electron system to become hot. In a semiclassical picture this can be visualized as a gas of fast moving particles (electrons) in a slow "cold" plasma (holes). The high energy electrons captured in the quantum well then lose energy not only through LO phonon emission but also through electron-hole interactions. At the typical carrier densities present in quantum well lasers the electron-hole interaction is often more effective than the electron-phonon scattering for the cooling of electrons.

For most laser materials, the effective mass of electrons is much smaller than that of holes and in order to obtain inversion in the population and stimulated emission the electron distribution must be degenerate with a value of approximately 1 at the lasing energy. Due to the Pauli exclusion principle, this results in a dramatic reduction of all scattering mechanisms in that region, except for stimulated photon emission, whose intensity continuously increases with the carrier concentration. The stimulated recombination then creates a *hole* in the electron distribution that is partially filled up by the various scattering processes. However, due to the reduced efficiency of the scattering mechanisms, this *hole* is never entirely destroyed and can seriously affect the laser performance. This effect has been observed experimentally.[10] In Figure 4 we show the electron distribution function obtained from the MC simulation (solid line). Apart from the relatively noisy result, which is unavoidable given the limited number of simulated particles, we can clearly see a significant difference with respect to the equilibrium distribution (dashed line). The carrier depletion does not appear in the hole distribution function due to the much higher values of all scattering rates and, most importantly, a much smaller distribution function, resulting in a less effective "degeneracy blockade." Even though the deviation from the equilibrium distribution by *hole* burning appears quite small, one must not forget that the gain of the laser is proportional to $(f_e + f_h - 1)$, and even a small decrease in the electron distribution results in a large loss of gain. In our case this loss is ~ 6% and at higher power can seriously limit the operating bandwidth of the device. Our result must be considered,however, as the upper limit for the gain reduction because, since we underestimate CC scattering by using static screening, we overestimate the *hole* burning. For a more correct analysis a full dynamical screening model must be used.

The tail of the distribution shows heating (i.e. a higher value with respect to the equilibrium distribution) created by the carriers entering the well. Note that the use of the term "carrier heating" can be misleading. The distribution is heated in the sense that there are fewer particles at low energies and more at the higher energies, but we are dealing with a non-equilibrium distribution that cannot simply be characterized by a higher temperature.

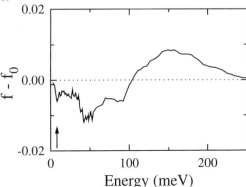

**Figure 4.** The difference between the electron distribution obtained by the MC calculation and a Fermi distribution, $f_0$, corresponding to the same carrier density. The arrow indicates the lasing energy.

In conclusion, we have shown that non-equilibrium distribution functions play an important role in semiconductor quantum well lasers and that carrier-carrier Auger like processes are as important for capture in the quantum well as are carrier-phonon interactions. Also, spectral *hole* burning in the electron distribution due to stimulated recombination can seriously degrade laser performance.

We wish to acknowledge financial support from the NATO Collaborative Research Grants Programme CRG 941281, and from the Office of Naval Research N00014-89-J-1470 (M.G., K.H.).

## REFERENCES

1 J.A. Brum and G. Bastard, Phys. Rev. **B33**, 1420 (1986).

2 J. Feldman, G. Peter, E.O. Göbel, K. Leo, H.J. Polland, K. Ploog, K. Fujiwara, and T. Nakayama, Appl. Phys. Lett. **51**, 226 (1987).

3 D.J. Westland, D. Mihailovic, J.F. Ryan, and M.D. Scott, Appl. Phys. Lett. **51**, 590 (1987).

4 B. Deveaud, J. Shah, T.C. Damen, W.T. Tsang, Appl. Phys. Lett. **52**, 1886 (1988).

5 P.W.M. Blom, C. Smit, J.E.M. Haverkort, and J.H. Wolter, Phys. Rev. **B47**, 2072 (1993).

6 The dephasing by scattering mechanisms has recently been investigated by F. Register and K. Hess; to be published in *Superlattices and Microstructures*

7 H. Rücker, E. Molinari, and P. Lugli, Phys. Rev. **B45**, 6747 (1992).

8 M. Grupen, K. Hess, and L. Rota in *Physics and Simulation of Optoelectronics Devices III*, edited by W.W. Chow and M.A. Osinski, SPIE vol. 2399, 468 (1995).

9 M. Grupen and K. Hess, Appl. Phys. Lett. **65**, 2454 (1994).

10 M.A. Alam and M.S. Lundstrom in *Physics and Simulation of Optoelectronics Devices III*, edited by W.W. Chow and M.A. Osinski, SPIE vol. 2399, 292 (1995).

10 F. Girardin, G. Duan, P. Galliol, A. Talneau, and A. Ougazzaden, Appl. Phys. Lett. **67**, 771 (1995).

# BLUE-GREEN QUANTUM WELL LASERS: COMPETITION BETWEEN EXCITONIC AND FREE-CARRIER LASING

R.Cingolani, L.Calcagnile, G.Coli'

Dipartimento Scienza dei Materiali, Universita' di Lecce, 73100 Lecce Italy

M.Lomascolo and M.DiDio

CNRSM-Pastis, Strada Prov. Mesagne, 72100 Brindisi, Italy

L.Vanzetti, L.Sorba and A.Franciosi

Laboratorio TASC-INFM, Padriciano 99, Trieste Italy

G.C.LaRocca

Scuola Normale Superiore, 56100 Pisa Italy

D.Campi

CSELT-Centro Studi e Laboratori Telecomunicazioni, 10148, Torino Italy

## Abstract

We investigated the excitonic or free-carrier nature of lasing in II-VI quantum wells by means of pump and probe transmission and stimulated emission measurements in magnetic field. We demonstrate that the phase transition between exciton and free-carrier gas determines the actual recombination mechanism.

Due to the strong exciton stability, lasing in II-VI quantum wells can be generated either by free-carrier or exciton stimulated recombination [1-4]. In this work we study the stimulated emission processes of ZnCdSe/ZnSe quantum wells, where exciton lasing and free-carrier lasing are strongly competing. The samples were grown by Molecular Beam Epitaxy following the procedure described in Ref.[5]. The Cd content was varied in the range $0.1< x <0.26$, and the well widths ranged between $3nm< L_z <20$ nm, allowing us to tune the exciton binding energy $(E_b)$ from about 20 meV (in wide and shallow quantum wells) to about 38 meV (in narrow and Cd-rich samples) [6]. Pump and probe non-linear transmission experiments, together with time resolved luminescence and stimulated emission measurements in high-magnetic field have been performed in order to clarify the origin of lasing in these heterostructures. In Figs.1a-1b, we display the intensity dependence of the pump and probe transmission spectra measured at 10 K in two samples with exciton binding energy of 25 meV (fig.1a) and 38 meV (Fig.1b). A clear exciton saturation is observed. The saturation intensity ranges between $I_{sat} \simeq 2kW\,cm^{-2}$ in the shallowest quantum wells and $I_{sat} \simeq 50kW\,cm^{-2}$ in the deep one, having a considerably stabler exciton. These data have to be correlated with the stimulated emission threshold, which is found to be $I_{lasing} \simeq 6 - 8kW\,cm^{-2}$ in all investigated samples, independent of the quantum well size and composition. We find that deep and narrow quantum wells $(E_b \geq 32$ meV) exhibit stimulated emission at power densities $I_{lasing} < I_{sat}$, suggesting

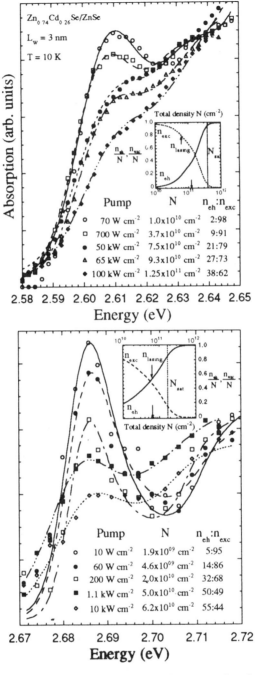

Fig.1a) Experimental (symbols) and calculated (curves) pump and probe transmission spectra of a 11 nm $Zn_{0.89}Cd_{0.11}Se/ZnSe$ quantum well samples measured at 10 K and under different pump power densities. The table indicates the pump power density (pump), the corresponding total density of photogenerated elementary excitations (N) obtained from the self-consistent line shape calculation (eqs.(1,2,6)), and the ratio of the free-carrier to exciton population ($n_{eh} : n_{exc}$) for each spectrum.

Inset: Phase-diagram of the exciton-free carrier. The dashed vertical line indicates the theoretical value of $N_{sat} \cdot n_{lasing}$ indicates the carrier density corresponding to the threshold intensity for stimulated emission. The white arrow indicates the density for total exciton saturation obtained from the line shape analysis of the pump and probe spectra.

Fig.1b - Same as in Fig.1a but for a 3nm quantum well with Cd-content x=0.26.

that exciton can play a role in the lasing process. On the contrary, in shallow and thick quantum wells ($E_b \leq 25$ meV) lasing occurs at $I_{lasing} > I_{sat}$, when the exciton resonance is saturated (Fig.2b). In this case a dominant free-carrier recombination is expected. In the wide range of intermediate size and composition $I_{sat}$ and $I_{lasing}$ are comparable, so that the exciton or free-carrier character of the emission depends on the photogeneration rate and sample parameters.

These experiments are interpreted assuming that a dense exciton gas and a carrier plasma coexist in the quantum well. Excitonic or free-carrier lasing can be obtained depending on the phase diagram of the exciton and free-carrier gas in thermal equilibrium. The total density of elementary excitations N is given by the sum of free-carrier ($n_{eh}$) and exciton density ($n_{exc}$). The equilibrium between the two phases is established by the mass-action law , including the exciton binding energy renormalized by the many-body interactions ($E_b'(n_{eh}, n_{exc})$ and the finite temperature of the distribution functions. This results in a set of coupled equations for the total density of photogenerated elementary excitations and for the relative balance of free-carriers and excitons:

$$N = n_{eh}(T) + n_{exc}(T) \qquad (1)$$

$$\frac{n_{eh}^2}{n_{exc}} = \frac{m_e m_h k_b T}{2\pi\hbar^2(m_e + m_h)} \cdot exp(\frac{-E_b'[n_{exc}(T), n_{eh}(T)]}{k_b T}) \qquad (2)$$

Due to the large density of excitons coexisting with the carrier plasma, both the band gap renormalization $\Delta E_g$ and the many-body correction to the exciton energy$\Delta E_{exc}$ depend on the free-carrier and exciton populations. The renormalized exciton binding energy $E_b'$ can thus be written as:

$$E_b'(N) = E_b - |\Delta E_g(n_{eh})| - |\Delta E_g(n_{exc})| - |\Delta E_{exc}(n_{eh})| - |\Delta E_{exc}(n_{exc})| \qquad (3)$$

In eq.(3) $\Delta E_g(n_{eh})$ and the hard core exchange repulsion among excitons $[\Delta E_{exc}(n_{exc})]$ are evaluated according to the analytical models or refs.[7] and [8], respectively. The excitonic contribution to the band gap renormalization is found to be

$$\Delta E_g(n_{exc}) = -\langle u_{1s}|V_s f_{e,h}|u_{1s}\rangle = -128\frac{\pi e^2}{\epsilon} n_{exc} \cdot \int_0^\infty \frac{k}{k+\chi} \frac{a_o^2}{[4+(a_o k)^2]^3} dk \qquad (4)$$

which can be handled analytically. The phase space filling effect on the exciton energy $\Delta E_{exc}(n_{eh})$ is obtained by evaluating analytically the expectation value of the many-body corrections to the Hamiltonian on the 1s exciton wavefunction, which gives:

$$\Delta E_{exc}(n_{eh}) \simeq \pi a_0^2[13.1 n_{eh}(T)] * E_b \qquad (5)$$

All these calculations are first order (perturbative) approximations strictly valid in the non degenerate free-carrier limit and for $E_b >> k_b T$. Both these assumptions are quite reasonable in II-VI heterostructures, by virtue of the large exciton binding energy and effective masses.The non-linear pump and probe spectra are thus given by :

$$\alpha(\hbar\omega, N) = C\cdot(\frac{1}{1+\frac{N}{N_{sat}}})\cdot G\{\hbar\omega - [E_g - E_b'(N)]\} + C'\cdot D[\hbar\omega - E_g'(N)]\cdot S(\hbar\omega - E_b)\cdot(f_e - f_h) \qquad (6)$$

The first factor of eq.(6) accounts for the saturation of the exciton oscillator strength, through the critical saturation density $N_{sat}$ [8]. G and D model the homogeneously [7] and inhomegeneously broadened density of states of the exciton (Gaussian) and of the continuum, respectively. $E'_g$ is the renormalized band gap and $S(\hbar\omega)$ is the Sommerfeld factor. The calculation of the non-linear absorption spectra must be performed self-consistently, by varying the relative exciton/plasma population for a given total density, until convergence is achieved. The calculated spectra reproduce quite well the experimental data (lines). The hot-carrier temperatures (between 2 and 5 meV) were directly obtained from the slope of the high-energy tails of the spontaneous luminescence. The corresponding phase diagrams are plotted in the insets of Figs.1a-1b. A smooth transition from exciton gas to free-carrier gas occurs with increasing the total photogeneration rate (N). The exciton resonance is *totally* bleached at N-values close to the transition from the exciton-rich to the free carrier-rich phase ($6 \cdot 10^{10} cm^{-2}$ and $1.3 \cdot 10^{11} cm^{-2}$ in the two samples, respectively) , well below the $N_{sat}$ value calculated in ref.[8]. Both the lasing threshold and the exciton saturation can occur in the free-carrier side or the exciton-side of the phase diagram depending on the exciton stability of the investigated sample.

These results are unambiguously confirmed by measuring the diamagnetic shift of the stimulated emission in high magnetic field [3]. The experiments performed in all samples show that deep and narrow quantum wells exhibit a stimulated emission with clear excitonic shift (of the order of $2~\mu eV \cdot T^{-2}$) , whereas shallow quantum wells are characterized by a Landau-type shift of the lasing line (of the order of 0.5 meV/T). This is shown in Fig.2 where we display the data of two representative heterostructures of well width 7 nm and Cd content 11 % and 23 %. The excitonic nature of the lasing observed in deep quantum wells is probably due to the relevant exciton localization occurring in Cd-rich quantum wells, due to compositional fluctuations [1]. This phenomenon provides an efficient three-level system for the operation of the excitonic laser. The occurrence of localization is indeed demonstrated by time resolved luminescence experiments, showing long rise and decay times in the low energy tail of the exciton resonance, due to trapping at compositional fluctuations (Fig.3). This temporal evolution changes with the excitation

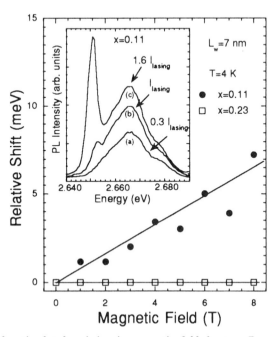

Fig.2 - Shift of the stimulated emission in magnetic field for two $Zn_{1-x}Cd_xSe/ZnSe$ multiple quantum well samples of well width $L_w = 7$nm and x =0.11 and x=.23. In the inset: stimulated emission spectra of the x=0.11 sample around threshold. $I_{lasing} \simeq 7kWcm^{-2}$.

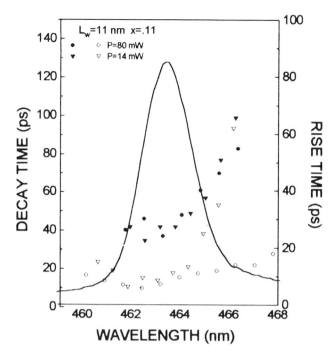

Fig.3 - Spectrally resolved decay time and rise time of a shallow ZnCdSe/ ZnSe quantum well at 10 K, measured at different excitation intensity.

intensity. In shallow quantum wells both the rise time and the decay time get shorter, with increasing the pump power, revealing the saturation of the available density of localization states and the recover of the intrinsic recombination (Fig.3). On the contrary, no recover of the intrinsic recombination is observed in deep quantum wells even at injection rates comparable of $I_{lasing}$, indicating a dominant localized exciton recombination even under stimulated emission conditions.

1 - J.Ding, H.Jeon, T.Ishiara, M.Hagerott, A.V.Nurmikko, H.Luo, N .Samarth, and J.Furdyna,Phys. Rev. Lett. **69**, 1707 (1992)
2 - Y.Kawakami, I.Hauksonn, H.Stewart, J.Simpson, I. Galbraith, K.A.Prior, and B.C.Cavenett, Phys. Rev. **B48**, 11994 (1993)
3 - R.Cingolani,R.Rinaldi, L.Calcagnile, P.Prete, P.Sciacovelli , L.Tapfer, L.Vanzetti, F.Bassani, L.Sorba,and A.Franciosi, Phys. Rev. **B49**, 16769 (1994)
4 - A.Diessel, W.Ebeling, J.Gutowski, B.Jobst, K.Schuell, D. Hommel, and K.Henneberger, Phys. Rev. **B52** in press (1995)
5 - G.Bratina, R.Nicolini, L.Sorba, L.Vanzetti, G.Mula, X.Yu, and A.Franciosi, J.Cryst. Growth **127**, 387 (1993)
6 - R.Cingolani, P.Prete, D.Greco, P.V.Giugno, M.Lomascolo, R. Rinaldi,L.Calcagnile, L.vanzetti, L.Sorba and A.Franciosi, Phys. Rev. **B51**, 5176 (1995)
7 - D.Campi and C.Coriasso, Phys. Rev. **B51**, 7985 (1995)
8 - S.Schmitt-Rink, D.S.Chemla, and D.A.B.Miller, Phys. Rev.**32**, 6601 (1985)

# ELECTRON TRANSPORT IN InGaAsP/InP QUANTUM WELL LASER STRUCTURES

S. Marcinkevičius,[1] U. Olin,[2] C. Silfvenius,[3] B. Stålnacke,[3]
J. Wallin,[3] and G. Landgren[3]

[1]Department of Physics II, Royal Institute of Technology,
 S-100 44 Stockholm, Sweden
[2]Institute of Optical Research, S-100 44 Stockholm, Sweden
[3]Semiconductor Laboratory, Royal Institute of Technology,
 Electrum 229, S-164 40 Kista, Sweden

## INTRODUCTION

Carrier transport in quantum well (QW) lasers is one of the factors affecting the high frequency performance of these devices. Usually the transport of holes is assumed to be of major significance. However, if the active region of a laser is shifted towards the $p$ contact of the device,[1] or this region is $p$-doped,[2] the electron transport may become an important factor.

Previously, most of investigations on the carrier transport have been performed for the GaAs/AlGaAs material system. Here we present an experimental study of electron transport processes for InGaAsP/InP laser structures which are of major importance for optical telecommunications. The transport processes studied include electron transport in the confinement region, capture into the QWs, and transport between the QWs. Time-resolved photoluminescence (PL) investigations using upconversion were performed. Experimental set-up, based on a fs Ti:sapphire laser, assured a temporal resolution of 150 fs. All experiments were carried out at room temperature.

## TRANSPORT IN THE CONFINEMENT REGION

The InGaAsP/InP QW structures were grown by low-pressure MOVPE on $n$-InP substrates. To study each of the above mentioned transport effects, different sets of structures were produced. The electron transport in the confinement region has been investigated using a uniformly $p$-doped structure ($p = 1 \times 10^{18}$ cm$^{-3}$) with a step-graded confinement region (figure 1). The steps of InGaAsP are each 22 nm thick and have room-temperature band-gap values of 1.15, 1.09 and 1.02 eV. The band gaps for the barriers and the QWs are at 0.95 eV and 0.79 eV, respectively. The top InP layer is 500 nm thick, so that absorption of the optical excitation at 770 nm would mainly occur in this layer ($\alpha^{-1} = 300$ nm). The photoexcited carrier density was $\sim 10^{16}$ cm$^{-3}$, much lower than the background doping. It has been shown,[3] that this is a sufficient condition to assure that it is electron and not ambipolar carrier transport which is studied in the experiment. We have checked it for our structures performing PL measurements at different excitation intensities.[4]

The dynamics of the electron movement is followed by measuring the temporal

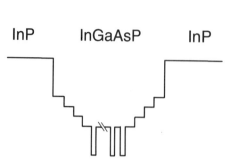

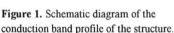

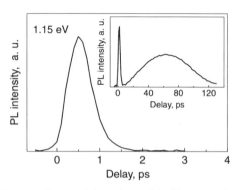

**Figure 1.** Schematic diagram of the conduction band profile of the structure.

**Figure 2.** Temporal dependence of the PL measured at 1.15 eV.

dependencies of the PL signal at the energies of the band gaps of the different layers. The PL decay at the InP band-gap energy, with a decay time of 9.7 ps, reflects electron escape from this layer, while the increase of the PL signal at the energy of the bottom of the confinement region, with the rise time of 10.3 ps, indicates electron arrival. A subpicosecond difference between these time values reflects a very fast electron transport across the steps. This influences the PL spectra: no signal can be detected at energies, corresponding to the band gaps of the steps.

The PL signal at energies of the step band gaps appears if the excitation intensity is somewhat increased. The PL transients then have a double peak structure (inset, figure 2). We assign the first peak to the electrons generated directly in the step layer, while the second peak originates from the carriers excited in the InP layer and arriving to the step layer with a considerable delay. The situation here is quite peculiar: in the InP layer the carriers are generated at a rather high density and diffuse towards the bottom of the graded region in the ambipolar manner, while, because of the finite penetration of the exciting light, the electrons in the step layer are generated at densities more appropriate for the unipolar transport.

The first PL peak exhibits a fast decay with a time constant of 250 fs which is determined by the electron escape from the step layer. Then, the three steps forming the graded region would be passed in no more than 750 fs. However, it should be noted that 250 fs is an upper time limit for the electron transport across the step, as it is the transfer time measured for the electrons which have cooled down close to the bottom of the conduction band and have little excess energy. Electrons, photoexcited in the step layer, have an excess energy of 0.39 eV, and it would take over a ps for most of them to cool via electron-hole and electron-LO phonon scattering. Consequently, most of the electrons would leave the layer remaining hot. Even electrons injected into the graded region from the InP layer would not have enough time to cool down at each step and follow the potential profile of the steps, as the time of LO-phonon emission (160 fs) is comparable to the electron transit time. Thus, the electron transport mechanism in the stepped region is hot electron diffusion.

## ELECTRON CAPTURE

To study electron capture into the QWs, a structure with a 25 nm thick InP cap layer has been used, so the exciting light is mainly absorbed in the confinement region. For the excitation energy of 1.61 eV, electrons are excited with much higher excess energy than the holes (0.60 and 0.07 eV, with respect to the bottom of the confinement region). Since the hole relaxation and capture are much faster processes than those of the electrons, presented PL data would reflect relaxation of the electron system. In these measurements $p$-doping and low excitation densities are not essential, thus, photoexcited carrier densities of the order of $10^{18}$ cm$^{-3}$ have been used.

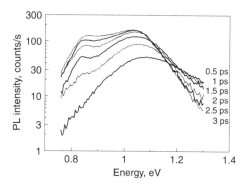

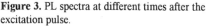

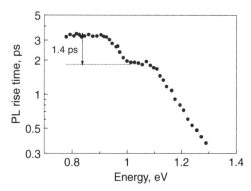

**Figure 3.** PL spectra at different times after the excitation pulse.

**Figure 4.** PL rise times as a function of the PL detection energy.

The electron relaxation and capture is studied by monitoring temporal evolution of the PL spectra and measuring PL transients at different detection energies. The PL spectra measured at different times after the excitation (figure 3) clearly show carrier relaxation in the confinement region and capture into the QWs. Measuring the PL dynamics at fixed detection energies, we concentrate on PL rise times which reflect electron arrival at a particular energy level. The energy dependence of the PL rise times is shown in figure 4. At high energies the PL decay is fast and determined by the electron down-scattering in the conduction band. With decrease of the PL energy the rise times continually increase until, close to the bottom of the confinement region, their values saturate at 1.8 ps. At still lower PL energies, corresponding to the transitions in the QWs, the PL rise time exhibits another plateau which continues down to the band gap of the wells. The difference between the two plateaus of 1.4 ps can be interpreted as the electron capture time into the QWs. The obtained value should be attributed to the local capture time, since in the studied multiple QW structure the barriers are thin enough to neglect carrier transport in the barriers.[5]

## TRANSPORT IN THE QW REGION

To study electron transport across the QWs, $p$-doped structures are used again ($p = 1 \times 10^{18}$ cm$^{-3}$), and the top InP layer is 500 nm thick. A number of shallow QWs is followed by one or several deeper wells, which serve as a marker (inset, figure 5). Then, the decay of the PL signal at the InP band-gap energy would reflect electron transfer from the top InP layer into the shallow QW region, and rise of the deep QW PL would be determined by the arrival of electrons which have passed the shallow wells. The photoexcited electron density has been kept low, at about $10^{16}$ cm$^{-3}$, to prevent deep wells from filling.

First, let us examine results obtained for structures with narrow (5 nm) barriers in the shallow QW region. The four samples have 4, 8, 16 and 32 shallow QWs, with a band-gap energy of 0.81 eV, followed by a few deep QWs ($E_g = 0.76$ eV). The barriers for the both types of wells have a band gap of 0.95 eV.

Figure 5 shows PL temporal dependencies measured at the band gaps of InP and deep QWs for different samples. The InP PL decay is exponential with a time constant of 14 ps. The increase of the PL signal for the deep wells can be approximated by a relation $I_{PL} \propto [1 - \exp(-(t/\tau)]$, where the $\tau$ values are 16, 18, 22 and 29 ps for the structures with 4, 8, 16 and 32 shallow QWs, respectively. The difference between the deep QW PL rise times and the decay time of the InP PL is the time for the electron transfer across the shallow QW region. These transport times scale linearly with the number of shallow QWs. This discrepancy with a diffusion model can be explained taking into account the dependence of

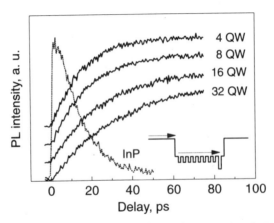

**Figure 5.** Normalised PL transients measured at the band-gap energies of InP and deep QWs.

the diffusion coefficient on the photoexcited carrier density. At the same photoexcited carrier density in the InP layer, the carrier densities in the shallow QW region for the different samples are different because of the different widths of this region. Since the mobility of the photoexcited carriers is determined by the relation $\mu = (n+p)/(n/\mu_p + p/\mu_n)$, where $n$ and $p$ are electron and background hole densities, and $\mu_n$ and $\mu_p$ are electron and hole mobilities, the carrier mobility is larger in the structures with a larger number of shallow QWs . When the experiment is performed at different photoexcitation intensities, but at similar electron densities in the shallow QW region, the dependence of the carrier transport time on the number of shallow QWs has a more quadratic character.

Estimations of the electron mobility in the shallow QW region obtained from the measured transfer times produce values close to 2400 cm/Vs$^2$, the mobility of the InGaAsP alloy.[7] This could be expected, since for 5 nm barriers the electron miniband extends up to the barrier energy, and all electrons in the shallow QW region should move freely, either they are in the QWs, or in continuum states above the wells.

A different picture appears if the barrier width is increased. The deep QW PL rise times for the samples with 4 and 16 shallow QWs (now the shallow wells are 8 nm and the barriers 12 nm wide) are 32 and 82 ps, significantly longer as in the previous case.

The electron ground state for the shallow QWs is only 30 meV below the barriers. Thus, at room temperature a considerable fraction of electrons would have the energy exceeding that of the barriers. Assuming that only these electrons are mobile, the overall electron mobility in the shallow QW region can be evaluated from the structural data.[8] For the studied structures with wide barriers, the effective electron mobility value is about 400 cm/Vs$^2$. The electron transport times across the shallow QW region, estimated using this value, agree fairy well with the experimental ones. The presented results clearly show the influence of barrier width on the electron transport in InGaAsP/InP multiple QW structures.

Nir Tessler is acknowledged for valuable discussions. This work has been supported by the The Swedish Research Council for Engineering Sciences (TFR)

1. F. Steinhagen *et al.*, *Electron. Lett.* 31:274 (1995).
2. P.A. Morton *et al.*, *Electron. Lett.* 28:2156 (1992).
3. B. Lambert *et al.*, *Semicond. Sci. Technol.* 4:513 (1989).
4. S. Marcinkevičius, U. Olin, J. Wallin, K. Streubel, and G. Landgren, *Appl. Phys. Lett.* 66:2098 (1995).
5. R. Kersting, R. Schwedler, K. Wolter, K. Leo, and H. Kurz, *Phys. Rev. B* 46:1639 (1992).
6. U. Hohenester, P. Supancic, P. Kocevar, X. Zhou, W. Kütt, and H. Kurz, *Phys. Rev. B* 47:13233 (1993).
7. J.R. Hayes, A.R. Adams, and P.D. Greene, *in* "GaInAsP Alloy Semiconductors," T. P. Pearsall, ed., Wiley, New York (1982).
8. N. Tessler and G. Eisenstein, *IEEE J. Quantum Electron.* QE-29: 1586 (1993).

# HOT CARRIER EFFECTS IN FEMTOSECOND GAIN DYNAMICS
# OF InGaAs/AlGaAs QUANTUM WELL LASERS

G. D. Sanders[1,*], C. J. Stanton[1,*] C.-K. Sun[2], B. Golubovic[2]
and J. G. Fujimoto[2]

[1]Mikroelektronik Centret
Danmarks Tekniske Universitet
DK2800 Lyngby, Denmark

[2]Department of Electrical Engineering and Computer Science
Massachusetts Institute of Technology
Cambridge, MA 02139

Ultrafast optical nonlinearities in semiconductors play a central role in determining transient amplification and pulse-dependent gain saturation in quantum-well diode lasers. Both carrier-phonon and carrier-carrier scattering are expected to influence the nonlinearities. In this paper, we investigate hot electron effects on the semiconductor gain dynamics of strained-layer quantum well lasers. We present a relaxation time approximation model for carrier-carrier scattering in strained layer lasers. The relaxation approximation makes the problem an effective *one dimensional problem* which can then be solved directly using an adaptive Runge Kutta routine[1]. This procedure requires substantially less computational resources than a full Monte Carlo simulation. Results show that the inclusion of carrier-carrier scattering improves previous results with only carrier-phonon scattering.

In our theoretical model, transient gain and differential transmission are computed using: 1) a multiband effective mass model including biaxial strain, with valence subband mixing, 2) polar optical phonon scattering both within and between subbands 3) carrier-carrier scattering, 4) transient photogeneration of electron-hole pairs. The calculations are based on a numerical (not Monte Carlo) solution of the Boltzmann transport equations. We follow the time evolution of the distribution functions and directly determine the time dependent gain/absorption of the laser diode. Corrections to the Boltzmann equation such as 1) band gap renormalization, 2) many-body enhancement of the gain and 3) broadening of the initial carrier distributions due to rapid dephasing are also included in the calculations.

Previous work [1] focused on only carrier-phonon scattering. Here we focus on the carrier-carrier scattering. We model carrier-carrier collision integral using an effective relaxation operator approach [2]. The collision term for both electrons and holes is described by an operator of the form

$$\left[\frac{\partial f_n(k,t)}{\partial t}\right]_{cc} = -\frac{\left(f_n(k,t) - f_n^0(k,\mu^*,T^*)\right)}{\tau_n(k)} \tag{1}$$

with the relaxation time given by

$$\frac{1}{\tau_n\left(\vec{k}\right)} = \frac{2\pi}{\hbar} \sum_{n'',\vec{k}'',\vec{q}} |V(q)|^2 \times \{f_n^0\left(\vec{k}+\vec{q}\right) f_{n''}^0\left(\vec{k}'' - \vec{q}\right)\left(1 - f_n^0\left(\vec{k}''\right)\right) +$$

$$f_{n''}^0\left(\vec{k}\right)\left(1 - f_n^0\left(\vec{k}+\vec{q}\right)\right)\left(1 - f_{n''}^0\left(\vec{k}'' - \vec{q}\right)\right)\} \times \tag{2}$$

$$\times \delta\left(E_n\left(\vec{k}+\vec{q}\right) + E_n\left(\vec{k}'' - \vec{q}\right) - E_n\left(\vec{k}\right) - E_n\left(\vec{k}''\right)\right).$$

$f_n^0(k,\mu^*,T^*)$ is a quasi-equilibrium Fermi-Dirac distribution function characterized by an effective temperature, $T^*$, and chemical potential, $\mu^*$. Both $T^*$ and $\mu^*$ are determined by the constraints of the conservation of particle number and energy. They are time dependent as the actual distribution functions change in time. Since $\tau$ is $k$ dependent, the actual distribution function $f$ and the relaxation function $f_o$ may be characterized by different effective temperatures and chemical potentials.

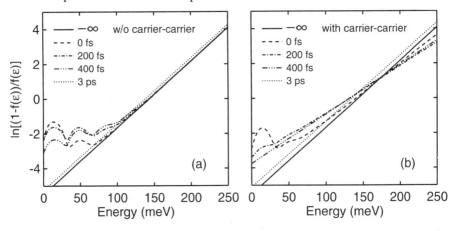

**Figure 1.** $\ln\left((1 - f(E))/f(E)\right)$ plotted as a function of energy for the time-dependent electron distribution function in the first conduction band without (a) and with (b) carrier-carrier scattering.

Results are shown in figures 1 and 2. In figure 1, we show $\ln\left((1 - f(E))/f(E)\right)$ for the first conduction band with a background carrier density of $5 \times 10^{12}$ $cm^{-2}$ and a 200 fs pulse centered 18 meV above the conduction band edge without (a) and with (b) carrier-carrier scattering. The different curves are: before the pulse (solid line), center of the pulse (dashed line), 200 fs (dash-dotted line), 400 fs (dash-multi dotted line) and 3ps (dotted line). This function should be a straight line with the slope equal to the inverse temperature and the zero value of energy being the chemical potential if the distribution function is a Fermi-Dirac distribution. As can be seen in (a), without carrier-carrier scattering, the distribution relaxes through sequential optic phonon emission. The temperature of the tail never heats up. When carrier-carrier scattering is included (b), the phonon replicas are smeared out and the tail heats up at intermediate times before finally relaxing back to the lattice temperature. The shift in the chemical potential results from the fact that carriers are taken out of the band due to the laser pulse. At carrier concentrations around $1 \times 10^{12}$ $cm^{-2}$, carrier-carrier scattering is negligable and electron-phonon scattering is the dominant mechanism.

The distribution functions themselves are not directly related to the experimentally measured quantity, the differential change in absorption. In figure 2, we show experimental and theoretical calculations of $\Delta T/T$ for 4.2 mA of injection current corresponding to a carrier density of $2.1 \times 10^{12}$ $cm^{-2}$ with the pump pulse at 935 nm and the transparency point at 925 nm (pumping in the gain region). $\Delta T/T$ as a function of probe wavelength at t=-400 fs (dotted line), t=-200 fs (short dashed line), t=80 fs (solid line), t=400 fs (long-dashed line) and t=3 ps (dash-dotted line) are shown for the experimental (a), and theoretical calculations without (b) and with (c) carrier-carrier scattering. $\Delta T/T$ as a function of delay time at $\lambda$= 905nm (dashed line), $\lambda$=920 nm (dash-dotted line), $\lambda$=935 nm (solid line), and $\lambda$=950nm (dotted line) for the experimental (d), and theoretical calculations without (e) and with (f) carrier-carrier scattering. While the calculations with only phonon scattering produce reasonable agreement with the experiments, carrier-carrier scattering is needed to get more accurate results. In future work, dynamic screening will be included.

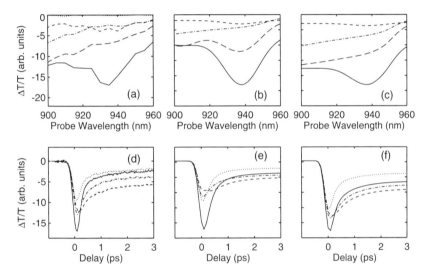

**Figure 2.** $\Delta T/T$ as a function of wavelength (a-c) and time (d-f) for experimental (a,d) and theoretical calculations without (b,e) and with (c,f) carrier-carrier scattering.

In conclusion, we have solved a relaxation time model for carrier-carrier scattering in quantum well lasers. Comparison to experiment shows that carrier-carrier scattering is needed to produce better agreement with the observed gain dynamics and to produce heating in the tails of the distributions.

### Acknowledgments

This work was supported by U.S. Office of Naval Research through Grant N00091–J–1956 and partially through NSF grant DMR 8957382 and the Alfred P. Sloan foundation. We also wish to thank the Danish Research Academy for support during part of this work.

### REFERENCES

*Permanent Address: Dept. of Physics, University of Florida, Gainesville, FL 32611.

[1] G. D. Sanders, C.-K. Sun, J. G. Fujimoto, H. K. Choi, C. A. Wang, and C. J. Stanton, Phys. Rev. B50, 8539 (1994).

[2] N. S. Wingreen, C. J. Stanton, and J. W. Wilkins, Phys. Rev. Lett. 57, 1084 (1986).

# CARRIER AND PHOTON DYNAMICS IN InAlGaAs/InP MQW LASER STRUCTURES:

## INFLUENCE OF CARRIER TRANSPORT ON HIGH-FREQUENCY MODULATION

H. Hillmer[1], A. Greiner[2], F. Steinhagen[3], R. Lösch[1], W. Schlapp[1],
E. Binder[2], T. Kuhn[2] and H. Burkhard[1]

1 Deutsche Telekom, Forschungszentrum, P.O. Box 100003, 64276 Darmstadt, Germany
2 Institut für Theoretische Physik, Universität Stuttgart, 70569 Stuttgart, Germany
3 Institut für Hochfrequenztechnik, Technische Hochschule, 64283 Darmstadt, Germany

## INTRODUCTION

Ultrafast semiconductor laser devices are key components for advanced lightwave transmission systems. Presently, for the third telecommunication window close to 1.55 μm, mainly InGaAsP/InP distributed feedback (DFB) lasers are investigated and applied. Due to larger technological difficulties, the alternative InAlGaAs/InP material system was by far less studied and used[1-8], although this system seems to be superior in many physical points such as band gap discontinuities, carrier quantum well (QW) tunneling, thermal carrier re-emission from QWs, differential gain, gain compression and compositional refractive index variation. We performed detailed experimental and theoretical studies of the carrier and photon dynamics in MQW InAlGaAs/InP laser structures focussing on the influence of carrier transport.

## LASER STRUCTURE DESIGN

We modelled and processed DFB laser structures having different geometrical, compositional and doping conditions in the optical confinement region close to the QW active layers. In our studies we compare lasers of mainly asymmetric waveguide structures, i.e. the p-sided and n-sided confinement layers (CL) have different widths $d_p$ and $d_n$. Fig. 1a shows the calculated band structure in vertical direction of a laser structure consisting of 10 compressively strained 3-nm InGaAs wells and unstrained InAlGaAs barriers. Between the p-doped InAlAs buffer ($\geq 2 \cdot 10^{18}$ cm$^{-3}$ Be) and this active MQW stack, the "p-sided" p-doped InAlGaAs ($5 \cdot 10^{17}$ cm$^{-3}$ Be ) CL of width $d_p$ is located. On the opposite side, between the DFB grating and the MQWs, the "n-sided" p-doped InAlGaAs ($5 \cdot 10^{17}$ cm$^{-3}$ Be ) CL is located. The DFB grating is covered by an n-doped InP top layer ($\geq 2 \cdot 10^{18}$ cm$^{-3}$ Si ). Since the InAlAs and InP bulk layers are highly doped, the respective dielectric relaxation times are very small leading to a very small influence of these layers on the modulation properties. However, the InAlGaAs

*Applications of Advanced Technology to Ash-Related Problems in Boilers*
Edited by L. Baxter and R. DeSollar, Plenum Press, New York, 1996

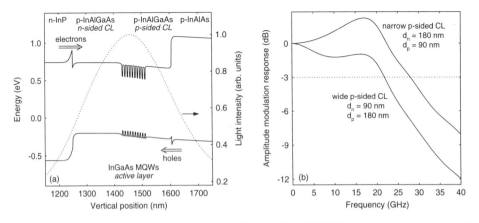

Fig. 1 (a) Band structure of an InGaAs/InAlGaAs/InP MQW laser structure under bias.
(b) Amplitude modulation response calculated for different transport geometries.

CLs are lower doped to reduce the optical free carrier reabsorption losses of the guided optical field, but this implies a decrease in the high-frequency properties due to transport influence[9-13] in the CL. Varying $d_n$ and $d_p$, these asymmetric structures are very useful to study the influence of carrier transport on the modulation bandwidth, since electrons and holes have considerably different mobilities.

## RESULTS OF AMPLITUDE MODULATION STUDIES

Among different laser structures, which we considered in our experiments and simulations, we now select and compare two having nearly the same waveguide thickness $d_n + d_p$, nearly the identical optical confinement factor of 6% but different positions of the MQW stack: either asymmetrically n-side or p-side shifted. The calculations are based on a drift diffusion model using a self-consistent solution of carrier continuity equations, Poisson's equation and photon rate equations including effective relaxation dynamics of the carriers in the QWs[10,11]. The calculations include electron and hole mobilities measured as a function of doping in the related quaternary and ternary material layers. For a photon density of $14 \cdot 10^{15} \, cm^{-3}$, Fig. 1 b demonstrates considerable differences between the two structures. The laser with the long p-sided CL shows a -3 dB frequency of 22 GHz and at these photon densities already a strong RC-like rolloff. The laser with the short p-sided CL provides a -3 dB frequency of 28 GHz. This laser still reveals a distinct resonance, i.e. an inherent potential to reach still higher bandwidths for increasing bias (rising photon densities). This can be mainly attributed to carrier transport, i.e. the difference in hole and electron mobilities, in agreement to Refs. 6 and 9. This result demonstrates the importance of geometry and doping of the laser structures for optimized high-frequency behaviour. Effects like the low and high frequency rolloffs in the linear response of the system are directly obtained and identified from the spatial transport description. These theoretical results were further confirmed for implemented laser devices of the same geometry showing also a difference between the two lasers in the -3 dB frequency of 5 GHz. However, the absolute experimental values are lower due to electrical parasitics. According to the theoreretical results obtained we have grown InAlGaAs/InP laser structures by MBE designed for high-speed operation: optimized doping profiles, reduced parasitics, improved technology and further reduced p-sided CL at the expense of the n-sided CL to enable a faster transfer of the less mobile holes to the QWs, leading to a faster dynamic optical response of the MQW laser devices under high-speed current modulation. Fig. 2 shows 26 GHz in modulation bandwidth which is a distinct record value for the InAlGaAs material system to the best of our

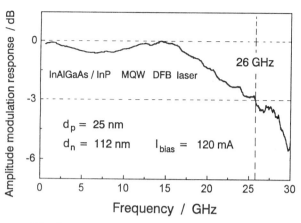

Fig. 2 Experimental amplitude modulation response of an InAlGaAs/InP laser including 10 QWs and asymmetric confinement layers.

knowledge[5,6]. Note that the record value in the widely applied InGaAsP/InP of 26 GHz has now also been reached. The calculated maximum bandwidths are still higher. This may, e.g., be due to remaining parasitics, leakage currents, capture and relaxation processes, non-isothermal transport and hot phonon effects.

We wish to thank R. Göbel and H. Janning for VPE, E. Kuphal for MOCVD, B. Hübner and B. Kempf for the grating definition and G. Mahler and S. Hansmann for stimulating discussions. The support by Deutsche Telekom and the European Community under RACE 2006 Welcome is gratefully acknowledged.

## REFERENCES

1. M. Blez, C. Kazmierski, M. Quillec, D. Robein et al., Electron. Lett. **27,** 94 (1991).
2. J. Thompson, R. M. Ash, N. Maung, A. J. Moseley, J. Electron. Mat. **19**, 349 (1990).
3. B. Stegmüller, B. Borchert, H. Hedrich, et al., Jap. J. Appl. Phys. **30** 2781 (1991).
4. M. J. Moudry, Z. M. Chuang, M. G. Peters, L. A. Coldren, Electr. Lett. **28**, 1471 (1992).
5. Ch. Zah, R. Bhat, T. P. Lee et al., IEEE J. Quantum Electron. **30**, 511 (1994).
6. F. Steinhagen, H. Hillmer, R. Lösch, W. Schlapp, H. Walter, R. Göbel, E. Kuphal, H. L. Hartnagel and H. Burkhard, Electron. Lett. **31**, 274 (1995).
7. H. Hillmer, R. Lösch and W. Schlapp, J. Appl. Phys. **77**, 5440 (1995).
8. H. Hillmer, R. Lösch, W. Schlapp, F. Steinhagen and H. Burkhard, Electron. Lett. **31** No. 16 (1995).
9. R. Nagarajan, R. P. Mirin, T. E. Reynolds, and J. E. Bowers, Electron. Lett. **29** 1688 (1993).
10. Special issue on "Carrier transport effects in QW lasers", e.g. H. Hillmer, T. Kuhn, A. Greiner, S. Hansmann, H. Burkhard, Optical and Quantum Electron. **26** S691 (1994) and other contributions in this volume.
11. A. Greiner, T. Kuhn, H. Hillmer, S. Hansmann and H. Burkhard, NATO ASI on "Quantum transport in ultrasmall devices" 1994, Ed. D. K. Ferry, in press (1995).
12. M. Grupen and K. Hess, Appl. Phys. Lett. **65** 2454 (1994).
13. M. Alam and M. Lundstrom, IEEE Photon. Technol. Lett. **6**, 1418 (1995).

# INFLUENCE OF AMPLIFIED SPONTANEOUS EMISSION ON ENERGY RELAXATION, RECOMBINATION AND ULTRAFAST EXPANSION OF HOT CARRIERS IN DIRECT-GAP SEMICONDUCTORS

Yu. D. Kalafati,[1] V.A. Kokin,[1] H.M. van Driel[2] and G.R. Allan[2]

[1]Institute of Radio Engineering and Electronics RAN, Moscow, Russia
[2]Department of Physics, University of Toronto,Toronto, Canada, M5S-1A7

## INTRODUCTION

The picosecond cooling and spatial expansion dynamics of high density ( $> 10^{18}$ cm$^{-3}$) photoexcited plasmas in direct gap semiconductors has commanded much attention over the past twenty years. However one aspect of the plasma evolution that has often been ignored is the role of amplified spontaneous emission (ASE). The common belief is that radiative recombination occurs on a nanosecond time scale and is therefore unimportant for plasma dynamics. Nonetheless, for a plasma which becomes degenerate as it cools, optical gain can occur and ASE can induce recombination and plasma reheating on a picosecond time scale, slowing down the cooling process. This can occur in density regimes where hot phonon effects[1] have previously been surmised to be the major factor in determining a reduced cooling rate. In addition, ASE can lead to an effective ultrafast carrier diffusion[2,3] through radiative transfer. Overall, for a plasma initially generated in a certain volume and with density above a certain critical value, we find that ASE can induce strong nonlocal behavior on a picosecond time scale whereby the density and temperature depend on position as well as excitation volume. To illustrate this behavior we present here results of simulations of plasma evolution in GaAs for a 1-D geometry, which is appropriate to many lasers or induced-grating experiments . Similar effects are expected in 2- and 3-D.

## NUMERICAL SIMULATIONS

Our model solves coupled partial differential equations for electron-hole density and temperature in the rate equation approximation, incorporating well-known components of hot carrier evolution such as carrier cooling (assuming instantaneous electron-hole thermalization and an energy relaxation time of 300 fs) and band-gap renormalization. In addition we explicitly include, radiative and auger recombination, ASE and free carrier and interband reabsorption processes. We have considered plasmas with initially homogeneous

*Hot Carriers in Semiconductors*
Edited by K. Hess *et al.*, Plenum Press, New York, 1996

characteristics over an excitation region between 10 and 50 $\mu$m wide. Because of the nonlocal effects induced by ASE, the plasma evolution depends on the excitation region; here we discuss only results for a 20 $\mu$m wide initial excitation region. We choose the initial plasma (lattice) temperature at t = 0 to be 2000K (300K).

We have identified three characteristic regimes of plasma evolution corresponding to initial carrier densities in the excitation region to be n < $2 \times 10^{18}$ cm$^{-3}$, $2 \times 10^{18}$ cm$^{-3}$ < n < $10^{19}$ cm$^{-3}$ and n > $2 \times 10^{19}$ cm$^{-3}$; the demarcation densities here are functions of the initial excitation region. In the low initial density regime, the plasma density does not change significantly on a picosecond time scale (spontaneous recombination occurs on a nanosecond time scale). The temperature is not spatially dependent and evolves only in time; cooling is governed by phonon emission processes alone. For higher densities, ASE induces recombination on a picosecond time scale. Since the ASE irradiance varies across the plasma, the density and temperature become spatially inhomogeneous with temporal evolution strongly influenced by emission and reabsorption of light. For n > $2 \times 10^{19}$ cm$^{-3}$, all parts of the initial excitation region are influenced by ASE. Fig. 1 illustrates how the plasma temperature evolves at the centre and edge of the initial excitation region.

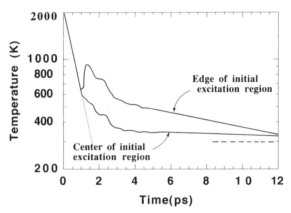

**Figure 1.** Variation of plasma temperature with time at the center and edge of the initial excitation region at times following excitation of a homogeneous plasma of density $2 \times 10^{19}$ cm$^{-3}$ over -20 < x < 20 $\mu$m and with initial electron-hole temperature of 2000 K. The dotted curve illustrates how cooling would occur at both locations in the absence of ASE.

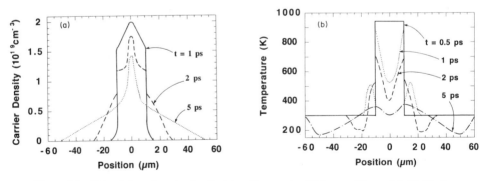

**Figure 2** Spatial variation of carrier density (a) and temperature(b) for conditions related to Fig.1.

For t < 0.8 ps optical gain does not exist, and the plasma cools similar to that of a low density plasma. For t > 0.8 ps, the onset of plasma degeneracy leads to significant gain[4] with the ASE being strongest at the edges of the initial excitation region. Peak irradiances exceed 3 GWcm$^{-2}$ at t= 1 ps and decrease thereafter. For t > 0.8 ps the center and edge temperature evolution differ significantly reflecting redistribution of energy through ASE. The temporal oscillations reflect details of the feedback between plasma heating (less plasma degeneracy, less gain) and cooling (more plasma degeneracy, more gain), however there is a general cooling trend, albeit at a rate which is much slower than that which occurs under nondegenerate plasma conditions (represented by dotted line in Fig. 1).

Figure 2 illustrates how the plasma density, temperature and ASE irradiance evolve with time. The strong stimulated emission also leads to ultra-rapid expansion of the plasma on a picosecond time scale with an effective diffusion coefficient as high as $10^7$ cm$^2$s$^{-1}$. The redistribution of plasma energy through ASE leads to a complex spatial and temporal evolution of the plasma temperature. For t > 2 ps it is interesting to note that outside the initial excitation region a plasma temperature below that of the lattice can exist for several picoseconds. This is caused by emission of photons in regions of high density where there is significant band-gap renormalization and subsequent reabsorption and carrier creation in regions of lower density and higher band-gap.

## CONCLUSIONS

We have illustrated in a 1-D geometry how amplified spontaneous emission can significantly influence the spatial and temporal evolution of hot carriers in a direct gap semiconductor. Much of the hot-carrier literature has focused on carrier energy dynamics, in particular cooling, and the reduction of the cooling rate at high densities. Screening and the hot phonon effect have long been considered the primary factors in explaining this reduction. However there are several reports of other types of "anomalous" behavior at high excitation density such as picosecond recombination rates, non-Maxwellian carrier distributions from time-resolved luminescence experiments, ultrafast expansion of plasmas and the apparent dependence of some of the plasma characteristics on excitation spot-size. To date many of these results have not been explained. Although stimulated emission and recombination are often cited to explain qualitatively features observed in certain experiments, there has been no systematic investigation of its role at a quantitative level for high excitation conditions. We have illustrated quantitatively some of these effects here and in future we will make direct comparison with experimental results and extend our analysis to higher spatial dimensions as well.

## ACKNOWLEDGMENTS

We gratefully acknowledge support of the Natural Sciences and Engineering Research Council of Canada and the Premier of Ontario's Technology Fund.

## REFERENCES

1) H.M. van Driel, Phys. Rev.B19: 5928 (1979); W. Pötz and P. Kocevar, Phys. Rev. B 28: 7040 (1983).

2) Yu. D. Kalafati and V.A. Kokin, Sov. Phys. JETP 72: 1003 (1991).

3) E.C. Fox and H.M. van Driel, Phys. Rev. B 47: 1663 (1993).

4) T. Gong, P. M. Fauchet, J.F. Young and P.J. Kelly, Phys. Rev. B, 44: 6542 (1991).

# EXPERIMENTAL DETERMINATION OF THE INTRABAND
# RELAXATION TIME IN STRAINED QUANTUM WELL LASERS

M. Zimmermann[1], S. Krämer[1], F. Steinhagen[2], H. Hillmer[2],
H. Burkhard[2], A. Ougazzaden[3], C. Kazmierski[3], and A. Hangleiter[1]

[1]4. Physikalisches Institut, Universität Stuttgart
70550 Stuttgart, Germany
[2]Deutsche Telekom AG, Forschungs und Technologiezentrum
64276 Darmstadt, Germany
[3]France TELECOM, CNET, Paris B,
92220 Bagneux Cedex, France

## INTRODUCTION

Intraband relaxation of electrons and holes in semiconductors occurs on a femtosecond
time scale. Nevertheless, it has a considerable impact on both the static and dynamic properties
of optoelectronic devices like semiconductor lasers. Sharp spectral features are broadened
by the lifetime broadening due to intraband relaxation and the spectral holeburning related to
the finite intraband relaxation time contributes to nonlinearities of the optical gain[1,2,3]. Since
intraband relaxation is mediated by electron-electron and electron-phonon scattering, a strong
influence of size quantization as well as of strain-induced changes of the band structure on the
intraband relaxation times is expected. There are by now a number of reports dealing with
this problem from a theoretical point of view [1,2,3,4]. However, there is only very little direct
experimental information on how the intraband relaxation times change in strained quantum
well structures.

## EXPERIMENT AND METHOD OF INVESTIGATION

In this paper, we report on our experimental studies of spectral holeburning in strained
and unstrained InGaAs/InP quantum well lasers. The spectral holeburning is observed when
these lasers are operated far above threshold, i.e. when there is a high photon density in
the laser cavity giving rise to fast stimulated recombination of electrons and holes. Due to
the finite intraband relaxation time, a hole in the optical gain spectrum is formed, which is
observed experimentally. So far, there has been only one further experimental approach,
which was performed in the small signal regime for unstrained laser samples[5].

In order to measure the shape of the optical gain spectrum in these lasers and the changes
due to spectral hole burning, we use the method of Hakki and Paoli, where the gain is basically

determined from the peak-to-valley ratio of the Fabry-Perot modes. In order to cope with the extremely large peak-to-valley ratio of up to 60 dB close to the lasing line, we use a double monochromator set-up which offers a dynamic range of up to 80 dB. The validity of this approach is checked by the value of the net cavity gain at the lasing line which should be zero above threshold, but would be negative if stray light had any influence. We typically find $g > -0.5$ cm$^{-1}$ indicating that our stray light suppression is sufficiently good.

The optical gain spectrum in the absence of any spectral holeburning would have a very smooth shape and would reach zero net cavity gain at exactly one point in the spectrum, which is the lasing mode. Experimentally, we observe the spectra develop a "flat top" far above threshold, which is obviously due to spectral hole burning. Since in a laser there can be no net cavity gain larger than zero, we can not expect to observe a real hole in the spectrum, but rather spectral holeburning leads to lasing of several modes and to the "flat top" of the spectrum, which is due the superposition of the holes associated with the individual modes.

By comparing the measured spectra with calculated spectra from a 6-band $k \cdot p$-model we are able to determine the "missing part" of the gain spectrum: Assuming a Lorentzian shape of the spectral holes associated with each lasing mode, this allows us to determine the width of the spectral holes. Following Agrawal[2,3] it is possible to calculate the gain compression factor $\epsilon_{hb}$ of the laser from the involved intraband relaxation times. By correcting the measured gain spectra with the fraction $\epsilon_{hb} \cdot S$ of the total gain and fitting them with our calculated gain spectra, we are able to directly extract the intraband relaxation times.

## RESULTS

Fig. 1 shows the measured optical gain spectra together with fitted model spectra for a 1% compressively strained laser. On the right hand side, for a bias level of about two and three times threshold, the flattening of the top of the gain spectra is clearly visible, indicating spectral holeburning.

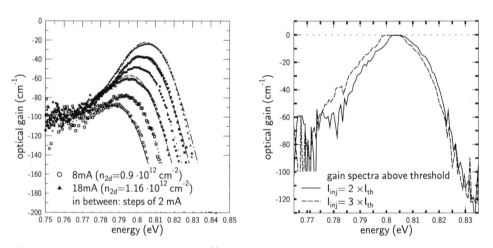

**Figure 1.** Left side: Gain spectra of a 1% compressively strained FP-MQW laser structure. The sub-threshold measurements are fitted by calculated gain curves, which help in the determination of the intraband relaxation time, when compared with spectra above threshold. Right side: Gain spectra above threshold, the flat top is caused by spectral holeburning.

Comparing strained with unstrained lasers, we find a value of the intraband relaxation time of about 300fs for electrons, 50fs for holes in the unstrained and 70fs for holes in the 1% compressively strained case (Fig. 2). Due to the large hole-to-electron mass ratio, the polarization relaxation time is supposed to be mainly determined by the hole intraband relaxation time. Qualitatively, our experimental result is in agreement with the change expected from a smaller heavy hole mass in the strained quantum well. Furthermore we are able to compare $\epsilon_{hb}$ with the total value of $\epsilon_{tot}$, derived from parasitic free high frequency measurements. The larger $\epsilon_{tot}$ is likely be explained by other gain compression effects like transport.

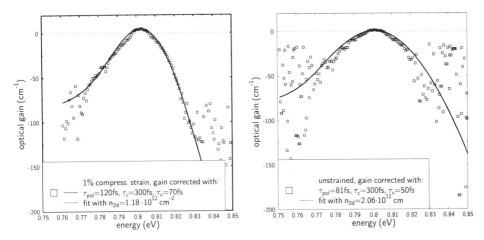

**Figure 2.** Left side: 1% compressively strained MQW laser structure. Right side: Nominally unstrained FP-MQW laser sample, showing less spectral holeburning and in consequence a faster hole relaxation time of about 50fs (measured at DC current of $6 \times I_{th}$)

## SUMMARY

We have determined the contribution of spectral holeburning to gain compression in strained and unstrained InGaAs/InP FP-MQW laser structures by gain measurements above laser threshold. We deduce hole intraband relaxation times of 50fs for unstrained and 70fs for 1% strained lasers.
This work has been supported in part by the EU under contract R2006 "WELCOME".

## REFERENCES

1. M. Willatzen, T. Takahashi, and Y. Arakawa, Nonlinear gain effects due to carrier heating and spectral holeburning in strained-quantum well lasers, *IEEE Photon. Technol. Lett.*, 4:682 (1992)
2. G.P. Agrawal, Spectral hole-burning and gain saturation in semiconductor lasers: Strong-signal theory, *IEEE J. Quantum Electron.* 23:860 (1987)
3. G.P. Agrawal, Modulation bandwidth of high-power single-mode semiconductor lasers: Effect of intraband gain saturation, *Appl. Phys. Lett.* 57:1 (1990)
4. S. Seki, P. Sotirelis, and K. Hess, Theoretical analysis of gain saturation coefficients in InGaAs/AlGaAs strained layer quantum well lasers, *Appl. Phys. Lett.* 61:2147 (1992)
5. R. Frankenberger and R. Schimpe, Measurement of the gain saturation spectrum in InGaAsP diode lasers, *Appl. Phys. Lett.* 57:2520 (1990)

# TRANSIENT PULSE RESPONSE OF $In_{0.2}Ga_{0.8}As$/GaAs MICROCAVITY LASERS

Peter Michler,[1] Gernold Reiner,[2] and Wolfgang W. Rühle[1]

[1]Max-Planck-Institut
für Festkörperforschung
70569 Stuttgart, Germany
[2]Universität Ulm
Abt. Optoelektronik
89069 Ulm, Germany

## INTRODUCTION

The large signal response of semiconductor lasers is important because it limits speed of digital optical communication. In the last few years the concept of microcavity vertical surface emitting lasers has found increasing interest, since it promises several advantages with respect to conventional edge-emitting lasers. However, little is known about their time response and chirp characteristics.

## EXPERIMENT

Two planar microcavity structures were grown by molecular beam epitaxy. The one wavelength thick central region contains either one (sample M1) or three (sample M3) 8 nm thick $In_{0.2}Ga_{0.8}As$ quantum wells (QW's). The QW's are embedded in 10 nm thick GaAs barriers and sandwiched between $Al_{0.33}Ga_{0.67}As$ cladding layers which are surrounded by top and bottom AlAs/GaAs distributed Bragg reflectors with 20.5 and 30 periods (theoretical reflectivities > 99.9 %). A thickness variation of the cavity causes a shift of the resonance from 1020 nm at the wafer center to 860 nm at the edge, while the QW band-gap wavelength is merely unchanged ($\Delta\lambda < 2$ nm) and is centered at 923 nm at 20 K. The stimulated emission, which occurs at the cavity resonance, can therefore be tuned within the gain spectrum of the QW's by probing different points on the wafer.

The quantum wells in the microcavities are directly excited with 821 nm pulses of a femtosecond Ti:sapphire, i.e., with a wavelength below the barrier and just below the 105 nm wide stopband of the cavity. Transient stimulated emission is measured using a streak camera (S1 cathode) or photoluminescence up-conversion for detection.

## RESULTS

The time dependence of the wavelength-integrated emission of M1 is displayed at 925 nm [Fig. 1(a)] and at 884 nm [Fig. 1(b)].

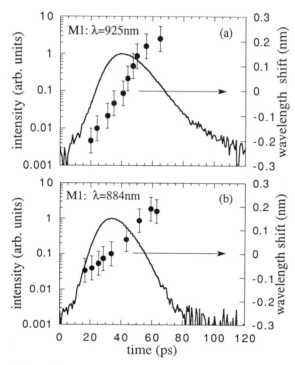

**Figure 1.** Time evolution of the measured emission (solid lines) and wavelength shift of the spectral maximum (solid circles) from $In_{0.2}Ga_{0.8}As/GaAs$ microcavity lasers with one QW at 925 nm (a) and at 884 nm (b).

Additionally, the instantaneous laser wavelength shift is shown (solid circles with error bars) where the wavelength at the temporal maximum of the pulse is set to zero. The transient pulse response at the two different emission wavelengths (see Fig. 1 (a), (b)) shows as reported[1] that the stimulated emission is faster when the cavity resonance is tuned to shorter wavelengths within the gain spectrum of the quantum well. Comparison of the wavelength chirps shows that the total chirp is with $\Delta\lambda = 0.26$ nm at 884 nm smaller than with $\Delta\lambda = 0.41$ nm at 925 nm.

In Fig. 2 the wavelength dependent measurements of pulse width and threshold are summarized. The wavelength dependences of the FWHM of the pulses of M1 (full triangles) and M3 (full circles) are similar, however, M3 is always faster. Starting at long wavelengths, the pulse widths first decrease, reach a minimum near 884 nm and increase again for shorter wavelengths. For comparison, we plotted also the wavelength dependences of the thresholds of M1 (open triangles) and M3 (open circles). Obviously, the fastest response and the minimum threshold are at different wavelengths.

The laser dynamics was modeled using a set of rate equations that has been described previously[1,2]. The model is based on a simple three level system where all the excited carriers are at t = 0 in the highest state with an energy corresponding to the pump energy. The carrier transfer into the upper laser level occurs with a time constant which phenomenologically includes all relaxation processes. The model attributes the decreasing pulse width with decreasing wavelength to an increasing differential gain.

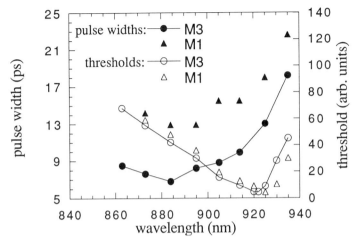

**Figure 2.** FWHM of the emission (closed symbols) and thresholds (open symbols) versus wavelength for two different microcavities with one (triangles) and three (circles) QW's, respectively. The lines are guides to the eye.

The laser dynamics becomes faster with increasing number of QW's inside the microcavity due to the increase of the longitudinal confinement factor, which increases with the size of the active gain region. A smaller total chirp is observed at shorter emission wavelengths since the linewidth broadening factor $\alpha$ is reduced at shorter emission wavelengths.

## SUMMARY

In conclusion, optimum high speed characteristics of a microcavity laser is not obtained if the cavity resonance is at the wavelength of lowest threshold but at shorter wavelengths. We obtain best performance for a detuning of -38 nm for the structure containing three QWs. The best values are a 1 ps rise time, a 6.8 ps pulse width (FWHM), and a time-bandwidth product of 0.68 at the expense of an increase of threshold by a factor of 5.

## REFERENCES

1. P. Michler, A. Lohner, G. Reiner, and W. W. Rühle, Transient pulse response of $In_{0.2}Ga_{0.8}As/$ GaAs microcavity lasers, *Appl. Phys. Lett. 66:1599 (1995)*.
2. P. Michler, W. W. Rühle, G. Reiner, K. J. Ebeling, and A. Moritz, Time-bandwidth product of gain-switched $In_{0.2}Ga_{0.8}As/$GaAs microcavity lasers, *to be published in Appl. Phys. Lett.*

# HOT ELECTRON LIGHT EMISSION FROM GUNN DOMAINS IN LONGITUDINALLY BIASED GaAs p - n JUNCTIONS AND IN n- GaAs EPILAYERS

N. Balkan[1], M. Hostut[1], T. de Kort[2], A. Straw[1]

[1]University of Essex, Department of Physics
Colchester, UK

[2] Eindhoven University of Technology
Department of Physics, Eindhoven, The Netherlands

## INTRODUCTION

Hot carrier Light emission from GaAs, biased in the Negative Differential Resistance region has been observed since the mid-sixties.[1]. Most earlier devices investigated were of dimensions in the 1 mm range. With the advent of sophisticated lithographic techniques it has been possible to fabricate devices with spatial dimensions reduced to the sub-micron scale. Over the last few years, there has been a growing interest in hot electron light emission in such devices. There are two reasons for this. Firstly, the study of hot-carrier luminescence is a powerful technique to understand impact ionisation phenomena that occurs in heavily biased devices with a high density of conduction electrons. Secondly, from an engineering point of view, in micron and sub-micron devices, operated at bias voltages above the NDR threshold, degradation of the performance characteristics or break-down can occur[2]. It is therefore, desirable to understand the phenomena for improved device reliability. A large number of functional devices to date have been investigated. These include GaAs based MESFET[2,3], HEMT[2,4], PM-HEMT[2,5] and, HBT[2,6] as well as similar devices based on other material structures[2]. The devices studied have channel lengths usually between 0.5 and 5 μm. Most studies involve the DC biasing of the devices above the NDR threshold and the measurement of spatially and temporally averaged electroluminescence. Spectral measurements provide information about the energy distribution of the associated electrons. Electron temperatures, as determined from the high energy tail of the spectra by assuming a Maxwellian distribution are reported to vary between a few 1000 K[2] to about 15000 K[7]. These temperatures are significantly higher than those reported in bulk GaAs[1]. In order to explain the observed light emission a number of mechanisms have been suggested. These include inter-band recombination of the impact ionised electron-hole pairs, Bremsstrahlung, electron- impurity transitions , and inter-conduction-band transitions ( References 2 - 7 ). However, there is no conclusive evidence to support fully any of the proposed mechanisms.

The aim of this work is to study the spectral and the temporal behaviour of the light emission in simple devices based on n-doped GaAs epilayers . The length of the samples investigated are chosen to be long enough to observe the transients of the Gunn Domains and the emitted light within the time resolution available to us (1 nsec). We also investigated the effect on the emission spectra of a p-layer adjacent to the active ( n-type ) layer by studying the phenomena in a GaAs p-n junction with electric fields applied parallel to the junction plane.

*Hot Carriers in Semiconductors*
Edited by K. Hess *et al.*, Plenum Press, New York, 1996

## EXPERIMENTAL RESULTS AND DISCUSSIONS

The samples used in the investigations were grown by the MBE ( n- GaAs) and MOCVD ( p-n GaAs) techniques. The structural parameters of the samples and the device dimensions are given in Table I.

Table I. Sample parameters

| Sample | Growth | Doping (cm$^{-3}$) | Epilayer thickness | Carrier concentration T=77K(cm$^{-2}$) | Mobility T=77K (cm$^2$/V.s) | Device dimensions length x width |
|--------|--------|--------------------|--------------------|-----------------------------------------|-----------------------------|----------------------------------|
| A1053 | MBE | Si: 1 10$^{17}$ | 2 μm | 1.24 10$^{13}$ | 6038 | 316 x 98μm |
| QT623 | MOCVD | Si: 7 10$^{17}$ C: 5 10$^{17}$ | 0.15 μm 0.15 μm | 3.3 10$^{12}$ 3.16 10$^{12}$ | 2280 1450 | 600 x     μm |

The samples were fabricated in the form of Hall bars ( for electrical characterisation) and simple bars ( for high speed I-V measurements). Ohmic contacts were formed by alloying Au/Ge/Ni. Optical characterisation was carried out using orthodox Photoluminescence techniques. Electric field pulses of duration between 5 and 100 nanosecond were applied along the layers with a duty cycle of less than 10$^{-3}$. I-V transients were recorded using a 1 GHz. oscilloscope. The spectral and integrated electroluminescence measurements were carried out by employing simple gating techniques as described by us elsewhere[8]. In the experiments the whole sample is focused onto a monochromator slit, and therefore, the EL is representative of emission from the whole surface . Fig. 1 shows the PL spectra of the 2 samples at around liquid nitrogen temperatures. The n- type epilayer has a single emission peak at $h\upsilon$ =1.51 eV corresponding to a band to band transition. The emission in the p-n structure, however, is dominated by a peak s at $h\upsilon$ =1.48 eV indicative of an electron - acceptor transition.

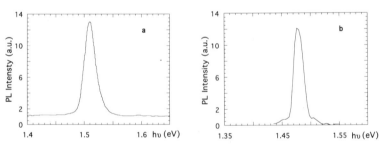

Fig. 1 PL Spectra of A1053 (a) and QT623 (b) at T=87 and 78 K respectively

Figure 2 shows the electroluminescence (EL) spectra from the samples. The EL spectra is measured at fields above the NDR region as indicated in figure 3. The arrows in figure3 indicate the threshold electric fields at which the NDR, current instabilities and light emission were observed.

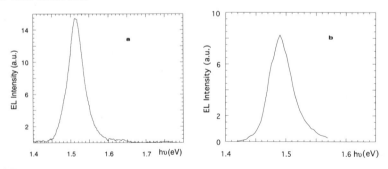

Fig. 2 EL Spectra of A 1053 (a) and QT 623 measred at electric fields F=3.3 kV/cm and 3.6 kV/cm.

A simple comparison of figures 1 and 2 indicate that the mechanism producing the EL is identical to that producing the PL results. Namely, band-to-band recombination and electron-acceptor recombination in A1053 and QT623 respectively. However, unlike the PL spectra, the EL spectra of both samples have a high energy tail. The field dependence of the peak EL intensity for A 1503 and electron temperature (as obtained from the high energy tails of the spectra, as described in reference 8) are shown in figure 4. It is evident that the electron temperatures are significantly lower than those reported in sub-micron and micron devices[2-7], but much higher than those in mm samples[1]. In view of the fact that the electric fields in the domains might exceed, $E=10^4$ V / cm, therefore, electron temperatures would be expected to be very high[9]. We believe that most of the luminescence is due to the recombination of the impact ionised holes with equilibrium electrons outside the domain where the electric field and hence, the electron temperature is low.

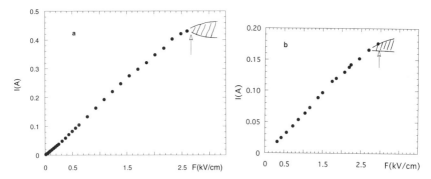

Fig.3 Current- Field characteristics in A1053 (a) and QT623 (b) at T= 87 K and 78K. The arrows indicate the fields above which instabilities and light emission occurs.

Further evidence for this comes from the temporal behaviour of the emitted light during single or consecutive transits of Gunn domains during the electric field pulse. This is shown in figure 5. The spikes in the current pulse in figure 5 (a) have a period of 2.5 nsec. corresponding to a domain transit velocity of 1. 3 x10 [7] cm/sec. When the applied electric field pulse width is 4.5 nsec, so that only a single domain transit occurs during the pulse duration, the EL pulse has the form as indicated in figure5 (b). There is a time delay between the current and the light pulse of 25 nsec which has no significance but merely represents the electron transit time of the photo multiplier. What is interesting, however, is that the light pulse persists after the collapse of the domain at the anode. This suggests that once the negative space charge disappears at the anode there is an accumulation of non-equilibrium impact ionised holes in the vicinity of the anode. These holes recombine with the equilibrium electrons, in accord with previous studies that the light emission is concentrated mainly in the drain region [2-7].

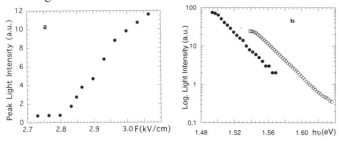

Fig.4 (a) Applied field dependence of the peak EL intensity in A1053 at T= 87 K. (b) Logarithim of the high energy tail of EL spectra for A1053 ( open circles) and for QT 623 closed circles. Electron temperatures are T= 250 and 230 K respectively. Electric fields are the same as in figure 2.

When the width of the electric field pulse is increased, the transit of 3 domains during the pulse duration can be observed as shown in figure 5 (c). This is accompanied by an increase in the amplitude of the light pulse. However, the latter does not follow the transit of the domains. i.e., the light pulse is not in the form of three individual peaks associated with the three domains in transit. When the pulse width is increased further to 40 nsec, figure 5 (d), only very small wriggles are seen in the current pulse after the initial 6 domain transits ( 15 nsec.). This suggests that the domains are either not forming at the cathode or not collapsing at the anode. The light pulse, however has a sharp increase at the point ( after 15 nsec.) where the current spikes disappear. This observation may be explained only if the accumulation of non-equilibrium holes increases after the disassociation of 6 consecutive domains at the anode and, therefore, the domains will now be no longer moving in a uniform background as they approach the anode where the positive charge has accumulated . As a result, at this point in time, the EL intensity is expected to increase rapidly due to the recombination of the high density non-equilibrium holes .

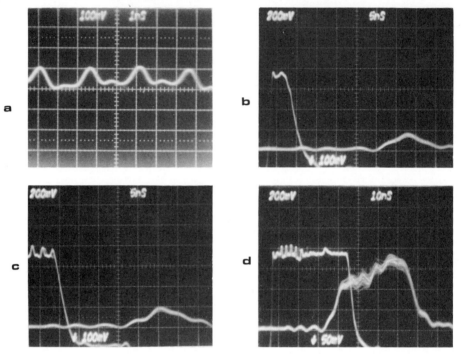

Fig. 5 The oscillograms of the current pulse and the emitted light. (a) 2.5 nsec.oscillations in the current pulse, (b), (c), (d) Current and the light pulses during the transits of one, three and six domains during the pulse duration. Light pulse is delayed 25nsec with respesct to the current pulse.

## CONCLUSIONS

In n- doped epilayer GaAs and p-n junction GaAs light is emitted when electric fields in the NDR region are applied along the layers. Emitted light is due to the radiative recombination of impact ionised holes with the hot equilibrium electrons. Most of the steady - state electroluminescence in devices with lengths much greater than the Gunn domain width are expected to originate from outside the domain.

## REFERENCES

1. P. D. Southgate, Journ. Appl. Phys. 18: 4589 (1967)
2. C. Canali, C. Tedesco, E. Zanoni, M. Manfredi, A. Paccagnella p 215 in" NDR and Instabilities in 2-D Semiconductors"edited by N. Balkan, B.K. Ridley andA.J. Vickers, Plenum Publ. New York (1993)
3. A. Paccagnella, E. zanoni, C. Tedesco, C. Lanzieri, A. Cetronio, IEEE, Electr. Dev.38: 2682 (1991)
4. H. P. Zappe, D. J. As, Appl. Phys. Lett. 59: 2257 (1991)
5. F. Aniel, P. Boucaud, A. Sylvestre, P. Crozat, F.H. Julien, R.Adde, J.Appl. Phys. 77: 2184 (1995)
6. E. Zanoni, L. Vendrame, P. Pavan, M. Manfredi, S. Bigliari, R. Malik, Appl. Phys. Lett. 62: 402 (1993)
7. R. Ostermeir, F. Koch, H. Brugger, P. Narozny, H. Dambkes, Semicon. Sci. Technol. 9: 659 (1994)
8. R. Gupta, N. Balkan, B.K. Ridley, Phys. Rev. 46: 7745 (1992)
9. B. K. Ridley, p 1 in " NDR and Instabilities in 2-D Semiconductors" edited by N. Balkan, B.K. Ridley andA.J. Vickers, Plenum Publ. New York (1993)

# HOT ELECTRON LIGHT EMITTING SEMICONDUCTOR HETEROJUNCTION DEVICES (HELLISH) - TYPE - 1 AND TYPE - 2

N. Balkan,[1] A. da Cunha,[1] A. O'Brien,[1] A. Teke,[1] R. Gupta,[1] A. Straw[1], M. Ç. Arikan[2]

1University of Essex, Department of Physics, Colchester, UK
2University of Istanbul, Department of Physics, Istanbul and TUBITAK, MRC, Kocaeli, Turkey

## INTRODUCTION

One of the draw-backs of the conventional light emitters appears to be that the light emission is confined to a small region of the facets of the devices[1]. Thus, the compatibility in generic integration technology remains a problem. The research on simple devices that emit light from the surface with good control of wavelength tunability, and which can be fabricated in large scale 2- dimensional arrays has been largely stimulated by potential applications in optical signal processing. One possible candidate for such a simple functional device is the light emitting charge injection transistor ( CHINT) [2]. Another light emitter, HELLISH-1 (Hot Electron Light Emission and Lasing in Semiconductor Heterostructures ) has been proposed by us[3-5]. In this paper we present a novel surface emitting device, HELLISH-2 and demonstrate its operation with a simple model. We also report the results of our recent studies on a heavily p-n doped HELLISH-1 device.

## RESULTS AND DISCUSSIONS

### Hellish - 1

The structure and the schematic band diagram of the device investigated are shown in figure 1. Ohmic contacts were formed by diffusing Au/Ge/Ni to all the layers. The quantum well is in the depletion region close to the n-side. Also shown in figure 1 is the model for the carrier injection and the recombination in the device. When the electric field is applied parallel to the layers electrons on the n-side and the holes on the p-side of the junction heat up. Hot electrons are injected into the quantum well via tunnelling and thermionic emission. The well acts, therefore, as a giant trap for the excess hot electrons. The accumulation of the negative charge in the well induces the change of the potential profile as to preserve charge neutrality. The potential barrier and the depletion region in the p-side is reduced, enabling excess hot holes to diffuse into the quantum well where they recombine with the excess electrons. The operation of the device is, therefore, independent of the polarity of the applied voltage. In figure 2 the PL spectra from the sample shows two peaks associated band to band recombination (1.505 eV) and a deep impurity recombination (1.417eV), originating from the GaAs substrate and buffer layers. The quantum well emission appears as a weak high energy shoulder in the PL spectra. The weak PL emission from the well is expected. Because the incident photons (the 647 nm line of the Kr laser) create electron-hole pairs across the band-gap of the GaAlAs layers these carrier pairs are swept out into respective majorty regions in the high built-in field, thus reducing the capture in the well.

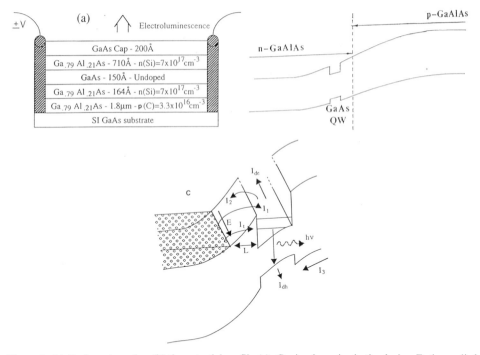

**Figure 1.** (a) Device schematics, (b) the potential profile (c) Carrier dynamics in the device. E: the applied longitudinal field. $I_1$, $I_t$ are the injection of excess electrons into the well by thermionic emission and by tunnelling. $I_2$, $I_{de}$, $I_{dh}$ are the reverse current from the well into the n- layer, electron and hole drift currents in the well respectively. $I_3$ diffusion current of hot holes into the well.

The EL, however, which originates from the trapping of excess hot carriers via tunnelling and thermionic emission, shows two peaks due to two subbands in the well ( 1.535 eV and 1.597 eV). ( In the figure the EL intensity is reduced by about two orders of magnitude). In Figure 3. The EL intensity is shown to have two broad peaks at electric fields as indicated.These peaks were also observed for a lightly doped sample[3-5]. When the occupation of the 3-D states, in resonance with the quantum well subbands, increses with increasing electric field or electron temperature, the tunnelling of hot electrons from the n-layer into the well gives rise to broad resonance peaks as observed. The lattice temperature dependence of the EL intensity, measured

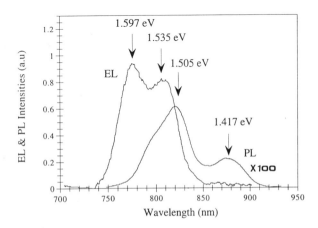

**Figure 2.** The PL and the EL spectra of the sample at TL= 77K and at F= 455 V/cm.

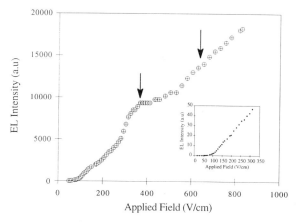

**Figure 3.** The electric field dependence of the EL intensity at TL= 77K

at a fixed electric field, also shows these resonance features as illustrated in figure 4. It is evident from figure 4 that, unlike the conventional devices, the efficiency of HELLISH-1 is very little affected by the change of lattice temperature due to efficient non-equilibrium carrier collection by the well via the tunnelling of hot electrons[3-5].

## Hellish - 2

The structure and the energy - band diagram of the device, designed for single or double wavelength operation, is shown in figure 5. Ohmic point contacts were formed by diffusing Au/Ge/Ni to all the layers. Figure 6. shows the surface EL spectra of the sample at T=77 K. As in HELLISH-1, the EL was independent of the polarity of the applied voltage, because the light emission is associated with the electric field heating of carriers.

At low electric fields the EL has a peak at $h\nu_1$. When the electric field is increased however, the spectra develops a high energy tail. At higher fields another peak at energy $h\nu_2$ appears. The intensity of $h\nu_2$ grows faster than that of the $h\nu_1$ peak with increasing field. At an electric field Feq=1.2 kV/cm the spectra has two peaks of equal intensity.

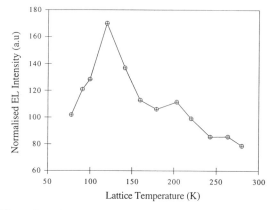

**Figure 4.** Lattice temperature dependence of EL at F= 353 V/cm

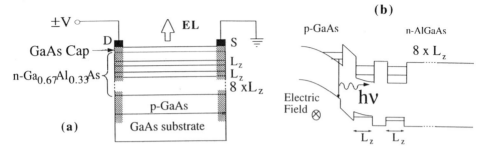

**Figure 5.** (a) The structure of HELLISH-2, (b) Energy - band diagram.

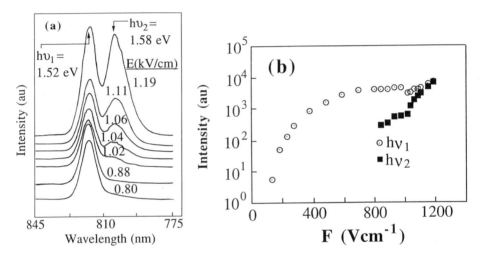

**Figure 6.** (a) EL spectra at T= 77K. (b) Peak EL intensity vs. applied electric field for both emission peaks

The carrier dynamics in the device are illustrated in figure 7(a). The potential profile of the device has been calculated as described elsewhere[6]. When the electric field is applied to the device in the plane of the layers, both the electrons in the wells and in the inversion layer and holes in the p-layer are heated up. Hot electrons in the well adjacent to the junction plane are injected into the inversion layer via phonon assisted tunnelling and thermionic emission currents at finite non-equilibrium temperatures[6,7]. The accumulation of excess negative charge in the inversion layer induces modification of the junction. The p-side of the depletion region decreases so as to preserve the charge neutrality. Thus the electron and the hole wave functions overlap in the vicinity of the inversion layer giving rise to emission at around the GaAs interband transition ($h\nu_1$) When the field is increased further, the hot electron current into the inversion layer increases. Also the non- equilibrium electrons in the inversion layer (which also see the same external field) heat up and occupy higher energy states . The inversion layer luminescence is therefore expected to develop a high energy tail. However, photons with energies high enough to be absorbed by the quantum wells are re-emitted at an energy corresponding to e1-hh1 transition ($h\nu_2$). With increasing field both the injected non-equilibrium electron density and the occupancy of the higher energy states in the inversion layer increase further. Therefore, more high energy photons become available for absorption in the quantum wells. As a result, the intensity of re-emission at ($h\nu_2$) increases rapidly with increasing field as observed in figure 6. The net current into the inversion layer is plotted against the inverse electron temperature in the quantum well in figure 7 (b). Also shown in the same figure is the integrated EL intensity versus electron temperature as determined from the high energy tail of the $h\nu_2$ emission. The agreement is excellent.The structural configuration of the device can be designed as to tune the wavelength for three or more emissions[7].

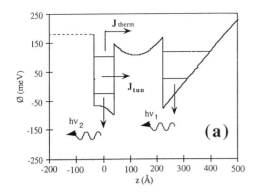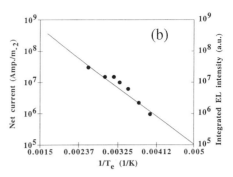

**Figure 7.** (a) Non-equilibrium hot electron injection from the quantum well into the inversion layer and the origin of the observed EL emissions. (b) Calculated net excess current into the inversion layer ( Line), and the EL intensity vs. inverse electron temperature ( circles).

## CONCLUSIONS

The devices have the following advantages over the conventional semiconductor light emitters:
i)   The operation of the devices utilise longitudinal transport of hot carriers.
ii)  Only two diffused in point contacts are required for the operation,
iii) Light is emitted from the surface, hence the fabrication and the quality assessment of 2-D array surface emitters can be achieved very easily,
iv) Light emission is independent of the polarity of the applied voltage, the light output is therefore an  XOR function of the input voltage,
v)   The HELLISH - 2 device can be tuned to emit light at single or multiple wavelengths.
vi) The HELLISH - 2  device, therefore, has a built-in multi- wavelength light logic.

Furthermore both light emitters are excellent candidates for surface emitting lasers, where the HELLISH structure is placed in the cavity between  a DBR's  a vertical cavity  laser structure. The work on HELLISH-VCSEL is underway and the results will be published in due course.

## ACKNOWLEDGEMENTS

The work is supported by EPSRC and NATO (Grant no: CRG 931510 ) .

## REFERENCES

1. M.Harberg, B.Jonsson, A.Larsson, J.Vac. Sci. Techn. B, 10, 2243 (1992)
2. S. Luryi, M.Mastrapasqua, in " NDR and Instabilities in 2D Semiconductors" edited by  N. Balkan, B.K.Ridley, A.J.Vickers  Plenum Press, Newyork  (1993)
3. A.da Cunha, R,Gupta, A.Straw, N.Balkan, B.K.Ridley, Semiconductor Science and Technology, 9, 677 (1994)
4. A.DaCunha, A.Straw, R,Gupta, N. Balkan B.K.Ridley. Electro Chem. Soc.  94-17, 313 (1995)
5. A.Straw, A.daCunha, R.Gupta, N.Balkan and B.K.Ridley Superlattices and Microstr. 16, 173 (1994)
6. N. Balkan, A. Teke, R.Gupta, A. Straw, J.H. Wolter, W. van Vleuten, Appl. Phys. Lett. 67: 7, 1 (1995)
7. R. Gupta, N. Balkan, A. Teke, A. Straw, To be published in the Proc. of ICSMM-8 (1995)

# INTRABAND POPULATION INVERSION BY ULTRAFAST RECOMBINATION IN PROTON BOMBARDED InP  (abstract)

R.A. Höpfel[1], Ch. Teissl[1], K.F. Lamprecht[1] and L. Rota[2]

[1]Institut für Experimentalphysik, Universität Innsbruck,
A-6020 Innsbruck, Austria
[2]Clarendon Laboratory, University of Oxford
Parks Road, Oxford OX1 3PU, United Kingdom

In a previous conference[1] we have reported on carrier distributions in radiation-damaged semiconductors with ultrashort lifetimes. The short existence of the electron-hole distribution has lead to luminescence spectra far from equilibrium due to the short carrier lifetime. Recombination within 100 fs has been observed, which is clearly faster than the energy relaxation of optically excited carriers. Thermalisation, however, can occur within the timescale of 100 fs (Ref. 2), strongly depending on the carrier concentration, so that the issue of intraband inversion was still an open physical question.

At this conference we have presented a combined experimental and theoretical study of the scenario at low carrier concentrations, where the recombination is faster than the thermalisation. Optically excited carriers recombine before complete thermalisation. The time-integrated luminescence over the lifetime of 100 fs is studied. We observe *strongly inverted luminescence spectra* at carrier densities of ~ $3x10^{16}$ cm$^{-3}$, indicating intraband population inversion over more than 100 meV. We have also investigated the effect of recombination on the energy relaxation in samples with lifetimes of the order of 1 ps. Reduced cooling is observed in the presence of fast recombination. We interpret this as a consequence of energy dependent recombination, which increases the degree of inversion.

Theoretical Monte-Carlo simulations reproduce the observed luminescence spectra and give quantitative information on the carrier distributions in the involved bands. The electron distribution is most strongly inverted, the distributions in the heavy-hole, light-hole and split-off valence band show interesting nonthermalised structures. Thus the evidence for intraband population inversion in a single continuous band of a solid is now clearly given. Since the inversion is present for the time-average of the carrier existence, it might be maintained also with stationary (cw) excitation. We propose the application for a solid-state free-electron laser. Details of the results are contained in ref. 3. This work has been supported by the Jubiläumsfonds der Österreichischen Nationalbank (project 5059).

[1]K.Lamprecht, S.Juen, L.Palmetshofer, R.A.Höpfel, Semicond.Sci.Technol.**7**, B151 (1992).
[2]L. Rota, P. Lugli, T. Elsässer, J. Shah, Phys. Rev. **B47**, 4226 (1993).
[3] R.A. Höpfel, Ch. Teissl, K.F. Lamprecht, L. Rota, submitted to Phys. Rev. B.

# ON FEL-LIKE STIMULATED FIR EMISSION IN SUPERLATTICES

A.Andronov,[1] E.Gornik,[2] W.Hess,[2] I.Nefedov,[1]

[1] Institute for Physics of Microstructures, Russian Academy of Sciences
Nizhny Novgorod 603600, Russia
[2] Inst. for Solid State Electronics Technich. University of Vienna,
Floragasse 7/1 Stock A-1040 Vienna Austria

## INTRODUCTION

In this work we show that the FIR emission of hot carriers due to weak super-lattice (SL) potential [1] - similar the one used in the Free Electron lasers (FEL) may be made stimulated in crossed electric E and magnetic H fields for lateral transport in 2d gas with weak lateral superlattice (SL) potential and for lateral transport in "standard" SL - MQW with narrow barriers - provided carriers drift in the fields across the SL with high enough velocity: $\vec{q}\vec{v}_c > \omega_c$ where $\vec{q}$ is the wavevector of the SL potential $u = u_0 * [exp(i\vec{q} * \vec{r}) + exp(-i\vec{q} * \vec{r})]$, $\vec{v}_c = c[\vec{E} * \vec{H}]/H^2$ is the drift velocity and $\omega_c = eH/m^*c$ is the cyclotron frequency.

## GENERAL CONSIDERATIONS

FEL-like emission may be described as simultaneous emission of a photon $\hbar\omega$ and the Bragg scattering in the SL potential and is governed by the conservation laws:

$$\epsilon - \epsilon^{'} = \hbar\omega \quad \vec{p} - \vec{p} = \hbar\vec{q} \tag{1}$$

Here $\epsilon$ and $\vec{p}$ are energy and momentum of an electron. The conductivity $\sigma$ of the system is proportional to difference of electron populations at $\epsilon$ and $\epsilon^{'}$. Because the SL potential consists of the two q-harmonics $\sigma < 0$ for electrons with drift velocity $v_d$ is possible for $\omega < qv_d$ only for small RMS velocity $v_T$ ($v_d > v_T$) because of absorption due to the harmonic with $\vec{q} * \vec{v}_d < 0$. On the other hand since work [2] it is known that it is possible to achieve drift of carriers with small RMS velocity in semiconductors with strong carrier-optical phonon interaction in the crossed E, H fields due to carrier accumulation below an optical phonon energy $\hbar\omega_0$. Figure 1, A gives simulated electron momentum distribution function in GaAs; we see that for distribution along the drift $v_d > v_T$ and one should expect $\sigma < 0$ if the drift is along a SL axis and $\omega < q * v_c$.

Such a drift can be easily established for hot electrons in 2d gas in perpendicular to the gas magnetic field (where similar distributions should occur). However it is difficult to achieve in "standard" SL's for the needed direction of $\vec{H}$ - parallel to the SL layers. However for the lateral transport in the latter case if the Hall field $E_H$ is not too high compared with applied field $E_a$ a group of electrons drifting across the SL (and the drift velocity $\vec{v}_d$) may exists which could provide $\sigma < 0$ (Fig 1,A,b).

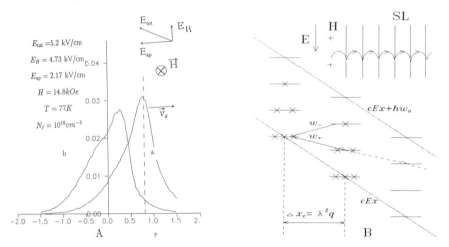

**Figure 1** A.Distribution function of electrons in GaAs along drift (a) and along the Hall field (b); B. Scheme of the Landau levels and emission processes in cross fields

This is the more so if we recall that crossed fields affect emission processes due to quantization of an electron energy which is now:

$$\epsilon = \hbar\omega_c(N + 1/2) + \hbar^2 * k_z^2/2m^* + m^* * c^2(E/H)^2 + eEx_c \qquad (2)$$

Here N is the Landau level number, $x_c = -\lambda^2 * k_y$, is wavefunction center position, $k_y = p_y/\hbar$, $\lambda$ is the magnetic length. Now the FEL-like emission involves change in N and change in $x_c$ to $\Delta x_c = -\lambda^2 q_y$ due to the Bragg scattering. From (2) for the two $q_y = \pm \mid q_y \mid = q_0$ we have the two sets of the emission frequencies:

$$\omega = \omega_+ = v_c * q_0 - M\omega_c \qquad \omega = \omega_- = -v_c * q_0 + L\omega_c \qquad (3)$$

Here M and L are integers and represent changes in the Landau level number during emission: $M, L = N - N'$ where N and N' are the numbers before and after emission (Fig 1,b). For $\omega_+$ with $M < 0$ transition takes place to the higher (less populated) Landau level - and $\sigma$ should be negative. Because $\omega_+ \neq \omega_-$ generally one should not care for the absorption process which involves $q_y < 0$ at small $v_c/v_T$ and which suppress $\sigma < 0$ in the case the effects of the fields excluded.

## ESTIMATE OF NEGATIVE CONDUCTIVITY

Conductivity $\sigma$ of the electron system in crossed fields with the SL potential may be calculated by second order perturbation theory (cf. [1]). We present here only the conductivity along $\vec{v}_c \parallel \vec{q}$ for the simplest (classical) case of small enough q and $\omega_c$ when one can express difference in population of the states responsible for conductivity via appropriate derivative, for approximation of the electron distribution by drifted with velocity $v_c$ maxwellian distribution with electronic temperature $T_e$ and ignoring effect of the fields. This calculations are appropriate for the case $\omega \gg \omega_c$ and $qr_c \gg 1$, $r_c$ is electron gyroradius. The result is:

$$\sigma = \sigma_0 * [F_+ + F_-], \qquad (4)$$

where $\sigma_0 = (\pi)^{1/2} * (eE_s/\omega)^2 * (e^2 N_0/m^*\omega) * (1/m^* * T_e)$, $\Omega = \omega/q * v_c$, $\kappa = v_c/v_T$, $\beta = \omega_c/q * v_c$, $E_s = q * U_0/e$ is electric field of SL the potential and $N_0$ is electron concentration; the functions $F_\pm = (\Omega \mp 1) * \exp(-\kappa^2 * (\Omega \mp 1))$ with $F_+ < 0$ for $\Omega < 1$ and eventually describe "envelops" of the conductivity spectrum determined by (3). From (4) it follows that $\sigma < 0$ only for $\kappa > 1/\sqrt{2}$. To estimate highest value of $\sigma < 0$ we should incorporate spectrum (3) and choose value of the SL potential because $\sigma \sim E_s^2$. By observing that in the crossed fields with the SL potential the electron drift picture is conserved unless $E_s < E$ and by putting $E_s \approx E$ we get the following estimate for the highest value of negative conductivity at $\kappa \simeq 1$, $\omega \simeq \omega_c$: $\sigma_{max} \simeq -(1/10 - 1/5)e^2 * N_0/(m^*\nu)$. Here $\nu$ is scattering rate at $\epsilon < \hbar\omega_0$ which account for linewidth at $\omega \simeq \omega_+$. For small $\nu/\omega_c \simeq 0, 1$ this value is substantially larger than the Drude conductivity $\sigma_D = e^2 * N * \nu/(m^*\omega)^2$ what shows a real possibility for creation of a FIR source. Efficiency of the FIR emission should be high enough because for electrons which continue the down the slope motion (Fig 1, B) and hit the $eEx + \hbar\omega_0$ line before being scattered the energy transferred to the FIR field is of about that transformed to Joule heat.

## STRUCTURES FOR OBSERVATION OF THE EFFECTS

The first step in establishing of the phenomena described should be observation of spontaneous emission and identification of emission spectrum (3) what can be also made by observation of FIR photocurrent in crossed fields.
The SL period d needed follows from the main condition for $\sigma < 0$: $\vec{q} * \vec{v}_c > \omega_c$. For the 2d gas in GaAs-based system $d \simeq 0.1 - 0.3\mu m$ for $H \sim 10 - 30$ Tesla may be quite appropriate. At the first stage - observation of spontaneous FIR emission and FIR photoresponse - the SL potential may be introduced by metallic gates ( which is unappropriated for the FIR source - due to absorption). For lateral transport in "standard" SL smaller $d \leq 0.05 - 0.07\mu m$ is needed. The main problem here is a way to produce high mobility weak potential SL. The FIR source should include resonator system which may be produced just by etching the substrate beneath the 2d gas or MQW to provide a waveguide with FIR field located in the region with $\sigma < 0$.

**Acknowledgments.** This work is supported by Russian Foundation for Fundamental Research, Russian Nonophysics Program and by INTAS

## REFERENCES

1. C.Wirner, C.Kiener, W.Boxleitner, M.Witzany, E.Gornik, P.Vogl, G.Bohm and G.Weimann. "Direct observation of hot electron distribution function in GaAs/AlGaAs heterostructures", Phys. Rev. Letters, v.70, n17, pp 2609-2612,(1993)
2. H.Maeda and T.Kurosawa. "Hot carrier populaton inversion in crossed electric and magnetic fields", Journ. Phys. Soc. Japan, v. 33, N 2, p 562-563 (1972).

# FAST MODULATION OF A LASER-PHOTOTRANSISTOR
# BY LONG-WAVELENGTH INFRARED RADIATION

Victor Ryzhii[1], Vladimir Mitin[2], Maxim Ershov[1],
Irina Khmyrova[1], Valerii Korobov[2] and Maxim Ryzhii[1]

[1]Computer Solid State Physics Laboratory, University of Aizu
Aizu-Wakamatsu City, 965-80, Japan
[2] Department of Electrical and Computer Engineering,
Wayne State University, Detroit, MI 48202

Quantum-well (QW) structures are utilized in a variety of modern semiconductor devices, for example, in QW lasers and intersubband QW infrared photodetectors.[1,2] A novel device – the photodetector with optical output has been proposed and studied theoretically.[3,4] This device consists of a QW photodetector integrated with a light emitting diode. Recently such integrated device was fabricated and measured by Liu et al.[5] The purpose of the present work is to propose and evaluate two-terminal and three-terminal QW laser-phototransistor structures, which can generate a laser radiation in short-wavelength infrared range of spectrum modulated by long-wavelength infrared radiation (see Figure 1). The operation of the QW laser-transistors in question is connected with the injection of hot electrons into a laser active region or their extraction from it controlled by the potential of the QW specially inserted in the emitter (Figure 1a) or collector (Figure 1b) region.

The analytical model of the two-terminal and three-terminal QW laser phototransistors is developed. This model includes equation for the current injected into the laser active region or extracted from it , balance equation of the electrons in the QW, and Poisson equation. In contrast with the model used previously[6] new model takes into account the nonstationary process of the filling of the QW by the electrons or its devastation under the influence of modulated long-wavelength infrared radiation. The developed analytical model gives for both structures the following formula for the variation of the power of laser radiation $\Delta P_\Omega$ versus the power of incident infrared long-wavelength radiation $P_\omega$:

$$\Delta P_\Omega = S \cdot \frac{e \, \Sigma \, \sigma}{\hbar \, \omega \, (1 - \beta)} \cdot P_\omega. \qquad (1)$$

*Hot Carriers in Semiconductors*
Edited by K. Hess *et al.*, Plenum Press, New York, 1996

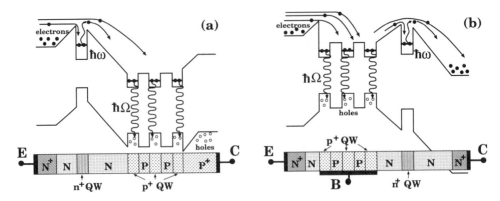

**Figure 1.** Sequences of layers and band diagrams of (a) two-terminal and (b) three-terminal QW laser phototransistor. Here $E$ - emitter, $C$ - collector, $B$ - base; arrows show the electron transitions.

Here $S$ is the differential responsivity of the laser, $e$ is the elementary charge, $\Sigma$ is the sheet density of the electrons in the absorbing QW, $\sigma$ is the cross-section of the photoescape of the electrons from the QW, $(1 - \beta)$ is the probability of the electron capture in the QW, and $\hbar\omega$ is the energy of the infrared photons. For the QW laser-phototransistor structure based on AlGaAs-GaAs with $S = 0.4$ W/A, $\sigma = 2 \times 10^{-15}$ cm$^{-2}$, $\hbar\omega = 0.1$ eV, $\Sigma = 10^{12}$ cm$^{-2}$ and $\beta = 0.9 - 0.99$ we have for the modulation efficiency $\mathcal{M} = \frac{\Delta P_\Omega}{P_\omega} \approx 0.08 - 0.8$.

Previously it has been shown by Gorfinkel and Luryi[7] that the interband absorption coefficient in a laser active region and output power can be controlled by long-wavelength radiation also due to to the heating of the electrons in the active laser region. Comparison of the modulation efficiency $\mathcal{M}$ with that corresponding to the heating mechanism $\mathcal{M}_\mathcal{H}$ results in the following estimation:

$$\frac{\mathcal{M}}{\mathcal{M}_\mathcal{H}} = \frac{1}{2}\left(\frac{\Sigma}{\Sigma_t}\right) \cdot \left(\frac{1+\beta}{1-\beta}\right) \cdot \left(\frac{\tau_R}{\tau_\varepsilon}\right) \cdot \left(\frac{kT}{\hbar\omega}\right). \tag{2}$$

Here $\Sigma_t$ is the sheet electron concentration in the active laser region required to just make the optical chanel transparent, $\tau_R$ and $\tau_\varepsilon$ are the life time and energy relaxation time of the electrons in the active laser region.

For real structures the following inequalities are valid: $\Sigma/\Sigma_t \ll 1$, $\tau_R/\tau_\varepsilon \gg 1$, $(1+\beta)/(1-\beta) \gg 1$. Because of this $\mathcal{M}/\mathcal{M}_\mathcal{H}$ can be less or larger than unity. For example, for typical values $\Sigma = 10^{12}$ cm$^{-2}$, $\Sigma_t = 10^{13} - 10^{14}$ cm$^{-2}$, $\tau_R = 10^{-9}$ s, $\tau_\varepsilon = 6 \times 10^{-12}$ s, $T = 300$ K, $\hbar\omega = 0.1$ eV we have $\mathcal{M}/\mathcal{M}_\mathcal{H} \approx 0.3 - 3$ for the structure with $\beta = 0.1$ and $\mathcal{M}/\mathcal{M}_\mathcal{H} \approx 4 - 40$ for $\beta = 0.9$. These estimates show that the integration of the laser with the QW intersubband infrared phototransistor can significantly enhance the modulation efficiency, especially in the case of structures with $(1 - \beta) \ll 1$. In the last case this effect is connected with large photoelectric gain exhibited by the phototransistor with high efficiency of hot electron transport via the QW (see, for example, ref. [8]).

The above formulae are valid if the modulation frequency $f \ll f_{max}$. The maximum modulation frequency $f_{max}$ of the laser-phototransistor under consideration is limited by the time of recharging of the QW $\tau_{QW}$ in the phototransistor part of the device and the electron energy relaxation time $\tau_\varepsilon$ in its laser domain. If $\tau_\varepsilon < \tau_{QW}$ our model results in the following expression

$$ f_{max} = \frac{2\,e^2\,W}{\ae\,k\,T} \cdot \left( \frac{j_T}{e} + \frac{\sigma\,\Sigma\,\overline{P_\omega}}{\hbar\omega} \right). \tag{3} $$

Here $W = W_1\,W_2/(W_1 + W_2)$, where $W_1$ and $W_2$ are the widths of the barriers on both sides of the QW, $j_T$ is the density of the leakage current from the QW due to thermionic excitation of the electrons. The second term in parentheses of expression (3) corresponds to the current of the electrons photoexcited by long-wavelength radiation with average power density $\overline{P_\omega}$. Normally, in the case of the two-terminal device, the quantity $j_T\,(1 - \beta)^{-1}$ should be of the order of the laser threshold current density $j_t = e\,\Sigma_t/\tau_R$ or larger. In the case of weak intensity of long-wavelength radiation $(P_\omega < \frac{\hbar\omega}{e\,\sigma\,\Sigma} j_T)$ we have from formula (3) the following estimation for the maximum modulation frequency

$$ f_{max} \geq \frac{2\,e^2\,W\,\Sigma_t}{\ae\,k\,T\,\tau_R} \cdot (1 - \beta). \tag{4} $$

If $W = 10^{-5}$ cm, $\Sigma_t = 10^{13}$ cm$^{-2}$, $T = 300K$, $\tau_R = 10^{-9}$ s, $\beta = 0.9$ the last relationship gives $f_{max} \geq 10^{10}$Hz. The maximum modulation frequency $f_{max}$ increases with the increase of current density through the device. But under high current density condition $f_{max}$ is limited by the processes connected with energy relaxation of the injected electrons in the laser region of the device. The above estimates show that the proposed QW laser-phototransistor can effectively operate in the range of modulation frequencies $f \geq 10$ GHz.

## REFERENCES

1. P. S. Zory, ed. "Quantum Well Lasers", Academic Press, San Diego (1993).

2. B. F. Levine, Quantum-well infrared photodetectors, *J. Appl. Phys.* 74:R1 (1993).

3. V. Ryzhii and M. Ryzhii, Device physics of quantum well infrared photodetector with optical output, *Abstr. of 7th Int. Conf. on Narrow Gap Semicond.*, Santa Fe, Jan. 8–12, 1995, p. H1.

4. V. Ryzhii, M. Ershov, M. Ryzhii and I. Khmyrova, Quantum well infrared photodetector with optical output, *Jpn. J. Appl. Phys.* 34:L38 (1995).

5. H. C. Liu, J. Li, Z. R. Wasilewski and M. Buchman, Integrated quantum well intersub-band photodetector and light emitting diode, *Electronics Lett.* 31:832 (1995).

6. V. Ryzhii, I. Khmyrova, M. Ershov, M. Ryzhii and T. Iizuka, Theoretical study of an infrared-to-visible wavelength quantum-well converter, *Semicond. Sci. Technol.* to appear (1995).

7. V. B. Gorfinkel and S. Luryi, Rapid modulation of interband optical properties of quantum wells by intersubband absorption, *Appl. Phys. Lett.* 60:314 (1992).

8. V. Ryzhii and M. Ershov, Electrical and optical properties of a quantum-well infrared photo-transistor , *Semicond. Sci. Technol.* 10:687 (1995).

# ROOM TEMPERATURE 10 μm INTERSUBBAND
# LASERS BASED ON CARRIER CAPTURE
# PROCESSES IN STEP QUANTUM WELLS

X.Zhang, and G.I.Haddad

Solid State Electronics Laboratory
Department of EECS
The University of Michigan
Ann Arbor, MI 48109

In this work, a new room temperature intersubband laser at 10 μm wavelength is proposed in which population inversion is achieved between subbands in a step quantum well based on carrier capture processes. It is known that in quantum well structures, emission of LO phonons is the main carrier capture mechanism. The LO-phonon scattering rate depends on the Froehlich matrix element and the overlap of the initial and final wave functions squared[1,2]. In a step quantum well with a high ratio of the step width to the well width the wave function overlap of the state in the step and the state in the well is reduced and thus a lower LO-phonon scattering rate and a longer capture time result. The carriers in the ground state in the well can be removed by a coupled quantum well by tunneling and thus population inversion can be realized.

Fig.1 shows the band diagram of the $Al_{.6}Ga_{.4}As/Al_{.25}Ga_{.75}As/GaAs$ asymmetrical step quantum well active region. The asymmetrical step quantum well has a 500 Å $Al_{.25}Ga_{.75}As$ step and a 40 Å GaAs well. The barriers are $Al_{.6}Ga_{.4}As$. A second GaAs well of 60 Å is coupled to the first well with a 30 Å $Al_{.6}Ga_{.4}As$ barrier in between. The dashed lines are the wave functions of the first three states. $E_1$, $E_2$, and $E_3$ are the energies of each state. ($E_3$ - $E_2$) = 124 meV, corresponding to the emission wavelength of 10 μm. ($E_2$ - $E_1$) = 36 meV, which is the optical phonon energy in GaAs. Time constants for a bias of 0.1 V are shown in the inset. Due to the reduced wave function overlap, the relaxation time $\tau_{32}$ between state 3 and state 2 is long ($\tau_{32}$ = 14 ps). However, the removal of carriers from state 2 to state 1 is fast because the rate of optical phonon assisted tunneling is high ($\tau_{21}$ = 2.1 ps). The escape time from state 3 is extremely long ($\tau_3$ = 152 ps), providing a high injection efficiency of 0.91. As the relaxation rate $1/\tau_{32}$ is lower than the removal rate $1/\tau_{21}$, population inversion arises[3].

Carrier capture processes in asymmetrical step quantum wells have been investigated by time resolved differential transmission spectroscopy. The samples for the experiment were grown by molecular beam epitaxy on semi-insulating GaAs substrate, consisting of 10 undoped asymmetrical step quantum wells. To allow transmission measurement the substrate was completely removed by chemical etching. Femtosecond white-light

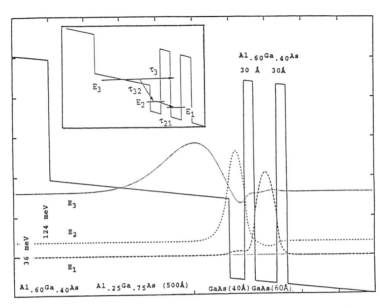

**Figure 1.** Conduction band diagram of the asymmetrical step quantum well active region.

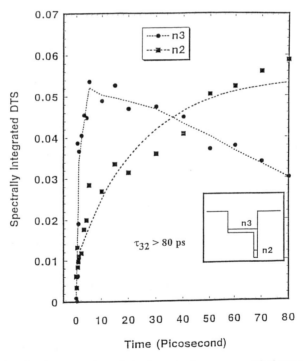

**Figure 2.** Relaxation time $\tau_{32}$ and population inversion in an asymmetrical step quantum well.

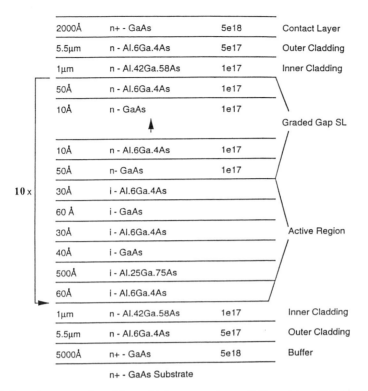

| | | | |
|---|---|---|---|
| 2000Å | n+ - GaAs | 5e18 | Contact Layer |
| 5.5µm | n - Al.6Ga.4As | 5e17 | Outer Cladding |
| 1µm | n - Al.42Ga.58As | 1e17 | Inner Cladding |
| 50Å | n - Al.6Ga.4As | 1e17 | |
| 10Å | n - GaAs | 1e17 | Graded Gap SL |
| 10Å | n - Al.6Ga.4As | 1e17 | |
| 50Å | n- GaAs | 1e17 | |
| 30Å | i - Al.6Ga.4As | | |
| 60 Å | i - GaAs | | |
| 30Å | i - Al.6Ga.4As | | Active Region |
| 40Å | i - GaAs | | |
| 500Å | i - Al.25Ga.75As | | |
| 60Å | i - Al.6Ga.4As | | |
| 1µm | n - Al.42Ga.58As | 1e17 | Inner Cladding |
| 5.5µm | n - Al.6Ga.4As | 5e17 | Outer Cladding |
| 5000Å | n+ - GaAs | 5e18 | Buffer |

n+ - GaAs Substrate

**Figure 3.** Schematic cross section of the multiple step quantum well 10 µm intersubband laser structure.

differential transmission spectroscopy (DTS) was measured in a standard pump-probe setup[4] to determine subband relaxation times and to demonstrate population inversion. The differential transmission is due to the carriers in subbands generated by the optical pump and the DTS in time domain shows carrier relaxation in quantum wells. When the pump is tuned to the wavelength which is longer than the wavelength corresponding to the band gap of the step, the generated carriers can only be populated in the wells. The relaxation time $\tau_{21}$ can thus be determined and we obtained $\tau_{21}$ = 2 - 3 ps. When the pump is tuned to the wavelength corresponding to the band gap of the step resonantly, we can measure the relaxation time $\tau_{32}$ and demonstrate population inversion. Fig.2 shows the time resolved spectrally integrated DTS in an asymmetrical step quantum well shown in the inset, which is proportional to the number of carriers in the states. It can be seen that $\tau_{32}$ is longer than 80 ps and population is inverted between state 3 and state 2. The calculated relaxation time $\tau_{32}$ at zero bias is 85 ps.

Fig.3 shows the 10 µm intersubband laser structure. It comprises 10 stages and the multiple active regions are connected by the graded gap superlattice[5,6]. It is a 9 period GaAs/Al.6Ga.4As superlattice with a constant period of 50 Å while the thicknesses of the GaAs and Al.6Ga.4As are gradually varied between 45 and 5 Å. The energy difference between the maximum and minimum effective conduction band edges of the graded gap superlattice is 403 meV. The flat band condition in the graded gap superlattice is obtained at an electric field of 90 kV/cm to transfer carriers between active regions. Optical transition and carrier transport are shown in Fig.4. Carriers are injected into the step quantum well and stimulated emission takes place between state 3 and state 2. Carriers in state 2 can efficiently be removed to state 1 by tunneling and escape from the active region through a 30 Å barrier. Then they are transferred to the next active region through the

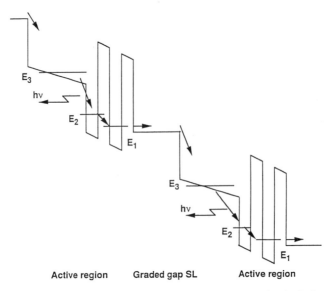

**Figure 4.** Optical transition and carrier transport under laser operation in the intersubband lasers.

graded gap superlattice to continue laser action. Inner and outer cladding layers confine the optical wave to the active regions. The confinement factor is 0.17. The maximum gain is determined by the oscillator strength $f_{32}$, population inversion $\Delta n$, intersubband scattering time $T_s$, and the spontaneous radiation time $\tau_s$. $f_{32}$ and $\tau_s$ are calculated to be 6.70 and 68.5 ns, respectively. $T_s$ is 0.082 ps, which is estimated from the FWHM in the FTIR absorption spectra of the active region at room temperature. The threshold gain $g_{th}$ is determined by optical losses. The total losses at room temperature including the mirror losses for a cavity length of 1 mm is 32.6 cm$^{-1}$. Then we obtain $g_{th} = 187$ cm$^{-1}$, $\Delta n = 1.4 \times 10^{11}$ cm$^{-2}$ and the threshold current density $J_{th} = 2$ kA/cm$^2$. We therefore conclude that room temperature realization of the intersubband laser at 10 $\mu$m wavelength is feasible. This work was supported by the ARO-URI program under grant No.DAA L03-92-G-0109.

## REFERENCES

1. P.W.M.Blom, C.Smit, J.E.M.Haverkort, and J.H.Wolter, Carrier capture into a semiconductor quantum well, Phys. Rev. B, 47: 2072 (1992)

2. D.Morris, B.Deveaud, A.Regreny, and P.Auvray, Electron and hole capture in multiple-quantum-well structures, Phys. Rev. B, 47:6819 (1992)

3. X.Zhang, G.I.Haddad, J.P.Sun, A.Afzali-Kushaa, C.Y.Sung and T.B.Norris, Population inversion in step quantum wells at 10 $\mu$m wavelength, 53rd Device Research Conference, Session VA, Charlottesville, Virginia, (1995)

4. C.Y.Sung, T.B.Norris, X.Zhang, M.Sneed and P.K.Bhattacharya, Directly time-resolved carrier capture process in semiconductor quantum well, CLEO/QELS '95, Baltimore Maryland, (1995)

5. F.Capasso, H.M.Cox, S.G.Hummel, Pseudo-quaternary GaInAsP semiconductors: A new GaInAs/Inp graded gap superlattice and its applications to avalanche photodiodes, Appl. Phys. Lett., 45:1193 (1984)

6. J.Faist, F.Capasso, D.L.Sivco, C.Sirtori, A.L.Hutchinson, A.Y.Cho, Quantum cascade laser, Science, 264:553 (1994)

# NONEQUILIBRIUM ELECTRON-HOLE PLASMAS UNDER CROSSED ELECTRIC AND MAGNETIC FIELDS AND STIMULATED LANDAU EMISSION OF n-InSb AT THE QUANTUM LIMIT

Takeshi Morimoto[1], Meiro Chiba[1], and Giyuu Kido[2]

[1]Institute of Atomic Energy, Kyoto University, Uji, Kyoto 611, Japan
[2]National Research Institute for Metals, Tukuba 305, Japan

In quantizing high magnetic fields, highly nonequilibrium electron-hole plasmas can be excited in bulk n-InSb samples at low temperatures, when a constant dc current $J$ ( $// \hat{x}$ ) is passed through the sample across the magnetic field $H$ ( $// \hat{z}$ ), resulting in stimulated interband Landau emission, as shown in Fig. 1.[1-3]

In the extrinsic region at low temperatures, there would be no holes in originally n-type materials at thermal equilibrium, in contrast to the intrinsic case.[4-8] However, at low temperatures highly nonequilibrium generation of electrons-hole pairs has been observed owing to the transverse excitation by hot electrons under high magnetic fields.[1-3] In such nonequilibrium state at high magnetic fields, electrons and (heavy) holes are driven to the -y-direction with almost equal drift velocity by the action of $J \times H$ force, or the equivalent effective transverse electric field $E^*_y$, as given by $E^*_y \sim JH / n^*|e|c$, $n^*$ being the effective concentration of electrons.

At high magnetic fields the value of $E^*_y$ becomes much higher than the longitudinal electric field $E_x$ originally applied.[5,9] Then excess electrons will be excited along the y-direction by the action of $E^*_y$ which acts during the travel time through the effective mean free path $l^*$ until the annihilation by recombination with excess holes, escaping from severe damping by collisions.[1,2,5] (As for the concept of such lucky electrons see Ref. 10). As a result of the radiative recombination of excess electrons and holes, interband Landau emission takes place (see, Fig. 1). At the quantum limit, thus excited electrons and holes will populate almost in the lowest Landau levels, $\ell = \ell' = 0$, $\ell$ and $\ell'$ being Landau's orbital quantum numbers of electrons and (heavy) holes, respectively. When they are highly excited, inverted population of electrons and holes will be realized near the band edge, $k_z = 0$, $k_z$ being the wave vector of electrons along the magnetic field direction.

For the peak frequencies, $\omega_\pm$, of the strongest emissions $\alpha_1$ ( $0_+ \to 0'$ ) and $\alpha_2$ ( $0_- \to 0'$ ), the following relation has been observed:[1-3]

$$\hbar\omega_\pm = \varepsilon_g + \frac{1}{2}\hbar\omega_{c1}[1 \pm \frac{m^*_1 g_1}{2m_0}] + \frac{1}{2}\hbar\omega_{c2,h} - |e|E^*_y l^*, \tag{1}$$

with $E^*_y \sim JH / n^*|e|c$, corresponding to the transitions between the lowest Landau levels of electrons and heavy holes which mainly contribute to the recombination. Here, the $\pm$ signs correspond to the up and down spin states of conduction electrons, $\omega_{c1}$ and $\omega_{c2,h}$ are the cyclotron frequencies of electrons and heavy holes, respectively, $\varepsilon_g$ is the energy gap, $m^*_1$ is the effective mass of electrons, $g_1$ is the effective g-factor, $m_0$ is the rest mass of electrons, and $l^*$ is the mean free path of lucky electrons as mentioned earlier.

*Hot Carriers in Semiconductors*
Edited by K. Hess *et al.*, Plenum Press, New York, 1996

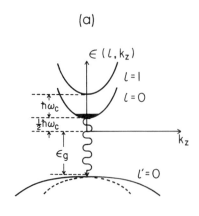

(a)

$\in (\ell, k_z)$

$\ell = 1$

$\ell = 0$

$\hbar\omega_c$

$\frac{1}{2}\hbar\omega_c$

$k_z$

$\varepsilon_g$

$\ell' = 0$

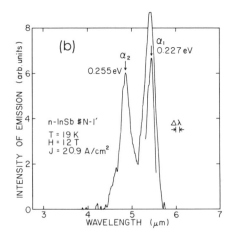

(b)

$\alpha_1$ 0.227 eV

$\alpha_2$ 0.255 eV

n-InSb #N-I'
T = 19 K
H = 12 T
J = 20.9 A/cm²

$\Delta\lambda$

INTENSITY OF EMISSION (arb units)

WAVELENGTH (μm)

**Figure 1.** (a) Population of electrons and Landau emission at the quantum limit, neglecting spin splitting. (b) Emission spectrum of n-InSb for $H$ = 12T at 19K. $J$ = 20.9A / cm². $H$ // [ 110 ] direction.

The last term of Eq. (1) is added in order to explain the recuction of virtual energy gap in crossed electric and magnetic fields.[1-3]

At the quantum limit, the gain, $g( \omega, H )$, of the stimulated emission becomes diversingly high value at the band edge, $k_z$ = 0, for $\alpha_1$ ( $0_+\rightarrow 0'$ ) and $\alpha_2$ ( $0_-\rightarrow 0'$ ) emissions, since it has singularities at $\omega = \omega_\pm$ of the form:[11]

$$g(\omega,H) = A\frac{eH}{c\omega}\sum_{\pm}(\hbar\omega - \hbar\omega_\pm)^{-1/2}. \qquad (2)$$

$A$ is a constant proportional to the square of the momentum matrix element $p_{12}$. Then, the stimulated emission will take place at extremely low value of the critical current density $J_c$ as observed here.

The high field experiments up to 22T were performed using a hybrid-type magnet at the high field laboratory of Tohoku University. The electron concentration $n_0$ and the mobility $\mu$ of the sample were $5\times10^{14}$ cm⁻³ and $3.7\times10^5$ cm² / Vs at 80K. The dimensions were 4.06 mm long, 3.18 mm wide and 0.22 mm thick. The surfaces were chemically etched. A constant dc current, chopped with a 50 % duty factor, was passed through the sample using a constant current supply.

Figure 2(a) shows an example of the current dependence of the total output of the IR emission from n- InSb for $H$ = 10T at 43K. The simultaneously measured $V$-$I$ curves are shown in Fig. 2(b). Beginning of the spontaneous emission was observed at low current density to be $J \sim 5$A / cm², as shown in Fig. 2(a), where the output is nearly proportional to $J$. However, at the critical value, $J_c \approx$ 17.1A / cm², the emitted intensity increases so steeply in proportion to $J^{21}$, suggesting the occurrence of the stimulated emission. The emission spectrum for $H$ = 12T at 19K is shown in Fig. 1(b). The spectrum shows line narrowing within the spectrum resolution $\Delta\lambda$ = 0.08 μm.

Corresponding to the sharp increase in the emitted intensity near the critical value $J_c$, abrupt increase in the voltage drop across the sample has been observed in the $V$-$I$ curves, as shown in Fig. 2(b), indicating characteristic of current saturation at the critical current density $J_c$. Such an abrupt increase in the voltage drop must be needed in order to keep the constant current operation against the rapid decrease in the number of electron-hole pairs accompanied by the stimulated emission.The constant current operation is essential to stabilize the emission in the negative resistance region.

The subsequent decrease in the voltage drop, as seen in Fig. 2(b) at the higher current region of $J > 20$A / cm² for $H$ = 10T, comes from the increase in the number of thermal electrons due to heat-up of the sample at higher currents. The increase in the number of thermal electrons $n_0$ and of thermal holes $p_0$ causes saturation of the emitted power, as seen in Fig. 2(b) in the output curve at $J$ > 20A / cm² for $H$ = 10T. Such a feature is enhanced with increasing magnetic field.

The critical concentration of electrons, $n_T$, for realizing the inverted population is calculated as $n_T = eH(2m^*_1 kT)^{1/2} / 2ch^2$, taking into account the Landau degeneracy of electrons at $\ell = 0$.

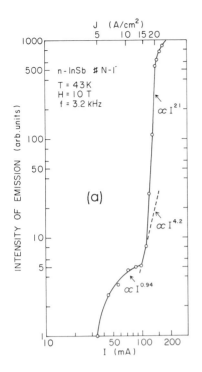

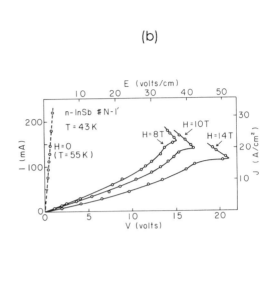

Figure 2. (a) Current variation of the total output of the emission from n-InSb for $H$ = 10T at 43K. (b) $V$-$I$ characteristics of n-InSb at 43K. $J \perp H$. All data were taken under the constant current operation.

The value is estimatated to be $n_T = 5.24 \times 10^{15}$ cm$^{-3}$ for $H$ = 10T at 20K, which is realizable value even for the cw excitation at low temperatures[12]. Thus, the inverse square root singularity of the gain $g(\omega, H)$ at $\omega = \omega_{\pm}$ will drastically reduce the value of $J_c$.

Finally it should be noted that the total system involving the emission and the background radiation fields constitutes a highly driven open-system, accompanying with generation and annihilation of electrons and holes. In this case, the symmetric properties expected for non-perturbed, equilibrium system seems to be broken in part, causing anomalous galvanomagnetic properties.[13]

## Acknowledgments

We are indebted to S. Ueda, M. Inoue and T. Hamamoto for their technical assistance. Thanks are also due to Dr. S. Akai for offering high quality InSb samples.

## References

1. T. Morimoto and M. Chiba, J.Phys. Soc. Jpn. **60**, 2446 (1991).
2. T. Morimoto, M. Chiba, G. Kido, and A. Tanaka, Semicond. Sci. Technol. **8**, S417 (1993).
3. T. Morimoto, M. Chiba, G. Kido, and M. Inoue, Physica B**184**, 123 (1993).
4. T. Morimoto and M. Chiba, Jpn. J. Appl. Phys. **23**, L821 (1984).
5. T. Morimoto and M. Chiba, Infrared Phys. **29**, 371 (1989).
6. V. K. Malyutenko, S. S. Volgov, and E. I. Yablonovsky, Infrared Phys. **25**, 115 (1985).
7. F. R. Kessler, P. Paul, and R. Nies, Phys. Stat. Sol. (b) **167**, 349 (1991).
8. P. Berdahl, J. Appl. Phys. 63. 5846 (1988).
9. M. Toda and M. Glicksman, Phys. Rev. **140**, A1317 (1965).
10. For a review see, F. Cappaso, Physics of Avalanche Photodiodes, in Semiconductors and Semimetals, Vol. 22, Part D, R. K. Willardson and A. C. Beer, ed., Academic Press, New York (1985).
11. T. Morimoto and M. Chiba, Semicond. Sci. Technol. 7, B652 (1992).
12. T. Morimoto, M. Chiba, and G. Kido, in Proc. 22nd Int. Conf. Physics of Semiconductors, D. J. Lockwood, ed., World Scientific, Singapore (1995).
13. T. Morimoto and S. Ueda, unpublished (1994).

# EFFECTS OF SPACE CHARGE ON THE ELECTRIC FIELD
# DISTRIBUTION IN *P*-GE HOT HOLE LASERS

R.C. Strijbos[1], A.V. Muravjov[2], and W.Th. Wenckebach[1]

[1] Faculty of Applied Physics, Delft University of Technology
P.O. Box 5046, 2600 GA Delft, The Netherlands
[2] Institute for Physics of Microstructures, Russian Academy of Sciences
Nizhny Novgorod 603600, Russia

The small-signal gain of $p$-Ge far-infrared lasers ($50\text{cm}^{-1} < \nu < 140\text{cm}^{-1}$), operated in crossed electric ($E = 0.3 - 5\text{kVcm}^{-1}$) and magnetic ($B = 0.3 - 5\text{T}$) fields at helium temperature, depends strongly on the magnitude and direction of the electric field.[1] In order to obtain a higher efficiency of these lasers, it is, therefore, important to have a homogeneous electric field in the active area of the laser. However, it has been pointed out that, due to the large Hall-field, the electric field distribution can be strongly inhomogeneous in the commonly used sample geometries: a rectangular block of $p$-Ge with electrical contacts extending along two opposite sides.[2,3] In this contribution we show that the occurrence of space charge in the p-Ge laser crystal has a remarkable influence on the electric field distribution. Using a simple model for the conductivity, it is found that the space charge is located near the electric contacts and causes a large voltage drop close to the contacts. It is concluded that, due to space-charge effects, long-sample geometries, where the distance $a$ between the contacts is larger than the length $b$ of the contact sides, are favourable for the operation of $p$-Ge lasers.

Since the problem is symmetric along the magnetic field direction ($z$-axis), the following 2D Poisson equation has to be solved:

$$\frac{\partial^2 \Phi}{\partial x^2} + \frac{\partial^2 \Phi}{\partial y^2} = -4\pi\rho, \tag{1}$$

where $\rho$ is determined by the equation:

$$\text{div}\mathbf{j} = \text{div}\hat{\sigma}\mathbf{E} = 0, \tag{2}$$

where $\mathbf{j}$ is the current vector and $\hat{\sigma}$ is the conductivity tensor. The boundary conditions for the electrical contacts are defined by the applied voltage $V_1 - V_2$ and are given by $\Phi = V_1$ and

*Hot Carriers in Semiconductors*
Edited by K. Hess *et al.*, Plenum Press, New York, 1996

$\Phi = V_2$ at $x = \mp a/2$, respectively. The condition for the lateral sides ($y = \pm b/2$) is found by considering that the current has to flow along the edges:

$$\left(\frac{\partial\Phi/\partial y}{\partial\Phi/\partial x}\right) = \tan\alpha_H, \tag{3}$$

where $\alpha_H$ is the Hall angle, i.e. the angle between current and electric field.

In order to calculate the space charge $\rho$, the conductivity tensor has to be defined. For $p$-Ge lasers several contributions to the conductivity have to be accounted for: a contribution from streaming heavy holes, i.e. heavy holes repeatedly accelerated by the electric field and scattered due to optical phonon emission, and contributions from both light and heavy holes that are accumulated in closed cyclotron orbits below the optical phonon energy $\varepsilon_{op}$. For such a complex system, the conductivity tensor can only be roughly estimated from experimental data[4] or Monte Carlo simulations,[5] especially since the degree of hole accumulation strongly depends on temperature and ionized impurity concentration.

However, in order to illustrate the space-charge effects, we can use the following simple model system which incorporates the most prominent feature of the current-field relation of $p$-Ge lasers, i.e. the current saturation due to streaming heavy holes, in a direct way:

$$\hat{\sigma} = E^{-f}\left(\begin{array}{cc} \cos\alpha_H & \sin\alpha_H \\ -\sin\alpha_H & \cos\alpha_H \end{array}\right), \tag{4}$$

where f ($0 \le f \le 1$) is a free parameter for setting the rate of current saturation, and $\alpha_H$ is assumed to be a constant parameter. Using Eqs. (2) and (4) the space charge at each point of the sample cross section can then be evaluated:

$$4\pi\rho = \nabla \cdot \mathbf{E} = \frac{f}{2E\cos\alpha_H}\left(\nabla E^2 \cdot \hat{\mathbf{e}}\right) \tag{5}$$

with $\hat{\mathbf{e}} = \mathbf{j}/|\mathbf{j}|$ the local unit current vector. In order to find the electric field distribution in the sample, Eq. (1) is solved by iteration: after each step $\rho$ is evaluated and put into the right-hand side of Eq. (1). It is clear that in the absence of current saturation ($f = 0$) there is no space charge. In Fig. 1, the equipotential lines are shown for the calculations with and without

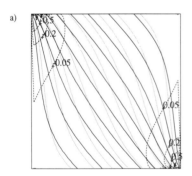 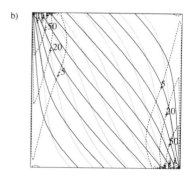

**Figure 1:** Equipotential lines for a square cross section with a) $f = 0.5$ and b) $f = 0.7$; $E_{appl} = 1\text{kVcm}^{-1}$, $\alpha_H = 45°$. Dashed lines: without space charge, solid lines: with space charge. The space charge (in units $10^7\text{C m}^{-3}$) is indicated by the the dash-dotted contour lines.

space-charge effects. The space-charge distribution that causes the differences is also indicated. The corresponding electric field distribution is shown in Fig. 2. The space charge is located near the electrical contacts and causes a large voltage drop close to the contacts. As a consequence, the electric field in the centre of the sample cross section becomes more homogeneous and its direction is almost determined by the Hall angle.

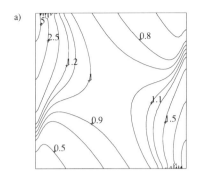 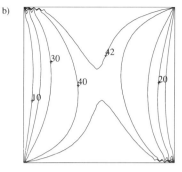

**Figure 2:** Lines of constant a) $E = |\mathbf{E}|$ and b) $\Theta = \arctan(E_y/E_x)$ for a square cross section; $E_{appl} = 1\text{kVcm}^{-1}$, $\alpha_H = 45^\circ$, $f = 0.7$.

Finally, we have calculated the electric field in the centre of the sample cross section as a function of the ratio $a/b$ of the sample sides, again with and without space-charge effects. It is concluded that the space charge forces the current to flow more perpendicular to the contacts, which results in a lower electric field in the centre of the sample.

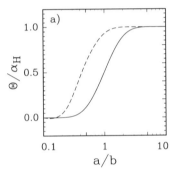 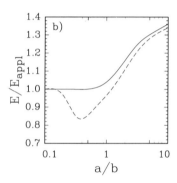

**Figure 3:** a) $\Theta = \arctan(E_y/E_x)$ and b) electric field strength in the centre of the sample cross section versus the ratio $a/b$ of the sample sides with (dashed) and without (solid) space-charge effects, $E_{appl} = 1\text{kVcm}^{-1}$, $\alpha_H = 45^\circ$, $f = 0.7$.

We conclude that space-charge effects can play an important role in p-Ge hot hole lasers. First, the internal reflection modes reflecting at the $b$-sides experience an average gain comparable to that of axial modes, contrary to those associated with the $a$-sides. Second, due to space-charge effects, the long sample-geometry ($a/b \geq 1$) seems to be a better choice for a p-Ge laser crystal than the short sample-geometry ($a/b < 1$) because of increased electric field homogeneity and a higher $E/E_{appl}$ in the centre of the sample cross-section. Both conclusions are in agreement with experimental observations.[1]

One of us (A.V. Muravjov) acknowledges the support by the Russian Basic Research Foundation (Project 93-02-14661).

## REFERENCES

1. *Opt. Quantum Electron.* 23, special issue far-infrared semiconductor lasers (1991).
2. R.C. Strijbos, J.E. Dijkstra, J.G.S. Lok, S.I. Schets, and W.Th Wenckebach, Effects of sample shape in p-Ge 'hot-hole' lasers, *Semicond. Sci. Technol.* 9:648 (1994).
3. A.V. Bespalov and K.F. Renk, Electric field dependence of the cyclotron p-germanium laser line, *Semicond. Sci. Technol.* 9:645 (1994).
4. S. Komiyama and R. Spies, Hot carrier population inversion in p-Ge, *J. Physique, Colloq.* C7–387 (1981).
5. V.A. Kozlov, Hot electron population inversion and bulk ndc in semiconductors, *J. Physique, Colloq.* C7–413 (1981).

# ACTIVE MODE LOCKING OF A *P*-GE LIGHT-HEAVY HOLE BAND LASER BY ELECTRICALLY MODULATING ITS GAIN: THEORY AND EXPERIMENT

R.C. Strijbos,[1] A.V. Muravjov,[2] J.H. Blok,[1] J.N. Hovenier,[1] J.G.S. Lok,[1] S.G. Pavlov,[2] R.N. Schouten,[1] V.N. Shastin,[2] and W.Th. Wenckebach[1]

[1] Faculty of Applied Physics, Delft University of Technology, P.O. Box 5046, 2600 GA Delft, The Netherlands
[2] Institute for Physics of Microstructures, Russian Academy of Sciences Nizhny Novgorod 603600, Russia

The *p*-Ge intervalenceband (IVB) far-infrared laser ($50 \text{cm}^{-1} < \nu < 140 \text{cm}^{-1}$), operating in crossed electric and magnetic ($\mathbf{E} \perp \mathbf{B}$) fields, is based on a population inversion between accumulated light holes and streaming heavy holes in the **k**-space region below the optical phonon energy $\varepsilon_{\text{op}}$.[1] Its broad amplification band ($\Delta \nu \approx 100 \text{cm}^{-1}$) allows amplification of far-infrared pulses at a picosecond timescale.[2] We have investigated the feasibility of achieving active mode locking in a *p*-Ge IVB laser by locally modulating the gain in a small part of the *p*-Ge crystal. By applying a radiofrequency (RF) electric field to small additional contacts along the magnetic field direction, the population inversion in the active part between these contacts is modulated strongly: at peak amplitudes of the RF cycle the light holes are accelerated out of the so-called passive region below $\varepsilon_{\text{op}}$, while at the RF nodes the light hole population is restored due to the fast pumping by optical phonon scattering from streaming heavy holes. Fig. 1a shows the expected response of the small-signal gain due to the direct IVB transition during one period of a 500MHz RF electric field, as it has been calculated using a Monte Carlo simulation program.[3] From Fig. 1b it is clear that a peak-to-peak modulation depth of 30 – 40 % can be obtained in the modulated part of the *p*-Ge crystal with only a small RF field amplitude $E_{\text{RF}}^{0}$ of approximately 20Vcm$^{-1}$.

When the RF frequency is chosen to be half the cavity roundtrip frequency of the total resonator, this way of local small-signal gain modulation may be used to achieve active mode locking of a *p*-Ge IVB laser: a pulse, which is passing the modulated area at maximum gain, will be built-up in the cavity and will be shortened during each roundtrip due to the modulation of the gain. Based on a simple ideal model,[4] it is estimated that when the gain of a *p*-Ge hot-hole laser is modulated at approximately 1GHz, this pulse might reach a steady-state pulsewidth of approximately 10ps after a few thousand roundtrips.[3]

In order to implement this scheme experimentally, a *p*-Ge sample is mounted in Voigt

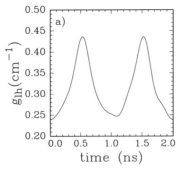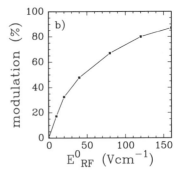

**Figure 1:** a) Small-signal gain $g_{lh}$ for far-infrared radiation ($h\nu = 12$meV , $\mathbf{e}_\nu \parallel \mathbf{B}$) in the modulated part of the $p$-Ge crystal due to the direct IVB transition during one (cosine) period of the RF field, as calculated from the Monte Carlo simulation; $\mathbf{E} = 2$kVcm$^{-1} \parallel$ [111] and $\mathbf{B} = 1.3$T $\parallel$ [0$\bar{1}$1] , $T = 10$K, $E_{RF}^0 = 40$Vcm$^{-1}$, $N_I = 2 \times 10^{14}$cm$^{-3}$. $g_{lh}$ is modulated considerably at the second harmonic of the RF frequency $f_m = 500$ MHz but is more 'peaked': at least the first 8 Fourier coefficients have to be sampled to represent $g_{lh}$. b) peak-to-peak gain modulation depth versus amplitude $E_{RF}^0$ of the RF field.

configuration with $\mathbf{E} \parallel$ [1$\bar{1}$0] and $\mathbf{B} \parallel$ [112]. The sample is cut from a 40$\Omega$cm Ge:Ga crystal with dimensions $d \times h \times l = 7 \times 5 \times 50$mm$^3$. Four ohmic contacts to the sample have been made: two contacts for the main electric field (high voltage (HV) contacts) covering the $5 \times 50$mm$^2$ sides and two additional contacts of $1 \times 10$mm$^2$ covering the central part of the $7 \times 50$mm$^2$ sides close to the outcouple mirror. An external resonator is constructed by pressing a plane copper mirror (separated by a 20$\mu$m thin Teflon film) and a gold capacitive mesh (grid constant $g = 30\mu$m, linewidth $2a = 1\mu$m on a 400$\mu$m pure silicon substrate) directly to the $7 \times 5$mm$^2$ sides of the $p$-Ge sample. The radiation is detected by a cryogenic $p$-Ge detector mounted approximately 20cm above the laser crystal and by a pyro-electric detector or a Schottky diode outside the cryostat. Normally, the laser is operated by supplying HV pulses of 2.5$\mu$s at a 1Hz repetition rate. The peak power outside the cryostat is estimated to be approximately 50mW.

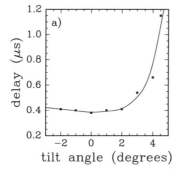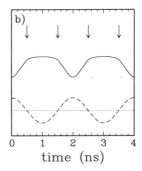

**Figure 2:** a) Delay of laser output pulse from the beginning of the HV pulse versus sample tilt angle. b) gain-modulation scheme for the tilted sample: the solid line shows the response of the small-signal gain to a strong 500MHz RF field along $\mathbf{B}$ (dashed line), as expected from a). The arrows indicate when a pulse travelling through the resonator at a cavity roundtrip frequency of 1GHz has to pass the modulator in order to be selectively amplified.

When the RF field is applied continuously to the additional contacts, the lasing is blocked by local sample heating from the RF field. This is confirmed by applying a gated 300$\mu$s RF pulse before the HV pulse; just when the delay between the two pulses is more than 700$\mu$s, the laser starts to oscillate again. The heating effect can be overcome by reducing the RF pulse to approximately 3$\mu$s. When the 3$\mu$s RF burst is applied during HV excitation, the RF field influences lasing only when the sample is slightly tilted around its long axis. This can be understood from Fig. 2a, where we have plotted the delay of laser output with respect to the HV excitation – which is a good measure of the small-signal gain – as a function of the angle by which the sample is tilted around its long axis. The small-signal gain remains constant up to $2^o$ and just decreases strongly if $\mathbf{E}$ and $\mathbf{B}$ are $4^o$ non-orthogonal. We assume that this behaviour is due to an electric field inhomogeneity in the sample caused by the inhomogeneity of impurity doping in the $p$-Ge crystal.[5]

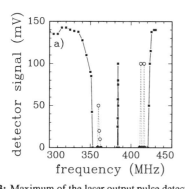 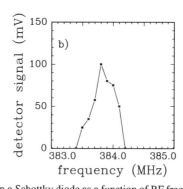

**Figure 3:** Maximum of the laser output pulse detected on a Schottky diode as a function of RF frequency. The sample has been tilted approximately 2 degrees around its long axis; $\mathbf{E}_{appl} = 1.15\text{kVcm}^{-1} \parallel [1\bar{1}0]$, $\mathbf{B} = 0.90\text{T} \parallel [111]$, $\tau_{HV} = 2.5\mu s$, $\tau_{RF} \approx 3\mu s$. Maximally 60W of RF power is dissipated in the $p$-Ge crystal. a) large frequency range; the dotted lines indicate unstable lasing, probably due to small resonances in the impedance matching circuit. b) small frequency range.

Since the RF field 'tilts' the total electric field only one or two degrees, it is clear that the RF field can only modulate the small-signal gain significantly if the sample is slightly tilted. The result of such an experiment is shown in Fig. 3: the lasing is blocked at full RF power in the frequency range $350 - 420$ MHz, corresponding to the resonance of the impedance matching circuit. However, the lasing does not disappear in a narrow frequency range around 383.8MHz. This resonance frequency corresponds very nicely with half the estimated cavity roundtrip frequency $\nu_{RT} = c/2L' \approx 765$MHz, with $L'$ the total optical path length of the resonator. Furthermore, we note that the spectral width of this resonance coincides with the spectral width of homodyne mixing products of the axial modes of the $p$-Ge laser, as measured by heterodyne mixing spectroscopy.[6]

Although the small-signal gain is now modulated at the RF frequency, it is still possible to selectively amplify pulses passing the modulator with $\nu_{RT} = 2\nu_{RF}$. This is due to the strongly non-linear diode-like response of the small-signal gain, as illustrated in Fig. 2b. Although we consider this result as a strong indication for mode locking, further investigations are necessary to provide decisive experimental demonstrations in the time- and frequency domain.

## REFERENCES

1. *Opt. Quantum Electron.* 23, special issue far-infrared semiconductor lasers (1991).
2. Keilmann and R. Till, Saturation spectroscopy of the $p$-Ge far-infrared laser, *Opt. Quantum Electron.* 23:S231–S246 (1991).
3. R.C. Strijbos, J.G.S. Lok, and W.Th. Wenckebach, A Monte Carlo simulation of mode-locked hot-hole laser operation, *J. Phys. Condens. Matter* 6:7461 (1994).
4. A. E. Siegman, Lasers, University Science Books, Mill Valley (1986).
5. A.V. Murav'ev, I.M. Nefedov, S.G. Pavlov, and V.N. Shastin, Tunable narrowband laser that operates on interband transitions of hot holes in germanium, *Quantum Electron.* 23:119 (1993).
6. E. Bründermann, H. P. Röser, A. V. Muravjov, S. G. Pavlov, and V. N. Shastin, Mode fine structure of the $p$-Ge intervalenceband laser measured by heterodyne mixing spectroscopy with an optically pumped ring gas laser, *Infrared Phys. Technol.* 36:59 (1995).

# INDUCED RADIATIVE TRANSITIONS BETWEEN
# STRAIN-SPLIT ACCEPTOR LEVELS IN GERMANIUM
# AT STRONG ELECTRIC FIELD

I.V.Altukhov, E.G.Chirkova, M.S.Kagan,
K.A.Korolev, and V.P.Sinis

Institute of Radioengineering & Electronics
of Russian Academy of Sciences
11, Mokhovaya
Moscow, 103907
RUSSIA

Earlier we have observed the induced far-infrared emission from uniaxially compressed p-Ge at strong electric field but with no magnetic field [1]. The possible reason for the induced emission was shown to be a population inversion of strain-split shallow acceptor levels [2]. In this report we present the first spectral investigations of the induced radiation carried out by means of far-infrared grating monochromator. The data obtained show that the induced emission in compressed p-Ge is due to the radiative transitions between strain-split impurity levels when the one of them lies in a continuous band spectrum.

Uniaxial deformation removes the degeneracy of the valence band at k=0 and splits this band into two subbands separated by an energy gap $\Delta$ proportional to the pressure P (the proportionality factor is about 4 meV/kbar for $P\|[111]$ and 6 meV/kbar for $P\|[100]$ crystallographic direction [3]). The degenerate ground state of an acceptor in germanium is split as well into two levels with an energy separation depending on pressure. Figure 1,a shows the energy splitting of these levels (curve 1) and the ionization energy of the lower level (2) as a function of $\Delta$ calculated from eq. (27.18) of [3]. Fig.1,b represents schematically the valence band structure and impurity level positions at corresponding pressures. (For convenience, the direction of growing hole energy is accepted here as for electrons). The upper level enters into the continuous energy spectrum at $\Delta \approx$ 16 meV (the cross point of curves in Fig.1,a and band scheme 2 in Fig.1,b). It is natural to suppose that the population of these two ground states can be inverted because the lower level located in the forbidden band is empty due to impact ionization in strong fields while the upper one is in the valence band and should be filled. Indeed, the onset of induced emission in some samples was observed just at the pressure corresponded to $\Delta \approx 16$ meV [2], the energy gap between the impurity levels being about 8 meV (Fig.1). The condition of turning-off the induced radiation can be seen in Fig.2. Shown in this figure is the signal from Ge<Ga> photodetector as a function of $\Delta$ at the pressure $P\|[100]$ for various electric fields E. The threshold stress for the onset of induced emission depends on E; corresponding values of the threshold band splitting for different E are shown by arrows in Fig.2. The pressure at which the induced emission turns off (shown by the dotted arrow) is independent of E. At this pressure, the energy of the upper impurity level counted from the bottom of the light-hole band is close to 36 meV, that is the optical phonon energy for Ge (see band scheme 4 in Fig.1b). The upper acceptor level becomes empty because of

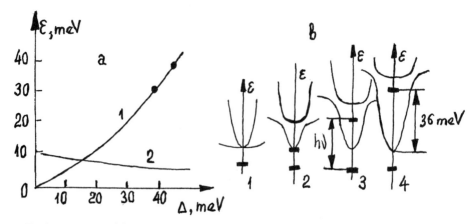

**Fig.1. a:** Energy splitting of the ground states of the shallow acceptor (1) and the ionization energy of the lower level (2) as a function of the energy gap (Δ) between the valence subbands.
**b:** The valence band structure and the positions of impurity levels for various values of Λ, in meV: 1-0; 2-16; 3-36; 4-48.

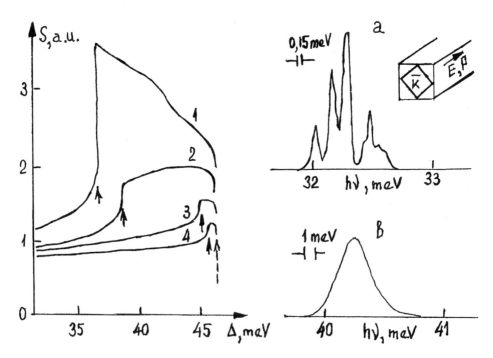

**Fig.2.** Far-infrared luminescence intensity S as a function of Δ for various values of E, in kV/cm: 1-4.1; 2-3.5; 3-3.1; 4-2.6. **E∥P∥[100].**

**Fig.3.** Induced emission spectra of uniaxially stressed p-Ge for the pressure P, in kbar: a-9.5; b-11.5. The insert schematically shows the ray path in the sample cross section plane.

636

optical phonon-assisted hole transitions from the upper level into the bottom of lower-energy valence subband. The population inversion should be weakened and this leads to turning off the stimulated emission. Thus, the range of the band splitting for induced emission is from 16 to 48 meV. Correspondingly, the energy of quantum of induced emission should change between 8 and 42 meV (Fig.1).

Shown in Fig.3 is the spectrum of induced radiation from compressed p-Ge at two values of applied pressure P≈9.5 kbar (Fig.3,a) and P≈11.5 kbar (Fig.3,b) and the electric field E=4 kV/cm ($\mathbf{E} \| \mathbf{P} \|$ [111]) measured by the grating monochromator. The maxima in the spectrum correspond to the energy of 32 meV for 9.5 kbar and 40 meV for 11.5 kbar and coincide with the energy separation between strain-split impurity levels at corresponding pressures (the points in Fig.1,a). Thus, the induced emission is due to the direct optical transitions of holes between these acceptor levels. The energy of quantum of induced radiation depends strongly on stress in accordance with the energy splitting of impurity ground states by uniaxial strain.

The spectrum of the induced emission is rather wide: of about 0.5 meV. The reason for wide spectrum may be the broadening of the upper impurity level being in a continuum [4].

The spectra of Fig.3,a and 3,b were measured with different resolution. The measurements with higher spectral resolution (Fig.3,a) show a mode structure of the induced radiation with a spacing of about 0.15 meV. In our case, as well as in [5], the optical resonator is formed due to total internal reflection from crystal planes (see insert in Fig.3,a). The resonance condition is Nλ=nL, where n is the refractive index, L is the optical path length, and N is an integer. This condition gives just the same mode spacing for the specimen with the cross section of 0.8 x 0.8 mm². The mode structure of the induced radiation gives the direct evidence for the laser action of uniaxially stressed p-Ge.

It should be noted, that the lines corresponding to transitions between excited and ground states of shallow acceptors have been observed [6] in the spectrum of stimulated emission from undeformed p-Ge in crossed electric and magnetic fields. However, as it was pointed out in [6], these transitions were initiated by a population inversion of the light- and heavy-hole subbands which is the primary cause of the stimulated emission in this case. In stressed p-Ge, the stimulated emission originates from the population inversion of acceptor levels split by deformation.

The data obtained show that the induced far-infrared emission in uniaxially compressed p-Ge is due to the population inversion of strain-split acceptor levels as the upper of them is in the valence band. The possibility of a strong frequency tuning by stress is also shown.

The research described in this publication was made possible in part by Grants NN. N7H000, N7H300 from the International Science Foundation and Russian Government.

References
1. I.V.Altukhov, M.S.Kagan, K.A.Korolev, V.P.Sinis, and F.A.Smirnov. JETP, V.74 (2) P.404 (1992).
2. I.V.Altukhov, M.S.Kagan, K.A.Korolev, and V.P.Sinis. JETP Lett., V.59, P.476 (1994).
3. G.L.Bir and G.E.Pikus. Symmetry and Strain-Induced Effects in Semiconductors. Wiley, NY (1974).
4. A.K.Ramdas, S.Rodrigues. Rep.Prog.Phys., V.44, P.1297 (1981).
5. A.A.Andronov, I.V.Zverev, V.A.Kozlov, Yu.N.Nozdrin, S.A.Pavlov, and V.N.Shastin. JETP Lett., V.40, P.804 (1984).
6. A.V.Murav'ov, S.G.Pavlov, and V.N.Shastin. JETP Lett., V.52, P.343 (1990).